STATIC AND DYNAMIC ELECTRICITY

INTERNATIONAL SERIES IN PURE AND APPLIED PHYSICS
LEONARD I. SCHIFF, CONSULTING EDITOR

Adler, Bazin, and Schiffer Introduction to General Relativity
Allis and Herlin Thermodynamics and Statistical Mechanics
Becker Introduction to Theoretical Mechanics
Bjorken and Drell Relativistic Quantum Fields
Bjorken and Drell Relativistic Quantum Mechanics
Chodorow and Susskind Fundamentals of Microwave Electronics
Clark Applied X-rays
Collin Field Theory of Guided Waves
Evans The Atomic Nucleus
Feynman and Hibbs Quantum Mechanics and Path Integrals
Ginzton Microwave Measurements
Hall Introduction to Electron Microscopy
Hardy and Perrin The Principles of Optics
Harnwell Electricity and Electromagnetism
Harnwell and Livingood Experimental Atomic Physics
Henley and Thirring Elementary Quantum Field Theory
Houston Principles of Mathematical Physics
Jammer The Conceptual Development of Quantum Mechanics
Kennard Kinetic Theory of Gases
Leighton Principles of Modern Physics
Lindsay Mechanical Radiation
Livingston and Blewett Particle Accelerators
Middleton An Introduction to Statistical Communication Theory
Morse Vibration and Sound
Morse and Feshbach Methods of Theoretical Physics
Morse and Ingard Theoretical Acoustics
Muskat Physical Principles of Oil Production
Newton Scattering Theory of Waves and Particles
Present Kinetic Theory of Gases
Read Dislocations in Crystals
Richtmyer, Kennard, and Lauritsen Introduction to Modern Physics
Schiff Quantum Mechanics
Schwartz Introduction to Special Relativity
Seitz The Modern Theory of Solids
Slater Introduction to Chemical Physics
Slater Quantum Theory of Atomic Structure, Vol. I
Slater Quantum Theory of Atomic Structure, Vol. II
Slater Quantum Theory of Matter
Slater Electronic Structure of Molecules: Quantum Theory of Molecules and Solids, Vol. 1
Slater Symmetry and Energy Bands in Crystals: Quantum Theory of Molecules and Solids, Vol. 2
Slater Insulators, Semiconductors, and Metals: Quantum Theory of Molecules and Solids, Vol. 3
Slater and Frank Introduction to Theoretical Physics
Slater and Frank Mechanics
Smythe Static and Dynamic Electricity
Stratton Electromagnetic Theory
Tinkham Group Theory and Quantum Mechanics
Townes and Schawlow Microwave Spectroscopy
Wang Solid-State Electronics
White Introduction to Atomic Spectra

The late F. K. Richtmyer was Consulting Editor of the series from its inception in 1929 to his death in 1939. Lee A. DuBridge was Consulting Editor from 1939 to 1946; and G. P. Harnwell from 1947 to 1954.

STATIC AND DYNAMIC ELECTRICITY

Third Edition, Revised Printing

WILLIAM R. SMYTHE

Professor Emeritus of Physics
California Institute of Technology

A SUMMA BOOK

HEMISPHERE PUBLISHING CORPORATION
A member of the Taylor & Francis Group

New York Washington Philadelphia London

STATIC AND DYNAMIC ELECTRICITY:
Third Edition, Revised Printing

1 2 3 4 5 6 7 8 9 0 BRBR 8 9 8 7 6 5 4 3 2 1 0 9 8

Library of Congress Cataloging-in-Publication Data

Smythe, William Ralph, date.
Static and dynamic electricity.

(A SUMMA book)
Includes bibliographies and index.
1. Electrostatics. 2. Electrodynamics. I. Title.
QC571.S59 1989 537′.2 88-24722
ISBN 0-89116-916-4 (hard)
ISBN 0-89116-917-2 (soft)

CONTENTS

TABLE OF SYMBOLS

Note: In this table bold-face symbols ($\boldsymbol{v}$, $\boldsymbol{u}$, $\boldsymbol{\phi}$, . . .) are space vectors. Phasors ($\check{I}$, $\check{\mathcal{E}}$, . . .), phasor space vectors ($\check{\boldsymbol{E}}$, $\check{\boldsymbol{B}}$, $\check{\boldsymbol{\Pi}}$, . . .), conjugate phasors ($\hat{I}$, $\hat{\mathcal{E}}$, . . .) and conjugate phasor space vectors ($\hat{\boldsymbol{E}}$, $\hat{\boldsymbol{B}}$, $\hat{\boldsymbol{\Pi}}$, . . .) are shown by an erect ($\smile$) or an inverted ($\frown$) flat vee above the symbol. Magnitudes of vectors and scalars, whether time-dependent or not, are written without designation.

$\boldsymbol{A}$, A_ϕ, A_x, etc.	Vector potential.
A^0	Normalized vector potential.
A, A_x, etc.	Quasi-vector potential
$\boldsymbol{B}$, B_ϕ, B_x, etc.	Magnetic induction
B	Susceptance.
B^0	Normalized or relative susceptance, $B\check{Z}_k$.
C	Capacitance. A constant.
C^0	Normalized or relative capacitance, $C\check{Z}_k$.
c	Velocity of light. A length.
c_{nn}	Self-capacitance.
c_{mn}	Mutual capacitance.
$\boldsymbol{D}$, D_ϕ, D_x, etc.	Electric displacement.
Dw	Dwight integral tables.
$d\boldsymbol{s}$	Differential element along s.
$d\boldsymbol{r}$	Differential change in $\boldsymbol{r}$.
$\boldsymbol{E}$, E, $\check{\boldsymbol{E}}$, $\check{E}$, etc.	Electric field intensity.
$E(k)$	Complete elliptic integral.
e	Electronic charge. 2.71828.
$\boldsymbol{\mathcal{E}}$, $\mathcal{E}$, $\check{\boldsymbol{\mathcal{E}}}$, $\check{\mathcal{E}}$, etc.	Electromotance.
$\mathcal{E}_e$	Effective or rms electromotance.
$\boldsymbol{F}$, F_x	Force
G	Conductance, $\check{Y} = G + jB$.
g	Acceleration of gravity.
$\boldsymbol{H}$, H, $\check{\boldsymbol{H}}$, $\check{H}$, etc.	Magnetic field intensity.
$H_n^{(1)}$, $H_n^{(1)}(v)$, $H_n^{(2)}$, $H_n^{(2)}(v)$	Hankel functions.
h	Planck's constant.
h_1, h_2, h_3	In orthogonal curvilinear coordinates. Length elements are h_1du_1, h_2du_2, h_3du_3.
$h_n^{(1)}$, $h_n^{(1)}(v)$, $h_n^{(2)}$, $h_n^{(2)}(v)$	Spherical Hankel functions.

$\mathbf{I}$, I, $\check{I}$, $\hat{I}$, etc.	Electric current.
I_e	Effective or rms current.
i_e	Effective or rms current density.
$\mathbf{i}$, $\check{\mathbf{i}}$, i, i_x, etc.	Current density. Current.
$\mathbf{i}, \mathbf{j}, \mathbf{k}$	Unit x, y, z vectors.
J_n, $J_n(v)$	Bessel functions.
j	$(-1)^{\frac{1}{2}}$.
j_n, $j_n(v)$	Spherical Bessel functions.
K	Relative capacitivity, ϵ/ϵ_v.
$K(k)$	Complete elliptic integral.
K_m	Relative permeability, μ/μ_v.
K_n, $K_n(v)$	Modified Bessel functions.
k_n, $k_n(v)$	Modified spherical Bessel functions.
k	Boltzman constant.
L, L_{nn}, L_n	Inductance.
L_{mn}	Mutual inductance.
L^0	Normalized or relative inductance, $L/\check{Z}_k$.
l, m, n	Direction cosines with x, y, z axis.
$\mathbf{M}$, M	Magnetization.
M	Mutual inductance.
M_z, $\mathbf{M}$, M, $\check{\mathbf{M}}$, $\check{M}$, etc.	Dipole or loop moment.
$\mathbf{M}'$, M'	Classical magnetic dipole moment (IX).
m	Mass. A number (usually integer).
N	Electric or magnetic flux.
$\mathbf{n}$	Unit normal vector.
n	Index of refraction. A number.
n_n, $n_n(v)$	Spherical Bessel function.
$2n!!$	$2 \cdot 4 \cdot 6 \cdots 2n$
$(2n + 1)!!$	$1 \cdot 3 \cdot 5 \cdots (2n + 1)$
$\mathbf{P}$	Polarization
P, $\check{P}$	Power.
$\bar{P}$	Average power.
P_n^m, $P_n^m(\mu)$	Associated Legendre function.
Pc	Peirce integral tables.
$\mathbf{p}$, p	Momentum.
p	A number. $\omega\mu\gamma$. ω.
Q	Electric charge. Quality of cavity.
Q	Quadrupole moment.
Q_n^m, $Q_n^m(\mu)$	Associated Legendre function.
q	Point or variable charge.
R, R_n, R_{nn}, R_{mn}	Resistance.
R, $R(r)$	Function of r only.
R_n, $R_n(v)$	Solution of Bessel's equation.

R_n^0, $R_n^0(v)$	Solution of Bessel's modified equation.
$\mathbf{R}$, R	Distance between two points.
$\mathbf{r}$, r	Distance from origin.
S	Area or surface.
S_c, S_o	Cavity cross-section areas.
S, S_n, S_n^m	Surface harmonic.
S, S_n, S_{mn}, S_{nn}	Elastance.
s_{nn}	Self-elastance.
s_{mn}	Mutual elastance.
s	Distance along curve. An integer.
$\mathbf{T}$, T	Torque.
T	Absolute temperature. Period.
t	Time.
TE	Transverse electric.
TM	Transverse magnetic.
te	Subscript for TE wave quantities.
tm	Subscript for TM wave quantities.
U	Stream or potential function.
$[U]$	$\oint dU$ around constant V curve.
$\mathbf{u}$, u	Velocity.
u	$\cos\theta$.
u_1, u_2, u_3	Orthogonal curvilinear coordinates.
V	Potential or steam function.
$[V]$	$\oint dV$ around constant U curve.
v	Volume.
$\mathbf{v}$, v	Velocity.
$\mathbf{W}$, W, $\check{\mathbf{W}}$, $\check{W}$	Solutions of scalar wave equation.
W	Energy. $U + jV$.
W_{te}	Solutions yielding TE waves.
W_{tm}	Solutions yielding TM waves.
X	Reactance.
x, y, z	Rectangular coordinates.
Y_n, $Y_n(v)$	Bessel functions.
$\mathbf{Y}$, $\check{Y}$	Admittance, $G + jB$.
$\check{Y}^0$	Normalized or relative admittance, $\check{Y}\check{Z}_k$.
$\mathbf{Z}$, $\check{\mathbf{Z}}$	Hertz vector.
$\check{Z}$, $\mathbf{Z}$, Z, $\mathbf{Z}_{nn}$, $\mathbf{Z}_n$	Impedance.
$\check{Z}_k$	Characteristic impedance.
$\check{Z}^0$	Normalized or relative impedance, $\check{Z}/\check{Z}_k$.
Z, $Z(z)$	Function of z only.
z	Complex variable, $x + jy$.
α, β, γ, δ, ϵ, θ, ϕ, χ, ψ	Often used for angles.

β	Ratio v/c. Ratio $(\mu - \mu_v)/(\mu + \mu_v)$.
β	Free space wave number, $\omega(\mu\epsilon)^{\frac{1}{2}}$.
β'_{mn}	Wave-guide wave number, $(\beta^2 - \beta^2_{mn})^{\frac{1}{2}}$.
β_{mn}	Cutoff wave number.
β_{mnp}	Cavity resonance wave number.
$\mathbf{\Gamma}$, $\check{\Gamma}$	Phasor propagation constant.
γ	Electrical conductivity.
Δ, Δ_{rs}	Determinant. A small part of.
δ	Skin depth. Phase difference.
δ	A small quantity. A small part of.
δ^m_n	Kronecker delta, zero if $m \neq n$, one if $m = n$.
ϵ	Capacitivity. A small quantity.
ϵ_v	Free space capacitivity.
ϵ, ϵ_n	Phase angle.
η	Intrinsic impedance.
Θ, $\Theta(\theta)$	Function of θ only.
θ	Colatitude angle.
$\boldsymbol{\theta}$	Unit vector in θ direction.
θ, θ', θ''	Angles of incidence, reflection, and refraction.
κ	Magnetic susceptibility. $(1 - \beta^2)^{-\frac{1}{2}}$.
λ	Wave length.
λ_{mn}	Cutoff wave length.
λ_g	Wave-guide wave length.
λ_{mnp}	Cavity resonance wave length.
μ	Permeability. $\cos\theta$.
μ_v	Free space permeability, $4\pi \times 10^{-7}$.
ν	Frequency in cycles per second.
ν_{mn}	Cutoff frequency.
ν_{mnp}	Cavity resonance frequency.
Ξ, $\Xi(\xi)$	Function of ξ only.
ξ, ζ, ϕ	Oblate spheroidal coordinates.
ξ, η, ϕ	Prolate spheroidal coordinates.
$\mathbf{\Pi}$, $\check{\mathbf{\Pi}}$, Π	Poynting vector.
$\bar{\mathbf{\Pi}}$	Rms Poynting vector.
ρ	Distance from z- or θ-axis. Charge density.
$\boldsymbol{\rho}_1$	Unit vector in ρ-direction.
σ	Surface electric charge density.
$\mathfrak{s}$	Area or surface resistivity.
τ	Volume resistivity. Density. Time.
Φ	Function of ϕ only.
$\boldsymbol{\phi}$	Unit vector in ϕ-direction.
ϕ	Longitude angle. Phase angle.
Ψ	Scalar potential.

Ω	Magnetomotance. Solid angle.
ω	Frequency in radians per second.
$\boldsymbol{\nabla}$	Vector operator, $\boldsymbol{i}\partial/\partial x + \boldsymbol{j}\partial/\partial y + \boldsymbol{k}\partial/\partial z$.
$\boldsymbol{\nabla}_2$	Two-dimensional vector operator.
$\boldsymbol{a} \times \boldsymbol{b}$	Vector product of $\boldsymbol{a}$ and $\boldsymbol{b}$.
$\boldsymbol{a} \cdot \boldsymbol{b}$	Scalar product of $\boldsymbol{a}$ and $\boldsymbol{b}$.
$\boldsymbol{\nabla}^2$	Laplace operator.
$[\boldsymbol{v}]$	Retarded $\boldsymbol{v}$.

PREFACE TO THE THIRD EDITION, REVISED PRINTING

It is a source of great satisfaction to my father on his 95th birthday that a revision of this book is being published. In the past 50 years it has been used as a textbook by countless students who have, sometimes painfully, learned the value of a rigorous problem course. Others have found it valuable as a reference that contains solutions to hundreds of difficult electromagnetic problems with widespread applicability.

WILLIAM RODMAN SMYTHE

PREFACE TO THE THIRD EDITION

The arrival of the digital computer since the appearance of the second edition has considerably altered the electromagnetic computation problem and made methods feasible that were probably discarded years ago as impractical. These devices iterate so efficiently that the summation of a simple, slowly convergent series is often faster than closed-form evaluations requiring interpolation in an inadequate function table. Solution of linear simultaneous algebraic equations is now easy and very useful for obtaining coefficients in the series solutions of mixed boundary and mixed coordinate problems such as those added to the text and problems of Chap. V. Many of these appear here for the first time.

About 25 useful elliptic function conformal transformations are added to text and problems of Chap. IV as well as minor alterations and additions to Chaps. VI, VII, and XI (former Chap. XIII). Old Chaps. IX and X have been deleted because for about 10 years all physics and electrical engineering graduate students at the California Institute of Technology, and probably elsewhere, have seemed to be already familiar with linear circuits. The coverage of eddy currents in Chap. X (former Chap. XI) is greatly extended. Little or none of this added material appears elsewhere. Chapter XII (former Chap. XIV) on wave guides and cavities is entirely rewritten, with the addition of new matter and use of the variational method. The treatment of radiation from accelerated particles in Chap. XIII (former Chap. XV) has been expanded.

The new material herein, unlike that in former editions, is unchecked by students. As much as possible of the almost error-free last edition has been left undisturbed. The author has checked the new material as well as he can but knows from past experience that some errors may survive. It is hoped that those finding them will inform him so that they may be corrected in future printings.

William R. Smythe

PREFACE TO THE SECOND EDITION

The wide use of rationalized mks units and the increased importance of microwaves made this radical revision of the first edition imperative. The units are changed throughout. The resultant extensive resetting of the text permits a modernization of nomenclature through such changes as "capacitor" for "condenser" and "electromotance" for "electromotive force," The original wording has been preserved only in the Cambridge problems. In static-field chapters, 40 problems of above-average difficulty have been added, usually covering boundary conditions omitted in the first edition. The expanded treatment of electromagnetic waves made necessary the rewriting of the parts of Chap. V dealing with Bessel functions and led to the introduction of vector surface harmonics, which greatly simplify some calculations. Much of Chap. XI on eddy currents has been rewritten, and two of the three electromagnetic-wave chapters are entirely new. Both the text and the 150 problems include methods and results not found in the literature. Two groups of advanced Ph.D. students worked over this material to get practice in attacking every type of wave-field problem. Many are too difficult for first-year graduate students, but every problem was solved by at least one of the advanced students. They can be worked either directly from the text or by fairly obvious extensions of it. Some useful results appear in the problems and are listed in the Index, which should be consulted by engineers with boundary value problems to solve. Chapter XV of the first edition is omitted because none of the remaining theory is based on it and because to bring it up to date would require an excessive amount of space.

None of the new topics appears to lie outside the scope of the mathematical preparation assumed for readers of the first edition. That the successful solution of electrical problems depends on physical rather than mathematical insight is borne out by the author's experience with the first edition, which shows that graduate students in electrical engineering and physics greatly excel those in mathematics.

It is believed that very few of the errors and obscure or ambiguous statements in the first edition escaped the scrutiny of the 375 students at the California Institute of Technology who worked it through. No infallible system for locating errors caused by the transposition of units has been found, and the author will appreciate letters from readers pointing them out.

WILLIAM R. SMYTHE

July, 1950

PREFACE TO THE FIRST EDITION

It has been found that, in most cases, the average graduate student, even though he seems to be thoroughly familiar with advanced electrical theory, is unable to solve the electrical problems he encounters in research if they fall outside the routine types and so must be worked out from first principles. The present book is the result of having taught, for the last twelve years, a course designed to train first-year graduate students in physics, electrical engineering, geophysics, and mathematics to apply the principles of electricity and magnetism. Students in these fields must show a proficiency equivalent to an average grade in this course to be admitted to candidacy for the Ph.D. degree at this institute. It is hoped that this book will also provide a reference where workers in these fields may find the methods of attack on those problems, common in research, for which the formulas in the handbooks are inadequate. It is assumed that the reader has the mathematical preparation usually acquired in a good undergraduate course in introduction to mathematical physics. This implies a reasonable familiarity with vector analysis, the calculus, and elementary differential equations. All the mathematical development beyond this point is done herein by methods that a reader with such training can follow. The author has succeeded, with some difficulty, in avoiding the use of contour integration but believes that anyone going further into the subject should acquire a working knowledge of this powerful mathematical tool.

As has already been implied, this book is written for the experimental research physicist and engineer rather than for the theoretical man. For this reason, only that theory which has applications is included, and this is developed by the most concise method compatible with the assumed preparation of the reader. No subject is included for its historical interest alone. More than the usual number of problems have been worked out in the text, and these have been selected for one or more of the following reasons: the result has important applications, it clarifies the theory, it illustrates some useful mathematical device, it proves the utility of some concept in the theory. At the end of each chapter is an extensive collection of problems involving nearly all the theory in the text. Many of these are taken from the Cambridge University examination questions as printed in Jeans's "The Mathematical Theory of Electricity and Magnetism," Cambridge, 1925, and are included with the permission of the Cambridge University Press to whom we express our thanks. Although the best students can work all the problems, the average student cannot. They provide an opportunity for the reader who is working up the subject by himself to test his proficiency. In many cases important results, which lack of space prevented working out in the text, are included as problems and are listed in the index.

The treatment and arrangement of the subject matter depart from the most common practice in several respects. In the first place, it will be noticed that all developments are based directly on the macroscopic experimental facts rather than on the hypothetical microscopic structure of conductors and dielectrics. There are two reasons for this. The first is that, although the microscopic theory meets the crucial test, namely, gives the observed macroscopic laws within the precision of observation, this does not imply that the theory is unique or that other deductions from it are correct. The second is that the development of the most satisfactory theories, those based on wave mechanics, require mathematical technique with which we have assumed the reader to be unfamiliar when starting this book, but with which he will be quite familiar after finishing two-thirds of it. The development of these theories, necessarily brief, is therefore postponed until the last chapter. The second departure from the most common practice consists in the complete omission of the concept of the isolated magnetic pole. All magnetic theory is therefore based on the interactions of electric currents or moving charges. This logically leads to the use of the magnetic vector potential rather than the scalar potential, so that the former is used extensively, although not exclusively, in all magnetic and electromagnetic theory. It will surprise many to find that this leads to considerable simplification in certain places, notably in the calculation of inductance and the treatment of eddy currents and electromagnetic radiation. Other minor departures from the usual practice include the more extensive use of Bessel functions and conformal transformations and the treatment of forces between moving charges exclusively by the methods of special relativity. The latter procedure, besides having a very firm experimental basis, makes hypotheses as to the shape or size of electric charges unnecessary and gives a clearer idea of the limits within which the standard formulas can be applied without using such hypotheses.

Several subjects usually included in books on electricity and magnetism are omitted here. The theories of electrolytic conduction, thermionic emission, photoelectric phenomena, thermoelectric effects, etc., of which the reader is assumed to have an elementary knowledge, are entirely omitted because it is believed that a treatment of them, on the same level as the remainder of the book, would require a background of physical chemistry, thermodynamics, and quantum theory not possessed by the average reader. The theory of electrical machines and instruments, including vacuum tubes, is also omitted because it is believed that such subjects are best treated in connection with laboratory courses. Certain alternative methods such, for example, as the Heaviside operational method and the dynamic method of circuit analysis have been omitted for lack of space.

Before starting this book the average reader will have become familiar with all the common systems of electrical units and will have his own preference. The actual system used will make little difference provided it is clearly indicated. For each section of the subject, the author has used that system which seemed simplest to work with. Thus the c.g.s. electrostatic units are used in Chaps. I to V, the c.g.s. electromagnetic units in Chaps. VII to XII, and the Gaussian system in Chaps. XIII, XIV, and XV. To avoid confusion, the units used are noted at the bottom of each page and a very complete system of conversion tables is given in the Appendix, enabling the results of any calculation to be expressed in any units. To see whether the use of rational units would have simplified calculations, all the numbered formulas occurring in the preliminary lithoprinted edition were scrutinized. It was found that the complexity of 1196 formulas would be unchanged, that of 169 would be decreased, and that of 123 would be increased. Thus in the theory there is little to choose between rationalized and unrationalized units. The results in problems were not investigated.

It has been the practice throughout, whenever an integration is performed or a mathematical transformation made, to refer, by number, to the appropriate formula in both Peirce's "A Short Table of Integrals," Ginn, 1929, and Dwight's "Tables of Integrals and Other Mathematical Data," Macmillan, 1934 [4th ed., 1961]. It is therefore advisable for the prospective reader to procure a copy of one of these inexpensive tables. The bibliographies at the ends of the chapters are not by any means complete but include merely those books that have come to the author's attention which contain useful additional material or instructive alternative treatments.

The author has used every device he can think of to eliminate mistakes but is perfectly certain that they still persist and will be grateful to anyone pointing them out.

The author wishes to express his gratitude to the hundred graduate students, especially Dr. Pasternack, who worked through the preliminary lithoprinted edition, zealously detecting errors and obscure spots, and checking the answers to problems. He also wishes to thank Professors Ira S. Bowen and William V. Houston for reading sections of the manuscript and for valuable discussions. In particular, he wishes to thank Dr. Charles H. Townes for checking all the derivations in the final manuscript and the answers to those problems not appearing in the preliminary edition.

WILLIAM R. SMYTHE

August, 1939

CHAPTER I

BASIC IDEAS OF ELECTROSTATICS

1.00. Electrification, Conductors, and Insulators.—The word "electricity" is derived from the Greek word for amber. Apparently Thales of Miletus first discovered, about 600 B.C. that amber, when rubbed, attracts light objects to itself. It is now known that most substances possess this property to some extent. If we rub with a piece of silk a glass rod or a metal rod supported by a glass handle, we find that both will attract small bits of paper, and we say they are electrified. We see that only in the case of the metal rod can we destroy the electrification by touching with the finger. Furthermore, we find when we touch the electrified metal rod with pieces of various materials held in the hand that metals and damp objects destroy the electrification whereas substances like glass and silk do not. We call substances that remove the electrification conductors and those which do not insulators. We can find materials that remove the electrification very slowly and that we might call poor conductors or poor insulators. Thus there is no definite division between the two classes.

1.01. Positive and Negative Electricity.—If we rub a glass rod with silk and touch some light conducting body, such as a gilded pith ball suspended by a silk string, with either the rod or the silk, we find it is electrified. If two balls are electrified by the glass or by the silk, they repel each other; but if one is electrified by the rod and one by the silk, they attract. Thus we conclude that there are two kinds of electrification and that bodies similarly electrified repel each other and those oppositely electrified attract each other. By experimenting with many substances, we conclude that there are only two kinds of electricity. Arbitrarily we define the electricity on the glass rod to be positive and that on the silk negative.

1.02. Coulomb's Law, Unit Charge, Dielectrics.—We next observe that the force between the balls decreases rapidly as they are separated. Coulomb investigated these forces with a torsion balance and found that the force between two small electrified bodies is directed along the line joining them, is proportional to the product of the charges, and is inversely proportional to the square of the distance between them. This is known as Coulomb's or Priestley's law. The statement of this law implies the definition of a quantity of electrification or of electric charge. We formulate this as follows: An electrostatic unit of charge is that charge which, when placed one centimeter away from a like charge in a vacuum, repels it with a force of one dyne. The practical unit of charge is the coulomb which is 3.0×10^9 electrostatic units.

In many homogeneous, isotropic, nonconducting mediums, the inverse-square law is found to hold, but the force between the same charges is reduced, so that for such a medium we may write this law, in mks units,

$$F = \frac{qq'}{4\pi\epsilon r^2}\boldsymbol{r}_1 \tag{1}$$

where F is the force on the charge of q' coulombs due to the charge q, $\boldsymbol{r}$ the vector from q to q' whose magnitude is r meters, $\boldsymbol{r}_1$ a unit vector along $\boldsymbol{r}$, and ϵ the capacitivity which is a constant characteristic of the medium. *In vacuo* the numerical value of ϵ is 8.85×10^{-12} farad per meter and is written ϵ_v. The specific inductive capacity or relative capacitivity K is the ratio ϵ/ϵ_v which is dimensionless and has the same numerical value for all systems of units.

1.021. Limitations of the Inverse-square Law.—The precision of Coulomb's measurement has been greatly exceeded by modern methods, and we may now state that the exponent of r in 1.02 (1) has been verified, recently, to about 1 part in 10^9 for measurable values of r. It should be borne in mind that we can apply Coulomb's law with certainty only to dimensions for which it has been verified. We shall endeavor throughout this book to avoid basing any macroscopic theory on the assumption of the validity of this law at atomic distances. The law applies strictly only to charged bodies whose dimensions are small compared with the distances between them. Their shape and composition are immaterial.

1.03. Electrical Induction.—The fact that the electrical charge on a conductor is mobile indicates that when an electric charge is brought near an uncharged conductor electricity of the opposite sign will move to the parts nearer the charge and that of the same sign to the parts more remote from it. From Coulomb's law, the force between nearer charges will be greater so that the charge will attract the conductor. These charges on the conductor are known as induced charges. Unless the more remote charge is removed, for example, by touching with the finger, the conductor will return to its neutral state when the inducing charge is removed. If, with the inducing charge in place, we separate the near and far parts of the conductor, being careful to keep them insulated, we find that the two parts are oppositely charged, as we would expect.

Many "static machines" have been devised that repeat this operation automatically and accumulate the separated charges.

When sufficiently sensitive methods are used, it is found that a charge also exerts a small attractive force on uncharged insulators. This seems to indicate that even in insulators charges are present and are not absolutely fixed but may suffer some displacement. The hypotheses concerning the actual behavior of the electric charges in conductors and insulators will not be discussed here. The theories are still imperfect but have greatly improved since 1930.

1.04. The Elementary Electric Charges.—It has been found that an electric charge cannot be subdivided indefinitely. The smallest known negative charge is that of the negative electron and mesotron, first determined with considerable precision by Millikan. The accepted value is now 1.60×10^{-19} coulomb. The smallest known positive charge is that on the positron, or positive electron, on the mesotron, and on the proton. To a high order of precision, all these elementary charges are the same in magnitude. The mass of the negative electron and also, probably, that of the positron is 9.1×10^{-31} kg. The mass of the proton is about 1837 times that of the electron.

In handling electrical problems, we shall treat electrical charges as infinitely divisible, using charge density as if this were so. Clearly such a procedure is justifiable only when we are dealing with quantities much greater than 1.60×10^{-19} coulomb and certainly is useless if we go to atomic dimensions. We have seen that the electricity in conductors is free to move, so that if it possess inertia we might expect it to lag behind when the body is accelerated, thus producing an electric current which could be detected by its magnetic field. Maxwell predicted such phenomena but their magnitude is so small that they were detected and measured by Tolman, Barnett, and others not until long after his death. The results show that the mobile electricity in conductors is negative and that the ratio of its charge to its mass equals, within experimental error, the ratio for the negative electron. It appears that almost all the phenomena with which we deal in this book involve the distribution or motion of such electrons, a positive charge appearing when there is a dearth of electrons. As far as the mathematics is concerned, it is immaterial whether the positive, the negative, or both take part in the transfer of electric charge.

1.05. Electric Field Intensity.—When an infinitesimal electric charge, placed in a region, experiences a force, we say an electric field exists there. We define the electric intensity to be a vector equal to the force per unit charge acting on a positive charge placed at that point, the charge being so minute that no redistribution of charge takes place due to its presence. The last qualification is necessary because of the phenomenon of electric induction.

Just as a number of mechanical forces acting on a body can be resolved into a single resultant force by taking their vector sum, so the resultant electric intensity due to any charge distribution can be obtained by taking the vector sum of the electric intensities due to each element of the distribution. Thus the electric intensity at point P due to n charges is

$$\boldsymbol{E}_p = -\frac{1}{4\pi\epsilon}\sum_{i=1}^{n}\frac{q_i}{r_i^3}\boldsymbol{r}_i \tag{1}$$

where $\boldsymbol{E}_p$ is the electric intensity in volts per meter, $\boldsymbol{r}_i$ is the vector of magnitude r_i directed from P to q_i, and ϵ is the capacitivity of the infinite homogeneous medium in which the system is immersed.

1.06. Electrostatic Potential.—Work is done against electric forces when a charge is moved in an electric field. The potential in volts at a point P in an electrostatic field is the work in joules per coulomb to bring a charge from the point of zero potential to the point P, the charge being so small that there is no redistribution of electricity due to its presence. The choice of the point of zero potential is a matter of convenience. It is frequently, but not always, chosen at infinity. To avoid other than electrostatic effects, it will be necessary to move the charge very slowly.

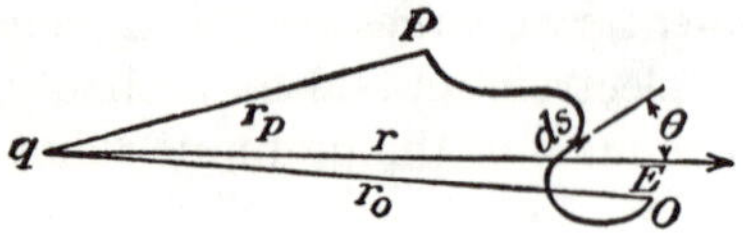

FIG. 1.06.

Let us compute the potential due to a point charge q. The work dV required to move a unit charge a distance $\boldsymbol{ds}$ in a field of intensity $\boldsymbol{E}$ is $-\boldsymbol{E}\cdot\boldsymbol{ds}$ or $-E\,ds\cos\theta$ where θ is the angle between $\boldsymbol{E}$ and $\boldsymbol{ds}$. In the case of the point charge, this becomes

$$dV = -\frac{q\cos\theta}{4\pi\epsilon r^2}\,ds$$

where $\boldsymbol{r}$ is the vector from the charge q to the element of path $\boldsymbol{ds}$ and θ is the angle between $\boldsymbol{r}$ and $\boldsymbol{ds}$ as shown in the figure. Evidently $dr = ds\cos\theta$, so that we have for the potential in volts

$$\int_0^{V_P} dV = \frac{-q}{4\pi\epsilon}\int_{r_o}^{r_p}\frac{dr}{r^2} \qquad \text{or} \qquad V_p = \frac{q}{4\pi\epsilon}\left(\frac{1}{r_p} - \frac{1}{r_o}\right) \tag{1}$$

If r_o is chosen at infinity, this becomes

$$V_p = \frac{q}{4\pi\epsilon r_p} \tag{2}$$

The electrostatic potential is a scalar point function and is independent of the path by which we bring our charge to the point. The potential at any point in an electrostatic field can be obtained by adding up the potentials due to the individual charges producing the field; thus

$$V_p = \frac{1}{4\pi\epsilon}\sum_{i=1}^{n}\frac{q_i}{r_i} \tag{3}$$

where r_i is the distance from P to q_i in meters.

Since it is usually much easier to take a scalar sum than a vector sum, it is evident why (3) is to be preferred to 1.05 (1) for purposes of computation. The field intensity at P can be obtained from (3) by taking the gradient; thus

$$\boldsymbol{E} = -\text{grad}\, V = -\nabla V \tag{4}$$

In rectangular coordinates, the components are

$$E_x = -\frac{\partial V}{\partial x}, \qquad E_y = -\frac{\partial V}{\partial y}, \qquad E_z = -\frac{\partial V}{\partial z} \tag{5}$$

The components of the gradient in any other fixed coordinate system can be obtained by expressing V and the components of the gradient in terms of that system. A method of doing this is given in 3.03 and 3.05.

If the elementary charges are very close together compared with the dimensions involved, as is always the case in ordinary practice, we may treat the distribution as continuous and speak of the charge per unit volume as volume charge density ρ and the charge per unit area as the surface charge density σ. The summation in (3) now becomes an integral

$$V_p = \frac{1}{4\pi\epsilon}\int_V \frac{\rho\, dv}{r} + \frac{1}{4\pi\epsilon}\int_S \frac{\sigma\, dS}{r} \tag{6}$$

where dv is the element of volume and dS is the element of surface. It should be noted that this formula applies only when all space, including the interior of any material bodies present, has the capacitivity ϵ. When this is not the case, we must use the methods of Chaps. IV and V.

1.07. Electric Dipoles and Multipoles.—Consider the potential at a point P at x, y, z due to charges $-q$ at x_0, y_0, z_0 and $+q$ at $x_0 + h, y_0, z_0$ for small h. If the distance $r_- = f(x_0, y_0, z_0, x, y, z)$ from $-q$ to the field point P is $[(x - x_0)^2 + (y - y_0)^2 + (z - z_0)^2]^{\frac{1}{2}}$, then from $+q$ to P it will be $r_+ = f(x_0 + h, y_0, z_0, x, y, z)$. Write down this potential from 1.06 (2) and expand r_+^{-1} in a Taylor series by *Dw* 39, which gives

$$V = \frac{q}{4\pi\epsilon}\left(\frac{1}{r_+} - \frac{1}{r_-}\right) = \frac{q}{4\pi\epsilon}\left[\frac{1}{r_-} + h\frac{\partial}{\partial x_0}\left(\frac{1}{r_-}\right) + \frac{h^2}{2!}\frac{\partial^2}{\partial x_0^2}\left(\frac{1}{r_-}\right) + \cdots - \frac{1}{r_-}\right]$$

If $h \to 0$ and $q \to \infty$ in such a way that $hq = M$ remains a finite constant, the result is an electric dipole of moment $\boldsymbol{M}$ in the x-direction. Only the h term in the expression above survives, so that, if θ is the angle between $\boldsymbol{h}$ and $\boldsymbol{r}$, which replaces r_- as $h \to 0$, there results

$$V = \frac{M}{4\pi\epsilon}\frac{\partial}{\partial x_0}\left(\frac{1}{r}\right) = \frac{-M}{4\pi\epsilon}\frac{\partial}{\partial x}\left(\frac{1}{r}\right) = \frac{M\cos\theta}{4\pi\epsilon r^2} = \frac{\boldsymbol{M}\cdot\boldsymbol{r}}{4\pi\epsilon r^3} \tag{1}$$

Evidently this can be extended so that if the potential V_p at the point P due to a set of n charges, the radius vector from q_i to P being r_i, is given by 1.06 (3), then the potential V'_p, due to a set of n dipoles of the same strength and sign located at the same points with axes parallel to x is

$$V'_p = -\frac{\partial V_p}{\partial x_p} = \sum_{i=1}^{n} \frac{q_i(x_p - x_i)}{4\pi\epsilon r_i^3} \tag{2}$$

By differentiating the expression for the potential of a unit electric dipole with respect to any of the rectangular coordinates, we get the potential due to a unit quadrupole whose dimensions are QL^2; thus

$$\frac{1}{4\pi\epsilon}\frac{\partial^2(1/r)}{\partial x^2}, \qquad \frac{1}{4\pi\epsilon}\frac{\partial^2(1/r)}{\partial x\,\partial y}, \qquad \text{etc.}$$

represent the potentials due to a linear quadrupole (Fig. 1.07a) and a square quadrupole (Fig. 1.07b), etc. Further differentiations give us the potentials due to more complicated multipole arrangements but always ones in which the total charge is zero. For other cases see 1.12.

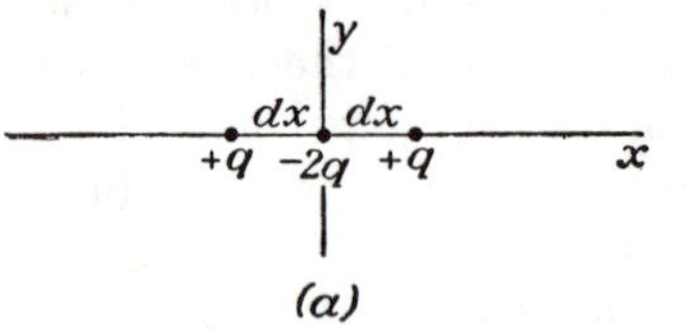

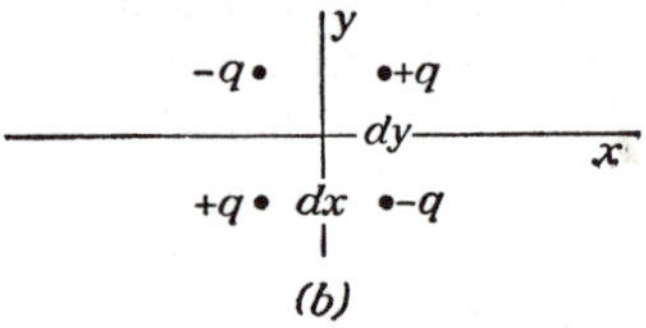

FIG. 1.07.

The translational force on a dipole M in a field E will be the vector sum of the forces on each charge. Since the charges are equal and opposite, this means the vector difference in the field strength $(\mathbf{ds} \cdot \nabla)E$, at the two ends, multiplied by q; thus

$$F = q(\mathbf{ds} \cdot \nabla)E = (M \cdot \nabla)E \tag{3}$$

In a uniform field, this force is zero.

In a uniform field, the charges are subject to forces $+qE$ and $-qE$ applied at a distance $ds \sin\theta$ apart where θ is the angle between $\mathbf{ds}$ and $\mathbf{E}$. Thus there is a torque acting on the dipole of amount

$$T = \mathbf{t}Eq\,ds \sin\theta = \mathbf{t}ME \sin\theta = M \times E \tag{4}$$

where $\mathbf{t}$ is a unit vector normal to M and E.

1.071. Interaction of Dipoles.—The potential energy of a dipole, in any field whose potential is V, equals the total work done, against this field, in bringing each charge separately into place. If the potential of the field at P_1 where $+q$ is placed is V_1 and at P_2 where $-q$ is placed it is V_2, then

$$W = q(V_1 - V_2) = q\overline{P_1P_2}\frac{\partial V}{\partial s} = M\frac{\partial V}{\partial s} \tag{1}$$

where s is in the direction of the dipole axis and M is its moment. In vector notation, this is

$$W = (M \cdot \nabla)V \tag{2}$$

If $\boldsymbol{M}_1$ and $\boldsymbol{M}_2$ are the vector moments of two dipoles A and B and if $\boldsymbol{r}$ is the vector from A to B, then the potential at B due to A is, from 1.07 (1),

$$4\pi\epsilon V = \frac{M_1 \cos\theta}{r^2} = \frac{\boldsymbol{M}_1 \cdot \boldsymbol{r}}{r^3} = -\boldsymbol{M}_1 \cdot \nabla\left(\frac{1}{r}\right) \tag{3}$$

Substituting this value for V in (2) gives

$$\begin{aligned} 4\pi\epsilon W &= +\boldsymbol{M}_2 \cdot \nabla(\boldsymbol{M}_1 \cdot \boldsymbol{r}r^{-3}) = \boldsymbol{M}_2 \cdot (r^{-3}\nabla \boldsymbol{M}_1 \cdot \boldsymbol{r} + \boldsymbol{M}_1 \cdot \boldsymbol{r}\,\nabla r^{-3}) \\ &= \boldsymbol{M}_2 \cdot (\boldsymbol{M}_1 r^{-3} - \boldsymbol{M}_1 \cdot \boldsymbol{r}3\boldsymbol{r}r^{-5}) = \boldsymbol{M}_1 \cdot \boldsymbol{M}_2 r^{-3} - 3\boldsymbol{M}_1 \cdot \boldsymbol{r}\boldsymbol{M}_2 \cdot \boldsymbol{r}r^{-5} \end{aligned}$$

If $\boldsymbol{M}_1$ and $\boldsymbol{M}_2$ make angles θ and θ' with $\boldsymbol{r}$ and ϕ with each other, we have

$$W = \frac{M_1 M_2}{4\pi\epsilon r^3}(\cos\phi - 3\cos\theta\cos\theta') \tag{4}$$

If ψ is the angle between the planes intersecting in $\boldsymbol{r}$ which contain $\boldsymbol{M}_1$ and $\boldsymbol{M}_2$, then, taking $\boldsymbol{r}$ in the direction of x and $\boldsymbol{M}_1$ in the xy-plane, the direction cosines are $l_1 = \cos\theta, l_2 = \cos\theta'$, $m_1 = \sin\theta$, $m_2 = \sin\theta'\cos\psi$, and $n_1 = 0$, so that $\cos\phi$ and W become, respectively,

$$\cos\phi = l_1 l_2 + m_1 m_2 + n_1 n_2 = \cos\theta\cos\theta' + \sin\theta\sin\theta'\cos\psi$$

$$W = \frac{M_1 M_2}{4\pi\epsilon r^3}(\sin\theta\sin\theta'\cos\psi - 2\cos\theta\cos\theta') \tag{5}$$

This gives the radial force between two dipoles, by differentiation, to be

$$F = -\frac{\partial W}{\partial r} = \frac{3M_1 M_2}{4\pi\epsilon r^4}(\sin\theta\sin\theta'\cos\psi - 2\cos\theta\cos\theta') \tag{6}$$

This has its maximum value when $\psi = 0$, $\theta = \theta' = 0$. The torque trying to rotate the dipole in the direction α is obtained similarly, giving

$$T = -\frac{\partial W}{\partial \alpha} \tag{7}$$

1.08. Lines of Force.—A most useful method of visualizing an electric field is by drawing the "lines of force" and "equipotentials." A line of force is a directed curve in an electric field such that the forward drawn tangent at any point has the direction of the electric intensity there. It follows that if $\boldsymbol{ds}$ is an element of this curve,

$$\boldsymbol{ds} = \lambda \boldsymbol{E} \tag{1}$$

where λ is a scalar factor. Writing out the components in rectangular coordinates and equating values of λ, we have the differential equation of the lines of force

$$\frac{dx}{E_x} = \frac{dy}{E_y} = \frac{dz}{E_z} \tag{2}$$

Analogous equations can be written in terms of other coordinate systems by using 3.03 and 3.05. Usually much easier methods of getting the equations of the lines of force exist than by integrating these equations.

We shall, however, give one example of the integration of this equation. Consider the field due to two charges $+q$ at $x = a$ and $\pm q$ at $x = -a$. Since, from symmetry, any section of the field by a plane including the x-axis will look the same, we may take the section made by the xy-plane. The sum of the x-components of the intensity at any point due to the two charges is E_x, where

$$4\pi\epsilon E_x = \frac{q(x-a)}{[y^2 + (x-a)^2]^{\frac{3}{2}}} \pm \frac{q(x+a)}{[y^2 + (x+a)^2]^{\frac{3}{2}}}$$

If we substitute,

$$u = \frac{x+a}{y} \qquad \text{and} \qquad v = \frac{x-a}{y} \tag{3}$$

this becomes

$$4\pi\epsilon E_x = \frac{qv}{y^2(1+v^2)^{\frac{3}{2}}} \pm \frac{qu}{y^2(1+u^2)^{\frac{3}{2}}}$$

Similarly,

$$4\pi\epsilon E_y = \frac{q}{y^2(1+v^2)^{\frac{3}{2}}} \pm \frac{q}{y^2(1+u^2)^{\frac{3}{2}}}$$

Equation (2) becomes

$$\frac{dy}{dx} = \frac{E_y}{E_x} = \frac{(1+v^2)^{\frac{3}{2}} \pm (1+u^2)^{\frac{3}{2}}}{u(1+v^2)^{\frac{3}{2}} \pm v(1+u^2)^{\frac{3}{2}}}$$

Solving (3) for y and x and taking the ratio of the differentials give

$$\frac{dy}{dx} = \frac{dv - du}{u\,dv - v\,du}$$

Comparing these two expressions for dy/dx, we see that

$$\frac{du}{dv} = \mp\left(\frac{1+u^2}{1+v^2}\right)^{\frac{3}{2}}$$

Separating variables and integrating, we have

$$u(1+u^2)^{-\frac{1}{2}} \pm v(1+v^2)^{-\frac{1}{2}} = C$$

In terms of x and y, this becomes

$$(x+a)[(x+a)^2 + y^2]^{-\frac{1}{2}} \pm (x-a)[(x-a)^2 + y^2]^{-\frac{1}{2}} = C \tag{4}$$

This is the equation giving the lines of force shown in Figs. 1.08*a* and 1.08*b* in which the value of C for each line is marked. An easier method for getting this equation, by Gauss's electric flux theorem, is given in 1.101.

We may write the left side of (4) in the form

$$(x+a)r^{-1}(1 + 2axr^{-2} + a^2r^{-2})^{-\frac{1}{2}} - (x-a)r^{-1}(1 - 2axr^{-2} + a^2r^{-2})^{-\frac{1}{2}}$$

where $r^2 = x^2 + y^2$. If we let $a \to 0$, expand the radicals by *Pc* 750 or *Dw* 9.03, neglect the square and higher powers of a, and write C' for $C/(2a)$, we obtain

$$\frac{y^2}{r^3} = C' = \frac{\sin^2 \theta}{r} \tag{5}$$

which is the equation of the lines of force for an electric dipole, shown in Fig. 1.08*c*.

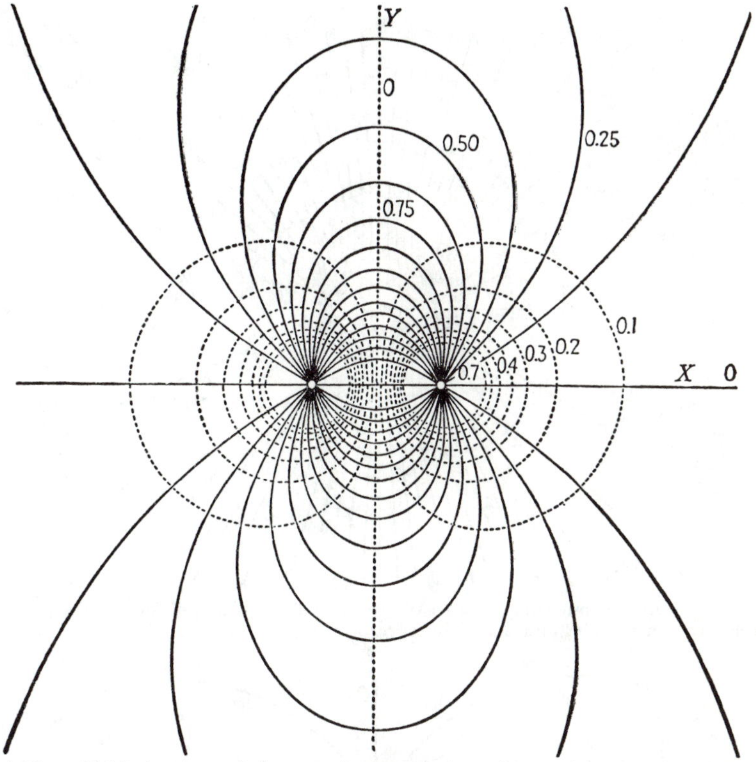

Fig. 1.08*a*.—Field about equal charges of opposite sign. Lines of force and equipotential lines are shown by solid and dotted lines, respectively.

1.09. Equipotential Surfaces.—An equipotential surface in an electric field is one at all points of which the potential is the same. The equation of an equipotential surface is therefore

$$V = C \tag{1}$$

where C is a constant. Numerous maps of electric fields showing equipotentials and lines of force will be found in subsequent chapters. It should be noted that since no work is done in moving charges over an equipotential surface, the lines of force and equipotentials must intersect orthogonally. As an example of the use of this equation, we shall take the case just considered. The potential at any point P will be C, where

$$q[(x - a)^2 + y^2]^{-\frac{1}{2}} \mp q[(x + a)^2 + y^2]^{-\frac{1}{2}} = 4\pi\epsilon C \tag{2}$$

This is the equation of the equipotential surfaces shown in section in Figs. 1.08*a* and 1.08*b* by broken lines. C values are for $q = 4\pi\epsilon$.

There are often points or lines in an electrostatic field where an equipotential surface crosses itself at least twice so that ∇V vanishes there.

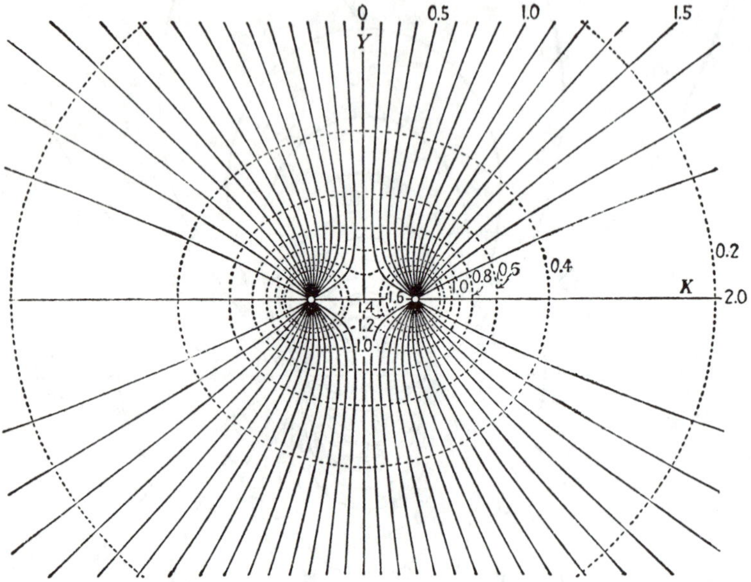

FIG. 1.08*b*.—Field about equal charges of the same sign. Lines of force and equipotential lines are shown by solid and dotted lines, respectively.

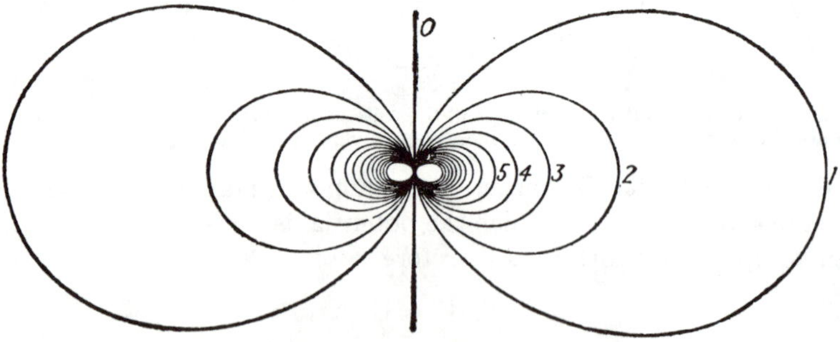

FIG. 1.08*c*.—Lines of force of electric dipole.

These are called neutral or equilibrium or singular points or lines. The origin in Fig. 1.08*b* is such a point. Some of the properties of such points are considered in 5.235.

1.10. Gauss's Electric Flux Theorem.—We shall derive this flux theorem from the inverse-square law on the assumption, to be modified later, that all space is filled with a uniform dielectric.

Consider a small element dS of a closed surface (Fig. 1.10), whose outward normal makes an angle α with the radius vector from a point charge q at P. Now draw a line from every point of the boundary of dS to the point P so that the small cone so formed cuts out an area $d\Sigma$ from the spherical surface, having P as a center and passing through Q. Then

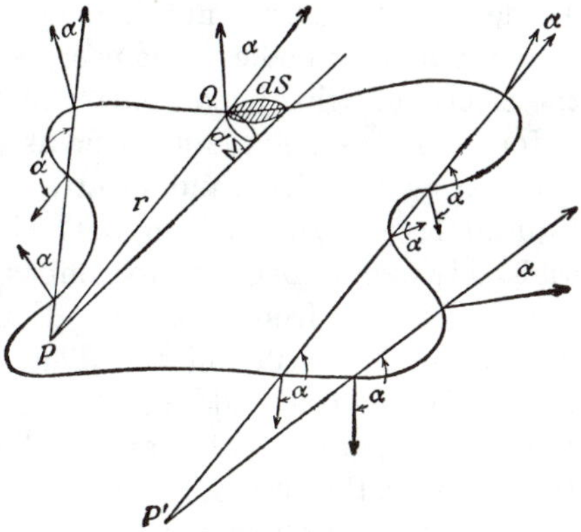

FIG. 1.10.

$d\Sigma = dS \cos \alpha$. The normal component of the intensity at Q due to a charge q at P is

$$E_n = \frac{q\mathbf{r} \cdot \mathbf{n}}{4\pi\epsilon r^3} = \frac{q \cos \alpha}{4\pi\epsilon r^2}$$

The normal component of the flux through dS is defined as

$$dN = \epsilon E_n \, dS = \frac{q \cos \alpha \, dS}{4\pi r^2} = \frac{q \, d\Sigma}{4\pi r^2}$$

But the solid angle subtended by dS at P is $d\Sigma r^{-2} = d\Omega$, so that

$$4\pi \, dN = q \, d\Omega$$

We notice that if the point is inside the surface the cone cuts the surface n times where n is an odd integer, and the angle α will be acute $\frac{1}{2}(n + 1)$ times and obtuse $\frac{1}{2}(n - 1)$ times so that the net value of the flux through the cone is $(q/4\pi) \, d\Omega$. But if the point is outside the surface, n is even and we have the same number of positive and negative values of $d\Omega$ so that the net contribution is zero. To obtain the total flux through the surface surrounding the charge, we integrate the normal component and obtain

$$4\pi \int_S dN = q \int_0^{4\pi} d\Omega \qquad \text{or} \qquad N = q$$

By adding up the flux due to all the charges inside, we arrive at Gauss's electric flux theorem which is: If any closed surface is taken in an electric field and if $\mathbf{E}$ is the electric intensity at any point on the surface and $\mathbf{n}$ the unit outward normal vector to the surface, then

$$\epsilon\int_S \boldsymbol{E}\cdot\boldsymbol{n}\,dS = q \tag{1}$$

where the integration extends over the whole surface which includes the charge q.

If the space outside this surface is not homogeneous but contains various dielectric and conducting bodies, it is necessary to make certain hypotheses as to the electrical behavior of matter in electrostatic fields. It is assumed therefore that, as regards such fields, matter is entirely electrical in nature consisting of positive and negative charges whose fields obey the inverse-square law. With this hypothesis the electrostatic effect of any material body is obtained by superimposing the fields of its constituent charges as calculated from this law. Equation (1) is valid, therefore, no matter what the nature of the dielectric or conducting material outside the surface considered may be because the fields of external charges were considered in its derivation. The above hypothesis is used, explicitly or implicitly, in most treatments of electrostatics.

1.101. Lines of Force from Collinear Charges.—To illustrate the application of this theorem, we shall use it to find the equation of the

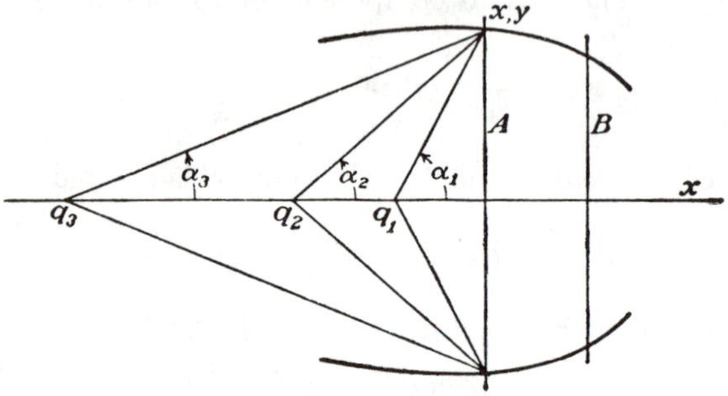

FIG. 1.101.

lines of force about any set of collinear electric charges q_1, q_2, q_3 . . . situated at x_1, x_2, x_3 . . . along the x-axis. From symmetry, no lines of force will pass through the surfaces of revolution generated by rotating the lines of force lying in the xy-plane about the x-axis. Gauss's electric flux theorem, applied to the charge free space inside such a surface between the planes $x = A$ and $x = B$ (Fig. 1.101), tells us that the total normal flux N entering through section A equals that leaving through section B since none passes through the surface. The equation of the surface is therefore obtained by setting N equal to a constant. From 1.05 (1), N equals the sum of the normal fluxes due to each charge and, as we have just found in proving Gauss's electric flux theorem, this is

$$4\pi N = q_1\Omega_1 + q_2\Omega_2 + q_3\Omega_3 + \cdot\cdot\cdot$$

where $\Omega_1, \Omega_2, \Omega_3 \ldots$ are the solid angles which section A subtends at $x_1, x_2, x_3 \ldots$ In terms of the angles $\alpha_1, \alpha_2, \alpha_3 \ldots$ in Fig. 1.101 this is

$$N = \sum_{i=1}^{n} \tfrac{1}{2} q_i (1 - \cos \alpha_i) = C' - \tfrac{1}{2} \sum_{i=1}^{n} q_i \cos \alpha_i$$

Collecting the constants on one side of the equation and substituting for the cosines in terms of the coordinates x, y of a line of force in the xy-plane, we have

$$C = \sum_{i=1}^{n} q_i (x - x_i)[(x - x_i)^2 + y^2]^{-\frac{1}{2}} \tag{1}$$

for the equation of a line of force. This agrees with 1.08 (4).

1.102. Lines of Force at Infinity.—If $\bar{r} = [(x - \bar{x})^2 + y^2]^{\frac{1}{2}}$ we neglect $[(\bar{x} - x_i)/\bar{r}]^n$ when $\bar{x} - x_i \ll \bar{r}$ and $n \geqslant 2$ and write

$$C = \frac{x - \bar{x}}{\bar{r}} \sum_{i=1}^{n} q_i + \left[\frac{1}{\bar{r}} - \frac{(x - \bar{x})^2}{\bar{r}^3}\right] \sum_{i=1}^{n} q_i (\bar{x} - x_i) = \frac{x - \bar{x}}{\bar{r}} \sum_{i=1}^{n} q_i \tag{1}$$

where $\bar{x}$ is the coordinate of the "center of gravity" of the charges. Thus at infinity, the field is the same as if the algebraic sum of the charges were concentrated at their center of gravity. We may extend this rule to noncollinear charges by pairing off members of a group of charges and taking the centers of gravity of each pair by (1), then pairing off these centers of gravity, etc., until we arrive at the center of gravity of the group.

1.11. Potential Maxima and Minima. Earnshaw's Theorem.—Consider a small spherical surface enclosing a point P in an electric field. The average value of the potential over this sphere is

$$\bar{V} = \frac{1}{4\pi r^2} \int_S V\, dS = \frac{1}{4\pi} \int_0^{\pi} \int_0^{2\pi} V \sin\theta \, d\theta \, d\phi$$

Taking derivatives and applying Gauss's electric flux theorem gives

$$\frac{d\bar{V}}{dr} = \frac{1}{4\pi} \int_0^{\pi} \int_0^{2\pi} \frac{dV}{dr} \sin\theta \, d\theta \, d\phi$$
$$= \frac{1}{4\pi r^2} \int_S \frac{dV}{dr} dS = -\frac{q}{4\pi\epsilon r^2}$$

where q is the charge inside the sphere, giving, on integration, the result

$$\bar{V} = \frac{q}{4\pi\epsilon r} + C$$

If $q = 0$, the average value of the potential over the small sphere enclosing P is the same as at P. Hence the theorem that the potential cannot have a maximum or a minimum value at any point in space not occupied by an electric charge. It follows from the definition of potential that to be in stable equilibrium a positive point charge must be at a point of minimum potential and a negative point charge must be at a point of maximum potential, the potential due to the charge itself being excluded. From the theorem just proved, this is impossible, giving us Earnshaw's theorem which states that a charge, acted on by electric forces only, cannot rest in stable equilibrium in an electric field. This shows that if we are to consider matter as purely electrical in nature and as made up of positive and negative charges controlled by electrical forces only, these forces must be different from those operating on a laboratory scale in electrostatics.

1.12. Potential of Electric Double Layer.—We saw in 1.07 that the potential due to a dipole may be obtained from that due to a single charge by differentiation in the direction of the dipole axis. From this, we see that if the potential at P due to an element dS of a surface having a charge density σ is

$$dV = \frac{\sigma}{4\pi\epsilon r}\,dS$$

where r is the distance from dS to P, then

$$\frac{\sigma}{4\pi\epsilon}\,dS\frac{\partial}{\partial n}\left(\frac{1}{r}\right)$$

is the potential at P due to a dipole of strength $\sigma\,dS$ which has the direction of n. Therefore, if we have an electric double layer of moment Φ per unit area, the potential at P due to it is

$$V = \frac{1}{4\pi\epsilon}\int_S \Phi\frac{\partial}{\partial n}\left(\frac{1}{r}\right)dS = \frac{1}{4\pi\epsilon}\int_S \Phi\frac{\boldsymbol{n}\cdot\boldsymbol{r}}{r^3}\,dS \tag{1}$$

But we saw in 1.10 that $\boldsymbol{n}\cdot\boldsymbol{r}\,r^{-3}\,dS = d\Omega$ where $d\Omega$ is the solid angle subtended at P by the element dS. Thus we have

$$V = \frac{1}{4\pi\epsilon}\int \Phi\,d\Omega \tag{2}$$

If the double layer is of uniform strength Ψ, this becomes

$$V = \frac{\Psi\Omega}{4\pi\epsilon} \tag{3}$$

where Ω is the total solid angle subtended by it on the positive side at P.

1.13. Electric Displacement and Tubes of Force.—The product of the capacitivity by the electric intensity occurs frequently and, in an isotropic dielectric, this product is designated as the electric displacement $\boldsymbol{D}$; thus

$$\boldsymbol{D} = \epsilon\boldsymbol{E} \tag{1}$$

In the mks system of units, $\boldsymbol{D}$ is measured in coulombs per square meter and $\boldsymbol{E}$ in volts per meter.

Lines of electric displacement, which are analogous to lines of electric intensity, will have the same direction as the latter in an isotropic dielectric, but if ϵ is greater than ϵ_v, they will be more closely spaced. Taking a small element of area normal to the displacement and drawing lines of displacement through its boundary, we cut out a tubular region in space known as a tube of force. Applying Gauss's electric flux theorem to the charge free space bounded by two normal cross sections of such a tube, we see that, since there is no contribution to the surface integral by the side walls of the tube, the flux entering one end must equal that leaving the other, so that if the areas of the sections are S_1 and S_2 the flux through the tube is

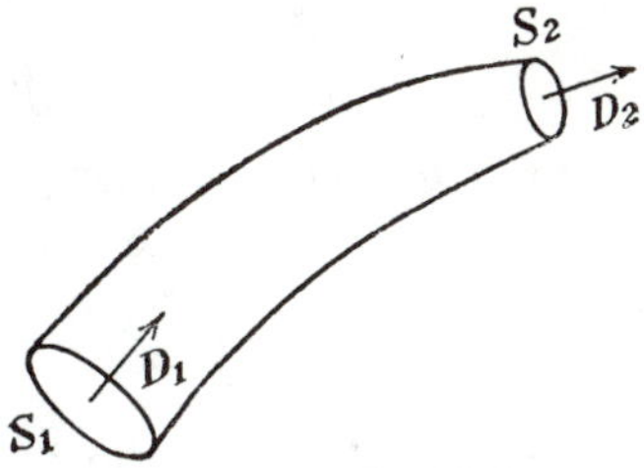

FIG. 1.13.

$$N = S_1D_1 = S_2D_2$$

Thus we define a unit tube of force to be one in which the flux through any section is unity. Many diagrams showing tubes of force will be found in subsequent chapters. Since there are 4π square meters surface on the sphere of unit radius surrounding a point charge q and since the displacement on this surface is $q/(4\pi)$, it follows that there are q unit tubes of force emerging from a charge q. Thus the charge on the end of a unit tube is 1 coulomb.

1.14. Stresses in an Electric Field.—We have introduced the conception of lines and tubes of force merely as an aid toward visualizing an electric field. It is possible to carry this idea considerably further, as Faraday did, and actually look upon these tubes as transmitting electrical forces. Since this is sometimes a most useful method of attacking a problem, let us see if a system of stresses can be postulated to account for observed electrical forces. Let us determine what function of the electric intensity the tension along the tubes of force must be in order to account for the observed Coulomb force between two equal charges of opposite sign at a distance $2a$ apart. Call this function $\Phi(E)$. From 1.05 (1), the intensity in the plane of symmetry is

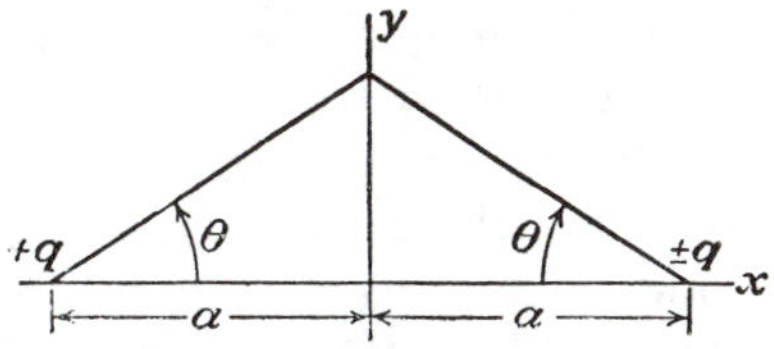

FIG. 1.14.

$$E = \frac{2aq}{4\pi\epsilon(a^2 + y^2)^{\frac{3}{2}}} = \frac{q\cos^3\theta}{2\pi\epsilon a^2}$$

The geometry is shown in Fig. 1.14. The ring element of area is

$$dS = \frac{2\pi a^2 \sin\theta}{\cos^3\theta}\, d\theta$$

Writing the Coulomb force on the left and the tension across the yz-plane on the right and dividing both sides by $2\pi a^2$ give

$$\frac{q^2}{32\pi^2\epsilon a^4} = \frac{1}{2\pi a^2}\int \Phi(E)\,dS = \int_0^{\frac{\pi}{2}} \Phi\left(\frac{q\cos^3\theta}{2\pi\epsilon a^2}\right)\frac{\sin\theta\,d\theta}{\cos^3\theta} \tag{1}$$

Let $x = q/(2\pi\epsilon a^2)$, and expand Φ in powers of E; then

$$\frac{\epsilon x^2}{8} = \frac{1}{2\pi a^2}\int\sum_{n=0}^{\infty} C_n E^n\,dS = \sum_{n=0}^{\infty} C_n x^n \int_0^{\frac{\pi}{2}} \cos^{3(n-1)}\theta\,\sin\theta\,d\theta$$

This equation must hold for all values of q and a and hence for all values of x. Thus $C_n = 0$ except when $n = 2$. For this we have, canceling out x,

$$\frac{\epsilon}{8} = C_2\int_0^{\frac{\pi}{2}} \cos^3\theta\,\sin\theta\,d\theta = \frac{C_2}{4}$$

Thus we find that

$$\Phi(E) = \frac{\epsilon E^2}{2} \tag{2}$$

This is the tension along the lines of force required to account for the Coulomb law of attraction between charges of opposite sign.

It is evident in the preceding case that if there were only a tension along the tubes of force they would shorten as much as possible and all would lie in the line joining the charges. Since we know that in equilibrium they fill the entire space surrounding the charges, there must be some repulsion between them to prevent this. To determine this pressure $\Psi(E)$, let us consider the force between two equal charges of the same sign. This differs from the case just considered because lines of force terminate at infinity. The tension per unit area across a spherical surface of large radius falls off as the inverse fourth power of its radius because of (2) and the inverse-square law. The area of the sphere increases only with the square of its radius so that no force is transmitted this way. Therefore the entire force may be considered as due to the repulsion of the lines of force across the plane of symmetry. From 1.05 (1), the intensity in this plane is

$$E = \frac{2qy}{4\pi\epsilon(a^2+y^2)^{\frac{3}{2}}} = \frac{q\cos^2\theta\,\sin\theta}{2\pi\epsilon a^2} \tag{3}$$

Proceeding as before, we obtain instead of (1)

$$\frac{-q^2}{32\pi^2\epsilon a^4} = \frac{1}{2\pi a^2}\int \Psi(E)\,dS = \int_0^{\frac{\pi}{2}} \Psi\left(\frac{q\cos^2\theta\,\sin\theta}{2\pi\epsilon a^2}\right)\frac{\sin\theta\,d\theta}{\cos^3\theta} \tag{4}$$

For the same reasons as before, $\Psi(E)$ must be of the form C_2E^2, and we evaluate C_2 in the same way which gives

$$-\frac{\epsilon}{8} = C_2 \int_0^{\frac{\pi}{2}} \sin^3 \theta \cos \theta \, d\theta = \frac{C_2}{4}$$

Thus we find that

$$\Psi(E) = -\frac{\epsilon E^2}{2} \tag{5}$$

This is the repulsive force per unit area between adjacent lines of force necessary to account for the Coulomb law of repulsion between charges of the same sign. These results may be put in the equivalent forms

$$\frac{\epsilon E^2}{2} = \frac{\boldsymbol{E} \cdot \boldsymbol{D}}{2} = \frac{D^2}{2\epsilon} \tag{6}$$

Since Φ and Ψ are functions of ϵ and E only, the origin or shape of the field is immaterial and they must have the same form in all fields.

1.15. Gauss's Electric Flux Theorem for Nonhomogeneous Mediums. We are now prepared to extend Gauss's electric flux theorem to isotropic mediums within which the capacitivity varies continuously from place to place. Let us suppose that a point charge q is situated at a point P somewhere inside a closed surface S in such a medium. We shall draw about P a sphere S' so small that ϵ' is constant throughout it. Now take an element dS of the surface S, so small that ϵ is constant over dS, and consider the tube of force, of which dS is a section, that cuts out an element dS' on S' and ends on q. Now apply Gauss's theorem to the uncharged dielectric inside the tube between dS and dS'. Since the normal component of $\boldsymbol{D}$ over the walls is zero, the only contribution to the surface integral is from dS and dS', so that

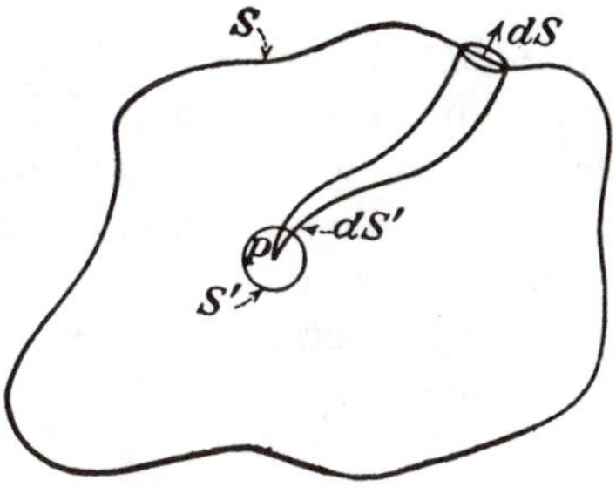

FIG. 1.15.

$$\epsilon' \boldsymbol{E}' \cdot \boldsymbol{n}' \, dS' = \epsilon \boldsymbol{E} \cdot \boldsymbol{n} \, dS$$

Integrating over the two surfaces, we have

$$\epsilon' \int_{S'} \boldsymbol{E}' \cdot \boldsymbol{n}' \, dS' = \int_S \epsilon \boldsymbol{E} \cdot \boldsymbol{n} \, dS$$

since ϵ' is the same for all elements dS'. But we have already proved in 1.10 that the left-hand integral is q, so that

$$\int_S \epsilon \boldsymbol{E} \cdot \boldsymbol{n} \, dS = q \tag{1}$$

where ϵ and $\boldsymbol{E}$ are both functions of position. This may be extended as before, so that q includes all the charges inside S.

A complicated field may result from simple sources. The application of (1) in such a case is often greatly simplified by calculating the flux from each source separately and adding the result. Thus

$$\int(\boldsymbol{E}_1 + \boldsymbol{E}_2 + \cdots + \boldsymbol{E}_n)\cdot\boldsymbol{n}\,dS = \int\boldsymbol{E}_1\cdot\boldsymbol{n}\,dS + \int\boldsymbol{E}_2\cdot\boldsymbol{n}\,dS + \cdots + \int\boldsymbol{E}_n\cdot\boldsymbol{n}\,dS \qquad (2)$$

This permits many problem answers to be written down by inspection.

1.16. Boundary Conditions and Stresses on the Surface of Conductors. When the electric charge on a conductor is in static equilibrium, there can be no fields in the interior or along the surface. Otherwise, since by definition charges are free to move on a conductor, there would be a movement of the charges, contradicting the postulated equilibrium. It follows that the whole conductor is at one potential and that lines of force meet its surface normally and end there.

The charge density on the surface is σ coulombs per square meter, and there is one unit tube of force, pointing outward if σ is positive, from each charge; so we have

$$\boldsymbol{D} = \epsilon\boldsymbol{E} = \sigma \qquad (1)$$

Since lines of force leave the conducting surface normally, they can intersect each other only at infinitely sharp points or edges. We have seen that this occurs at a mathematical point or edge. The converse also holds. At the bottom of a sharp V-shaped groove or conical depression, $\boldsymbol{D}$ and σ are zero.

We have seen in 1.14 that there is a tension along these lines of force of the amount

$$F = \frac{D^2}{2\epsilon} = \frac{\sigma^2}{2\epsilon} \qquad (2)$$

and so this is the outward force per square meter on a charged conducting surface. This force is independent of the sign of the charge.

It should be noted that we have neglected the hydrostatic forces that may be present in the dielectric owing to its tendency to expand or contract in an electric field. An expression including such forces will be derived later in 2.09.

1.17. Boundary Conditions and Stresses on the Surface of a Dielectric.—Let us apply Gauss's electric flux theorem to the small disk-shaped volume whose flat surfaces of area dS lie on opposite sides of the plane boundary between two dielectrics ϵ' and ϵ'' (Fig. 1.17*a*). The disk is so thin that the area of the edges is negligible compared with that of the faces. If there is no free charge on the surface and the normal components of displacement are D'_n and D''_n, we find from 1.15 that

$$D'_n\,dS = D''_n\,dS \qquad \text{or} \qquad D'_n = D''_n \qquad (1)$$

The stress on this surface due to the normal components of the displacement must be the difference between the stresses on the two sides, so that, from 1.14 (6), we have

$$T_n = \frac{D_n'^2}{2\epsilon'} - \frac{D_n''^2}{2\epsilon''} = -\frac{D_n'^2(\epsilon' - \epsilon'')}{2\epsilon'\epsilon''} = -\frac{D_n'^2}{2\epsilon_v}\frac{K' - K''}{K'K''} \tag{2}$$

Let us consider the work required to take a unit charge around the path shown in Fig. 1.17*b* in which the length of path normal to the boundary is vanishingly small. Starting at any point, the work done in taking

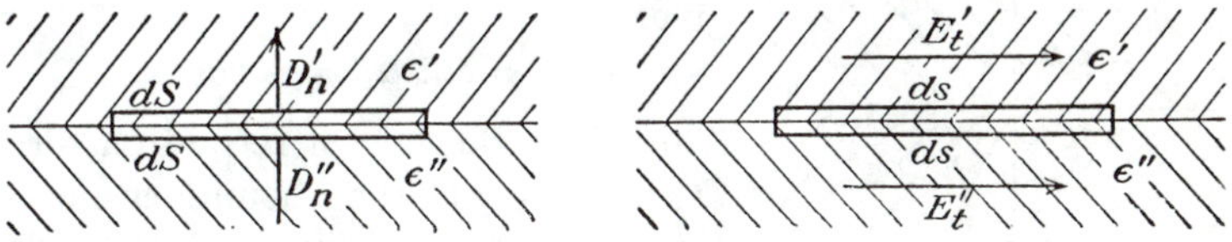

FIG. 1.17*a*. FIG. 1.17*b*.

a unit charge around the path is, if energy is to be conserved, zero, so that $E_t'\,ds = E_t''\,ds$ or

$$E_t' = E_t'' \tag{3}$$

The pressure on the interface will be the difference in the pressure on the two sides; thus, from 1.14 (5), we have

$$P_n = \tfrac{1}{2}\epsilon' E_t'^2 - \tfrac{1}{2}\epsilon'' E_t''^2 = \tfrac{1}{2}E_t'^2(\epsilon' - \epsilon'') = \tfrac{1}{2}E_t'^2\epsilon_v(K' - K'') \tag{4}$$

We may state then that at the uncharged interface between two dielectrics, the normal component of the displacement and the tangential component of the intensity are continuous. In terms of the potential these boundary conditions may be written

$$\epsilon'\frac{\partial V'}{\partial n} = \epsilon''\frac{\partial V''}{\partial n} \qquad \text{or} \qquad K'\frac{\partial V'}{\partial n} = K''\frac{\partial V''}{\partial n} \tag{5}$$

$$V' = V'' \tag{6}$$

where V' and V'' are the potentials in ϵ' and ϵ''. Equation (6) implies that the zero of potential is chosen in both mediums, so that at some point on the boundary $V' = V''$. It then follows, by integration of (3), that (6) holds for all points on the boundary.

The normal stresses directed from ϵ'' to ϵ' on the interface between two dielectrics may also be written, from (2) and (4),

$$F_n = T_n - P_n = -\frac{K' - K''}{2\epsilon_v K'}\left\{\frac{D_t'^2}{K'} + \frac{D_n'^2}{K''}\right\} = -\frac{\epsilon' - \epsilon''}{2\epsilon'}\left(\frac{D_t'^2}{\epsilon'} + \frac{D_n'^2}{\epsilon''}\right) \tag{7}$$

The foregoing formulas do not take account of the fact that some dielectrics tend to contract or expand in the presence of an electric field. In such a medium, there will be an additional force of a hydrostatic nature acting on the surface. An expression including this force will be derived later, in 2.09.

At the boundary between two isotropic dielectrics, the lines of force and the lines of displacement will be refracted in the same way. Let the angle between $\boldsymbol{E}_1$ or $\boldsymbol{D}_1$, in the dielectric ϵ_1, and the normal to the boundary be α_1, and let the corresponding angle in ϵ_2 be α_2 (Fig. 1.17*c*). Then from (1) and (3), we have

$$D_1 \cos \alpha_1 = \epsilon_1 E_1 \cos \alpha_1 = D_2 \cos \alpha_2 = \epsilon_2 E_2 \cos \alpha_2$$
$$D_1 \epsilon_1^{-1} \sin \alpha_1 = E_1 \sin \alpha_1 = D_2 \epsilon_2^{-1} \sin \alpha_2 = E_2 \sin \alpha_2$$

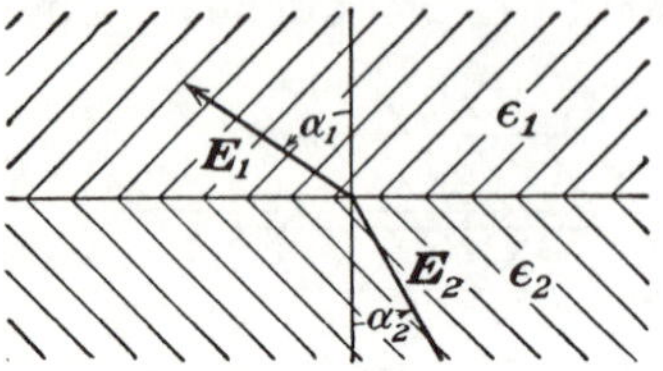

Fig. 1.17*c*.

Dividing the first equation by the second gives

$$\epsilon_1 \cot \alpha_1 = \epsilon_2 \cot \alpha_2 \qquad (8)$$

This gives the law of refraction for both $\boldsymbol{D}$ and $\boldsymbol{E}$ at the boundary between two isotropic mediums with different capacitivities.

1.18. Displacement and Intensity in Solid Dielectric.—The capacitivity was first introduced in the statement of Coulomb's law [(1.02 (1)] as a factor that depends on the medium in which the electric force is measured. This hypothetical measurement is difficult to conceive in a solid dielectric, but by using the boundary conditions just derived we may devise a method of determining the displacement and intensity and hence the capacitivity in such a medium.

To determine the displacement and intensity in a solid dielectric, let us excavate a small evacuated disk-shaped cavity whose thickness is very small compared with its radius. The intensity inside the cavity

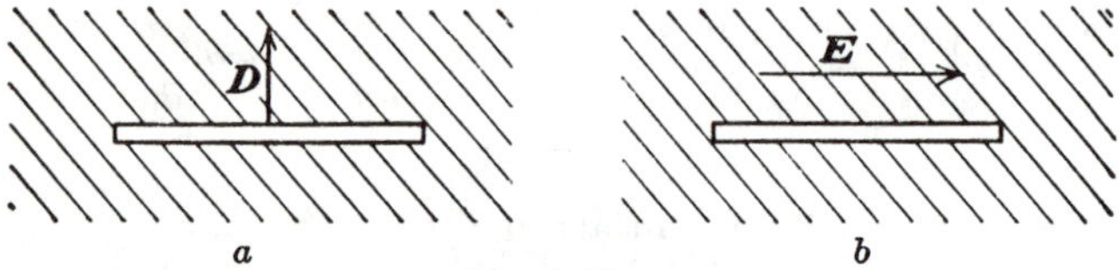

Fig. 1.18.

far from the edges will be determined entirely by the boundary conditions over the flat surfaces, as shown in Fig. 1.18*a*.

To determine the displacement, we orient the cavity so that the intensity inside is normal to the flat faces in Fig. 1.18*a*. From 1.17 (1), we know that the displacement in the dielectric equals that in the cavity. Thus by measuring the intensity in the cavity and multiplying by ϵ_v, we determine the displacement in the dielectric.

To determine the intensity in the solid dielectric, we orient a long thin cylindrical cavity so that the intensity inside is parallel to the axis as in Fig. 1.18*b*. Then from 1.17 (3), we know that the intensity inside is the same as that in the solid dielectric. The ratio of the displacement to the intensity, the cavities being small compared with all dimensions of the dielectric mass and the external field remaining constant, gives the capacitivity of the medium.

The values of displacement and intensity found in this way certainly do not represent the actual fields on a molecular scale found inside a dielectric but are some type of average. Other types of average might give results contradicting macroscopic observations.

1.19. Crystalline Dielectrics.—We shall now use the experimental tests of the last article to find the relation between displacement $\boldsymbol{D}$ and intensity $\boldsymbol{E}$ in a homogeneous crystalline dielectric. Let us cut three plane parallel plates of thickness d from the positive x, y, and z faces of a large cube of this material. We now deposit a conducting layer on each face of each plate and apply a difference of potential V. If we consider areas of the plate sufficiently far from its edges we see that the potential boundary conditions on all such areas of all plates are identical so that the potential distribution in the central areas of all plates is the same. Thus the equipotentials near the center of each plate are parallel to its faces and from 1.06 (4) the intensity $\boldsymbol{E}$ equals V/d. By experimenting with our disk-shaped cavity which is taken small compared with d so that it does not disturb the charge distribution on the conducting faces we find that $\boldsymbol{D}$ is proportional to $\boldsymbol{E}$ but not in the same direction. Thus we have in the x, y and z plates, respectively,

$$\begin{aligned} (D_x)_x &= \epsilon_{11}E_x, & (D_y)_x &= \epsilon_{12}E_x, & (D_z)_x &= \epsilon_{13}E_x \\ (D_x)_y &= \epsilon_{21}E_y, & (D_y)_y &= \epsilon_{22}E_y, & (D_z)_y &= \epsilon_{23}E_y \\ (D_x)_z &= \epsilon_{31}E_z, & (D_y)_z &= \epsilon_{32}E_z, & (D_z)_z &= \epsilon_{33}E_z \end{aligned} \tag{1}$$

Even when $\boldsymbol{E}$ is the same in all plates the normal components of $\boldsymbol{D}$ will in general differ. However in all cases it is found experimentally that

$$(D_x)_yE_x = (D_y)_xE_y, \qquad (D_x)_zE_x = (D_z)_xE_z, \qquad (D_y)_zE_y = (D_z)_yE_z \tag{2}$$

Superimposing the results in (1) gives

$$\begin{aligned} D_x &= \epsilon_{11}E_x + \epsilon_{21}E_y + \epsilon_{31}E_z \\ D_y &= \epsilon_{12}E_x + \epsilon_{22}E_y + \epsilon_{32}E_z \\ D_z &= \epsilon_{13}E_x + \epsilon_{23}E_y + \epsilon_{33}E_z \end{aligned} \tag{3}$$

Comparing (2) and (1), we see that

$$\epsilon_{12} = \epsilon_{21}, \qquad \epsilon_{13} = \epsilon_{31}, \qquad \epsilon_{23} = \epsilon_{32} \tag{4}$$

Thus the single simple factor ϵ connecting the quantities $\boldsymbol{D}$ and $\boldsymbol{E}$ in the case of an isotropic medium becomes, for a crystal, a quantity known as a symmetrical tensor, having nine components of which six are different.

Let us see if it is possible to orient our axes so as to simplify (3). The product $\boldsymbol{E} \cdot \boldsymbol{D}$, being a scalar quantity, must be independent of the choice of axes. Writing out in terms of the components of $\boldsymbol{E}$, we have, from (3), using (4),

$$\boldsymbol{E} \cdot \boldsymbol{D} = \epsilon_{11}E_x^2 + \epsilon_{22}E_y^2 + \epsilon_{33}E_z^2 + 2\epsilon_{12}E_xE_y + 2\epsilon_{13}E_xE_z + 2\epsilon_{23}E_yE_z \tag{5}$$

This is the equation of a quadric surface in E_x, E_y, and E_z. We can vary these coordinates in any way we please, keeping $E_x^2 + E_y^2 + E_z^2$ constant, by rotating our axes. We may therefore orient our axes so that the cross products E_xE_y, E_xE_z, and E_yE_z disappear. The equation of the quadric referred to the new axes is

$$\boldsymbol{E} \cdot \boldsymbol{D} = \epsilon_1 E_x^2 + \epsilon_2 E_y^2 + \epsilon_3 E_z^2 \tag{6}$$

The components of the displacement referred to these axes are

$$D_x = \epsilon_1 E_x \qquad D_y = \epsilon_2 E_y \qquad D_z = \epsilon_3 E_z \tag{7}$$

The directions of the coordinate axes in (7) are called the electrical axes of the crystal. If ϵ_1, ϵ_2, ϵ_3 are all the same, we have an isotropic medium; if only two are the same, we have a uniaxial crystal; and if all three are different, we have a biaxial crystal.

Problems

Problems marked *C* are taken from the Cambridge examination questions as reprinted by Jeans by permission of the Cambridge University Press.

1. The work required to bring a point charge q up to the centers, which are a distance b apart, of two thin parallel conducting coaxial rings, each of radius a, is W_1 and W_2, respectively. Show that the charges on the rings are

$$Q_{1,2} = \frac{4\pi\epsilon a}{b^2 q}(a^2 + b^2)^{\frac{1}{2}}[(a^2 + b^2)^{\frac{1}{2}}W_{1,2} - aW_{2,1}]$$

2. Four equal parallel line charges lie at the corners of a square prism, those at the ends of one diagonal being positive and the others negative. Find the fraction of the total flux that enters the prism.

3. There is a charge q at $x = +a$, $y = 0$, $z = 0$. Find how large a charge must be placed at $x = -a$, $y = 0$, $z = 0$ so that a flux N passes in the positive direction through the circle $x = 0$, $y^2 + z^2 = a^2$.

4. Two thin concentric coplanar rings of radii 1 and 2 carry charges $-Q$ and $+(27)^{\frac{1}{2}}Q$, respectively. Show that the only neutral points in the field are at $x = 0$ and $\pm 2^{-\frac{1}{2}}$.

5. Show that the equation of the lines of force between two parallel line charges, of strength q and $-q$ per unit length, at $x = a$ and $x = -a$, in terms of the flux per unit length N, between the line of force and the positive x-axis is

$$\left[y - a\cot\left(\frac{2\pi N}{q}\right)\right]^2 + x^2 = a^2 \csc^2\left(\frac{2\pi N}{q}\right)$$

6*C*. Charges $+4q$, $-q$ are placed at the points A, B, and C is the point of equilibrium. Prove that the line of force that passes through C meets AB at an angle of 60° at A and at right angles at C. Find the angle at A between AB and the line of force that leaves B at right angles to AB.

(Write potential for small region about C in polar coordinates with origin at C.)

7*C*. Two positive charges q_1 and q_2 are placed at the points A and B, respectively. Show that the tangent at infinity to the line of force, which starts from q_1 making an angle α with BA produced, makes an angle

$$2\sin^{-1}\left(q_1^{\frac{1}{2}}(q_1 + q_2)^{-\frac{1}{2}} \sin\frac{\alpha}{2}\right)$$

with BA and passes through the point C in AB such that $AC{:}CB = q_2{:}q_1$.

8C. Point charges $+q$, $-q$ are placed at the points A, B. The line of force that leaves A making an angle α with AB meets the plane that bisects AB at right angles in P. Show that

$$\sin\frac{\alpha}{2} = 2^{\frac{1}{2}} \sin\frac{1}{2}\angle PAB$$

9C. If any closed surface is drawn not enclosing a charged body or any part of one, show that at every point of a certain closed line on the surface it intersects the equipotential surface through the point at right angles.

10C. Charges $3q$, $-q$, $-q$ are placed at A, B, C, respectively, where B is the middle point of AC. Draw a rough diagram of the lines of force, show that a line of force that starts from A making an angle α with $AB > \cos^{-1}(-\frac{1}{3})$ will not reach B or C, and show that the asymptote of the line of force for which $\alpha = \cos^{-1}(-\frac{2}{3})$ is at right angles to AC.

11C. If there are three electrified points A, B, C in a straight line, such that $AC = f$, $BC = a^2/f$, and the charges are q, $-qa/f$, and $4\pi\epsilon Va$, respectively, show that there is always a spherical equipotential surface, and discuss the position of the equilibrium points on line ABC if $4\pi\epsilon V = q(f + a)/(f - a)^2$ and if $4\pi\epsilon V = q(f - a)/(f + a)^2$.

12C. A and C are spherical conductors with charges $q + q'$ and $-q$, respectively. Show that there is either a point or a line of equilibrium depending on the relative size and positions of the spheres and on q'/q. Draw a diagram for each case, giving the lines of force and the sections of the equipotentials by a plane through the centers.

13C. An electrified body is placed in the vicinity of a conductor in the form of a surface of anticlastic curvature. Show that at that point of any line of force passing from the body to the conductor, at which the force is a minimum, the principal curvatures of the equipotential surface are equal and opposite.

14C. If two charged concentric shells are connected by a wire, the inner one is wholly discharged. If the force law were $r^{-(2+p)}$, prove that there would be a charge B on the inner shell such that if A were the charge on the outer shell and f, g the sum and difference of the radii $2gB = -Ap[(f - g)\ln(f + g) - f\ln f + g\ln g]$, approximately.

15C. Three infinite parallel wires cut a plane perpendicular to them in the angular points A, B, C of an equilateral triangle and have charges q, q, $-q'$ per unit length, respectively. Prove that the extreme lines of force which pass from A to C make at starting angles $(2q - 5q')\pi(6q)^{-1}$ and $-(2q + q')\pi(6q)^{-1}$ with AC, provided that $q' \ngtr 2q$.

16C. A negative point charge $-q_2$ lies between two positive point charges q_1 and q_3 on the line joining them and at distances α, β from them, respectively. Show that if the magnitudes of the charges are given by $q_1\beta^{-1} = q_3\alpha^{-1} = q_2\lambda^3(\alpha + \beta)^{-1}$ and if $1 < \lambda^2 < (\alpha + \beta)^2(\alpha - \beta)^{-2}$, there is a circle at every point of which the force vanishes. Determine the general form of the equipotential surface on which this circle lies.

17C. Charges of electricity q_1, $-q_2$, q_3, $(q_3 > q_1)$ are placed in a straight line, the negative charge being midway between the other two. Show that, if $4q_2$ lie between $(q_3^{\frac{1}{3}} - q_1^{\frac{1}{3}})^3$ and $(q_3^{\frac{1}{3}} + q_1^{\frac{1}{3}})^3$, the number of unit tubes of force that pass from q_1 to q_2 is $\frac{1}{2}(q_1 + q_2 - q_3) + 3(2^{\frac{1}{3}}4)^{-1}(q_3^{\frac{2}{3}} - q_1^{\frac{2}{3}})(q_1^{\frac{2}{3}} - 2^{\frac{1}{3}}q_2^{\frac{2}{3}} + q_3^{\frac{2}{3}})^{\frac{1}{2}}$.

18C. An infinite plane is charged to surface density σ, and P is a point distant $\frac{1}{2}$ in. from the plane. Show that of the total intensity $\frac{1}{2}\sigma/\epsilon$ at P half is due to the charges at points that are within 1 in. of P and half to the charges beyond.

19C. A disk of vulcanite (nonconducting), of radius 5 in., is charged to a uniform surface density by friction. Find the electric intensities at points on the axis of the disk distant, respectively, 1, 3, 5, 7 in. from the surface.

20. Two parallel coaxial circular rings of radii a and b carry uniformly distributed charges Q_1 and Q_2. The distance between their planes is c. Show that the force between them is

$$F = \frac{ck^3Q_1Q_2}{16\pi^2\epsilon(ab)^{\frac{3}{2}}}\left(\frac{E}{1-k^2}\right) \qquad \text{where } k^2 = \frac{4ab}{c^2 + (a+b)^2}$$

and E is a complete elliptic integral of modulus k.

21. Show that at a great distance the field of a ring charge $-Q$ of radius b concentric and coplanar with a ring charge Q of radius c is identical with that of a linear quadrupole in which the end charges $-Q$ are at a distance a from the center charge $2Q$, provided that $b^2 - c^2 = 4a^2$.

References

Useful treatments of the material of this chapter occur in the following books:

ABRAHAM, M., and R. BECKER: "Classical Electricity and Magnetism," Blackie, 1932. Uses vector notation and treats simple cases.

CORSON, D. R., and P. LORRAIN: "Introduction to Electromagnetic Fields and Waves," Freeman, 1962. Well-illustrated dielectric discussion.

DURAND, E.: "Electrostatique et Magnetostatique," Masson et Cie, 1953. Complete and well-illustrated treatment.

GEIGER-SCHEEL: "Handbuch der Physik," Vol. XII, Berlin, 1927.

JACKSON, J. D.: "Classical Electrodynamics," Wiley, 1962. Chapters I and IV are especially pertinent.

JEANS, J. H.: "The Mathematical Theory of Electricity and Magnetism," Cambridge, 1925. Uses long notation but gives comprehensive treatment.

MASON, M., and W. WEAVER: "The Electromagnetic Field," University of Chicago Press, 1929, and Dover. Gives an excellent treatment using vectors.

MAXWELL, J. C.: "Electricity and Magnetism," 3d ed. (1891), Dover, 1954. Gives extensive treatment using long notation with many good field drawings.

OWEN, G. E.: "Electromagnetic Theory," Allyn and Bacon, 1963. Has clear and well-illustrated discussion of multipoles.

PANOFSKY, W. K. H., and MELBA PHILLIPS: "Classical Electricity and Magnetism," Addison-Wesley, 1962. Chapter I discusses clearly the material in this chapter.

STRATTON, J. A.: "Electromagnetic Theory," McGraw-Hill, 1941. Gives extensive mathematical treatment based on Maxwell's equations.

WEBSTER, A. G.: "Electricity and Magnetism," Macmillan, 1897. Gives clear treatment using long notation.

WIEN-HARMS: "Handbuch der Experimentalphysik," Vol. X, Leipzig, 1930.

CHAPTER II

CAPACITORS, DIELECTRICS, SYSTEMS OF CONDUCTORS

2.00. Uniqueness Theorem.—Before trying to find the potentials in a system of conductors when the charges are given or vice versa, it is well to be sure that there is only one correct solution.

First let us assume that a surface density σ on the conductors produces the same potentials as a different surface density σ'. The potential at a point P on the surface of one of the conductors due to the surface density $\sigma - \sigma'$ will be, from 1.06 (6),

$$V_P = \int_S \frac{\sigma - \sigma'}{4\pi\epsilon r}\,dS = \frac{1}{4\pi\epsilon}\int_S \frac{\sigma}{r}\,dS - \frac{1}{4\pi\epsilon}\int_S \frac{\sigma'}{r}\,dS$$

where r is the distance from P to the surface element dS and the integration covers the surface of all conductors. The two integrals on the right were initially assumed equal, so that V_P is zero. The surface density $\sigma - \sigma'$ therefore makes all conductors at zero potential, so that there is no electric field and $\sigma - \sigma'$ is zero. Therefore $\sigma = \sigma'$, and the distribution is unique. We have proved then that only one distribution of electric charge will give a specified potential to every conductor in an electric field.

Now let us assume that a surface density σ on the conductors gives the same total charge Q in each conductor as a different surface density σ'. The surface density $\sigma - \sigma'$ will then give a total charge zero on each conductor. The density $\sigma - \sigma'$ may be zero over the whole conductor or negative on some areas and positive on others. If the latter is the case, the tubes of force ending on negative areas must originate at points of higher potential and those ending on positive areas at points of lower potential. This reasoning applies to every conductor in the field so that, with the density $\sigma - \sigma'$, no conductor can be at the maximum or the minimum of potential. Therefore, all of them must be at the same potential, and the field vanishes, which gives $\sigma = \sigma'$. We have now proved that there is only one surface charge distribution which will give a specified total charge to each conductor in an electric field.

2.01. Capacitance.—One consequence of the fact (1.14) that all electrical stresses in a medium in an electric field depend on the intensity in the same way is that the equilibrium is undisturbed if the intensity everywhere in the field is changed by the same factor. Thus if we double the surface density at every point in a system of charged conductors, the configuration of the field is unchanged but the intensity is doubled, and hence the work required to take a small charge from one conductor to another is also doubled. This constant charge to potential ratio of an isolated conductor is its capacitance, and the reciprocal ratio is its elastance. These terms are not precise when other conductors are in the field unless all are *both* earthed and uncharged. If Q is the charge in coulombs, C the capacitance in farads, S the elastance in darafs, and V the potential in volts, then the definitive equations are

$$Q = CV \qquad V = SQ \tag{1}$$

Two insulated conductors near together constitute a simple capacitor. If these two conductors are given equal and opposite charges, the capacitance of the capacitor is the ratio of the charge on one to the difference of potential between them. The ratio is always taken so as to make the capacitance a positive quantity. Thus, for a capacitor we have

$$Q = C(V_1 - V_2) \qquad V_1 - V_2 = SQ \tag{2}$$

2.02. Capacitors in Series and Parallel.—If we take n simple uncharged capacitors, connect one member of each capacitor to a terminal A and the other member to a terminal B, as shown in Fig. 2.02a, and then apply a potential V between A and B, we have

$$Q_1 = C_1V, \qquad Q_2 = C_2V, \quad . \ . \ . \ , \qquad Q_n = C_nV$$

where $Q_1, Q_2, \ldots, Q_n$ are the charges on the capacitors whose capacitances are $C_1, C_2, \ldots, C_n$, respectively. The total charge is then

$$Q = Q_1 + Q_2 + \cdots + Q_n = V(C_1 + C_2 + \cdots + C_n)$$

Thus these capacitors, which are said to be connected in parallel or "multiple arc," behave like a single capacitor whose capacitance is

$$C = C_1 + C_2 + \cdots + C_n \tag{1}$$

Let us take n simple uncharged capacitors, connecting the first member of capacitor 1 to A and the second member to the first member of capacitor 2, then connecting the second member of 2 to the first member of 3, and so forth, finally connecting the second member of the nth capacitor to B as shown in Fig. 2.02b. These capacitors are now said to be connected in series or "cascade." The application of a potential between A and B gives

$$V = V_1 + V_2 + \cdots + V_n = S_1Q_1 + S_2Q_2 + \cdots + S_nQ_n$$

where V_i is the potential across the ith capacitor. Since each pair of connected conductors has remained insulated, its net charge is zero. Hence, if all the tubes of force leaving one member of a capacitor end on the other member of the same capacitor, we have

$$Q_1 = Q_2 = \cdots = Q_n$$

so that the charge factors out of the right side of the equation. Thus these capacitors, connected in series, behave like a single capacitor of capacitance C or elastance S where

$$S = S_1 + S_2 + \cdots + S_n, \qquad \frac{1}{C} = \frac{1}{C_1} + \frac{1}{C_2} + \cdots + \frac{1}{C_n} \tag{2}$$

We observe that, in general, it will not be possible to confine the fields in the manner assumed so that this formula is an approximation. The additional "distributed" capacitance due to stray fields is usually negligible in cases where the members of a capacitor are close together, and when the capacitivity of the region between is high compared with that outside the capacitor. In the case of air capacitors with widely separated members, this formula may be worthless

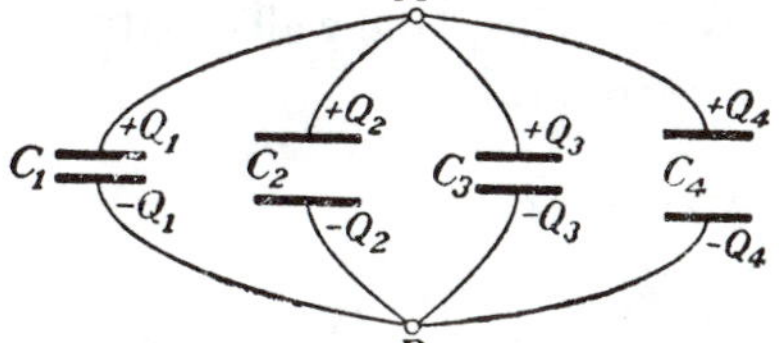

Fig. 2.02*a*.

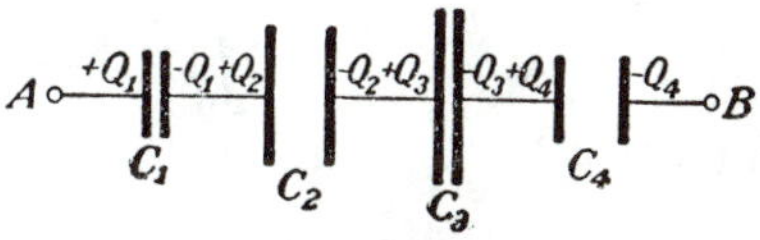

Fig. 2.02*b*.

2.03. Spherical Capacitors.—Consider a pair of concentric spherical conducting shells, the inner, of radius a, carrying a charge $+Q$ and the outer, of radius b, carrying a charge $-Q$, the space between being filled by a homogeneous isotropic dielectric of capacitivity ϵ. From symmetry, the electric displacement must be directed radially and its magnitude can depend only on r. Hence, applying Gauss's electric flux theorem to a concentric spherical surface of radius r where $b > r > a$, we have

$$\int_S \epsilon \mathbf{E} \cdot \mathbf{n}\, dS = 4\pi r^2 \epsilon E = Q$$

so that

$$E = -\frac{\partial V}{\partial r} = \frac{Q}{4\pi\epsilon r^2}$$

The potential difference between the spheres is therefore

$$V_a - V_b = -\frac{Q}{4\pi\epsilon}\int_b^a \frac{dr}{r^2} = \frac{Q}{4\pi\epsilon}\left(\frac{1}{a} - \frac{1}{b}\right) = \frac{b-a}{4\pi\epsilon ab}Q$$

This gives the capacitance of a spherical capacitor to be

$$C = \frac{4\pi\epsilon ab}{b-a} \tag{1}$$

The capacitance of a single sphere of radius a immersed in a dielectric of capacitivity ϵ can be obtained from (1) by letting $b \to \infty$ giving

$$C = 4\pi\epsilon a \tag{2}$$

It should be noted that in deriving (1) it is assumed that there is no charge on the outside of b, which requires that b be at zero potential. If this is not the case, the additional capacitance between the outside of b and infinity, computed from (2), must be considered.

2.04. Cylindrical Capacitors.—Consider a pair of concentric circular conducting cylinders of infinite length, the inner, of radius a, carrying a charge Q per unit length and the outer, of radius b, carrying a charge $-Q$ per unit length, the space between being filled with a homogeneous isotropic medium of capacitivity ϵ. From symmetry, the electric displacement must be directed radially outward from the axis and lie in a plane normal to the axis, and its magnitude must depend only on r. Apply Gauss's electric flux theorem to the volume enclosed by two planes, set normal to the axis and one meter apart, and the concentric circular cylinder of radius r when $b > r > a$. The plane walls contribute nothing to the surface integral; we have therefore

$$\int_S \epsilon \mathbf{E} \cdot \mathbf{n}\, dS = 2\pi r \epsilon E = Q$$

so that

$$E = -\frac{\partial V}{\partial r} = \frac{Q}{2\pi\epsilon r} \qquad (1)$$

The potential difference between the cylinders is therefore

$$V_a - V_b = -\frac{Q}{2\pi\epsilon}\int_b^a \frac{dr}{r} = -\frac{Q}{2\pi\epsilon}\ln\frac{a}{b} \qquad (1.1)$$

Thus the capacitance per unit length of a long cylindrical capacitor is

$$C = \frac{2\pi\epsilon}{\ln (b/a)} \qquad (2)$$

If we let $b \to \infty$, we see that $C \to 0$. Therefore, if there is a finite charge per unit length on a circular cylinder of finite radius and infinite length, the potential difference between its surface and infinity is infinite. Since physically we deal only with cylinders of finite length, this difficulty does not arise, but it indicates that the results of this article apply only where the distance to the surface of the cylinder is small compared with the distance to the ends.

2.05. Parallel-plate Capacitors.—When two infinite parallel conducting planes, carrying charges $+Q$ and $-Q$, are a distance a apart, the space between being filled with a homogeneous isotropic dielectric, we see from symmetry that the field between them must be uniform and normal to the plates. If σ is the charge per square meter, then there must be, from 1.13, σ unit tubes leaving every square meter of the plates. Thus we have for the displacement and intensity between the plates the relation

$$D = \epsilon E = -\epsilon \frac{\partial V}{\partial x} = \sigma$$

The difference of potential between the plates is

$$V_2 - V_1 = \frac{\sigma}{\epsilon} \int_0^a dx = \frac{\sigma a}{\epsilon} \tag{1.1}$$

Therefore the capacitance per unit area is ϵ/a, and the capacitance of an area A is

$$C = \frac{\epsilon A}{a} \tag{1}$$

In practice, the field will be uniform only far from the edges of the plate, so that this formula is an approximation which is good if a is small compared with all surface dimensions of the plate and still better if, in addition, the capacitivity of the region between the plates is higher than that of the region beyond the edges.

2.06. Guard Rings.—The derivation of formula 2.04 (2) for the capacitance per unit length of a cylindrical capacitor and that of 2.05 (1) for the capacitance of a parallel-plate capacitor both involve the hypothesis of conductors of infinite dimensions. To permit the application of these formulas to actual capacitors, a device known as a guard ring is used. For the cylindrical capacitors, shown in Fig. 2.06a, the end sections of one of the members are separated from the center section by narrow cracks but are maintained at the same potential. Thus the distorted field near the edges does not affect the center section, except for a very small effect produced by the cracks, so that the charge on this section is related to the potential difference by 2.04 (1.1).

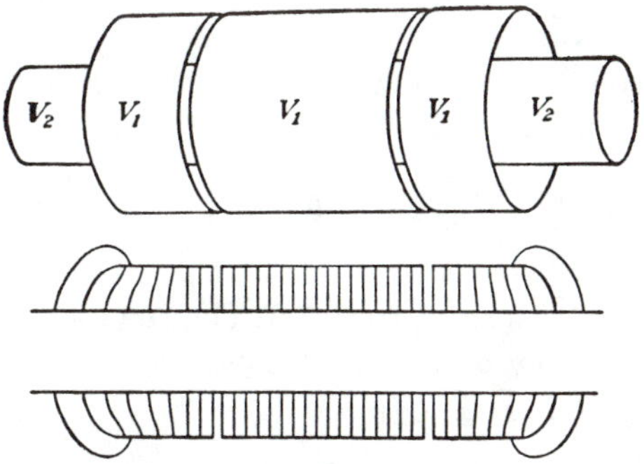

Fig. 2.06a.

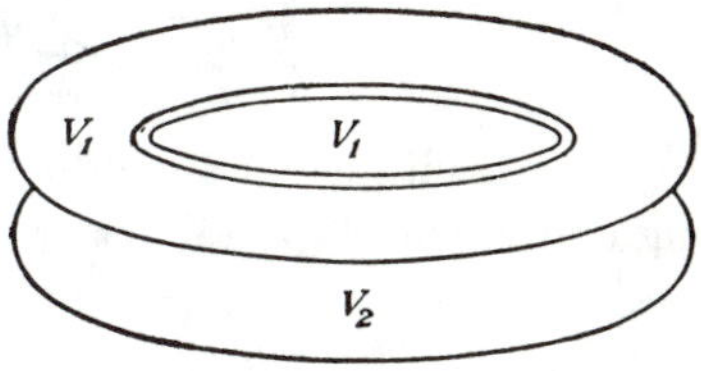

Fig. 2.06b.

A similar arrangement is used for the parallel-plate capacitor by leaving a narrow gap between the central section of one plate and the area surrounding it, maintaining both at the same potential as shown in Fig. 2.06b. The field between the central areas is now uniform except for the negligible effect of the small gap, and the charge on this section is related to the potential difference by 2.05 (1.1).

2.07. Energy of a Charged Capacitor.—We can compute the mutual energy of any system of charges directly from the definition of potential. The work in joules to put the jth charge in place will be, from 1.06 (3),

$$W_j = q_jV_j = \frac{q_j}{4\pi\epsilon}\sum_{i=1}^{n}\frac{q_i}{r_{ij}} \qquad \text{where } i \neq j$$

The total work to put all charges in place is

$$W = \frac{1}{8\pi\epsilon}\sum_{i=1}^{n}\sum_{j=1}^{n}\frac{q_iq_j}{r_{ij}} \qquad \text{where } i \neq j \tag{1}$$

The factor $\frac{1}{2}$ is necessary because the summation includes not only the work done in bringing the ith charge into its position in the field of the jth charge but also that done in bringing the jth charge into the field of the ith charge, which is the same. If V_i is the potential at the spot where the ith charge is situated, this may be written, from 1.06 (3),

$$W = \tfrac{1}{2}\sum_{i=1}^{n}q_iV_i \tag{2}$$

When all charges lie on the same conductor a, they are at the same potential and if their sum is Q_a, we may write

$$W_a = \frac{1}{2}\sum_{\text{over } a}q_iV_i = \frac{V_a}{2}\sum_{\text{over } a}q_i = \frac{1}{2}Q_aV_a \tag{3}$$

If the capacitance of a conductor is C, we have, from 2.01 (1), the following equivalent expressions for the energy of a charged conductor:

$$W = \tfrac{1}{2}QV = \tfrac{1}{2}Q^2/C = \tfrac{1}{2}CV^2 \tag{4}$$

For a capacitor, the two members of which carry charges Q and $-Q$ at potentials V_1 and V_2, respectively, the energy becomes

$$W = \tfrac{1}{2}QV_1 - \tfrac{1}{2}QV_2 = \tfrac{1}{2}Q(V_1 - V_2) \tag{5}$$

From 2.01 (2), (5) has the same equivalent forms as (4).

2.08. Energy in an Electric Field.—We have seen that visualizing electric forces as transmitted by stresses in the region occupied by an electric field gives results consistent with the observed laws of electrostatics. Where stresses exist, potential energy must be stored. We shall now compute this energy density. Consider an infinitesimal disk-shaped element of volume oriented in such a way that the two faces are equipotential surfaces. If this element is taken sufficiently small, the faces of the disk will be flat and parallel and the field inside uniform and identical with that in an infinitesimal parallel-plate capacitor. Let the thickness of the disk be ds; then the difference of potential between the faces is $(\partial V/\partial s)\,ds = -E\,ds$. Let $\boldsymbol{n}$ be the unit vector normal to the faces so that $\boldsymbol{E} = E\boldsymbol{n}$. The charge on an area dS of the face is

$$\boldsymbol{D} \cdot \boldsymbol{n}\,dS = \frac{\boldsymbol{D} \cdot \boldsymbol{E}}{E}\,dS$$

and the capacitor volume is $dv = ds\,dS$, so that 2.07 (4) gives

$$dW = \frac{\boldsymbol{D} \cdot \boldsymbol{E}}{2}\,dv$$

This gives, for the energy density in the field,

$$\frac{dW}{dv} = \frac{\boldsymbol{D} \cdot \boldsymbol{E}}{2} \tag{1}$$

In an isotropic dielectric, $\boldsymbol{D} \cdot \boldsymbol{E} = DE$, giving

$$\frac{dW}{dv} = \frac{\epsilon E^2}{2} = \frac{DE}{2} = \frac{D^2}{2\epsilon} \tag{2}$$

In a crystalline dielectric, we have, from 1.19 (5),

$$\frac{dW}{dv} = \tfrac{1}{2}(\epsilon_{11}E_1^2 + \epsilon_{22}E_2^2 + \epsilon_{33}E_3^2 + 2\epsilon_{12}E_1E_2 + 2\epsilon_{13}E_1E_3 + 2\epsilon_{23}E_2E_3) \tag{3}$$

or if the coordinate axes are chosen to coincide with the electric axes of the crystal, we have, from 1.19 (6),

$$\frac{dW}{dv} = \tfrac{1}{2}(\epsilon_1E_x^2 + \epsilon_2E_y^2 + \epsilon_3E_z^2) \tag{4}$$

2.081. Parallel-plate Capacitor with Crystalline Dielectric.—Let us now calculate the capacitance per square meter of a parallel-plate capacitor where the dielectric consists of a slab of crystal of thickness d. Let the capacitivities along the crystal axes x, y, and z be ϵ_1, ϵ_2, and ϵ_3, respectively, and let the direction cosines of the angle made by the normal to the capacitor plates with these axes be l, m, and n. Since, electrically, one square meter section is like any other, the equipotentials must be parallel to the plates and equally spaced and the electric intensity must lie along the normal. Thus we have

$$E = (X^2 + Y^2 + Z^2)^{\frac{1}{2}} = \frac{V}{d}$$

where V is the potential across the capacitor. Thus

$$X = \frac{lV}{d}, \qquad Y = \frac{mV}{d}, \qquad Z = \frac{nV}{d} \tag{1}$$

Substituting in 2.08 (4), multiplying by the volume d of a square meter section, and using 2.07 (4), we have

$$\frac{1}{2}C_1V^2 = \frac{d}{2}\left[\epsilon_1 l^2\left(\frac{V}{d}\right)^2 + \epsilon_2 m^2\left(\frac{V}{d}\right)^2 + \epsilon_3 n^2\left(\frac{V}{d}\right)^2\right]$$

so that the capacitance per square meter is

$$C_1 = \frac{l^2\epsilon_1 + m^2\epsilon_2 + n^2\epsilon_3}{d} \tag{2}$$

2.09. Stresses When the Capacitivity Is a Function of Density.—In considering stresses in a dielectric heretofore in 1.16 and 1.17, we have ignored the possibility that the capacitivity may actually change with density τ so that there may be a hydrostatic stress tending to expand or contract the dielectric. By working with a volume element of the shape and orientation used in 2.08, we can simplify the investigation to that of a small parallel-plate capacitor of area ΔS and spacing δ in which the charge on the plates is considered as embedded in the dielectric at the boundary. Combining 2.07 (4) with 2.05 (1) and assuming an isotropic dielectric of capacitivity ϵ, we have for the energy of our capacitor

$$\Delta W = \frac{\delta}{2\epsilon\,\Delta S}Q^2 = \frac{m}{2\epsilon\tau\,\Delta S}Q^2 = \frac{mD^2}{2\epsilon\tau}\,\Delta S$$

where we have let m be the mass of the dielectric per unit area between the plates so that $m = \tau\delta$. If m is assumed constant and ϵ is taken as a function of τ, the force on an area ΔS of the plate is

$$\Delta F = -\frac{\partial(\Delta W)}{\partial\delta} = -\frac{\partial(\Delta W)}{\partial\tau}\frac{\partial\tau}{\partial\delta} = -\frac{D^2}{2\epsilon^2}\frac{\partial(\epsilon\tau)}{\partial\tau}\,\Delta S$$

Thus the stress or force per unit area pulling on the surface of the conductor is

$$-\frac{\Delta F}{\Delta S} = \frac{D^2}{2\epsilon^2}\frac{\partial(\epsilon\tau)}{\partial\tau} \tag{1}$$

Carrying out the differentiation and comparing with 1.16 (2), we see that the additional hydrostatic stress is

$$-\frac{D^2}{2\epsilon^2}\tau\frac{\partial\epsilon}{\partial\tau} = -\frac{E^2}{2}\tau\frac{\partial\epsilon}{\partial\tau} = -\frac{\epsilon_v E^2}{2}\tau\frac{\partial K}{\partial\tau} \tag{2}$$

At the surface between two dielectrics, we shall now have to add to those stresses already considered the difference in this hydrostatic pressure giving, in place of 1.17 (7), for the total stress directed from K' to K'' the value

$$F_n = \frac{1}{2\epsilon_v}\left[\frac{K' - K''}{K'}\left(\frac{D_t'^2}{K'} + \frac{D_n'^2}{K''}\right) - \frac{D'^2\tau'}{K'^2}\frac{\partial K'}{\partial\tau'} + \frac{D''^2\tau''}{K''^2}\frac{\partial K''}{\partial\tau''}\right] \tag{3}$$

At the surface between a dielectric and a vacuum set $K'' = 1$ and $\partial K''/\partial\tau'' = 0$, we have

$$F_n = \frac{1}{2\epsilon_v}\left[\frac{K' - 1}{K'}\left(\frac{D_t'^2}{K'} + D_n'^2\right) - \frac{D'^2\tau'}{K'^2}\frac{\partial K'}{\partial\tau'}\right]$$

$$= \frac{\epsilon_v E'^2}{2}\left[(K' - 1) - \tau'\frac{\partial K'}{\partial\tau'}\right] + \frac{\epsilon_v E_n'^2}{2}(K' - 1)^2 \tag{4}$$

where

$$E'^2 = E_t'^2 + E_n'^2 = \left(\frac{D_t'}{\epsilon'}\right)^2 + \left(\frac{D_n'}{\epsilon'}\right)^2$$

The sign of $\partial K'/\partial\tau'$ may be either positive or negative, and if this term is large enough to predominate in (4) the dielectric may either expand or contract in the field. The phenomenon is known as electrostriction and has been observed by Quincke and others.

2.10. Electrostriction in Liquid Dielectrics.—There is a relation known as the Clausius-Mossotti formula which connects the relative capacitivity with the density for a given liquid.

$$\frac{K-1}{K+2} = C\tau \tag{1}$$

C is a constant characteristic of the liquid. This formula, although theoretically derived on incomplete assumptions, has been extremely well confirmed in many cases experimentally (see Geiger-Scheel, "Handbuch der Physik," Band XII, p. 518). Differentiating, it gives

$$\frac{\partial K}{\partial\tau} = \frac{(K+2)^2}{3}C = \frac{(K+2)(K-1)}{3\tau}$$

Substituting in 2.09 (2), we find that the hydrostatic pressure tending to contract the liquid is

$$P = \frac{\epsilon_v E^2}{2}\,\frac{(K+2)(K-1)}{3} \tag{2}$$

We have assumed throughout that the liquid is nearly incompressible so that τ is nearly constant. Thus we see that if we had a charged sphere immersed in a large body of liquid dielectric the pressure given by (2) would vary inversely as the fourth power of the distance from the center of the sphere, and if the liquid were slightly compressible it would be most dense at the sphere's surface, when other effects such as gravitation are neglected.

2.11. Force on Conductor in Dielectric.—So far we have considered the force at the charged boundary between a dielectric and a conductor as if the charge were on the dielectric side of the boundary so that the entire field lies in the dielectric. This gives, from 1.16 (2), the force per unit area to be $D^2/(2\epsilon)$, neglecting electrostriction. We know from considering the energy of a charged capacitor that this result is correct. Now let us investigate the force if the charge is taken to be on the conducting side of the boundary, the capacitivity of the conductor being ϵ'. The force then becomes $D^2/(2\epsilon')$. The surface of the dielectric is now in the field so that there will be a tension on it pressing against the conductor whose amount is given by 2.09 (3). Neglecting electrostriction and remembering that the field is normal to the surface we obtain for the sum of these forces

$$F_2 = \frac{K - K'}{2\epsilon_v KK'} D^2 = \frac{\epsilon - \epsilon'}{2\epsilon\epsilon'} D^2$$

The total force acting on the conductor will now be

$$F = F_1 - F_2 = \frac{K - K + K'}{2\epsilon_v KK'} D^2 = \frac{D^2}{2\epsilon} \tag{1}$$

Thus either point of view as to the location of the charge gives the same resultant force.

If, instead of being discontinuous, the relative capacitivity is considered to change rapidly but continuously in crossing the boundary, the stresses must be determined by integrating. The total resultant stresses come out the same as before although their distribution near the boundary is different.

2.12. Green's Reciprocation Theorem.—We shall now prove that if charges $Q_1, Q_2, \ldots, Q_n$ on the conductors of a system give rise to potentials $V_1, V_2, \ldots, V_n$ and if charges $Q'_1, Q'_2, \ldots, Q'_n$ give rise to potentials $V'_1, V'_2, \ldots, V'_n$, then

$$\sum_{i=1}^{n} Q_i V'_i = \sum_{i=1}^{n} Q'_i V_i \tag{1}$$

Let us first consider a system of point charges and write down in the form of a matrix n^2 terms, each of which comprises the product of one point charge by the potential of a second point charge. Using 1.06 (3), we shall write the sum of the columns in the bottom row and the sum of the horizontal rows in the right-hand column. Thus we obtain

$$\begin{array}{ccccccccccc}
0 & + & \dfrac{q'_2 q_1}{4\pi\epsilon r_{21}} & + & \dfrac{q'_3 q_1}{4\pi\epsilon r_{31}} & + & \cdots & + & \dfrac{q'_n q_1}{4\pi\epsilon r_{n1}} & = & q_1 V'_1 \\
\dfrac{q'_1 q_2}{4\pi\epsilon r_{12}} & + & 0 & + & \dfrac{q'_3 q_2}{4\pi\epsilon r_{32}} & + & \cdots & + & \dfrac{q'_n q_2}{4\pi\epsilon r_{n2}} & = & q_2 V'_2 \\
\dfrac{q'_1 q_3}{4\pi\epsilon r_{13}} & + & \dfrac{q'_2 q_3}{4\pi\epsilon r_{23}} & + & 0 & + & \cdots & + & \dfrac{q'_n q_3}{4\pi\epsilon r_{n3}} & = & q_3 V'_3 \\
\cdots & & \cdots & & \cdots & & \cdots & & \cdots & & \\
\dfrac{q'_1 q_n}{4\pi\epsilon r_{1n}} & + & \dfrac{q'_2 q_n}{4\pi\epsilon r_{2n}} & + & \dfrac{q'_3 q_n}{4\pi\epsilon r_{3n}} & + & \cdots & + & 0 & = & q_n V'_n \\
\hline
q'_1 V_1 & & q'_2 V_2 & & q'_3 V_3 & & \cdots & & q'_n V_n & &
\end{array}$$

The order of summation being immaterial, the sum may be obtained by adding the terms of the last row or last column. Equating these gives

$$\sum_{s=1}^{n} q'_s V_s = \sum_{s=1}^{n} q_s V'_s \tag{2}$$

It should be noted that for these point charges V_s is the potential at q_s due to all unprimed charges except q_s itself. All charges located on the same conductor are multiplied by the same potential and so may be collected; thus

$$\Sigma q_i' V = V \Sigma q_i' = Q' V$$

giving (1). One important application of this theorem is obtained by putting Q_1', Q_3', . . . , $Q_n' = 0$, Q_2, Q_3, . . . , $Q_n = 0$, and $Q_1 = Q_2'$ in (1). This gives $V_1' = V_2$. Thus the potential to which an uncharged conductor A is raised by putting a charge Q on B is the same as that to which B, when uncharged, will be raised by putting the charge Q on A. This theorem is still valid if dielectric boundaries are present as is proved in 3.07.

2.13. Superposition of Fields.—By adding $\Sigma Q_i V_i$ or $\Sigma Q_i' V_i'$ to both sides of 2.12 (1) and comparing with 2.12 (1), we see that if charges $Q_1, Q_2, \ldots, Q_n$ produce potentials $V_1, V_2, \ldots, V_n$ etc., then charges $Q_1 + Q_1', Q_2 + Q_2', \ldots, Q_n + Q_n'$ give $V_1 + V_1', V_2 + V_2', \ldots, V_n + V_n'$. A great many new problems can be solved by superimposing the solutions of known problems. As an example of this, suppose we are given the charges $Q_1, Q_2, \ldots, Q_n$ on n concentric conducting spheres of radii $r_1, r_2, \ldots, r_n$ and wish to find the potential of any sphere, say the sth. We may superimpose the potential produced by each of these charged shells separately, giving

$$4\pi\epsilon V_s = (Q_1 + Q_2 + \cdots + Q_s) r_s^{-1} + Q_{s+1} r_{s+1}^{-1} + \cdots + Q_n r_n^{-1} \quad (1)$$

since the potential outside a charged shell at a distance r from its center is independent of its radius and equals $Q(4\pi\epsilon r)^{-1}$, and the potential inside a conducting spherical shell of radius a is equal to $Q(4\pi\epsilon a)^{-1}$.

2.14. Induced Charges on Earthed Conductors.—If we place a point charge q at P in the neighborhood of a group of earthed conductors, induced charges will appear on them. We can find the induced charge Q on any one of these conductors if we know the potential V_p' to which the point P would be raised, the charge q being absent, by raising that conductor to a potential V'. For, from 2.12 (1),

$$QV' + qV_p' + q_1 \cdot 0 + q_2 \cdot 0 + \cdots = Q' \cdot 0 + 0 \cdot V_p + q_1' \cdot 0 + q_2' \cdot 0 + \cdots$$

so that

$$Q = -\frac{V_p'}{V'} q \quad (1)$$

If a conducting sphere is the only conductor and P is at a distance r from its center, we have, from 2.03 (2), $V' = q_1'(4\pi\epsilon a)^{-1}$ and $V_p' = q_1'(4\pi\epsilon r)^{-1}$, so that the charge induced on it by a charge q at P is

$$Q = -\frac{aq}{r} \quad (2)$$

If a point P lies between two conductors, one of which encloses the other, and if the potential at P is known when the conductors are at potentials V_1' and V_2', respectively, then the charge induced on either, when both are earthed, by a charge q at P can be found by 2.12 (1). Let the induced charges be Q_1 and Q_2; then

$$Q_1V_1' + Q_2V_2' + qV_p' = 0$$

Also since all tubes of force from q end on these conductors, we have the relation $Q_1 + Q_2 = -q$. Solving gives

$$Q_1 = \frac{V_2' - V_p'}{V_1' - V_2'}q \qquad \text{and} \qquad Q_2 = \frac{V_1' - V_p'}{V_2' - V_1'}q \tag{3}$$

Thus from 2.03 if the point P lies between two earthed spheres, the charge induced on the inner and outer spheres are, respectively,

$$Q_1 = -\frac{r_1(r_2 - r)}{r(r_2 - r_1)}q \qquad \text{and} \qquad Q_2 = -\frac{r_2(r - r_1)}{r(r_2 - r_1)}q \tag{4}$$

where r is the distance of P from the center and $r_1 < r < r_2$.

From 2.04 if P lies between two earthed cylinders, the charge induced on the inner and outer are, respectively,

$$Q_1 = -\frac{\ln(r_2/r)}{\ln(r_2/r_1)}q \qquad \text{and} \qquad Q_2 = -\frac{\ln(r_1/r)}{\ln(r_1/r_2)}q \tag{5}$$

From 2.05 if P lies between two earthed plates at a distance a from the first plate and b from the second, we have

$$Q_1 = -\frac{bq}{a+b} \qquad \text{and} \qquad Q_2 = -\frac{aq}{a+b} \tag{6}$$

2.15. Self- and Mutual Elastance.—Consider n initially uncharged conductors, fixed in position and shape. We have seen that putting a charge on any conductor of the group will affect the potential of all other conductors in a definite way which depends only on the geometrical configuration of the system and the capacitivity. The ratio of the rise in potential V_r of conductor r to the charge Q_s placed on conductor s to produce this rise is called the coefficient of potential or mutual elastance s_{sr}. The first application of Green's reciprocation theorem (2.12) shows that $s_{sr} = s_{rs}$. A superposition of solutions for charges Q_r, Q_s, Q_t, etc., on conductors r, s, t gives

$$\begin{aligned} V_1 &= s_{11}Q_1 + s_{21}Q_2 + \cdots + s_{n1}Q_n \\ V_2 &= s_{12}Q_1 + s_{22}Q_2 + \cdots + s_{n2}Q_n \\ &\cdots\cdots\cdots\cdots\cdots\cdots \\ V_n &= s_{1n}Q_1 + s_{2n}Q_2 + \cdots + s_{nn}Q_n \end{aligned} \tag{1}$$

Thus s_{sr} is the potential to which the rth conductor is raised when a unit charge is placed on the sth conductor, all the other conductors being present but uncharged. Putting a positive charge on a conductor always raises the potential of neighboring insulated conductors so s_{rs} is always positive. The coefficient s_{rr} is called self-elastance.

2.16. Self- and Mutual Capacitance.—We may solve the set of equations 2.15 (1) and obtain the charge on a conductor in terms of the potentials of neighboring conductors. The solutions are

$$\begin{aligned} Q_1 &= c_{11}V_1 + c_{21}V_2 + \cdots + c_{n1}V_n \\ Q_2 &= c_{12}V_1 + c_{22}V_2 + \cdots + c_{n2}V_n \\ &\cdots\cdots\cdots\cdots\cdots\cdots \\ Q_n &= c_{1n}V_1 + c_{2n}V_2 + \cdots + c_{nn}V_n \end{aligned} \qquad (1)$$

where

$$c_{11} = \frac{1}{\Delta}\begin{vmatrix} s_{22}s_{32} & \cdots & s_{n2} \\ s_{23}s_{33} & \cdots & s_{n3} \\ \cdots & \cdots & \cdots \\ s_{2n}s_{3n} & \cdots & s_{nn} \end{vmatrix} \qquad c_{12} = c_{21} = -\frac{1}{\Delta}\begin{vmatrix} s_{21}s_{31} & \cdots & s_{n1} \\ s_{23}s_{33} & \cdots & s_{n3} \\ \cdots & \cdots & \cdots \\ s_{2n}s_{3n} & \cdots & s_{nn} \end{vmatrix}$$

and

$$\Delta = \begin{vmatrix} s_{11}s_{21} & \cdots & s_{n1} \\ s_{12}s_{22} & \cdots & s_{n2} \\ \cdots & \cdots & \cdots \\ s_{1n}s_{2n} & \cdots & s_{nn} \end{vmatrix} \qquad (2)$$

so that c_{rs} is the cofactor of s_{rs} in Δ divided by Δ.

The quantity c_{rr} is called the coefficient of capacitance or the self-capacitance and is defined as the charge to potential ratio on the rth conductor when the other conductors are present but earthed. Since the potential has the same sign as the charge, c_{rr} is always positive.

The quantity c_{sr} is called the coefficient of induction or the mutual capacitance and may be defined as the ratio of the induced charge on the rth conductor to the potential of the sth conductor when all conductors, except the sth, are grounded. The induced charge is always opposite in sign to the inducing charge so c_{rs} is always negative or zero.

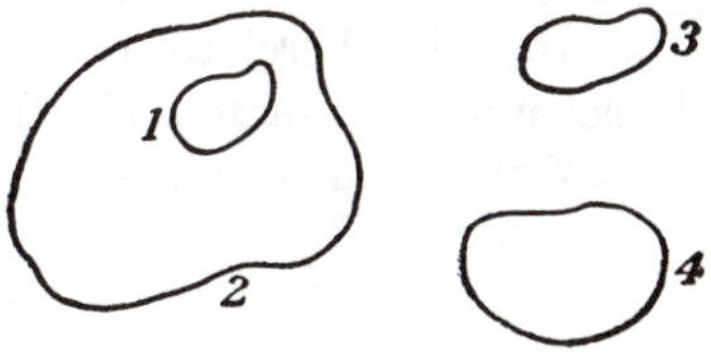

Fig. 2.17.

2.17. Electric Screening.—Suppose conductor 2 surrounds conductor 1 as in Fig. 2.17. All tubes of force from 1 end on 2 so that, if $V_2 = 0$, the charge on 1 depends only on its own potential, which means that, in 2.16 (1),

$$c_{31} = c_{41} = \cdots = c_{n1} = 0 \qquad (1)$$

Thus there is no inductive effect between 1 and any of the conductors outside of 2. In this case, we say that they are electrically screened from

1. We notice also that, since $Q_2 = -Q_1$ when all but 1 are earthed, $c_{11} = -c_{12}$.

Because of the complete similarity of 2.15 (1) and 2.16(1), we see that we can express the elastances in terms of the capacitances by an interchange of s and c in 2.16 (2). When 1 is screened by 2, we put in the values given by (1) and obtain

$$s_{1r} = s_{2r} \qquad 1 < r \tag{2}$$

Substitution of s_{2r} for s_{1r} in the first equation of 2.15 (1) and subtraction of the second from it give

$$V_1 - V_2 = (s_{11} - s_{12})Q_1$$

If $V_1 > V_2$, tubes of force pass from V_1 to V_2 and Q_1 is positive so that $s_{11} - s_{12}$ is positive, and since, from (2), $s_{12} = s_{22}$, we have

$$s_{11} > s_{12} \qquad \text{and} \qquad s_{11} > s_{22} \tag{3}$$

2.18. Elastances and Capacitances for Two Distant Conductors.—Let two conductors 1 and 2 have capacitances C_1 and C_2 when alone. When 1 is uncharged, let us bring up 2, which has a charge Q_2, to a mean distance r from 1, r being large compared with the linear dimensions, of magnitude a, of either conductor. The potential to which 1 is raised is $Q_2(4\pi\epsilon r)^{-1}$, if we neglect the variation of this potential over the region occupied by 1 which is of the order $a(4\pi\epsilon r^2)^{-1}$. Comparison of this with the first equation of 2.15 (1) shows that $s_{21} = (4\pi\epsilon r)^{-1}$. A charge of opposite sign to Q_2 and of order of magnitude $Q_2a(4\pi\epsilon r)^{-1}$ will be induced on the nearer parts of 1 and an equal charge of the same sign on the more remote parts. Here we have equal and opposite charges, separated by a distance smaller than a which is small compared with r, so that the field at the distance r is essentially a dipole field, and the potential [1.07 (1)] at 2 due to the presence of 1 uncharged is at most of order $Q_2a^2(4\pi\epsilon r^3)^{-1}$. To this order of magnitude, therefore, 1 does not affect the potential at 2 so that, from the second equation in 2.15 (1), we have $s_{22} = V_2Q_2^{-1} = C_2^{-1}$ and similarly $s_{11} = C_1^{-1}$. Thus to a first approximation,

$$s_{11} = \frac{1}{C_1}, \qquad s_{12} = s_{21} = \frac{1}{4\pi\epsilon r}, \qquad s_{22} = \frac{1}{C_2} \tag{1}$$

Solving for the self- and mutual capacitances by 2.16 (2), we have, neglecting r^{-3} terms,

$$c_{11} = \frac{16\pi^2\epsilon^2r^2C_1}{16\pi^2\epsilon^2r^2 - C_1C_2}, \quad c_{12} = c_{21} = -\frac{C_1C_2}{4\pi\epsilon r}, \quad c_{22} = \frac{16\pi^2\epsilon^2r^2C_2}{16\pi^2\epsilon^2r^2 - C_1C_2} \tag{2}$$

2.19. Energy of a Charged System.—If the electrical intensity and the displacement were known at all points outside a set of charged conductors, it would be possible to get the energy of the system by integration from

2.08 (1); thus

$$W = \tfrac{1}{2}\int_V \boldsymbol{D}\cdot\boldsymbol{E}\,dv \tag{1}$$

where the integration extends over the whole volume outside the conductors.

Frequently we do not know the field at all points but do know the potentials and charges on conductors and the self- or mutual capacitances. To determine the energy, suppose we add to the charges $Q_1\alpha$, $Q_2\alpha$, . . . , $Q_n\alpha$ on the conductors, by infinitesimal steps, $Q_1\,d\alpha$, $Q_2\,d\alpha$, . . . , $Q_n\,d\alpha$, starting with the initial uncharged state where $\alpha = 0$ and ending with the final state where $\alpha = 1$. Since the potentials are $V_1\alpha$, $V_2\alpha$, . . . , $V_n\alpha$ when the charges are $Q_1\alpha$, $Q_2\alpha$, . . . , $Q_n\alpha$, the work done in a single step is

$$dW = V_1Q_1\alpha\,d\alpha + V_2Q_2\alpha\,d\alpha + \cdots + V_nQ_n\alpha\,d\alpha$$

The energy in the final state is then

$$W = \sum_{i=1}^{n} V_iQ_i\int_0^1 \alpha\,d\alpha = \tfrac{1}{2}\sum_{i=1}^{n} V_iQ_i \tag{2}$$

If we substitute for Q_i from 2.16 (1), we have

$$W_V = \tfrac{1}{2}(c_{11}V_1^2 + 2c_{12}V_1V_2 + c_{22}V_2^2 + \cdots) \tag{3}$$

If we substitute for V_i from 2.15 (1), we have

$$W_Q = \tfrac{1}{2}(s_{11}Q_1^2 + 2s_{12}Q_1Q_2 + s_{22}Q_2^2 + \cdots) \tag{4}$$

2.20. Forces and Torques on Charged Conductors.—If the electrical intensity and displacement are known at all points on the surface of a conductor, the total resultant force on the conductor in the direction of the unit vector $\boldsymbol{p}$ can be obtained from 2.11 (1) by integration over the surface of the conductor, giving

$$F_p = \int_S \frac{\boldsymbol{D}\cdot\boldsymbol{E}}{2}\boldsymbol{p}\cdot\boldsymbol{n}\,dS \tag{1}$$

where $\boldsymbol{n}$ is the unit vector normal to the surface element dS.

When the charges and the elastances of a set of conductors are known, the potential energy of the system is given by 2.19 (4) in which s_{11}, s_{12}, etc., are functions of the configuration of the system. As in mechanics, we determine the force or torque tending to produce a change in any element of this configuration by taking the negative derivative of the potential energy with respect to this element.

$$-\frac{\partial W}{\partial \eta} = -\frac{1}{2}\left(\frac{\partial s_{11}}{\partial \eta}Q_1^2 + 2\frac{\partial s_{12}}{\partial \eta}Q_1Q_2 + \cdots\right) \tag{2}$$

This gives either a force or a torque tending to increase η depending on whether η is a length or an angle. In this case where the charge is kept constant, the change in electrical energy equals the mechanical work done. If, on the other hand, we work with 2.19 (3), keeping the potentials constant by a battery or equivalent device, we may supply additional energy to the system. The force in both cases must be the same, since it depends only on the initial state of the system which may be expressed in terms of either charge or potential.

To determine the force in terms of the potential, let us combine 2.19 (4), (2), and (3) in the equation

$$\Psi = W_Q + W_V - \Sigma V_i Q_i = 0 \tag{3}$$

Taking the total differential, we have

$$d\Psi = \sum_{i=1}^{n} \frac{\partial \Psi}{\partial Q_i} dQ_i + \sum_{i=1}^{n} \frac{\partial \Psi}{\partial V_i} dV_i + \sum \frac{\partial \Psi}{\partial \eta_s} d\eta_s = 0 \tag{4}$$

But from (3), 2.19 (4), and 2.15 (1)

$$\frac{\partial \Psi}{\partial Q_i} = \frac{\partial W_Q}{\partial Q_i} - V_i = \sum_{j=1}^{n} s_{ij} Q_j - V_i = 0$$

and from (3), 2.19 (3), and 2.16 (1)

$$\frac{\partial \Psi}{\partial V_i} = \frac{\partial W_V}{\partial V_i} - Q_i = \sum_{j=1}^{n} c_{ij} V_j - Q_i = 0$$

Substituting in (4) gives

$$\sum \frac{\partial \Psi}{\partial \eta_s} d\eta_s = 0$$

This sum is zero for any values of $d\eta_s$ we choose; so each term must be separately equal to zero. Substituting for Ψ from (3) gives

$$\frac{\partial \Psi}{\partial \eta_s} d\eta_s = \frac{\partial W_Q}{\partial \eta_s} d\eta_s + \frac{\partial W_V}{\partial \eta_s} d\eta_s = 0 \tag{5}$$

But we know that $-\partial W_Q/\partial \eta_s$ is the force or torque tending to increase η_s, so that in terms of the potentials this is

$$+\frac{\partial W_V}{\partial \eta_s} = \frac{1}{2}\left(\frac{\partial c_{11}}{\partial \eta_s} V_1^2 + 2\frac{\partial c_{12}}{\partial \eta_s} V_1 V_2 + \cdots\right) \tag{6}$$

When the charges are kept constant, the work done making a small displacement is given by (2), and when the potentials are kept constant it is given by (6). The difference must represent the work done by the device maintaining the potential giving

$$\left(\frac{\partial W_V}{\partial \eta_s} - \frac{\partial W_Q}{\partial \eta_s}\right) d\eta_s = 2\frac{\partial W_V}{\partial \eta_s} d\eta_s \tag{7}$$

Problems

Problems marked C are taken from the Cambridge examination questions as reprinted by Jeans by permission of the Cambridge University Press.

1. Three identical spheres of radius a are placed with their centers in line, the intervals being r_1 and r_2, and are numbered in order. At first, only the central one 2 carries a charge Q. It is now connected with 1. This connection is then broken, and it is connected with 3. Show that, if the intervals are large compared with a, the final charge on 3 is

$$Q_3 = \frac{Q}{4}\left[\frac{ar_2^2}{r_1(r_1 + r_2)(r_2 - a)} + 1\right]$$

2. Four identical uncharged conducting spheres are placed at the corners of a square and numbered in rotation. A charge is given to 1, which is then connected for an instant by a thin wire to 2, 3, and 4 in turn. Show that finally

$$Q_4 = \frac{Q}{8}\frac{s_{11} - s_{24}}{s_{11} - s_{14}} \qquad Q_1 = \frac{Q}{8}\frac{s_{11} - 2s_{14} + s_{24}}{s_{11} - s_{14}}$$

3. If, initially, spheres 1 and 2 in the last problem had charges $+Q$ and $-Q$, respectively, and if, as described, 1 is connected for an instant to 3 and 4 in turn, and if the side of the square r is large compared with the radius a of the spheres, show that, finally,

$$Q_4 = 2^{-\frac{5}{2}}Q[2^{\frac{1}{2}}r - (2^{\frac{5}{2}} - 3)a]/(r - a)$$

approximately, and find Q_3 and Q_1.

4. Each of three similar spheres, situated a considerable distance apart at the corners of an equilateral triangle, has initially a charge Q. Each in turn is now earthed for an instant and then insulated. Show that the final charge on 3 is

$$a^2r^{-2}Q(3 - 2a/r)$$

and find the final charges on 1 and 2.

5. Four similar conductors are arranged at the corners of a regular tetrahedron in such a way that each is perfectly symmetrical with respect to the other three. All are initially uncharged. One is now given a charge Q by means of a battery of voltage V and is then insulated. It is now connected for an instant to each of the other three in turn and then to earth. The charge is now $-Q_1$. Show that all the mutual capacitances are $56Q^2Q_1/[V(24Q_1 + 7Q)(8Q_1 - 7Q)]$ and find the elastances.

6. Two conductors are mirror images of each other and are initially uncharged. A sphere with a charge Q is now brought in contact with a certain point on one conductor and then with the image point on the other. If, in each case, it shares its charge equally with the conductor, show that after a large number of alternate contacts the charge will be equally distributed between the three conductors.

7. Three similar insulated conductors are so arranged that each is perfectly symmetrical with respect to the other two. A wire from a battery of unknown voltage is touched to each in turn. The charges on the first two are found to be Q_1 and Q_2. What is the charge on the third?

8. The inner and outer of three concentric spherical conducting shells of radii a, b, and c are grounded. The intervening sphere is split into halves and charged. Find how large a must be in order just to prevent the halves from separating if $a < b < c$.

9. Three conductors are connected by fine wires and charged. Find how the charge is distributed between them if

$$s_{11} = s_{33} \neq s_{22}, \qquad s_{12} = s_{23} \neq s_{13}, \qquad s_{11} + s_{13} = s_{12} + s_{22}$$

10. A pair of concentric spherical conductors of radii a and b are connected by a wire. A point charge q is detached from the inner one and moved radially with uniform velocity v to the outer one. Show that the rate of transfer of the induced charge from the inner to the outer sphere is $dQ/dt = -qab(b-a)^{-1}v(a+vt)^{-2}$.

11. A ring carrying a total charge Q and of radius a lies inside of an earthed sphere of radius b, its axis coinciding with a diameter and its plane at a distance c from the center. Find the potential at the center.

12. A charge q when placed on the axis of a thin earthed conducting ring of radius a at a distance b from the center is found to induce a charge $-Q$ on it. Show that the capacitance of the ring is $4\pi\epsilon Qq^{-1}(a^2+b^2)^{\frac{1}{2}}$.

13. A conducting sphere of radius a is embedded in the center of a sphere of dielectric of radius b and relative capacitivity K. Show that the capacitance of the conducting sphere is $4\pi\epsilon_v Kab(Ka+b-a)^{-1}$.

14. The conductor in the last problem is earthed and a point charge q is placed at a distance r from its center, where $r > b > a$. Show that the charge induced on the sphere is $-Kabq/\{r[b+(K-1)a]\}$.

15. A spherical capacitor with inner radius a and outer radius b is filled with two spherical layers of capacitivities ϵ_1 and ϵ_2, the boundary between being given by $r = \frac{1}{2}(a+b)$. If, when both shells are earthed, a point charge on the dielectric boundary induces equal charges on the inner and outer shells, find the ratio ϵ_1/ϵ_2.

16. An uncharged conducting spherical shell of mass M floats with one quarter of its volume submerged in a liquid of capacitivity ϵ. To what potential must it be charged to float half submerged?

17*C*. If the algebraic sum of the charges on a system of conductors is positive, then on one at least the surface density is everywhere positive.

18*C*. There are a number of insulated conductors in given fixed positions. The capacities of any two of them in their given positions are C_1 and C_2, and their mutual coefficient of induction is B. Prove that if these conductors are joined by a thin wire, the capacity of the combined conductor is $C_1 + C_2 + 2B$.

19*C*. In a system of insulated conductors, having been charged in any manner, charges are transferred from one conductor to another, till they are all brought to the same potential V. Show that $V = E/(s_1 + 2s_2)$, where s_1, s_2 are the algebraic sums of the coefficients of capacity and induction, respectively, and E is the sum of the charges.

20*C*. Prove that the effect of the operation described in the last question is a decrease of the electrostatic energy equal to what would be the energy of the system if each of the original potentials were diminished by V.

21*C*. Two equal similar condensers, each consisting of two spherical shells, radii a, b, are insulated and placed at a great distance r apart. Charges e, e' are given to the inner shells. If the outer surfaces are now joined by a wire, show that the loss of energy is, approximately $(16\pi\epsilon)^{-1}(e-e')^2(b^{-1}-r^{-1})$.

22*C*. A condenser is formed of two thin concentric shells, radii a, b. A small hole exists in the outer sheet through which an insulated wire passes connecting the inner sheet with a third conductor of capacity c at a great distance r from the condenser. The outer sheet of the condenser is put to earth, and the charge on the two connected conductors is E. Prove that the force on the third conductor is, approximately,

$$ac^2E^2(4\pi\epsilon r^3)^{-1}[4\pi\epsilon ab(a-b)^{-1}+c]^{-2}$$

23*C*. Two closed equipotentials V_1, V_0 are such that V_1 contains V_0, and V_p is the potential at any point P between them. If now a charge E is put at P and both equipotentials are replaced by conducting shells and earth connected, then the charges

E_1, E_0 induced on the two surfaces are given by

$$E_1(V_0 - V_p)^{-1} = E_0(V_p - V_1)^{-1} = E(V_1 - V_0)^{-1}$$

24*C*. A conductor is charged from an electrophorus by repeated contacts with a plate, which after each contact is recharged with a quantity E of electricity from the electrophorus. Prove that if e is the charge of the conductor after the first operation, the ultimate charge is $Ee(E - e)^{-1}$.

25*C*. Four equal uncharged insulated conductors are placed symmetrically at the corners of a regular tetrahedron and are touched in turn by a moving spherical conductor at the points nearest to the center of the tetrahedron, receiving charges e_1, e_2, e_3, e_4. Show that the charges are in geometrical progression.

26*C*. In 25*C*, replace "tetrahedron" by "square" and prove that

$$(e_1 - e_2)(e_1e_3 - e_2^2) = e_1(e_2e_3 - e_1e_4)$$

27*C*. Two insulated fixed conductors are at given potentials when alone in the electric field and charged with quantities E_1, E_2 of electricity. Their coefficients of potential are s_{11}, s_{12}, s_{22}. But if they are surrounded by a spherical conductor of very large radius R at zero potential with its center near them, the two conductors require charges E_1', E_2' to produce the given potentials. Prove, neglecting R^{-2}, that

$$(E_1' - E_1)(E_2' - E_2)^{-1} = (s_{22} - s_{12})(s_{11} - s_{12})^{-1}$$

28*C*. Show that the locus of the positions, in which a unit charge will induce a given charge on a given uninsulated conductor, is an equipotential surface of that conductor supposed freely electrified.

29*C*. Prove (1) that if a conductor, insulated in free space and raised to unit potential, produces at any external point P a potential denoted by (P), then a unit charge placed at P in the presence of this conductor uninsulated will induce on it a charge $-(P)$; (2) that if the potential at a point Q due to the induced charge is denoted by (PQ), then (PQ) is a symmetrical function of the positions of P and Q.

30*C*. Two small uninsulated spheres are placed near together between two large parallel planes, one of which is charged, and the other connected to earth. Show by figures the nature of the disturbance so produced in the uniform field, when the line of centers is (1) perpendicular, (2) parallel to the planes.

31*C*. A hollow conductor A is at zero potential and contains in its cavity two other insulated conductors B and C, which are mutually external. B has a positive charge, and C is uncharged. Analyze the different types of lines of force that are possible within the cavity, classifying with respect to the conductor from which the line starts and the conductor at which it ends, and proving the impossibility of the geometrically possible types that are rejected. Hence, prove that B and C are at positive potentials, the potential of C being less than that of B.

32*C*. A portion P of a conductor, the capacity of which is C, can be separated from the conductor. The capacity of this portion, when at a long distance from other bodies, is c. The conductor is insulated, and the part P when at a considerable distance from the remainder is charged with a quantity e and allowed to move under the mutual attraction up to it. Describe and explain the changes that take place in the electrical energy of the system.

33*C*. A conductor having a charge Q_1 is surrounded by a second conductor with charge Q_2. The inner is connected by a wire to a very distant uncharged conductor. It is then disconnected, and the outer conductor connected. Show that the charges Q_1', Q_2' are now

$$Q_1' = \frac{mQ_1 - nQ_2}{m + n + mn}, \qquad Q_2' = \frac{(m + n)Q_2 + mnQ_1'}{m + n}$$

where C, $C(1 + m)$ are the coefficients of capacity of the near conductors, and Cn is the capacity of the distant one.

34C. If one conductor contains all the others, and there are $n + 1$ in all, show that there are n relations between either the coefficients of potential or the coefficients of induction, and if the potential of the largest is V_0 and that of the others $V_1, V_2, \ldots, V_n$, then the most general expression for the energy is $\frac{1}{2}CV_0^2$ increased by a quadratic function of $V_1 - V_0, V_2 - V_0, \ldots, V_n - V_0$, where C is a definite constant for all positions of the inner conductors.

35C. The inner sphere of a spherical condenser (radii a, b) has a constant charge E, and the outer conductor is at potential zero. Under the internal forces, the outer conductor contracts from radius b to radius b_1. Prove that the work done by the electric forces is $E^2(b - b_1)/(8\pi\epsilon_v bb_1)$.

36C. If, in the last question, the inner conductor has a constant potential V, its charge being variable, show that the work done is $2\pi\epsilon V^2a^2(b - b_1)/[(b_1 - a)(b - a)]$ and investigate the quantity of energy supplied by the battery.

37C. With the usual notation, prove that $(s_{11} + s_{23}) > (s_{12} + s_{13})$, $s_{11}s_{23} > s_{12}s_{13}$.

38C. Show that if s_{rr}, s_{rs}, s_{ss} are three coefficients before the introduction of a new conductor and s'_{rr}, s'_{rs}, s'_{ss} the same coefficients afterward, then

$$(s_{rr}s_{ss} - s'_{rr}s'_{ss}) \nless (s_{rs} - s'_{rs})^2$$

39C. A system consists of $p + q + 2$ conductors $A_1, A_2, \ldots, A_p, B_1, B_2, \ldots, B_q, C, D$. Prove that when the charges on the A's and on C and the potentials of the B's and of C are known, there cannot be more than one possible distribution in equilibrium, unless C is electrically screened from D.

40C. A, B, C, D are four conductors, of which B surrounds A and D surrounds C. Given the coefficients of capacity and induction (1) of A and B when C and D are removed, (2) of C and D when A and B are removed, (3) of B and D when A and C are removed, determine those for the complete system of four conductors.

41C. Two equal and similar conductors A and B are charged and placed symmetrically with regard to each other; a third movable conductor C is carried so as to occupy successively two positions, one practically wholly within A, the other within B, the positions being similar and such that the coefficients of potential of C in either position are p, q, r in ascending order of magnitude. In each position, C is in turn connected with the conductor surrounding it, put to earth, and then insulated. Determine the charges on the conductors after any number of cycles of such operations, and show that they ultimately lead to the ratios $1 : -\beta : \beta^2 - 1$, where β is the positive root of $rx^2 - qx + p - r = 0$.

42C. Two conductors are of capacities C_1 and C_2, when each is alone in the field. They are both in the field at potentials V_1 and V_2, respectively, at a great distance r apart. Prove that the repulsion between the conductors is

$$C_1C_2(4\pi\epsilon rV_1 - C_2V_2)(4\pi\epsilon rV_2 - C_1V_1)(4\pi\epsilon)/[(16\pi^2\epsilon^2r^2 - C_1C_2)^2]$$

As far as what power of r^{-1} is this result accurate?

43C. Two equal and similar insulated conductors are symmetrically placed with regard to each other, one of them being uncharged. Another insulated conductor is made to touch them alternately in a symmetrical manner, beginning with the one that has a charge. If e_1, e_2 are their charges when it has touched each once, show that their charges, when it has touched each r times, are, respectively,

$$\frac{e_1^2}{2e_1 - e_2}\left[1 + \left(\frac{e_1 - e_2}{e_1}\right)^{2r-1}\right] \quad \text{and} \quad \frac{e_1^2}{2e_1 - e_2}\left[1 - \left(\frac{e_1 - e_2}{e_1}\right)^{2r}\right]$$

(For 43C and 44C see difference equations, 5.081.)

44C. Three conductors A_1, A_2, and A_3 are such that A_3 is practically inside A_2. A_1 is alternately connected with A_2 and A_3 by means of a fine wire, the first contact being with A_3. A_1 has a charge E, initially, A_2 and A_3 being uncharged. Prove that the charge on A_1 after it has been connected n times with A_3 is

$$\frac{E\beta}{\alpha+\beta}\left[1+\frac{\alpha(\gamma-\beta)}{\beta(\alpha+\gamma)}\left(\frac{\alpha+\beta}{\alpha+\gamma}\right)^{n-1}\right]$$

where α, β, γ stand for $s_{11} - s_{12}$, $s_{22} - s_{12}$, and $s_{33} - s_{12}$.

45C. A spherical condenser, radii a, b, has air in the space between the spheres. The inner sphere receives a coat of paint of uniform thickness t and of a material of which the inductive capacity is K. Find the change produced in the capacity of the condenser.

46C. A conductor has a charge e, and V_1, V_2 are the potentials of two equipotential surfaces completely surrounding it ($V_1 > V_2$). The space between these two surfaces is now filled with a dielectric of inductive capacity K. Show that the change in the energy of the system is $\frac{1}{2}e(V_1 - V_2)(K - 1)/K$.

47C. The surfaces of an air condenser are concentric spheres. If half the space between the spheres is filled with solid dielectric of specific inductive capacity K, the dividing surface between the solid and the air being a plane through the center of the spheres, show that the capacity will be the same as though the whole dielectric were of uniform specific inductive capacity $\frac{1}{2}(1 + K)$.

48C. The radii of the inner and outer shells of two equal spherical condensers, remote from each other and immersed in an infinite dielectric of inductive capacity K, are, respectively, a and b, and the inductive capacities of the dielectric inside the condensers are K_1, K_2. Both surfaces of the first condenser are insulated and charged, the second being uncharged. The inner surface of the second condenser is now connected to earth, and the outer surface is connected to the outer surface of the first condenser by a wire of negligible capacity. Show that the loss of energy is

$$\frac{Q^2[2(b-a)K + aK_2]}{8\pi b\epsilon[(b-a)K + aK_2]}$$

where Q is the quantity of electricity that flows along the wire.

49C. The outer coating of a long cylindrical condenser is a thin shell of radius a, and the dielectric between the cylinders has inductive capacity K on one side of a plane through the axis and K' on the other side. Show that when the inner cylinder is connected to earth and the outer has charge q per unit length, the resultant force on the outer cylinder is

$$\frac{q^2(K-K')}{\pi^2\epsilon_v a(K+K')^2} \text{ per unit length}$$

50C. A heterogeneous dielectric is formed of n concentric spherical layers of specific inductive capacities $K_1, K_2, \ldots, K_n$, starting from the innermost dielectric, which forms a solid sphere; also, the outermost dielectric extends to infinity. The radii of the spherical boundary surfaces are $a_1, a_2, \ldots, a_{n-1}$, respectively. Prove that the potential due to a quantity Q of electricity at the center of the spheres at a point distant r from the center in the dielectric K_s is

$$\frac{1}{4\pi\epsilon_v}\left[\frac{Q}{K_s}\left(\frac{1}{r}-\frac{1}{a_s}\right)+\frac{Q}{K_{s+1}}\left(\frac{1}{a_s}-\frac{1}{a_{s+1}}\right)+\cdots+\frac{Q}{K_n}\frac{1}{a_{n-1}}\right]$$

51C. A condenser is formed by two rectangular parallel conducting plates of breadth b and area A at distance d from each other. Also, a parallel slab of a dielectric of thickness t and of the same area is between the plates. This slab is pulled along its length from between the plates so that only a length x is between the plates. Prove that the electric force sucking the slab back to its original position is

$$\frac{E^2\, dbt'(d - t')}{2\epsilon_v[A(d - t') + xbt']^2}$$

where $t' = t(K - 1)/K$. K is the specific inductive capacity of the slab, E is the charge, and the disturbances produced by the edges are neglected.

52*C*. Three closed surfaces 1, 2, 3 are equipotentials in an electric field. If the space between 1 and 2 is filled with a dielectric ϵ, and that between 2 and 3 is filled with a dielectric ϵ', show that the capacity of a condenser having 1 and 3 for faces is C, given by $1/C = \epsilon_v/(\epsilon A) + \epsilon_v/(\epsilon' B)$ where A, B are the capacities of air condensers having as faces the surfaces 1, 2 and 2, 3, respectively.

53*C*. The surface separating two dielectrics (K_1, K_2) has an actual charge σ per unit area. The electric intensities on the two sides of the boundary are F_1, F_2 at angles c_1, c_2 with the common normal. Show how to determine F_2, and prove that $K_2 \cot c_2 = K_1 \cot c_1[1 - \sigma/(\epsilon_v K_1 F_1 \cos c_1)]$.

54*C*. The space between two concentric spheres, radii a, b, which are kept at potentials A, B is filled with a heterogeneous dielectric of which the inductive capacity varies as the nth power of the distance from their common center. Show that the potential at any point between the surfaces is

$$[(Aa^{n+1} - Bb^{n+1})/(a^{n+1} - b^{n+1})] - (ab/r)^{n+1}(A - B)/(a^{n+1} - b^{n+1})$$

55*C*. A condenser is formed of two parallel plates, distant h apart, one of which is at zero potential. The space between the plates is filled with a dielectric whose inductive capacity increases uniformly from one plate to the other. Show that the capacity per unit area is $\epsilon_v(K_2 - K_1)/[h \ln (K_2/K_1)]$ where K_1 and K_2 are the values of the inductive capacity at the surfaces of the plate. The inequalities of distribution at the edges of the plates are neglected.

56*C*. A spherical conductor of radius a is surrounded by a concentric spherical conducting shell whose internal radius is b, and the intervening space is occupied by a dielectric whose specific inductive capacity at a distance r from the center is $(c + r)/r$. If the inner sphere is insulated and has a charge E, the shell being connected with the earth, prove that the potential in the dielectric at a distance r from the center is $[E/(4\pi\epsilon c)] \ln \{[b/r][(c + r)/(c + b)]\}$.

57*C*. A spherical conductor of radius a is surrounded by a concentric spherical shell of radius b, and the space between them is filled with a dielectric of which the inductive capacity at distance r from the center is $\mu e^{-p^2}p^{-3}$ where $p = ra^{-1}$. Prove that the capacity of the condenser so formed is $8\pi\epsilon_v\mu a/(e^{b^2/a^2} - e)$.

58*C*. Show that the capacity of a condenser consisting of the conducting spheres $r = a$, $r = b$, and a heterogeneous dielectric of inductive capacity $K = f(\theta, \phi)$ is $\epsilon_v ab(b - a)^{-1} \int\int f(\theta, \phi) \sin \theta \, d\theta \, d\phi$.

59*C*. In an imaginary crystalline medium, the molecules are disks placed so as to be all parallel to the plane of xy. Show that the components of intensity and polarization (displacement in our notation) are connected by equations of the form

$$f = \epsilon_{11}X + \epsilon_{21}Y, \qquad g = \epsilon_{12}X + \epsilon_{22}Y, \qquad h = \epsilon_{33}Z$$

60*C*. A slab of dielectric of inductive capacity K and of thickness x is placed inside a parallel-plate condenser so as to be parallel to the plates. Show that the surface of the slab experiences a tension $(\frac{1}{2}\sigma^2/\epsilon_v)[1 - K^{-1} - x\, d(K^{-1})/dx]$.

61*C*. For a gas $K = 1 + \theta\rho$, where ρ is the density and θ is small. A conductor is immersed in the gas. Show that if θ^2 is neglected the mechanical force on the conductor is $\frac{1}{2}\sigma^2/\epsilon_v$ per unit area. Give a physical interpretation of this result.

62*C*. The curve

$$\frac{1}{(x^2+y^2)^{\frac{1}{2}}} - \frac{9a}{16}\left\{\frac{a+x}{[(x+a)^2+y^2]^{\frac{3}{2}}} + \frac{a-x}{[(x-a)^2+y^2]^{\frac{3}{2}}}\right\} = \frac{1}{a}$$

when rotated round the axis of x generates a single closed surface, which is made the bounding surface of a conductor. Show that its capacity will be $4\pi\epsilon_v a$ and that the surface density at the end of the axis will be $e/(3\pi a^2)$, where e is the total charge.

63. The potential ratios, P_{rs}, of a system of n conductors may be defined by

$$\begin{aligned} V_1 &= c_{11}^{-1}Q_1 + P_{12}V_2 + \cdots + P_{1n}V_n \\ V_2 &= P_{21}V_1 + c_{22}^{-1}Q_2 + \cdots + P_{2n}V_n \\ &\cdots\cdots\cdots\cdots\cdots\cdots\cdots\cdots \\ V_n &= P_{n1}V_1 + P_{n2}V_2 + \cdots + c_{nn}^{-1}Q_n \end{aligned}$$

Show that P_{rs} is given in terms of the capacitances, if $s \neq r$, by

$$P_{rs} = -c_{rs}c_{rr}^{-1}$$

References

The author has found helpful treatments of the subject matter of this chapter in the following books:

ABRAHAM, M., and R. BECKER: "Classical Electricity and Magnetism," Blackie,

CORSON, D., and P. LORRAIN: "Introduction to Electromagnetic Fields and Waves," Freeman, 1962. Second chapter includes much of this material.

DURAND, E.: "Electrostatique et Magnetostatique," Masson et Cie, 1953. Includes a well-illustrated treatment of this material.

JEANS, J. H.: "The Mathematical Theory of Electricity and Magnetism," Cambridge, 1925. Gives comprehensive treatment using long notation.

MAXWELL, J. C.: "Electricity and Magnetism," 3d ed. (1891), Dover, 1954. Gives extensive treatment using long notation and contains many good scale drawings of fields.

PLANCK, M. K. E. L.: "Theory of Electricity and Magnetism," Macmillan, 1932. Uses vector notation.

RAMSEY, A. S.: "Electricity and Magnetism," Cambridge, 1937. Gives clear elementary treatment with problems. Uses vectors.

RUSSELL, A.: "Alternating Currents," Cambridge, 1914. Gives applications to multiple conductors.

THOMSON, J. J.: "Mathematical Theory of Electricity and Magnetism," Cambridge, 1921.

THOMSON, W.: "Papers on Electrostatics and Magnetism," Macmillan, 1884. Solves many problems and gives numerical results on spheres and spherical segments.

WEBSTER, A. G.: "Electricity and Magnetism," Macmillan, 1897. Gives a clear treatment using long notation.

WIEN-HARMS: "Handbuch der Experimentalphysik," Vol. X, Leipzig, 1930.

CHAPTER III

GENERAL THEOREMS

3.00. Gauss's Theorem.—We now proceed to find a relation between the integral over a closed surface S and over $m - 1$ other closed surfaces, inside it, of the normal component of a vector $\boldsymbol{A}$, which is a continuous function of position in space, and the integral of its divergence defined as $\nabla \cdot \boldsymbol{A}$, throughout the volume v between these surfaces. Let the components of $\boldsymbol{A}$ in rectangular coordinates be A_x, A_y, A_z so that

$$\nabla \cdot \boldsymbol{A} = \frac{\partial A_x}{\partial x} + \frac{\partial A_y}{\partial y} + \frac{\partial A_z}{\partial z} \tag{1}$$

Let us suppose v divided up into slender prisms of rectangular cross section $dy\,dz$. One of these (Fig. 3.00) cuts elements dS_j', dS_j'' from S_j,

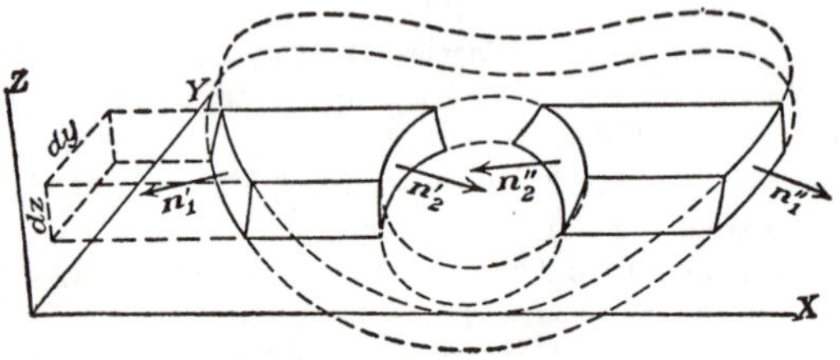

FIG. 3.00.

the coordinates of which are x_j' and x_j'' and the outward drawn normal unit vectors to which are $\boldsymbol{n}_j'$, $\boldsymbol{n}_j''$. The contributions of the elements dS_j' and dS_j'' to the surface integral are

$$A_{x_j'}\boldsymbol{i} \cdot \boldsymbol{n}_j'\,dS_j' + A_{x_j''}\boldsymbol{i} \cdot \boldsymbol{n}_j''\,dS_j''$$

But from Fig. 3.00, $\boldsymbol{i} \cdot \boldsymbol{n}_j'\,dS_j' = -\boldsymbol{i} \cdot \boldsymbol{n}_j''\,dS_j'' = dy\,dz$ so that the total contribution to all sections of that prism, cut from v, which penetrates surfaces p to q is

$$\sum_{j=p}^{q} dy\,dz(A_{x_j'} - A_{x_j''}) = \sum_{j=p}^{q} dy\,dz \int_{x_j'}^{x_j''} \frac{\partial A_x}{\partial x}\,dx$$

We now sum over all the prisms into which v is divided, and we have

$$\sum_{j=1}^{m} \int_{S_j} A_x\boldsymbol{i} \cdot \boldsymbol{n}_j\,dS_j = \int_{\mathrm{v}} \frac{\partial A_x}{\partial x}\,dx\,dy\,dz$$

Working out similar expressions for A_y and A_z and adding, we obtain, after substituting $\nabla \cdot A$ from (1) and writing $dv = dx\,dy\,dz$, the expression

$$\sum_{j=1}^{m} \int_{S_j} A \cdot n_j \, dS_j = \int_V \nabla \cdot A \, dv \tag{2}$$

This formula, apparently first given by Green, states a law widely known as Gauss's theorem.

3.01. Stokes's Theorem.—We can derive another important theorem directly from Gauss's theorem. Let us apply 3.00 (2) to an infinitesimal right cylinder of height h, base area S, curved area S', and peripheral length s. Let $\boldsymbol{n}$ be the unit vector normal to the base and $\boldsymbol{n}_1$ that normal to the side. Let $\boldsymbol{F}$ be some vector function and replace $\boldsymbol{A}$ by $\boldsymbol{n} \times \boldsymbol{F}$. Since $\boldsymbol{n}$ is a constant in the following expression $\boldsymbol{F} \cdot \nabla \times \boldsymbol{n}$ vanishes.

$$\nabla \cdot n \times F = F \cdot \nabla \times n - n \cdot \nabla \times F$$

But $[n \times F] \cdot n$ is zero on the flat faces so that the surface integral over S vanishes and 3.00 (2) becomes

$$\int_{S'} n \times F \cdot n_1 \, dS' = \int_{S'} F \cdot n_1 \times n \, dS' = -\int_V n \cdot \nabla \times F \, dv$$

But $dv = h\,dS$ and $dS' = h\,ds$, so that the surface and volume integral become line and surface integrals, respectively, giving

$$\oint F \cdot n_1 \times n \, ds = -\int_S n \cdot \nabla \times F \, dS$$

Now $n \times n_1 = -n_1 \times n$ is a unit vector directed along the boundary, so that if we choose it for the positive direction of $\boldsymbol{s}$ we have $n_1 \times n \, ds = -\boldsymbol{ds}$, and summing up for all the infinitesimal cylinders composing a surface we have

$$\sum \oint F \cdot ds = \sum \int_S n \cdot \nabla \times F \, dS$$

In summing up the line integrals, every interior contour is covered twice in opposite directions, so that they cancel each other leaving only the integral around the outer edge of the whole area. The sum of the surface integrals is of course the integral over the whole surface, so that

$$\oint F \cdot ds = \int_S n \cdot \nabla \times F \, dS \tag{1}$$

This is Stokes's theorem and may be stated as follows: The line integral of a vector function $\boldsymbol{F}$ around any closed contour equals the surface integral of its curl over any surface bounded by this contour.

3.02. Equations of Poisson and Laplace.—In Gauss's theorem, let the vector $\boldsymbol{A}$ be the electric displacement $\boldsymbol{D} = \epsilon \boldsymbol{E}$ and apply Gauss's electric flux theorem [1.15 (1)] to the surface integral, and we obtain

$$\int_V \nabla \cdot D \, dv = Q = \int_V \rho \, dv$$

where ρ is the electric charge density. Thus we have, if dv becomes vanishingly small,

$$\text{div } D = \nabla \cdot D = \frac{dQ}{dv} = \rho \tag{1}$$

If we write $D = \epsilon E = -\epsilon \text{ grad } V = -\epsilon \nabla V$, this becomes

$$\text{div } (\epsilon \text{ grad } V) = \nabla \cdot (\epsilon \nabla V) = -\rho \tag{2}$$

This is Poisson's equation for a nonhomogeneous dielectric. If the dielectric is homogeneous, ϵ is a constant and may be factored out, giving

$$\text{div grad } V = \nabla \cdot \nabla V = \nabla^2 V = -\frac{\rho}{\epsilon} \tag{3}$$

If $\rho = 0$, we obtain Laplace's equation which may then be written, in rectangular coordinates, for a nonhomogeneous but isotropic dielectric

$$\frac{\partial}{\partial x}\left(\epsilon \frac{\partial V}{\partial x}\right) + \frac{\partial}{\partial y}\left(\epsilon \frac{\partial V}{\partial y}\right) + \frac{\partial}{\partial z}\left(\epsilon \frac{\partial V}{\partial z}\right) = 0 \tag{4}$$

If the dielectric is homogeneous but not isotropic and if the coordinate axes lie on the electric axes of the crystal as in 1.19 (7), this becomes

$$\epsilon_1 \frac{\partial^2 V}{\partial x^2} + \epsilon_2 \frac{\partial^2 V}{\partial y^2} + \epsilon_3 \frac{\partial^2 V}{\partial z^2} = 0 \tag{5}$$

If the dielectric is both homogeneous and isotropic, the equation becomes

$$\frac{\partial^2 V}{\partial x^2} + \frac{\partial^2 V}{\partial y^2} + \frac{\partial^2 V}{\partial z^2} = 0 \tag{6}$$

3.03. Orthogonal Curvilinear Coordinates.—In most electrostatic problems, the data given are the charges on, or the potentials of, all conductors involved, the size and location of all other charges, and the capacitivity at all field points. The problem is considered solved when the potential at all points has been determined. To do this, it is necessary to find a solution of Laplace's equation that can be made to fit the given boundary conditions. There is usually one system of coordinates in terms of which these conditions can be most simply expressed, and it is therefore desirable to solve Laplace's equation in this coordinate system. All commonly used coordinate systems can be classed as orthogonal curvilinear coordinates.

Consider three families of mutually orthogonal surfaces, one member of each family passing through every point of the region under consideration. Any member of the first family can be specified by giving the proper numerical value to u_1 and similarly for the other families by u_2 or u_3. An infinitesimal rectangular parallelepiped will be cut out of space by the six surfaces u_1, $u_1 + du_1$, u_2, $u_2 + du_2$, u_3, $u_3 + du_3$. Since the quantities u_1, u_2, and u_3 will represent distances in a few cases only, it will, in general, be necessary to multiply du_1, du_2, and du_3 by certain factors* h_1, h_2, and h_3 to get the actual length of its edges. These factors may be variable from point to point in the field, and so they are functions of u_1, u_2, and u_3. The lengths of the edges of our parallelepiped are therefore

$$ds_1 = h_1\,du_1, \qquad ds_2 = h_2\,du_2, \qquad ds_3 = h_3\,du_3 \quad (1)$$

These are shown in Fig. 3.03.

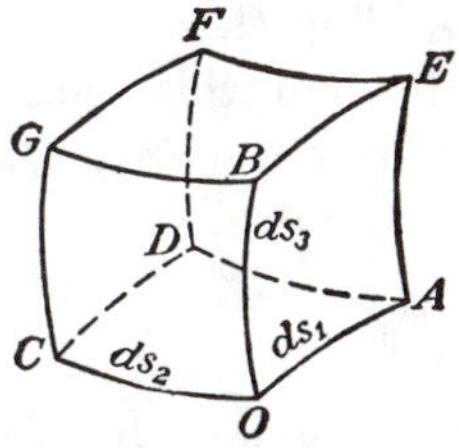

FIG. 3.03.

If V is a scalar, then, by definition, components of the gradient are

$$\frac{\partial V}{\partial s_1} = \frac{\partial V}{h_1\,\partial u_1}, \qquad \frac{\partial V}{\partial s_2} = \frac{\partial V}{h_2\,\partial u_2}, \qquad \frac{\partial V}{\partial s_3} = \frac{\partial V}{h_3\,\partial u_3} \quad (2)$$

To calculate the divergence in this coordinate system apply Gauss's theorem [3.00 (2)] to the infinitesimal volume shown in Fig. 3.03. The outward normal component of the flux of the vector A through the faces $OCGB$ and $ADFE$ is

$$\frac{\partial}{\partial s_1}(A_1\,ds_2\,ds_3)\,ds_1 = \frac{\partial}{\partial u_1}(h_2h_3A_1)\,du_1\,du_2\,du_3 = \frac{1}{h_1h_2h_3}\frac{\partial}{\partial u_1}(h_2h_3A_1)\,dv$$

Adding the two corresponding expressions for the other four faces and comparing with 3.00 (2), we see that

$$\operatorname{div} A = \nabla\cdot A = \frac{1}{h_1h_2h_3}\left[\frac{\partial}{\partial u_1}(h_2h_3A_1) + \frac{\partial}{\partial u_2}(h_3h_1A_2) + \frac{\partial}{\partial u_3}(h_1h_2A_3)\right] \quad (3)$$

Now letting $A = \epsilon\nabla V$ and substituting in Poisson's equation [3.02 (2)], we have

$$\frac{1}{h_1h_2h_3}\left[\frac{\partial}{\partial u_1}\left(\frac{h_2h_3\epsilon}{h_1}\frac{\partial V}{\partial u_1}\right) + \frac{\partial}{\partial u_2}\left(\frac{h_3h_1\epsilon}{h_2}\frac{\partial V}{\partial u_2}\right) + \frac{\partial}{\partial u_3}\left(\frac{h_1h_2\epsilon}{h_3}\frac{\partial V}{\partial u_3}\right)\right] = -\rho$$

$$\nabla\cdot(\epsilon\nabla V) = -\rho \quad (4)$$

Putting $\rho = 0$ gives us Laplace's equation for a nonhomogeneous isotropic dielectric. If the dielectric is homogeneous and isotropic, ϵ as well as $(h_1h_2h_3)^{-1}$ may be factored out of the equation.

3.04. Curl in Orthogonal Curvilinear Coordinates.—Let us apply Stokes's theorem to the face $OADC$ of the elementary curvilinear cube of Fig. 3.03. Let F_1, F_2, and F_3 be the components of a vector along u_1, u_2, and u_3. Then, from 3.03 (2), the line integral along OA and DC is

$$[h_1(u_1, u_2)F_1(u_1, u_2) - h_1(u_1, u_2 + du_2)F_1(u_1, u_2 + du_2)]\,du_1$$
$$= -\frac{\partial(h_1F_1)}{\partial u_2}\,du_1\,du_2$$

* Our notation is that used by Houston, "Principles of Mathematical Physics," and Abraham-Becker, "Classical Electricity and Magnetism." The Peirce and Smithsonian Tables and Jeans, "Electricity and Magnetism," use h_1, h_2, and h_3 for the reciprocal quantities.

and along AD and CO it is

$$[h_2(u_1 + du_1, u_2)F_2(u_1 + du_1, u_2) - h_2(u_1, u_2)F_2(u_1, u_2)]\, du_2 = \frac{\partial(h_2F_2)}{\partial u_1}\, du_1\, du_2$$

Adding these gives the line integral around this face, and, by Stokes's theorem, this equals the integral of the normal component of the curl of $\boldsymbol{F}$ over the area $h_1h_2\, du_1\, du_2$ of the face. Canceling out the factor $du_1\, du_2$, this gives

$$(\nabla \times F)_3 = \frac{1}{h_1h_2}\left[\frac{\partial(h_2F_2)}{\partial u_1} - \frac{\partial(h_1F_1)}{\partial u_2}\right] \tag{1}$$

and similarly for the other faces, we have

$$(\nabla \times F)_1 = \frac{1}{h_2h_3}\left[\frac{\partial(h_3F_3)}{\partial u_2} - \frac{\partial(h_2F_2)}{\partial u_3}\right] \tag{2}$$

$$(\nabla \times F)_2 = \frac{1}{h_3h_1}\left[\frac{\partial(h_1F_1)}{\partial u_3} - \frac{\partial(h_3F_3)}{\partial u_1}\right] \tag{3}$$

3.05. $\nabla \cdot (\epsilon \nabla V)$ in Other Coordinate Systems.—In spherical polar coordinates where $r = u_1$ is the distance from the origin, $\theta = u_2$ is the

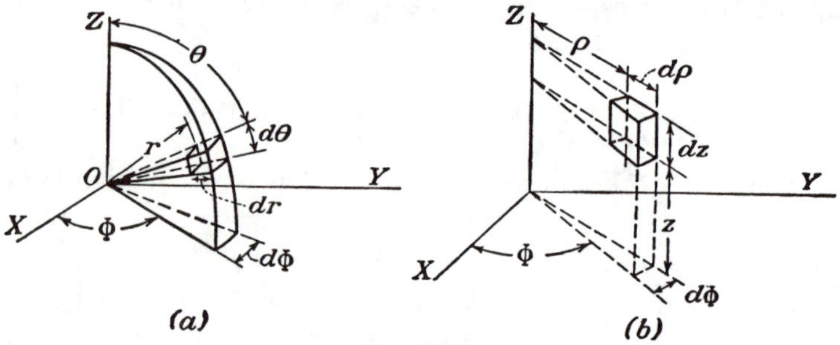

FIG. 3.05.

colatitude angle, and $\phi = u_3$ is the longitude angle, we have

$$\begin{aligned} ds_1 &= h_1\, du_1 = dr & h_1 &= 1 \\ ds_2 &= h_2\, du_2 = r\, d\theta & h_2 &= r \\ ds_3 &= h_3\, du_3 = r \sin\theta\, d\phi & h_3 &= r\sin\theta \end{aligned}$$

From 3.03 (4), Laplace's equation becomes

$$\frac{1}{r^2}\frac{\partial}{\partial r}\left(\epsilon r^2\frac{\partial V}{\partial r}\right) + \frac{1}{r^2 \sin\theta}\frac{\partial}{\partial\theta}\left(\epsilon \sin\theta\frac{\partial V}{\partial\theta}\right) + \frac{1}{r^2 \sin^2\theta}\frac{\partial}{\partial\phi}\left(\frac{\epsilon\, \partial V}{\partial\phi}\right) = 0 \tag{1}$$

In cylindrical coordinates where $\rho = u_1$ is the distance from the z-axis, $\phi = u_2$ is the longitude angle, and $z = u_3$ is the distance from the xy-plane, we have

$$\begin{aligned} ds_1 &= h_1\, du_1 = d\rho & h_1 &= 1 \\ ds_2 &= h_2\, du_2 = \rho\, d\phi & h_2 &= \rho \\ ds_3 &= h_3\, du_3 = dz & h_3 &= 1 \end{aligned}$$

This arrangement is shown in Fig. 3.05*b*. From 3.03 (4), Laplace's equation becomes

$$\nabla \cdot (\epsilon \nabla V) = \frac{1}{\rho}\frac{\partial}{\partial \rho}\left(\epsilon\rho\frac{\partial V}{\partial \rho}\right) + \frac{1}{\rho^2}\frac{\partial}{\partial \phi}\left(\epsilon\frac{\partial V}{\partial \phi}\right) + \frac{\partial}{\partial z}\left(\epsilon\frac{\partial V}{\partial z}\right) = 0 \tag{2}$$

We shall have occasion to use still other coordinate systems, such as the confocal system. These will be taken up later in connection with special problems.

3.06. Green's Theorems.—In 3.00 (2), let $A = (\epsilon \operatorname{grad} \Phi)\Psi = \Psi\epsilon\nabla\Phi$ where Ψ and Φ are scalar quantities which are finite and continuous in the region of integration and can be differentiated twice, and ϵ is a scalar quantity which may be differentiated once and which may be discontinuous at certain boundaries in the region. We shall exclude these discontinuities by drawing surfaces around them which fit them closely on both sides. Let n'_p and n''_p be the unit normal vectors drawn into the pth boundary from the two sides, and let the values of A on the two sides be A'_p and A''_p. Then if there are q such surfaces enclosing q discontinuities, the integral over them is

$$\sum_{p=1}^{q}\int_{S_p}(A'_p \cdot n'_p + A''_p \cdot n''_p)\,dS_p$$

where dS_p is an element of the original surface of discontinuity. Adding these terms to 3.00 (2) and substituting $A = \Psi\epsilon\nabla\Phi$, we have

$$\sum_{j=1}^{m}\int_{S_j}\epsilon\Psi\frac{\partial\Phi}{\partial n_j}\,dS_j + \sum_{p=1}^{q}\int_{S_p}\left(\epsilon'_p\Psi'_p\frac{\partial\Phi'_p}{\partial n'_p} + \epsilon''_p\Psi''_p\frac{\partial\Phi''_p}{\partial n''_p}\right)dS_p$$
$$= \int_V \epsilon\nabla\Psi\cdot\nabla\Phi\,dv + \int_V \Psi\nabla\cdot(\epsilon\nabla\Phi)\,dv \tag{1}$$

If we write a similar equation with Ψ and Φ interchanged and subtract the two, we obtain

$$\sum_{p=1}^{q}\int_{S_p}\left[\epsilon'_p\left(\Psi'_p\frac{\partial\Phi'_p}{\partial n'_p} - \Phi'_p\frac{\partial\Psi'_p}{\partial n'_p}\right) + \epsilon''_p\left(\Psi''_p\frac{\partial\Phi''_p}{\partial n''_p} - \Phi''_p\frac{\partial\Psi''_p}{\partial n''_p}\right)\right]dS_p$$
$$+ \sum_{j=1}^{m}\int_{S_j}\epsilon\left(\Psi\frac{\partial\Phi}{\partial n_j} - \Phi\frac{\partial\Psi}{\partial n_j}\right)dS_j = \int_V[\Psi\nabla\cdot(\epsilon\nabla\Phi) - \Phi\nabla\cdot(\epsilon\nabla\Psi)]\,dv \tag{2}$$

When ϵ is a constant, without discontinuities, (1) becomes

$$\sum_{j=1}^{m}\int_{S_j}\Psi\frac{\partial\Phi}{\partial n_j}\,dS_j = \int_V(\nabla\Psi\cdot\nabla\Phi + \Psi\nabla^2\Phi)\,dv \tag{3}$$

and (2) becomes

$$\sum_{j=1}^{m}\int_{S_j}\left(\Psi\frac{\partial\Phi}{\partial n_j} - \Phi\frac{\partial\Psi}{\partial n_j}\right)dS_j = \int_{V}(\Psi\,\nabla^2\Phi - \Phi\,\nabla^2\Psi)\,dv \qquad (4)$$

A useful vector analogue of (3) and (4) has been proved by Stratton as follows. In 3.00 (2) let $A = \mathbf{\Psi} \times (\nabla \times \mathbf{\Phi})$ where $\mathbf{\Psi}$ and $\mathbf{\Phi}$ are vector quantities that are finite and continuous in the region of integration and can be differentiated twice. Then

$$\sum_{j=1}^{m}\int_{S_j}[\mathbf{\Psi}\times(\nabla\times\mathbf{\Phi})]\cdot \mathbf{n}_j\,dS_j = \int_{V}\nabla\cdot[\mathbf{\Psi}\times(\nabla\times\mathbf{\Phi})]\,dv$$

$$\sum_{j=1}^{m}\int_{S_j}\mathbf{\Psi}\cdot[(\nabla\times\mathbf{\Phi})\times n]\,dS_j = \int_{V}[(\nabla\times\mathbf{\Psi})\cdot(\nabla\times\mathbf{\Phi}) - \mathbf{\Psi}\cdot\nabla\times(\nabla\times\mathbf{\Phi})]\,dv \qquad (5)$$

Subtraction of a similar equation with $\mathbf{\Psi}$ and $\mathbf{\Phi}$ interchanged gives

$$\sum_{j=1}^{m}\int_{S_j}[\mathbf{\Psi}\times(\nabla\times\mathbf{\Phi}) - \mathbf{\Phi}\times(\nabla\times\mathbf{\Psi})]\cdot n\,dS_j$$
$$= \int_{V}\{\mathbf{\Phi}\cdot[\nabla\times(\nabla\times\mathbf{\Psi})] - \mathbf{\Psi}\cdot[\nabla\times(\nabla\times\mathbf{\Phi})]\}\,dv \qquad (6)$$

3.07. Green's Reciprocation Theorem for Dielectrics.—In 3.06 (2), let $\Psi = V$ be the potential of one distribution of electricity $\Phi = V'$ that of another and let ϵ be the capacitivity. If the contact surfaces of the different dielectrics are uncharged, we have, from 1.17 (5),

$$\epsilon'_p\frac{\partial\Phi'}{\partial n'_p} = \epsilon''_p\frac{\partial\Phi''}{\partial n'_p} = -\epsilon''_p\frac{\partial\Phi''}{\partial n''_p}$$

and, from 1.17 (6), $\Psi'_p = \Psi''_p$ and $\Phi'_p = \Phi''_p$ and a similar relation for Ψ, so that the integrals over the surfaces of discontinuity disappear; and if there are no charges throughout the volume, $\nabla\cdot(\epsilon\,\nabla\Psi) = 0$ and $\nabla\cdot(\epsilon\,\nabla\Phi) = 0$, so that the volume integral vanishes, leaving

$$\sum_{j=1}^{m}\int_{S_j}\left(V_j\epsilon\frac{\partial V'}{\partial n_j} - V'_j\epsilon\frac{\partial V}{\partial n_j}\right)dS_j = 0$$

or

$$\sum_{j=1}^{m}\left(V_j\int_{S_j}\sigma'\,dS_j - V'_j\int_{S_j}\sigma\,dS_j\right) = 0$$

or

$$\sum_{j=1}^{m}Q_jV'_j = \sum_{j=1}^{m}Q'_jV_j \qquad (1)$$

Thus we have proved Green's reciprocation theorem 2.12 (1) to hold when dielectrics are present.

3.08. Green's Function.—Let Ψ be the potential due to a unit charge at the point P, and let Φ be the potential due to the induced surface density σ on some closed surface S at zero potential, which surrounds P. We may consider the unit charge at P as consisting of charge density ρ which is everywhere zero except in an infinitesimal volume dv enclosing P. This element is so small that Φ has a constant value Φ_p throughout it. On the surface $\Phi = -\Psi$, at a distance r from P, $\Psi = (4\pi\epsilon r)^{-1}$; also, $\nabla^2\Phi$ is zero throughout v, whereas $\nabla^2\Psi = -\rho/\epsilon$ in dv and is zero elsewhere. Substituting in 3.06 (4) gives

$$\int_S \frac{1}{4\pi r}\left\{\frac{\partial[\Phi + (4\pi\epsilon r)^{-1}]}{\partial n}\right\} dS = \Phi_p \int_V \rho\, dv = \Phi_p$$

where, as in 3.00 and 3.06, the positive direction of n is from v into S. Writing G for $\Phi + (4\pi\epsilon r)^{-1}$, this becomes

$$\Phi_p = \frac{1}{4\pi}\int_S \frac{1}{r}\frac{\partial G}{\partial n}\, dS = \frac{1}{4\pi\epsilon}\int_S \frac{\sigma}{r}\, dS \tag{1}$$

We shall refer to G as Green's function although some authors designate Φ by this name. It is evident that G is a solution of Laplace's equation which is zero over a given boundary and has a simple pole at a point P inside. Electrically, it represents the potential inside an earthed conducting surface under the influence of a unit charge at a point P within it. Equation (1), which is identical with 1.06 (6), is usually worthless for determining G as we rarely know σ on the conductor. Many methods of determining G for various types of boundaries will appear in subsequent chapters.

The force F_1 acting along u_1 on a charge q at u_1', u_2', u_3' in the coordinate system of 3.03 can be found from $G(u_1, u_2, u_3)$ by the formulas

$$\frac{q}{2}\lim_{\delta\to 0}\left\{\left[\frac{\partial G(u_1, u_2', u_3')}{-h_1\partial u_1}\right]_{u_1'+\delta} + \left[\frac{\partial G(u_1, u_2', u_3')}{-h_1\partial u_1}\right]_{u_1'-\delta}\right\} = \frac{q}{2}\frac{\partial G(u_1', u_2', u_3')}{-h_1\partial u_1'} \tag{2}$$

The first is valid because, when δ is small, the fields of the induced charges at $u_1' + \delta$ and $u_1' - \delta$ are the same, but those of the charge itself are equal and opposite and cancel out. The second form follows from 2.07 (2) because the field energy of the charge itself is unaltered by a change in u_1' so only the induced charge energy is affected.

3.09. Solution of Poisson's Equation.—In a vacuum, the potential at the point P due to a charge density ρ in the element of volume dv is, from 1.06 (2), $dV = \rho(4\pi\epsilon r)^{-1}\, dv$, where r is the distance from the element dv to P. Thus the potential at P due to all charges in space is

$$V_p = \frac{1}{4\pi\epsilon}\int_V \frac{\rho\, dv}{r} \tag{1}$$

But we know from 3.02 (3) that this potential satisfies Poisson's equation

$$\nabla^2 V = -\frac{\rho}{\epsilon} \tag{2}$$

Thus (1) is a solution of (2).

We may also solve (2) directly by means of Green's theorem. Let us apply this theorem to the region v between a very small sphere of area σ and a larger sphere of area Σ and radius R, both centers being at P. Write $\Psi = r^{-1}$ and $\Phi = V$ in 3.06 (4), and we have

$$\int_\sigma\left[\frac{1}{r}\frac{\partial V}{\partial n} - V\frac{\partial}{\partial n}\left(\frac{1}{r}\right)\right] dS + \int_\Sigma\left[\frac{1}{r}\frac{\partial V}{\partial n} - V\frac{\partial}{\partial n}\left(\frac{1}{r}\right)\right] dS = \int_{\mathrm{v}}\left[\frac{1}{r}\nabla^2 V - V\nabla^2\left(\frac{1}{r}\right)\right] d\mathrm{v} \quad (3)$$

Consider the first of these integrals. On the small sphere, $\partial/\partial n = -\partial/\partial r$ and the solid angle subtended by any area dS is $d\Omega = r^{-2}\,dS$, so that, since $\partial V/\partial r$ is finite, the integral is

$$-r\int\frac{\partial V}{\partial r}\,d\Omega - \int V\,d\Omega \underset{r\to 0}{\longrightarrow} -4\pi V_p \quad (4)$$

If the total charge Q producing the potential V lies in the finite region, then $V \to Q(4\pi\epsilon R)^{-1}$ as $R \to \infty$ so on the surface Σ, both terms in the integrand approach $Q(4\pi\epsilon R^3)^{-1}$ whereas the surface area is $4\pi R^2$. Thus the second integral becomes zero as $Q(\epsilon R)^{-1}$. Since r^{-1} is a solution of Laplace's equation, $\nabla^2(r^{-1}) = 0$ and (3) becomes, if (2) is used,

$$-4\pi V_p = \int_{\mathrm{v}}\frac{\nabla^2 V}{r}\,d\mathrm{v} = -\frac{1}{\epsilon}\int_{\mathrm{v}}\frac{\rho}{r}\,d\mathrm{v} \quad (5)$$

which gives us (1) again.

3.10. Uniqueness Theorem with Dielectrics Present.—If there were several solutions to Laplace's equation satisfying the same boundary conditions so that we had to resort to experiment to choose the correct one, there would be little use for potential theory. We shall now prove that if the location and size of all fixed charges inside a certain region are specified, as well as the value of the potential over all its boundaries, with the possible exception of certain closed conducting surfaces on which only the total charge is given; then the value of V is uniquely determined in this region. Suppose there are two values V and V', both of which satisfy the boundary conditions. Since both satisfy Poisson's equation, their difference is, from 3.02 (2),

$$\nabla\cdot[\epsilon\nabla(V - V')] = 0$$

everywhere, since fixed charges cancel. Put $\Phi = \Psi = V - V'$ in 3.06 (1), and remember that, as in 3.07, the integrals over the boundaries between dielectrics vanish if these are uncharged. The substitution gives

$$\sum_{j=1}^{m}\int_{S_j}\epsilon(V - V')\frac{\partial}{\partial n_j}(V - V')\,dS_j = \int_{\mathrm{V}}\epsilon[\nabla(V - V')]^2\,d\mathrm{v} \quad (1)$$

The surface integrals vanish in those terms on the left representing boundaries on which the potentials are specified, since $V = V'$ at every point. All surfaces on which the total charge is specified, but not the actual potential, are conducting surfaces so that V and V' are constant and may be taken out of the integrals, giving, for the sth such surface, by 1.15 (1),

$$(V_s - V'_s)\int_{S_s}\epsilon\frac{\partial}{\partial n_s}(V_s - V'_s)\,dS_s = (V_s - V'_s)(Q_s - Q'_s)$$

By hypothesis the charge on this surface is fixed so that $Q_s - Q'_s$ is zero. The whole left side of (1) is, therefore, zero and, since the integrand on the right is always positive, $\nabla(V - V')$ must vanish everywhere. Thus $V - V'$ must be a constant and since it is zero on the boundaries it must be zero everywhere and the theorem is proved.

3.11. Introduction of New Conductor.—We shall now prove that if a new uncharged or earthed conductor is introduced into the electric field produced by a system of charged conductors, the charge on all conductors remaining unchanged, the energy of the system is decreased. Let the energy, electric field, and the space occupied by this field be designated by W, E, and v, respectively, in the original system and by W', E', and v′ in the final system; then we have

$$W - W' = \frac{\epsilon}{2}\int_{\mathrm{v}} E^2\,d\mathrm{v} - \frac{\epsilon}{2}\int_{\mathrm{v}'} E'^2\,d\mathrm{v} \qquad (1)$$

$$= \frac{\epsilon}{2}\int_{\mathrm{v}-\mathrm{v}'} E^2\,d\mathrm{v} + \frac{\epsilon}{2}\int_{\mathrm{v}'}[(E - E')^2 - 2E'\cdot(E' - E)]\,d\mathrm{v}$$

Now putting $\Psi = V'$ and $\Phi = V - V'$ in 3.06 (3) and remembering that $\nabla^2\Phi = 0$ throughout the volume give

$$\int_{\mathrm{v}'}\nabla V'\cdot\nabla(V - V')\,d\mathrm{v} = \int_{\mathrm{v}'}E'\cdot(E - E')\,d\mathrm{v} = \sum_{j=1}^{m}\int_{S_j'}V'\left(\frac{\partial V}{\partial n_j} - \frac{\partial V'}{\partial n_j}\right)dS_j$$

$$= \frac{1}{\epsilon}\sum_{j=1}^{m}V'_j\int_{S_j'}(\sigma - \sigma')\,dS_j = \frac{1}{\epsilon}\sum_{j=1}^{m}V'_j(Q_j - Q'_j) = 0$$

since $Q'_j = Q_j$, so that the last term in the last integral (1) vanishes, and $W - W'$ is a positive quantity. Since we do no work in making an earth connection, any motion of electric charge caused thereby must be at the expense of the electric field and so cause a further decrease in energy.

3.12. Green's Equivalent Stratum.—Let V_P be the potential at some point P, outside a closed surface S, due to an electric charge density $\rho(x, y, z)$, throughout the volume v, enclosed by S. In 3.06 (4), let $\Psi = r^{-1}$ and $\Phi = V$, so that $\nabla^2\Phi = -\rho/\epsilon$ by 3.02 (3). This gives

$$\int_S\frac{1}{r}\frac{\partial V}{\partial n}\,dS - \int_S V\frac{\partial}{\partial n}\left(\frac{1}{r}\right)dS = -\frac{1}{\epsilon}\int_{\mathrm{v}}\frac{\rho\,d\mathrm{v}}{r} = -4\pi V_P \qquad (1)$$

by 1.06 (6), where r is the distance from P to $d\text{v}$. Thus we see that the potential V_P at P can be expressed either as a volume integral of the charge density in v or by two surface integrals over the surface, S, enclosing v. Referring to 1.06 (6) and 1.12 (1), we see that this means that V_P will be unchanged if we remove the volume distribution ρ and replace it by a surface stratum which coincides with S and carries a charge $-\epsilon(\partial V/\partial n)$ per unit area and an outward pointing dipole distribution of moment ϵV per unit area.

If S is an equipotential surface, the second integral becomes, by 3.00 (2), when V is taken out,

$$V\int_S \boldsymbol{n}\cdot\nabla\left(\frac{1}{r}\right)dS = V\int_{\text{v}}\nabla^2\left(\frac{1}{r}\right)d\text{v} \tag{2}$$

This is zero since $1/r$ is the potential of a point charge $4\pi\epsilon$ at P, and therefore $\nabla^2(1/r) = 0$ inside v. In this case, no dipole layer is required. This shows that an electric field is unchanged when any area of an equipotential surface is replaced by a very thin uncharged conducting sheet, because such a sheet may be regarded as two equipotential surfaces infinitely close together enclosing this area. By (1), they must have equal and opposite densities on the exterior faces to leave V_P unchanged.

3.13. Energy of a Dielectric Body in an Electric Field.—From the last article the forces exerted on external fixed charges by a given volume distribution and by its equivalent stratum are the same. It follows from Newton's laws that the reverse is also true. This fact facilitates the calculation of the work done in placing an uncharged dielectric body in the field produced by fixed sources in a region of capacitivity ϵ_0. If the stratum is brought into the field fully formed, 1.06 and 1.071 (2) give the energy of an area dS to be

$$dW = [\sigma V + (\boldsymbol{M}\cdot\nabla)V]\,dS \tag{1}$$

where σ is the charge density, $\boldsymbol{M}$ the dipole moment density, and V the potential of the external fixed charges. The dielectric polarization and hence the strength of the equivalent stratum are proportional in this case to the external field and are built up by it from zero to its final value so that the total work done is one-half that given by (1). Substitution for σ and $\boldsymbol{M}$ from the last article and integration over the surface of the stratum give

$$W = \tfrac{1}{2}\epsilon_0\int_S(-V\boldsymbol{n}\cdot\nabla V_o + V_o\boldsymbol{n}\cdot\nabla V)\,dS \tag{2}$$

where V_o is the final potential just outside the equivalent stratum or the dielectric surface. By 1.17 (5) and (6) the potential just inside the dielectric surface is related to V_o by the equations

$$V_o = V_i \qquad \text{and} \qquad \epsilon_0\boldsymbol{n}\cdot\nabla V_o = \epsilon\boldsymbol{n}\cdot\nabla V_i \tag{3}$$

where ϵ is the capacitivity of the body. Substitution of (3) in (2) and application of Gauss's theorem 3.00 (2) to the result give

$$W = \tfrac{1}{2}\int_S(\epsilon_o V_i \boldsymbol{n}\cdot\nabla V - \epsilon V \boldsymbol{n}\cdot\nabla V_i)\, dS = \tfrac{1}{2}\int_{\mathrm{V}}(\epsilon_o - \epsilon)\nabla V\cdot\nabla V_i\, d\mathrm{v} \quad (4)$$

since $\nabla^2 V = \nabla\cdot\epsilon\nabla V_i = 0$. Thus if the electric intensity produced in the volume v of a uniform isotropic medium of capacitivity ϵ_o by a fixed charge distribution is $\boldsymbol{E}$, and when v is occupied by a uniform isotropic body ϵ it is $\boldsymbol{E}_i$, then the energy difference is

$$W = \tfrac{1}{2}\int_{\mathrm{V}}(\epsilon_o - \epsilon)\boldsymbol{E}\cdot\boldsymbol{E}_i\, d\mathrm{v} \quad (5)$$

The torque or force acting on the body in the direction θ is

$$F = -\frac{\partial W}{\partial\theta} \quad (6)$$

3.14. Effect of an Increase of Capacitivity.—If the capacitivity at any point in the electric field produced by a system of charged conductors is increased, the energy of the system is decreased provided the charges on the conductors are kept constant. To prove this, let W be the energy of the system, Q_j the charge on the jth conductor, ρ the volume density of charge, V the potential at any point, and ϵ the capacitivity. We assume that Q_j and ρ remain constant when ϵ is varied but that V and W vary. Thus,

$$W = \tfrac{1}{2}\int_{\mathrm{V}}\epsilon E^2\, d\mathrm{v} = \tfrac{1}{2}\int_{\mathrm{V}}\epsilon(\nabla V)^2\, d\mathrm{v}$$

$$\delta W = \tfrac{1}{2}\int_{\mathrm{V}}\delta\epsilon(\nabla V)^2\, d\mathrm{v} + \int_{\mathrm{V}}\epsilon\nabla V\cdot\nabla(\delta V)\, d\mathrm{v} \quad (1)$$

Substitute δV for Ψ and V for Φ in 3.06 (1), set $\nabla\cdot(\epsilon\nabla V) = -\rho$, and assume uncharged dielectric surfaces over which the integral vanishes.

$$\int_{\mathrm{V}}\epsilon\nabla(\delta V)\cdot\nabla V\, d\mathrm{v} = \sum_{j=1}^{m}\int_{S_j}\delta V\epsilon\frac{\partial V}{\partial n_j}\, dS_j + \int_{\mathrm{V}}\delta V\,\rho\, d\mathrm{v}$$

$$= \sum_{j=1}^{m}\delta V_j\int_{S_j}\sigma\, dS_j + \int_{\mathrm{V}}\delta V\,\rho\, d\mathrm{v}$$

$$= \sum_{j=1}^{m}Q_j\,\delta V_j + \int_{\mathrm{V}}\delta V\,\rho\, d\mathrm{v} = 2\delta W$$

since we also have

$$W = \frac{1}{2}\sum_{j=1}^{m}Q_jV_j + \frac{1}{2}\int_{\mathrm{V}}\rho V\, d\mathrm{v}$$

Substituting in (1) and transferring $2\delta W$ to the left side, we have

$$\delta W = -\frac{1}{2}\int_V \delta\epsilon(\nabla V)^2\, dv \tag{2}$$

Thus δW is negative if $\delta\epsilon$ is positive.

3.15. Potential of Axially Symmetrical Field.—It can be verified by substitution in 3.05 (2) and integration by parts that a solution of Laplace's equation, if ϵ is constant and V does not depend on ϕ, is

$$V(z, \rho) = \frac{1}{2\pi}\int_0^{2\pi} \Phi(z + j\rho \sin \theta)\, d\theta \tag{1}$$

where $\Phi(z)$ is a real function of z whose Taylor expansion is, by *Dw* 39,

$$\Phi(z, \rho) = \Phi(z) + \Phi'(z) j\rho \sin\theta + (2!)^{-1}\Phi''(z)(j\rho \sin\theta)^2 + \cdots \tag{2}$$

Substitution of (2) in (1) and integration from 0 to 2π give

$$V(z, \rho) = \sum_{n=0}^{\infty} \frac{(-1)^n}{(n!)^2} \frac{\partial^{2n}\Phi(z)}{\partial z^{2n}} \left(\frac{\rho}{2}\right)^{2n} \tag{3}$$

It is evident that $V(z, 0)$ is identical with $\Phi(z)$. If no axial charges exist, (3) gives V uniquely at all points attainable from the axis without crossing a charged surface. The proof resembles that used for 7.07 (5).

Problems

Problems marked *C* are taken from the Cambridge examination questions as reprinted by Jeans by permission of the Cambridge University Press.

1. Show that the components of the curl in cylindrical coordinates are

$$\text{curl}_\rho A = \frac{1}{\rho}\frac{\partial A_z}{\partial \phi} - \frac{\partial A_\phi}{\partial z}$$

$$\text{curl}_\phi A = \frac{\partial A_\rho}{\partial z} - \frac{\partial A_z}{\partial \rho}$$

$$\text{curl}_z A = \frac{1}{\rho}\left[\frac{\partial(\rho A_\phi)}{\partial \rho} - \frac{\partial A_\rho}{\partial \phi}\right]$$

2. Show that the components of the curl in spherical polar coordinates are

$$\text{curl}_r A = \frac{1}{r \sin\theta}\left[\frac{\partial(\sin\theta A_\Phi)}{\partial\theta} - \frac{\partial A_\theta}{\partial\phi}\right]$$

$$\text{curl}_\theta A = \frac{1}{r\sin\theta}\left[\frac{\partial A_r}{\partial\phi} - \sin\theta\frac{\partial(rA_\Phi)}{\partial r}\right]$$

$$\text{curl}_\Phi A = \frac{1}{r}\left[\frac{\partial(rA_\theta)}{\partial r} - \frac{\partial A_r}{\partial\theta}\right]$$

3. Show that for ellipsoidal coordinates, as defined in 5.01, we write in 3.03 (1) $4h_1^2 = (u_1 - u_2)(u_1 - u_3)D_1$, $4h_2^2 = (u_2 - u_3)(u_2 - u_1)D_2$, $4h_3^2 = (u_3 - u_1)(u_3 - u_2)D_3$, where $D_{1,2,3} = [(a^2 + u_{1,2,3})(b^2 + u_{1,2,3})(c^2 + u_{1,2,3})]^{-1}$, $c > b > a$, $-b^2 < u_2 < -a^2$, and $-a^2 < u_1 < \infty$. For special cases of oblate and prolate spheroids, see 5.27 and 5.28.

4. If three sets of orthogonal surfaces are defined by the concentric spheres $u_1^2 = x^2 + y^2 + z^2$ and by the two cones $x^2u_2^{-2} + y^2(u_2^2 - b^2)^{-1} + z^2(u_2^2 - c^2)^{-1} = 0$ and $x^2u_3^{-2} + y^2(u_3^2 - b^2)^{-1} + z^2(u_3^2 - c^2)^{-1} = 0$, show that we must choose $h_1 = 1$, $h_2^2 = u_1^2(u_2^2 - u_3^2)[(u_2^2 - b^2)(c^2 - u_2^2)]^{-1}$, and $h_3^2 = u_1^2(u_2^2 - u_3^2)[(b^2 - u_3^2)(c^2 - u_3^2)]^{-1}$.

5. If the three sets of orthogonal surfaces are $x^2c^{-2}u_1^{-2} + y^2c^{-2}(u_1^2 - 1)^{-1} = 1$, $x^2c^{-2}u_2^{-2} - y^2c^{-2}(1 - u_2^2)^{-1} = 1$, and $z = u_3$, show that $h_1^2 = c^2(u_1^2 - u_2^2)(u_1^2 - 1)^{-1}$, $h_2^2 = c^2(u_1^2 - u_2^2)(1 - u_2^2)^{-1}$, and $h_3 = 1$.

6. If the orthogonal surfaces are $y^2 = 4cu_1x + 4c^2u_1^2$, $y^2 = -4cu_2x + 4c^2u_2^2$, and $z = u_3$, show that $h_1^2 = c^2u_1^{-1}(u_1 + u_2)$, $h_2^2 = c^2u_2^{-1}(u_1 + u_2)$, $h_3 = 1$.

7. If the sets of orthogonal surfaces are $z = c(u_1 - u_2)$, $x^2 + y^2 = 4c^2u_1u_2$, and $y = x \tan u_3$, show that $h_1 = c[u_1^{-1}(u_1 + u_2)]^{\frac{1}{2}}$, $h_2 = c[u_2^{-1}(u_1 + u_2)]^{\frac{1}{2}}$, $h_3 = 2c(u_1u_2)^{\frac{1}{2}}$.

8. If the orthogonal surfaces are $(x^2 + y^2)^{\frac{1}{2}} = c \sinh u_1(\cosh u_1 - \cos u_2)^{-1}$, $y = x \tan u_3$, and $z = c \sin u_2(\cosh u_1 - \cos u_2)^{-1}$, prove $h_1 = c(\cosh u_1 - \cos u_2)^{-1}$, $h_2 = h_1$, and $h_3 = c \sinh u_1(\cosh u_1 - \cos u_2)^{-1}$. These are known as toroidal coordinates and give orthogonal anchor rings, spheres, and planes.

9. If the orthogonal surfaces are $(x^2 + y^2)^{\frac{1}{2}} = c \sin u_2(\cosh u_1 - \cos u_2)^{-1}$, $y = x \tan u_3$, and $z = c \sinh u_1(\cosh u_1 - \cos u_2)^{-1}$, prove $h_1 = c(\cosh u_1 - \cos u_2)^{-1}$, $h_2 = h_1$, and $h_3 = c \sin u_2(\cosh u_1 - \cos u_2)^{-1}$. These are called bipolar coordinates, because if r_1 and r_2 are the vectors from a point P to the points $z = +c$ and $z = -c$, respectively, we have $u_1 = \ln (r_2r_1^{-1})$ and $\cos u_2 = (r_1 \cdot r_2)r_1^{-1}r_2^{-1}$.

10*C*. If the specific inductive capacity varies as $e^{-r/a}$, where r is the distance from a fixed point in the medium, verify that the differential equation solution satisfied by the potential is $a^2r^{-2}[e^{r/a} - 1 - ra^{-1} - r^2(2a^2)^{-1}] \cos \theta$, and hence determine the potential at any point of a sphere, whose inductive capacity is the above function of the distance from the center, when placed in a uniform field of force.

11*C*. If the electricity in the field is confined to a given system of conductors at given potentials and the inductive capacity of the dielectric is slightly altered according to any law such that at no point is it diminished and such that the differential coefficients of the increment are also small at all points, prove that the energy of the field is increased.

12. Find the condition where a set of two dimensional equipotentials $V_2 = f(z, y)$ can generate a set of equipotentials when rotated about the z axis. Show that if this is possible the potential is

$$V = A \int e^{-\int F(V_2)\,dV_2}\, dV_2 + B \qquad \text{where } F(V_2) = \frac{1}{y(\nabla V_2)^2} \frac{\partial V_2}{\partial y}$$

References

The subject matter of this chapter is treated in the following books:

ABRAHAM, M., and R. BECKER: "Classical Electricity and Magnetism," Blackie, 1932. Gives lucid derivations of general theorems using vector notation.

BUCHHOLZ, H.: "Electrische und Magnetische Potentialfelder," Springer, 1957. Gives a very complete treatment.

COLLIN, R. E.: "Field Theory of Guided Waves," McGraw-Hill, 1960. Chapter 2 includes a fine detailed treatment of Green's functions.

DURAND, E.: "Electrostatique et Magnetostatique," Masson et Cie, 1953. Extensive treatment in first four chapters.

GEIGER-SCHEEL: "Handbuch der Physik," Vols. III and XII, Berlin, 1928 and 1927.

JACKSON, J. D.: "Classical Electrodynamics," Wiley, 1962. Chapter 1 treats this material. Static Green's functions are applied neatly in Chap. 3.

JEANS, J. H.: "The Mathematical Theory of Electricity and Magnetism," Cambridge, 1925. Gives comprehensive treatment using long notation.

MASON, M., and W. WEAVER: "The Electromagnetic Field," University of Chicago Press, 1929, and Dover. Gives an excellent treatment using vector notation.

MAXWELL, J. C.: "Electricity and Magnetism," 3d ed. (1891), Dover, 1954. Gives extensive treatment using long notation.

PANOFSKY, W. K. H., and MELBA PHILLIPS: "Classical Electricity and Magnetism," Addison-Wesley, 1962. First two chapters cover this material.

STRATTON, J. A.: "Electromagnetic Theory," McGraw-Hill, 1941. Gives a very extensive and rigorous treatment.

TRALLI, N.: "Classical Electromagnetic Theory," McGraw-Hill, 1963. Chapter 1 contains clear and detailed treatment.

WEBSTER, A. G.: "Electricity and Magnetism," Macmillan, 1897. Clear treatment using long notation.

WIEN-HARMS: "Handbuch der Experimentalphysik," Vol. X, Leipzig, 1930.

CHAPTER IV

TWO-DIMENSIONAL POTENTIAL DISTRIBUTIONS

4.00. Field and Potential in Two Dimensions.—A potential problem is said to be two-dimensional when all equipotential surfaces in the field are cylindrical, which means that each can be generated by moving an infinite straight line parallel to some fixed straight line. The unit of charge is now a uniform line charge parallel to this axis and having a strength of 1 coulomb per meter length. We have already seen in 2.04 (1) that in a homogeneous isotropic dielectric the field intensity at a distance r from such a charge is radial and its magnitude is

$$E = \frac{q}{2\pi\epsilon r} \tag{1}$$

The potential obtained by integrating this is

$$V = -\frac{q}{2\pi\epsilon} \ln r + C \tag{2}$$

Clearly the zero of potential cannot be chosen, conveniently, at infinity as this would make C infinite. We usually give C a value that makes the computation as simple as possible.

Theoretically, a two-dimensional electrostatic problem can never occur as all conductors are finite. However, there are a vast number of important cases in which the lengths of the parallel cylindrical conductors are so great compared with the intervening spaces that the end effects are negligible, and the problem then becomes two-dimensional.

4.01. Circular Harmonics.—In its most general sense, the term "harmonic" applies to any solution of Laplace's equation. In the more usual but restricted sense, it applies to a solution of Laplace's equation in a specified coordinate system, which has the form of a product of three terms, each of which is a function of one coordinate only. The solution required to fit a given set of boundary conditions is then constructed by adding up a number of such harmonics which have been multiplied by suitable coefficients.

In the ordinary cylindrical coordinates described in 3.05, we should therefore have cylindrical harmonics of the form

$$V = R(\rho)\Phi(\phi)Z(z) \tag{1}$$

In the special case where $Z(z)$ is a constant, this reduces to a two-dimensional problem and the harmonics are called circular harmonics. For this case, we usually write r for ρ and θ for ϕ so that for a uniform isotropic dielectric Laplace's equation, 3.05 (2), becomes, when we multiply through by r^2,

$$r\frac{\partial}{\partial r}\left(r\frac{\partial V}{\partial r}\right) + \frac{\partial^2 V}{\partial \theta^2} = 0 \tag{2}$$

We now wish to find solutions of this equation of the form

$$V = R(r)\Theta(\theta) \tag{3}$$

Substituting in (2) and dividing through by (3) give

$$\frac{1}{R}\left(r\frac{\partial R}{\partial r} + r^2\frac{\partial^2 R}{\partial r^2}\right) + \frac{1}{\Theta}\frac{\partial^2\Theta}{\partial\theta^2} = 0$$

This equation will evidently be satisfied by solutions of the equations

$$\frac{d^2\Theta}{d\theta^2} = -n^2\Theta \tag{4}$$

$$\frac{d^2R}{dr^2} + \frac{1}{r}\frac{dR}{dr} = n^2\frac{R}{r^2} \tag{5}$$

A solution of (4), the simple harmonic motion equation, is

$$\Theta = A\cos n\theta + B\sin n\theta \tag{6}$$

and it is easily verified that the solution of (5) is

$$R = Cr^n + Dr^{-n} \tag{7}$$

where $n \neq 0$. If $n = 0$, we have the solutions

$$\Theta = A\theta + B \tag{8}$$

$$R = C\ln r + D \tag{9}$$

The number n is called the degree of the harmonic. The circular harmonics then are

Degree zero	$V = (A\theta + B)(C\ln r + D)$	(10)
Degree not zero	$V = (A\cos n\theta + B\sin n\theta)(Cr^n + Dr^{-n})$	(11)

A sum of such terms, with different constants for each n, or an integral with respect to n is also a solution of (2). Thus

$$V = \sum_n \Theta_n R_n \qquad \text{or} \qquad V = \int f(n)\Theta_n R_n\,dn \tag{12}$$

It should be noticed that we have placed no restrictions on n. If the equation

$$\Sigma(A_n\cos n\theta + B_n\sin n\theta) = \Sigma(C_n\cos n\theta + D_n\sin n\theta) \tag{13}$$

holds for all values of θ where n is an integer, then we have the relations

$$A_n = C_n \qquad \text{and} \qquad B_n = D_n \tag{14}$$

To prove this, it is necessary only to multiply both sides of (13) by $\cos m\theta$ and integrate from 0 to 2π, giving

$$\sum\left(A_n\int_0^{2\pi}\cos n\theta\cos m\theta\,d\theta + B_n\int_0^{2\pi}\sin n\theta\cos m\theta\,d\theta\right)$$
$$= \sum\left(C_n\int_0^{2\pi}\cos n\theta\cos m\theta\,d\theta + D_n\int_0^{2\pi}\sin n\theta\cos m\theta\,d\theta\right)$$

If $m \neq n$, using *Pc* 361 and 360 or *Dw* 445 and 465, we see that all terms on both sides drop out. If $m = n$, we use *Pc* 489 and 364 or *Dw* 450.11 and 858.4 and obtain

$$\tfrac{1}{2}\pi A_n = \tfrac{1}{2}\pi C_n$$

A similar procedure using the sine proves the second part of (14).

4.02. Harmonic Expansion of Line Charge Potential.—In using circular harmonics, it is frequently necessary to have the expansion due to a line charge whose coordinates are r_0, θ_0. If R is the distance from the line charge q to the field point P (Fig. 4.02), then, from 4.00 (2) when C is zero, the potential V at P is given by

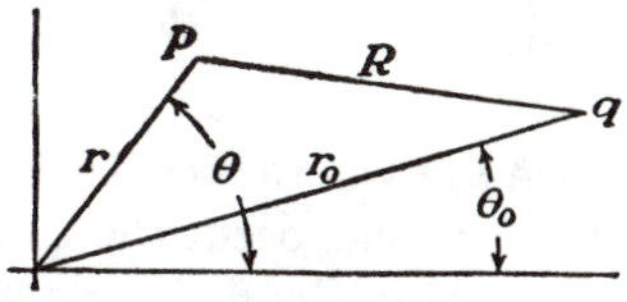

FIG. 4.02.

$$4\pi\epsilon V = -2q\ln R = -q\ln\left[r^2 + r_0^2 - 2rr_0\cos(\theta-\theta_0)\right]$$
$$= -2q\ln r - q\ln\left[1 - \frac{r_0}{r}e^{j(\theta-\theta_0)}\right]\left[1 - \frac{r_0}{r}e^{-j(\theta-\theta_0)}\right]$$

Using *Pc* 768 or *Dw* 601 where r_0 is less than r, we obtain

$$-2q\ln r + q\left\{\frac{r_0}{r}[e^{j(\theta-\theta_0)} + e^{-j(\theta-\theta_0)}] + \frac{1}{2}\left(\frac{r_0}{r}\right)^2[e^{j2(\theta-\theta_0)} + e^{-j2(\theta-\theta_0)}] + \cdots\right\}$$
$$= -2q\left[\ln r - \frac{r_0}{r}\cos(\theta-\theta_0) - \frac{1}{2}\left(\frac{r_0}{r}\right)^2\cos 2(\theta-\theta_0) - \cdots\right]$$

Writing as a summation and expanding $\cos n(\theta-\theta_0)$ by *Pc* 592 or *Dw* 401.04 give

$$V = \frac{q}{2\pi\epsilon}\left[\sum_{n=1}^{\infty}\frac{1}{n}\left(\frac{r_0}{r}\right)^n(\cos n\theta_0\cos n\theta + \sin n\theta_0\sin n\theta) - \ln r\right] \qquad (1)$$

This holds when $r > r_0$. When $r < r_0$, the same procedure gives, factoring out $\ln r_0$ instead of $\ln r$,

$$V = \frac{q}{2\pi\epsilon}\left[\sum_{n=1}^{\infty}\frac{1}{n}\left(\frac{r}{r_0}\right)^n(\cos n\theta_0\cos n\theta + \sin n\theta_0\sin n\theta) - \ln r_0\right] \qquad (2)$$

These are the required expansions.

4.03. Conducting or Dielectric Cylinder in Uniform Field.—As an example involving boundary conditions at both dielectric and conducting surfaces, let us find the field at all points when an infinite conducting cylinder of radius a surrounded by a layer, of relative capacitivity K and radius b, is set with its axis perpendicular to a uniform field of strength E.

The original potential outside the cylinder is of the form

$$V = Ex = Er \cos \theta$$

At infinity, the potential superimposed on this, due to the induced charges on the cylinder, must vanish so that no terms of the form r^n can occur in it. Since this potential must be symmetrical about the x-axis, no terms involving $\sin n\theta$ can occur. The final potential outside must therefore be of the form

$$V_o = Er \cos \theta + \sum_{n=1}^{\infty} A_n r^{-n} \cos n\theta \tag{1}$$

Since the values $r = 0$ and $r = \infty$ are excluded from the dielectric, both r^n and r^{-n} can occur there but $\sin n\theta$ terms are thrown out as before. The potential in the dielectric must therefore be of the form

$$V_i = \sum_{n=1}^{\infty} (B_n r^n + C_n r^{-n}) \cos n\theta \tag{2}$$

If we take the center of the cylinder at the origin, then $V = 0$ throughout the conductor. We have therefore found solutions of Laplace's equation that satisfy the boundary conditions at infinity and the symmetry conditions. It is now necessary only to determine A_n, B_n, and C_n so that the boundary conditions at the dielectric and conducting surfaces are satisfied. From 1.17 (5) and (6), at $r = b$ the boundary conditions are

$$\epsilon_v \frac{\partial V_o}{\partial r} = \epsilon \frac{\partial V_i}{\partial r} \quad \text{or} \quad \frac{\partial V_o}{\partial r} = K \frac{\partial V_i}{\partial r} \quad \text{and} \quad V_o = V_i \tag{3}$$

Substituting (1) and (2) in (3) gives

$$E \cos \theta - \Sigma n A_n b^{-n-1} \cos n\theta = K \Sigma n (B_n b^{n-1} - C_n b^{-n-1}) \cos n\theta$$
$$Eb \cos \theta + \Sigma A_n b^{-n} \cos n\theta = \Sigma (B_n b^n + C_n b^{-n}) \cos n\theta$$

From 4.01 (14), we have, if $n \neq 1$,

$$-A_n b^{-n-1} = K B_n b^{n-1} - K C_n b^{-n-1} \tag{4}$$
$$A_n b^{-n} = B_n b^n + C_n b^{-n} \tag{5}$$

At the conducting surface, $r = a$, $V_i = 0$, so that

$$0 = B_n a^n + C_n a^{-n} \tag{6}$$

Adding (4) $\times$ b to (5) gives

$$(K + 1) B_n = (K - 1) C_n b^{-2n} \tag{7}$$

The only way (6) and (7) can be satisfied is to have $B_n = C_n = 0$ or $(K + 1)/(K - 1) = -(a/b)^{2n}$. The latter condition is impossible since the left side is positive and the right side negative so that the first holds, and substituting in (5) we have

$$A_n = B_n = C_n = 0 \tag{8}$$

When $n = 1$ we have, instead of (4), (5), and (6),

$$\begin{aligned} E - A_1 b^{-2} &= KB_1 - KC_1 b^{-2} \\ E + A_1 b^{-2} &= B_1 + C_1 b^{-2} \\ 0 &= B_1 a + C_1 a^{-1} \end{aligned}$$

Solving these equations for A_1, B_1, and C_1 gives

$$\begin{aligned} A_1 &= -Eb^2 \frac{(K+1)a^2 + (K-1)b^2}{(K+1)b^2 + (K-1)a^2} \\ B_1 &= \frac{2Eb^2}{(K+1)b^2 + (K-1)a^2} \\ C_1 &= \frac{-2Ea^2b^2}{(K+1)b^2 + (K-1)a^2} \end{aligned} \tag{9}$$

The potentials (1) and (2) then become

$$V_o = \left(Er + \frac{A_1}{r}\right)\cos\theta, \qquad V_i = \left(B_1 r + \frac{C_1}{r}\right)\cos\theta \tag{10}$$

The lines of electric displacement are shown in Fig. 4.03a.

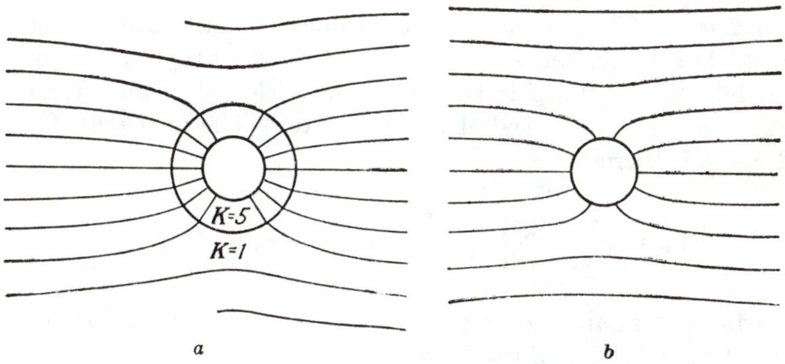

FIG. 4.03.

We may get the case of a conducting cylinder of radius a by letting $K = 1$ in (9), so that

$$A_1 = C_1 = -Ea^2 \qquad \text{and} \qquad B_1 = E$$

and the potential is

$$V_o = E\left(r - \frac{a^2}{r}\right)\cos\theta \tag{11}$$

The field is shown in Fig. 4.03b.

We get the case of a dielectric cylinder of radius b by letting $a = 0$ in (9), so that

$$A_1 = -\frac{K-1}{K+1}b^2E, \qquad B_1 = \frac{2E}{K+1}, \qquad C_1 = 0$$

and the potentials are

$$V_o = E\left(r - \frac{K-1}{K+1}\frac{b^2}{r}\right)\cos\theta, \qquad V_i = \frac{2E}{K+1}r\cos\theta \tag{12}$$

We notice that the field inside the dielectric is uniform. The displacement is shown, for $K = 5$, in Fig. 4.03c.

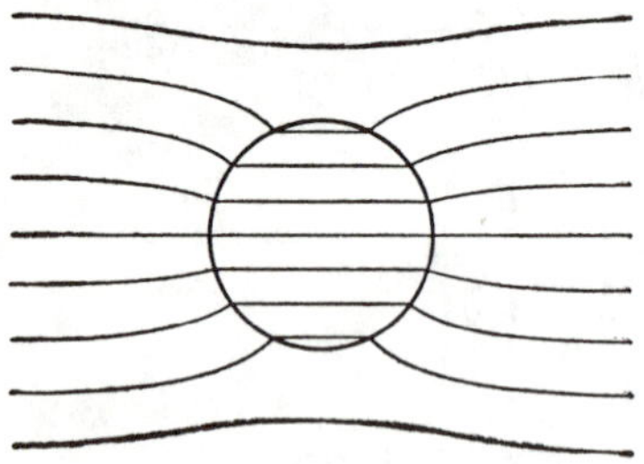

FIG. 4.03c.

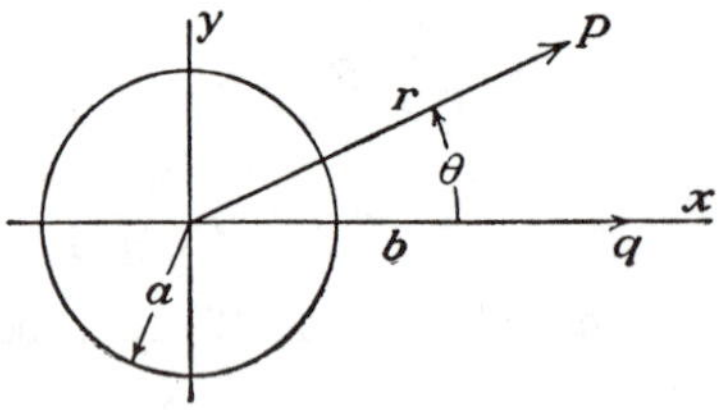

FIG. 4.04.

4.04. Dielectric Cylinder. Method of Images.—Let us consider a dielectric cylinder of radius a under the influence of a line charge at $r = b, \theta = 0$. The potential due to the line charge alone may be obtained from 4.02 (2) by putting $r_0 = b, \theta_0 = 0$. We must superimpose on this a potential, due to the polarization of the dielectric, which vanishes at infinity and is symmetrical about the x-axis. The final potential when $a < r < b$ is therefore

$$V_o = \frac{q}{2\pi\epsilon_v}\left\{\sum_{n=1}^{\infty}\left[\frac{1}{n}\left(\frac{r}{b}\right)^n + \frac{A_n}{r^n}\right]\cos n\theta - \ln b + C_1\right\} \tag{1}$$

Since the potential inside must be finite when $r = 0$ and symmetrical about the x-axis, it is of the form

$$V_i = \frac{q}{2\pi\epsilon_v}\left(\sum_{n=1}^{\infty} B_n r^n \cos n\theta + C_2\right) \tag{2}$$

Setting $V_o = V_i$ when $r = a$, we have

$$\frac{1}{n}\left(\frac{a}{b}\right)^n + \frac{A_n}{a^n} = a^n B_n \qquad \text{and} \qquad C_2 = -\ln b + C_1 \tag{3}$$

Setting $\epsilon_v \partial V_o/\partial r = \epsilon \partial V_i/\partial r$ or $\partial V_o/\partial r = K\partial V_i/\partial r$ when $r = a$, gives

$$\frac{a^{n-1}}{b^n} - \frac{n}{a^{n+1}}A_n = nKa^{n-1}B_n \tag{4}$$

Solving (3) and (4) for A_n and B_n gives

$$A_n = \frac{1-K}{1+K}\frac{a^{2n}}{nb^n}, \qquad B_n = \frac{2}{(1+K)nb^n}$$

The potential outside then becomes

$$V_o = \frac{q}{2\pi\epsilon_v}\left\{\sum_{n=1}^{\infty}\frac{1}{n}\left[\left(\frac{r}{b}\right)^n + \frac{1-K}{1+K}\left(\frac{a^2}{b}\right)^n\frac{1}{r^n}\right]\cos n\theta - \ln b + C_1\right\} \tag{5}$$

and that inside becomes

$$V_i = \frac{q}{\pi\epsilon_v(1+K)}\sum_{n=1}^{\infty}\frac{1}{n}\left(\frac{r}{b}\right)^n \cos n\theta - \frac{q}{2\pi\epsilon_v}(\ln b - C_1) \tag{6}$$

If we let

$$C_1 = 0 = -\frac{1-K}{1+K}\ln r + \frac{1-K}{1+K}\ln r$$

then (5) becomes exactly the expansion given by 4.02 (1) and (2) for the potential due to three line charges on the x-axis, one of strength q' at $x = a^2/b$, one of strength $-q'$ at $x = 0$, and one of strength q at $x = b$. Also, (6) gives the expansion due to a charge q'' situated at $x = b$, where

$$q' = \frac{1-K}{1+K}q \qquad \text{and} \qquad q'' = \frac{2}{1+K}q \tag{7}$$

Thus when an uncharged dielectric cylinder of radius a is brought into the neighborhood of a line charge q with its axis parallel to the charge and at a distance b from it, the additional potential in the region outside the cylinder due to its presence is the same as if it were replaced by a parallel "image" line charge q' located between q and the axis at a distance a^2/b from the latter, plus a line charge $-q'$ at the origin. The potential inside the cylinder is the same, except for the additive constant, as if it were absent and q were replaced by a charge q''. Some authors write $2K$ instead of 2 in the numerator of q'', in which case the potential inside the cylinder must be calculated as if all space were filled with the dielectric K.

When a dielectric cylinder is introduced into any electric field produced by a two-dimensional charge distribution parallel to its axis, it follows from the last paragraph, since such a distribution can always be built up from line charges, that the form of the field inside the cylinder is unchanged but its intensity is reduced by the factor $2/(K+1)$.

4.05. Image in Conducting Cylinder.—It follows from 1.17 (8) that as $K \to \infty$ the lines of force become incident normally at the dielectric surface. This is the condition at the boundary of a conductor, so that we may get the law of images in an uncharged conducting cylinder by letting $K \to \infty$ in 4.04 (7), giving $q' = -q$. Thus when a line charge of strength q is placed parallel to the axis of an uncharged conducting cylinder of radius a and at a distance b from it, the additional field outside the cylinder due to its presence is the same as if it were replaced by a parallel line charge $-q$ located between q and the axis at a distance a^2/b from the latter plus a charge q on the axis.

As will be noted from Fig. 4.05, the image position can be located by drawing a tangent to the dielectric or conducting cylinder from the point q. The line through this point, normal to $-qq'$, intersects $-qq'$ at q'.

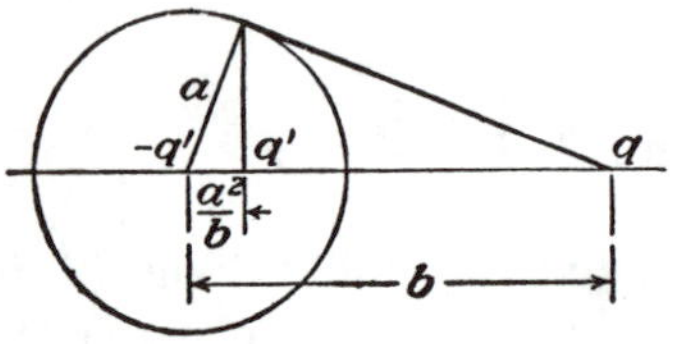

FIG. 4.05.

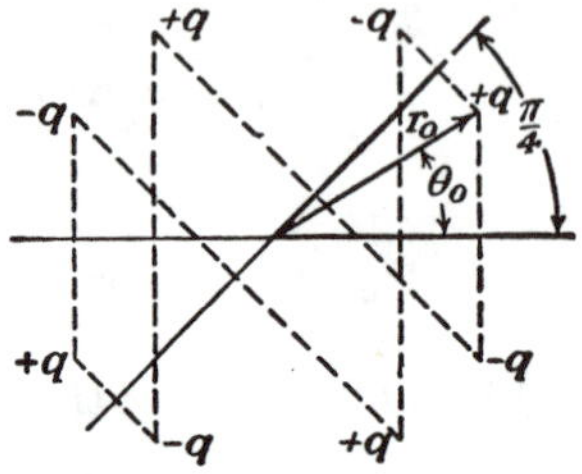

FIG. 4.06.

4.06. Image in Plane Face of Dielectric or Conductor. Intersecting Conducting Planes.—If we let the radius of the cylinder become infinite, keeping the distance $d = b - a$ of the line charge from the face constant, then the distance of the image q' from the face is

$$a - a^2b^{-1} = ab^{-1}(b - a) \to d$$

Thus the image laws derived in the last two sections apply to a line charge lying parallel to the faces of a semi-infinite dielectric or conducting block. The image charge is the same distance back of the face as the actual charge is in front. For a conductor, $q' = -q$, and for a dielectric, q' and q'' are given by 4.04 (7).

It is evident from Fig. 4.06 that if two planes intersect at the origin at an angle π/m, where m is an integer, both planes will be equipotentials under the influence of line charges parallel to the intersection lying in the cylinder $r = r_0$ and arranged with $+q$ at θ_0, $2\pi m^{-1} + \theta_0$, $4\pi m^{-1} + \theta_0$, $\ldots$, $2(m-1)\pi m^{-1} + \theta_0$ and $-q$ at $2\pi m^{-1} - \theta_0$, $4\pi m^{-1} - \theta_0$, $\ldots$, $2\pi - \theta_0$.

4.07. Dielectric Wedge.—Another solution of 4.01 (2) is obtained if $n \neq 0$ by writing jn for n in 4.01 (4), (5), (6), and (7), so that

$$\Theta_n = A \cosh n\theta + B \sinh n\theta \qquad (1)$$

$$R_n = C'_{1,2}r^{\pm jn} = C'_{1,2}e^{\pm jn \ln r} = C \cos (n \ln r) + D \sin (n \ln r) \qquad (2)$$

This solution is periodic in $\ln r$ instead of θ so that in 4.01 (12) R_n not Θ_n is now the orthogonal function. These harmonics can be used to solve the problem of a dielectric wedge of capacitivity ϵ_2 bounded by the planes $\theta = -\alpha$ and $\theta = \alpha$ under the influence of a line charge q at $\theta = \gamma$, $r = a$ in a medium of capacitivity ϵ_1 as shown in Fig. 4.07. There are no cylindrical boundaries where the sine and cosine terms in (2) must vanish so n is not restricted to discrete values but may vary continuously which indicates the integral form of 4.01 (12). We write β for $(\epsilon_1 - \epsilon_2)/(\epsilon_1 + \epsilon_2)$ and choose potentials of the forms

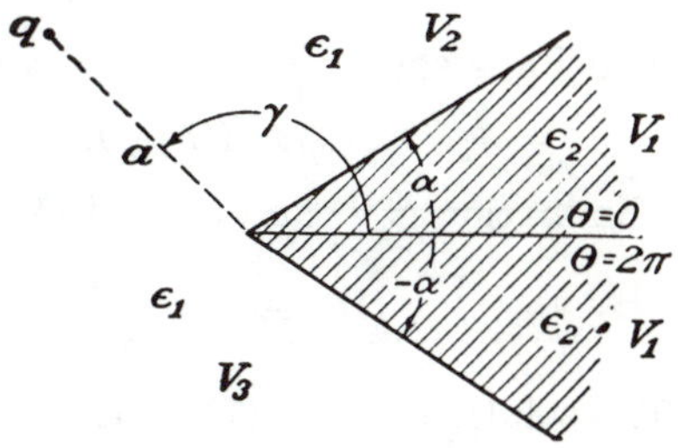

FIG. 4.07.

$$V_1 = (1+\beta)\int_0^\infty [A(k)e^{k\theta} + B(k)e^{-k\theta}] \cos [k \ln (r/a)]\, dk + C_0 \tag{3}$$

$$V_2 = \int_0^\infty [C(k)e^{k\theta} + D(k)e^{-k\theta}] \cos [k \ln (r/a)]\, dk + C_0 \tag{4}$$

$$V_3 = \int_0^\infty [E(k)e^{k\theta} + F(k)e^{-k\theta}] \cos [k \ln (r/a)]\, dk + C_0 \tag{5}$$

according to whether $-\alpha < \theta < \alpha$ (3), $\alpha < \theta < \gamma$ (4), or $\gamma < \theta < 2\pi - \alpha$ (5). The constant C_0 can be chosen to make V zero at any specified point. For this point at $r = a$, $\theta = 0$, it has the value

$$-(1 + \beta)\int_0^\infty [A(k) + B(k)]\, dk$$

The circle $r = a$ passing through q is a line of force because $\partial V/\partial r$ is zero there. Thus half the flux from q goes to $r = \infty$ and half to $r = 0$ where there is a charge $-\frac{1}{2}q$. By Fourier's integral theorem, if two of the integrals in (3), (4), and (5) are equal to each other over the whole range of $\ln (r/a)$, then their integrands are equal. Application of 1.17 (5) and (6) to the integrands of V_1 and V_2 at $\theta = \alpha$ gives, after rearrangement,

$$C = A + \beta e^{-2k\alpha}B, \qquad D = \beta e^{2k\alpha}A + B \tag{6}$$

Similarly joining V_1 at $\theta = -\alpha$ to V_3 at $\theta = 2\pi - \alpha$ gives

$$E = (A + \beta B e^{2k\alpha})e^{-2k\pi}, \qquad F = (\beta A e^{-2k\alpha} + B)e^{2k\pi} \tag{7}$$

To meet the remaining boundary conditions at $\theta = \gamma$, we write down an expression for the flux density originating anywhere in the $\theta = \gamma$ plane and evaluate it by Gauss's theorem, thus

$$\frac{\epsilon_1}{r}\left(\frac{\partial V_2}{\partial\theta} - \frac{\partial V_3}{\partial\theta}\right) = \frac{\epsilon_1}{r}\int_0^\infty k[Ce^{k\gamma} - De^{-k\gamma} - Ee^{k\gamma} + Fe^{-k\gamma}] \cos\left(k \ln \frac{r}{a}\right) dk$$

When both sides are multiplied by $\cos [t \ln (r/a)]\, dr$ and integrated from $r = 0$ to $r = \infty$, the left side is simply q by Gauss's theorem because the integrand vanishes except near $r = a$ and the right side is found by Fourier's integral theorem so that

$$q = \tfrac{1}{2}\pi\epsilon_1 k[(C - E)e^{k\gamma} - (D - F)e^{-k\gamma}]$$

Also V_2 is equal to V_3 at $\theta = \gamma$ so that

$$(C - E)e^{k\gamma} = -(D - F)e^{-k\gamma}$$

Elimination of D and F or of C and E from these equations gives

$$C = E + \frac{q}{\pi\epsilon_1 k}e^{-k\gamma}, \qquad D = F - \frac{q}{\pi\epsilon_1 k}e^{k\gamma} \tag{8}$$

The solution of (6), (7), and (8) for A and B gives

$$A,B = \frac{q[e^{\mp k(\gamma-\pi)} \sinh k\pi - e^{\pm k(\gamma-\pi)}\, \beta \sinh k(\pi - 2\alpha)]}{2\pi\epsilon_1 k[\sinh^2 k\pi - \beta^2 \sinh^2 k(\pi - 2\alpha)]} \tag{9}$$

where the upper signs go with A and the lower with B. In the special symmetrical case where $\gamma = \pi$, the potential in the dielectric wedge is

$$V_1 = \frac{q(1 + \beta)}{\pi\epsilon_1}\int_0^\infty \frac{\cosh k\theta \cos [k \ln (r/a)] - 1}{k[\sinh k\pi + \beta \sinh k(\pi - 2\alpha)]}\,dk \tag{10}$$

The integrand, which is finite at $k = 0$ and falls off exponentially as k increases, can be plotted as a function of k and the integral then evaluated with a planimeter. If the only charge is to be at $r = a$, $\theta = \gamma$, then the charge $-\frac{1}{2}q$ at the origin may be canceled by addition to V_1, V_2, and V_3 of the term

$$\tfrac{1}{4}q \ln \frac{r}{a}[\alpha(\epsilon_1 - \epsilon_2) - \pi\epsilon_1]^{-1} \tag{11}$$

If the cylinder $r = b$ is at zero potential, then we must superimpose two solutions of the form just derived, one for q at $r = a$, $\theta = \gamma$ and the other for $-q$ at $r = b^2/a$, $\theta = \gamma$. If the cylinders $r = b$ and $r = c$ are at zero potential, then we shall need discrete values of n in (2) and the potential function will be a series instead of an integral.

4.08. Complex Quantities.—Before taking up the general subjects of conjugate functions and conformal transformations, it will be well to review some of the more important properties of complex quantities. If $z = x + jy$, it is clear that to every point in the xy-plane, usually called, in this connection, the z-plane, there corresponds one value of z. In polar coordinates, we have, using *Pc* 609 or *Dw* 408.04,

$$z = x + jy = r \cos \theta + jr \sin \theta = re^{j\theta} \tag{1}$$

The magnitude of the vector r is known as the modulus of z and written $|z|$. The angle θ is known as the argument, amplitude,* phase or angle

* This term, universally understood by mathematicians to represent the angle in this case, is also, unfortunately, in general use to specify the maximum fluctuation of any alternating quantity in connection with alternating currents. Thus, when the complex notation is used for such quantities, it becomes practically identical with the term "modulus" as used here.

of z. When a complex number z is raised to a power n, we have

$$z^n = r^n e^{jn\theta} \tag{2}$$

so we may say that the modulus of z^n is the nth power of the modulus of z and the argument of z^n is n times the argument of z. When we take the product of two complex numbers, we get

$$zz_1 = rr_1 e^{j(\theta+\theta_1)} \tag{3}$$

so that the modulus of the product of two complex numbers equals the product of their moduli and the argument is the sum of their arguments. By putting z_1^{-1} for z_1 in (3), we see that the modulus of the quotient of two complex numbers equals the quotient of their moduli and the argument equals the difference of their arguments.

If we have $z_1 = x_1 + jy_1 = f(z) = f(x + jy)$, then the quantity $z_1^* = x_1 - jy_1 = f(x - jy)$ is called the complex conjugate of z. That the latter relation holds if $f(z)$ is an analytic function can be shown by expanding $f(x \pm jy)$ in a power series with real coefficients, for wherever $\pm j$ occurs to an even power the term is real and the choice of sign disappears, whereas whenever $\pm j$ occurs to an odd power the term is imaginary and the choice of sign remains as before. Thus we have

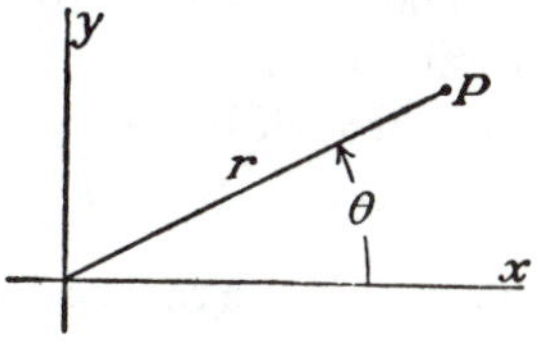

FIG. 4.08.

$$|z_1|^2 = x_1^2 + y_1^2 = z_1 z_1^* = f(x + jy) \cdot f(x - jy)$$

This gives us the rule that to obtain the modulus of a function of a complex variable we multiply it by its complex conjugate and take the square root of the product.

4.09. Conjugate Functions.—Laplace's equation in two dimensions and rectangular coordinates is written

$$\frac{\partial^2 U}{\partial x^2} + \frac{\partial^2 U}{\partial y^2} = 0 \tag{1}$$

Since this is a partial differential equation of the second order, the general solution must contain two arbitrary functions and it is easily verified by differentiation that such a solution is

$$U = \Phi(x + jy) + \Psi(x - jy)$$

It should be noted that to be a solution of (1), Φ and Ψ must possess definite derivatives where (1) holds, and hence are analytic functions in this region and capable of expansion in a power series (see Whittaker and Watson, "Modern Analysis," Chap. V). Since U is to be the electrostatic potential, it must be a real quantity and so the imaginary part of Φ must be equal and opposite to that of Ψ, so that, if $\Phi(x + jy) = u + jv$, then $\Psi(x - jy) = w - jv$, where u, v, and w are real quantities. Since both Φ and Ψ are analytic functions, we may expand them in series; thus we have

$$\Phi(x + jy) = \Phi(re^{j\theta}) = \sum_{n=0}^{\infty} A_n r^n e^{jn\theta} = \sum_{n=0}^{\infty} A_n r^n \cos n\theta + j\sum_{n=0}^{\infty} A_n r^n \sin n\theta$$

$$\Psi(x - jy) = \Psi(re^{-j\theta}) = \sum_{n=0}^{\infty} B_n r^n e^{-jn\theta} = \sum_{n=0}^{\infty} B_n r^n \cos n\theta - j\sum_{n=0}^{\infty} B_n r^n \sin n\theta$$

Since the imaginary parts of Φ and Ψ are equal and opposite over the entire range of θ, we see from 4.01 (14) that $A_n = B_n$ so that the real parts of Φ and Ψ are equal or $u = w$. Thus we have $U = 2u$.

Let V be another real quantity such that $V = 2v$, and we have

$$U + jV = 2(u + jv) = 2\Phi(x + jy) = f(x + jy) \tag{2}$$

The function V also satisfies Laplace's equation as may be shown by noting that the above expansion is in circular harmonics or by multiplying through by $-j$, giving

$$V - jU = -jf(x + jy) = F(x + jy)$$

Thus V is the real part of $F(x + jy)$, just as U is the real part of $f(x + jy)$. We shall write W for $U + jV$ and z for $x + jy$, giving

$$W = f(z) \tag{3}$$

The functions $U(x, y)$ and $V(x, y)$ are called conjugate functions.

4.10. The Stream Function.—Differentiating (3) with respect to x and y gives

$$\frac{\partial W}{\partial x} = \frac{\partial U}{\partial x} + j\frac{\partial V}{\partial x} = f'(z)\frac{\partial z}{\partial x} = f'(z)$$

$$\frac{\partial W}{\partial y} = \frac{\partial U}{\partial y} + j\frac{\partial V}{\partial y} = f'(z)\frac{\partial z}{\partial y} = jf'(z)$$

Multiply the second equation by j, add to the first, equate real and imaginary parts, and we have

$$\frac{\partial V}{\partial x} = -\frac{\partial U}{\partial y} \quad \text{and} \quad \frac{\partial V}{\partial y} = \frac{\partial U}{\partial x} \tag{1}$$

This is the condition that the two families of curves $U(x, y) =$ constant and $V(x, y) =$ constant intersect each other orthogonally. As we have seen, we may choose either set to represent equipotentials, in which case we call this the potential function. The other set which intersects this set everywhere orthogonally then represents the lines of force and is known as the stream function.

4.11. Electric Field Intensity. Electric Flux.—Let us consider the derivative

$$\frac{dW}{dz} = \frac{dU + j\,dV}{dx + j\,dy} = \frac{(\partial U/\partial x)dx + (\partial U/\partial y)dy + j[(\partial V/\partial x)dx + (\partial V/\partial y)dy]}{dx + j\,dy}$$

Substitute for $\partial U/\partial x$ and $\partial U/\partial y$ from 4.10 (1).

$$\frac{dW}{dz} = \frac{(\partial V/\partial y)(dx + j\,dy) + j(\partial V/\partial x)(dx + j\,dy)}{dx + j\,dy}$$

$$= \frac{\partial V}{\partial y} + j\frac{\partial V}{\partial x} = \frac{\partial U}{\partial x} - j\frac{\partial U}{\partial y} \qquad (1)$$

Thus if V is the potential function, the imaginary part of $-dW/dz$ gives the x-component of the electric field intensity and the real part gives the y-component. Regardless of whether U or V is the potential function, the absolute value of dW/dz at any point gives the magnitude of the electric field intensity at that point. If dn is an element of length in the direction of maximum increase of potential and ds is the element of length obtained by rotating dn counterclockwise $\frac{1}{2}\pi$ radians, then we get from (1)

$$\left|\frac{dW}{dz}\right| = \frac{\partial U}{\partial n} = \frac{\partial V}{\partial s} \quad \text{or} \quad \left|\frac{dW}{dz}\right| = \frac{\partial V}{\partial n} = -\frac{\partial U}{\partial s} \qquad (2)$$

according as U or V is the potential function. In the latter case, if we wish to know the flux through any section of an equipotential surface between the curves U_1 and U_2, we integrate, giving by 1.10 (1),

$$\text{Flux} = -\epsilon\int_{U_1}^{U_2}\frac{\partial V}{\partial n}\,ds = \epsilon\int_{U_1}^{U_2}\frac{\partial U}{\partial s}\,ds = \epsilon(U_2 - U_1) \qquad (3)$$

Thus just as the difference of potential between any two points in the field is given by the difference in the values of the potential function at the two points, so the total flux passing through a line joining any two points in a field equals the product of the capacitivity by the difference in the value of the stream function at the two points.

If the surfaces V_1 and V_2 are closed and all lines of force in the region they bound pass from one to the other, then, from 2.01, they form a capacitor. The charge Q on either is the total flux per unit length. From (3), this flux is the product of ϵ by the increment $[U]$ in U going once around a V-curve. Since the potential difference is $|V_2 - V_1|$, the capacitance and field energy per unit length are

$$C = \frac{[Q]}{|V_2 - V_1|} = \frac{\epsilon[U]}{|V_2 - V_1|} \qquad (4)$$

$$\text{Field energy} = \tfrac{1}{2}C|V_2 - V_1|^2 = \tfrac{1}{2}\epsilon|U_2 - U_1|\,|V_2 - V_1| \qquad (5)$$

4.12. Functions for a Line Charge.—Before taking up the methods available for finding the required $f(x + jy)$ to fit a given problem, let us consider a simple case where the form of this function is evident by inspection. In polar coordinates by *Pc* 609 or *Dw* 408.04, we have

$$z = x + jy = r\cos\theta + jr\sin\theta = re^{j\theta} \qquad (1)$$

The potential of a line charge at the origin is $U = -\frac{1}{2}q(\pi\epsilon)^{-1}\ln r$ from 4.00 (2). Clearly from (1) this is the real part of $-\frac{1}{2}q(\pi\epsilon)^{-1}\ln z$, so

$$W = U + jV = -\frac{q \ln z}{2\pi\epsilon} = -\frac{q \ln r}{2\pi\epsilon} - \frac{jq\theta}{2\pi\epsilon}$$
$$= -\frac{q \ln (x + jy)}{2\pi\epsilon} = -\frac{q \ln (x^2 + y^2)^{\frac{1}{2}}}{2\pi\epsilon} - \frac{jq \tan^{-1} (y/x)}{2\pi\epsilon} \quad (2)$$

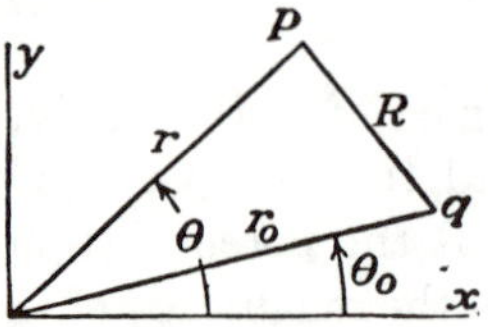

FIG. 4.12.

is the desired function. To check this we observe that, as θ goes from 0 to -2π, the function ϵV goes from 0 to q and so is the flux emerging from a charge q as required by 4.11.

We may now write down $f(z)$ when there is a line charge at r_0, θ_0, or z_0. The potential is

$$U = -\tfrac{1}{2}q(\pi\epsilon)^{-1} \ln R = -\tfrac{1}{4}q(\pi\epsilon)^{-1} \ln [r^2 + r_0^2 - 2rr_0 \cos (\theta - \theta_0)]$$
$$= -\tfrac{1}{4}q(\pi\epsilon)^{-1} \ln [(r \cos \theta - r_0 \cos \theta_0)^2 + (r \sin \theta - r_0 \sin \theta_0)^2]$$
$$= -\tfrac{1}{2}q(\pi\epsilon)^{-1} \ln (A^2 + B^2)^{\frac{1}{2}}$$

But from (2), substituting A for x and B for y, we see that $\ln (A^2 + B^2)^{\frac{1}{2}}$ is the real part of $\ln (A + jB)$ so that the required function is

$$W = \frac{-q}{2\pi\epsilon} \ln (A + jB) = \frac{-q}{2\pi\epsilon} \ln (re^{j\theta} - r_0e^{j\theta_0}) = \frac{-q}{2\pi\epsilon} \ln (z - z_0) \quad (3)$$

The function for n line charges situated at $z_1, z_2, \ldots, z_n$ is therefore

$$W = -\frac{1}{2\pi\epsilon}\sum_{s=1}^{n} q_s \ln (z - z_s) \quad (4)$$

4.13. Capacitance between Two Circular Cylinders.—We have already seen in 4.05 that the equipotentials about two equal and opposite line charges are circular cylinders. Let us therefore superimpose the fields due to charges $2\pi\epsilon$ at $y = a$ and $-2\pi\epsilon$ at $y = -a$. This choice of the charge q simplifies the coefficients. From 4.12 (3), the expression for W becomes, using *Pc* 645 or *Dw* 601.2 and 505.1,

$$W = \ln \frac{z + ja}{z - ja} = 2j \tan^{-1} \frac{a}{z} = 2j \cot^{-1} \frac{z}{a} \quad (1)$$

Solving for z and using *Pc* 601 or *Dw* 408.19, we have

$$z = a \cot \frac{U + jV}{2j} = \frac{-a \sin (U/j) + a \sin V}{\cos (U/j) - \cos V}$$

Using *Pc* 606 and 607, we can now separate real and imaginary parts, giving

$$x = \frac{a \sin V}{\cosh U - \cos V} \qquad y = \frac{a \sinh U}{\cosh U - \cos V} \quad (2)$$

Eliminating V from these equations gives

$$x^2 + y^2 - 2ay \coth U + a^2 = 0 \quad (2.1)$$

This may be written

$$x^2 + (y - a \coth U)^2 = a^2 \operatorname{csch}^2 U \tag{3}$$

Thus, as anticipated, the equipotentials are a set of circles with centers on the y-axis, positive potentials being above the x-axis and negative

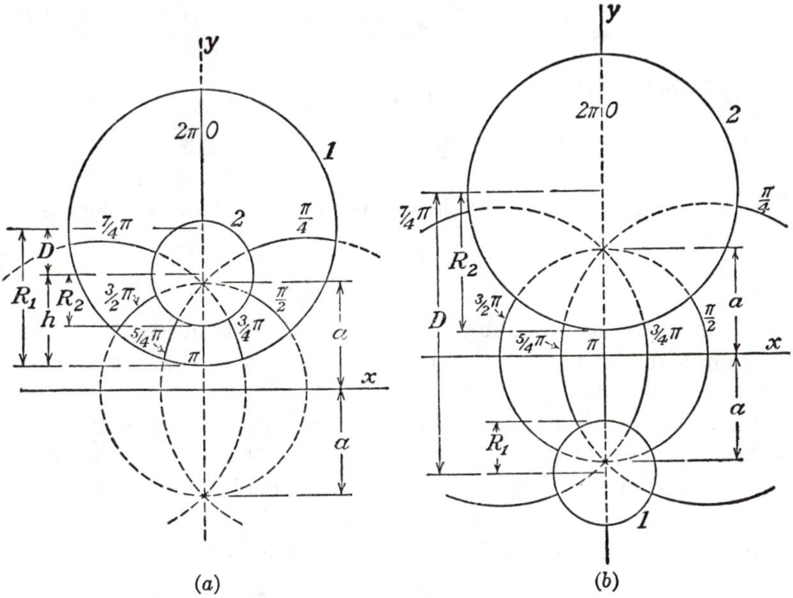

FIG. 4.13.

below. Eliminating U from (2) gives

$$x^2 - 2ax \cot V + y^2 - a^2 = 0$$

This may be written

$$(x - a \cot V)^2 + y^2 = a^2 \csc^2 V \tag{4}$$

Thus the lines of force are also a set of circles, all of which pass through the points $y = +a$ and $y = -a$ on the y-axis.

To determine the capacitance per unit length between the cylinders $U = U_1$ and $U = U_2$, it is necessary, from 4.11 (4), only to divide the charge $2\pi\epsilon$ by the difference of potential $U_2 - U_1$. We are given the radii R_1 and R_2 of the two cylinders and the distance D between their axes, from which we must determine a, U_1, and U_2. From (3), $R_1 = a|\operatorname{csch} U_1|$, $R_2 = a|\operatorname{csch} U_2|$ and $D = a(|\coth U_1| \pm |\coth U_2|)$, taking the lower sign if U_1 and U_2 are both positive, giving one cylinder inside the other, and taking the upper sign if U_1 is negative, giving one cylinder outside the other. We now write by *Pc* 661 or *Dw* 651.02, using the same sign rule,

$$\cosh (U_2 - U_1) = \cosh U_2 \cosh U_1 \pm |\sinh U_2 \sinh U_1|$$
$$= (|\coth U_2 \coth U_1| \pm 1)|\sinh U_2 \sinh U_1|$$

Now apply *Pc* 659 or *Dw* 650.08 to the $\pm 1 = \pm\frac{1}{2} \pm \frac{1}{2}$ term, giving

$$\cosh (U_2 - U_1) = [|\coth U_1 \coth U_2| \pm \tfrac{1}{2}(\coth^2 U_2 - \operatorname{csch}^2 U_2) \pm \tfrac{1}{2}(\coth^2 U_1 - \operatorname{csch}^2 U_1)]|\sinh U_1 \sinh U_2|$$
$$= \frac{\pm(|\coth U_2| \pm |\coth U_1|)^2 \mp \operatorname{csch}^2 U_1 \mp \operatorname{csch}^2 U_2}{2|\operatorname{csch} U_1 \operatorname{csch} U_2|}$$

Putting in values of D, R_1, and R_2 gives

$$\cosh (U_2 - U_1) = \pm\frac{D^2 - R_1^2 - R_2^2}{2R_1R_2}$$

Thus the capacitance per unit length between the two cylinders is

$$C = 2\pi\epsilon\left[\cosh^{-1}\left(\pm\frac{D^2 - R_1^2 - R_2^2}{2R_1R_2}\right)\right]^{-1} \tag{5}$$

where the lower sign is taken when one cylinder is inside the other and the upper sign when they are external to each other. The two cases are shown in Figs. 4.13*a* and 4.13*b*.

4.14. Capacitance between Cylinder and Plane and between Two Similar Cylinders.—In Fig. 4.13*a*, let $R_1 = D + h \to \infty$ so that the outer cylinder circles through infinity and coincides, in the finite region, with the x-axis. Then we have, neglecting R_2^2 compared with R_1^2 and D^2, and h compared with $2R_1$, $R_1 + D = 2R_1 - h \approx 2R_1$,

$$\frac{R_1^2 + R_2^2 - D^2}{2R_1R_2} \to \frac{(R_1 - D)(R_1 + D)}{2R_1R_2} \to \frac{h}{R_2}$$

Thus the capacitance per unit length of a conducting cylinder of radius R with its axis parallel to and at a distance h from an infinite conducting plane is

$$C = 2\pi\epsilon\left(\cosh^{-1}\frac{h}{R}\right)^{-1} \tag{1}$$

Two similar cylinders with their centers at a distance $D = 2h$ apart have half the capacitance per unit length given by (1) since they consist of two such capacitors in series. The resultant expression is in somewhat simpler form than that obtained by putting $R_1 = R_2$ and $D = 2h$ in 4.13 (5)

$$C = \pi\epsilon\left(\cosh^{-1}\frac{D}{2R}\right)^{-1} \tag{2}$$

4.15. Conformal Transformations.—Evidently conjugate functions furnish a powerful means for solving two-dimensional potential problems, provided we can find the proper function. Before taking up methods of doing this, we shall investigate certain special properties of functions of a complex variable. Suppose we plot values of $z = x + jy$ on one plane and values of $z_1 = x_1 + jy_1$ on a second plane and let z be an analytic function of z_1 so that at least one point in the z-plane corresponds to each point in the z_1-plane. Then as we move the latter along some curve, the corresponding point in the z-plane will also describe a curve, provided $z = f(z_1)$ is continuous, otherwise the second point may jump from place to place. If the z curve returns to the starting point when the z_1 curve does, then $f(z_1)$ is said to be single valued in this region of the z-plane. From the rule for quotients of complex numbers 4.08, we have

$$\left|\frac{dz}{dz_1}\right| = \frac{|dz|}{|dz_1|} = \frac{ds}{ds_1} = h \tag{1}$$

where ds is the length of an element dz of a curve in the z-plane and ds_1 is the length of the corresponding element dz_1 of the corresponding curve in the z_1-plane. Thus the modulus of dz/dz_1 measures the magnification of an element from a certain point in the z-plane when transformed to the corresponding point in the z_1-plane.

Let us draw the infinitesimal triangle made by the intersection of three curves in the z_1-plane, and let the lengths of the sides be ds_1, ds_1', ds_1''. Then the lengths of the sides of the transformed triangle will be $ds = h\,ds_1$, $ds' = h\,ds_1'$, $ds'' = h\,ds_1''$. Thus $ds_1{:}ds_1'{:}ds_1'' = ds{:}ds'{:}ds''$, so that the two triangles are similar and the angles of the intersections of corresponding curves in the two planes are the same. Such a transformation is said to be conformal.

Since the argument of the quotient of two complex numbers is the difference of their arguments, we see that the argument of (dz/dz_1) is the angle through which the element has been rotated in transforming from one plane to the other.

4.16. Given Equations of Boundary in Parametric Form.—If $f(x, y) = 0$ is the equation of one of the desired equipotential boundaries and x and y can be expressed as real analytic functions of a real parameter t whose total range just covers the conductor, then there is a simple method of obtaining a solution of Laplace's equation fitting this boundary. Let

$$x = f_1(t) \qquad y = f_2(t) \tag{1}$$

Then the desired solution is

$$x + jy = f_1(bW) + jf_2(bW) \tag{2}$$

This gives $V = 0$ as the conductor, for substituting this value gives exactly the parametric equation of the conductor with bU in place of t.

Unfortunately, the number of cases where this method is useful is very limited. Among them may be mentioned the confocal conics and the various cycloidal curves. As an example, let us find the field on one side of a corrugated sheet of metal whose equation is that of a cycloid and which forms one boundary of a uniform field. The equation of the surface is

$$x = a(\theta - \sin\theta) \qquad y = a(1 - \cos\theta) \tag{3}$$

so that $z = a(bW - \sin bW) + aj(1 - \cos bW) = a(bW + j - je^{-jbW})$ giving $x = a(bU - e^{bV}\sin bU)$, $y = a(bV + 1 - e^{bV}\cos bU)$. When V is large and negative $x = abU$ and $y = +abV$, so that we have a uniform field in the $-y$-direction of strength $E = +\partial V/\partial y = a^{-1}b^{-1}$. Substituting for b gives

$$z = a\left(\frac{W}{aE} + j - je^{-j\frac{W}{aE}}\right)$$

To find the field at any point, we differentiate, and

$$1 = \frac{1}{E}\frac{dW}{dz} - \frac{1}{E}e^{-j\frac{W}{aE}}\frac{dW}{dz}$$

$$\left|\frac{dW}{dz}\right| = \left|E\left(1 - e^{-j\frac{W}{aE}}\right)^{-1}\right| = E\left(1 - 2e^{\frac{V}{aE}}\cos\frac{U}{aE} + e^{\frac{2V}{aE}}\right)^{-\frac{1}{2}}$$

To get the surface density on the sheet, we note that on the conductor $V = 0$ and $y = a[1 - \cos(U/aE)]$, and

$$\sigma = \epsilon\left|\frac{dW}{dz}\right|_{V=0} = \left(\frac{a}{2y}\right)^{\frac{1}{2}}\epsilon E$$

This result gives the field on the side of the sheet with sharp ridges. The solution on the other side gives lines of discontinuity so that it is of no value there.

4.17. Determination of Required Conjugate Functions.—In most cases, the search for a function W which fits the given boundary conditions in the z-plane begins by looking for a transformation which reduces the boundaries to simpler shapes. If the new boundary conditions are unfamiliar, we endeavor to find a second transformation which will still further simplify the boundary conditions. Eventually, we should arrive at a situation where the solution can be written down by inspection. We then proceed backward along the steps we have come to the solution of the original problem. It is frequently possible to jump these steps and write $f(W, z) = 0$ by eliminating the intermediate complex variables. If this is impossible, those variables serve as parameters connecting W and z.

In manipulating these transformations, it is often helpful to visualize that portion of the z_1-plane between the boundaries as an elastic membrane which possesses the property that no matter how we distort the boundaries the angle of intersection of any lines drawn on the membrane remains unchanged. The membrane may not separate from a boundary but may slide along it and be expanded or contracted indefinitely.

For example, suppose the boundaries in our problem are two nonconcentric nonintersecting circles or two intersecting circles or two of one type and one or two of the other intersecting orthogonally. The region with such boundaries may be transformed into a rectangle by 4.13 (1), the relation being

$$z_1 = \ln \frac{z + ja}{z - ja} \tag{1}$$

Here we have written $z_1 = x_1 + jy_1$ for $W = U + jV$, since we are attaching no electrical significance to x_1 and y_1 at present. From 4.13 (3) and (4), we see that, when $-\infty < x < \infty$ and $-\infty < y < \infty$, then $-0 < y_1 < 2\pi$ and $-\infty < x_1 < \infty$. Thus (1) transforms a horizontal

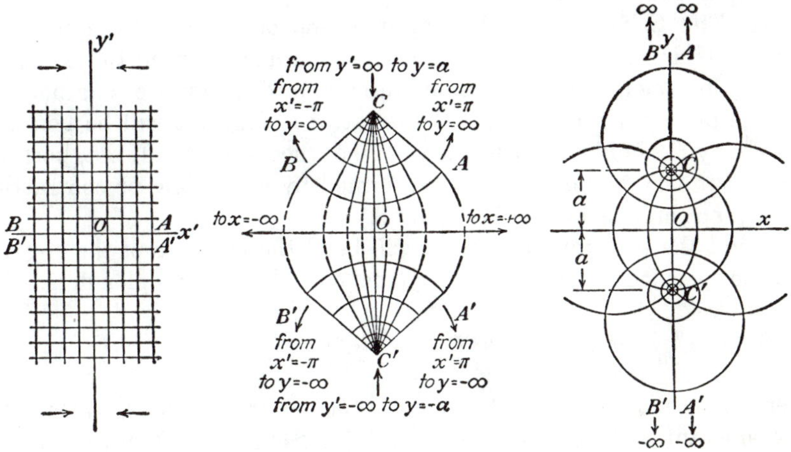

FIG. 4.17.

strip of width 2π in the z_1-plane into the whole of the z-plane. Vertical lines in this strip, from 4.13 (3), go into circles given by

$$x^2 + (y - a \coth x_1)^2 = a^2 \operatorname{csch}^2 x_1 \tag{2}$$

and horizontal lines go into circles passing through $y = \pm a$, $x = 0$ given by 4.13 (4) to be

$$(x - a \cot y_1)^2 + y^2 = a^2 \csc^2 y_1 \tag{3}$$

This transformation can be visualized by imagining an infinite horizontal strip of elastic membrane of width 2π being rotated counterclockwise to a vertical position in a z'-plane, so that points $x_1 = 0$, $y_1 = 0$, and $x_1 = 0$, $y_1 = 2\pi$ go to AA' and BB' and then pinched together at $y' = +\infty$ and $y' = -\infty$. These points C and C' are then brought toward each other along the y-axis while the center of the strip is expanded horizontally. The lines CA, CB and $C'A'$, $C'B'$ are opened out, as a fan is opened about the point C and C', respectively, until CA coincides with CB and $C'A'$ with $C'B'$. The membrane thus expands to fill the whole z-plane, the infinitesimal arcs AA' and BB' being stretched into infinite arcs bisected by the x-axis.

The operation, after the first rotation, is pictured in Fig. 4.17. If there are to be no discontinuities where CA folds against CB and $C'A'$ against $C'B'$, then the potential used in the horizontal strip in the z_1-plane must be periodic in y_1 with a period 2π.

Problems involving line charges and rectangular boundaries or ones where different sections of such boundaries are at different potentials can now be solved by images. In other cases, it may be necessary to unfold the rectangle into a half plane by a Schwarz transformation, giving, in general, elliptic functions.

4.18. The Schwarz Transformation.—One of the most useful transformations is that which transforms the upper half of the z_1-plane, bounded by the real axis and an infinite arc, into the interior of a polygon in the z-plane, or vice versa. If the latter is finite, it will be bounded entirely by the deformed real axis $y_1 = 0$ of the z_1-plane. If not, portions of its boundary at infinity may be formed by expanding or contracting the original infinite arc of the z_1-plane.

To find the transformation that will bend the real axis of the z_1-plane into the specified polygon in the z-plane, consider the complex derivative

$$\frac{dz}{dz_1} = C_1(z_1 - u_1)^{\beta_1}(z_1 - u_2)^{\beta_2} \cdot \cdot \cdot (z_1 - u_n)^{\beta_n} \tag{1}$$

where $u_1, u_2, \ldots, u_n$ and $\beta_1, \beta_2, \ldots, \beta_n$ are real numbers, C_1 is a complex constant and $u_n > u_{n-1} > \cdot \cdot \cdot > u_2 > u_1$. As we have seen, the argument of the product of several factors raised to certain powers equals the sum of the products of the argument of each factor by its exponent, so that

$$\arg \frac{dz}{dz_1} = \arg C_1 + \beta_1 \arg (z_1 - u_1) + \cdot \cdot \cdot + \beta_n \arg (z_1 - u_n) \tag{2}$$

When dz_1 is an element of the real axis in the z_1-plane, we may write it dx_1 and we have

$$\arg \frac{dz}{dz_1} = \arg \frac{dx + j\,dy}{dx_1} = \tan^{-1} \frac{dy}{dx} \tag{3}$$

This is the angle that the element dz, into which dz_1 is transformed, makes with the real axis $y = 0$ in the z-plane. When z_1 is real and lies between u_r and u_{r+1}, then $(z_1 - u_1), (z_1 - u_2), \ldots, (z_1 - u_r)$ are real positive numbers whose arguments are zero, and $(z_1 - u_{r+1}), (z_1 - u_{r+2}), \ldots, (z_1 - u_n)$ are real negative numbers whose arguments are π. This gives, from (2) and (3)

$$\theta_r = \tan^{-1}\frac{dy}{dx} = \arg C_1 + (\beta_{r+1} + \beta_{r+2} + \cdots + \beta_n)\pi \qquad (4)$$

Thus all elements of the x_1-axis which lie between u_r and u_{r+1} (Fig. 4.18) have the same direction after transformation and form a straight line, whose slope is given by (4), as shown. Similarly, the elements lying between u_{r+1} and u_{r+2} form a straight line having the slope

$$\theta_{r+1} = \tan^{-1}\frac{dy}{dx} = \arg C_1 + (\beta_{r+2} + \beta_{r+3} + \cdots + \beta_n)\pi \qquad (5)$$

The angle between these two lines is given by

$$\theta_{r+1} - \theta_r = -\pi\beta_{r+1}$$

We have now formed two sides of a polygon, and in a similar fashion we can evaluate the remaining β's and u's to give us the desired vertex angles and lengths of sides of our polygon.

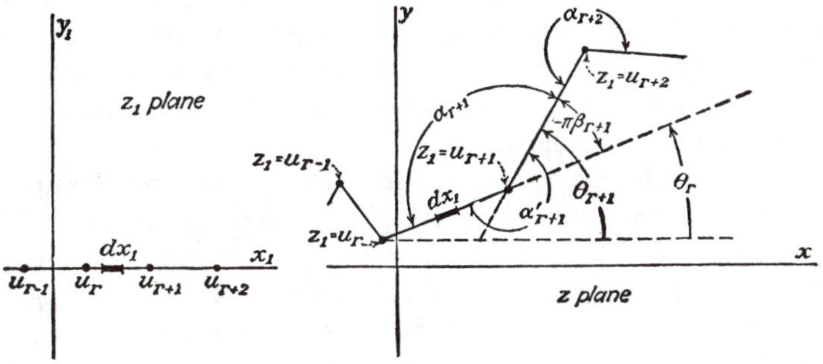

FIG. 4.18.

Suppose that the field is to be on the upper side of the broken line in the z-plane shown in Fig. 4.18. Then any angle, such as α_{r+1}, measured between two adjacent sides of the polygon on the side of the field is called an interior angle of the polygon, and since $\pi - \alpha_{r+1} = -\pi\beta_{r+1}$ we have, to get this angle, to choose

$$\beta_{r+1} = \frac{\alpha_{r+1}}{\pi} - 1 \qquad (6)$$

Substituting in (1) gives

$$\frac{dz}{dz_1} = C_1(z_1 - u_1)^{\frac{\alpha_1}{\pi}-1}(z_1 - u_2)^{\frac{\alpha_2}{\pi}-1} \cdots (z_1 - u_n)^{\frac{\alpha_n}{\pi}-1} \qquad (7)$$

Integrating this expression gives

$$z = C_1\int[(z_1 - u_1)^{\frac{\alpha_1}{\pi}-1}(z_1 - u_2)^{\frac{\alpha_2}{\pi}-1} \cdots]\,dz_1 + C_2 = C_1 f(z_1) + C_2 \qquad (8)$$

This is the desired transformation.

From (4), we see that we may orient our polygon in the z-plane any way we please by giving C_1 the proper argument. The size of the polygon is determined by the modulus of C_1. The polygon may be displaced in any direction without rotation by choosing the proper value of C_2.

To show that we have measured α_r on the correct side of the boundary, let us set $z_1 = W$ so that the real axis $y_1 = V = 0$ represents an equipotential line. If $\alpha_r < \pi$, the field $dW/dz = dz_1/dz$ must be zero in the angle; and if $\alpha_r > \pi$, it must be infinite. Setting $x_1 = U = u_r$ in (7), we see that this is true which verifies the result.

4.19. Polygons with One Positive Angle.—When the real axis is bent at only one point, no generality is lost by taking this point at the origin, so that $u_1 = 0$ in 4.18 (8) giving, if $\alpha > 0$,

$$z = C_1 z_1^{\frac{\alpha}{\pi}} + C_2 \tag{1}$$

If we choose $C_2 = 0$ and C_1 real, the polygon formed has its apex at the origin and is bounded by $\theta = 0$, $\theta = \alpha$ and by an arc at infinity. From 4.08, the modulus of z^n is the nth power of the modulus of z, so that circles given by $r_1 = a_1$ in the z_1-plane transform into circles $r = a$ in the z-plane. Thus a problem involving the region bounded by two radii of a circle and the included arc can be reduced by (1) to one involving a diameter and a semicircular arc.

When $\alpha = 2\pi$, the real axis is folded back on itself and the upper half of the z_1-plane opened into the full z-plane. Separating real and imaginary parts gives $y = 2C_1x_1y_1$ and $x = C_1(x_1^2 - y_1^2)$. Eliminating y_1 and x_1 in turn gives $y^2 = -4C_1x_1^2(x - C_1x_1^2)$ and $y^2 = 4C_1y_1^2(x + C_1y_1^2)$ which are the equations of two orthogonal families of confocal parabolas. Thus, if we plot the uniform field $W = z_1$ on the z_1-plane, we find it transformed in the z-plane into the field about a charged semi-infinite conducting plane. If we plot the field between a line charge and an earthed horizontal conducting plane passing through the origin in the z_1-plane, then we find it transformed in the z-plane into the field between a semi-infinite conducting plane and a line charge parallel to its edge.

When $\alpha = 3\pi/2$, if we set $W = z_1$, we have, in the z-plane, the field near the edge of a charged conducting 90° wedge whose sides coincide with the positive x-axis and the negative y-axis. The field between such a wedge and a line charge can be found in the same way as when $\alpha = 2\pi$.

When $\alpha = \pi$, there is, of course, no change in the coordinate system except that it is magnified by the factor C_1.

When $\alpha = \frac{1}{2}\pi$, if we set $W = z_1$, we find, in the z-plane that W represents the field in a 90° notch in a conductor, the positive x-axis and the positive y-axis forming the sides of the notch. In this case, $C_1^2U = x^2 - y^2$ and $C_1^2V = 2xy$, so that the equipotentials and lines of force are orthogonal families of equilateral hyperbolas. The field due to a charged filament parallel to the edge of the notch is obtained as when $\alpha = 2\pi$.

4.20. Polygon with Angle Zero.—In this important case, the x_1-axis is folded back parallel to itself and the upper half plane is compressed into the space between. In order that two parallel intersecting lines have a finite distance between them, it is necessary that the point of intersection be at infinity. Instead of 4.19 (1), we now have

$$z = C_1 \ln z_1 + C_2 \tag{1}$$

If, as before, we take $C_2 = 0$ and C_1 real, the origin $z_1 = 0$ is transformed to $z = -\infty$ and the new origin is at $z_1 = 1$. Writing $z_1 = r_1 e^{j\theta_1}$, we see that when $\theta_1 = 0$, $z = C_1 \ln r_1$, which is the real axis in the z-plane, and when $\theta_1 = \pi$, $x = C_1 \ln r_1$ and $y = C_1\pi$, which is a line parallel to the real axis and at a distance $C_1\pi$ above it. Thus the upper half of the z_1-plane is transformed into a horizontal strip in the z-plane. Radial lines $\theta_1 =$ constant are transformed into horizontal lines $y =$ constant; the semicircles $r_1 =$ constant go into vertical lines, of length $C_1\pi$. Frequently, we have problems in which a configuration is periodic; *i.e.*,the field may be split in identical strips. In such a case, the transformation (1) is useful. Suppose, for example, we have a charged filament between two parallel earthed conducting planes. Going back to the z_1-plane, we see that this is reduced to a filament parallel to a single earthed plane. By images 4.06 and 4.12 (4), the required function is

$$W = -\frac{q}{2\pi\epsilon}\left(\ln \frac{z_1 - e^{j\theta_0}}{z_1 - e^{-j\theta_0}} + C\right)$$

To make $W = 0$ when $z_1 = +1$, take $C = -\ln(-e^{j\theta_0})$ so that we have

$$W = -\frac{q}{2\pi\epsilon} \ln \frac{z_1 - e^{j\theta_0}}{1 - z_1 e^{j\theta_0}} \tag{2}$$

Transforming into the z-plane taking $C_2 = 0$, $C_1\pi = a$, and $C_1\theta_0 = b$, we have, substituting for z_1 from (1),

$$W = -\frac{q}{2\pi\epsilon} \ln \frac{e^{\pi z/a} - e^{j\pi b/a}}{1 - e^{\pi(z+jb)/a}} = -\frac{q}{2\pi\epsilon} \ln \frac{e^{\frac{1}{2}\pi(z-jb)/a} - e^{-\frac{1}{2}\pi(z-jb)/a}}{e^{-\frac{1}{2}\pi(z+jb)/a} - e^{\frac{1}{2}\pi(z+jb)/a}}$$

Now using *Pc* 652, 660, and 645 or *Dw* 654.1, 655.1, and 702, we have

$$\begin{aligned} W &= -\frac{q}{2\pi\epsilon} \ln -\frac{\sinh[\frac{1}{2}\pi(z - jb)/a]}{\sinh[\frac{1}{2}\pi(z + jb)/a]} \\ &= -\frac{q}{2\pi\epsilon} \ln \frac{1 + j\tanh(\frac{1}{2}\pi z/a)\cot(\frac{1}{2}\pi b/a)}{1 - j\tanh(\frac{1}{2}\pi z/a)\cot(\frac{1}{2}\pi b/a)} \\ &= -\frac{jq}{\pi\epsilon} \tan^{-1}[\tanh(\tfrac{1}{2}\pi z/a)\cot(\tfrac{1}{2}\pi b/a)] \end{aligned} \tag{3}$$

We can separate real and imaginary parts and get

$$\tan \frac{2\pi\epsilon V}{q} = \frac{-\sin(\pi b/a)\sinh(\pi x/a)}{-\cos(\pi b/a)\cosh(\pi x/a) + \cos(\pi y/a)} \tag{4}$$

$$\tanh \frac{2\pi\epsilon U}{q} = \frac{\sin(\pi b/a)\sin(\pi y/a)}{-\cos(\pi b/a)\cos(\pi y/a) + \cosh(\pi x/a)} \tag{5}$$

The z_1- and z-planes are shown in Fig. 4.20. If the plates are to be at a difference of potential U_0, we need superimpose on this only a vertical field given by $W' = -jU_0z/a$. The final solution will then be

$$W'' = W + W' = -[jq/(\pi\epsilon)] \tan^{-1} [\tanh (\tfrac{1}{2}\pi z/a) \cot (\tfrac{1}{2}\pi b/a)] - jU_0z/a \quad (6)$$

where U is the potential function.

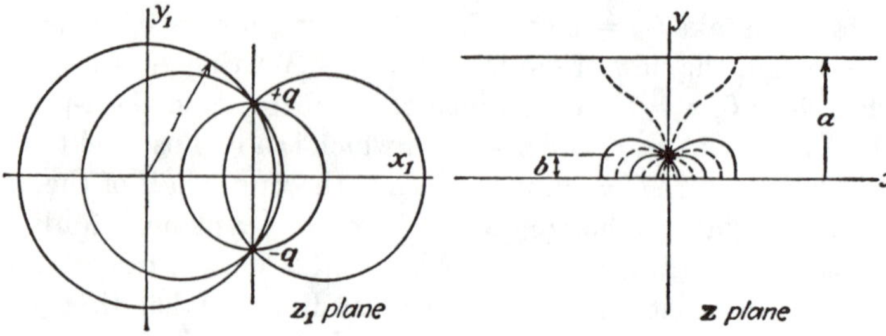

FIG. 4.20.—Line charge between earthed plates.

When a plane grating of parallel wires, spaced a distance a apart, forms one boundary of a uniform field, we can find the conjugate functions in the same way. It is necessary only to put $+q$ at both $z_1 = j$ and $z_1 = -j$ and proceed as before. We obtain, in the z-plane, a section of the field including one wire and bounded by lines of force proceeding from the wire to $x = +\infty$.

It may be noted that if we plot a uniform field in the z-plane and transform back into the z_1-plane, we get the field about a line charge already worked out in 4.12.

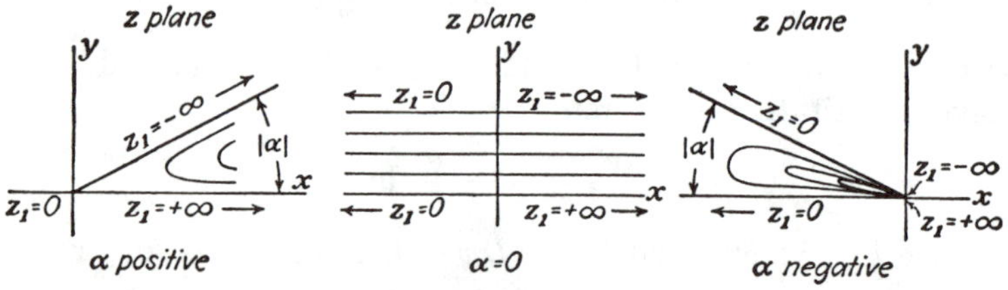

FIG. 4.21.

4.21. Polygons with One Negative Angle. Doublet. Inversion. It is now natural to investigate the meaning of Schwarz transformations with negative angles. The transition from positive to negative angles is clearly illustrated in Fig. 4.21. The ends of the real axis of the z_1-plane are now brought together at an angle α at the origin in the z-plane, if $C_2 = 0$ in 4.19 (1). The approximate form of the lines $y_1 =$ constant in the z-plane is shown in the figure.

The most important cases, by far, in this category are those in which $\alpha = -\pi$. Suppose, for example, we start with a uniform field $W = z_1$. If U is the potential function, we may think of this field as produced by an infinite positive charge at $x_1 = +\infty$ and an infinite negative charge at $x_1 = -\infty$. Inspection of Fig. 4.21 shows that the transformation in question brings these two charges infinitely close together on opposite sides of the y-axis in the z-plane. By definition, a two-dimensional dipole consists of two equal and opposite parallel infinite line charges, infinitely close together, the product of the charges per unit length by the distance between them being finite and equal to the dipole strength per unit length M. We can get the transformation directly from 4.13 (1) by writing a for ja and using *Pc* 769 or *Dw* 601.2; thus

$$W = -\frac{q}{2\pi\epsilon}\ln\frac{1-\frac{a}{z}}{1+\frac{a}{z}} \underset{a\to 0}{\longrightarrow} \frac{aq}{\pi\epsilon z} = \frac{M}{2\pi\epsilon z} \tag{1}$$

whence

$$U = \frac{M\cos\theta}{2\pi\epsilon r}, \qquad V = \frac{-M\sin\theta}{2\pi\epsilon r}$$

also

$$x^2 + y^2 - \frac{Mx}{2\pi\epsilon U} = 0, \qquad x^2 + y^2 + \frac{My}{2\pi\epsilon V} = 0 \tag{2}$$

Thus the equipotentials are circles tangent to the y-axis at the origin, and the lines of force are also circles but tangent to the x-axis at the origin.

Another important case where $\alpha = -\pi$ is obtained by letting $C_1 = a^2$ in 4.19 (1) giving

$$z = \frac{a^2}{z_1} \tag{3}$$

Write this in polar coordinates, and separate real and imaginary parts, and we have

$$rr_1 = a^2 \qquad \text{and} \qquad \theta = -\theta_1$$

Thus every point outside the circle $r_1 = a$ in the z_1-plane transforms into a point inside the circle $r = a$ in the z-plane. If $W = f(z_1)$ is a solution of Laplace's equation, then $W^* = f(z_1^*)$, where z_1^* is the complex conjugate of z_1, is also a solution and this field is the mirror image of the other, giving $r_1 = r_1^*$ and $\theta_1 = -\theta_1^*$. Comparing the z-plane and the z_1^*-plane, we have

$$rr_1^* = a^2 \qquad \text{and} \qquad \theta = \theta_1^* \tag{4}$$

When such a relation holds, r, θ and r_1^*, θ_1^* are said to be inverse points and a is the radius of inversion. If the sum of the charges in the z-plane is not zero, then there must be an equal and opposite charge at infinity, on which the excess lines of force end, and which will appear at the origin on the z_1- or z_1^*-plane. Hence we have the rule for inversion in two dimensions.

If a surface S is an equipotential under the influence of line charges q', q'', etc., at z', z'', etc., then the inverse surface will be an equipotential under the influence of charges q', q'', etc., at $z_1'^*$, $z_1''^*$, etc., plus a charge $-\Sigma q$ at the origin. This method is useful for obtaining, from the solution of a problem involving intersecting plane boundaries, a solution of a problem with intersecting circular cylindrical boundaries. In polar coordinates, the equation of the circle in the z-plane is

$$r^2 - 2ur\cos(\theta - \alpha) = R^2 - u^2 \tag{5}$$

where the center of the circle is at u, α and R is its radius. Multiplying through by r_1^{*2} and writing θ_1^* for θ and α_1^* for α from (4), we have

$$(rr_1^*)^2 - 2rr_1^*ur_1^*\cos(\theta_1^* - \alpha_1^*) = (R^2 - u^2)r_1^{*2}$$

Writing a^2 for rr_1^* from (4) and rearranging give

$$r_1^{*2} + 2\frac{a^2u}{R^2 - u^2}r_1^*\cos(\theta_1^* - \alpha_1^*) = \frac{a^4}{R^2 - u^2}$$

or

$$r_1^{*2} - 2u_1r_1^*\cos(\theta_1^* - \alpha_1^*) = R_1^2 - u_1^2 \tag{6}$$

where

$$u_1 = -\frac{a^2u}{R^2 - u^2} \quad \text{and} \quad R_1 = \frac{a^2R}{|R^2 - u^2|} \tag{7}$$

Thus the circle inverts into a circle. If the original circle passed through the origin, then $|u| = R$ so that $|u_1| \to R_1 = \infty$. Thus the inverted circle is of infinite radius with its center at infinity, which means a straight line. The nearest approach of this line to the origin is

$$|u_1| - |R_1| = \frac{a^2}{u + R} = \frac{a^2}{2R}$$

and the equation of the radius vector normal to it is $\theta_1^* = \alpha_1^*$. Conversely, a straight line inverts into a circle passing through the origin.

4.211. Images by Two-dimensional Inversion.—At this point, we shall pause in the discussion of Schwarz transformations long enough to give an example of two-dimensional inversion. We shall use the rules just derived to find an expression, in terms of images, for the field between an infinite cylindrical conductor, carrying a charge $-q$ per unit length and bounded by the external surfaces of two circular cylinders intersecting orthogonally and a parallel infinite line charge of strength $+q$ per unit length. In Fig. 4.211a, representing the z-plane, the charge is at P and the conducting surface is indicated by the solid line. From 4.21, we see that, if we invert about the point O, the circles will become straight lines, and since this process is conformal, these lines will intersect orthogonally. For simplicity, we take the circle of inversion, shown dotted, tangent to the larger cylinder. Figure 4.211b shows the inverse system in the z_1^*-plane. The latter system presents the familiar problem, treated in 4.06, of a line charge parallel to the line of intersection of two orthogonal conducting planes. From 4.06, we know the field in the angle between the planes to be identical with that produced in this region when all conductors are removed, but we have, in addition to q at P', line charges $-q$, $+q$, and $-q$ at P'_1, P'_2, and P'_3, respectively.

The rules of inversion tell us that the surface inverse to these planes, which is the surface of our conductor, coincides with a natural equipotential in the field of charges $+q$, $-q$, $+q$, and $-q$ at P, P_1, P_2, and P_3, respectively, where P, P_1, P_2, and P_3 are inverse points to P', P'_1, P'_2, and P'_3. We see that if $CP = r_c$ and $BP = r_b$, then P_1 is on BP at a distance b^2/r_b from B, P_3 is on CP at a distance c^2/r_c from C, and P_2

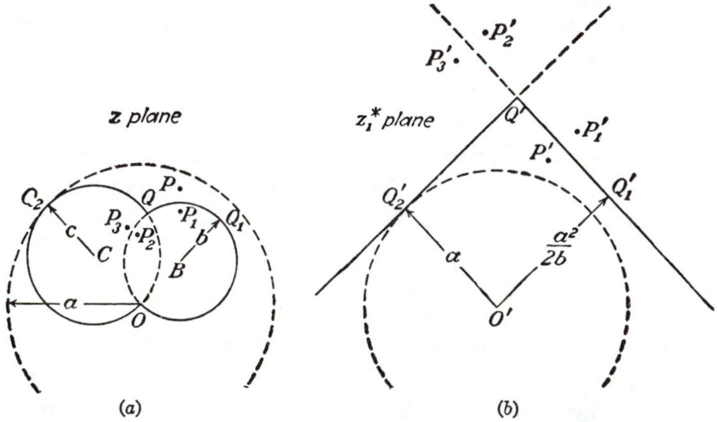

FIG. 4.211.—Two-dimensional inversion.

is at the intersection of BP_3 and CP_1. Since P', P'_1, P'_2, and P'_3 lie on a circle orthogonal to the lines $Q'Q'_1$ and $Q'Q'_2$ in the z_1^* plane, then P, P_1, P_2, and P_3 lie on a circle orthogonal to the circles in the z-plane.

4.22. Polygon with Two Angles.—From the large number of cases that come under this class, we shall select as an important example that in which the real axis is bent into a rectangular trough of width $2a$. Taking $\alpha_1 = \alpha_2 = \frac{1}{2}\pi$ in 4.18 (7), $u_1 = +a_1$, $u_2 = -a_1$, we obtain

$$\frac{dz}{dz_1} = \frac{A}{(z_1^2 - a_1^2)^{\frac{1}{2}}} \tag{1}$$

Integrating by *Pc* 126*b* or 127 or *Dw* 260.01 or 320.01 gives

$$z = A\cosh^{-1}\left(\frac{z_1}{a_1}\right) + C_1 = jA\sin^{-1}\left(\frac{z_1}{a_1}\right) + C_2$$

Taking $C_2 = 0$ gives, if the second form is used, $z = 0$ when $z_1 = 0$. In order that $z = \pm a$ when $z_1 = \pm a_1$, we must have $a = \frac{1}{2}jA\pi$ or $jA = 2a/\pi$, giving

$$z = \frac{2a}{\pi}\sin^{-1}\frac{z_1}{a_1} \qquad \text{or} \qquad z_1 = a_1 \sin\frac{\pi}{2a}z \tag{2}$$

The most useful application of this is obtained by applying (2) to a uniform field in the z-plane. Taking $a = \frac{1}{2}\pi$ and writing $W = z$, we have

$$z_1 = a_1 \sin W \qquad (3)$$

The vertical strip of the uniform field is unfolded as shown in Fig. 4.22a. Thus, if V is the potential function, the field in the z_1-plane is the upper half of that due to a charged strip of width $2a_1$, and if U is the potential, it is the upper half of that due to two semi-infinite coplanar sheets at a difference of potential π with a gap of width $2a_1$ between them. Separating real and imaginary parts of (3), if *Pc* 615 or *Dw* 408.16 is used, gives

$$x_1 = a_1 \sin U \cosh V \qquad y_1 = a_1 \cos U \sinh V$$

Divide the first equation by $a_1 \cosh V$ and the second by $a_1 \sinh V$, square, and add, and we have

$$\frac{x_1^2}{a_1^2 \cosh^2 V} + \frac{y_1^2}{a_1^2 \sinh^2 V} = 1 \qquad (4)$$

This shows the curves on which V is constant to be confocal ellipses with major and minor axes $2a_1 \cosh V$ and $2a_1 \sinh V$, respectively. In the same way, dividing by $a_1 \sin U$ and $a_1 \cos U$, squaring, and subtracting, we have

$$\frac{x_1^2}{a_1^2 \sin^2 U} - \frac{y_1^2}{a_1^2 \cos^2 U} = 1 \qquad (5)$$

The curves on which U is constant are confocal hyperbolas. We note that by giving V values from 0 to ∞ and U values from $-\pi$ to $-\frac{1}{2}\pi$ and π to $\frac{1}{2}\pi$ in the lower half plane, the field due to the strip is represented everywhere in the plane. The hyperbolic lines of force are then discontinuous in passing through the conducting strip. On the other hand, by giving U values between $-\frac{1}{2}\pi$ and $\frac{1}{2}\pi$ and V values from 0 to $-\infty$ in the lower half plane, we represent completely the field due to the two planes. The elliptic lines of force now are discontinuous in passing through the conducting planes.

Suppose that in place of a single charged strip we have a great number of similar strips lying uniformly spaced and parallel to each other in the same plane. It is clear that we may obtain a typical cell of such a field by folding up the x_1-axis in Fig. 4.22a at $u_1 = +b_1$ and $u_2 = -b_1$ where $b_1 > a_1$ and by applying (2), with b_1 substituted for a_1 and b for a, to (3), giving

$$z_1 = a_1 \sin W = b_1 \sin\left(\tfrac{1}{2}\pi z/b\right)$$

where $2b$ is the width of the cell in the grating. This is shown in Fig. 4.22b. Since $x = \pm a$, $y = 0$ when $V = 0$, $U = \pm\frac{1}{2}\pi$ we must have $a_1 = b_1 \sin\left(\frac{1}{2}\pi a/b\right)$, so that the transformation is

$$W = \sin^{-1}\left[\frac{\sin\left(\frac{1}{2}\pi z/b\right)}{\sin\left(\frac{1}{2}\pi a/b\right)}\right] \qquad (6)$$

Although derived for the region above the x-axis, we notice that this formula represents the field equally well when $0 > y > -\infty$, where choosing positive values of V requires that, when $0 < x < b$, $\frac{1}{2}\pi < U < \pi$ and, when $0 > x > -b$, $-\pi < U < -\frac{1}{2}\pi$. The strength of the field at $y = \infty$ is $-\frac{1}{2}\pi/b$, so that to get the transformation for a field of strength E' we must multiply by the factor $2bE'/\pi$. If we used this grating as the boundary of a uniform field of strength E extending to $y = +\infty$, then we need only superimpose a vertical field of strength E' when $E' = \frac{1}{2}E$,

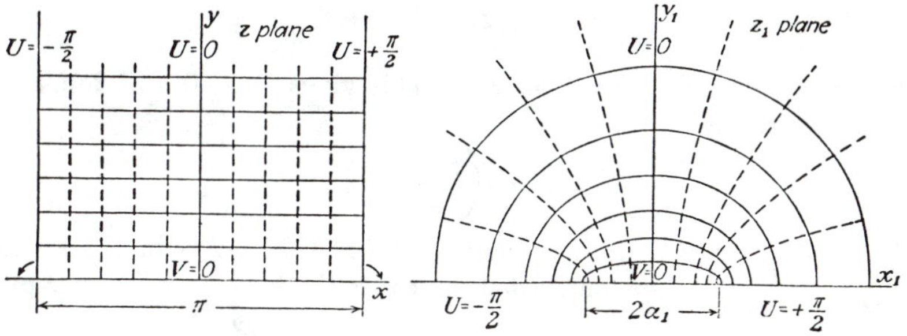

FIG. 4.22*a*.—Transformation of semi-infinite vertical strip into upper half plane.

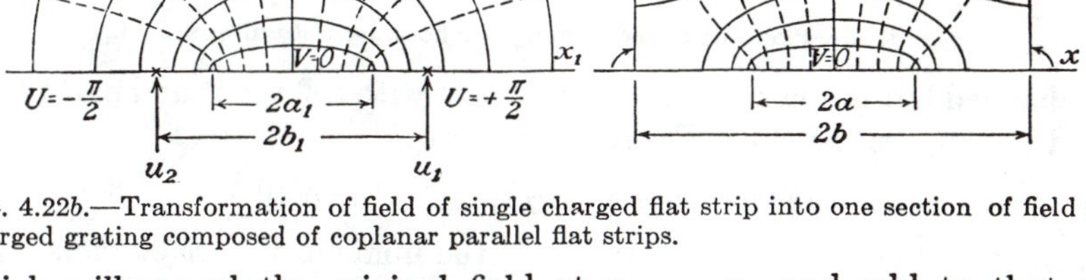

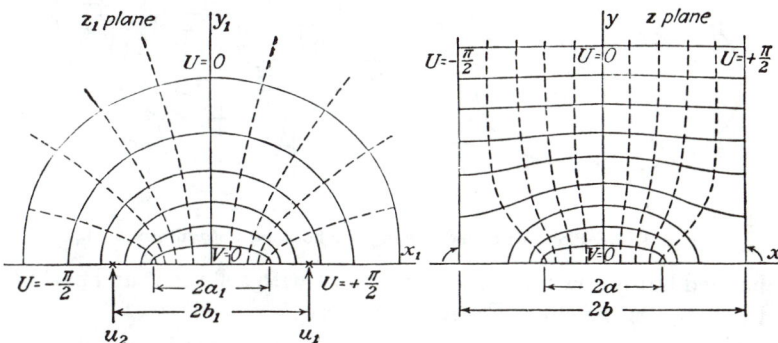

FIG. 4.22*b*.—Transformation of field of single charged flat strip into one section of field of charged grating composed of coplanar parallel flat strips.

which will cancel the original field at $y = -\infty$, and add to that at $y = +\infty$ giving

$$W' = \frac{E}{2}\left\{z + \frac{2b}{\pi}\sin^{-1}\left[\frac{\sin(\frac{1}{2}\pi z/b)}{\sin(\frac{1}{2}\pi a/b)}\right]\right\} \tag{7}$$

The field intensity at any point is then given by

$$\left|\frac{dW}{dz}\right| = \frac{E}{2}\left|1 \pm \frac{\cos(\frac{1}{2}\pi z/b)}{[\cos^2(\frac{1}{2}\pi z/b) - \cos^2(\frac{1}{2}\pi a/b)]^{\frac{1}{2}}}\right| \tag{8}$$

4.23. Slotted Plane.—To obtain the field in the neighborhood of a plane conducting sheet with a slit of width $2a$ in it, we may bend down that portion of the real axis lying between $x = +a$ and $x = -a$ and expand it to enclose the whole lower half of the z-plane, at the same time drawing that area of the z_1-plane near the origin through the gap. By examining Fig. 4.23, we see that $\alpha_1 = 2\pi$, $\alpha_2 = -\pi$, $\alpha_3 = 2\pi$, $u_1 = +a_1$, $u_2 = 0$, $u_3 = -a_1$, giving, from 4.18 (7),

$$\frac{dz}{dz_1} = C_1\frac{z_1^2 - a_1^2}{z_1^2}$$

Integrating gives 4.18 (8) to be

$$z = C_1\left(z_1 + \frac{a_1^2}{z_1}\right) + C_2$$

When $z_1 = \pm a_1$, we wish $z = \pm a$. This gives $C_2 = 0$ and $2a_1C_1 = a$, so that the result is

$$z = \frac{a}{2}\left(\frac{z_1}{a_1} + \frac{a_1}{z_1}\right) \tag{1}$$

If we start with a uniform vertical field $W = -Ez_1$ in the z_1-plane and take $2a_1 = a$, the transformation leaves the field at ∞ in the z-plane

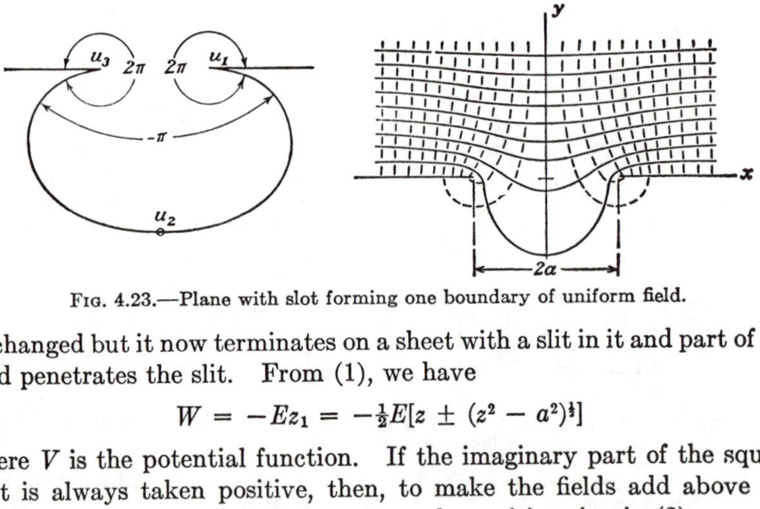

FIG. 4.23.—Plane with slot forming one boundary of uniform field.

unchanged but it now terminates on a sheet with a slit in it and part of the field penetrates the slit. From (1), we have

$$W = -Ez_1 = -\tfrac{1}{2}E[z \pm (z^2 - a^2)^{\frac{1}{2}}] \tag{2}$$

where V is the potential function. If the imaginary part of the square root is always taken positive, then, to make the fields add above the x-axis and subtract below, we must use the positive sign in (2).

The surface density on the sheet is given by

$$\sigma = \epsilon\left|\frac{dW}{dz}\right|_{y=0} = \frac{\epsilon E}{2}\left(\frac{x}{(x^2 - a^2)^{\frac{1}{2}}} \pm 1\right) \tag{3}$$

where the sign is chosen so that the two terms add on the upper surface and subtract on the lower.

The field between a charged filament and an earthed slotted sheet in the z-plane is easily obtained by (1) from the field between a charged filament and an earthed plane in the z_1-plane.

Writing the right side of (1) in polar coordinates and solving for x and y, we have

$$\frac{x}{r_1^2 + a_1^2} = \frac{a}{2r_1a_1} \cos \theta_1 \tag{4}$$

$$\frac{y}{r_1^2 - a_1^2} = \frac{a}{2r_1a_1} \sin \theta_1 \tag{5}$$

Squaring and adding give

$$\frac{x^2}{(r_1^2 + a_1^2)^2} + \frac{y^2}{(r_1^2 - a_1^2)^2} = \frac{a^2}{4r_1^2a_1^2} \tag{6}$$

Thus, if $r_1 > a_1$, the semicircle of radius r_1 and that of radius a_1^2/r_1 transform into the upper and lower halves, respectively, of the ellipse in the z-plane given by (6). The semicircle $r_1 = a_1$ flattens into the real axis between $x = +a$ and $x = -a$.

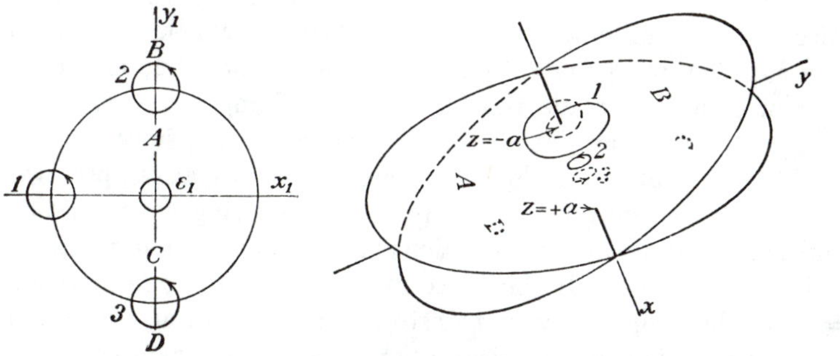

FIG. 4.24.—Riemann surface.

4.24. Riemann Surfaces.—To visualize completely the possibilities of a given transformation, it is often valuable to use the concept of a Riemann surface. This is well illustrated by the example in the last article. Although we have already used up all points in the z-plane to represent positive values of y_1, equation 4.23 (1) gives us a value of z for every negative value of y_1, *i.e.*, when $-\pi < \theta_1 < 0$ as well. From 4.23 (5), we see that if $0 < r_1 < a_1$ and $-\pi < \theta_1 < 0$ we arrive in the upper half of the z-plane, and if $a_1 < r_1 < \infty$ and $-\pi < \theta_1 < 0$ we arrive in the lower half. Thus there are two points in the z_1-plane corresponding to each point in the z-plane. We can eliminate this double value by making the z-plane a double sheet. We must be careful, however, to connect these two sheets in such a way that, in tracing a continuous circuit in one plane, we trace a continuous circuit through corresponding points in the other. The central portion of such a surface, known as a Riemann surface, is shown in Fig. 4.24 at the right. Lines in the lower surface are dotted, and the sheets are spread apart to clarify the diagram When $x^2 < a^2$, it is possible to pass only between A and B or C and D, and when $x^2 > a^2$ it is possible to pass only between A and C or between

B and D. The circle $r_1 = \epsilon_1$ where $\epsilon_1 \to 0$ becomes a circle at infinity lying in and surrounding the AC sheet. It is possible to construct Riemann surfaces for a great many transformations.

4.25. Circular Cylinder into Elliptic Cylinder.—The last two articles show that the region outside the circle $r_1 = a_1$ in the z_1-plane can be transformed into the region outside that part of the real axis lying between $x = +a$ and $x = -a$, in the BD sheet of the z-plane, by means of the relation

$$z = \frac{a}{2}\left(\frac{z_1}{a_1} + \frac{a_1}{z_1}\right) \tag{1}$$

Since, in the z_1-plane we are restricted to a region outside the circle $r_1 = a_1$, we cannot cross the x_1-axis where $-a_1 < x_1 < +a_1$. Consequently in the z-plane, we cannot cross the x-axis where $-a < x < +a$ and the line joining $+a$ and $-a$ is said to be a cut in the z-plane.

Electrically, this enables us to transform any field which, plotted in the z_1-plane, makes $r_1 = a_1$ an equipotential or a line of force into a field, plotted in the z-plane, in which the line joining $x = +a$ and $x = -a$ is an equipotential or a line of force. More generally, since from 4.23 (6) any circle $r_1 = b_1$ where $b_1 > a_1$ transforms into an ellipse, we may obtain from the solution of any problem involving concentric circular cylindrical boundaries the solution of a problem involving confocal elliptical boundaries. If the circle in the z_1-plane is taken eccentric to $r_1 = a_1$ in the proper way, it transforms into an airfoil. This is therefore the famous "airfoil transformation" used in aeronautics.

4.26. Dielectric Boundary Conditions.—With the aid of 1.17 (5) and (6) and the relations 4.10 (1), we shall now determine the boundary conditions that must be satisfied by the potential and stream functions when $W_1 = U_1 + jV_1 = f_1(z)$ and $W_2 = U_2 + jV_2 = f_2(z)$ are functions representing the electrostatic fields on the two sides of a boundary where the capacitivities are ϵ_1 and ϵ_2, respectively. Let $\partial/\partial n$ and $\partial/\partial s$ represent differentiation normal to and parallel to the boundary, respectively. Then, from 1.17 (5) and 4.10 (1), if U is the potential function,

$$\epsilon_1\frac{\partial U_1}{\partial n} = \epsilon_2\frac{\partial U_2}{\partial n} \qquad \text{or} \qquad \epsilon_1\frac{\partial V_1}{\partial s} = \epsilon_2\frac{\partial V_2}{\partial s} \tag{1}$$

If we take the zero stream line of W_1 and W_2 to join at the boundary, then we may integrate (1) along the boundary from this line to any other point V_1, V_2 and obtain in terms of the capacitivity

$$\epsilon_1 V_1 = \epsilon_2 V_2 \qquad \text{or} \qquad K_1V_1 = K_2V_2 \tag{2}$$

in terms of the relative capacitivity. From 1.17 (6), we have

$$U_1 = U_2 \tag{3}$$

These are the boundary conditions when U is the potential function.

If V is the potential function, the boundary conditions are

$$V_1 = V_2 \qquad \text{and} \qquad \epsilon_1 U_1 = \epsilon_2 U_2 \qquad \text{or} \qquad K_1 U_1 = K_2 U_2 \tag{4}$$

4.261. Elliptic Dielectric Cylinder.—Conformal transformations can be applied not only to cases involving equipotential or line of force boundaries, but also to many cases involving dielectric boundaries. The fact that such an operation preserves angles means that the law of refraction of the lines of force [1.17 (8)] will still be satisfied. Suppose, for example, our problem is to find the conjugate functions giving the field at all points when an elliptic cylinder is placed in a uniform field of force making an angle α with its major axis. Let the equation of the elliptic dielectric boundary be

$$\frac{x^2}{m^2} + \frac{y^2}{n^2} = 1 \tag{1}$$

The transformation of 4.25 immediately suggests itself as a means for deriving this elliptic boundary, represented in the z-plane, from a circular boundary $r_1 = b$, represented in the z_1-plane. For simplicity, let us take $a_1 = 1$ in 4.23 and 4.25, so that the z-plane represents the area outside the unit circle in the z_1-plane. Writing 4.23 (6) in the form of (1) and equating coefficients, we see that

$$a^2 = m^2 - n^2 \qquad \text{and} \qquad b^2 = \frac{m+n}{m-n} \tag{2}$$

It is often convenient, when elliptical boundaries are involved, to use confocal coordinates u and v instead of rectangular coordinates x and y. Such a system is shown in Figs. 4.22*a* and *b*. From 4.22 (3), we see that the relations are

$$z = a \sin w, \qquad x = a \sin u \cosh v, \qquad y = a \cos u \sinh v \tag{3}$$

The transformation of 4.25 (1) then becomes $z = \frac{1}{2}a(z_1 + z_1^{-1})$ or

$$z_1 = je^{-jw} = je^{-ju+v} = \frac{1}{a}[z + (z^2 - a^2)^{\frac{1}{2}}] = a[z - (z^2 - a^2)^{\frac{1}{2}}]^{-1} \tag{4}$$

This equation shows us that at a great distance from the origin, where $r_1 \to \infty$, $z = \frac{1}{2}az_1$, so that a uniform field $W = \frac{1}{2}aEz_1$ transforms into the uniform field $W = Ez$ in this region. It is clear that the desired field may be formed by superimposing a vertical field of strength $E \sin \alpha$ on a horizontal field of strength $E \cos \alpha$. These component fields are shown in Figs. 4.261*a* and *b*. Since the axis between the foci is an equipotential in *a* and a line of force in *b*, we see that in the z_1-plane the situation is as shown in Figs. 4.261*c* and *d*, respectively.

The field of Fig. 4.261*c* has already been solved in 4.03, from which we have, taking U' for the potential function and $U'_i = 0$ on the unit circle,

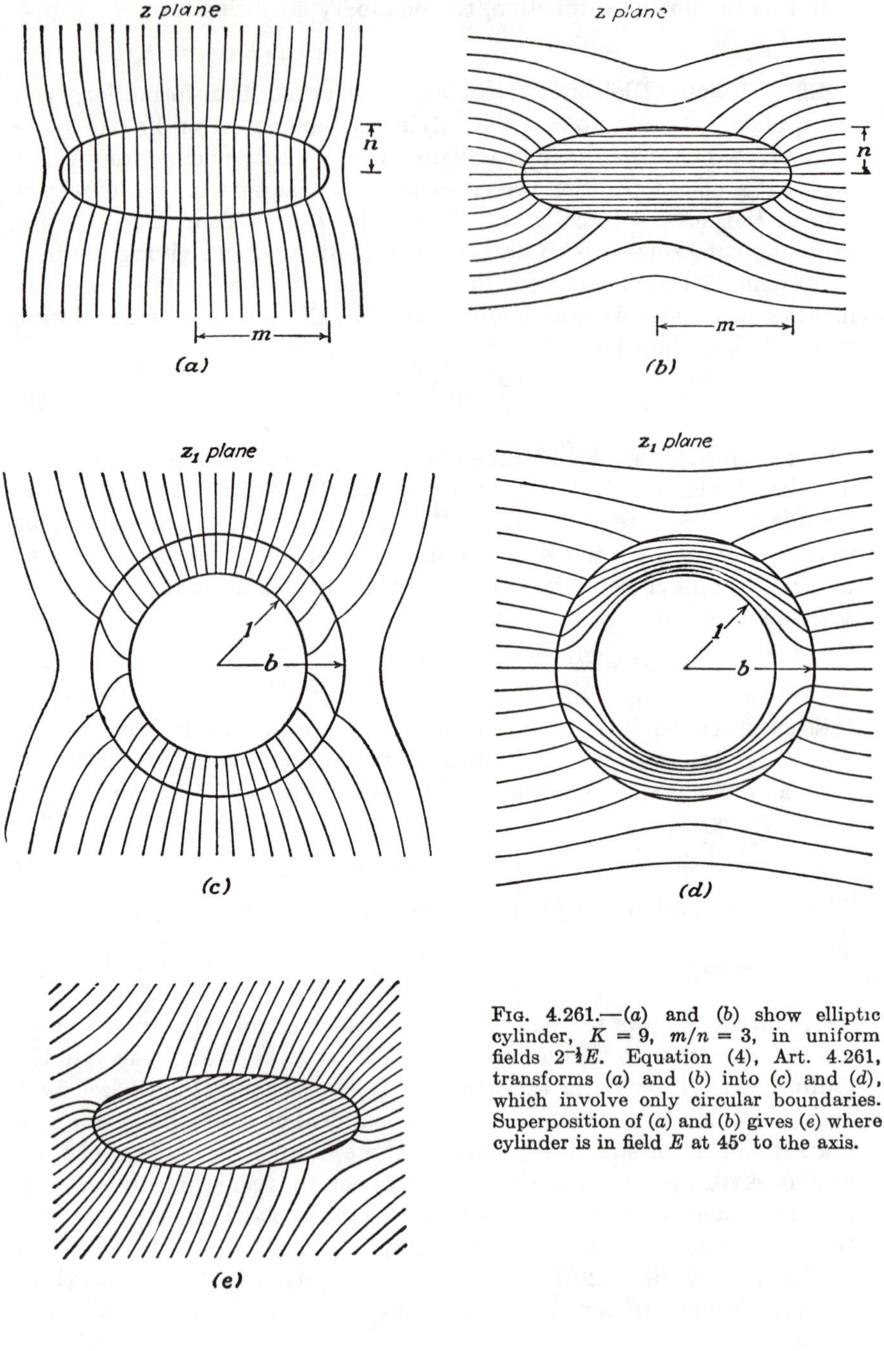

FIG. 4.261.—(a) and (b) show elliptic cylinder, $K = 9$, $m/n = 3$, in uniform fields $2^{-\frac{1}{2}}E$. Equation (4), Art. 4.261, transforms (a) and (b) into (c) and (d), which involve only circular boundaries. Superposition of (a) and (b) gives (e) where cylinder is in field E at 45° to the axis.

$$W'_o = \tfrac{1}{2}jaE \sin \alpha(-z_1 + A'z_1^{-1}) \tag{5}$$
$$W'_i = -\tfrac{1}{2}jaB'E \sin \alpha(z_1 + z_1^{-1}) \tag{6}$$

where, using (2), we have

$$A' = \frac{-b^2[(K+1) + (K-1)b^2]}{(K-1) + (K+1)b^2} = \frac{-(m+n)(Km-n)}{(m-n)(Km+n)} \tag{7}$$

$$B' = \frac{2b^2}{(K-1) + (K+1)b^2} = \frac{m+n}{Km+n} \tag{8}$$

For Fig. 4.261*d*, we see similar harmonics must occur. Since $V''_i = 0$ when $z_1 = e^{j\theta}$, the solutions must be of the form

$$W''_o = \tfrac{1}{2}aE \cos \alpha(z_1 + A''z_1^{-1}) \tag{9}$$
$$W''_i = \tfrac{1}{2}aB''E \cos \alpha(z_1 + z_1^{-1}) \tag{10}$$

Since, from 4.26, we must have $U''_o = U''_i$ and $V''_o = KV''_i$ when $r = b$, we have

$$b^2 + A'' = B''(b^2 + 1) \qquad \text{and} \qquad b^2 - A'' = KB''(b^2 - 1)$$

Solving for A'' and B'' gives

$$A'' = \frac{b^2[(K+1) - (K-1)b^2]}{b^2(K+1) - (K-1)} = \frac{(m+n)(m-Kn)}{(m-n)(m+Kn)} \tag{11}$$

$$B'' = \frac{2b^2}{b^2(K+1) - (K-1)} = \frac{m+n}{m+Kn} \tag{12}$$

Superimposing and using the transformation (4), we have, outside the cylinder,

$$\begin{aligned} W_o &= \tfrac{1}{2}aE[e^{-j\alpha}z_1 + (A'' \cos \alpha + jA' \sin \alpha)z_1^{-1}] \\ &= \tfrac{1}{2}ajE[e^{-j(\alpha+w)} - (A'' \cos \alpha + jA' \sin \alpha)e^{jw}] \\ &= \tfrac{1}{2}E\{e^{-j\alpha}[z+(z^2-a^2)^{\frac{1}{2}}]+(A'' \cos \alpha+jA' \sin \alpha)[z-(z^2-a^2)^{\frac{1}{2}}]\} \end{aligned} \tag{13}$$

Similarly, inside the cylinder, we have

$$\begin{aligned} W_i &= \tfrac{1}{2}aE(B'' \cos \alpha - jB' \sin \alpha)(z_1 + z_1^{-1}) \\ &= E(m+n)\left(\frac{\cos \alpha}{m+Kn} - j\frac{\sin \alpha}{Km+n}\right)z \end{aligned} \tag{14}$$

We notice that the field inside the cylinder is still uniform. The lines of electric displacement when the external field makes an angle of 45° with either axis of the ellipse are shown in Fig. 4.261*e*.

4.262. Torque on Dielectric Cylinder.—The torque T per unit length on an infinite dielectric cylinder perpendicular to a uniform electric field with which its major axis makes an angle α may be found from the last article and from 3.13 (5) and (6). Thus

$$T = -\frac{\partial}{\partial \alpha}\left[\frac{1}{2}\epsilon_v(1-K)\int_V EE_i \cos \phi \, dv\right] \tag{1}$$

where ϕ is the angle between the original field E and the final inside field E_i. E, E_i, and $\cos\phi$ are constant in v and may be taken outside the integral leaving $\int dv = \pi mn$ per meter. The relative directions and the magnitudes of E and E_i are given by 4.261 (14) with 1 or K replacing K.

$$E = \left|\frac{\partial W_i}{\partial z}\right|_1 = |E|e^{j\theta}, \qquad E_i = \left|\frac{\partial W_i}{\partial z}\right|_K = |E_i|e^{j\theta_i} \tag{2}$$

Simplification of the 4.261 (14) product and substitution in (1) give

$$|E||E_i|\cos(\theta - \theta_i) = EE_i^*(\text{real part}) = |E^{(r)}E_i^{(r)}| + |E^{(i)}E_i^{(i)}| \tag{3}$$

$$T = \tfrac{1}{2}\epsilon_v E^2(K-1)\pi mn(m+n)\frac{\partial}{\partial\alpha}\left(\frac{\cos^2\alpha}{m+Kn} + \frac{\sin^2\alpha}{Km+n}\right)$$
$$= \frac{\pi\epsilon_v E^2(K-1)^2 mn(n^2-m^2)\sin 2\alpha}{2(Km+n)(m+Kn)} \tag{4}$$

This torque acts to align the major axis with the field.

4.27. Polygon with Rounded Corner.—There are several methods of replacing the sharp corner in a Schwarz transformation with a rounded one. By one method, we replace the factor $z_1^{(\alpha/\pi)-1}$ in

$$\frac{dz}{dz_1} = z_1^{\frac{\alpha}{\pi}-1}(z_1 - u_2)^{\frac{\alpha}{\pi}-1} \cdots$$

by $[(z_1 + \lambda(z_1^2 - 1)^{\frac{1}{2}})]^{(\alpha/\pi)-1}$, where $|u_n| > |u_{n-1}| > \cdots |u_2| \gtrless 1$ and $\lambda < 1$. By another method, we replace the factor $z_1^{(\alpha/\pi)-1}$ by

$$(z_1+1)^{\frac{\alpha}{\pi}-1} + \lambda(z_1-1)^{\frac{\alpha}{\pi}-1}$$

In both cases, the argument of the new factor is zero when $z_1 \gtrless 1$ and is $\alpha - \pi$ when $z_1 \lessgtr -1$, so that the faces of the polygon on either side of the region $-1 < z_1 < +1$ make the same angle with each other as if the factor $z_1^{(\alpha/\pi)-1}$ were used. Between $z_1 = +1$ and $z_1 = -1$, we now have a curve whose shape can be adjusted somewhat by means of the factor λ.

4.28. Plane Grating of Large Cylindrical Wires.—We have already seen in 4.20 how the problem of a plane grating formed by cylindrical wires of small radius may be solved by taking for the cylindrical surfaces the natural equipotentials surrounding a grating of line charges. When the diameters of the wires become comparable with the spacing, this approximation breaks down completely. We may solve this case however by the method of the last article. Let us take the "cell" of the grating, outlined in Fig. 4.28*a*. From the figure, we see that the differential expression connecting z_1 and z is

$$\frac{dz}{dz_1} = C_1\frac{(z_1+1)^{\frac{1}{2}} + \lambda(z_1-1)^{\frac{1}{2}}}{[(z_1-1)(z_1+1)(z_1+a_1)]^{\frac{1}{2}}}$$
$$= \frac{C_1}{[(z_1-1)(z_1+a_1)]^{\frac{1}{2}}} + \frac{\lambda C_1}{[(z_1+1)(z_1+a_1)]^{\frac{1}{2}}} \tag{1}$$

We illustrate here a method we have not hitherto used for evaluating the constant C_1. Since $z_1 = r_1 e^{j\theta_1}$ if we keep r_1 constant, we have

$$dz_1 = jr_1 e^{j\theta_1}\, d\theta_1 = jz_1\, d\theta_1$$

We notice that when $r_1 \to \infty$ and θ_1 goes from 0 to π then z goes from $y = 0$ to $y = b$. Substituting $jz_1\, d\theta_1$ for dz_1 in (1) and letting $z_1 \to \infty$, we have

$$\int_0^{jb} dz = jC_1(1 + \lambda)\int_0^{\pi} d\theta_1 \qquad \text{or} \qquad jb = jC_1\pi(1 + \lambda)$$

$$C_1 = \frac{b}{\pi(1 + \lambda)} \tag{2}$$

Frequently, this method can be used to evaluate a constant by an elementary integration of a special case, usually $r_1 \to 0$ or $r_1 \to \infty$, before

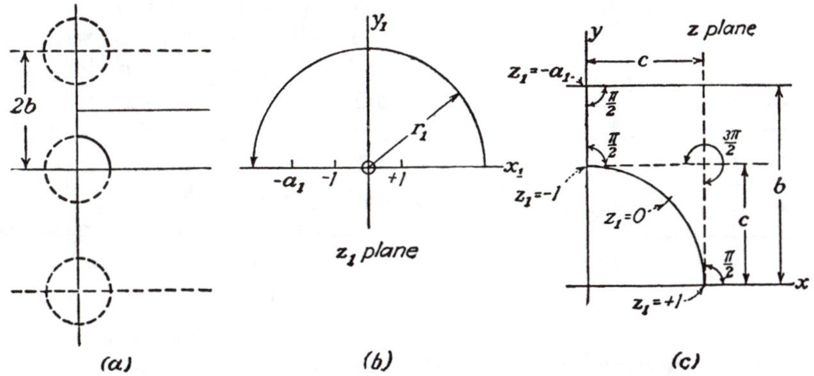

FIG. 4.28*a*, *b*, *c*.

performing the general integration. Integrating (1) using *Pc* 113 or letting $z_1 = (a_1 u^2 \pm 1)(1 - u^2)^{-1}$ and using *Dw* 140.02 gives

$$z = \frac{2b}{\pi(1 + \lambda)}\left[\tanh^{-1}\left(\frac{z_1 - 1}{z_1 + a_1}\right)^{\frac{1}{2}} + \lambda \tanh^{-1}\left(\frac{z_1 + 1}{z_1 + a_1}\right)^{\frac{1}{2}}\right] + C_2 \tag{3}$$

When $z = c$, $z_1 = +1$, so that C_2 is not real and

$$c = \frac{2b\lambda}{\pi(1 + \lambda)} \tanh^{-1}\left(\frac{2}{1 + a_1}\right)^{\frac{1}{2}} \quad \text{or} \quad \frac{a_1 + 1}{2} = \coth^2\left[\frac{\pi c(1 + \lambda)}{2b\lambda}\right] \tag{4}$$

When $z = jc$, $z_1 = -1$, so that C_2 is not imaginary and

$$c = \frac{+2b}{\pi(1 + \lambda)} \tan^{-1}\left(\frac{2}{a_1 - 1}\right)^{\frac{1}{2}} \quad \text{or} \quad \frac{a_1 - 1}{2} = \cot^2\left[\frac{\pi c(1 + \lambda)}{2b}\right] \tag{5}$$

Solving (4) and (5) for $(a_1 - 1)/(a_1 + 1)$ or eliminating a_1 gives

$$\frac{a_1 - 1}{a_1 + 1} = \cot^2 \frac{\pi c(1 + \lambda)}{2b} \tanh^2 \frac{\pi c(1 + \lambda)}{2b\lambda} = \cos^2 \frac{\pi c(1 + \lambda)}{2b} \tag{6}$$

$$\csc \frac{\pi c(1 + \lambda)}{2b} = \coth \frac{\pi c(1 + \lambda)}{2b\lambda} \tag{7}$$

To determine λ in terms of b and c, we can plot the ratio of the left to the right side of this equation as a function of λ and choose that value of λ for which the ratio is unity. Having determined λ, we now add (4) and (5) and obtain

$$a_1 = \coth^2\left[\frac{\pi c(1+\lambda)}{2b\lambda}\right] + \cot^2\left[\frac{\pi c(1+\lambda)}{2b}\right] \tag{8}$$

which determines a_1.

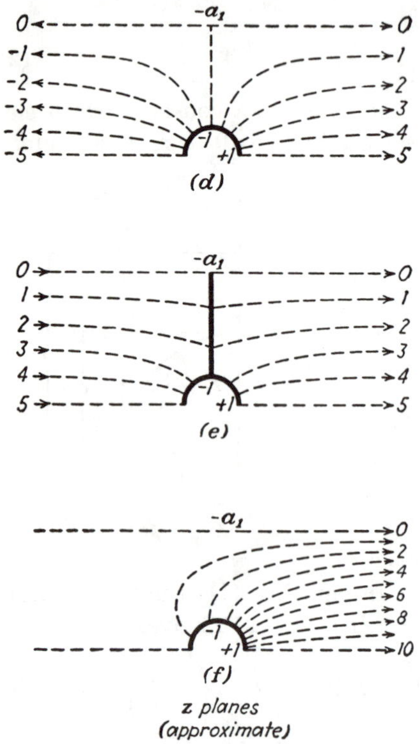

FIG. 4.28*d*, *e*, *f*.

The question of how close the curve approximates to $r = c$ has been investigated by Richmond,* and its distance from the origin has been found to vary from c by less than 2 per cent if $2c < b$.

To obtain the solution when the grating forms one boundary of a uniform field (Fig. 4.28*f*), we can superimpose Fig. 4.28*d* and Fig. 4.28*e* where lines of force are dotted and equipotentials solid. For case *d*, the desired function in the z_1-plane is evidently, from 4.22 (3),

$$W = A \sin^{-1} z_1$$

where V is the potential function. Since we wish $\frac{1}{2}Eb$ lines of force in the strip of Fig. 4.22*a* instead of π, we must take A equal to $\frac{1}{2}Eb/\pi$, giving

$$W = \frac{Eb}{2\pi} \sin^{-1} z_1 \tag{9}$$

Similarly for case *e*, we need the x_1-axis between $-a_1$ and $+1$ at potential zero; so, shifting the origin, we have

$$W = \frac{Eb}{2\pi} \sin^{-1} \frac{2z_1 + a_1 - 1}{a_1 + 1} \tag{10}$$

When these fields are superimposed, we get case *f* with a uniform field of strength E at $x = +\infty$ and strength 0 at $x = -\infty$. The potential function, as in 4.22 (3), is V. Accurate drawings of such fields may be seen in the article mentioned.

4.29. Elliptic Function Transformations. Two Coplanar Strips.—The integral in 4.18 (8) of the Schwarz transformation often leads to Jacobi

* *Proc. London Math. Soc.*, Ser. 2, Vol. 22, p. 389, 1924.

elliptic functions when the polygon angles are not integral multiples of π or when more than two $\frac{1}{2}\pi$ bends are involved. Examples of these appear in Kober and other references. The functions for two similar coplanar parallel conducting strips with equal and opposite charges is a starting point for the solution of several important practical problems. The data given are the strip boundaries $\pm a$, $\pm b$ and the potential difference $2U_0$. Figure 4.29 shows that four $\frac{1}{2}\pi$ real axis bends in (a) form a rectangular box in (b) which encloses an area of the uniform field bounded by $V = 0$, $V = V_1$, $U = +U_0$, and $U = -U_0$. Thus 4.18 (8) takes the form *Dw* 781.01.

$$W = \int \frac{C_1\,dz}{[(z^2 - b^2)(z^2 - a^2)]^{\frac{1}{2}}} = \frac{1}{C_2} \operatorname{sn}^{-1}\left(\frac{z}{b}, \frac{b}{a}\right) \tag{1}$$

There is no additive constant if $z = 0$ when $W = 0$, so

$$z = b \operatorname{sn}\left(C_2 W, \frac{b}{a}\right) \tag{2}$$

Special values of the elliptic sines are (HTF II, page 346)

$$\operatorname{sn}(K + jK', k) = k^{-1} \qquad \operatorname{sn}(K + \tfrac{1}{2}jK', k) = k^{-\frac{1}{2}} \tag{3}$$

where $k = b/a$ is the modulus, $k' = (1 - k^2)^{\frac{1}{2}}$ is the complementary modulus, and K and K' are the corresponding complete elliptic integrals

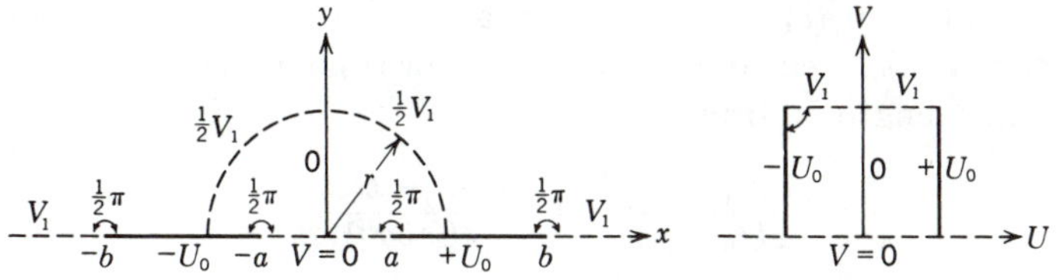

(a) FIG. 4.29a, b. (b)

(*Dw* 773.1). Thus when $z = a$ in Fig. 4.29a

$$C_2(U_0 + jV_1) = K + jK', \qquad C_2 = \frac{K}{U_0}, \qquad V_1 = \frac{K'U_0}{K} \tag{4}$$

The final transformation may then be written

$$z = b \operatorname{sn}\left(\frac{KW}{U_0}, \frac{b}{a}\right) \tag{5}$$

When $y = 0$, on the strips U_0 and $-U_0$ or the strip $V = 0$ and the plane V_1,

$$\left|\frac{\partial W}{\partial z}\right| = \frac{U_0}{K}\left|\frac{\partial}{\partial z} \operatorname{sn}^{-1}\left(\frac{z}{b}, \frac{b}{a}\right)\right| = \frac{U_0}{K}|[(x^2 - a^2)(x^2 - b^2)]^{-\frac{1}{2}}| \tag{6}$$

For the two strips at potential $\pm U_0$, $b < |x| < a$ so that, as in 4.23 (3),

$$\sigma = \pm U_0(\epsilon K)^{-1}[(x^2 - b^2)(a^2 - x^2)]^{-\frac{1}{2}} \tag{7}$$

For a strip at potential 0 coaxial with a slit in a coplanar plane at V_1,

$$\begin{aligned} \sigma_s &= -V_1(\epsilon K')^{-1}[(b^2 - x^2)(a^2 - x^2)]^{-\frac{1}{2}} \\ \sigma_p &= V_1(\epsilon K')^{-1}[(x^2 - b^2)(x^2 - a^2)]^{-\frac{1}{2}} \end{aligned} \tag{8}$$

It is evident that a and b are inverse points for the circle of radius $r = (ab)^{\frac{1}{2}}$ so that inversion of Fig. 4.29a about this circle leaves the equipotential strips U_0 and $-U_0$, and hence their fields, unaltered. This is impossible if a field line crosses the circle diagonally, so the circle itself must be a field line of value $\frac{1}{2}V_1$.

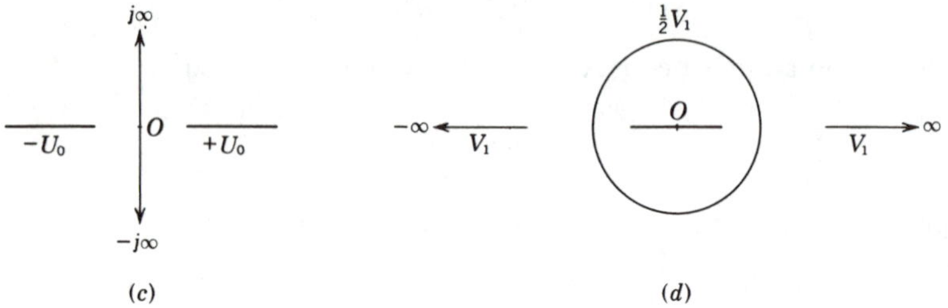

FIG. 4.29c, d.

The transformation (5) also solves the problems of two coplanar strips or a plane and strip (c) and a strip in a coaxial cylinder or a cylinder or strip coaxial with a slit in a plane (d). For coplanar strips the capacitance per unit length is, from 4.11 (4),

$$C = \frac{\epsilon[V]}{2U_0} = \frac{\epsilon K'}{K} = \frac{\epsilon K[(1 - b^2a^{-2})^{\frac{1}{2}}]}{K(ba^{-1})} \tag{9}$$

For strip and plane it is twice this, and for strip in slit it is

$$C = \frac{\epsilon[U]}{V_1} = \frac{4\epsilon U_0}{V_1} = \frac{4\epsilon K(ba^{-1})}{K[(1 - b^2a^{-2})^{\frac{1}{2}}]} \tag{10}$$

For a strip ($2b$) in a cylinder $[r_0 = (ab)^{\frac{1}{2}}]$ or a cylinder $[r_0 = (ab)^{\frac{1}{2}}]$ in a slit ($2a$), the capacitance per unit length is twice (10). A logarithmic transformation, $z_1 = \ln z$, applied to (5) will give the fields of an infinite stack of strips charged alternately to potentials $-U_0$ and $+U_0$, and includes a "strip line," which is a strip at potential U_0 halfway between and parallel to two infinite earthed planes. Useful formulas for Jacobi elliptic functions of a complex variable appear in HMF, Dwight, Whittaker and Watson, Erdelyi et al., and elsewhere. Jahnke and Emde give some excellent graphs.

4.30. Unequal Coplanar Strips by Inversion.—The inversion method of Art. 4.21, applied to the results of the last article, can solve directly the unsymmetrical case of two unequal and oppositely charged strips, but the solution takes a much simpler form if we first solve the problem of a strip of width B at potential zero separated by a gap of $A - B$ from a semi-infinite plane at potential V_1 as shown in Fig. 4.30*a*. This is done by folding the real axis in Fig. 4.29*a* back on itself at 0 so that in 4.18 (7) $\alpha_1 = 2\pi$, $u_1 = 0$, and $\alpha_2 = \alpha_3 = \cdots = \pi$. Then if the

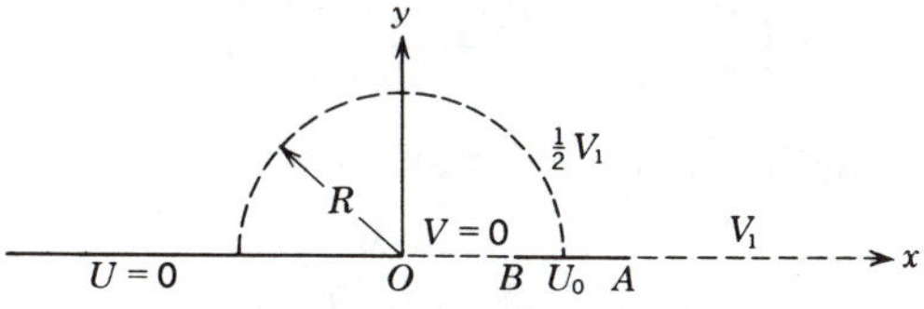

(*a*)

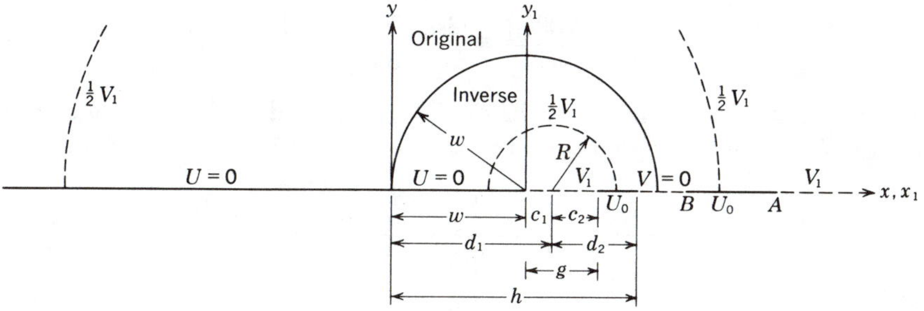

(*b*)

FIG. 4.30*a*, *b*.

origin is kept at $W = 0$, 4.18 (8) becomes $z = Cz_1^2$. If $z = B$ when $z_1 = b$, then C is B/b^2. Substitution of z_1 for z in 4.29 (5) gives

$$z = B\,\mathrm{sn}^2\left[\frac{KW}{U_0}, \left(\frac{B}{A}\right)^{\frac{1}{2}}\right] \tag{1}$$

The modulus is $(B/A)^{\frac{1}{2}}$ from 4.29 (3), for when $W = U_0 + jV_1$, then KW/U_0 is $K + jK'$ and $z = A$. From 4.11 (4) and 4.29 (4) the capacitance per unit length between strip and plane is

$$C = \frac{2\epsilon U_0}{V_1} = \frac{2\epsilon K[(B/A)^{\frac{1}{2}}]}{K\{[1-(B/A)]^{\frac{1}{2}}\}} \tag{2}$$

Since the cylinder of radius $R = (AB)^{\frac{1}{2}}$ is at potential $\frac{1}{2}V_1$, the capacitances between this cylinder and an enclosed strip of width B with one edge on its axis and between it and a semi-infinite plane whose edge is a distance A from its axis both equal $2C$ or twice that given by (2).

Inversion of Fig. 4.30*a* about point $z = w$ with an inversion radius $w(2w < B)$ will solve the problem of two coplanar strips at potentials $U = 0$ and $U = U_0$ when one strip is of width w, the inside gap is g, and the distance between outside edges is h. That part of Fig. 4.30a lying outside the circle of inversion is shown in Fig. 4.30b, along with that part of the inverse system that lies inside it. From the inversion law

$$z_1 = \frac{w^2}{z - w} \qquad A = \frac{w(w+g)}{g} \qquad B = \frac{wh}{h-w}$$

$$k = \left(\frac{B}{A}\right)^{\frac{1}{2}} = \left[\frac{hg}{(w+g)(h-w)}\right]^{\frac{1}{2}} \tag{3}$$

So that the transformation is, from (1) and Dw 755.1,

$$z_1 = \frac{w(h-w)}{w - h\,\mathrm{cn}^2\,(KW/U_0,k)} \tag{4}$$

From 4.29 (4) the capacitance between the strips is

$$C = \frac{2\epsilon[V_1]}{U_0} = \frac{2\epsilon K[(1-k^2)^{\frac{1}{2}}]}{K(k)} \tag{5}$$

Let the distance from the center of the $\frac{1}{2}V_1$ cylinder of radius R to the edges of $V = V_1$ be c_1 and c_2 and to the near edges of $V = 0$ be d_1 and d_2.

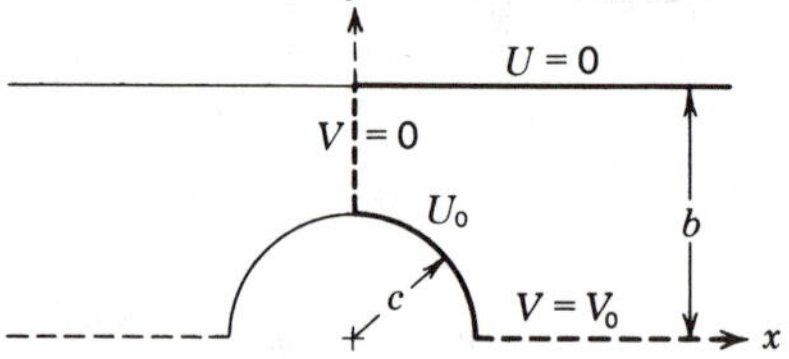

FIG. 4.31

Then $c_1d_1 = R^2$, $c_2d_2 = R^2$, $d_1 = c_1 + w$, $d_2 = c_2 + h - g - w$, and $c_1 + c_2 = g$. With these relations h, w, and g may be eliminated in (3) so that

$$k = \frac{R(c_1+c_2)}{R^2 + c_1c_2} = \frac{R(d_1+d_2)}{R^2 + d_1d_2} \tag{6}$$

The capacitance between the cylinder $\frac{1}{2}V_1$ and the strip V_1 equals that between $\frac{1}{2}V_1$ and the two semi-infinite planes $V = 0$ and is given by

$$C = \frac{2\epsilon U_0}{\frac{1}{2}V_1} = \frac{4\epsilon K(k)}{K[(1-k^2)^{\frac{1}{2}}]} \tag{7}$$

4.31. Charged Circular Cylinder between Plates.—The problem in which a conducting cylinder of radius c at potential U_0 lies midway between two infinite parallel earthed conducting planes at a distance $2b$ apart can be solved by using 4.30 (1) and 4.28 (3). The solution is almost exact for $c < \frac{1}{2}b$ but fails as $c \rightarrow b$. First Fig. 4.30a must be adjusted so that U_0 lies between $+1$ and -1 in Fig. 4.28b and $W = 0$ is at $-a$. Evidently the origin in Fig. 4.30a must be shifted a distance $a_1 = \frac{1}{2}(A + B)$ to the right where $B = a_1 - 1$ and A is $a_1 + 1$. Thus 4.30 (1) becomes

$$z_1 + a_1 = (a_1 - 1)\,\mathrm{sn}^2\left(\frac{KW}{U_0}, k\right) \tag{1}$$

The modulus, from 4.28 (6), is then

$$k = \left(\frac{B}{A}\right)^{\frac{1}{2}} = \left(\frac{a_1 - 1}{a_1 + 1}\right)^{\frac{1}{2}} = \cos\left[\frac{\pi c(1 + \lambda)}{2b}\right] \tag{2}$$

Substitute (1) in 4.28 (3), use the relations between $\mathrm{cn}^2\, u$, $\mathrm{dn}^2\, u$, and $\mathrm{sn}^2\, u$ in *Dw* 755.1, 755.2, and 755.3, and note that $\tanh^{-1}(jy) = j\tan^{-1} y$ from *Dw* 722.3 and HMF 5.07.32 and that $C_2 = 0$. Then 4.28 (3) becomes

$$\begin{aligned} z &= \frac{2jb}{\pi(1 + \lambda)}\left(\tan^{-1}\frac{\mathrm{dn}\, u}{k\,\mathrm{sn}\, u} + \lambda\tan^{-1}\frac{\mathrm{cn}\, u}{\mathrm{sn}\, u}\right) \\ &= 2jb[\cos^{-1}(k\,\mathrm{sn}\, u) + \lambda\sin^{-1}(\mathrm{cn}\, u)][\pi(1 + \lambda)]^{-1} \end{aligned} \tag{3}$$

where $u = (KW/U_0, k)$. The parameter λ must be chosen to satisfy 4.28 (6). From Fig. 4.31 and 4.11 (4) the capacitance per unit length is

$$C = \frac{\epsilon[V]}{U_0} = \frac{4\epsilon K[(1 - k^2)^{\frac{1}{2}}]}{K(k)} \tag{4}$$

This is a very close upper limit if $c < \frac{1}{2}b$, and the error is less than 1 per cent when $c = \frac{1}{2}b$. Numerical tables and formulas for the real and imaginary parts of $\mathrm{cn}\, u$, $\mathrm{dn}\, u$, and $\mathrm{sn}\, u$ are given in HMF, pages 574, 575.

Problems

Problems marked C are taken from the Cambridge examination questions as reprinted by Jeans, by permission of the Cambridge University Press.

1. That half of the xz-plane for which x is positive is at zero potential except for the strip between $x = 0$ and $x = a$ which is at potential V_0. The whole of the yz-plane is at zero potential. Show that the potential at any point for which x and y are positive is

$$\pi V = V_0\left(\tan^{-1}\frac{y}{x - a} - 2\tan^{-1}\frac{y}{x} + \tan^{-1}\frac{y}{x + a}\right)$$

2. The positive halves of the x and y planes are conducting sheets. All the surface is earthed except the region near the intersection, bounded by the lines $x = a$ and $y = b$, which is insulated and raised to a potential V. Find the surface density at any point on the sheet.

3. By inversion, establish the image law for a line charge in a parallel circular conducting cylinder.

4. A conductor is formed by the outer surface of two similar circular cylinders, of radius a, intersecting orthogonally. It carries a charge q per unit length. Show that $q(2r^2 - a^2)(4\pi a r^2)^{-1}$ is the surface charge density, where r is the distance from the axis.

5. Show that when an elliptic conducting cylinder is charged, the ratio of the maximum to the minimum surface densities equals the ratio of the major to the minor axis.

6. A filament carrying a charge q per unit length passes vertically through a vertical hole of radius a in a block of relative capacitivity K. The filament is a distance c from the center of the hole. Show that the force per unit length pulling it toward the wall is $[(K - 1)/(K + 1)]cq^2/[2\pi\epsilon_v(a^2 - c^2)]$.

7. Two fine fiber leads are to be run vertically through an earthed conducting circular cylindrical tube of radius a. Show that the spacing of the fibers must be $2(5^{\frac{1}{2}} - 2)^{\frac{1}{2}}a$ in order that, when charged to equal and opposite potentials, there will be no force on them.

8. An infinite circular conducting cylindrical shell of radius a is divided longitudinally into quarters. One quarter is charged to a potential $+V_1$ and the one diagonally opposite to $-V_1$. The other two are earthed. Show that the potential at any point inside is

$$\frac{V_1}{\pi}\left(\tan^{-1}\frac{2ay}{a^2 - r^2} + \tan^{-1}\frac{2ax}{a^2 - r^2}\right)$$

9. Consider the region of space between the cylinder $x^2 + y^2 = b^2$ and the xz-plane. All the curved boundary and that portion of the plane boundary for which $a < |x| < b$ is at zero potential. The part of the latter for which $-a < x < +a$ is at the potential V_0. Show that the equation of the lines of force inside is, when $a \lessgtr r \lessgtr b$

$$U = \frac{2V_0}{\pi}\sum\frac{a^n}{n}\left[\frac{1}{r^n} + \left(\frac{r}{b^2}\right)^n\right]\cos n\theta$$

where only odd values of n are taken.

10*C*. Three long thin wires, equally electrified, are placed parallel to each other so that they are cut by a plane perpendicular to them in the angular points of an equilateral triangle of side $(3)^{\frac{1}{2}}c$. Show that the polar equation of an equipotential curve drawn on the plane is

$$r^6 + c^6 - 2r^3c^3\cos 3\theta = \text{constant}$$

the pole being at the center of the triangle and the initial line passing through one of the wires.

11*C*. A long hollow cylindrical conductor is divided into two parts by a plane through the axis, and the parts are separated by a small interval. If the two parts are kept at potentials V_1 and V_2, the potential at any point within the cylinder is

$$\frac{1}{2}(V_1 + V_2) + \frac{V_1 - V_2}{\pi}\tan^{-1}\frac{2ar\cos\theta}{a^2 - r^2}$$

where r is the distance from the axis, and θ is the angle between the plane joining the point to the axis and the plane through the axis normal to the plane of separation.

12*C*. An electrified line with charge e per unit length is parallel to a circular cylinder of radius a and inductive capacity K; the distance of the wire from the center of the cylinder being c. Show that the force on the wire per unit length is

$$\frac{K - 1}{K + 1}\,\frac{a^2e^2}{2\pi\epsilon_v c(c^2 - a^2)}$$

13*C*. A cylindrical conductor of infinite length whose cross section is the outer boundary of three equal orthogonal circles of radius a has a charge e per unit length.

Prove that the electric density at distance r from the axis is

$$\frac{e}{6\pi a}\,\frac{(3r^2 + a^2)(3r^2 - a^2 - 6^{\frac{1}{2}}ar)(3r^2 - a^2 + 6^{\frac{1}{2}}ar)}{r^2(9r^4 - 3a^2r^2 + a^4)}$$

14. A horizontal plane at potential zero has its edge parallel to and at a distance c from an infinite vertical plane at potential $\frac{1}{2}\pi$. Show that the charge density on the vertical plane is $\epsilon(y^2 + c^2)^{-\frac{1}{2}}$ and on the horizontal plane $-\epsilon(x^2 - c^2)^{-\frac{1}{2}}$, where x and y are measured from the line in the vertical plane nearest the edge of the horizontal plane.

15. Show that the capacitance per unit length between a flat strip of width $2c$ and an elliptic cylinder whose foci coincide with the edges of the strip and whose semimajor axis is a is $2\pi\epsilon[\cosh^{-1}(a/c)]^{-1}$.

16. Show that the force of attraction per unit length between two similar parallel wires of radius a and carrying a charge $+q$ and $-q$ per unit length, respectively, is $q^2/[2\pi\epsilon(c^2 - 4a^2)^{\frac{1}{2}}]$ where c is the distance between centers.

17. A plane grating is formed of parallel flat coplanar strips of width $2a$ with their center lines at a distance $2b$ apart. If the grating is charged, show that the equipotential surface which is at a mean distance b from the plane of the grating deviates from a true plane by approximately $.028b \cos^2(\frac{1}{2}\pi a/b)$.

18. Find, approximately, the field about a grating composed of parallel wires of radius a at a distance 2π apart, charged to a potential U_0 and a parallel earthed plate at a distance b from the center of the wire, if $a \ll b$ and $a \ll 2\pi$.

$$U = \frac{-U_0}{U_1}\ln\frac{e^{2x} - 2ce^x\cos y + c^2}{c^2e^{2x} - 2ce^x\cos y + 1}$$

where

$$U_1 = -\ln\frac{e^{2b} - 2ce^b\cos a + c^2}{c^2e^{2b} - 2ce^b\cos a + 1}$$

and $c = \cosh b \cos a + (\cosh^2 b \cos^2 a - 1)^{\frac{1}{2}}$. Here we have taken $2a$ for the maximum thickness, measured parallel to the earthed plate, of a closed equipotential about a line charge, the distance from $2a$ to the plate being b.

19. Using the results of Arts. 4.03 and 4.25, find the transformation which gives the field about a conducting earthed strip of width $2c$ whose plane lies in the direction of a uniform electric field. Let U be the potential function.

$$W = \pm E(z^2 - c^2)^{\frac{1}{2}}$$

20. Superimposing a uniform field on the result of the last problem, find the transformation giving the field about a flat conducting strip of width $2c$, lying with its axis perpendicular to a uniform electric field and its plane inclined at an angle α to it. Find the torque per unit length acting on it.

$$W = E[\pm(z^2 - c^2)^{\frac{1}{2}}\cos\alpha - jz\sin\alpha], \qquad T = \tfrac{1}{2}\pi\epsilon c^2E^2\sin 2\alpha$$

21. By an inverse Schwarz transformation, find the field about two semi-infinite coplanar planes charged to potentials $+U_0$ and $-U_0$, whose edges are parallel and at a distance $2c$ apart. Let U be the potential function. Show that $z = c\sin(\frac{1}{2}\pi W/U_0)$.

22. By rotating the transformation of problem 21 through 90° and applying a Schwarz transformation, find the field about a freely charged horizontal grating composed of similar vertical strips of width $2a$ parallel to each other and at a distance $2b$

apart. Let V be the potential function. The charge on the strip is ϵU_0.

$$W = \frac{U_0}{2\pi} \sin^{-1} \frac{\sinh [\pi z/(2jb)]}{\sinh (\frac{1}{2}\pi a/b)}$$

23. By shifting the origin of problem 21 to the left and applying a logarithmic transformation, find the transformation which gives the field about the grating of problem 22 when it forms one boundary of a uniform field. Let the potential function be V. The charge on the strip is ϵU_0.

$$W = \frac{U_0}{2\pi} \sin^{-1} \frac{e^{\pi z/b} - \cosh (\pi a/b)}{\sinh (\pi a/b)}$$

24. Taking U for the potential function and taking $c = 1$ in problem 21, apply a logarithmic transformation and obtain the field about a set of semi-infinite conducting planes spaced a distance b apart whose edges lie on the y-axis and which are charged alternately to potentials $+U_0$ and $-U_0$.

$$z = \frac{b}{\pi} \ln \sin \frac{\pi W}{2U_0}$$

25. Show that it is possible to shape the edges of the plates in a many-plate parallel-plate capacitor so that the field over the whole plate surface including the edges is constant. Show that this surface is the surface $U = U_0/2$ in problem 24 and that its equation is

$$x = -\frac{b}{2\pi} \ln \left(2 \cos \frac{2\pi y}{b}\right)$$

26. A variable air capacitor consists of thin plates which slide between other fixed plates. Show, using problem 24, that the additional capacitance due to the bulging of the lines of force at the edges is equivalent to adding a strip of width $(b/\pi) \ln 2$ to the edges of the plate, where the spacing between fixed plates is b.

27. By rotating the transformation of 4.22 through 90°, letting $b = \pi$, and applying a logarithmic transformation, find the field about a cylindrical trough of radius 1 which subtends an angle 2α at the center of curvature and carries a charge $-Q$, in the presence of a charge $+\frac{1}{2}Q$ at the origin.

$$W = \frac{Q}{2\pi\epsilon} \sin^{-1} \left[\frac{j(r-1)\cos(\frac{1}{2}\theta) - (r+1)\sin(\frac{1}{2}\theta)}{2r^{\frac{1}{2}} \sin(\frac{1}{2}\alpha)}\right]$$

28. By superimposing a potential due to $-\frac{1}{2}Q$ at the origin in the last problem, get the potential due to a freely charged circular cylindrical trough of unit radius, subtending an angle 2α at the origin.

$$W = \frac{Q}{4\pi\epsilon}\left[2 \sin^{-1} \frac{j(r-1)\cos(\frac{1}{2}\theta) - (r+1)\sin(\frac{1}{2}\theta)}{2r^{\frac{1}{2}} \sin(\frac{1}{2}\alpha)} + j \ln r - \theta\right]$$

29. Show that the ratio of the charge on the convex surface to that on the concave surface in the last problem is $(\pi + \alpha)/(\pi - \alpha)$.

30. Show that the charge density on the surface in the last problem is

$$\mp\frac{1}{4}(Q/\pi)[1 \pm \cos \tfrac{1}{2}\theta(\cos^2 \tfrac{1}{2}\theta - \cos^2 \tfrac{1}{2}\alpha)^{-\frac{1}{2}}]$$

where the positive sign applies to the convex surface.

31. Two semi-infinite straight-edged thin conducting sheets are arranged with edges parallel, so that one is the image of the other in the y-plane. They are at zero potential and form one boundary of the field. Find the transformations for the following cases:

a. Sheets parallel, field outside,

$$z = -\left(\frac{b_0}{\pi}\right)(jW^2a^{-2} - 2j \ln W - \pi - j + 2j \ln a)$$

b. Sheet at 90°, field outside,

$$z = (1 - j)\left(\frac{3b_0a^{-\frac{3}{2}}}{8}\right)\left(\frac{2W^{\frac{3}{2}}}{3} + 2a^2W^{-\frac{1}{2}}\right) + jb_0$$

c. Sheets at 90°, field inside,

$$z = (1 + j)\tfrac{1}{2}b_0a^{-\frac{1}{2}}W^{-\frac{1}{2}}(W^2 + a^2) - jb_0$$

32. Find the transformation that gives field at all points when a filament carrying a charge q per unit length is placed at a distance d in front of the center of a slit of width $2b$ in an earthed conducting plane.

$$W = -\frac{q}{2\pi\epsilon} \ln \frac{z \pm (z^2 - b^2)^{\frac{1}{2}} - j[d \pm (d^2 + b^2)^{\frac{1}{2}}]}{z \pm (z^2 - b^2)^{\frac{1}{2}} + j[d \pm (d^2 + b^2)^{\frac{1}{2}}]}$$

33. A Wolff electroscope consists of two similarly charged fibers of radius c at a distance $2d$ apart placed symmetrically inside and parallel to the axis of an earthed cylinder of radius b. Show that the capacitance per unit length between the two fibers and the case lies between

$$C = 8\pi\epsilon\left\{\ln\left[\frac{b^4}{4c^2d^2}\right]\right\}^{-1} \quad \text{and} \quad C = 8\pi\epsilon\left\{\ln\left[\frac{b^4}{4c^2(d^2 + c^2)}\right]\right\}^{-1}$$

neglecting c^4 and d^4 compared with b^4.

34. Two similar parallel conducting cylinders are in contact and carry a total charge q per unit length. Show that the capacitance per unit length between these cylinders and a third large cylinder whose axis is their line of contact is approximately $2\pi\epsilon/\ln(2b/\pi a)$, where $b > a$.

35. One plate, of thickness $2A$, of a variable air capacitor moves midway between two other plates at a distance $2B$ apart. Show that the transformation for the field in the neighborhood of the edge is

$$z = A - \frac{2jB}{\pi}\cosh^{-1}\frac{B \tanh(\frac{1}{2}\pi W)}{[A(2B - A)]^{\frac{1}{2}}} + j\frac{2(B - A)}{\pi}\cosh^{-1}\frac{(B - A)\sinh(\frac{1}{2}\pi W)}{[A(2B - A)]^{\frac{1}{2}}}$$

and show that the apparent increase in width of the plate due to the bulging of the field at the edge is approximately

$$\Delta y = \frac{2}{\pi}\left\{B \ln\frac{2B - A}{B - A} - A \ln\frac{[A(2B - A)]^{\frac{1}{2}}}{B - A}\right\}$$

36. A thin infinite plate at potential $V = 1$ has in it an infinitely long slit of width $2K$. It lies parallel to and at a distance h from another plate at potential $V = 0$.

Show that the field at any point is given by

$$x = h\left[\frac{2a^2}{\pi(1 - a^2)} \frac{\sinh \pi U}{\cosh \pi U + \cos \pi V} + U\right]$$

$$y = h\left[\frac{2a^2}{\pi(1 - a^2)} \frac{\sin \pi V}{\cosh \pi U + \cos \pi V} + V\right]$$

where U is the stream function and a is determined from the relation

$$\pi K = +2h[\tanh^{-1} a + a/(1 - a^2)]$$

37. It is desired to find the field near a charged plate of thickness $2h$ having a rounded edge of radius h. Show that a suitable transformation is

$$z = h[W^2 + 2\pi^{-1}W(W^2 - 1)^{\frac{1}{2}} + 2j\pi^{-1} \sin^{-1} W]$$

where $V = 0$ on the plate whose surface is given by

$$y = (2h/\pi)[(x/h)^{\frac{1}{2}}(1 - x/h)^{\frac{1}{2}} + \sin^{-1} (x/h)^{\frac{1}{2}}]$$

when $0 < x < h$ and by $y = \pm h$ when $x > h$. Show that this curve deviates from the circle of radius h by less than $.14h$.

38. Show that when a conducting cylinder whose cross section is a hypocycloid of n points is at the potential $V = 0$, and carries a charge Q per unit length, the potential at any point outside is given by the transformation

$$z = \frac{a}{n} e^{\frac{2j\pi\epsilon W}{Q}} \left[(n - 1) + e^{-\frac{2nj\pi\epsilon W}{Q}}\right]$$

where a is the distance of an edge from the axis.

39. Show that a regular prism of n sides each of width

$$2a \sin (\tfrac{1}{2}\pi/n)[\cot (\tfrac{1}{2}\pi/n)]^{1/n}[(1 - 2/n) \sec (\pi/n) - 1]$$

may be circumscribed on the equipotential $V = Q/(2\pi\epsilon n) \ln \tan (\frac{1}{2}\pi/n)$ in the preceding problem in such a way that each side touches it along two lines at a distance

$$2a \sin (\tfrac{1}{2}\pi/n)[\cot (\tfrac{1}{2}\pi/n)]^{1/n}$$

apart. Show that the two surfaces come nearer coincidence as n increases and that when $n = 5$ their maximum axial distances differ by less than 6 per cent and their minimum by less than 1 per cent.

40*C*. A cylinder whose cross section is one branch of a rectangular hyperbola is maintained at zero potential under the influence of a line charge parallel to its axis and on the concave side. Prove that the image consists of three such line charges, and hence find the density of the induced distribution.

41*C*. A cylindrical space is bounded by two coaxial and confocal parabolic cylinders, whose *latera recta* are $4a$ and $4b$, and a uniformly electrified line which is parallel to the generators of the cylinder intersects the axes which pass through the foci in points distant c from them ($a > c > b$). Show that the potential throughout the space is

$$A \ln \frac{\cosh \dfrac{\pi r^{\frac{1}{2}} \cos \frac{1}{2}\theta}{a^{\frac{1}{2}} - b^{\frac{1}{2}}} - \cos \dfrac{\pi(r^{\frac{1}{2}} \sin \frac{1}{2}\theta - c^{\frac{1}{2}})}{a^{\frac{1}{2}} - b^{\frac{1}{2}}}}{\cosh \dfrac{\pi r^{\frac{1}{2}} \cos \frac{1}{2}\theta}{a^{\frac{1}{2}} - b^{\frac{1}{2}}} + \cos \dfrac{\pi(r^{\frac{1}{2}} \sin \frac{1}{2}\theta + c^{\frac{1}{2}} - a^{\frac{1}{2}} - b^{\frac{1}{2}})}{a^{\frac{1}{2}} - b^{\frac{1}{2}}}}$$

where r, θ are polar coordinates of a section, the focus being a pole. Determine A in terms of the electrification per unit length of the line.

42*C*. An infinitely long elliptic cylinder of inductive capacity K, given by $\xi = \alpha$ where $x + jy = c \cosh (\xi + j\eta)$, is in a uniform field P parallel to the major axis of any section. Show that the potential at any point inside the cylinder is

$$-Px(1 + \coth \alpha)/(K + \coth \alpha)$$

43*C*. Two insulated uncharged circular cylinders outside each other, given by $\eta = \alpha$ and $\eta = -\beta$ where $x + jy = c \tan \frac{1}{2}(\xi + j\eta)$, are placed in a uniform field of force of potential Fx. Show that the potential due to the distribution on the cylinders is

$$-2Fc\sum_{1}^{\infty}(-1)^n \frac{e^{n(\eta-\alpha)} \sinh n\beta + e^{-n(\eta+\beta)} \sinh n\alpha}{\sinh n(\alpha + \beta)} \sin n\xi$$

44*C*. Two circular cylinders outside each other, given by $\eta = \alpha$ and $\eta = -\beta$ where $x + jy = c \tan \frac{1}{2}(\xi + j\eta)$, are put to earth under the influence of a line charge q on the line $x = 0$, $y = 0$. Show that the potential of the induced charge outside the cylinders is

$$-\frac{q}{\pi\epsilon}\sum \frac{1}{n} \frac{e^{-n\alpha} \sinh n(\eta + \beta) + e^{-n\beta} \sinh n(\alpha - \eta)}{\sinh n(\alpha + \beta)} \cos n\xi + C$$

the summation being taken for all odd positive integral values of n.

45*C*. The cross sections of two infinitely long metallic cylinders are the curves

$$(x^2 + y^2 + c^2)^2 - 4c^2x^2 = a^4 \qquad \text{and} \qquad (x^2 + y^2 + c^2)^2 - 4c^2x^2 = b^4$$

where $b > a > c$. If they are kept at potentials V_1 and V_2, respectively, the intervening space being filled with air, prove that the surface densities per unit length of the electricity on the opposed surfaces are

$$\frac{\epsilon(V_1 - V_2)(x^2 + y^2)^{\frac{1}{2}}}{a^2 \ln (b/a)} \qquad \text{and} \qquad \frac{\epsilon(V_2 - V_1)(x^2 + y^2)^{\frac{1}{2}}}{b^2 \ln (b/a)}$$

respectively.

46*C*. What problems are solved by the transformation

$$\frac{d(x + jy)}{dt} = \frac{c(t^2 - 1)^{\frac{1}{2}}}{a^2 - t^2}, \quad \pi(\Psi + j\Phi) = \ln \left(\frac{a + t}{a - t}\right)$$

where $a > 1$?

47*C*. What problem in electrostatics is solved by the transformation

$$x + jy = \operatorname{cn} (\Phi + j\Psi)$$

where Ψ is taken as the potential function, Φ being the function conjugate to it?

48. A uniform field E_0 is bounded by the upper face ($y = 0$) of an infinite plate conductor of thickness b. The portion between $x = a$ and $x = -a$ is removed leaving an infinite slit. If U is the potential function, show that the transformation for the field may be written in the alternative forms:

$$E_0z = -(1 - W^2)^{\frac{1}{2}}\left(\frac{m^2}{W^2} - 1\right)^{\frac{1}{2}} + (1 - m^2)F\left(\sin^{-1}\frac{W}{m}, m\right) - 2E\left(\sin^{-1}\frac{W}{m}, m\right) + C_1E_0$$

$$jE_0z = (1 - W^2)^{\frac{1}{2}}\left(1 - \frac{m^2}{W^2}\right)^{\frac{1}{2}} + (1 + m^2)F\left[\sin^{-1}\left(\frac{1 - W^2}{1 - m^2}\right)^{\frac{1}{2}}, (1 - m^2)^{\frac{1}{2}}\right]$$

$$- 2E\left[\sin^{-1}\left(\frac{1 - W^2}{1 - m^2}\right)^{\frac{1}{2}}, (1 - m^2)^{\frac{1}{2}}\right] + jaE_0$$

$$E_0z = \left(1 - \frac{m^2}{W^2}\right)^{\frac{1}{2}}(W^2 - 1)^{\frac{1}{2}} - (1 - m^2)F\left(\sin^{-1}\frac{1}{W}, m\right) + 2E\left(\sin^{-1}\frac{1}{W}, m\right) + C_2E_0$$

where $F(p, q)$ and $E(p, q)$ are elliptic integrals of modulus q of the first and second kinds, respectively. The constants C_1, C_2, and m satisfy the equations

$$C_2 - a = a + jb - C_1 = [(1 - m^2)K(m) - 2E(m)]/E_0$$
$$bE_0 = -(1 + m^2)K[(1 - m^2)^{\frac{1}{2}}] + 2E[(1 - m^2)^{\frac{1}{2}}]$$

where K and E are complete elliptic integrals of modulus m.

49. An infinite conducting sheet at potential zero occupies the $x = 0$ plane except for a groove or ridge whose section is a circular arc of radius b with its center at $x = c$. This surface forms one boundary of a field E that extends to $x = \infty$ and is uniform except near the groove or ridge. Show that the potential is the imaginary part of

$$W = -\frac{(b^2 - c^2)^{\frac{1}{2}}\pi E}{\alpha}\,\frac{[z + j(b^2 - c^2)^{\frac{1}{2}}]^{\pi/\alpha} + [z - j(b^2 - c^2)^{\frac{1}{2}}]^{\pi/\alpha}}{[z + j(b^2 - c^2)^{\frac{1}{2}}]^{\pi/\alpha} - [z - j(b^2 - c^2)^{\frac{1}{2}}]^{\pi/\alpha}}$$

where $\cos \alpha = c/b$ and for a ridge $0 < \alpha < \pi$ and for a groove $\pi < \alpha < 2\pi$.

50. Show that in two-dimensional rectangular harmonics a solution of Laplace's equation is $V = (A \sin mx + B \cos mx)(C \sinh my + D \cosh my)$ when m is not equal to zero and $V = (A + Bx)(C + Dy)$ when m is zero.

51. The walls of an infinitely long conducting prism are given by $x = 0$, $x = a$ and $y = 0$, $y = b$. A line charge of strength q per unit length lies at $x = c$, $y = d$, where $0 < c < a$ and $0 < d < b$. Show that the potential inside the prism is

$$V' = \frac{2q}{\pi\epsilon}\sum_{m=1}^{\infty}\frac{1}{m}\operatorname{csch}\frac{m\pi b}{a}\sinh\frac{m\pi}{a}(b - d)\sinh\frac{m\pi y}{a}\sin\frac{m\pi c}{a}\sin\frac{m\pi x}{a}, \quad \text{where } 0 < y < d$$

$$V'' = \frac{2q}{\pi\epsilon}\sum_{m=1}^{\infty}\frac{1}{m}\operatorname{csch}\frac{m\pi b}{a}\sinh\frac{m\pi d}{a}\sinh\frac{m\pi}{a}(b - y)\sin\frac{m\pi c}{a}\sin\frac{m\pi x}{a}, \quad \text{where } d < y < b$$

Show by 3.08 (2) that the force per unit length is

$$\frac{q^2}{\epsilon}\sum_{m=1}^{\infty}\left[\frac{i}{b}\operatorname{csch}\frac{m\pi a}{b}\sinh\frac{m\pi(2c - a)}{b}\sin^2\frac{m\pi d}{b} + \frac{j}{a}\operatorname{csch}\frac{m\pi b}{a}\sinh\frac{m\pi(2d - b)}{a}\sin^2\frac{m\pi c}{a}\right]$$

Note that, unless c and d are very small, this series converges rapidly.

52. Show that the set of three line charges q at z_{10}, q at z_{10}^{-1}, and $-q$ at the origin in the z_1-plane of Art. 4.261 transform into the undistorted field of a single line charge q, at z_0 in the z-plane where $z_0 = z_{10} + z_{10}^{-1}$.

53. Using the image law of 4.04 for a circular dielectric cylinder and the image law established in the last problem for the unit circle, find the conjugate functions giving the field when a line charge q is placed parallel to the axis of an elliptic dielectric cylinder K. In the confocal coordinates of 4.261, the charge is at u_0, v_0 outside the cylinder whose surface is given by v_1. Obtain the result, inside the dielectric.

$$W_i = \frac{-q}{\pi\epsilon_v(K+1)}\sum_{n=0}^{\infty}\left(\frac{K-1}{K+1}\right)^n \ln\left(2\{\sin(u+jv) - \sin[u_0 \pm j(2nv_1 + v_0)]\}\right)$$

and outside, due to polarization alone,

$$W'_o = \frac{q}{2\pi\epsilon_v}\left\{\frac{K-1}{K+1}\ln\left[1 - e^{j(u-u_0)+2v_1-v_0-v}\right] - \frac{4K}{(K+1)^2}\sum_{n=1}^{\infty}\left(\frac{K-1}{K+1}\right)^n \ln\left[1 \pm e^{j(u\pm u_0)-(v+v_0+2nv_1)}\right]\right\}$$

where the upper sign is taken when n is even and the lower when n is odd.

54. A line charge q is placed parallel to the generators of a parabolic dielectric cylinder. In terms of the parabolic coordinates $y = 2\xi\eta$ and $x = \xi^2 - \eta^2$, where $-\infty < \xi < +\infty$ and $0 < \eta < \infty$, the coordinates of the charge are ξ_0 and η_0 and of the surface of the cylinder η_1. Show that inside the dielectric the transformation giving the field is

$$W_i = \frac{-q}{\pi\epsilon_v(K+1)}\sum_{n=0}^{\infty}\left(\frac{K-1}{K+1}\right)^n \ln\{(\xi+j\eta)^2 - [\xi_0 \pm j(\eta_0 + 2n\eta_1)]^2\}$$

and that outside, due to the polarization alone, it is

$$W'_o = \frac{q}{2\pi\epsilon_v}\left(\frac{K-1}{K+1}\ln[\xi - \xi_0 - j(2\sigma_1 - \eta_0 - \eta)] - \frac{4K}{(K+1)^2}\sum_{n=1}^{\infty}\left(\frac{K-1}{K+1}\right)^n \ln\{\xi \mp \xi_0 + j[\eta + (\eta_0 + 2n\eta_1)]\}\right)$$

where the upper sign is taken when n is even and the lower when n is odd.

55. The cylinder $(x/a)^{2n} + (y/b)^{2n} = 1$ carries a charge $-Q$ per unit length. Show by applying Gauss's theorem to the surface $V = 0$ that

$$z = a\left(\cos\frac{2\pi\epsilon W}{Q}\right)^{\frac{1}{n}} + jb\left(\sin\frac{2\pi\epsilon W}{Q}\right)^{\frac{1}{n}}$$

Then show by applying it to the cylinder $r = \infty$ that the enclosed charge per unit length is nQ. Hence the space outside the cylinder is charge free only when n is 1.

56. A conductor is bounded by $\theta = \alpha$, $\theta = -\alpha$, and $r = 1$, and carries a charge $\pi\epsilon$ per unit length. Show that V is the potential function in the transformation

$$\ln z = -2j\left\{\sin^{-1}\frac{\pi\sin W}{[\alpha(2\pi-\alpha)]^{\frac{1}{2}}} + \frac{n-\alpha}{\pi}\tan^{-1}\frac{(\pi-\alpha)\sin W}{[\alpha(2\pi-\alpha) - \pi^2\sin^2 W]^{\frac{1}{2}}}\right\}$$

57. An infinite plane conducting plate with a thin ridge of uniform height a at $x = 0$ forms one boundary of a field which is uniform and of strength E far from the ridge. By a simple Schwarz transformation show that $W = E(a^2 + z^2)^{\frac{1}{2}}$.

58. A conductor of star-shaped section is formed by $2m$ perfectly conducting strips of width a with one edge on the axis and with uniform angular spacing. The total charge per unit length is q. Show that the complex potential function is $W = [q/(2m\pi\epsilon)] \sin^{-1} (z/a)^m$.

59. One plate at potential V_0 of a capacitor lies at $y = a$, $-\infty < x < 0$, and the whole x-axis is at zero potential. Show that the complex potential function is given by

$$z = \frac{a}{\pi}\left(\frac{\pi W}{V_0} + e^{-\pi W/V_0}\right) + 1$$

60. The surface $V = \pi$ is the negative x- and y-axis, and the surface $V = 0$ is $x = h$ when $y < k$ and $y = k$ when $x < h$. Show that the transformation giving the field is

$$\tfrac{1}{2}\pi z = h \tan^{-1} [k(e^W + 1)^{\frac{1}{2}}(h^2e^W - k^2)^{-\frac{1}{2}}] - k \tanh^{-1} [h(e^W + 1)^{\frac{1}{2}}(h^2e^W - k^2)^{-\frac{1}{2}}]$$

Show that the additional capacitance per unit length over what would exist if the only fields between $V = 0$ and $V = \pi$ were uniform and in the regions $x < 0$ and $y < 0$ is

$$\frac{2\epsilon}{\pi}\left(\ln \frac{h^2 + k^2}{4hk} + \frac{k}{h} \tan^{-1} \frac{h}{k} + \frac{h}{k} \tan^{-1} \frac{k}{h}\right)$$

61. A conducting flat strip of width $2a$ lies on the real axis midway between and parallel to two infinite earthed conducting planes at $y = b$ and $y = -b$. Apply a logarithmic transformation to 4.29 (5) and thus show that the complex potential function and its derivative are

$$W = \frac{U_0}{K(e^{-\pi a/b})} \operatorname{sn}^{-1} (e^{\frac{1}{2}\pi(z+a)/b}, e^{-\pi a/b})$$

$$\frac{dW}{dz} = \frac{\frac{1}{2}\pi U_0}{bK(e^{-\pi a/b})}\left(2 \cosh \frac{\pi z}{b} - 2 \cosh \frac{\pi a}{b}\right)^{-\frac{1}{2}}$$

where $K(k)$ is a complete elliptic integral. Show from 4.29 (9) that the capacitance per unit length of this strip line is

$$C = \frac{2\epsilon K[(1 - e^{-2\pi a/b})^{\frac{1}{2}}]}{K(e^{-\pi a/b})}$$

The origin is at the center of the strip.

62. An infinite plane at potential V_1 occupies the $y = 0$ plane except for a gap bounded by $x = d$ and $x = -d$. A flat strip at potential 0 occupies the y-axis between $y = -h$ and $y = h$. By application of a Schwarz transformation to Fig. 4.29a, show that

$$z = jh \operatorname{cn}\left[\frac{KW}{U_0}, \frac{h}{(d^2 + h^2)^{\frac{1}{2}}}\right]$$

and that the capacitance per unit length is

$$C = \frac{\epsilon K[h/(d^2 + h^2)^{\frac{1}{2}}]}{K[d/(d^2 + h^2)^{\frac{1}{2}}]}$$

63. Two right-angle bends in the real axis of Fig. 4.29a solve the problem of a horizontal conducting strip at potential $V = 0$ with its edges at $x = c$ and $x = -c$ between two infinite vertical conducting planes at potential V_2 located at $x = d$ and $x = -d$. Show that the transformation is

$$z = \frac{2d}{\pi} \cos^{-1}\left[\operatorname{dn}\left(\frac{KW}{U_0}, \sin\frac{c\pi}{2d}\right)\right]$$

and that the capacitance per unit length is, if $K(k)$ is a complete elliptic integral,

$$C = \frac{4\epsilon K[\sin(\frac{1}{2}c\pi/d)]}{K[\cos(\frac{1}{2}c\pi/d)]}$$

64. Apply a Schwarz transformation to Fig. 4.30a with $\frac{1}{2}\pi$ bends at B and A and a 2π bend at $\frac{1}{2}(A + B)$. Thus show that the complex potential function giving the field when the negative x-axis is at potential zero and a flat strip with one edge at $x = b$, $y = h$ and the other at $x = b$, $y = -h$ is at potential U_0 is

$$z = b\left[1 + \operatorname{dn}\left(\frac{KW}{U_0}, k\right) \operatorname{cn}\left(\frac{KW}{U_0}, k\right)\right]$$

where k is $[(h^2 + b^2)^{\frac{1}{2}} - h]/b$. Show that the capacitance per unit length is

$$C = \frac{2\epsilon K[(1 - k^2)^{\frac{1}{2}}]}{K(k)}$$

65. Write z_1 for z in 4.30 (1) and invert the result about $z_1 = 2R(\cot\gamma - j)$ with a radius of inversion $2R$. Thus solve the problem in which that part of the surface of a cylinder of radius R which lies between $\theta = 0$ and $\theta = \alpha$ is at potential V_1 and that part between $\theta = \beta$ and $\theta = \gamma$ at potential zero. Let $\alpha < \beta < \frac{1}{2}\pi$ and $\beta < \gamma < \pi$. Take $z = 0$ at the center of inversion and let the x ($\theta = 0$)-axis parallel the x_1-axis. Show that the complex potential function is

$$z = 2R\left\{M \operatorname{sn}^2\left[\frac{KW}{U_0}, \left(\frac{M}{N}\right)^{\frac{1}{2}}\right] - \cot\beta + j\right\}^{-1}$$

where M is $\cot\gamma + \cot\beta$ and N is $\cot\gamma + \cot\alpha$. Show that the capacitance per unit length is

$$C = (\epsilon + \epsilon_v)K(M^{\frac{1}{2}}N^{-\frac{1}{2}})\{K[(1 - MN^{-1})^{\frac{1}{2}}]\}^{-1}$$

if the cylinder is of dielectric constant ϵ.

66. The centers of two parallel flat conducting strips lie at $x = d$ and $x = -d$ and the edges at $x = d$, $y = \pm h$ and $x = -d$, $y = \pm h$. Show, by use of a Schwarz transformation on Fig. 4.29a, that the capacitance is given by 4.29 (9) where b/a is found from k in

$$E(k)F(\phi,k) - K(k)E(\phi,k) = \frac{\frac{1}{2}\pi h}{d}, \qquad k^2 = 1 - \left(\frac{b}{a}\right)^2$$

$$\sin\phi = \left[\frac{K(k) - E(k)}{k^2K(k)}\right]^{\frac{1}{2}}$$

and a is then given by $a = 2K(k)d/\pi$. Use Legendre's relation HTF II (15), page 320, to simplify relations.

67. A rectangular bar with sides $2a$ and $2b$ carries a charge $-2\pi\epsilon$ per unit length. By the application of a Schwarz transformation in which the real axis bends are $\frac{3}{2}\pi$ at $x_1 = \pm k$ and $\frac{1}{2}\pi$ at $x_1 = \pm 1$ to 4.22 (3) with $a = 1$, derive the transformation

$$z = \frac{E[\sin^{-1}(k^{-1}\sin W),k] - (1-k^2)F[\sin^{-1}(k^{-1}\sin W),k]}{E(k) - (1-k^2)K(k)} + jb$$

where the origin is at the center of the rectangle and k is found from

$$\frac{a}{b} = \frac{E(k) - (1-k^2)K(k)}{k^2K[(1-k^2)^{\frac{1}{2}}] - E[(1-k^2)^{\frac{1}{2}}]}$$

On the top surface $0 < \sin U < k$ and on the side surface $k < \sin U < 1$ so that the charge densities are

$$\sigma_a = \frac{E(k) - (1-k^2)K(k)}{a(k^2 - \sin^2 U)^{\frac{1}{2}}}\epsilon \qquad \sigma_b = \frac{E(k) - (1-k^2)K(k)}{a(\sin^2 U - k^2)^{\frac{1}{2}}}\epsilon$$

68. A rectangular conducting tube bounded by $x = -b$, $x = b$, $y = 0$, and $y = b$ contains a line charge of strength q per unit length at x_0, y_0. Show that the complex potential is potential is

$$W = -\frac{q}{2\pi\epsilon}\ln\frac{\operatorname{sn}(z/C,k) - c - jd}{\operatorname{sn}(z/C,k) - c + jd}$$

where C and k are determined by $b = CK(k)$ and $a = CK[(1-k^2)^{\frac{1}{2}}]$, and c and d are given by

$$c + jd = \frac{\operatorname{sn}(u,k)\operatorname{dn}(v,k') + j\operatorname{cn}(u,k)\operatorname{dn}(u,k)\operatorname{sn}(v,k')\operatorname{cn}(v,k')}{\operatorname{cn}^2(v,k') - k^2\operatorname{sn}^2(u,k)\operatorname{sn}^2(v,k')}$$

where $u = x_0/C$, $v = y_0/C$, and $k' = (1-k^2)^{\frac{1}{2}}$.

69. A slit electron lens consists of two earthed planes at $y = h$ and $y = -h$, each of which has a slit with edges at $x = k$ and $x = -k$ and a plane at potential V_0 occupying the x-axis except for a slit with edges at $x = l$ and $x = -l$. Show that the complex potential function is given implicitly by

$$z = \left[b^2 - \frac{b^2 - a^2}{\sin^2(\frac{1}{2}\pi W/jV_0)}\right]^{\frac{1}{2}} + \frac{c^2 - b^2}{b}\cosh^{-1}\left[\frac{b\sin(\frac{1}{2}\pi W/jV_0)}{(b^2 - a^2)^{\frac{1}{2}}}\right]$$

where b, a, and c are given by

$$k = \left[b\left(b + \frac{2h}{\pi}\right)\right]^{\frac{1}{2}} + \frac{2h}{\pi}\sinh^{-1}\left(\frac{\frac{1}{2}\pi b}{h}\right)^{\frac{1}{2}}$$

$$l = a + \frac{2h}{\pi}\tanh\frac{a}{b}, \qquad c = \left[b\left(b + \frac{2h}{\pi}\right)\right]^{\frac{1}{2}}$$

70. An ion source consists of a thin flat earthed strip opposite a thin infinite plate at potential V_1 with a slit in it. The slit lies on the real axis between $x = a$ and $x = -a$ and the plane is at $y = -b$ with the slit between $x = g$ and $x = -g$. Apply a Schwarz transformation to Fig. 4.29a with 2π bends at $x = c$, $x = -c$, $x = d$, $x = -d$; $\frac{1}{2}\pi$ bends at $x = b$, $x = -b$; and $-\frac{1}{2}\pi$ bends at $x = a$ and $x = -a$ where $0 < c < b < a < d$. Thus show that the field is given everywhere by

$$z_1 = \frac{2h}{\pi}\left[K(k)Z(u) + \frac{E(k)k^2AB\operatorname{sn}u\operatorname{cn}u}{(A-B)(1-k^2)\operatorname{dn}u}\right]$$

where $u = KW/U_0$, $Z(u)$ is a Jacobi zeta function, and A, B, and the modulus k are found from

$$[1 - (1 - k^2)^{-1}AB]E(k) = (A - B)K(k)$$

$$f = \frac{2h}{\pi}\left[K(k)Z(u_1) + \frac{E(k)A^{\frac{1}{2}}B(1 - A)^{\frac{1}{2}}(A - 1 + k^2)^{\frac{1}{2}}}{(A - B)(1 - k^2)}\right]$$

$$g = \frac{2h}{\pi}\left[K(k)Z(u_2) + \frac{E(k)AB^{\frac{1}{2}}(1 + B)^{\frac{1}{2}}(B + 1 - k^2)^{\frac{1}{2}}}{(A - B)(1 - k^2)}\right]$$

where $u_1 = \text{sn}^{-1}\,[(1 - A)^{\frac{1}{2}}/k]$ and $u_2 = \text{sn}^{-1}\,(1 + B)^{-\frac{1}{2}}$. Show that the capacitance per unit length is $K(k)/K[(1 - k^2)^{\frac{1}{2}}]$. Milne-Thompson gives tables of all functions used and all formulas are in Whittaker and Watson.

71. Show that the Green's function for a line charge q per unit length at z_0 in an equilateral triangular conducting tube whose base lies between $x = 1$ and $x = -1$ is

$$W = \frac{-q}{2\pi}\ln\frac{f(cz) - f(cz_0)}{f(cz) - f(cz_0^*)} \qquad \text{where } f(cz) = \frac{\text{sn}\,(cz,k)\,\text{dn}\,(cz,k)}{[1 + \text{cn}\,(cz,k)]^2}$$

The modulus k is $\frac{1}{2}(1 + 3^{\frac{1}{2}})2^{-\frac{1}{2}}$ or 0.96593, and c must be chosen so that sn (c,k) is $2(1 + 3^{\frac{1}{2}})^{-1}3^{\frac{1}{4}}$, which gives $c = 1.9138$.

72. A cylindrical conductor has a lens-shaped section bounded by two circular arcs and carries a charge q per unit length. The exterior angle of intersection of the arcs is α and the exterior angle made by the upper convex surface A with the plane passing through the edges is β. Let the distance between the edges be $2c$, the distances between a point on the A surface and the two edges a and a_1 and between a point on the B surface and the edges b and b_1. Show that the charge densities on the two surfaces are

$$\sigma_A = \frac{2qc(aa_1)^{(\pi/\alpha)-1}\sin\,(\pi\beta/\alpha)}{\alpha[a^{2\pi/\alpha} - 2(aa_1)^{\pi/\alpha}\cos\,(\pi\beta/\alpha) + a_1^{2\pi/\alpha}]}$$

$$\sigma_B = \frac{2qc(bb_1)^{(\pi/\alpha)-1}\sin\,(\pi\beta/\alpha)}{\alpha[b^{2\pi/\alpha} + 2(bb_1)^{\pi/\alpha}\cos\,(\pi\beta/\alpha) + b_1^{2\pi/\alpha}]}$$

This result can be found by inverting about the charge the Green's function for a line charge parallel to the edge of a conducting wedge. The latter is found either by circular harmonics or by a conformal transformation.

73. Solve the preceding problem by a conformal transformation taking the edges of the lens-shaped section at $x = c$ and $x = -c$. Thus show that the complex potential function outside the cylinder is

$$W = \frac{-q}{2\pi\epsilon}\ln\frac{(z + c)^{\pi/\alpha} - (z - c)^{\pi/\alpha}}{(z + c)^{\pi/\alpha} - (z - c)^{\pi/\alpha}e^{2j\pi\beta/\alpha}}$$

Show that the potential U may be written

$$U = \frac{-q}{4\pi\epsilon}\ln\frac{r_1^{2\pi/\alpha} + r_2^{2\pi/\alpha} - 2(r_1r_2)^{\pi/\alpha}\cos\,(\pi\psi/\alpha)}{r_1^{2\pi/\alpha} + r_2^{2\pi/\alpha} - 2(r_1r_2)^{\pi/\alpha}\cos\,[\pi(2\beta - \psi)/\alpha]}$$

where r_1 and r_2 are the distances from the field point to $x = c$ and $x = -c$ and ψ is the angle between them. The surfaces of the cylinder $U = 0$ are given by $\psi = \beta$ and $\psi = \beta - \alpha$.

74. The cylindrical conductors of lens-shaped sections of the last two problems are earthed in the presence of a charge q per unit length at z_0 outside the cylinder. Using the symbols of the previous problems, show that the complex potential is

$$W = \frac{-q}{2\pi\epsilon}\ln\frac{[(z + c)(z_0 - c)]^{\pi/\alpha} - [(z - c)(z_0 + c)]^{\pi/\alpha}}{[(z + c)(z_0^* - c)]^{\pi/\alpha} - [(z - c)(z_0^* + c)]^{\pi/\alpha}e^{2j\beta\pi/\alpha}}$$

Show that the electrostatic potential is

$$U = \frac{-q}{4\pi\epsilon}\ln\frac{(r_1r_{20})^{2\pi/\alpha} - 2(r_1r_2r_{10}r_{20})^{\pi/\alpha}\cos[\pi(\psi - \psi_0)/\alpha] + (r_2r_{20})^{2\pi/\alpha}}{(r_1r_{20})^{2\pi/\alpha} - 2(r_1r_2r_{10}r_{20})^{\pi/\alpha}\cos[\pi(2\beta - \psi - \psi_0)/\alpha] + (r_2r_{20})^{2\pi/\alpha}}$$

where r_{10} and r_{20} are the distances from q to $x = c$ and $x = -c$ and ψ_0 is the angle between them. When $\psi = \beta$ or $\psi = \beta - \alpha$, $U = 0$.

75. A conductor at potential zero lies at $y = 0$ from $x = h$ to $x = \infty$ and at $x = h$ from $y = 0$ to $y = -\infty$. A second at potential V_0 lies at $y = k$ from $x = -h$ to $x = -\infty$ and at $x = -h$ from $y = k$ to $y = -\infty$. Show that the complex potential function is

$$z = (z_1 - a_1)^{\frac{1}{2}}(z_1 + b_1)^{\frac{1}{2}} + (b_1 - a_1)\ln\frac{(z_1 - a_1)^{\frac{1}{2}} + (z_1 + b_1)^{\frac{1}{2}}}{(a_1 + b_1)^{\frac{1}{2}}} - (a_1b_1)^{\frac{1}{2}}\sin^{-1}\frac{(b_1 - a_1)z_1 + 2a_1b_1}{(a_1 + b_1)z_1}$$

where $z_1 = e^{\pi W/V_0}$, $a_1 = 2[(h^2 + k^2)^{\frac{1}{2}} - k]/\pi$, $b_1 = 2[(h^2 - k^2)^{\frac{1}{2}} + k]/\pi$.

76. With a Schwarz transformation, bend the real axis in the z_1-plane through angles $3\pi/2$ at the origin, 2π at $x_1 = 2^{-\frac{1}{2}}$, $\pi/4$ at $x_1 = 3^{\frac{1}{2}}2$ and $x_1 = -3^{\frac{1}{2}}2$ and require that $z_1 = 0$ and $z_1 = \pm 3^{\frac{1}{2}}2$ all fall at $x = y = 0$. Thus show that the transformation $z = bz_1^{\frac{3}{2}}(3z_1^2 - 4)^{\frac{1}{4}}$ forms the x_1-axis into an arrow with the point at $x = y = 0$, the shaft on the positive x-axis, and the barbs of length b at 45° with the shaft.

77. Solve the problem of a thick cylindrical plate with a rounded edge by the first method of Art. 4.27 with $\alpha = 2\pi$ so that

$$z - -z_1^2 - \lambda z_1(z_1^2 - 1)^{\frac{1}{2}} + \lambda\cosh^{-1}z_1$$

When $x_1 > 1$, $-\infty < x < -1$ and $y = 0$; when $x_1 < -1$, $-\infty < x < -1$ and $y = 2$; when $x_1 = 0$, $x = 0$ and $y = 1$. Show that if $y = 3.735$, points $x = -1, -\frac{1}{2}, 0, +\frac{1}{2}, +1$ all lie on a circle of radius 1 centered at $x = -1$, $y = 1$ in the z-plane.

78. A semi-infinite conducting sheet at potential U_0 occupies the plane $x = 0$ from $y = 0$ to $y = \infty$. The plane $y = 0$ is a conducting sheet at potential $V = 0$ except for a gap between $x = -a$ and $x = +a$. Show that the force per unit length sucking the semi-infinite sheet into the gap is $2\epsilon U_0^2/\pi a$.

79. A circular cylinder of radius a centered at the origin is at potential $U = U_0$. A set of four rectangular hyperbolic cylinders given by $xy = \pm b^2$ is at potential $U = 0$. Show from 4.31 that the transformation giving the field is

$$z = 2b\left\{j\cos^{-1}\left[\operatorname{sn} k\left(\frac{KW}{U_0}, k\right)\right] + j\lambda\sin^{-1}\left[\operatorname{cn}\left(\frac{KW}{U_0}, k\right)\right]\right\}^{\frac{1}{2}}[\pi(1 + \lambda)]^{-\frac{1}{2}}$$

where the modululus k is $\cos[\frac{1}{2}b^{-2}\pi c^2(1 + \lambda)]$ and the parameter λ is found from 4.28 (6) with a^2 in place of $2c$ and b^2 in place of $2b$. Show that the capacitance per unit length is

$$C = 8\epsilon K[(1 - k^2)^{\frac{1}{2}}][K(k)]^{-1}$$

For good accuracy $c/b \leqslant 2^{-\frac{1}{2}}$.

80. A circular cylinder of radius a centered at the origin is at potential $U = U_0$. A parabolic cylinder given by $y^2 = 4p(x + p)$ is at potential $U = 0$. By folding the x-axis back on itself in 4.31 show that the transformation giving the field is

$$z = -8p\left[\cos^{-1}\left(k\operatorname{sn}\frac{KW}{U_0}\right) + \lambda\sin^{-1}\left(\operatorname{cn}\frac{KW}{U_0}\right)\right]^2[\pi(1 + \lambda)]^{-2}$$

where the modulus k is $\cos\,[\frac{1}{2}\pi p^{-\frac{1}{2}}a^{\frac{1}{2}}(1 + \lambda)]$ and the parameter λ is found from 4.28 (6) with $2a^{\frac{1}{2}}$ in place of c and $2b^{\frac{1}{2}}$ in place of b. Show that the capacitance per unit length is $C = 2\epsilon K[(1 - k^2)^{\frac{1}{2}}][K(k)]^{-1}$. For good accuracy $a/p \leqslant \frac{1}{4}$.

81. The equation of a parabolic cylinder at potential $U = 0$ is $y^2 = 4p(x + p)$. An infinite thin strip at potential U_0 lies between $x = 0$ and $x = a$. Show by folding the x-axis in problem 61 back on itself that the complex potential function for the field between parabola and strip is

$$W = U_0[K(k)]^{-1}\,\mathrm{sn}^{-1}\,\{\exp\,[\tfrac{1}{2}\pi p^{\frac{1}{2}}(z^{\frac{1}{2}} - a^{\frac{1}{2}})],k\}, \qquad k = \exp\,(-\pi a^{\frac{1}{2}}/p^{\frac{1}{2}})$$

Show that the capacitance per unit length is $\epsilon K[(1 - k^2)^{\frac{1}{2}}][K(k)]^{-1}$.

82. A rectangular bar at potential U_0 is bounded by $x' = \pm b$, $y' = \pm c$ and lies between earthed conducting planes at $y = \pm a$. Form one-quarter of this system by applying a Schwarz transformation to Fig. 4.30a with $\frac{1}{2}\pi$ bends at $x = 0$, $x = B$, $x = A$ and a $\frac{3}{2}\pi$ bend at $x = 1$, where $B < 1 < A$. Show that C_1 in 4.18 (7) is a/π by the method used for 4.28 (2), and for the main integration substitute a new variable u where z is $(B - u)/(1 - u)$. The resultant integrals are in Milne-Thompson, page 29. Thus show that the relations needed to evaluate B and A in terms of a, b, and c are

$$(a - c)K(k) = bK[(1 - k^2)^{\frac{1}{2}}], \qquad A = \tfrac{1}{2}k\{(1 + p^2)k + [(1 + p^2)^2k^2 - 4p^2]^{\frac{1}{2}}\}$$
$$\pi bp = 2akK(k), \qquad A = (1 - B)p^2$$

The first equation is solved for k by noting the intersection of the plotted curves for left and right sides as a function of k. The full accuracy of the $K(k)$ table may be attained by successive plots with increasing scales. Now p, A, and B are found in terms of a,b,c. The capacitance per unit length is, from 4.29 and 4.30,

$$C = \frac{4\epsilon V_1}{U_0} = \frac{4\epsilon K\,[1 - (B/A)]^{\frac{1}{2}}}{K[(B/A)^{\frac{1}{2}}]}$$

83. One edge of a conducting strip at potential U_0 lies at a distance b from the edge of a conducting wedge at potential zero and the other at a distance a, where $a > b$. The plane of the strip bisects the wedge angle making an angle α with each face. From Art. 4.30 show that the transformation for the fields is

$$z = b\left\{\mathrm{sn}\left[\frac{KW}{U_0}, \left(\frac{b}{a}\right)^{\frac{1}{2}\pi/\alpha}\right]\right\}^{2\alpha/\pi}$$

Show that the capacitance per unit length between wedge and strip is

$$C = \frac{2\epsilon V_1}{U_0} = \frac{2\epsilon K\{[1 - (b/a)^{\pi/\alpha}]^{\frac{1}{2}}\}}{K[(b/a)^{\frac{1}{2}\pi/\alpha}]}$$

84. The sharp edges of two identical infinite conducting wedges at potential zero, split by the x_1-axis, face each other at a distance $2a$ apart. A strip at potential U_0 occupies the x_1-axis midway between them leaving a gap of width b on either side. Apply a Schwarz transformation to Fig. 5.30a with a bend α, between $\frac{1}{2}\pi$ and π, at the origin and $\frac{1}{2}\pi$ at $z = A$ to form one-quarter of the field, so that if $k^2 = B/A$

and $B_x\ (M,N)$ is an incomplete beta function,

$$\frac{b}{a} = \int_0^{k^2} \frac{z^{(\alpha/\pi)-1}}{(1-z)^{\frac{1}{2}}}\,dz \left\{ \int_0^1 \frac{z^{(\alpha/\pi)-1}}{(1-z)^{\frac{1}{2}}}\,dz \right\}^{-1} = \frac{B_{k^2}[(\alpha/\pi),(\frac{3}{2})]}{B[(\alpha/\pi),(\frac{3}{2})]}$$

$$= \frac{\pi k^{2\alpha/\pi}(1-k^2)^{\frac{1}{2}}}{\alpha B[(\alpha/\pi),(\frac{3}{2})]} \left\{ 1 + \sum_{n=0}^{\infty} \frac{B[1+(\alpha/\pi),n+1]}{B[(\frac{3}{2})+(\alpha/\pi),n+1]} k^{2n+2} \right\}$$

by HMF 6.6.1, 26.5.1, and 26.5.4. This gives k^2 implicitly in terms of b/a. Show that the capacitance per unit length between strip and wedges is

$$C = \frac{4\epsilon V_1}{U_0} = \frac{4\epsilon K[(l-k^2)^{\frac{1}{2}}]}{K(k)}$$

References

The following books contain material pertinent to this chapter.

BATEMAN, H.: "Partial Differential Equations," Cambridge, 1932. Gives some interesting bipolar transformations.

BUCHHOLZ, H.: "Electrische und Magnetische Potentialfelder," Springer, 1957. Has very complete and detailed treatment of all methods.

DURAND, E.: "Electrostatique et Magnetostatique," Masson et Cie, 1953. Gives extensive, well-illustrated treatment in Chap. X.

ERDELYI, A., W. MAGNUS, F. OBERHETTINGER, and F. G. TRICOMI: "Higher Transcendental Functions," 3 vols., McGraw-Hill, 1953. Vol. II has extremely useful elliptic function formulas. (HTF)

FLUGGE, S., "Handbuch der Physik," Vol. XVI, Springer, 1958. Pages 40 to 82 give two-dimensional fields including many conformal transformations.

GEIGER-SCHEEL: "Handbuch der Physik," Vol. XII, Berlin, 1927.

JAHNKE, E., F. EMDE, and F. LOSCH.: "Tables of Higher Functions," McGraw-Hill, 1960. Numerical tables and graphs of many functions.

JEANS, J. H.: "The Mathematical Theory of Electricity and Magnetism," Cambridge, 1925. Follows Maxwell and gives examples.

KOBER, H.: "Dictionary of Conformal Transformations," Dover, 1952. The most extensive collection.

MAXWELL, J. C.: "Electricity and Magnetism," 3d ed. (1891), Dover, 1954. Considers images inversion and conjugate functions and gives good field drawings.

MILNE-THOMPSON, L. M.: "Jacobian Elliptic Function Tables," Gives all needed manipulations formulas and numerical tables of sn u, cn u, dn u, and $Z(u)$.

MORSE, P. M., and H. FESHBACH: "Methods of Theoretical Physics," 2 vols., McGraw-Hill, 1953. Many conformal transformations scattered through both volumes.

NATIONAL BUREAU OF STANDARDS: "Handbook of Mathematical Functions," U.S. Bureau of Commerce, 1964. Good for all functions. Has both formulas and numerical tables. (HMF)

ROTHE, R., F. OLLENDORF, and K. POHLHAUSEN: "Theory of Functions," Technology Press, 1933. Gives theory with important practical applications.

RYSHIK, I. M., and I. S. GRADSTEIN: "Tafelin," Deutscher Verlag der Wissenschaften, 1957. Formulas for most functions. Very good on elliptic functions.

WHITTAKER, E. T., and G. N. WATSON, "Modern Analysis," Cambridge, 1920. Gives very lucid treatment of elliptic functions.

WEBER, E.: "Electromagnetic Fields," Vol. I, Wiley, 1950. Gives many examples and references.

CHAPTER V

THREE-DIMENSIONAL POTENTIAL DISTRIBUTIONS

5.00. When Can a Set of Surfaces Be Equipotentials?—At first glance, one might think that the class of three-dimensional potential distributions in which there is symmetry about an axis could be obtained by rotation of a section of a two-dimensional distribution provided that, in so doing, the boundaries in the latter case generated the boundaries in the former. This is not true in general. We shall now find the condition that a set of nonintersecting surfaces in space must satisfy in order to be a possible set of equipotential surfaces. Let the equation of the surfaces be

$$F(x, y, z) = C \tag{1}$$

Since one member of the family corresponds to each value of C, if it is to be an equipotential, we must have one value of V for each value of C, so that

$$V = f(C)$$

must satisfy Laplace's equation. Differentiating results in

$$\frac{\partial V}{\partial x} = f'(C)\frac{\partial C}{\partial x}, \text{ etc.,} \qquad \frac{\partial^2 V}{\partial x^2} = f''(C)\left(\frac{\partial C}{\partial x}\right)^2 + f'(C)\frac{\partial^2 C}{\partial x^2}, \text{ etc.}$$

Substituting in Laplace's equation gives

$$\nabla^2 V = \frac{\partial^2 V}{\partial x^2} + \frac{\partial^2 V}{\partial y^2} + \frac{\partial^2 V}{\partial z^2} = f''(C)(\nabla C)^2 + f'(C)\nabla^2 C = 0$$

giving

$$\frac{\nabla^2 C}{(\nabla C)^2} = -\frac{f''(C)}{f'(C)} = \Phi(C) \tag{2}$$

The condition then that the surface $F(x, y, z) = C$ can be an equipotential is that $\nabla^2 C/(\nabla C)^2$ can be a function of C only.

By integration of (2), we can now obtain the actual potential. Since $f''(C)/f'(C) = d[\ln f'(C)]/dC$, we have

$$\int \Phi(C)\, dC = -\ln [f'(C)] + A'$$

or

$$f'(C) = Ae^{-\int \Phi(C)\, dC}$$

Integrating again gives

$$V = f(C) = A\int e^{-\int \Phi(C)\, dC}\, dC + B \tag{3}$$

The constants A and B can be determined by specifying the values of the potential on any two of the surfaces given by (1).

5.01. Potentials for Confocal Conicoids.—As an application of the formula just derived, we shall now show that any one of the three sets of nonintersecting confocal conicoids, given by the equation

$$\frac{x^2}{a^2+\theta} + \frac{y^2}{b^2+\theta} + \frac{z^2}{c^2+\theta} = 1 \tag{1}$$

where $c > b > a$ and $-c^2 < \theta < \infty$, is a possible set of equipotential surfaces. To get a picture of these surfaces, let us vary θ over the given range. For the range $-a^2 < \theta < \infty$, every term in (1) is positive so that it represents an ellipsoid. When $\theta = \infty$, we have a sphere of infinite radius, and when $\theta = -a^2$, the ellipsoid is flattened to an elliptical disk lying in the yz-plane. When θ passes from $-a^2 + \delta$ to $-a^2 - \delta$, we pass from the region of the yz-plane inside the disk to that outside. The latter is one limiting case of the hyperboloid of one sheet which (1) represents when $-b^2 < \theta < -a^2$. When $\theta = -b^2$, the hyperboloid of one sheet is flattened into that region of the xz-plane which includes the x-axis and lies between the hyperbolas cutting the z-axis at

$$z = \pm(c^2 - b^2)^{\frac{1}{2}}.$$

When θ passes from $-b^2 + \delta$ to $-b^2 - \delta$, we pass to the region of the xz-plane on the other side of these hyperbolas which is the limiting case in which the hyperboloid of two sheets, represented by (1) when $-c^2 < \theta < -b^2$, is flattened into the xz-plane. When $\theta = -c^2$, we have the other limiting case in which the two sheets of this hyperboloid coalesce in the xy-plane. Thus one curve of each set passes through each point in space; and, since it can be shown that the three sets are orthogonal, we can apply to them the theory developed in 3.03 for orthogonal curvilinear coordinates which leads to ellipsoidal harmonics. The latter are too complicated to be treated here, although later we shall treat the special cases of spheroidal harmonics.

To return to our problem: Let

$$M_n = \frac{x^2}{(a^2+\theta)^n} + \frac{y^2}{(b^2+\theta)^n} + \frac{z^2}{(c^2+\theta)^n}$$

and

$$N = \frac{1}{a^2+\theta} + \frac{1}{b^2+\theta} + \frac{1}{c^2+\theta}$$

With this notation, (1) becomes $M_1 = 1$, and differentiating this we have

$$\frac{2x}{a^2+\theta} - M_2\frac{\partial\theta}{\partial x} = 0 \qquad \text{or} \qquad \frac{\partial\theta}{\partial x} = \frac{2x}{M_2(a^2+\theta)}, \text{ etc.}$$

so that

$$(\nabla\theta)^2 = \left(\frac{\partial\theta}{\partial x}\right)^2 + \left(\frac{\partial\theta}{\partial y}\right)^2 + \left(\frac{\partial\theta}{\partial z}\right)^2 = \frac{4M_2}{M_2^2} = \frac{4}{M_2} \tag{1.1}$$

Differentiating again gives

$$\frac{\partial^2\theta}{\partial x^2} = \frac{2}{M_2(a^2+\theta)} - \frac{2x}{M_2(a^2+\theta)^2}\frac{\partial\theta}{\partial x} - \frac{2x}{a^2+\theta}\frac{1}{M_2^2}\left[\frac{2x}{(a^2+\theta)^2} - 2M_3\frac{\partial\theta}{\partial x}\right]$$

$$= \frac{2}{M_2(a^2+\theta)} - \frac{4x^2}{M_2^2(a^2+\theta)^3} - \frac{4x^2}{M_2^2(a^2+\theta)^3} + \frac{8x^2M_3}{(a^2+\theta)^2M_2^3}, \text{ etc.}$$

Adding similar expressions for y and z gives

$$\nabla^2\theta = \frac{2N}{M_2} - \frac{8M_3}{M_2^2} + \frac{8M_2M_3}{M_2^3} = \frac{2N}{M_2}$$

Substituting in 5.00 (2), we have

$$\frac{\nabla^2\theta}{(\nabla\theta)^2} = \frac{2N}{M_2}\cdot\frac{M_2}{4} = \frac{N}{2} \quad \text{so} \quad \Phi(\theta) = \frac{1}{2}\left(\frac{1}{a^2+\theta} + \frac{1}{b^2+\theta} + \frac{1}{c^2+\theta}\right) \tag{2}$$

This proves that such a set of equipotentials is possible. We now find the potential by 5.00 (3) to be

$$V = A\int^{\theta}[(a^2+\theta)(b^2+\theta)(c^2+\theta)]^{-\frac{1}{2}}\,d\theta + B \tag{3}$$

This is an elliptic integral given by Peirce 542 to 549 with $x = -\theta$. The constants A and B may be taken real or imaginary, whichever makes V real.

5.02. Charged Conducting Ellipsoid.—If we choose $V = 0$ when $\theta = \infty$, 5.01 (3) takes the form

$$V = -A\int_{\theta}^{\infty}[(a^2+\theta)(b^2+\theta)(c^2+\theta)]^{-\frac{1}{2}}\,d\theta \tag{1}$$

If we choose $V = V_0$ when $\theta = 0$ then, substituting in (1) gives

$$-A = V_0\left\{\int_0^{\infty}[(a^2+\theta)(b^2+\theta)(c^2+\theta)]^{-\frac{1}{2}}\,d\theta\right\}^{-1} \tag{2}$$

The field at infinity due to this ellipsoid, if its total charge is Q, will be $Q/(4\pi\epsilon r^2)$. We see from 5.01 (1) that as $\theta \to \infty$, $x^2 + y^2 + z^2 = r^2 \to \theta$, and so $\partial\theta/\partial r \to 2r$ giving

$$\frac{\partial V}{\partial r}\xrightarrow[r\to\infty]{}\frac{\partial V}{\partial\theta}\frac{\partial\theta}{\partial r} = \frac{A}{r^3}\cdot 2r = \frac{2A}{r^2} = -\frac{Q}{4\pi\epsilon r^2} \tag{3}$$

Hence

The capacitance of the ellipsoid is, from (2),

$$C = \frac{Q}{V_0} = -\frac{8\pi\epsilon A}{V_0} = 8\pi\epsilon\left\{\int_0^{\infty}[(a^2+\theta)(b^2+\theta)(c^2+\theta)]^{-\frac{1}{2}}\,d\theta\right\}^{-1}$$

$$= 4\pi\epsilon(a^2-b^2)^{\frac{1}{2}}F[(a^2-b^2)^{\frac{1}{2}}(a^2-c^2)^{\frac{1}{2}}, \sin^{-1}(1-c^2a^{-2})^{\frac{1}{2}}] \tag{4}$$

The surface density is given by

$$\sigma = -\epsilon(\nabla V)_{\theta=0} = -\epsilon\left(\frac{\partial V}{\partial\theta}|\nabla\theta|\right)_{\theta=0}$$

From (1), $(\partial V/\partial \theta)_{\theta=0} = A(abc)^{-1}$ and, from 5.01 (1.1), $|\nabla\theta| = 2M_2^{-\frac{1}{2}}$ so that

$$\sigma = \frac{Q}{4\pi abc}\left(\frac{x^2}{a^4} + \frac{y^2}{b^4} + \frac{z^2}{c^4}\right)^{-\frac{1}{2}} \tag{5}$$

5.03. Elliptic and Circular Disks.—The capacitance of an elliptic disk, obtained by putting $a = 0$ in 5.02 (4), is still an elliptic integral. To get the surface density, we write 5.02 (5) in the form

$$\sigma = \frac{Q}{4\pi bc}\left(\frac{x^2}{a^2} + \frac{a^2y^2}{b^4} + \frac{a^2z^2}{c^4}\right)^{-\frac{1}{2}}$$

Now let $a \to 0$, and the terms involving y and z can be neglected. Since both x and a are zero, the first term must be evaluated from 5.01 (1) where θ is put equal to zero, giving

$$\sigma = \frac{Q}{4\pi bc}\left(1 - \frac{y^2}{b^2} - \frac{z^2}{c^2}\right)^{-\frac{1}{2}} \tag{1}$$

The capacitance of a circular disk is obtained by putting $a = 0$ and $b = c$ in 5.02 (4), giving by *Pc* 114 or *Dw* 186.11

$$C = 8\pi\epsilon\left[\int_0^\infty \theta^{-\frac{1}{2}}(b^2 + \theta)^{-1}d\theta\right]^{-1} = 8\pi\epsilon\left(\frac{2}{b}\tan^{-1}\frac{\theta^{\frac{1}{2}}}{b}\bigg|_0^\infty\right)^{-1} = 8\epsilon b \tag{2}$$

Letting $\rho^2 = y^2 + z^2$ and $b = c$ in (1), the surface density on each side is

$$\sigma = \frac{Q}{4\pi b(b^2 - \rho^2)^{\frac{1}{2}}} \tag{3}$$

The potential due to such a disk given by 5.02 (1) with $a = 0$ and $b = c$ is

$$V = \frac{2V_0}{\pi}\left(\frac{1}{2}\pi - \tan^{-1}\frac{\theta^{\frac{1}{2}}}{b}\right) = \frac{2V_0}{\pi}\tan^{-1}\frac{b}{\theta^{\frac{1}{2}}}$$

Putting in the value of θ obtained by letting $r^2 = x^2 + y^2 + z^2$, $a = 0$, and $b = c$ in 5.01 (1) gives

$$V = \frac{2V_0}{\pi}\tan^{-1}\left(2^{\frac{1}{2}}b\{r^2 - b^2 + [(r^2 - b^2)^2 + 4b^2x^2]^{\frac{1}{2}}\}^{-\frac{1}{2}}\right) \tag{4}$$

This problem can also be solved by oblate spheroidal harmonics (5.271).

5.04. Method of Images. Conducting Planes.—An application of the test of 5.00 shows that in no case involving more than one point charge can we obtain the potential from the analogous two-dimensional case. Nevertheless, two of the methods used in such cases can also be applied to three-dimensional problems. One of these is the method of images. Any case in which the equation of a closed conducting surface under the influence of a point charge can be expressed in the form

$$0 = \frac{q}{r} + \sum_{s=1}^{n}\frac{q_s}{r_s}$$

where r is the distance from q to any point P on the surface and r_s is the distance from some point s on the other side of the surface to P can be solved by the method of images. We shall consider only simple spherical and plane surfaces. It is evident from symmetry that a single infinite conducting plane or two such planes intersecting in the z-axis at an angle π/m under the influence of a single point charge q in the xy-plane can be solved by locating point images in the xy-plane at the same places as in the two-dimensional case shown in Fig. 4.06. Adding up the potentials due to the point charge q and its point images gives exactly the same potential V in the region between the intersecting planes as would be obtained by adding up the potentials due to the charge q and the equal and opposite induced charge distributed over the planes. Hence we can determine this induced surface density σ on the conductor by finding $-\epsilon\partial V/\partial n$ on the plane to be replaced by the conductor. In the case of a point charge q at a distance a from an earthed conducting plane from 1.14, and 1.16 (1), the induced density at P is

$$\sigma = \frac{-aq}{2\pi r^3} \tag{1}$$

where r is the distance from q to P. From 2.17, we know that whatever the actual distribution of charges in the space separated from q by the earthed plane this cannot affect σ on the side facing q.

5.05. Plane Boundary between Dielectrics.—Since the case of a uniform line charge parallel to the plane face bounding two dielectrics may be considered as built up of equal point charges uniformly spaced along the line and since the images can be built up in the same way, it is plausible to suppose that the same image law holds in both cases.

Let the relative capacitivities of regions of positive and negative z be K_1 and K_2, respectively. Consider any configuration of charges in dielectric K_1 whose potential, with no dielectric present, would be given by

$$V_{vac} = f(x, y, z)$$

so that, if only K_1 were present, it would be, from 1.06 (3),

$$V_1 = \frac{1}{K_1} f(x, y, z)$$

The potential of an image in the $z = 0$ plane would be

$$V_i = C_1 f(x, y, -z)$$

The law of images of 4.04 suggests that when K_2 is present the fields in K_1 and K_2 should have the forms

$$V_1 = \frac{1}{K_1} f(x, y, z) + C_1 f(x, y, -z) \tag{1}$$

and

$$V_2 = C_2 f(x, y, z) \tag{2}$$

When $z = 0$, $V_1 = V_2$ and $K_1 \partial V_1/\partial z = K_2 \partial V_2/\partial z$ from 1.17 (5) so that

$$1 + K_1C_1 = K_1C_2 \qquad \text{and} \qquad 1 - K_1C_1 = K_2C_2$$

solving and substituting in (1) and (2), we have

$$V_1 = \frac{1}{K_1}\left[f(x, y, z) + \frac{K_1 - K_2}{K_1 + K_2} f(x, y, -z)\right] \tag{3}$$

and

$$V_2 = \frac{2}{K_1 + K_2} f(x, y, z) \tag{4}$$

Referring again to 1.06 (3), we see that if we have an actual charge q at P_1 in K_1, then the field in K_1 is the same as if the whole region were occupied by K_1 and we had an additional charge q' at the image point P_2, and the field in K_2 is that which would exist if the whole region were occupied by dielectric K_n and there were only a charge q'' at P_1, where

$$q' = \frac{K_1 - K_2}{K_1 + K_2} q \qquad \text{and} \qquad q'' = \frac{2K_n}{K_1 + K_2} q \tag{5}$$

For calculating the field the choice of K_n is immaterial but all authors take K_1, 1, or K_2. We use K_1.

5.06. Image in Spherical Conductor.—We saw in 4.05 that the cylinder $\rho = a$ is an equipotential under the influence of a line charge q at $\rho = b$ and a second line charge $-q$ at $\rho = +a^2/b$. Likewise we shall now see that it is possible to have the sphere $r = a$ at zero potential with a point charge q at $r = b$ and a point charge q' at $r = a^2/b$, and we shall determine q', which does not, as in the two-dimensional case, equal $-q$. The potential V' at $r = a$ due to q' is, from Fig. 5.06,

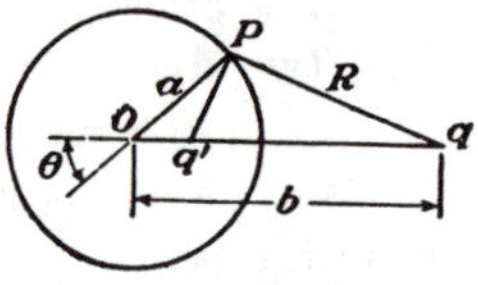

Fig. 5.06.

$$\begin{aligned} 4\pi\epsilon V' &= q'(a^4b^{-2} + a^2 - 2a^3b^{-1} \cos \theta)^{-\frac{1}{2}} \\ &= ba^{-1}q'(a^2 + b^2 - 2ab \cos \theta)^{-\frac{1}{2}} = ba^{-1}R^{-1}q' \end{aligned}$$

The potential at $r = a$ due to q is $q/(4\pi\epsilon R)$. In order that the sphere be at zero potential, these must be equal and opposite, giving

$$q' = -\frac{aq}{b} \tag{1}$$

The potential at any point is then

$$V = (4\pi\epsilon)^{-1}q[(r^2 + b^2 - 2br \cos \theta)^{-\frac{1}{2}} - a(b^2r^2 + a^4 - 2a^2br \cos \theta)^{-\frac{1}{2}}] \tag{2}$$

The surface density on the sphere will be

$$\sigma = -\epsilon\left(\frac{\partial V}{\partial r}\right)_{r=a} = \frac{q}{4\pi}\left[\frac{a - b \cos \theta}{R^3} - \frac{b(b - a \cos \theta)}{aR^3}\right] = \frac{a^2 - b^2}{4\pi a R^3} q \tag{3}$$

We can evidently raise the potential of the sphere to any desired value V by adding the potential due to a point charge q at the center where

$$q = 4\pi\epsilon a V \tag{4}$$

If we wish the field between a point charge q and a conducting sphere having a total charge Q, we need only add to (2) the potential due to a charge $Q + aq/b$ at the center.

We find that there is not, as in the analogous two-dimensional case, a simple image solution for the dielectric sphere and the point charge. We shall have to use spherical harmonics to solve this problem.

5.07. Example of Images of Point Charge.—As an example illustrating the last three articles, let us locate the images in the case where the earthed conductor consists of a plane sheet, lying in the yz-plane with a spherical boss of radius a centered at the origin, and the region below the xz-plane is filled with a material of relative capacitivity K. The point charge q lies at x_0, y_0, z_0 and $x_0^2 + y_0^2 + z_0^2 = b^2$. To locate the required images, we complete the boundaries with dotted lines as shown in Fig. 5.07. For the field above the xz-plane, the potential can be found by superimposing that due to the original charge plus seven images located as follows:

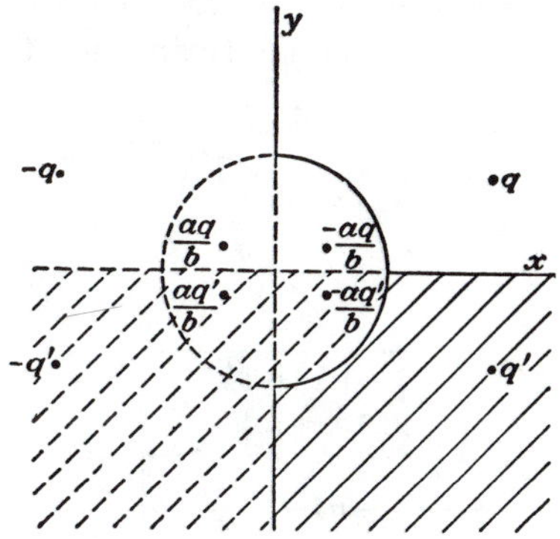

FIG. 5.07.

q at x_0, y_0, z_0	q' at $x_0, -y_0, z_0$
$-q$ at $-x_0, y_0, z_0$	$-q'$ at $-x_0, -y_0, z_0$
$-\dfrac{aq}{b}$ at $\left(\dfrac{a}{b}\right)^2 x_0, \left(\dfrac{a}{b}\right)^2 y_0, \left(\dfrac{a}{b}\right)^2 z_0$	$-\dfrac{aq'}{b}$ at $\left(\dfrac{a}{b}\right)^2 x_0, -\left(\dfrac{a}{b}\right)^2 y_0, \left(\dfrac{a}{b}\right)^2 z_0$
$\dfrac{aq}{b}$ at $-\left(\dfrac{a}{b}\right)^2 x_0, \left(\dfrac{a}{b}\right)^2 y_0, \left(\dfrac{a}{b}\right)^2 z_0$	$\dfrac{aq'}{b}$ at $-\left(\dfrac{a}{b}\right)^2 x_0, -\left(\dfrac{a}{b}\right)^2 y_0, \left(\dfrac{a}{b}\right)^2 z_0$

where, from 5.05 (5),

$$q' = \frac{1 - K}{1 + K} q$$

Below the xz-plane, in the dielectric, the potential can be obtained using only the images above the xz-plane and substituting q'' for q where, from 5.05 (5),

$$q'' = \frac{2q}{1 + K}$$

A section of the conductor by the xy-plane together with a projection of the images upon this plane is shown in Fig. 5.07.

5.08. Infinite Set of Images. Two Spheres.—Frequently, it is impossible to get equipotentials of the desired shape by any arrangement of a finite number of point charges. In some cases, however, one surface can be made an equipotential by some interior configuration of point charges. Then, by a suitable arrangement of image charges, the other surfaces may be made equipotentials, at the same time, however, distorting the first surface. By a third set of images, the latter is restored to its original shape at the expense of the other surfaces, and so forth. If, because they become more remote or grow smaller or tend to cancel each other, the effect of each successive group of images diminishes, we may get as close an approximation to the exact solution as we wish by taking enough of them.

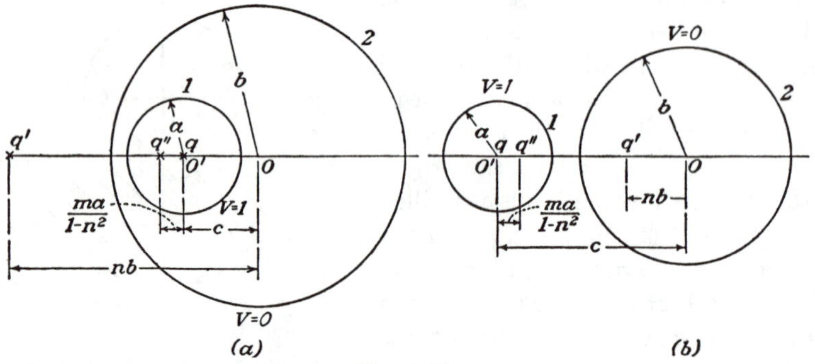

Fig. 5.08.

This method can be used to find the self- and mutual capacitances for two spheres. Let the radii of spheres 1 and 2 be a and b, and let the distances between their centers be c. From 2.16, the coefficient c_{11} is the charge on 1 and c_{12} that on 2, when 2 is grounded and 1 is raised to unit potential. By interchanging a and b in c_{11}, we can obtain c_{22}. The two cases, when $c < (b - a)$ and when $c > (b + a)$, are shown in Figs. 5.08a and 5.08b, respectively. In these figures, $m = a/c$ and $n = b/c$.

First, we put sphere 1 at unit potential by placing a charge $q = 4\pi\epsilon a$ at its center O'. We then put 2 at zero potential by placing an image $q' = -4\pi\epsilon ab/c = -4\pi\epsilon na$ [see 5.06 (1)] at a distance $b^2/c = nb$ to the left of O. Next, we restore 1 to unit potential with the image

$$q'' = \frac{\mp aq'}{(c - nb)} = \frac{\pm 4\pi\epsilon mna}{(1 - n^2)}$$

at a distance $a^2/(c - nb) = ma/(1 - n^2)$ to the right of O' and then restore 2 to zero potential by placing an image

$$q''' = \frac{-bq''}{c - ma/(1 - n^2)} = \frac{\mp 4\pi\epsilon mn^2 a}{(1 - m^2 - n^2)}$$

at the proper distance from O, and so forth. We note that the size of each successive image decreases so that we are approaching the exact solution. Adding up the charges on 1 gives

$$c_{11} = 4\pi\epsilon a\left[1 \pm \frac{mn}{1-n^2} + \frac{m^2n^2}{(1-n^2)^2 - m^2} + \cdots\right] \tag{1}$$

where the upper sign refers to case b. Adding up the charge on 2 gives, in case b,

$$c_{12} = 4\pi\epsilon\left[-na - \frac{mn^2a}{1-n^2-m^2} - \cdots\right] \tag{2}$$

In case a, from 2.17, $c_{12} = -c_{11}$. From symmetry, we can write down, in case b,

$$c_{22} = 4\pi\epsilon b\left[1 + \frac{mn}{1-m^2} + \frac{m^2n^2}{(1-m^2)^2 - n^2} + \cdots\right] \tag{3}$$

In case a, c_{22} is of no importance.

From 2.20 (6), the force of attraction between the two spheres will be, since $V_2 = 0$,

$$F = \frac{V_1^2}{2}\frac{\partial c_{11}}{\partial c} = -\frac{4\pi\epsilon a V_1^2}{c}\left\{\frac{mn}{(1-n^2)^2} + \frac{m^2n^2[2(1-n^2)-m^2]}{[(1-n^2)^2 - m^2]^2} + \cdots\right\} \tag{4}$$

If we wish the potential at any point P, we must add up the potentials at P due to each of the image charges. If the spheres are at potentials V_1 and V_2 where $V_2 \neq 0$, we shall have, in case b, a second set of images, for we must now start with a charge $q_1 = 4\pi\epsilon a V_1$ at the center of 1 and $q_2 = 4\pi\epsilon b V_2$ at the center of 2. The sizes and positions of the second set of images can be obtained by interchanging O and O' and a and b in the values for the first set.

5.081. Difference Equations. Two Spheres.—The formulas of the last article are not very convenient for precise calculation unless m and n are small. Better formulas, involving hyperbolic functions, can be obtained by getting the general relation between successive images and then solving the resulting difference equation. Let us designate the nth image in 1 by q_n so that the initial charge at its center is $q_1 = 4\pi\epsilon a$ and the image of q_n in 2 by p_n. In Fig. 5.08, the distance from O' to q_n is s_n and that from O to p_n is r_n. Then, using lower signs for case a and upper for case b, we have

$$p_n = \frac{-bq_n}{c \mp s_n}, \qquad r_n = \frac{b^2}{c \mp s_n} \tag{1}$$

$$q_{n+1} = \frac{\mp ap_n}{c - r_n} = \frac{\pm abq_n}{c(c \mp s_n) - b^2} \tag{2}$$

$$s_{n+1} = \frac{\pm a^2}{c - r_n} = \frac{\pm a^2(c \mp s_n)}{c(c \mp s_n) - b^2} = \frac{a}{b}\frac{q_{n+1}}{q_n}(c \mp s_n) \tag{3}$$

Eliminating $c \mp s_n$ from (2) and (3), we have

$$\pm \frac{q_n}{q_{n+1}} = \frac{c}{a^2}\frac{q_n}{q_{n+1}}s_{n+1} - \frac{b}{a}$$

so that

$$\pm \frac{q_{n-1}}{q_n} = \frac{c}{a^2}\frac{q_{n-1}}{q_n}s_n - \frac{b}{a} \tag{4}$$

Eliminating s_n from (2) and (4) rearranging give

$$\frac{1}{q_{n+1}} + \frac{1}{q_{n-1}} = \pm\frac{c^2 - a^2 - b^2}{ab}\frac{1}{q_n} \tag{5}$$

This is known as a difference equation of the second order with constant coefficients. The general method of solution is to substitute $1/q_n = u^n$, divide through by u^{n-1}, and solve the resultant quadratic equation for u algebraically. If the two solutions are u_1 and u_2, the solution of the difference equation is

$$\frac{1}{q_n} = Au_1^n + Bu_2^n \tag{6}$$

where A and B are to be determined by the terminal conditions. The present equation is specially simple because the coefficients of $1/q_{n+1}$ and $1/q_{n-1}$ are the same. Comparing with *Dw* 651.03 and 651.04 or *Pc* 669 and 671, which give

$$\sinh (n+1)\theta + \sinh (n-1)\theta = 2 \cosh \theta \sinh n\theta$$
$$\cosh (n+1)\theta + \cosh (n-1)\theta = 2 \cosh \theta \cosh n\theta$$

we see that a solution of (5) is

$$\frac{1}{q_n} = A \cosh n\alpha + B \sinh n\alpha \tag{7}$$

if we choose

$$\cosh \alpha = \pm\frac{c^2 - a^2 - b^2}{2ab} \tag{8}$$

To evaluate A and B, let us write down for the first two images

$$\frac{1}{q_1} = \frac{1}{4\pi\epsilon a} = A \cosh \alpha + B \sinh \alpha$$

$$\frac{1}{q_2} = \pm\frac{c^2 - b^2}{ab}\frac{1}{4\pi\epsilon a} = \left(2 \cosh \alpha \pm \frac{a}{b}\right)\frac{1}{4\pi\epsilon a} = A \cosh 2\alpha + B \sinh 2\alpha$$

Multiply the first of these equations by 2 cosh α so that it can be written in terms of cosh 2α and sinh 2α, and solve for A and B, and we have

$$A = \mp\frac{1}{4\pi\epsilon b} \qquad B = \frac{b \pm a \cosh \alpha}{4\pi\epsilon ab \sinh \alpha} \tag{9}$$

Substituting in (7) and using *Dw* 651.01 or *Pc* 660, we have

$$\frac{1}{q_n} = \frac{b \sinh n\alpha \pm a \sinh (n-1)\alpha}{4\pi\epsilon ab \sinh \alpha} \tag{10}$$

Adding the charges on 1 then gives

$$c_{11} = 4\pi\epsilon ab \sinh \alpha \sum_{n=1}^{\infty} [(b \sinh n\alpha \pm a \sinh (n-1)\alpha]^{-1} \tag{11}$$

where the lower sign refers to case a (Fig. 5.08a) and the upper to case b (Fig. 5.08b). In case a, from 2.17, $c_{12} = -c_{11}$. In case b, to get c_{12}, we must determine p_n. Taking the upper signs and eliminating $c - s_n$ from (1) and (2), we have

$$-\frac{1}{p_n} = \frac{a}{c}\frac{1}{q_{n+1}} + \frac{b}{c}\frac{1}{q_n}$$

Substituting for $1/q_{n+1}$ and $1/q_n$ from (10), using (8) and *Dw* 651.06, we have

$$\frac{1}{p_n} = -\frac{c \sinh n\alpha}{4\pi\epsilon ab \sinh \alpha} \tag{12}$$

so that adding up the charges on 2 gives

$$c_{12} = -\frac{4\pi\epsilon ab \sinh \alpha}{c} \sum_{n=1}^{\infty} \operatorname{csch} n\alpha \tag{13}$$

We can write down c_{22} from (11) for case b, by symmetry, thus:

$$c_{22} = 4\pi\epsilon ab \sinh \alpha \sum_{n=1}^{\infty} [a \sinh n\alpha + b \sinh (n-1)\alpha]^{-1} \tag{14}$$

5.082. Sphere and Plane and Two Equal Spheres.—A special case of some interest is that of a plane and a sphere. We can get this case by letting $d + c = b \to \infty$ in case a or $c - d = b \to \infty$ in case b, where d is the distance from the plane to the center of sphere 1. In either case, $n \to 1$, $m \to 0$, and $m/|1 - n| = a/d$ so that we have

$$c_{11} = 4\pi\epsilon a\left(1 + \frac{a}{2d} + \frac{a^2}{4d^2 - a^2} + \cdots\right) = 4\pi\epsilon a \sinh \alpha \sum_{n=1}^{\infty} \operatorname{csch} n\alpha \tag{1}$$

where $d = a \cosh \alpha$ and a is negligible compared with b. In either case, $c_{12} = -c_{11}$. In the case of the sphere and the plane, the force is

$$F = \frac{V_1^2}{2}\frac{\partial c_{11}}{\partial d} = -2\pi\epsilon a^2 V_1^2 \left\{\frac{1}{2d^2} + \frac{8ad}{(4d^2 - a^2)^2} + \cdots\right\}$$

$$= 2\pi\epsilon V_1^2 \sum_{n=1}^{\infty} [\operatorname{csch} n\alpha(\coth \alpha - n \coth n\alpha)] \tag{2}$$

In the case of two equal spheres of radius a at a distance c apart, the formulas of 5.081 are somewhat simplified by writing in terms of

$\beta = \frac{1}{2}\alpha$ where, setting $a = b$, we have, by *Dw* 652.6 or *Pc* 667,

$$\cosh \beta = \left[\frac{1}{2}(1 + \cosh \alpha)\right]^{\frac{1}{2}} = \frac{c}{2a} \tag{3}$$

In terms of β, 5.081 (11) becomes, setting $a = b$,

$$c_{11} = c_{22} = 4\pi\epsilon a \sinh \beta \sum_{n=1}^{\infty} \operatorname{csch} (2n - 1)\beta \tag{4}$$

and 5.081 (13) becomes

$$c_{12} = -4\pi\epsilon a \sinh \beta \sum_{n=1}^{\infty} \operatorname{csch} 2n\beta \tag{5}$$

5.09. Inversion in Three Dimensions. Geometrical Properties.—We have already seen in 4.21 that if, in a plane, we draw a circle of radius K about the origin and then draw a radial line outward from its center, any two points on this line, distant r and r' from the origin, are said to be inverse points if

$$rr' = K^2 \tag{1}$$

In 4.21, the circle represented a section of a cylinder, but it can just as well represent a principal section of a sphere. Thus for every point of any surface in space there corresponds an inverse point, and the surface formed by these inverse points is said to be the inverse surface. If the equation, in spherical polar coordinates, of the original surface was $f(r, \theta, \phi)$, then the equation of the inverse surface is $f(K^2/r, \theta, \phi)$. We proved in 4.21 that straight lines invert into circles lying in the same plane and passing through the center of inversion, and vice versa, whereas circles not passing through the origin invert into circles. Hence, in the present case, planes invert into spheres passing through the center of inversion and spheres not passing through the center of inversion invert into spheres. Since, in 4.21, we arrived at the laws of inversion by a conformal transformation, we know that angles of intersection are not altered by inversion. It is evident from this that when a small cone of solid angle $d\Omega$ with its vertex at the origin cuts out area elements dS and dS' from the surface S and its inverse surface S', the angle θ between the axis of the cone and the area elements is the same, so that

$$\frac{dS}{dS'} = \frac{r^2\, d\Omega \cos \theta}{r'^2\, d\Omega \cos \theta} = \frac{r^2}{r'^2} \tag{2}$$

5.10. Inverse of Potential and Image Systems.—We shall now show that it is possible to formulate laws for the inversion of electrical quantities such that we can obtain from the solution of a problem with a certain boundary surface the solution of a second problem with the inverse boundary surface.

Consider Fig. 5.10 in which P', R', and Q' are the inverse points of P, R, and Q, respectively, and O is the center of inversion. Then a charge q at P gives a potential $V = q/(4\pi\epsilon PQ)$ at Q, and a charge q' at P' will produce a potential $V' = q'/(4\pi\epsilon P'Q')$ at Q'. Triangles $OQ'P'$ and OPQ are similar since $K^2 = OP \cdot OP' = OQ \cdot OQ'$, and the included angle of both is α. The necessary relation between the potential V at Q before inversion to the potential V' at Q' after inversion is then

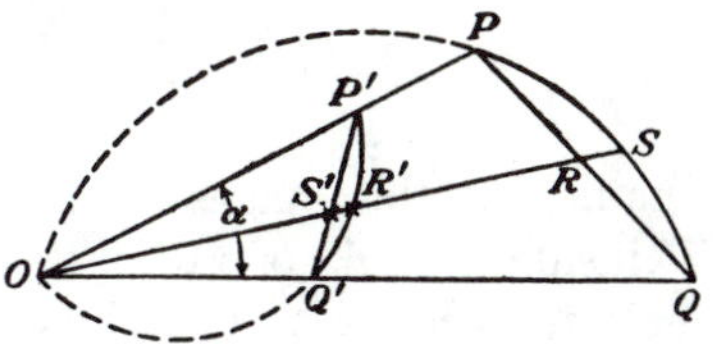

FIG. 5.10.

$$\frac{V'}{V} = \frac{q'}{q}\frac{PQ}{P'Q'} = \frac{q'}{q}\frac{OP}{OQ'}$$

In order to make this relation useful, we must find a suitable law for the inversion of charges. It was shown in 5.06 that a sphere of radius K is at zero potential under the influence of a charge q at $r = b$ and a charge $|q'| = +K|q|/b$ at $r' = K^2/b$ which is the inverse point. If the sphere is to remain at zero potential after inversion about itself, which will interchange the charges, the law of inversion of charges must be

$$\frac{q'}{q} = \frac{K}{b} = \frac{K}{OP} = \frac{OP'}{K} \tag{1}$$

We give the ratio the positive sign because we wish the inversion of charge to leave its sign unchanged. Substituting this value in the equation for the inversion of potential, we have the relation

$$\frac{V'}{V} = \frac{K}{OP}\cdot\frac{OP}{OQ'} = \frac{K}{OQ'} = \frac{OQ}{K} \tag{2}$$

This shows that $V' = 0$ if $V = 0$ provided OQ is finite, which means that if any surface is at zero potential under the influence of point charges $q_1, q_2, \ldots, q_n$ at $P_1, P_2 \ldots, P_n$, which are at finite distances, not zero, from the center of inversion, then the inverse surface will be at zero potential under the influence of the inverse charges $q'_1, q'_2, \ldots, q'_n$ at $P'_1, P'_2, \ldots, P'_n$. The qualification is necessary since (1) does not provide for inverting charges when OP is zero or infinite. It also follows from this rule that if any problem can be solved by images, the inverse problem can also be solved by images.

5.101. Example of Inversion of Images.—Let us now compute the force acting on a point charge q placed in the plane of symmetry at P at a distance b from the point of contact of two earthed spheres of radius a. Clearly, the two spheres can be obtained by inverting the system of planes shown in Fig. 5.101a. Taking the circle of inversion, shown dotted, tangent to the planes simplifies the computation. Before inversion, the potential at P' due to the images alone is

$$V' = \frac{1}{2\pi\epsilon}\sum_{n=1}^{\infty}(-1)^n\frac{q'}{4na} = -\frac{q'}{8\pi\epsilon a}\ln 2$$

The field of the images has a neutral point at P' so that the value V'_P of V' inside a circle of radius δ' around P' is constant if terms in δ'^n are neglected when $n > 1$. Thus inversion by 5.10 (2) gives the potentials at

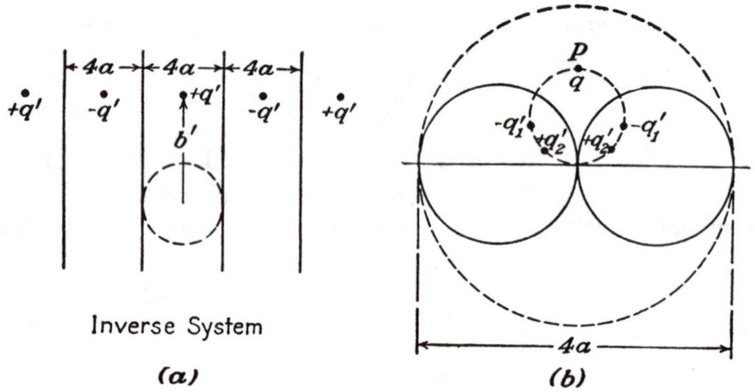

FIG. 5.101.

P and at a distance δ above P of the images that lie in the spheres on a circle through P and the point of contact to be, respectively,

$$V_b = \frac{2a}{b}V'_P, \qquad V_{b+\delta} \approx \frac{2a}{b+\delta}V'_P \approx V_b - \frac{2a\delta}{b^2}V'_P$$

From 5.10 (1), q' inverts into $2aq/b$ so that the force on q is

$$F = qE = \lim_{\delta\to 0}\frac{q(V_b - V_{b+\delta})}{\delta} = -\frac{aq^2}{2\pi\epsilon b^3}\ln 2 \tag{1}$$

5.102. Inversion of Charged Conducting Surface.—Let us now consider a conducting surface S charged to potential $1/K$, which produces a surface density σ, and let Q be any point on this surface. From 5.10 (2), the potential at the corresponding point Q' on the inverse surface S' is $V'_Q = 1/OQ'$. Since the potential at Q' due to a negative charge $4\pi\epsilon$ at O is $-1/OQ'$, it is clear that we can make the potential over the inverse surface everywhere zero by superimposing the potential due to such a charge. Reversing the procedure, we have the useful rule that, if we can solve the problem of a conductor at potential zero under the influence of a positive charge $4\pi\epsilon$, then we obtain by inversion, with this charge as a center, the solution of the problem of the inverse conducting surface raised to a potential $-K^{-1}$.

For the inversion of the surface density at P, we have, from 5.09 (1) and (2) and 5.10 (1), the relation

$$\frac{\sigma'}{\sigma} = \frac{q'}{dS'} \cdot \frac{dS}{q} = \frac{K}{OP} \cdot \frac{(OP)^2}{(OP')^2} = \frac{(OP)^3}{K^3} = \frac{K^3}{(OP')^3} \tag{1}$$

5.103. Capacitance by Inversion.—As an application of the rule of 5.102, we shall compute the capacitance of a conductor formed by two spheres of radii a and b which intersect orthogonally. Clearly, the inverse system is that shown in Fig. 5.103a, in which two earthed planes,

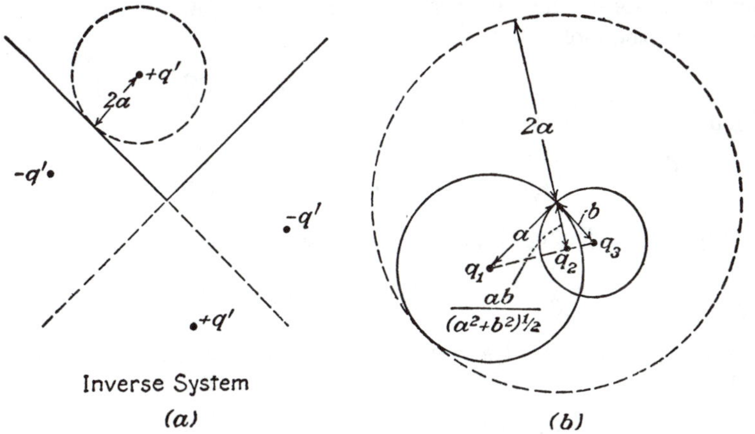

FIG. 5.103.

intersecting at right angles, are under the influence of the point charge $q' = 4\pi\epsilon$. This field can be replaced by that due to the image charges shown. The desired conductor, obtained by inversion, will be at potential $V = -1/K = -1/2a$. By 1.10 (1), the charge on this surface equals the sum of the image charges, q_1, q_2, and q_3, giving, by 5.10 (1),

$$Q = q_1 + q_2 + q_3 = 4\pi\epsilon\left[\frac{-a}{2a} + \frac{ab}{2a(a^2+b^2)^{\frac{1}{2}}} - \frac{b}{2a}\right]$$

and the capacitance is

$$C = \frac{Q}{V} = 4\pi\epsilon\left[\frac{(a+b)(a^2+b^2)^{\frac{1}{2}} - ab}{(a^2+b^2)^{\frac{1}{2}}}\right] \tag{1}$$

The potential around the conductor can be computed directly from the image charges q_1, q_2, q_3 or by inversion of the potential due to image charges alone in the region in the angle of the planes.

5.11. Three-dimensional Harmonics.—We saw in 5.00 that it is not, in general, possible, by rotating the orthogonal set of curves representing a normal section of a two-dimensional field, to obtain a set of three-dimensional equipotential surfaces. It is, however, possible to generate in this way a set of surfaces which, together with the planes intersecting on the axis and defined by the longitude angle, form a complete set of orthogonal curvilinear coordinates which can be treated by the method of 3.03. When the axial section of our three-dimensional boundary problem gives a two-dimensional boundary solvable by a conformal transformation, this provides a method of finding a coordinate system in which the original boundary conditions are very simple. The problem then becomes that of obtaining a general solution of Laplace's equation in this coordinate system.

Let us suppose that $u_1 = f_1(x, y)$ and $u_2 = f_2(x, y)$ are conjugate functions in the z-plane, given by

$$z = x + jy = f(u_1 + ju_2) = f(u) \tag{1}$$

Then, by 4.11 (1), we have

$$\frac{dz}{du} = f'(u) \qquad \text{or} \qquad dx + j\,dy = f'(u)(du_1 + j\,du_2)$$

Multiply this by the conjugate complex, and we get

$$ds_1^2 = dx^2 + dy^2 = \left|\frac{dz}{du}\right|^2 (du_1^2 + du_2^2)$$

If this system is rotated about the y-axis, we have, for the element of length,

$$ds^2 = ds_1^2 + x^2(d\phi)^2 = \left|\frac{dz}{du}\right|^2 (du_1^2 + du_2^2) + x^2(d\phi)^2 \tag{2}$$

Comparing with 3.03 (1), where $d\phi = du_3$, we see that

$$h_1 = h_2 = \left|\frac{dz}{du}\right| \qquad \text{and} \qquad h_3 = x$$

so that, from 3.03 (4), Laplace's equation is

$$x\left|\frac{dz}{du}\right|^{-2}\left[\frac{\partial}{\partial u_1}\left(x\frac{\partial V}{\partial u_1}\right) + \frac{\partial}{\partial u_2}\left(x\frac{\partial V}{\partial u_2}\right)\right] + \frac{\partial^2 V}{\partial \phi^2} = 0 \tag{3}$$

We can immediately split off the last term by the method of 4.01 by assuming a solution of the form

$$V = U(u_1, u_2)\Phi(\phi) \tag{4}$$

Dividing through by V and setting the last term of (3) equal to $-m^2$, we have for Φ the form, as in 4.01 (6),

$$\Phi = A_1 e^{jm\phi} + B_1 e^{-jm\phi} = A\cos m\phi + B\sin m\phi \tag{5}$$

The partial differential equation to be solved for $U(u_1, u_2)$ then becomes

$$\frac{1}{x}\frac{\partial}{\partial u_1}\left(x\frac{\partial U}{\partial u_1}\right) + \frac{1}{x}\frac{\partial}{\partial u_2}\left(x\frac{\partial U}{\partial u_2}\right) - \frac{m^2}{x^2}\left|\frac{dz}{du}\right|^2 U = 0 \tag{6}$$

The difficulty of solving this equation depends on the form of $x(u_1, u_2)$ and of $|dz/du|$. In many important coordinate systems, including all those treated in this chapter, x has the form

$$x = g_1(u_1)g_2(u_2) \tag{7}$$

and, when 4.11 (1) is used,

$$\left|\frac{dz}{du}\right|^2 = \left|\frac{\partial x}{\partial u_1} - j\frac{\partial x}{\partial u_2}\right|^2 = \left(\frac{\partial x}{\partial u_1}\right)^2 + \left(\frac{\partial x}{\partial u_2}\right)^2 = g_1'^2 g_2^2 + g_1^2 g_2'^2 \tag{8}$$

Assuming a solution of the harmonic form

$$U(u_1, u_2) = U_1(u_1)U_2(u_2) \tag{9}$$

and substituting (7), (8), and (9) in (6), we find that, when divided through by $U_1(u_1)U_2(u_2)$, each term contains only one variable so that, putting terms containing u_1 equal to a constant $+(s + m^2c)$ and those containing u_2 equal to $-(s + m^2c)$, we obtain the total differential equations

$$\frac{1}{g_1U_1}\frac{d}{du_1}\left(g_1\frac{dU_1}{du_1}\right) - m^2\left[\left(\frac{g_1'}{g_1}\right)^2 + c\right] - s = 0 \tag{10}$$

$$\frac{1}{g_2U_2}\frac{d}{du_2}\left(g_2\frac{dU_2}{du_2}\right) - m^2\left[\left(\frac{g_2'}{g_2}\right)^2 - c\right] + s = 0 \tag{11}$$

The remainder of this chapter will be concerned with the solving of these equations and with applying these solutions and (5) to potential problems.

We obtain spherical polar coordinates from 4.12 (2) by interchanging x and y so that

$$g_1 = e^{u_1} = r \qquad \text{and} \qquad g_2 = \sin u_2 = \sin\theta \tag{12}$$

If we put $c = -1$ and $s = n(n + 1)$ in (10) and (11) and write in terms of r and θ, these equations become 5.12 (2.1) and 5.14 (2).

We get oblate spheroidal harmonics from 4.22 (3) giving

$$g_1 = a_1 \sin u_1 = c_1(1 - \xi^2)^{\frac{1}{2}} \qquad \text{and} \qquad g_2 = \cosh u_2 = (1 + \zeta^2)^{\frac{1}{2}} \tag{13}$$

If we put $c = +1$ and $s = -n(n + 1)$ in (10) and (11) and write in terms of ξ and ζ, these equations become exactly 5.271 (2).

We obtain prolate spheroidal harmonics from 4.22 (3) with x and y interchanged, giving

$$g_1 = a_1 \cos u_1 = c_2(1 - \xi^2)^{\frac{1}{2}} \qquad \text{and} \qquad g_2 = \sinh u_2 = (\eta^2 - 1)^{\frac{1}{2}} \tag{14}$$

If we put $c = +1$ and $s = -n(n + 1)$ in (10) and (11) and write in terms of ξ and η, we get the standard form of Laplace's equation in prolate spheroidal coordinates referred to in 5.28.

We obtain cylindrical coordinates by rotating the transformation $z = u$ so that we have $g_1 = u_1 = \rho = k^{-1}v$ and $g_2 = 1$. If, in (10), we put $c = 0$, $s = -k^2$, $m = n$ and divide through by k^2, we obtain

Bessel's equation [5.291 (3)]. The same substitution in (11) gives 5.291 (2).

5.111. Surface of Revolution and Orthogonal Wedge.—In the particular case where $m \neq 0$ and Eq. 5.11 (7) is satisfied, we can obtain the solutions of 5.11 (10) and 5.11 (11) in a simple form. Let the conjugate functions to be rotated be $w_1 = f_1(x, y)$ and $w_2 = f_2(x, y)$. Then let us choose new orthogonal coordinates u_1 and u_2, such that

$$g_1(u_1) = (A_1 e^{u_1} + B_1 e^{-u_1}) = F_1(w_1) \qquad (1)$$

$$g_2(u_2) = (A_2 \cos u_2 + B_2 \sin u_2) = F_2(w_2) \qquad (2)$$

We have from 5.11 (7), by hypothesis, using ρ for the radius of rotation of any point,

$$\rho = g_1(u_1)g_2(u_2) = F_1(w_1)F_2(w_2) \qquad (3)$$

If we set $mc = 1$ and $s = 0$ in 5.11 (10) and 5.11 (11), we can easily verify by substitution that solutions of these equations are

$$U_1 = C_1 g_1^m \qquad (4)$$

and

$$U_2 = C_2 g_2^m \qquad (5)$$

The differential equations are of the second order so that there must be a second solution of each. Since the method of finding this from the known solution will be useful in other cases besides the present one, we shall work out the case with general coefficients here. Suppose that $y = v$ is a particular solution of

$$\frac{d^2y}{dx^2} + M\frac{dy}{dx} + Ny = 0 \qquad (6)$$

where M and N are function of x. Substitute $y = vz$ in this equation, and write z' for dz/dx, and we have, after eliminating v by (6), the result

$$v\frac{dz'}{dx} + \left(2\frac{dv}{dx} + Mv\right)z' = 0$$

Multiply by dx, divide by vz', and integrate

$$\ln z' + \ln v^2 + \int M\,dx = C$$

or

$$z' = \frac{dz}{dx} = Bv^{-2}e^{-\int M\,dx}$$

Integrating gives

$$z = A + B\int v^{-2}e^{-\int M\,dx}\,dx$$

so that

$$y = v\left(A + B\int v^{-2}e^{-\int M\,dx}\,dx\right) \qquad (7)$$

In the present case, putting 5.11 (10) and 5.11 (11) in the form of (6) by carrying out the first differentiation and multiplying through by

$U_{1,2}$, we see that $M_{1,2}$, has the form

$$M_{1,2} = \frac{g'_{1,2}}{g_{1,2}} = \frac{d \ln g_{1,2}}{du_{1,2}}$$

Using for v the solutions of (4) and (5), we obtain the second solutions from (7), setting $A = 0$, to be

$$U_1 = D_1 g_1^m \int \frac{du_1}{g_1^{2m+1}} \tag{8}$$

and

$$U_2 = D_2 g_2^m \int \frac{du_2}{g_2^{2m+1}} \tag{9}$$

Thus a solution of Laplace's equation is given by

$$V = g_1^m\left(C_1 + D_1 \int \frac{du_1}{g_1^{2m+1}}\right) g_2^m\left(C_2 + D_2 \int \frac{du_2}{g_2^{2m+1}}\right) \cos(m\phi + \beta) \tag{10}$$

This form of solution is particularly useful when the surface of a charged conductor is formed by a figure of rotation and a wedge with its edge on the axis of rotational symmetry.

A simple example will illustrate an application of this solution. A charged infinite conducting wedge of exterior angle α has a spherical conducting boss of radius a intersecting both faces orthogonally. Here, taking the center of the sphere as the origin and the faces of the wedge to be $\phi = \pm\frac{1}{2}\alpha$, we have $\rho = r \sin\theta$, $g_1 = e^{u_1} = r$, $g_2 = \sin u_2 = \sin\theta$. All the boundary conditions are satisfied by taking

$$m = \pi/\alpha, \qquad D_1 = \left(1 + \frac{2\pi}{\alpha}\right) C_1 a^{1+\frac{2\pi}{\alpha}}, \qquad C_2 = 1,\ D_2 = 0, \text{ and } \beta = 0$$

Thus we obtain

$$V = C_1 r^{\frac{\pi}{\alpha}}\left[1 - \left(\frac{a}{r}\right)^{1+\frac{2\pi}{\alpha}}\right](\sin\theta)^{\frac{\pi}{\alpha}} \cos\frac{\pi\phi}{\alpha} \tag{11}$$

This gives $V = 0$ on the surface of the wedge and sphere and agrees with 4.19 when we are far from the origin.

5.12. Spherical Harmonics.—When the boundary conditions of a potential problem are simply expressed in spherical polar coordinates, it is useful to have a general solution of Laplace's equation in this system. To get this solution, we proceed in exactly the same way as in 4.01. In terms of the distance from the origin r, the colatitude angle θ measured from the z-axis, and the longitude angle ϕ measured about the z-axis from the zx-plane, Laplace's equation, from 3.05 (1), takes the form

$$\frac{\partial}{\partial r}\left(r^2 \frac{\partial V}{\partial r}\right) + \frac{1}{\sin\theta}\frac{\partial}{\partial\theta}\left(\sin\theta \frac{\partial V}{\partial\theta}\right) + \frac{1}{\sin^2\theta}\frac{\partial^2 V}{\partial\phi^2} = 0 \tag{1}$$

We wish to find solutions of the form

$$V = R\Theta\Phi = RS \tag{2}$$

where R is a function of r only, Θ a function of θ only, and Φ a function of ϕ only. The function $S = \Theta\Phi$ is called a surface harmonic, and the function Θ, when Φ is a constant, is called a zonal surface harmonic. Substituting RS for V in (1) and dividing through by RS, we have

$$\frac{1}{R}\frac{d}{dr}\left(r^2\frac{dR}{dr}\right) + \frac{1}{S\sin\theta}\frac{\partial}{\partial\theta}\left(\sin\theta\frac{\partial S}{\partial\theta}\right) + \frac{1}{S\sin^2\theta}\frac{\partial^2 S}{\partial\phi^2} = 0$$

The first term is a function of r only, and the other ones involve only the angles. For all values of the coordinates, therefore, the equation can be satisfied only if

$$\frac{1}{S\sin\theta}\frac{\partial}{\partial\theta}\left(\sin\theta\frac{\partial S}{\partial\theta}\right) + \frac{1}{S\sin^2\theta}\frac{\partial^2 S}{\partial\phi^2} = -K$$

$$\frac{1}{R}\frac{d}{dr}\left(r^2\frac{dR}{dr}\right) = K \qquad (2.1)$$

The solution of the second equation is easily seen to be

$$R = Ar^n + Br^{-n-1} \qquad (3)$$

where $K = n(n + 1)$. Substituting this value of K in the first equation and multiplying through by S give

$$\frac{1}{\sin\theta}\frac{\partial}{\partial\theta}\left(\sin\theta\frac{\partial S}{\partial\theta}\right) + \frac{1}{\sin^2\theta}\frac{\partial^2 S}{\partial\phi^2} + n(n+1)S = 0 \qquad (4)$$

Equation (2) therefore takes the form

$$V = (Ar^n + Br^{-n-1})S_n \qquad (5)$$

It is evident that this is a solution of Laplace's equation only if the same value of n is used in both terms, hence the subscript on S_n. Any sum or integral with respect to n of terms like (5) is also a solution.

In the special case where $n = 0$, (4) becomes

$$\sin\theta\frac{\partial}{\partial\theta}\left(\sin\theta\frac{\partial S_0}{\partial\theta}\right) + \frac{\partial^2 S_0}{\partial\phi^2} = 0 \qquad (6)$$

In 6.20 it is shown that either U or V satisfies this equation, provided

$$U + jV = F[(\cos\phi + j\sin\phi)\tan\tfrac{1}{2}\theta] \qquad (7)$$

Thus every conjugate function of the last chapter yields two solutions of Laplace's equation in three dimensions by substitution of $\cos\phi\tan\frac{1}{2}\theta$ for x, and $\sin\phi\tan\frac{1}{2}\theta$ for y, and multiplication by $A + Br^{-1}$. Especially useful solutions obtained by choosing $W = z^{\pm m}$ and $W = \ln z$ are, respectively,

$$V = (A + Br^{-1})(C\cot^m\tfrac{1}{2}\theta + D\tan^m\tfrac{1}{2}\theta)\cos(m\phi + \delta_m) \qquad (8)$$

$$V = (A + Br^{-1})(C\ln\tan\tfrac{1}{2}\theta + D\phi) \qquad (9)$$

5.13. General Property of Surface Harmonics.—Before obtaining solutions of (4), we may derive, by Green's theorem, a useful property of

S_n. Let us again write down 3.06 (4), giving

$$\int_V (\Psi\nabla^2\Phi - \Phi\nabla^2\Psi)\, dv = \int_S \left(\Psi\frac{\partial\Phi}{\partial n} - \Phi\frac{\partial\Psi}{\partial n}\right) dS \tag{1}$$

Let us take $\Psi = r^m S_m$ and $\Phi = r^n S_n$ so that $\nabla^2\Phi = \nabla^2\Psi = 0$ and the volume integral vanishes. Taking the surface to be a unit sphere and writing $d\Omega$ for the element of solid angle, we have

$$\frac{\partial\Psi}{\partial n} = \frac{\partial(r^m S_m)}{\partial r} = mr^{m-1}S_m = mS_m$$

and similarly $\partial\Phi/\partial n = nS_n$. Substituting in (1) gives

$$\int_\Omega (nS_nS_m - mS_nS_m)\, d\Omega = (n - m)\int_\Omega S_nS_m\, d\Omega = 0$$

so that if $n \neq m$ we have the result

$$\int_\Omega S_nS_m\, d\Omega = 0 \tag{2}$$

5.131. Potential of Harmonic Charge Distribution.—Let us suppose that, on the surface of a sphere, we have an electric charge density, σ_n, which is finite, continuous, and of such a nature that we can choose a small area, ΔS, anywhere on it so that, over ΔS, $\sigma_n - \bar{\sigma}_n$ is negligible compared with the mean value, $\bar{\sigma}_n$, of σ_n over ΔS. This charge gives rise to a potential, V_o, outside the sphere and the potential, V_i, inside it. Applying Gauss's electric flux theorem [1.10 (1)] to a small box fitting closely this element of the shell, we have

$$\sigma_n = \epsilon\left(\frac{\partial V_i}{\partial r} - \frac{\partial V_o}{\partial r}\right)_{r=a} \tag{1}$$

Let us consider that σ_n is such that

$$\epsilon V_0 = \frac{1}{2n+1}\frac{a^{n+2}}{r^{n+1}}S_n \tag{2}$$

Then, since $V_o = V_i$ when $r = a$, we must have

$$\epsilon V_i = \frac{1}{2n+1}\frac{r^n}{a^{n-1}}S_n \tag{3}$$

Substituting (2) and (3) in (1) gives

$$\sigma_n = S_n \tag{4}$$

In studying S_n we shall see that when θ enters as $P_n^m(\cos\theta)$, it meets the conditions imposed on σ_n at the beginning of this article. By using these equations and 1.06 (6) we can evaluate two useful integrals.

$$\int_S \frac{S_n}{R_i}\, dS = \frac{4\pi}{2n+1}\frac{r^nS_n}{a^{n-1}}, \qquad \int_S \frac{S_n}{R_o}\, dS = \frac{4\pi}{2n+1}\frac{a^{n+2}S_n}{r^{n+1}} \tag{5}$$

The angles in S_n are the coordinates of the terminal point of R_i or R_o. Thus by superposition, the potential due to any surface distribution which can be expanded in the form

$$\sigma = S_0 + S_1 + S_2 + \cdots \tag{6}$$

is given by

$$V_i = \frac{a}{\epsilon}\left[S_0 + \left(\frac{r}{a}\right)\frac{S_1}{3} + \left(\frac{r}{a}\right)^2\frac{S_2}{5} + \cdots\right] \quad \text{if } r < a$$

$$V_0 = \frac{a}{\epsilon}\left[\frac{a}{r}S_0 + \left(\frac{a}{r}\right)^2\frac{S_1}{3} + \left(\frac{a}{r}\right)^3\frac{S_2}{5} + \cdots\right] \quad \text{if } r > a \tag{7}$$

5.14. Differential Equations for Surface Harmonics.—The variables θ and ϕ, in the differential equation, 5.12 (4), for the surface harmonics $S = \Theta\Phi$, may be separated by the process already used. Substitute $\Theta\Phi$ for S in 5.12 (4), and divide through by $\Theta\Phi/\sin^2\theta$ giving

$$\frac{\sin\theta}{\Theta}\frac{d}{d\theta}\left(\sin\theta\frac{d\Theta}{d\theta}\right) + \frac{1}{\Phi}\frac{d^2\Phi}{d\phi^2} + n(n+1)\sin^2\theta = 0$$

For all values of θ and ϕ, this equation can be satisfied only if

$$\frac{\sin\theta}{\Theta}\frac{d}{d\theta}\left(\sin\theta\frac{d\Theta}{d\theta}\right) + n(n+1)\sin^2\theta = K_1 \quad \text{and} \quad \frac{1}{\Phi}\frac{d^2\Phi}{d\phi^2} = -K_1$$

If we put $K_1 = m^2$, the solution of the second equation is easily seen to be

$$\Phi = C\cos m\phi + D\sin m\phi \tag{1}$$

except when $m = 0$, when it becomes

$$\Phi = M\phi + N \tag{1.1}$$

Putting $K_1 = m^2$ in the first equation and multiplying through by $\Theta/\sin^2\theta$ give

$$\frac{1}{\sin\theta}\frac{d}{d\theta}\left(\sin\theta\frac{d\Theta}{d\theta}\right) + \left[n(n+1) - \frac{m^2}{\sin^2\theta}\right]\Theta = 0 \tag{2}$$

This is the differential equation for Θ.

5.15. Surface Zonal Harmonics. Legendre's Equation.—Before considering a more general solution of 5.14 (2), let us solve the most important special case in which V is independent of ϕ so that Φ is constant and, from 5.14 (1), m is zero. Equation 5.14 (2) then becomes, when μ is written for $\cos\theta$,

$$\frac{d}{d\mu}\left[(1-\mu^2)\frac{d\Theta_n}{d\mu}\right] + n(n+1)\Theta_n = 0 \tag{1}$$

This is Legendre's equation, and its solutions are surface zonal harmonics.

5.151. Series Solution of Legendre's Equation.—To obtain a solution of 5.15 (1) in series, let us assume the solution

$$\Theta_n = \Sigma a_r\mu^r \tag{1}$$

Substituting this in 5.15 (1) gives

$$\Sigma\{r(r-1)a_r\mu^{r-2} + [n(n+1) - r(r+1)]a_r\mu^r\} = 0$$

To be satisfied for all values of μ, the coefficient of each power of μ must equal zero separately so that

$$(r+1)(r+2)a_{r+2} + [n(n+1) - r(r+1)]a_r = 0$$

$$a_r = -\frac{(r+1)(r+2)}{n(n+1) - r(r+1)}a_{r+2} = -\frac{(r+1)(r+2)}{(n-r)(n+r+1)}a_{r+2} \qquad (2)$$

To expand in increasing powers of μ, we notice that if $a_r = 0$ then $a_{r-2} = a_{r-4} = \cdots = 0$ and from (2) a_{-1} and a_{-2} are zero if a_0 and a_1 are finite, so that all negative powers of μ disappear. Hence, if we choose $a_0 = 1$ and use the even powers, we have the solution

$$p_n = 1 - \frac{n(n+1)}{2!}\mu^2 + \frac{n(n-2)(n+1)(n+3)}{4!}\mu^4 - \cdots \qquad (3)$$

If we choose $a_1 = 1$ and use the odd powers, we have

$$q_n = \mu - \frac{(n-1)(n+2)}{3!}\mu^3 + \frac{(n-1)(n-3)(n+2)(n+4)}{5!}\mu^5 - \cdots \qquad (4)$$

A complete solution of 5.15 (1), if $-1 < \mu < +1$, is

$$\Theta_n = A_n p_n + B_n q_n$$

regardless of whether n is an integer or a fraction, real or complex, provided the series converges. Recurrence formulas for p_n and q_n may be obtained by subtracting p_{n+1} from p_{n-1}, giving

$$\begin{aligned} p_{n-1} - p_{n+1} &= \left[\frac{(n+1)(n+2) - n(n-1)}{2!}\mu^2 \right. \\ &\quad \left. - \frac{(n+1)(n+4) - n(n-3)}{4}\,\frac{(n-1)(n+2)}{3!}\mu^4 + \cdots\right] \\ &= (2n+1)\mu\left(\mu - \frac{(n-1)(n+2)}{3!}\mu^3 + \cdots\right) \\ &= (2n+1)\mu q_n \end{aligned} \qquad (5)$$

By a similar procedure, we obtain

$$(n+1)^2 q_{n+1} - n^2 q_{n-1} = (2n+1)\mu p_n \qquad (6)$$

Differentiating (4) and adding $n\,dq_{n-1}/d\mu$ and $(n+1)\,dq_{n+1}/d\mu$ give

$$\begin{aligned} nq'_{n-1} + (n+1)q'_{n+1} &= \left[2n + 1 - \frac{(n-2+n+3)n(n+1)}{2!}\mu^2 \right. \\ &\quad \left. + \frac{(n-4+n+5)n(n-2)(n+1)(n+3)}{4!}\mu^4 - \cdots\right] \\ &= (2n+1)\left(1 - \frac{n(n+1)}{2!}\mu^2 + \frac{n(n-2)(n+1)(n+3)}{4!}\mu^4 - \cdots\right) \\ &= (2n+1)p_n \end{aligned} \qquad (7)$$

By a similar procedure, we obtain

$$(n + 1)p'_{n-1} + np'_{n+1} = -n(n + 1)(2n + 1)q_n \tag{8}$$

5.152. Legendre Polynomials. Rodrigues's Formula.—If n is a positive even integer, the series in 5.151 (3) evidently terminates and has $\frac{1}{2}(n + 2)$ terms and may be written

$$p_n = (-1)^{\frac{n}{2}}2^n\frac{[(\frac{1}{2}n)!]^2}{n!}\sum_{r=0}^{\frac{1}{2}n}(-1)^{r-\frac{1}{2}n}\frac{(n + 2r)!\,\mu^{2r}}{2^n[\frac{1}{2}(n - 2r)]!\,[\frac{1}{2}(n + 2r)]!(2r)!}$$

In this case, we define the polynomials $P_n(\mu)$ to be

$$P_n(\mu) = \frac{(-1)^{\frac{1}{2}n}n!}{2^n[(\frac{1}{2}n)!]^2}p_n \tag{1}$$

If n is a positive odd integer, the series in 5.151 (4) evidently terminates and has $\frac{1}{2}(n + 1)$ terms and may be written

$$q_n = (-1)^{\frac{1}{2}(n-1)}2^{n-1}\frac{\{[\frac{1}{2}(n - 1)]!\}^2}{n!}\sum_{r=0}^{\frac{1}{2}(n-1)}(-1)^{r-\frac{1}{2}(n-1)}$$
$$\frac{(n + 2r + 1)!\,\mu^{2r+1}}{2^n[\frac{1}{2}(n - 2r - 1)]!\,[\frac{1}{2}(n + 2r + 1)]!(2r + 1)!}$$

In this case, we define the polynomial $P_n(\mu)$ to be

$$P_n(\mu) = \frac{(-1)^{\frac{1}{2}(n-1)}n!}{2^{n-1}\{[\frac{1}{2}(n - 1)]!\}^2}q_n \tag{2}$$

Legendre's polynomial, given by $P_n(\mu)$ where n is a positive integer, in ascending powers of μ by (1) and (2) may be written in reverse order by substituting $s = \frac{1}{2}n - r$ in (1) and $s = \frac{1}{2}(n - 1) - r$ in (2), both giving the result

$$P_n(\mu) = \sum_{s=0}^{m}(-1)^s\frac{(2n - 2s)!}{2^n(s!)(n - s)!(n - 2s)!}\mu^{n-2s} \tag{3}$$

where $m = \frac{1}{2}n$ or $\frac{1}{2}(n - 1)$, whichever is an integer.

An expression for $P_n(\mu)$, known as Rodrigues's formula, may be obtained by writing (3) in the form

$$P_n(\mu) = \frac{1}{2^n n!}\sum_{s=0}^{m}(-1)^s\frac{n!}{s!(n - s)!}\,\frac{(2n - 2s)!}{(n - 2s)!}\mu^{n-2s}$$
$$= \frac{1}{2^n n!}\frac{d^n}{d\mu^n}\sum_{s=0}^{n}(-1)^s\frac{n!}{s!(n - s)!}\mu^{2n-2s}$$

The last summation is the binomial expansion of $(\mu^2 - 1)^n$ so that

$$P_n(\mu) = \frac{1}{2^n n!}\frac{d^n}{d\mu^n}(\mu^2 - 1)^n \tag{4}$$

Equations (3) and (4) are valid solutions of Legendre's equation [5.15 (1)] whatever the range of the variable μ. In prolate spheroidal harmonics, we have $0 < \mu < \infty$. For very large values of μ, the highest power outweighs all others; so we have

$$P_n(\mu) \xrightarrow[\mu\to\infty]{} \frac{(2n)!}{2^n(n!)^2}\mu^n \tag{5}$$

5.153. Legendre Coefficients. Inverse Distance.—The polynomials of 5.152 are also known as Legendre's coefficients, the reason being evident from the following considerations. The reciprocal of the distance between two points at distances a and b from the origin, where $b > a$, when the angle between a and b is θ and $\mu = \cos\theta$, can be written:

$$\begin{aligned}\frac{1}{R} &= (a^2 + b^2 - 2ab\mu)^{-\frac{1}{2}} = \frac{1}{b}\left(1 + \frac{a^2 - 2ab\mu}{b^2}\right)^{-\frac{1}{2}} \\ &= \frac{1}{b}\left[1 - \frac{1}{2}\frac{a^2 - 2ab\mu}{b^2} + \frac{1\cdot 3}{2\cdot 4}\left(\frac{a^2 - 2ab\mu}{b^2}\right)^2 - \cdots\right] \\ &= \frac{1}{b}\left[1 + \mu\frac{a}{b} + \frac{3\mu^2 - 1}{2}\left(\frac{a}{b}\right)^2 + \frac{5\mu^3 - 3\mu}{2}\left(\frac{a}{b}\right)^3 + \cdots\right]\end{aligned}$$

We see that the coefficient of $(a/b)^n$ is exactly the expression for $P_n(\mu)$ in 5.152 (3) so that we may write

$$\frac{1}{R} = \frac{1}{b}\left[P_0(\mu) + \left(\frac{a}{b}\right)P_1(\mu) + \left(\frac{a}{b}\right)^2 P_2(\mu) + \cdots\right] \tag{1}$$

We shall refer to this expansion many times in solving problems.

5.154. Recurrence Formulas for Legendre Polynomials.—If n is an odd integer, we may substitute for p_{n-1}, p_{n+1}, and q_n in 5.151 (5) from 5.152 (1) and (2) and obtain, after dividing out the factor

$$\frac{2^{n-1}\{[\frac{1}{2}(n-1)]!\}^2}{n!(-1)^{\frac{1}{2}(n+1)}}$$

the result

$$nP_{n-1} + (n+1)P_{n+1} = (2n+1)\mu P_n \tag{1}$$

An identical expression is obtained, when n is even, from 5.151 (6). We omit the argument of these polynomials when there is no ambiguity in doing so.

If n is an even integer, we may substitute in 5.151 (7) for p_n, q'_{n-1}, and q'_{n+1} from 5.152 (1) and (2), divide out the factor

$$\frac{2^n(2n+1)[(\frac{1}{2}n)!]^2}{(-1)^{\frac{1}{2}n}n!}$$

and obtain

$$P'_{n+1} - P'_{n-1} = (2n + 1)P_n \tag{2}$$

An identical expression is obtained, when n is odd, from 5.151 (8).

The integral of $P_n(\mu)$ is given by integrating (2), the result being

$$\int P_n(\mu)\, d\mu = \frac{P_{n+1} - P_{n-1}}{2n + 1} \tag{3}$$

The derivative of $P_n(\mu)$ is given, by adding successive equations of the type of (2), to be

$$P'_n(\mu) = (2n - 1)P_{n-1} + (2n - 5)P_{n-3} + \cdots \tag{4}$$

Another useful expression for the derivative may be obtained by differentiating (1) and eliminating P'_{n-1} by (2), giving

$$P'_{n+1} = \mu P'_n + (n + 1)P_n \qquad \text{or} \qquad P'_n = \mu P'_{n-1} + nP_{n-1} \tag{4.1}$$

Eliminating P'_{n-1} and P'_{n+1} between these equations and (2) and combining with (1), (2), or (3) give the following equivalent forms:

$$\begin{aligned} P'_n &= \frac{(n + 1)(\mu P_n - P_{n+1})}{1 - \mu^2} = \frac{-n(\mu P_n - P_{n-1})}{1 - \mu^2} \\ &= \frac{n(n + 1)}{2n + 1}\,\frac{P_{n-1} - P_{n+1}}{1 - \mu^2} = \frac{-n(n + 1)}{1 - \mu^2}\int P_n(\mu)\, d\mu \end{aligned} \tag{5}$$

5.155. Integral of Product of Legendre Polynomials.—In using Legendre polynomials, the integral of their product over the range $\theta = 0$ to π, or $\mu = -1$ to $+1$, is important. We saw in 5.13 (2) that

$$\int_{-1}^{+1} P_n(\mu)P_m(\mu)\, d\mu = 0 \qquad \text{if } m \neq n \tag{1}$$

If $m = n$, we substitute for one P_n from 5.152 (4)

$$\int_{-1}^{+1} [P_n(\mu)]^2\, d\mu = \frac{1}{2^n n!}\int_{-1}^{+1} P_n(\mu)\frac{d^n}{d\mu^n}(\mu^2 - 1)^n\, d\mu$$

Integrate the right side n times by parts, letting u be the first term and dv the second each time. Since $\dfrac{d^r(\mu^2 - 1)^n}{d\mu^r}$ always contains the factor $(\mu^2 - 1)^{n-r}$, v will always be zero when the limits are inserted and so the product uv drops out and we have finally

$$\int_{-1}^{+1} [P_n(\mu)]^2\, d\mu = \frac{(-1)^n}{2^n n!}\int_{-1}^{+1} \frac{d^n P_n(\mu)}{d\mu^n}(\mu^2 - 1)^n\, d\mu$$

But from 5.152 (3),

$$\frac{d^n P_n(\mu)}{d\mu^n} = \frac{(2n)!n!}{2^n n!n!} = \frac{(2n)!}{2^n n!} \tag{2}$$

so that

$$\int_{-1}^{+1}[P_n(\mu)]^2\,d\mu = \frac{(2n)!}{2^{2n}(n!)^2}\int_{-1}^{+1}(1-\mu^2)^n\,d\mu$$

$$= \frac{(2n-1)!!}{(2n)!!}\int_0^\pi \sin^{2n+1}\theta\,d\theta \qquad (2.1)$$

Integration by *Pc* 483 or *Dw* 854.1 gives

$$\int_{-1}^{+1}[P_n(\mu)]^2\,d\mu = \frac{2}{2n+1} \qquad (3)$$

5.156. Expansion of Function in Legendre Polynomials.—Any function that can be expanded in Fourier's series in the interval $-1 < \mu < +1$ can also be expanded in a series of Legendre polynomials in the same interval and by a similar method. Assume the expansion

$$f(\mu) = a_0P_0(\mu) + a_1P_1(\mu) + \cdots + a_nP_n(\mu) + \cdots \qquad (1)$$

Multiply by $P_m(\mu)$, and integrate from $\mu = -1$ to $+1$. By **5.13 (2)**, all terms vanish except the mth term; so we have, from 5.155 (3).

$$a_m = \tfrac{1}{2}(2m+1)\int_{-1}^{+1}f(\mu)P_m(\mu)\,d\mu \qquad (2)$$

Note that if $f(\mu) = 0$ when $-1 \leqslant \mu \leqslant +1$ then $a_m = 0$. This means that if we have an expansion in Legendre's polynomials equal to zero, the coefficient of each term must be separately equal to zero. As in the case of a Fourier's series, at a discontinuity this expansion gives half the sum of the values of $f(\mu)$ on each side. A formula for a_m which is frequently more convenient than (2) can be obtained by substituting Rodrigues's formula [5.152 (4)] in (2). Thus

$$a_m = (-1)^m\frac{2m+1}{2^{m+1}m!}\int_{-1}^{+1}f(\mu)\frac{d^m(1-\mu^2)^m}{d\mu^m}\,d\mu$$

Integrating this by parts repeatedly, always letting the first member be u and the second dv, we find that $|uv|_{-1}^{+1}$ is zero and $\int_{-1}^{+1}u\,dv$ alternates in sign until finally we are left with

$$a_m = \frac{2m+1}{2^{m+1}m!}\int_{-1}^{+1}\frac{d^mf(\mu)}{d\mu^m}(1-\mu^2)^m\,d\mu \qquad (3)$$

If the derivatives of $f(\mu)$ are simple, this gives usually an easy integration.

5.157. Table of Legendre Polynomials.—A table of values of $P_n(\mu)$ can be computed from 5.152 (3) or (4). The values of $n < 9$ are

$$P_0(\mu) = 1, \qquad P_1(\mu) = \mu, \qquad P_2(\mu) = \tfrac{1}{2}(3\mu^2 - 1)$$

$$P_3(\mu) = \frac{1}{2}(5\mu^3 - 3\mu), \qquad P_4(\mu) = \frac{(35\mu^4 - 30\mu^2 + 3)}{8}$$

$$P_5(\mu) = \frac{(63\mu^5 - 70\mu^3 + 15\mu)}{8}, \qquad P_6(\mu) = \frac{(231\mu^6 - 315\mu^4 + 105\mu^2 - 5)}{16}$$

$$P_7(\mu) = \frac{(429\mu^7 - 693\mu^5 + 315\mu^3 - 35\mu)}{16}$$

$$P_8(\mu) = \frac{(6435\mu^8 - 12{,}012\mu^6 + 6930\mu^4 - 1260\mu^2 + 35)}{128}$$

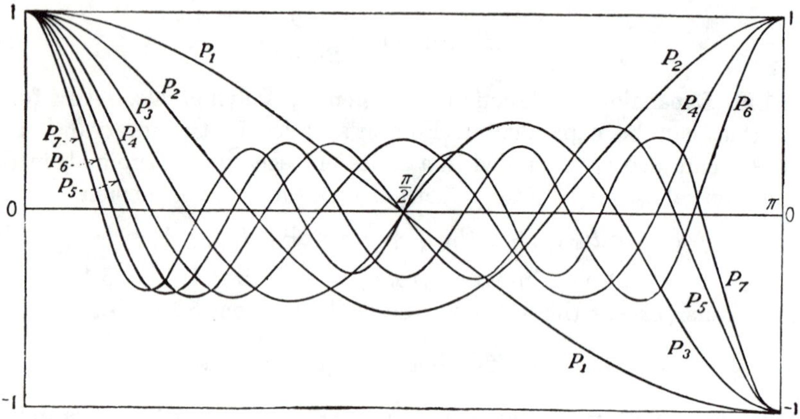

FIG. 5.157.—Legendre polynomials of orders 1 to 7.

Other useful values of $P_n(\mu)$ are

$$\begin{array}{lll}
(n \text{ odd}) & P_n(0) = 0 & \\
(n \text{ even}) & P_n(0) = (-1)^{\frac{1}{2}n}\dfrac{1 \cdot 3 \cdot 5 \cdots (n-1)}{2 \cdot 4 \cdot 6 \cdots n} & \\
(\text{any } n) & P_n(1) = 1 & \\
(\text{any } n) & P_n(-\mu) = (-1)^n P_n(\mu) & \\
(\text{any } n) & P'_n(0) = -(n+1)P_{n+1}(0) & [\text{from 5.154 (5)}] \\
(\text{any } n) & P'_n(1) = \frac{1}{2}n(n+1) & [\text{from 5.15 (1)}]
\end{array}$$

The values of $P_n(\mu)$ for $0 \leqslant n \leqslant 7$ are shown in Fig. 5.157.

5.158. Legendre Polynomial with Imaginary Variable.—We shall have occasion, in oblate spheroidal harmonics, to deal with $P_n(j\zeta)$ where $j = (-1)^{\frac{1}{2}}$ and $0 \leqslant \zeta < \infty$. Substituting $j\zeta$ for μ in 5.152 (3), we have

$$P_n(j\zeta) = (-1)^{\frac{1}{2}n} \sum_{s=0}^{m} \frac{(2n-2s)!}{2^n(s)!(n-s)!(n-2s)!}\zeta^{n-2s} \tag{1}$$

where $m = \frac{1}{2}n$ or $\frac{1}{2}(n-1)$, whichever is an integer. A similar substitution in 5.152 (5) gives

$$P_n(j\zeta) \xrightarrow[\zeta \to \infty]{} \frac{(2n)!}{2^n(n!)^2}(-1)^{\frac{1}{2}n}\zeta^n \tag{2}$$

5.16. Potential of Charged Ring.—If V is symmetrical about the x-axis and its value is known at all points on this axis and if this value can be expressed by a finite or convergent infinite series involving only integral powers of x, then the potential at any point can be obtained by multiplying the nth term by $P_n(\cos\theta)$ and writing r for x. The result holds for the same range of values of r as the range of x in the original expansion.

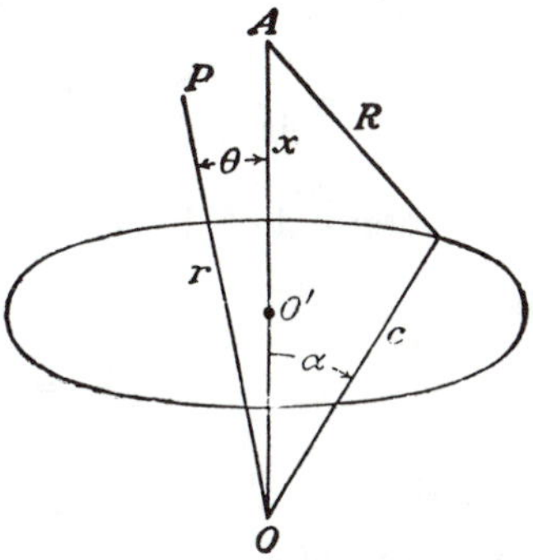

FIG. 5.16.

Let us apply this to a ring carrying a total charge Q (Fig. 5.16) giving

$$4\pi\epsilon V_A = Q(c^2 + x^2 - 2cx\cos\alpha)^{-\frac{1}{2}}$$

Expanding this by 5.153 (1) gives

$$x > c \qquad V_A = \frac{Q}{4\pi\epsilon c}\sum_{n=0}^{\infty}\left(\frac{c}{x}\right)^{n+1} P_n(\cos\alpha)$$

$$x < c \qquad V_A = \frac{Q}{4\pi\epsilon c}\sum_{n=0}^{\infty}\left(\frac{x}{c}\right)^{n} P_n(\cos\alpha)$$

The potential at any point P at r, θ is given by

$$r > c, \text{ or } \theta \neq \alpha,\ r = c \qquad V_P = \frac{Q}{4\pi\epsilon c}\sum_{n=0}^{\infty}\left(\frac{c}{r}\right)^{n+1} P_n(\cos\alpha)P_n(\cos\theta)$$

$$r < c, \text{ or } \theta \neq \alpha,\ r = c \qquad V_P = \frac{Q}{4\pi\epsilon c}\sum_{n=0}^{\infty}\left(\frac{r}{c}\right)^{n} P_n(\cos\alpha)P_n(\cos\theta) \qquad (1)$$

Other examples of this method will appear at the end of this and subsequent chapters.

5.17. Charged Ring in Conducting Sphere.—If the value of the potential due to a given fixed charge distribution is known in a certain region, then the values of the potential when an earthed conducting spherical shell is placed there can be found. Expand the original potential in spherical harmonics, and superimpose a second potential, similarly expanded, due to the induced charge such that the sum is zero over the sphere. The latter should vanish at infinity if the original distribution is outside the sphere and be finite at the center if it is inside.

As an example, let us find the potential at any point inside a spherical ionization chamber of radius b, if the collecting electrode is a thin concentric circular ring of radius a. Let us set $\alpha = \frac{1}{2}\pi$ and take $r > a$ in the last problem, inserting the value for $P_n(0)$ from 5.157 and writing $2n$

for n, since only even powers remain, we have for the potential due to the ring alone

$$V_r = \frac{Q}{4\pi\epsilon a}\sum_{n=0}^{\infty}(-1)^n\frac{1\cdot 3\cdot 5\cdots(2n-1)}{2\cdot 4\cdot 6\cdots 2n}\left(\frac{a}{r}\right)^{2n+1}P_{2n}(\cos\theta)$$

The potential of the induced charge must be finite at the origin; so it has the form

$$V_i = \frac{Q}{4\pi\epsilon a}\sum_{n=0}^{\infty}A_{2n}r^{2n}P_{2n}(\cos\theta)$$

But $V_i + V_r = 0$ when $r = b$, and from 5.156 (2) we may equate the coefficient of each $P_n(\cos\theta)$ separately to zero, so that

$$A_{2n} = -(-1)^n\frac{1\cdot 3\cdot 5\cdots(2n-1)}{2\cdot 4\cdot 6\cdots 2n}\frac{1}{b^{2n}}\left(\frac{a}{b}\right)^{2n+1}$$

and if $a < r \leqslant b$ or $r = a$, $\theta \neq \frac{1}{2}\pi$, the potential becomes

$$V = \frac{Q}{4\pi\epsilon a}\sum_{n=0}^{\infty}(-1)^n\frac{(2n-1)!!}{(2n)!!}\left[\left(\frac{a}{r}\right)^{2n+1} - \left(\frac{a}{b}\right)^{2n+1}\left(\frac{r}{b}\right)^{2n}\right]P_{2n}(\cos\theta) \quad (1)$$

If $r < a$ or $r = a$, $\theta \neq \frac{1}{2}\pi$, we have

$$V = \frac{Q}{4\pi\epsilon a}\sum_{n=0}^{\infty}(-1)^n\frac{(2n-1)!!}{(2n)!!}\left[\left(\frac{r}{a}\right)^{2n} - \left(\frac{a}{b}\right)^{2n+1}\left(\frac{r}{b}\right)^{2n}\right]P_{2n}(\cos\theta) \quad (2)$$

Care should be taken in the application of this method to fields produced by a distribution of charge on conductors. In general, in an actual case, the field of the induced charges will cause the inducing charges to redistribute themselves and invalidate the result.

5.18. Dielectric Shell in Uniform Field.—We now compute the field inside a dielectric shell of internal and external radii a and b, placed in a uniform electrostatic field of strength E. As in the last problem, we may consider the potential outside as that due to the original field $Er\cos\theta$ plus that due to the polarization of the dielectric. The latter must vanish at infinity so that it can contain only reciprocal powers of r. Furthermore, the only surface harmonic appearing in our boundary condition at infinity is $P_1(\mu) = \cos\theta$ so that the potential outside must be of the form

$$V_1 = \left(Er + \frac{A}{r^2}\right)\cos\theta$$

In the dielectric of relative capacitivity K, since r is neither zero nor infinite, we include both terms, giving

$$V_2 = \left(Br + \frac{C}{r^2}\right) \cos \theta$$

In the cavity, the potential must be finite so our only choice is

$$V_3 = Dr \cos \theta$$

The four boundary conditions necessary to determine A, B, C, and D are

$$r = b, \qquad V_1 = V_2 \qquad \text{or} \qquad Eb + \frac{A}{b^2} = Bb + \frac{C}{b^2} \tag{1}$$

$$\frac{\partial V_1}{\partial r} = K\frac{\partial V_2}{\partial r} \qquad \text{or} \qquad E - \frac{2A}{b^3} = KB - \frac{2KC}{b^3} \tag{2}$$

$$r = a, \qquad V_2 = V_3 \qquad \text{or} \qquad Ba + \frac{C}{a^2} = Da \tag{3}$$

$$K\frac{\partial V_2}{\partial r} = \frac{\partial V_3}{\partial r} \qquad \text{or} \qquad KB - \frac{2KC}{a^3} = D \tag{4}$$

Solving these equations, we obtain

$$D = \frac{9KE}{9K - 2(1 - K)^2[(a/b)^3 - 1]} \tag{5}$$

Looking at the expression for V_3, we see that this is the strength of the electric field inside the shell.

5.19. Off-center Spherical Capacitor.—As an example of boundary conditions involving surface harmonics, let us compute, approximately, the charge distribution on the inner shell of a slightly off-center spherical capacitor. The formulas derived by images in 5.08 require many terms to give accuracy in this case if the inner radius a is nearly equal to the outer radius b. Let us choose the origin at the center of the inner surface; then the approximate equation of the outer surface is

$$r = b + cP_1(\mu) \tag{1}$$

where c is the distance between centers and $\mu = \cos \theta$. This is obtained from 5.153 (1), the reciprocal of which may be written

$$b = r[1 + cr^{-1}P_1(\mu)]^{-1} = r - cP_1(\mu)$$

when terms in c^n are neglected if $n > 1$. Since here the boundary conditions involve both $P_0(\mu)$ and $P_1(\mu)$ and since both $r = 0$ and $r = \infty$ are excluded from the field, the potential must be of the form

$$V = A + \frac{B}{r} + \left(Cr + \frac{D}{r^2}\right)P_1(\mu) \tag{2}$$

where C and D are small correction terms of the order of c. The boundary conditions are

$$r = a \qquad V_1 = A + \frac{B}{a} + \left(Ca + \frac{D}{a^2}\right)P_1(\mu)$$

$$r = b + cP_1(\mu) \qquad V_2 = A + \frac{B}{b}\left[1 - \frac{c}{b}P_1(\mu)\right] + \left(Cb + \frac{D}{b^2}\right)P_1(\mu)$$

where the products Cc and Dc have been neglected. From 5.156 (2), we may equate the coefficients of $P_0(\mu)$ and $P_1(\mu)$ separately to zero, giving

$$A + \frac{B}{a} - V_1 = 0 \qquad Ca + \frac{D}{a^2} = 0$$

$$A + \frac{B}{b} - V_2 = 0 \qquad -\frac{Bc}{b^2} + Cb + \frac{D}{b^2} = 0$$

Solving, we have

$$B = \frac{ab(V_1 - V_2)}{b - a}, \qquad C = \frac{abc(V_1 - V_2)}{(b-a)(b^3 - a^3)}, \qquad D = -\frac{a^4bc(V_1 - V_2)}{(b-a)(b^3 - a^3)}$$

We may now obtain the surface density

$$\sigma = -\epsilon\left(\frac{\partial V}{\partial r}\right)_{r=a} = \frac{\epsilon ab(V_1 - V_2)}{b - a}\left(\frac{1}{a^2} - \frac{3c}{b^3 - a^3}\cos\theta\right)$$

where terms in c^n are neglected. We notice that if we integrate this over the surface of the sphere, the correction term drops out so that the capacitance is the same as if the spheres were concentric, if only terms in c are included.

5.20. Simple Conical Boundary.—We have already seen in 5.131 how the potential due to any charge distribution on the surface of a sphere can be expressed in spherical harmonics. It might also be pointed out that if the value of the potential V on the surface of any cone of revolution, whose equation is $\theta = \alpha$, can be expressed in the form

$$V = \Sigma(A_n r^n + B_n r^{-n-1})$$

where n is an integer, then, at any point inside the cone

$$V = \sum \frac{(A_n r^n + B_n r^{-n-1})P_n(\cos\theta)}{P_n(\cos\alpha)} \tag{1}$$

for it can be seen by inspection that this solution satisfies Laplace's equation and the boundary conditions.

5.21. Zonal Harmonics of the Second Kind.—The second solution of Legendre's equation given by the infinite series in 5.151 (3) or (4) is called a zonal harmonic of the second kind and is denoted by $Q_n(\mu)$. We shall define it by expressions similar to 5.152 (1) and (2) to be

$$(n \text{ odd}) \qquad Q_n(\mu) = (-1)^{\frac{1}{2}(n+1)}\frac{\{[\frac{1}{2}(n-1)]!\}^2 2^{n-1}}{n!}p_n \tag{1}$$

$$(n \text{ even}) \qquad Q_n(\mu) = (-1)^{\frac{1}{2}n}\frac{[(\frac{1}{2}n)!]^2 2^n}{n!}q_n \tag{2}$$

These hold for $-1 < \mu < +1$.

Although at present we are dealing with solutions of Legendre's equation in which $\mu = \cos\theta$, we shall find later, in using spheroidal harmonics, that we need solutions in which $\mu^2 > 1$. This requires the expansion 5.151 (1) with negative values of r. We may write 5.151 (2) in the form

$$a_{r+2} = -\frac{(n-r)(n+r+1)}{(r+1)(r+2)}a_r \tag{3}$$

We observe that if $a_{r+2} = 0$, $a_{r+4} = a_{r+6} = \cdots = 0$ and if $a_{-r+1} = 0$, $a_{-r+3} = a_{-r+5} = \cdots = 0$. But a_{n+2} is zero if a_n is finite, and a_{-n+1} is zero if a_{-n-1} is finite. If we let $a_n = 2n!/[2^n(n!)^2]$, we get 5.152 (3), a polynomial defining $P_n(\mu)$, but if we take $a_{-n-1} = 2^n(n!)^2/(2n+1)!$ we obtain, by writing $r = -n-3$, $r = -n-5$, etc., in (3), the coefficients in the series

$$Q_n(\mu) = \frac{2^n(n!)^2}{(2n+1)!}\left[\frac{1}{\mu^{n+1}} + \frac{(n+1)(n+2)}{2(2n+3)}\frac{1}{\mu^{n+3}} + \cdots\right]$$

$$= \frac{2^n(n!)^2}{(2n+1)!}\sum_{r=0}^{\infty}\frac{(n+r)!(n+2r)!(2n+1)!}{r!(n!)^2(2n+2r+1)!}\mu^{-n-2r-1}$$

Writing s for r to agree with the notation of 5.152 (3), we have

$$Q_n(\mu) = 2^n\sum_{s=0}^{\infty}\frac{(n+s)!(n+2s)!}{s!(2n+2s+1)!}\mu^{-n-2s-1} \tag{4}$$

This series evidently converges when $\mu^2 > 1$ and defines $Q_n(\mu)$ in this range. When μ is very large, the smallest negative power in (4) outweighs the remainder so that we have

$$Q_n(\mu) \xrightarrow[\mu\to\infty]{} \frac{2^n(n!)^2}{(2n+1)!\mu^{n+1}} \tag{5}$$

5.211. Recurrence Formulas for Legendre Functions of the Second Kind.—The following formulas connecting different orders of $Q_n(\mu)$ can be obtained from 5.151 (5) and (6) by substituting for p_n and q_n the values of 5.21 (1) and (2) in exactly the same way as 5.154 (1), (2), (3), (4), and (5) were derived.

$$nQ_{n-1} + (n+1)Q_{n+1} = (2n+1)\mu Q_n \tag{1}$$

$$Q'_{n+1} - Q'_{n-1} = (2n+1)Q_n \tag{2}$$

$$\int Q_n(\mu)\,d\mu = \frac{Q_{n+1} - Q_{n-1}}{2n+1} \tag{3}$$

$$Q'_n = (2n-1)Q_{n-1} + (2n-5)Q_{n-3} + \cdots + \mu(1-\mu^2)^{-1} \tag{4}$$

$$Q'_n = \frac{(n+1)(\mu Q_n - Q_{n+1})}{1-\mu^2} \tag{5}$$

5.212. Legendre Functions of the Second Kind in Terms of Legendre Polynomials.—A useful expression for $Q_n(\mu)$ may be obtained from Legendre's equation if we know that $P_n(\mu)$ is a solution. In 5.111 (6) and (7), we saw that if v is one solution of the differential equation

$$\frac{d^2y}{dx^2} + M\frac{dy}{dx} + Ny = 0 \tag{1}$$

where M and N are functions of x, then a second solution is

$$y = v(A + B\int v^{-2}e^{-\int M\,dx}\,dx) \tag{2}$$

In Legendre's equation [5.15 (1)] $v = P_n(\mu)$ and $M = -2\mu(1 - \mu^2)^{-1}$ so that $\int M\,d\mu = \ln(1 - \mu^2)$ and $e^{-\int M\,d\mu} = (1 - \mu^2)^{-1}$ giving (2) the form

$$Q_n(\mu) = P_n(\mu)\left\{A + B\int \frac{d\mu}{(1 - \mu^2)[P_n(\mu)]^2}\right\}$$

To determine the constants A and B, let $n = 0$ and $n = 1$, and integrate by *Dw* 140 or *Pc* 48 and by *Dw* 152.1 or *Pc* 57, and expand the logarithm by *Dw* 601.2 or *Pc* 769. This gives

$$Q_0(\mu) = A + \frac{1}{2}B\ln\frac{1+\mu}{1-\mu} = A + B\left(\mu + \frac{1}{3}\mu^3 + \frac{1}{5}\mu^5 + \cdots\right)$$

$$Q_1(\mu) - A\mu = \frac{\mu B}{2}\ln\frac{1+\mu}{1-\mu} - B = B\left(-1 + \mu^2 + \frac{\mu^4}{3} + \frac{\mu^6}{5} + \cdots\right)$$

From 5.21 (2) and (1), we see that $Q_0(\mu) = q_0$ and $Q_1(\mu) = -p_1$, and comparing the values of q_0 and p_1 in 5.151 (4) and 5.151 (3) we see that

$$A = 0 \qquad \text{and} \qquad B = 1$$

so that the general expression for $Q_n(\mu)$ is

$$Q_n(\mu) = P_n(\mu)\int \frac{d\mu}{(1 - \mu^2)[P_n(\mu)]^2} \tag{3}$$

and in particular

$$Q_0(\mu) = \frac{1}{2}\ln\frac{1+\mu}{1-\mu} \tag{4}$$

and

$$Q_1(\mu) = \frac{1}{2}\mu\ln\frac{1+\mu}{1-\mu} - 1 \tag{5}$$

Applying 5.211 (1), we obtain

$$Q_2(\mu) = \frac{3}{2}\mu Q_1(\mu) - \frac{1}{2}Q_0(\mu) = \frac{1}{4}(3\mu^2 - 1)\ln\frac{1+\mu}{1-\mu} - \frac{3\mu}{2}$$

$$= \frac{1}{2}P_2(\mu)\ln\frac{1+\mu}{1-\mu} - \frac{3}{2}P_1(\mu) \tag{6}$$

By repeated application of 5.211 (1), we get

$$Q_n(\mu) = \frac{1}{2}P_n(\mu)\ln\frac{1+\mu}{1-\mu} - \frac{2n-1}{1\cdot n}P_{n-1}(\mu) - \frac{2n-5}{3(n-1)}P_{n-3}(\mu) - \cdots \tag{7}$$

This holds for $\mu^2 < 1$. The general solution of Legendre's equation is

$$\Theta = AP_n(\mu) + BQ_n(\mu)$$

If we write $A' = A - \frac{1}{2}B \ln(-1)$ and substitute in (7), we have

$$\Theta = A'P_n(\mu) + B\left[\frac{1}{2}P_n(\mu)\ln\frac{\mu+1}{\mu-1} - \frac{2n-1}{1\cdot n}P_{n-1}(\mu) - \cdots\right]$$

So that when $\mu^2 > 1$, we may define $Q_n(\mu)$ by the equation

$$Q_n(\mu) = \frac{1}{2}P_n(\mu)\ln\frac{\mu+1}{\mu-1} - \frac{2n-1}{1\cdot n}P_{n-1}(\mu) - \frac{2n-5}{3(n-1)}P_{n-3}(\mu) - \cdots \quad (8)$$

Putting $n = 0$ and $n = 1$, expanding the logarithm by *Dw* 601.3 or *Pc* 770, and comparing the values of $Q_0(\mu)$ and $Q_1(\mu)$ with 5.21 (4), we see that the formulas agree.

5.213. Special Values of Legendre Functions of the Second Kind. The actual values of $Q_n(\mu)$ may now be found easily from 5.212 (7) or (8), but it is frequently convenient to know certain important special values for real arguments as follows:

$$(n \text{ even}) \qquad Q_n(0) = 0 \quad (1)$$

$$(n \text{ odd}) \qquad Q_n(0) = (-1)^{\frac{1}{2}(n+1)}\frac{2\cdot 4\cdot 6\cdots(n-1)}{1\cdot 3\cdot 5\cdots n} \quad (2)$$

$$(\text{any } n) \qquad Q_n(-\mu) = (-1)^{n+1}Q_n(\mu) \quad (3)$$

$$Q_n(1) = \infty \qquad Q_n(\infty) = 0 \quad (4)$$

5.214. Legendre Function of the Second Kind with Imaginary Variable.—In connection with oblate spheroidal harmonics, we shall have to deal with $Q_n(j\zeta)$ where $j = (-1)^{\frac{1}{2}}$ and $0 \leqslant \zeta < \infty$. No difficulty is experienced when $\zeta > 1$, for substituting $j\zeta$ for μ in 5.21 (4) we have

$$Q_n(j\zeta) = (-j)^{n+1}2^n\sum_{s=0}^{\infty}\frac{(-1)^s(n+2s)!(n+s)!}{s!(2n+2s+1)!}\zeta^{-n-2s-1} \quad (1)$$

If we make the same substitution in 5.212 (8), the result is ambiguous owing to the fact that the logarithm term is multiple valued. Using *Pc* 770 and 780 or *Dw* 601.2 and 506.2, we may write

$$\ln\frac{j\zeta+1}{j\zeta-1} = 2j\left(-\frac{1}{\zeta} + \frac{1}{3\zeta^3} - \frac{1}{5\zeta^5} + \cdots\right) = -2j\cot^{-1}\zeta$$

If we expand the coefficient in 5.212 (8) by 5.158 (1) and substitute the above series for the logarithm, we find the result is identical with (1) so that a consistent value has been chosen and we may write

$$Q_n(j\zeta) = -j\cot^{-1}\zeta P_n(j\zeta) - \frac{2n-1}{1\cdot n}P_{n-1}(j\zeta) - \frac{2n-5}{3(n-1)}P_{n-3}(j\zeta) - \cdots \quad (2)$$

where the arc cotangent ranges from 0 to π. We may use this expression to define $Q_n(j\zeta)$ over the whole range $-\infty < \zeta < \infty$ since it has no peculiarities at $\zeta = 1$. We note that $Q_n(j\zeta)$ is finite when ζ is finite and that

$$(n \text{ even})\ Q_n(j\cdot 0) = -\frac{1}{2}j\pi P_n(j\cdot 0) = (-1)^{\frac{1}{2}n}\frac{-j\pi}{2}\,\frac{1\cdot 3\cdot 5\cdots(n-1)}{2\cdot 4\cdot 6\cdots n} \tag{3}$$

$$(n \text{ odd})\ Q_n(j\cdot 0) = (-1)^{\frac{1}{2}(n+1)}\frac{2\cdot 4\cdot 6\cdots(n-1)}{1\cdot 3\cdot 5\cdots n} \tag{4}$$

When ζ is very large, the smallest negative power in (1) outweighs the remainder so that we have

$$Q_n(j\zeta) \xrightarrow[\zeta\to\infty]{} \frac{2^n(n!)^2}{(2n+1)!}(-j)^{n+1}\frac{1}{\zeta^{n+1}} \tag{4.1}$$

An expression that is sometimes useful is obtained by differentiating 5.212 (3) and eliminating the integral from the resultant equation by 5.212 (3). Substituting $j\zeta$ for μ gives

$$P_n(j\zeta)Q_n'(j\zeta) - P_n'(j\zeta)Q_n(j\zeta) = (1+\zeta^2)^{-1} \tag{5}$$

5.215. Use of Legendre Function of the Second Kind in Potential Problems.—The most important applications of zonal harmonics of the second kind occur in connection with spheroidal harmonics. Owing to the fact that $Q_n(\mu)$ is infinite when $\mu = 1$, they are used as spherical harmonics chiefly in problems where conical boundaries exclude the axis from the region of the electrostatic field.

Let us consider the case of two coaxial cones. Let the potential be zero over the cone $\theta = \beta$, and let $V_\alpha = \Sigma(A_n r^n + B_n r^{-n-1})$ over the cone $\theta = \alpha$. Then, at any point in between,

$$V = \sum \frac{(A_n r^n + B_n r^{-n-1})[Q_n(\cos\beta)P_n(\mu) - P_n(\cos\beta)Q_n(\mu)]}{P_n(\cos\alpha)Q_n(\cos\beta) - P_n(\cos\beta)Q_n(\cos\alpha)} \tag{1}$$

where $\mu = \cos\theta$. It is evident by inspection that this solution satisfies the boundary conditions.

A particular case of some interest is that in which one cone is at potential zero and the other at potential V_1. We then have, since $P_0(\mu) = 1$,

$$V_1 = A_0\frac{Q_0(\cos\beta) - Q_0(\cos\alpha)}{Q_0(\cos\beta) - Q_0(\cos\alpha)} = A_0$$

Also $Q_0(\cos\theta) = \frac{1}{2}\ln[(1+\cos\theta)/(1-\cos\theta)] = -\ln\tan\frac{1}{2}\theta$ from *Dw* 406.2 or *Pc* 578 so that the potential has the form

$$V = V_1\left(\ln\frac{\tan\frac{1}{2}\beta}{\tan\frac{1}{2}\theta}\right)\Big/\left(\ln\frac{\tan\frac{1}{2}\beta}{\tan\frac{1}{2}\alpha}\right) \tag{2}$$

5.22. Nonintegral Zonal Harmonics.—For many problems involving conical boundaries, we find harmonics with integral values of n inadequate. In such cases, we must expand in terms of harmonics with values of n so chosen that the roots of $P_n(\mu)$ or $Q_n(\mu)$ are zero on the cones in question. Many of the equations already used, such as the recurrence formulas for $P_n(\mu)$ and $Q_n(\mu)$ are equally valid for nonintegral values of n but the definitions and derivations need modification to fit this case. A convenient expression for $P_n(\mu)$ good for all values of n is the series

$$\sum_{r=0}^{\infty}\frac{(n+1)(n+2)\cdots(n+r)(-n)(1-n)\cdots(r-1-n)}{(r!)^2}\left(\frac{1-\mu}{2}\right)^r \tag{1}$$

This series converges for $\mu = \cos\theta$ except at $\theta = \pi$. An example of this type of harmonic is given in 5.26, 5.261, and 5.262, if we take the charge on the axis of the box so that $m = 0$.

When ν is not an integer, $P_\nu(\mu)$ and $P_\nu(-\mu)$ are independent solutions of Legendre's equation and are connected with $Q_\nu(\mu)$ by the relation

$$Q_\nu(\mu) = \frac{\pi[\cos\nu\pi P_\nu(\mu) - P_\nu(-\mu)]}{2\sin\nu\pi} \tag{2}$$

When n is integral, the $Q_n(\mu)$ of 5.21 is the limit of $Q_\nu(\mu)$ as $\nu \to n$.

5.23. Associated Legendre Functions.—We have seen in 5.12 and 5.14 that a solution of Laplace's equation, in spherical coordinates, of the form $R\Theta\Phi$ is possible when

5.12 (3) $$R = Ar^n + Br^{-n-1} \tag{1}$$

5.14 (1) $$\Phi = C\cos m\phi + D\sin m\phi \tag{2}$$

and Θ is a solution of 5.14 (2) which becomes, when μ is written for $\cos\theta$,

$$\frac{d}{d\mu}\left[(1-\mu^2)\frac{d\Theta}{d\mu}\right] + \left[n(n+1) - \frac{m^2}{1-\mu^2}\right]\Theta = 0 \tag{3}$$

To obtain a solution of this equation, we start with Legendre's equation, obtained by putting $m = 0$. Differentiating the product composing the first term, we may write Legendre's equation

$$(1-\mu^2)\frac{d^2y}{d\mu^2} - 2\mu\frac{dy}{d\mu} + n(n+1)y = 0$$

We know solutions of this to be $y = P_n(\mu)$ and $y = Q_n(\mu)$. Differentiate this m times, and write v for $d^my/d\mu^m$, and we obtain

$$(1-\mu^2)\frac{d^2v}{d\mu^2} - 2\mu(m+1)\frac{dv}{d\mu} + (n-m)(n+m+1)v = 0 \tag{4}$$

Write $w = (1-\mu^2)^{\frac{1}{2}m}v$ or $v = (1-\mu^2)^{-\frac{1}{2}m}w$, and this is

$$(1-\mu^2)\frac{d^2w}{d\mu^2} - 2\mu\frac{dw}{d\mu} + \left[n(n+1) - \frac{m^2}{1-\mu^2}\right]w = 0 \tag{4.1}$$

This equation is identical with (3), when the first differentiation in (3) is carried out, so that a solution of (3) is

$$\Theta = w = (1 - \mu^2)^{\frac{1}{2}m}v = (1 - \mu^2)^{\frac{1}{2}m}\frac{d^my}{d\mu^m}$$

Since y is a solution of Legendre's equation, a complete solution of (3) is

$$\Theta = A'P_n^m(\mu) + B'Q_n^m(\mu) \tag{5}$$

where if $-1 < \mu < +1$, $P_n^m(\mu)$ and $Q_n^m(\mu)$ are defined by

$$P_n^m(\mu) = (1 - \mu^2)^{\frac{1}{2}m}\frac{d^mP_n(\mu)}{d\mu^m} \tag{6}$$

$$Q_n^m(\mu) = (1 - \mu^2)^{\frac{1}{2}m}\frac{d^mQ_n(\mu)}{d\mu^m} \tag{7}$$

For $-1 < \mu < +1$ Hobson introduces the factor $(-1)^m$ on the right. When μ is imaginary or real, but greater than unity, we define these functions by the equations

$$P_n^m(\mu) = (\mu^2 - 1)^{\frac{1}{2}m}\frac{d^mP_n(\mu)}{d\mu^m} \tag{8}$$

$$Q_n^m(\mu) = (\mu^2 - 1)^{\frac{1}{2}m}\frac{d^mQ_n(\mu)}{d\mu^m} \tag{9}$$

These functions are known as associated Legendre functions of the first and second kind. We can compute them from the expressions for $P_n(\mu)$ and $Q_n(\mu)$ already obtained by using (6) and (7).

For μ real but less than unity, (6) and (7) give

$$P_1^1(\mu) = (1 - \mu^2)^{\frac{1}{2}} \qquad P_3^3(\mu) = 15(1 - \mu^2)^{\frac{3}{2}}$$

$$P_2^1(\mu) = 3(1 - \mu^2)^{\frac{1}{2}}\mu \qquad P_4^1(\mu) = \tfrac{5}{2}(1 - \mu^2)^{\frac{1}{2}}(7\mu^3 - 3\mu)$$

$$P_2^2(\mu) = 3(1 - \mu^2) \qquad P_4^2(\mu) = \tfrac{15}{2}(1 - \mu^2)(7\mu^2 - 1)$$

$$P_3^1(\mu) = \tfrac{3}{2}(1 - \mu^2)^{\frac{1}{2}}(5\mu^2 - 1) \qquad P_4^3(\mu) = 105(1 - \mu^2)^{\frac{3}{2}}\mu$$

$$P_3^2(\mu) = 15(1 - \mu^2)\mu \qquad P_4^4(\mu) = 105(1 - \mu^2)^2$$

$$Q_1^1(\mu) = (1 - \mu^2)^{\frac{1}{2}}\left(\frac{1}{2}\ln\frac{1+\mu}{1-\mu} + \frac{\mu}{1-\mu^2}\right)$$

$$Q_2^1(\mu) = (1 - \mu^2)^{\frac{1}{2}}\left(\frac{3}{2}\mu\ln\frac{1+\mu}{1-\mu} + \frac{3\mu^2 - 2}{1-\mu^2}\right) \tag{10}$$

$$Q_2^2(\mu) = (1 - \mu^2)\left(\frac{3}{2}\ln\frac{1+\mu}{1-\mu} + \frac{5\mu - 3\mu^3}{(1-\mu^2)^2}\right)$$

For higher values of m and n, use the recurrence formulas of 5.233.

For μ real but greater than unity, the values given by (8) and (9) can be obtained by substituting in the above formulas for the first factor $(\mu^2 - 1)^{\frac{1}{2}m}$ instead of $(1 - \mu^2)^{\frac{1}{2}m}$ and by writing, in addition, in the logarithm term of $Q_n^m(\mu)$ the factor $\mu - 1$ instead of $1 - \mu$. For higher values of m and n, use the recurrence formulas of 5.233.

For the imaginary values of the argument, we have, from 5.157, 5.214, and (8) and (9),

$$
\begin{aligned}
P_1(j\zeta) &= j\zeta & Q_1(j\zeta) &= \zeta \cot^{-1} \zeta - 1 \\
P_1^1(j\zeta) &= j(1+\zeta^2)^{\frac{1}{2}} & Q_1^1(j\zeta) &= (1+\zeta^2)^{\frac{1}{2}}\left(\cot^{-1}\zeta - \frac{\zeta}{1+\zeta^2}\right) \\
P_2(j\zeta) &= -\tfrac{1}{2}(3\zeta^2+1) & Q_2(j\zeta) &= \tfrac{1}{2}j[(3\zeta^2+1)\cot^{-1}\zeta - 3\zeta] \\
P_2^1(j\zeta) &= -3(1+\zeta^2)^{\frac{1}{2}}\zeta & Q_2^1(j\zeta) &= j(1+\zeta^2)^{\frac{1}{2}}\left(3\zeta\cot^{-1}\zeta - \frac{3\zeta^2+2}{1+\zeta^2}\right) \\
P_2^2(j\zeta) &= -3(1+\zeta^2) & Q_2^2(j\zeta) &= j(1+\zeta^2)\left[3\cot^{-1}\zeta - \frac{5\zeta+3\zeta^3}{(1+\zeta^2)^2}\right]
\end{aligned} \tag{11}
$$

In the formulas for $Q_n^m(j\zeta)$, the arc cotangent ranges from 0 to π as ζ ranges from $+\infty$ to $-\infty$.

When $\mu \to \infty$, we have, using 5.152 (5) and 5.21 (5) in (8) and (9),

$$P_n^m(\mu) \xrightarrow[\mu\to\infty]{} \frac{(2n)!}{2^n n!(n-m)!}\mu^n \tag{12}$$

$$Q_n^m(\mu) \xrightarrow[\mu\to\infty]{} (-1)^m \frac{n!(n+m)!2^n}{(2n+1)!\mu^{n+1}} \tag{13}$$

When $m = n$, insertion of 5.152 (3) in (6) leaves only the $s = 0$ term and yields the solution used in 5.111 (11).

$$P_m^m(\cos\theta) = (2m-1)!!\sin^m\theta \tag{14}$$

An integral, useful for all values of n when $x \gtrless 1$, is

$$\pi P_n^m(x) = (n+1)(n+2)\cdots(n+m)\int_0^\pi [x + (x^2-1)^{\frac{1}{2}}\cos\phi]^n \cos m\phi \, d\phi \tag{15}$$

Substitution in (3) and two integrations by parts prove that it satisfies Legendre's equation. The constant factor is verified by letting $x \to \infty$ so that the integral can be evaluated and the result compared with (12).

5.231. Integrals of Products of Associated Functions.—Equation 5.13 (2) shows us that, if $n \neq n'$,

$$\int_{-1}^{+1}\int_0^{2\pi} [P_n^m(\mu)P_{n'}^{m'}(\mu)(A\cos m\phi + B\sin m\phi)(A'\cos m'\phi + B'\sin m'\phi)]\,d\mu\,d\phi = 0 \tag{1}$$

Because of the trigonometric products occurring, this integral is also zero if m is an integer, and $m \neq m'$, by *Pc* 359, 360, and 361 or *Dw* 435, 445, and 465, regardless of the value of n and n'.

To get the result when $n = n'$ and $m = m'$ we must determine the integral of the square of $P_n^m(\mu)$ over a unit sphere.

Substitute from 5.23 (6) and 5.152 (4) for $P_n^m(\mu)$ and rearrange the factors and we have, integrating by parts,

$$\int_{-1}^{+1} [P_n^m(\mu)]^2\,d\mu = \int_{-1}^{+1} u\,dv$$
$$= \frac{(-1)^m}{2^{2n}(n!)^2}\int_{-1}^{+1}\left[(\mu^2-1)^m \frac{d^{n+m}}{d\mu^{n+m}}(\mu^2-1)^n\right] d\left[\frac{d^{n+m-1}}{d\mu^{n+m-1}}(\mu^2-1)^n\right]$$
$$= \frac{-(-1)^m}{2^{2n}(n!)^2}\int_{-1}^{+1}\left\{\frac{d}{d\mu}\left[(\mu^2-1)^m \frac{d^{n+m}}{d\mu^{n+m}}(\mu^2-1)^n\right]\right\} d\left[\frac{d^{n+m-2}}{d\mu^{n+m-2}}(\mu^2-1)^n\right]$$

We now integrate repeatedly by parts, the general form of u and v being

$$u = \frac{d^{s-1}}{d\mu^{s-1}}\left[(\mu^2-1)^m \frac{d^{n+m}}{d\mu^{n+m}}(\mu^2-1)^n\right], \qquad v = \frac{d^{m+n-s}}{d\mu^{m+n-s}}(\mu^2-1)^n$$

The product uv always vanishes at both limits since u contains the factor $\mu^2 - 1$ when $m \geqslant s$ and v when $m < s$. Thus after $m + n$ integrations by parts we obtain

$$\int_{-1}^{+1} [P_n^m(\mu)]^2\,d\mu = \int_{-1}^{+1} \frac{(1-\mu^2)^n}{2^{2n}(n!)^2}\left\{\frac{d^{n+m}}{d\mu^{n+m}}\left[(\mu^2-1)^m \frac{d^{n+m}}{d\mu^{n+m}}(\mu^2-1)^n\right]\right\} d\mu \quad (2)$$

The second factor in the integrand is evidently a constant, μ being eliminated by the differentiations. Keeping only the highest power of μ, by substituting μ^2 for $\mu^2 - 1$, we see by inspection that its value is

$$[2n(2n-1)(2n-2)\cdots(n-m+1)](n+m)! = \frac{(2n)!(n+m)!}{(n-m)!}$$

The equation now becomes, using 5.155 (2.1) and 5.155 (3),

$$\int_{-1}^{+1} [P_n^m(\mu)]^2\,d\mu = \frac{(n+m)!}{(n-m)!}\frac{(2n)!}{2^{2n}(n!)^2}\int_{-1}^{+1}(1-\mu^2)^n\,d\mu$$
$$= \frac{(n+m)!}{(n-m)!}\int_{-1}^{+1}[P_n(\mu)]^2\,d\mu = \frac{2}{2n+1}\frac{(n+m)!}{(n-m)!} \quad (3)$$

From (1) when $n \neq n'$,

$$\int_{-1}^{+1} P_n^m(\mu)P_{n'}^m(\mu)\,d\mu = 0 \quad (4)$$

When $m \neq m'$, another integral relation that is sometimes useful is

$$\int_{-1}^{+1} \frac{P_n^m(\mu)P_n^{m'}(\mu)}{1-\mu^2}\,d\mu = 0 \quad (5)$$

To obtain this formula, write 5.23 (3) with $\theta = y$, $m = m$ and with $\theta = y'$, $m = m'$, multiply the first by y', the second by y, subtract and integrate from -1 to $+1$. To integrate when $m = m'$, transfer the middle term of 5.233 (4) to the right side, square, write $n - 1$ for n throughout, multiply by $n + m$, and eliminate $\Theta_{n-1}^{m+1}\Theta_{n-1}^{m-1}$ from the result by using the square of 5.233 (7) multiplied by $n - m$. When integrated from -1 to $+1$, all terms not involving $1 - \mu^2$ are integrated by (3) and cancel out, leaving

$$\int_{-1}^{+1}\frac{(\Theta_n^m)^2\,d\mu}{1-\mu^2}=\frac{n+m}{n-m}\int_{-1}^{+1}\frac{(\Theta_{n-1}^m)^2\,d\mu}{1-\mu^2}=\frac{(n+m)!}{(2m)!(n-m)!}\int_{-1}^{+1}\frac{(\Theta_m^m)^2\,d\mu}{1-\mu^2}$$

Substitution of 5.23 (14) for Θ_m^m and integration by *Dw* 854.1 give

$$\int_{-1}^{+1}\frac{[P_m^m(\mu)]^2}{1-\mu^2}\,d\mu=\frac{(n+m)!}{(n-m)!}\frac{(2m-1)!!}{(2m)!!}\int_0^\pi \sin^{2m-1}\theta\,d\theta=\frac{1}{m}\frac{(n+m)!}{(n-m)!}\tag{6}$$

In dealing with vector potentials, we shall need to use the orthogonal properties of the surface vector function of cos θ defined by

$$\mathbf{P}_n^m(\mu)=\overset{\boldsymbol{\theta}}{\underset{\boldsymbol{\phi}}{}}(1-\mu^2)^{\frac{1}{2}}P_n^{m\prime}(\mu)\pm\overset{\boldsymbol{\phi}}{\underset{\boldsymbol{\theta}}{}}m(1-\mu^2)^{-\frac{1}{2}}P_n^m(\mu)\tag{7}$$

Integration of the scalar product of two such functions gives

$$\int_{-1}^{+1}\mathbf{P}_n^m(\mu)\cdot\mathbf{P}_p^m(\mu)\,d\mu=\int_{-1}^{+1}[(1-\mu^2)P_n^{m\prime}(\mu)P_p^{m\prime}(\mu)$$
$$+\,m^2(1-\mu^2)^{-1}P_n^m(\mu)P_p^m(\mu)]\,d\mu$$

Substitution for the second term from 5.23 (3) and rearrangement give

$$\int_{-1}^{+1}\left\{[(1-\mu^2)P_n^{m\prime}]P_p^{m\prime}+\frac{d}{d\mu}[(1-\mu^2)P_n^{m\prime}]P_p^m\right\}d\mu+n(n+1)\int_{-1}^{+1}P_n^mP_p^m\,d\mu$$
$$=\Big[(1-\mu^2)P_n^{m\prime}P_p^m\Big]_{-1}^{+1}+n(n+1)\int_{-1}^{+1}P_n^mP_p^m\,d\mu=n(n+1)\int_{-1}^{+1}P_n^mP_p^m\,d\mu$$

By (1) this is zero if $n\neq p$ and by (3) if $n=p$ it is

$$\int_{-1}^{+1}\mathbf{P}_n^m(\mu)\cdot\mathbf{P}_n^m(\mu)\,d\mu=\frac{2n(n+1)}{2n+1}\frac{(n+m)!}{(n-m)!}\tag{8}$$

5.232. Associated Functions with Imaginary Argument.—In connection with oblate spheroidal harmonics, the functions $P_n^m(j\zeta)$ and $Q_n^m(j\zeta)$ occur, where ζ ranges from 0 to ∞. Series for these functions can be found by applying 5.23 (8) and (9) to 5.158 (1) and 5.214 (1). It will be noted that the resultant series for $P_n^m(j\zeta)$ has only zero or positive powers of ζ so that $P_n^m(j\cdot 0)$ is finite and $P_n^m(j\cdot\infty)$ is infinite. The series for $Q_n^m(j\zeta)$ contains only negative powers of ζ so that $Q_n^m(j\cdot\infty)$ is zero.

We shall obtain an expression analogous to 5.214 (5) which is sometimes useful. From 5.212 (2) we have

$$Q_n^m(\mu)=P_n^m(\mu)\left\{A+B\int\frac{d\mu}{(1-\mu^2)[P_n^m(\mu)]^2}\right\}$$

Eliminating the integral from this equation and the equation obtained by differentiating it and substituting $j\zeta$ for μ, we obtain

$$P_n^m(j\zeta)\frac{d}{d\zeta}Q_n^m(j\zeta)-Q_n^m(j\zeta)\frac{d}{d\zeta}P_n^m(j\zeta)=\frac{jB}{1+\zeta^2}\tag{1}$$

Letting $\zeta \to \infty$ and using 5.23 (12) and (13) we find that

$$B = (-1)^m \frac{(n+m)!}{(n-m)!} \tag{2}$$

5.233. Recurrence Formulas for Legendre Associated Functions. We have seen in 5.154 and 5.211 that $P_n(\mu)$ and $Q_n(\mu)$ have identical recurrence formulas. The signs in the corresponding formulas for $\Theta_n^m = AP_n^m(\mu) + BQ_n^m(\mu)$ depend on whether $-1 < \mu < +1$ or whether $\mu > 1$ or is imaginary. We shall now obtain these formulas, the upper sign referring to the case $\mu = \cos\theta$. Differentiate 5.154 (4.1) m times, multiply through by $\sin^{m+1}\theta$ or $(\mu^2 - 1)^{\frac{1}{2}(m+1)}$, and then by 5.23 (6) and (7) or 5.23 (8) and (9) we have

$$\Theta_{n+1}^{m+1} = (m + n + 1)[\pm(1-\mu^2)]^{\frac{1}{2}}\Theta_n^m + \mu\Theta_n^{m+1} \tag{1}$$

Now differentiate 5.154 (2) m times, multiply by $[\pm(1-\mu^2)]^{\frac{1}{2}(m+1)}$, and we have, by 5.23 (6) and (7) or 5.23 (8) and (9),

$$\Theta_{n+1}^{m+1} - \Theta_{n-1}^{m+1} = (2n+1)[\pm(1-\mu^2)]^{\frac{1}{2}}\Theta_n^m \tag{2}$$

Subtract this from (1), and we obtain

$$\Theta_{n-1}^{m+1} = (m - n)[\pm(1-\mu^2)]^{\frac{1}{2}}\Theta_n^m + \mu\Theta_n^{m+1} \tag{3}$$

Write $n - 1$ for n in (1), and eliminate Θ_{n-1}^{m+1} and Θ_{n-1}^m by using (3) and (3) with $m - 1$ substituted for m, respectively. Dividing the result through by $-(1 - \mu^2)$, we get

$$\Theta_n^{m+1} \mp 2m\mu[\pm(1-\mu^2)]^{-\frac{1}{2}}\Theta_n^m \pm (m+n)(n-m+1)\Theta_n^{m-1} = 0 \tag{4}$$

This is the recurrence formula in m.

Now multiply (1) by $m - n$ and (3) by $m + n + 1$, and subtract. Writing m for $m + 1$ in the result, we have

$$(m - n - 1)\Theta_{n+1}^m + (2n+1)\mu\Theta_n^m - (m+n)\Theta_{n-1}^m = 0 \tag{5}$$

This is the recurrence formula in n.

Differentiating 5.23 (6) and (7) or 5.23 (8) and (9) and using the above formulas, we have

$$\begin{aligned}[\pm(1-\mu^2)]^{\frac{1}{2}}\Theta_n^{m\prime} &= \mp m\mu[\pm(1-\mu^2)]^{-\frac{1}{2}}\Theta_n^m + \Theta_n^{m+1} \\ &= \mp\tfrac{1}{2}(m+n)(n-m+1)\Theta_n^{m-1} + \tfrac{1}{2}\Theta_n^{m+1} \\ &= \pm m\mu[\pm(1-\mu^2)]^{-\frac{1}{2}}\Theta_n^m \mp (m+n)(n-m+1)\Theta_n^{m-1}\end{aligned} \tag{6}$$

We sometimes wish to express $[\pm(1-\mu^2)]^{-\frac{1}{2}}\Theta_n^m$ in terms of Legendre functions. To do this, write $m - 1$ for m in (1), divide through by $[\pm(1-\mu^2)]^{\frac{1}{2}}$, substitute for $\mu[\pm(1-\mu^2)]^{-\frac{1}{2}}\Theta_n^m$ from (4), and write $n - 1$ for n in the result, giving

$$2m[\pm(1-\mu^2)]^{-\frac{1}{2}}\Theta_n^m = \pm\Theta_{n-1}^{m+1} + (m+n-1)(m+n)\Theta_{n-1}^{m-1} \tag{7}$$

Writing $m - 1$ for m in (2) and substituting in the last form of (6) give

$$(1-\mu^2)\Theta_n^{m\prime} = \pm m\mu\Theta_n^m \mp (m+n)(n-m+1)(2n+1)^{-1}(\Theta_{n+1}^m - \Theta_{n-1}^m)$$

Substitution of (5) in the last, first, or middle term gives, respectively,

$$\begin{aligned}(1-\mu^2)\Theta_n^{m\prime} &= \pm(n+1)\mu\Theta_n^m \mp (n-m+1)\Theta_{n+1}^m \\ &= \pm(2n+1)^{-1}[(m-n-1)n\Theta_{n+1}^m + (n+1)(m+n)\Theta_{n-1}^m] \\ &= \mp n\mu\Theta_n^m \pm (m+n)\Theta_{n-1}^m \end{aligned} \tag{8}$$

5.234. Special Values of Associated Legendre Functions.—With the aid of the recurrence formulas of the last article, especially 5.233 (4) and 5.233 (6), combined with 5.157 and 5.213, we obtain the following relations

$$(n+m \text{ even}) \qquad P_n^m(0) = (-1)^{\frac{1}{2}(n-m)}\frac{1\cdot 3\cdot 5\cdots(n+m-1)}{2\cdot 4\cdot 6\cdots(n-m)}$$

$$(n+m \text{ odd}) \qquad P_n^m(0) = 0$$

$$(n+m \text{ odd}) \qquad Q_n^m(0) = (-1)^{\frac{1}{2}(n-m+1)}\frac{2\cdot 4\cdot 6\cdots(n+m-1)}{1\cdot 3\cdot 5\cdots(n-m)}$$

$$(n+m \text{ even}) \qquad Q_n^m(0) = 0$$

$$\left[\frac{d^r}{d\mu^r}P_n^m(\mu)\right]_{\mu=0} = P_n^{m+r}(0)$$

$$\left[\frac{d^r}{d\mu^r}Q_n^m(\mu)\right]_{\mu=0} = Q_n^{m+r}(0)$$

$$P_n^m(-\mu) = (-1)^{n+m}P_n^m(\mu)$$

$$Q_n^m(-\mu) = (-1)^{n+m+1}Q_n^m(\mu)$$

5.235. Neutral Points and Lines.—The nature of neutral points is clarified by taking one as the origin for a spherical harmonic potential expansion. Inside a sphere so small that it includes no charge, the potential must be everywhere finite and so of the form

$$V = \sum_{n=0}^{\infty}\sum_{m=0}^{n} A_{mn}r^nP_n^m(\cos\theta)\cos m(\phi-\delta_m) \tag{1}$$

At a neutral point, ∇V is zero so that as $r \to 0$, $\partial V/\partial r \to 0$, and A_{m1} is zero. Neutral points may be classified according to the highest p value for which all A_{mn}'s ($n \leqslant p$) vanish at $r = 0$. Thus for a neutral point of the pth order, p is the largest integer for which, for all values of θ and ϕ,

$$\frac{\partial V}{\partial r}\xrightarrow[r\to 0]{}\frac{\partial^2 V}{\partial r^2}\xrightarrow[r\to 0]{}\cdots\xrightarrow[r\to 0]{}\frac{\partial^p V}{\partial r^p}\to 0 \tag{2}$$

Sufficiently close to a neutral line or point of the pth order ,

$$V = A_{00} + r^{p+1}\sum_{m=0}^{p+1} A_{m,p+1}P_{p+1}^m(\cos\theta)\cos(m\phi+\delta_m) \tag{3}$$

The potential at the neutral point is A_{00}. The equipotential surfaces intersecting there are given when r is small, by equating the summation to zero. For an axially symmetrical field, m is zero so the surfaces are circular cones of half angles α given by $P_{p+1}(\cos\alpha) = 0$. From 5.157, for a first-order point, α is 54°44′ and 125°16′; for a second-order point, α is 39°14′, 90°, and 140°46′, etc. The simple neutral line occurs when all but the $m = p + 1$ term are zero. In this neutral line $p + 1$ planes intersect, the angles between adjacent planes being $\pi/(p+1)$.

5.24. Biaxial Harmonics.—It is sometimes convenient to be able to express a surface zonal harmonic $P_n(\cos\theta')$ in terms of general surface harmonics $S_n(\theta, \phi)$ referred to another axis. Let the two axes intersect at the origin and let the coordinates of the θ'-axis, referred to the θ, ϕ system, be $\theta = \Theta$ and $\phi = 0$. We must then determine the coefficients in the expansion

$$P_n(\cos\theta') = \sum_{m=0}^{m=n} A_m P_n^m(\cos\theta)\cos m\phi \tag{1}$$

Multiply both sides by $P_n^s(\cos\theta)\cos s\phi$, and integrate over the surface of a unit sphere. On the right side, by 5.231, all terms go out except the S_n terms. By 5.231 (3), this becomes

$$A_m\int_S [P_n^m(\cos\theta)]^2\cos^2 m\phi\, dS = \frac{2\pi}{2n+1}\frac{(n+m)!}{(n-m)!}A_m \tag{2}$$

on the left side, we must evaluate the integral

$$\int_S P_n(\cos\theta')S_n(\theta, \phi)\, dS$$

This identical integral appears as the coefficient of b^n when we find the potential V_p at the point $\theta' = 0$, $r = b$ due to a distribution of charge density $S_n(\theta, \phi)$ on the surface of a unit sphere. We have solved this problem in 5.131 so that we can evaluate the integral by equating it to the coefficient of b^n in that solution which must hold for all values of $b \lessgtr 1$. Thus we have

$$4\pi\epsilon V_P = \int_S\left(\frac{\sigma_Q}{PQ}\right)dS = \int_S S_n(\theta, \phi)(1 + b^2 - 2b\cos\theta')^{-\frac{1}{2}}\, dS$$

When expanded by 5.153 (1), the integrals of the products of the form $P_r(\cos\theta')S_n(\theta, \phi)$ go out by 5.13 (2) leaving only the term

$$4\pi\epsilon V_P = b^n\int_S P_n(\cos\theta')S_n(\theta, \phi)\, dS$$

But we have already evaluated V_P in 5.131 (5) in the form

$$4\pi\epsilon V_P = \frac{4\pi}{2n+1}b^n[S_n(\theta, \phi)]_P$$

In the present case, the coordinates of P, referred to the θ-axis, are $\theta = \Theta$, $\phi = 0$ and so $S_n(\theta, \phi) = P_n^m(\cos\Theta)$ and equating coefficients of b^n in the two expressions for V_P gives

$$\int_S P_n(\cos\theta')P_n^m(\cos\theta)\cos m\phi\,dS = \frac{4\pi}{2n+1}P_n^m(\cos\Theta) \tag{3}$$

Equating (2) and (3) and solving for A_m, we get

$$A_m = 2\frac{(n-m)!}{(n+m)!}P_n^m(\cos\Theta) \tag{4}$$

when $m = 0$, (2) and (3) become by 5.155 (3) and 5.131 (5)

$$A_0\int_S[P_n(\cos\theta)]^2\,dS = \frac{4\pi A_0}{2n+1} = \int_S P_n(\cos\theta')P_n(\cos\theta)\,dS = \frac{4\pi P_n(\cos\Theta)}{2n+1}$$

Substituting in (1) gives

$$P_n(\cos\theta') = \sum_{m=0}^{m=n}(2-\delta_m^0)\frac{(n-m)!}{(n+m)!}P_n^m(\cos\Theta)P_n^m(\cos\theta)\cos m\phi \tag{5}$$

where $\delta_m^0 = 1$ if $m = 0$ and $\delta_m^0 = 0$ if $m \neq 0$. This symbol is called the Kronecker delta and is more generally written δ_n^m where $\delta_n^m = 1$ if $m = n$ and $\delta_n^m = 0$ if $m \neq n$.

5.25. Conical Boundaries.—The potential inside a cone produced by an arbitrary potential distribution over its surface is found by choosing separation constants in 5.12 (1) so that R and Φ are orthogonal functions. Thus, if n is $jp - \frac{1}{2}$, then K is $-p^2 - \frac{1}{4}$ and R becomes, from 5.12 (3),

$$R_p = A'r^{jp-\frac{1}{2}} + B'r^{-jp-\frac{1}{2}} = r^{-\frac{1}{2}}A\cos(p\ln r + \delta_p) \tag{1}$$

The product $R_pR_{p'}\,dr$ then becomes $\cos(p\psi + \delta_p)\cos(p'\psi + \delta_{p'})\,d\psi$ where ψ is $\ln r$ and leads to Fourier's series or integrals in $\ln r$. There now appear in the Θ-factor Legendre functions of order $jp - \frac{1}{2}$ known as cone functions and treated by Hobson, Heine, and others. Hobson gives the series

$$P_{jp-\frac{1}{2}}(\pm\mu) = 1 + \frac{4p^2+1^2}{2^2}\left(\frac{1\mp\mu}{2}\right) + \frac{(4p^2+1^2)(4p^2+3^2)}{2^2\cdot 4^2}\left(\frac{1\mp\mu}{2}\right)^2 + \cdots \tag{2}$$

The upper sign solution, infinite at $\mu = 1$, is useful outside a cone, and the lower sign one is useful inside it. The function is not periodic. No assumptions regarding n were made in deriving 5.23 (6) so that

$$P_{jp-\frac{1}{2}}^m(\pm\mu) = (1-\mu^2)^{\frac{1}{2}m}\frac{\partial^m P_{jp-\frac{1}{2}}(\pm\mu)}{\partial\mu^m} \tag{3}$$

5.26. Nonintegral Associated Legendre Functions.—As mentioned in the last article, when working with conical boundaries we may use $P_n^m(\mu)$ or $Q_n^m(\mu)$ where n is not an integer. To use these functions, we need expansion in series of Legendre functions that include all values of n for which $\Theta_n^m(\mu) = 0$. For this purpose we must have formulas analogous to 5.13 (2) and 5.231 (3). Let $\Theta_n^m(\mu) = y$ and $\Theta_{n'}^m(\mu) = y'$ be solutions of 5.23 (3) such that

$$\Theta_{n'}^m(\mu_0) = \Theta_n^m(\mu_0) = 0 \tag{1}$$

Thus we have

$$(1-\mu^2)\frac{d^2y}{d\mu^2} - 2\mu\frac{dy}{d\mu} + \left[n(n+1) - \frac{m^2}{1-\mu^2}\right]y = 0$$

$$(1-\mu^2)\frac{d^2y'}{d\mu^2} - 2\mu\frac{dy'}{d\mu} + \left[n'(n'+1) - \frac{m^2}{1-\mu^2}\right]y' = 0$$

Multiply the first by y' and the second by y, and subtract, and we obtain

$$\frac{d}{d\mu}\left[(1-\mu^2)\left(y'\frac{dy}{d\mu} - y\frac{dy'}{d\mu}\right)\right] + (n-n')(n+n'+1)yy' = 0$$

Integrate from μ_0 to 1, and we get

$$\int_{\mu_0}^1 \Theta_n^m(\mu)\Theta_{n'}^m(\mu)\,d\mu = +\left|\frac{(1-\mu^2)\left(\Theta_n^m\dfrac{d\Theta_{n'}^m}{d\mu} - \Theta_{n'}^m\dfrac{d\Theta_n^m}{d\mu}\right)}{(n-n')(n+n'+1)}\right|_{\mu_0}^1 \tag{2}$$

From (1), we have, if $n \neq n'$,

$$\int_{\mu_0}^1 \Theta_n^m(\mu)\Theta_{n'}^m(\mu)\,d\mu = 0 \tag{3}$$

If $n = n'$, we must proceed as follows. Let $n - n' = \Delta n'$. Substitute $y = y' + (\partial y'/\partial n')\Delta n'$ in (2), and we get

$$y\frac{dy'}{d\mu} - y'\frac{dy}{d\mu} = y'\frac{\partial y'}{\partial\mu} + \frac{\partial y'}{\partial\mu}\frac{\partial y'}{\partial n'}\Delta n' - y'\frac{\partial y'}{\partial\mu} - y'\frac{\partial^2 y'}{\partial\mu\,\partial n'}\Delta n'$$

Since at $\mu = \mu_0$, $y' = 0$, we have, as $n \to n'$,

$$\int_{\mu_0}^1 [\Theta_n^m(\mu)]^2\,d\mu = -\frac{(1-\mu_0^2)}{2n+1}\left(\frac{\partial\Theta_n^m}{\partial\mu}\frac{\partial\Theta_n^m}{\partial n}\right)_{\mu=\mu_0} \tag{4}$$

To calculate values of $\partial\Theta_n^m/\partial n$, one may use series such as 5.151 (1) or definite integral expressions which may be found in works cited at the end of this chapter.

5.261. Green's Function for a Cone.—We shall now solve the problem of an earthed cone $\theta = \alpha$ under the influence of a point charge q at $r = a$, $\theta = \beta$, $\phi = \phi_0$. By a point charge, we mean one having dimensions too small to measure physically but mathematically not zero so that the field-intensity and potential functions are mathematically bounded. Our boundary condition $V = 0$ on the cone will be satisfied automatically if we use a series of Legendre functions in which we choose such values of n that $P_n^m(\mu_0)$ is zero where $\mu_0 = \cos\alpha$. Clearly, from

5.23 (1), (2), and (6), with such a choice of n, solutions of Laplace's equation which are finite at $r = 0$, $r = \infty$, agree at $r = a$, and have the proper symmetry about ϕ_0 are

$$r < a \qquad V_i = \sum_n \sum_{m=0}^{\infty} A_{mn}\left(\frac{r}{a}\right)^n P_n^m(\mu) \cos m(\phi - \phi_0) \tag{1}$$

$$r > a \qquad V_0 = \sum_n \sum_{m=0}^{\infty} A_{mn}\left(\frac{a}{r}\right)^{n+1} P_n^m(\mu) \cos m(\phi - \phi_0) \tag{2}$$

To determine the coefficients A_{mn}, we use a new variable $\phi' = \phi - \phi_0$ and proceed, as in 4.07, to write $\partial V_i/\partial r - \partial V_o/\partial r$, multiply both sides by $P_s^p(\mu) \cos p\phi'\, d\mu\, d\phi'$, and integrate from $\phi' = 0$ to $\phi' = 2\pi$ and from $\mu = \mu_0$ to $\mu = 1$. Then, by *Pc* 361 or *Dw* 445, all terms on the right are out unless $m = p$ and, by 5.26 (3), they are also out unless $n = s$. Thus we have, multiplying through by a^2,

$$a^2 \int_0^{2\pi} \int_{\mu_0}^{+1} P_s^p(\mu) \cos p\phi' \left(\frac{\partial V_i}{\partial r} - \frac{\partial V_o}{\partial r}\right)_{r=a} d\mu\, d\phi'$$
$$= aA_{ps}(2s + 1) \int_{\mu_0}^{+1} [P_s^p(\mu)]^2\, d\mu \int_0^{2\pi} \cos^2 p\phi'\, d\phi' \tag{3}$$

The integrals on the right are evaluated by 5.26 (4) and, except when $p = 0$, by *Pc* 363 or *Dw* 440.20. To evaluate the left side, we note that $a^2\, d\mu\, d\phi' = -a^2 \sin\theta\, d\theta\, d\phi' = -dS$. The field is continuous across the sphere $r = a$, except over the infinitesimal area ΔS at $\theta = \beta$, $\phi = \phi_0$ or $\phi' = 0$ where the charge is located. Thus $\partial V_i/\partial r = \partial V_0/\partial r$ except over ΔS and the integral vanishes elsewhere. We take ΔS so small that we may write $P_s^p(\mu_1)$ for $P_s^p(\mu)$, where $\mu_1 = \cos\beta$, and 1 for $\cos p\phi'$. We note that on the inner face of ΔS, $\partial V_i/\partial r = -\partial V/\partial n$ and that on the outer face of ΔS, $\partial V_o/\partial r = \partial V/\partial n$ so that the left side of (3) becomes, by Gauss's electric flux theorem [1.10 (1)]

$$-P_s^p(\mu_1) \int_{\Delta S} \frac{\partial V}{\partial n}\, dS = \frac{q}{\epsilon} P_s^p(\mu_1) \tag{4}$$

Solving (3) for A_{ps} and writing m for p and n for s to fit the notation of (1) and (2), we have

$$A_{mn} = \frac{-(2 - \delta_m^0)q}{2\pi\epsilon a(1 - \mu_0^2)} P_n^m(\mu_1) \Big/ \left[\frac{\partial P_n^m(\mu)}{\partial \mu} \frac{\partial P_n^m(\mu)}{\partial n}\right]_{\mu=\mu_0} \tag{5}$$

where $\delta_m^0 = 1$ if $m = 0$ and $\delta_m^0 = 0$ if $m \neq 0$.

5.262. Green's Function for a Conical Box.—Suppose that the charge q lies between the earthed spheres $r = d$ and $r = c$ and inside the earthed cone $\theta = \alpha$, so that $c < a < d$ and $0 < \beta < \alpha$. We must then superimpose on the Green's function for the cone a potential that will be zero on the cone and will give a resultant zero potential when $r = d$ and $r = c$. Since both $r = 0$ and $r = \infty$ are excluded, such a potential will have the form of 5.261 (1) or 5.261 (2) where the terms $(r/a)^n$ or $(a/r)^{n+1}$ are replaced by $C_n r^n + D_n r^{-n-1}$. When this is added to 5.261 (1) and we put $r = c$, the result must be zero so that

$$\left(\frac{c}{a}\right)^n + C_n c^n + \frac{D_n}{c^{n+1}} = 0 \tag{1}$$

When we add it to 5.261 (2) and put $r = d$, the result must also be zero so that

$$\left(\frac{a}{d}\right)^{n+1} + C_n d^n + \frac{D_n}{d^{n+1}} = 0 \tag{2}$$

Solving (1) and (2) for C_n and D_n and adding the new potential to 5.261 (1) give, if $r < a$,

$$V_i = \sum_n \sum_{m=0}^{\infty} A_{mn} \frac{a^{2n+1} - d^{2n+1}}{a^n(c^{2n+1} - d^{2n+1})}\left(r^n - \frac{c^{2n+1}}{r^{n+1}}\right) P_n^m(\mu) \cos m(\phi - \phi_0) \tag{3}$$

Similarly, if $r > a$, we have, from 5.261 (2),

$$V_o = \sum_n \sum_{m=0}^{\infty} A_{mn} \frac{a^{2n+1} - c^{2n+1}}{a^n(c^{2n+1} - d^{2n+1})}\left(r^n - \frac{d^{2n+1}}{r^{n+1}}\right) P_n^m(\mu) \cos m(\phi - \phi_0) \tag{4}$$

If, in addition, the planes $\phi = +\gamma$ and $\phi = 0$, where $\gamma > \phi_0 > 0$, are at zero potential, and if $\gamma = \pi/s$ where s is an integer, we can obtain the potential by the method of images, superimposing solutions of the type (3) and (4). If $\gamma \neq \pi/s$, we would need to use nonintegral harmonics of the form $\sin(m\pi\phi/\gamma)$. This would introduce in A_{mn} the factor $2\pi/\gamma$, and in (3) and (4) the factor $\sin(m\pi\phi_0/\gamma)\sin(m\pi\phi/\gamma)$ would replace $\cos m(\phi - \phi_0)$.

5.27. Oblate Spheroidal Coordinates.—Common geometrical forms occurring in electrical apparatus are the thin circular disk and the thin sheet with a circular hole. None of the coordinate systems so far studied gives these forms as natural boundaries except the confocal system of 5.01 and 5.02. In this system, we can specify a single value of one coordinate that gives exactly the desired surface, and no more, for the whole range of the remaining coordinates. Our problem is greatly simplified by the axial symmetry which gives oblate spheroids instead of the general ellipsoids. We shall now investigate the solution of Laplace's equation in such a system that introduces the functions called oblate spheroidal harmonics.

In 5.01 (1), let the longer semiaxes b and c be equal and let $y = \rho\cos\phi$ and $z = \rho\sin\phi$. We then have the equation

$$\frac{x^2}{a^2+\theta}+\frac{\rho^2}{b^2+\theta}=1 \tag{1}$$

In this equation, let $a^2 + \theta = (b^2 - a^2)\theta_1^2 = c_1^2\theta_1^2$. Then we have confocal oblate spheroids when $-a^2 < \theta < \infty$ or $0 < \zeta^2 < \infty$, where $\theta_1^2 = \zeta^2$, and when $-b^2 < \theta < -a^2$ or $0 < \xi^2 < 1$, where $-\theta_1^2 = \xi^2$ we have confocal hyperboloids of one sheet. Evidently the third coordinate is the longitude angle ϕ. The equation of the spheroids is

$$\frac{x^2}{c_1^2\zeta^2}+\frac{\rho^2}{c_1^2(\zeta^2+1)}=1 \tag{2}$$

and that of the hyperboloids is

$$\frac{x^2}{-c_1^2\xi^2}+\frac{\rho^2}{c_1^2(1-\xi^2)}=1 \tag{3}$$

Eliminating ρ from these equations gives

$$x = c_1\zeta\xi \tag{4}$$

and eliminating x gives

$$\rho = c_1[(1+\zeta^2)(1-\xi^2)]^{\frac{1}{2}} \tag{5}$$

The coordinate ρ is always positive, but x goes from $-\infty$ to $+\infty$. Therefore if we choose $0 < \zeta < \infty$, we must take $-1 \leqslant \xi \leqslant +1$ as shown in Fig. 5.273, and if we choose $-\infty < \zeta < +\infty$ we must take $0 \leqslant \xi \leqslant 1$ as shown in Fig. 5.272.

To write Laplace's equation in this system, we must determine h_1, h_2, and h_3 in 3.03 (4). From 3.03 (1),

$$\frac{1}{h_1}=\frac{\partial\xi}{\partial n}=j\frac{\partial\theta_1}{\partial n} \qquad \text{and} \qquad \frac{1}{h_2}=\frac{\partial\zeta}{\partial n}=\frac{\partial\theta_1}{\partial n}$$

With the aid of 5.01 (1.1), we have

$$\frac{\partial\theta_1}{\partial n}=\frac{1}{2c_1^2\theta_1}\frac{\partial\theta}{\partial n}=\frac{|\nabla\theta|}{2c_1^2\theta_1}=\frac{1}{c_1^2\theta_1 M_2^{\frac{1}{2}}}$$

Writing the coordinates in the order ξ, ζ, ϕ and substituting for x and ρ from (4) and (5) give

$$h_1 = c_1^2\xi\left[\frac{x^2}{(-c_1^2\xi^2)^2}+\frac{\rho^2}{c_1^4(1-\xi^2)^2}\right]^{\frac{1}{2}} = c_1\left(\frac{\xi^2+\zeta^2}{1-\xi^2}\right)^{\frac{1}{2}} \tag{6}$$

$$h_2 = c_1^2\zeta\left[\frac{x^2}{c_1^4\zeta^4}+\frac{\rho^2}{c_1^4(1+\zeta^2)^2}\right]^{\frac{1}{2}} = c_1\left(\frac{\xi^2+\zeta^2}{1+\zeta^2}\right)^{\frac{1}{2}} \tag{7}$$

$$h_3 = \frac{\rho\, d\phi}{d\phi} = \rho = c_1[(1+\zeta^2)(1-\xi^2)]^{\frac{1}{2}} \tag{8}$$

We may now write Laplace's equation by 3.03 (4).

$$\frac{\partial}{\partial\xi}\left[(1-\xi^2)\frac{\partial V}{\partial\xi}\right]+\frac{\partial}{\partial\zeta}\left[(1+\zeta^2)\frac{\partial V}{\partial\zeta}\right]+\frac{\xi^2+\zeta^2}{(1+\zeta^2)(1-\xi^2)}\frac{\partial^2 V}{\partial\phi^2}=0 \tag{9}$$

5.271. Oblate Spheroidal Harmonics.—A solution of 5.27 (9) in the form $V = \Xi Z \Phi$ where Ξ, Z, and Φ are functions of ξ, ζ, and ϕ only, respectively, is called an oblate spheroidal harmonic. Multiplying 5.27 (9) through by $(1 + \zeta^2)(1 - \xi^2)[\Xi Z \Phi(\xi^2 + \zeta^2)]^{-1}$, we find that the last term involves ϕ only. We can get a solution, as in 5.12, by setting this term equal to $-m^2$ and the other terms equal to $+m^2$, giving

$$\frac{1}{\Phi}\frac{d^2\Phi}{d\phi^2} = -m^2 \tag{1}$$

and

$$\frac{1}{\Xi}\frac{d}{d\xi}\left[(1 - \xi^2)\frac{d\Xi}{d\xi}\right] + \frac{1}{Z}\frac{d}{d\zeta}\left[(1 + \zeta^2)\frac{dZ}{d\zeta}\right] = \frac{m^2(\xi^2 + \zeta^2)}{(1 + \zeta^2)(1 - \xi^2)}$$
$$= \frac{m^2}{1 - \xi^2} - \frac{m^2}{1 + \zeta^2}$$

As before, this can be satisfied by taking

$$\frac{d}{d\xi}\left[(1 - \xi^2)\frac{d\Xi}{d\xi}\right] - \frac{m^2\Xi}{1 - \xi^2} + n(n + 1)\Xi = 0 \tag{2}$$

and

$$\frac{d}{d\zeta}\left[(1 + \zeta^2)\frac{dZ}{d\zeta}\right] + \frac{m^2 Z}{1 + \zeta^2} - n(n + 1)Z = 0$$

Putting $\zeta' = j\zeta$ in the last equation gives

$$\frac{d}{d\zeta'}\left[(1 - \zeta'^2)\frac{dZ}{d\zeta'}\right] - \frac{m^2 Z}{1 - \zeta'^2} + n(n + 1)Z = 0 \tag{3}$$

The solution of (1) is, from 5.23 (2),

$$\Phi_m = C \cos m\phi + D \sin m\phi \tag{4}$$

Since (2) and (3) are identical with 5.23 (3), the solution is given by 5.23 (5) to be

$$\Xi_{mn} = AP_n^m(\xi) + BQ_n^m(\xi) \tag{5}$$
$$Z_{mn} = A'P_n^m(\zeta') + B'Q_n^m(\zeta') = A'P_n^m(j\zeta) + B'Q_n^m(j\zeta) \tag{6}$$

The general solution is then

$$V = \sum_m \sum_n \Xi_{mn} Z_{mn} \Phi_m \tag{7}$$

Since a sphere is a special case of a spheroid, we should expect spherical harmonics to be a special case of spheroidal harmonics. Let us see how the solution given by (4), (5), (6), and (7) passes over into 5.23 (2), 5.23 (5), and 5.23 (1) as the eccentricity of our ellipsoid becomes zero. By letting $c_1 \to 0$, the disk given by $\zeta = 0$ shrinks to the point given by $r = 0$. From 5.27 (4), we see that, since $1 \geqslant \xi \geqslant 0$, $\zeta \to \infty$ as $c_1 \to 0$, we may neglect 1 compared with ζ^2 in 5.27 (5) and have

$$\rho \underset{c_1 \to 0}{\longrightarrow} c_1\zeta(1 - \xi^2)^{\frac{1}{2}} = \frac{x}{\xi}(1 - \xi^2)^{\frac{1}{2}}$$

Solving for ξ gives

$$\xi \underset{c_1 \to 0}{\longrightarrow} x(x^2 + \rho^2)^{-\frac{1}{2}} = \cos\theta = \mu \tag{8}$$

Thus (5) becomes 5.23 (5).

From 5.27 (4), we now see that

$$\zeta \underset{c_1 \to 0}{\longrightarrow} \frac{x}{c_1 \cos\theta} = \frac{r}{c_1} \tag{9}$$

Referring to 5.23 (12) and 5.23 (13), we have

$$A_1 P_n^m(j\zeta) \underset{c_1 \to 0}{\longrightarrow} A_2 r^n \qquad \text{and} \qquad B_1 Q_n^m(j\zeta) \underset{c_1 \to 0}{\longrightarrow} \frac{B_2}{r^{n+1}} \tag{10}$$

Thus (6) becomes 5.23 (1).

By means of (8) and (10), we can frequently choose, at once, the proper oblate spheroidal harmonics to use in a given problem from a knowledge of the form of the solution for the corresponding problem in spherical harmonics.

5.272. Conducting Sheet with Circular Hole.—Let us now consider the case of an infinite thin plane conducting sheet, with a circular hole, which either is freely charged or forms one boundary of a uniform field.

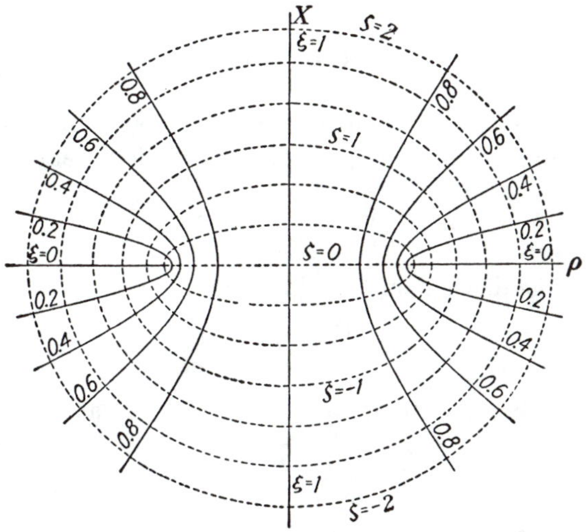

FIG. 5.272.

In order to have our coordinate continuous in the region of the field, we take $0 \leqslant \xi \leqslant 1$ and $-\infty < \zeta < +\infty$, ζ having the same sign as x. The equation of the sheet is $\xi = 0$, as shown in Fig. 5.272. Let us work out the case where such a sheet at potential zero forms one boundary of a uniform field. At $x = \infty$, the field will still be uniform and undisturbed by the hole, and at $x = -\infty$ the field will be zero. Equations 5.27 (4) and (5) simplify when $\zeta^2 \to \infty$, since 1 can then be neglected compared with ζ, giving

$$r^2 = x^2 + \rho^2 \to c_1^2\zeta^2\xi^2 + c_1^2\zeta^2 - c_1^2\zeta^2\xi^2 = c_1^2\zeta^2$$

so that $\zeta \to \pm r/c_1$ and $\xi = x/(c_1\zeta) \to |x|/r = |\cos\theta|$. Now the equation of the uniform field at $x = \infty$ is $V = Er\cos\theta$ so that ξ can enter as $\cos\theta$ only which permits only $P_1(\xi)$, and hence only $m = 0, n = 1$. Then the potential from 5.271 (7), (5), and (6) is

$$V = P_1(\xi)[A'P_1(j\zeta) + B'Q_1(j\zeta)] \tag{1}$$

Writing in the values of P_1 and Q_1 from 5.23 (11), we have

$$V = \xi[jA'\zeta + B'(\zeta\cot^{-1}\zeta - 1)] \tag{2}$$

To evaluate A' and B', consider $\zeta = \pm\infty$. When $\zeta = +\infty$, the coefficient of B' is negligible and

$$Er\cos\theta = \frac{jA'r}{c_1}\cos\theta \qquad \text{giving} \qquad Ec_1 = jA' \tag{3}$$

when $\zeta = -\infty$ the constant term is negligible and

$$0 = \cos\theta\left(\frac{jA'r}{c_1} + \frac{\pi B'r}{c_1}\right) \qquad \text{giving} \qquad \pi B' = -jA' \tag{4}$$

or

$$B' = -\frac{Ec_1}{\pi} \tag{5}$$

If the edge of the hole is given by $\rho = a$, we have, from 5.27 (5), since both ζ and ξ are zero there, $c_1 = a$, giving for the potential

$$V = aE\xi\left[\zeta - \frac{1}{\pi}(\zeta\cot^{-1}\zeta - 1)\right] \tag{6}$$

To get the charge density on the plate, we have

$$\sigma = -\epsilon\frac{\partial V}{\partial x}\underset{x\to 0}{\longrightarrow} \mp\epsilon\frac{\partial V}{a\zeta\,\partial\xi} \tag{7}$$

From 5.27 (5), when $\xi = 0$, $a\zeta = \pm(\rho^2 - a^2)^{\frac{1}{2}}$ and also

$$\cot^{-1}\left[\pm\frac{(\rho^2 - a^2)^{\frac{1}{2}}}{a}\right] = \frac{1}{2}\pi \mp \cos^{-1}\left(\frac{a}{\rho}\right),$$

so that from (6) and (7)

$$\sigma = -\epsilon E\left\{\pm\frac{1}{2} + \frac{1}{\pi}\left[\cos^{-1}\frac{a}{\rho} + \frac{a}{(\rho^2 - a^2)^{\frac{1}{2}}}\right]\right\} \tag{8}$$

where we take the plus sign on the upper side of the plate and the minus sign on the lower side.

5.273. Torque on Disk in Uniform Field.—The case of an uncharged conducting disk of radius a whose normal makes an angle α with an electric field E which was uniform before the introduction of the disk may be considered as the superposition of two cases, the first being a field $E \cos \alpha$ normal to the plane of the disk and the second a field $E' = E \sin \alpha$ parallel to the plane of the disk. To solve the latter case, we may use oblate spheroidal harmonics. In this case, in order that the coordinates will be continuous in the region of the field, we shall take $0 \leqslant \zeta < \infty$ and $-1 \leqslant \xi \leqslant +1$. The coordinates are then as shown in Fig. 5.273. As in the last article at $r = \infty$, $\zeta \to r/a$ and $\xi \to \pm \cos \theta$,

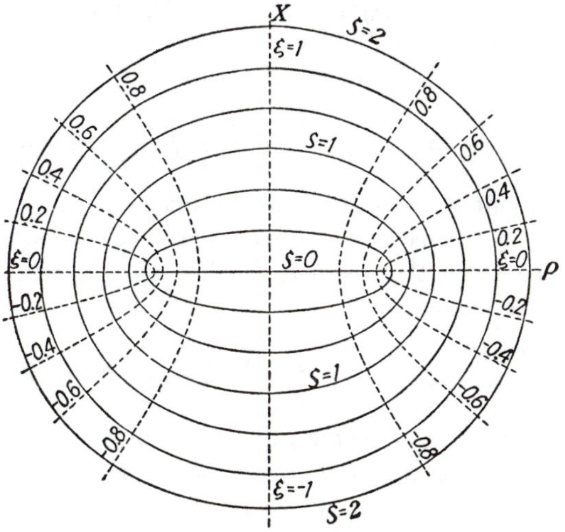

FIG. 5.273.

having the same sign as x. At infinity, where the field is still undisturbed, we have

$$V_\infty = E'\rho \cos \phi = E'a[(1 - \xi^2)(1 + \zeta^2)]^{\frac{1}{2}} \cos \phi \tag{1}$$

Since ϕ enters only as $\cos \phi$, we must have $m = 1$ in 5.271 (4), (5), (6), and (7). Since $n \geqslant m$, the smallest value of n that can occur is $n = 1$. Furthermore, when $r = \infty$, ξ and ζ can enter only as in (1) which is given by the factor $(1 - \mu^2)^{\frac{1}{2}m}$ in 5.23 (6) and (7). We can therefore not have $n > 1$ since the differentiation in 5.23 (6) and (7) would then bring in ξ and ζ as factors. From 5.23 (11),

$$Q_1^1(j\zeta) = (1 + \zeta^2)^{\frac{1}{2}}\left(\cot^{-1} \zeta - \frac{\zeta}{1 + \zeta^2}\right)$$

and from 5.23 (10),

$$P_1^1(\xi) = (1 - \xi^2)^{\frac{1}{2}}$$

Since the potential is finite when $\xi = \pm 1$, the factor $Q_1^1(\xi)$ is excluded and the potential must be

$$V' = P_1^1(\xi)[AP_1^1(j\zeta) + BQ_1^1(j\zeta)] \cos \phi$$
$$= (1 - \xi^2)^{\frac{1}{2}}\left[jA(1 + \zeta^2)^{\frac{1}{2}} + B(1 + \zeta^2)^{\frac{1}{2}}\left(\cot^{-1} \zeta - \frac{\zeta}{1 + \zeta^2}\right)\right] \cos \phi \tag{2}$$

When $\zeta = 0$ $V' = 0$, therefore $jA = -\frac{1}{2}\pi B$. When $\zeta = \infty$, the last two terms in ζ are negligible compared with the first term, and equating to (1)

$$E'a = -\frac{1}{2}\pi B \qquad \text{or} \qquad B = -\frac{2E'a}{\pi}$$

so that the final expression for the potential is

$$V' = 2\pi^{-1}E'a[(1 - \xi^2)(1 + \zeta^2)]^{\frac{1}{2}}\left(\tan^{-1}\zeta + \frac{\zeta}{1 + \zeta^2}\right)\cos\phi \tag{3}$$

Substituting ρ from 5.27 (5), this may be written

$$V' = 2\pi^{-1}E'\rho\left(\tan^{-1}\zeta + \frac{\zeta}{1 + \zeta^2}\right)\cos\phi \tag{4}$$

Since the component $E \cos \alpha$ of the field perpendicular to the disk is not disturbed by the presence of the uncharged disk, its potential is given by $V'' = Ex \cos \alpha$, so that, superimposing on the solution (4) gives $V = V' + V''$ the form

$$V = E\left[\frac{2\rho}{\pi}\left(\tan^{-1}\zeta + \frac{\zeta}{1 + \zeta^2}\right)\sin\alpha\cos\phi + x\cos\alpha\right] \tag{5}$$

where ζ is given in terms of ρ and x by 5.27 (4), (5).

We could get the torque from (5) by finding the surface density σ whence the force per unit area would be $\frac{1}{2}\sigma^2/\epsilon$, etc. An easier method, however, is to observe that the surface density induced by $E \cos \alpha$ produces no torque when acted on by $E \sin \alpha$ because the lever arm is zero. Hence, the entire torque must be that due to σ', induced by $E \sin \alpha$, and acted upon by $E \cos \alpha$. From (4),

$$\sigma' = -\epsilon\left[\frac{\partial V'}{\partial x}\right]_{\zeta=0} = -\frac{2\epsilon E'\rho}{\pi}\left\{\left[\frac{1}{1 + \zeta^2} + \frac{1 - \zeta^2}{(1 + \zeta^2)^2}\right]\frac{\partial\zeta}{\partial x}\right\}_{\zeta=0}\cos\phi$$

But from 5.27 (4), $\zeta = x/(a\xi)$ so that $\partial\zeta/\partial x = (a\xi)^{-1}$. When $\zeta = 0$, from 5.27 (5), $a^2\xi^2 = a^2 - \rho^2$ so that

$$\sigma' = -\frac{4\epsilon E \sin\alpha}{\pi}\frac{\rho\cos\phi}{(a^2 - \rho^2)^{\frac{1}{2}}} \tag{6}$$

For the element of area $\rho\, d\rho\, d\phi$, the lever arm is $\rho \cos \phi$, and the field is $E \cos \alpha$ so that the total torque is, including charge on both sides,

$$T = -\frac{8\epsilon E^2 \sin\alpha\cos\alpha}{\pi}\int_0^a\int_{-\pi}^{+\pi}\frac{\rho^3\cos^2\phi}{(a^2 - \rho^2)^{\frac{1}{2}}}\,d\phi\,d\rho$$

Since $2 \sin \alpha \cos \alpha = \sin 2\alpha$, we have upon integrating by *Dw* 858.3 or *Pc* 489

$$T = -4\epsilon E^2 \sin 2\alpha\int_0^a\frac{\rho^3\,d\rho}{(a^2 - \rho^2)^{\frac{1}{2}}}$$

Integrating again by *Dw* 323.01 or *Pc* 159 gives

$$T = -4\epsilon E^2 \sin 2\alpha\left[\frac{1}{3}(a^2-\rho^2)^{\frac{3}{2}} - a^2(a^2-\rho^2)^{\frac{1}{2}}\right]_0^a = -\frac{8\epsilon a^3E^2\sin 2\alpha}{3} \tag{7}$$

5.274. Potential of Charge Distribution on Spheroid.—Let us suppose that, on the surface of an oblate spheroid, $\zeta = \zeta_0$, we have an electric charge density σ_n of the nature described in the first paragraph of 5.131. This charge gives rise to a potential V_o outside the spheroid and V_i inside it. Applying Gauss's electric flux theorem, 1.10 (1), to a small box fitting closely an element of the spheroid gives

$$\frac{\sigma_n}{\epsilon} = \left\{\frac{\partial V_i}{\partial n} - \frac{\partial V_o}{\partial n}\right\}_S = \left[\frac{1}{h_2}\left(\frac{\partial V_i}{\partial \zeta} - \frac{\partial V_o}{\partial \zeta}\right)\right]_{\zeta=\zeta_0} \tag{1}$$

Let us consider that σ_n is such that

$$\epsilon V_o = j(-1)^m\frac{(n-m)!}{(n+m)!}(1+\zeta_0^2)P_n^m(j\zeta_0)Q_n^m(j\zeta)S_n^m \tag{2}$$

where

$$S_n^m = C_{mn}P_n^m(\xi)\cos m(\phi-\phi_m) \tag{3}$$

This choice makes V_o finite at $\zeta = \infty$. In order that ∇V_i be finite when $\zeta = 0$, $\xi = 0$, and V_i equal V_o when $\zeta = \zeta_0$, V_i must be of the form

$$\epsilon V_i = j(-1)^m\frac{(n-m)!}{(n+m)!}(1+\zeta_0^2)Q_n^m(j\zeta_0)P_n^m(j\zeta)S_n^m \tag{4}$$

Substituting (2) and (4) in (1) and using 5.232 (1) we obtain

$$\sigma_n = \left\{\frac{1}{h_2}\right\}_{\zeta_0} S_n^m \tag{5}$$

Setting

$$A_{mn} = \frac{j(-1)^m}{\epsilon}\frac{(n-m)!}{(n+m)!}(1+\zeta_0^2) \tag{6}$$

we can, with the aid of (2), (4), (5) and 1.06 (6), evaluate two useful integrals, which are

$$\int_{\zeta_0}\frac{S_n^m}{h_2R_o}\,dS = A_{mn}P_n^m(j\zeta_0)Q_n^m(j\zeta)S_n^m, \qquad \int_{\zeta_0}\frac{S_n^m}{h_2R_i}\,dS = A_{mn}Q_n^m(j\zeta_0)P_n^m(j\zeta)S_n^m \tag{7}$$

Now consider a surface density which is a superposition of densities of the form σ_{sr} so that the potentials are superpositions of potentials of the form of (7). Thus we have

$$\sigma = \Sigma\Sigma h_2^{-1}S_r^s = \Sigma\Sigma h_2^{-1}C_{sr}P_r^s(\xi)\cos s(\phi-\phi_s)$$

To determine the coefficients C_{mn}, we multiply by

$$P_n^m(\xi)\cos m(\phi-\phi_m)h_1h_3\,d\xi\,d\phi$$

and integrate over the spheroid. Exactly as with (5), all terms go out except those for which $m = s$ and $n = r$. For these terms, we use 5.24 (2) and so obtain, when $m \neq 0$,

$$C_{mn} = \frac{2n+1}{2\pi c_1}\frac{(n-m)!}{(n+m)!}\frac{1}{1+\zeta_0^2}\int_{-1}^{+1}\int_0^{2\pi} \sigma P_n^m(\xi)\cos m(\phi-\phi_m)h_1h_3\,d\xi\,d\phi \quad (8)$$

The potentials due to the distribution σ are then given by

$$\zeta < \zeta_0 \qquad V_i = \sum_{n=0}^{\infty}\sum_{m=0}^{n} M_{mn}P_n^m(j\zeta)P_n^m(\xi)\cos m(\phi - \phi_m) \quad (9)$$

$$\zeta > \zeta_0 \qquad V_o = \sum_{n=0}^{\infty}\sum_{m=0}^{n} N_{mn}Q_n^m(j\zeta)P_n^m(\xi)\cos m(\phi - \phi_m) \quad (10)$$

where, by combining (4) and (8),

$$M_{mn} = j(-1)^m\frac{(2-\delta_m^0)(2n+1)}{4\pi\epsilon c_1}\left[\frac{(n-m)!}{(n+m)!}\right]^2 Q_n^m(j\zeta_0)$$
$$\int_{-1}^{+1}\int_0^{2\pi}\sigma P_n^m(\xi)\cos m(\phi - \phi_m)h_1h_3\,d\xi\,d\phi \quad (11)$$

where $\delta_m^0 = 1$ if $m = 0$ and $\delta_m^0 = 0$ if $m \neq 0$. From (4) we have

$$N_{mn} = M_{mn}\frac{P_n^m(j\zeta_0)}{Q_n^m(j\zeta_0)} \quad (12)$$

The $2 - \delta_m^0$ factor is required because if $m = 0$ the integration with respect to ϕ used to obtain (8) introduces a factor 2π instead of π.

5.275. Potential of Point Charge in Oblate Spheroidal Harmonics. We can use the results of the last article to obtain the potential due to a point charge at ξ_0, ζ_0, ϕ_0. By a point charge, we mean one whose dimensions, although too small to measure physically, are mathematically not zero so that the field-intensity function and the potential function are everywhere bounded. We take the charge density σ everywhere zero except in an area S, at $\phi = \phi_0$, $\xi = \xi_0$, which is so small that $P_n^m(\xi)$ has the constant value $P_n^m(\xi_0)$ and $\cos m(\phi - \phi_m) = \cos m(\phi - \phi_0) = 1$. The integral in 5.274 (11) then becomes

$$P_n^m(\xi_0)\int\int \sigma h_1h_3\,d\xi\,d\phi = P_n^m(\xi_0)\int_s \sigma\,dS = qP_n^m(\xi_0)$$

The coefficient of 5.274 (11) now becomes

$$M_{mn} = j(2-\delta_m^0)q(-1)^m\frac{2n+1}{4\pi\epsilon c_1}\left[\frac{(n-m)!}{(n+m)!}\right]^2 Q_n^m(j\zeta_0)P_n^m(\xi_0) \quad (1)$$

By 5.274 (12) we have

$$N_{mn} = j(2-\delta_m^0)q(-1)^m\frac{2n+1}{4\pi\epsilon c_1}\left[\frac{(n-m)!}{(n+m)!}\right]^2 P_n^m(j\zeta_0)P_n^m(\xi_0) \quad (2)$$

The potentials due to q are then given by

$$\zeta < \zeta_0 \qquad V_i = \sum_{n=0}^{\infty} \sum_{m=0}^{n} M_{mn} P_n^m(j\zeta) P_n^m(\xi) \cos m(\phi - \phi_0) \tag{3}$$

$$\zeta > \zeta_0 \qquad V_o = \sum_{n=0}^{\infty} \sum_{m=0}^{n} N_{mn} Q_n^m(j\zeta) P_n^m(\xi) \cos m(\phi - \phi_0) \tag{4}$$

With the aid of these formulas, we can now get Green's function for regions bounded by surfaces of the oblate spheroidal coordinate system, using the same methods as with the spherical harmonics.

5.28. Prolate Spheroidal Harmonics.—A common form of spark gap consists of two coaxial rods with rounded ends which can be well approxi-

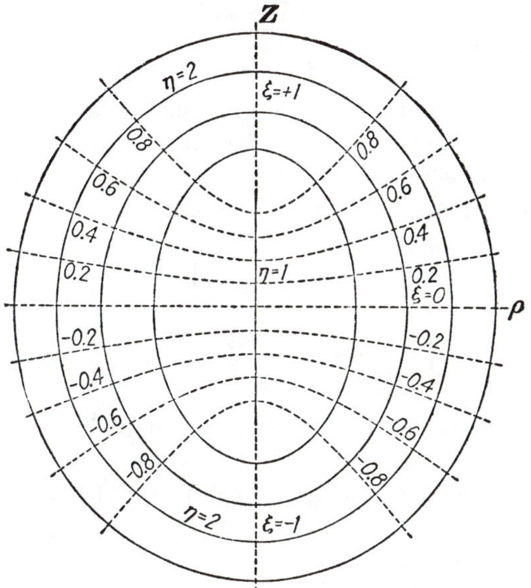

Fig. 5.28.

mated by the two sheets of an hyperboloid of revolution. We also find cases involving elongated conductors which resemble prolate spheroids. These surfaces are natural boundaries in the prolate spheroidal system of coordinates since a specified value of one coordinate will give one of these surfaces, and no more, for the entire range of the remaining coordinates. The solution of Laplace's equation in this system leads to functions called the prolate spheroidal harmonics.

We proceed exactly as in 5.27 except that we now let the two shorter semiaxes a and b in 5.01 (1) be equal and let $x = \rho \sin \phi$ and $y = \rho \cos \phi$. We then have the equation

$$\frac{\rho^2}{b^2 + \theta} + \frac{z^2}{c^2 + \theta} = 1 \tag{1}$$

Let $c^2 + \theta = (c^2 - b^2)\theta_2^2 = c_2^2\theta_2^2$. We have now confocal prolate spheroids when $-b^2 < \theta < \infty$ or $1 < \eta^2 < \infty$, where $\theta_2^2 = \eta^2$, and we have confocal hyperboloids of two sheets when $-c^2 < \theta < -b^2$ or $0 < \xi^2 < 1$ where $\theta_2^2 = \xi^2$. Figure 5.28 shows an axial section of this system. The third coordinate is the longitude angle ϕ. The equation of the spheroids is

$$\frac{\rho^2}{c_2^2(\eta^2 - 1)} + \frac{z^2}{c_2^2\eta^2} = 1 \tag{2}$$

and that of the hyperboloids is

$$\frac{\rho^2}{-c_2^2(1 - \xi^2)} + \frac{z^2}{c_2^2\xi^2} = 1 \tag{3}$$

Eliminating ρ from these equations gives

$$z = c_2\eta\xi \tag{4}$$

and eliminating z gives

$$\rho = c_2[(1 - \xi^2)(\eta^2 - 1)]^{\frac{1}{2}} \tag{5}$$

As in 5.27, if we take $1 \leqslant \eta < \infty$, we must take $-1 \leqslant \xi \leqslant +1$, but there is no reason to give negative values to η, since these coordinates are everywhere continuous. Taking the coordinates in the order ξ, η, ϕ, we determine h_1, h_2 and h_3 as in 5.27, giving

$$h_1 = c_2^2\xi\left[\frac{\rho^2}{c_2^4(1 - \xi^2)^2} + \frac{z^2}{c_2^4\xi^4}\right]^{\frac{1}{2}} = c_2\left(\frac{\eta^2 - \xi^2}{1 - \xi^2}\right)^{\frac{1}{2}} \tag{6}$$

$$h_2 = c_2^2\eta\left[\frac{\rho^2}{c_2^4(\eta^2 - 1)^2} + \frac{z^2}{c_2^4\eta^4}\right]^{\frac{1}{2}} = c_2\left(\frac{\eta^2 - \xi^2}{\eta^2 - 1}\right)^{\frac{1}{2}} \tag{7}$$

$$h_3 = \rho = c_2[(1 - \xi^2)(\eta^2 - 1)]^{\frac{1}{2}} \tag{8}$$

Laplace's equation is obtained exactly as in 5.27 (9) except that we have $\eta^2 - 1$ for $1 + \zeta^2$ and $\eta^2 - \xi^2$ for $\xi^2 + \zeta^2$. Proceeding as in 5.271, we find that, assuming a solution of the form $\Xi\mathrm{H}\Phi$, we have the same differential equation 5.271 (2) for both Ξ and H, and Φ as before is given by 5.271 (4) so that

$$\Xi_{mn} = AP_n^m(\xi) + BQ_n^m(\xi) \tag{9}$$

$$\mathrm{H}_{mn} = A'P_n^m(\eta) + B'Q_n^m(\eta) \tag{10}$$

$$\Phi_m = C\cos m\phi + D\sin m\phi \tag{11}$$

The general solution is given by

$$V = \sum_m\sum_n \Xi_{mn}\mathrm{H}_{mn}\Phi_m \tag{12}$$

5.281. Prolate Spheroid in Uniform Field.—As an application of the results of the last article, let us find the field where an earthed plane, having a conducting spheroidal boss of height c and base radius b, forms one boundary of a uniform field. This situation might arise, for example, in the field between a flat horizontal electrified storm cloud situated above a wet haystack or fir tree on the earth's surface. Let the field at $z = \infty$ be $Er\cos\theta$. As before, when

$$\eta \to \infty \qquad \eta \to \frac{r}{c_2} \qquad \text{and} \qquad \xi \to \frac{z}{r} = \cos\theta$$

Thus, referring to 5.23 (12), we see that ξ can enter only as $P_1(\xi)$; so the potential is, using 5.212 (8),

$$V = P_1(\xi)[(AP_1(\eta) + BQ_1(\eta)] = \xi\eta\left[A + B\left(\frac{1}{2}\ln\frac{\eta+1}{\eta-1} - \frac{1}{\eta}\right)\right]$$

$$= \frac{z}{c_2}\left[A + B\left(\coth^{-1}\eta - \frac{1}{\eta}\right)\right]$$

at $z = \infty$, $\eta = \infty$, $\eta^{-1} = 0$ and $\coth^{-1}\eta = 0$ so that

$$Ez = \frac{Az}{c_2} \qquad \text{or} \qquad A = c_2 E$$

The spheroid with semiaxes c and b is obtained by putting $\theta = 0$ in 5.28 (1) or $c_2^2\eta_0^2 = c^2$. Thus

$$0 = c_2E + B\left(\coth^{-1}\eta_0 - \frac{1}{\eta_0}\right) \qquad \text{or} \qquad B = -c_2E\left(\coth^{-1}\eta_0 - \frac{1}{\eta_0}\right)^{-1}$$

Thus the final expression for the potential is

$$V = Ez\left(1 - \frac{\coth^{-1}\eta - \dfrac{1}{\eta}}{\coth^{-1}\eta_0 - \dfrac{1}{\eta_0}}\right) \tag{1}$$

where

$$\eta_0 = \frac{c}{c_2} = \frac{c}{(c^2 - b^2)^{\frac{1}{2}}} \tag{2}$$

and η, ξ, and z are connected by 5.28 (4) and (5).

5.29. Laplace's Equation in Cylindrical Coordinates.—For potential problems where the boundary conditions are most simply expressed in a cylindrical coordinate system, we need a solution of Laplace's equation in such a system, where it takes the form

$$\frac{\partial^2 V}{\partial \rho^2} + \frac{1}{\rho}\frac{\partial V}{\partial \rho} + \frac{1}{\rho^2}\frac{\partial^2 V}{\partial \phi^2} + \frac{\partial^2 V}{\partial z^2} = 0 \tag{1}$$

Here $x = \rho\cos\phi$ and $y = \rho\sin\phi$. We have already in 4.01 investigated the special case in which V is independent of z and found it gives rise to circular harmonics. We now try a solution of the form $V = R\Phi Z$ where R, Φ, and Z are, respectively, functions of ρ, ϕ, and z alone. This solution will be called a cylindrical harmonic. Substituting in (1) and dividing through by $R\Phi Z\rho^{-2}$, we have

$$\frac{\rho}{R}\frac{d}{d\rho}\left(\rho\frac{dR}{d\rho}\right) + \frac{1}{\Phi}\frac{d^2\Phi}{d\phi^2} + \frac{\rho^2}{Z}\frac{d^2 Z}{dz^2} = 0 \tag{2}$$

5.291. Bessel's Equation and Bessel Functions.—In 5.29 (2), we set

$$\frac{1}{\Phi}\frac{d^2\Phi}{d\phi^2} = -n^2 \tag{1}$$

and

$$\frac{1}{Z}\frac{d^2Z}{dz^2} = k^2 \tag{2}$$

Then R satisfies the equation

$$\rho\frac{d}{d\rho}\left(\rho\frac{dR}{d\rho}\right) + (k^2\rho^2 - n^2)R = 0$$

If we set $v = k\rho$, this equation takes the form

$$\frac{d^2R}{dv^2} + \frac{1}{v}\frac{dR}{dv} + \left(1 - \frac{n^2}{v^2}\right)R = 0 \tag{3}$$

Complete solutions of 1 and 2 are

$$\Phi = Ae^{jn\phi} + Be^{-jn\phi} = A' \cos n\phi + B' \sin n\phi \tag{4}$$
$$Z = Ce^{kz} + De^{-kz} = C' \cosh kz + D' \sinh kz \tag{5}$$

A solution of (3), which is known as Bessel's equation, is called a Bessel function of the nth order. The cylindrical harmonic may now be written

$$V = R_n(k\rho)\Phi(n\phi)Z(kz) \tag{6}$$

except when $k = 0$ when, from 4.01 (5) and 4.07 (2), solutions are

$$V_0 = (M\rho^n + N\rho^{-n})(Cz + D)(A \cos n\phi + B \sin n\phi) \tag{7}$$
$$V_0 = [M \cos (n \ln \rho) + N \sin (n \ln \rho)](Cz + D)(A \cosh n\phi + B \sinh n\phi) \tag{8}$$

When k and n are both zero, we have

$$V_{00} = (M \ln \rho + N)(Cz + D)(A\phi + B) \tag{9}$$

5.292. Modified Bessel Equation and Functions.—A somewhat different equation is obtained by substituting jk_1 for k in 5.291 (2). We then get in place of 5.291 (3)

$$\frac{d^2R^0}{dv^2} + \frac{1}{v}\frac{dR^0}{dv} - \left(1 + \frac{n^2}{v^2}\right)R^0 = 0 \tag{1}$$

which is called the modified Bessel equation and whose solutions are called modified Bessel functions. In place of 5.291 (5), we now have

$$Z = Ce^{jk_1z} + De^{-jk_1z} = C' \cos k_1z + D' \sin k_1z \tag{2}$$

The cylindrical harmonic is now written

$$V = R_n^0(k_1\rho)e^{\pm jn\phi}e^{\pm jk_1z} \tag{3}$$

Solutions of this type will be considered in Arts. 5.32 to 5.36.

5.293. Solution of Bessel's Equation.—Let $R_n = \Sigma a_s v^{n+s}$ in 5.291 (3), and we obtain

$$v^n \Sigma[s(2n + s)a_s v^{s-2} + a_s v^s] = 0 \tag{1}$$

Whence,

$$a_s = -[s(2n + s)]^{-1}a_{s-2} \tag{2}$$

If a_0 is finite, then a_{-2}, a_{-4}, etc. are zero. Let a_0 be $[2^n\Gamma(n + 1)]^{-1}$, which reduces to $(2^n n!)^{-1}$ when n is an integer, then we have, by repeated use of (2), writing $J_n(v)$ for $R_n(v)$,

$$J_n(v) = \frac{v^n}{2^n\Gamma(n+1)}\left[1 - \frac{v^2}{2^2(n+1)} + \frac{v^4}{2^4 2!(n+1)(n+2)} - \cdots\right]$$

$$= \sum_{r=0}^{\infty}(-1)^r\frac{(\frac{1}{2}v)^{n+2r}}{r!\Gamma(n+r+1)} \tag{3}$$

The function $J_n(v)$ is called a Bessel function of the nth order of the first kind. Evidently $J_n(0) = 0$ when $n \neq 0$, and it will be shown in 5.303 that $J_n(\infty) = 0$ also.

Since 5.291 (3) is a second-order differential equation, there must be a second solution. Where n is not an integer, this is $J_{-n}(v)$; but when n is an integer, $J_{-n}(v)$ and $J_n(v)$ are not independent solutions. To show this, write $-n$ for n in (3) then, since $[\Gamma(-n)]^{-1}$ is zero, the series for $J_{-n}(v)$ and $(-1)^nJ_n(v)$ are identical. When n is not integral, the formula

$$Y_\nu(v) = \frac{J_\nu(v)\cos\nu\pi - J_{-\nu}(v)}{\sin\nu\pi} \tag{4}$$

defines a second solution. When ν is integral, this becomes 0/0. To find its value, write $-\nu$ for n in (3), split the 0 to ∞ r-summation into one from n to ∞ and one from 0 to $n - 1$, write $n + s$ for r in the former, and $\pi^{-1}\Gamma(\nu - r)\sin\nu\pi$ for $(-1)^r/\Gamma[1 - (\nu - r)]$ by *Dw* 850.3 in the latter. When $J_{-\nu}(v)$ in (4) is replaced by this result and $\cos\nu\pi$ by $(-1)^n$, it becomes

$$\frac{(-1)^n}{\sin\nu\pi}\left[J_\nu(v) - \sum_{s=0}^{\infty}\frac{(-1)^s(\frac{1}{2}v)^{\nu+2s}}{(n+s)!\Gamma(n-\nu+s+1)}\right] - \sum_{r=0}^{n-1}\frac{(\frac{1}{2}v)^{2r-\nu}\Gamma(\nu - r)}{\pi r!}$$

The bracket is zero when $\nu = n$ from (3) so the first term still has the form 0/0. To evaluate this, differentiate each factor with respect to ν and write n for ν. By referring to page 19 of Jahnke and Emde we find

$$\frac{d}{dz}\left[\frac{1}{\Gamma(z)}\right] = \frac{-1}{\Gamma(z)}\frac{d\ln\Gamma(z)}{dz} = \frac{-1}{\Gamma(z)}\left(-C + \sum_{m=1}^{z}\frac{1}{m}\right)$$

where $C = .5772157$ is Euler's number. With this formula and *Dw* 563.3 or *Pc* 828, the 0/0 form is evaluated and writing $\ln\alpha$ for $C - \ln 2$ gives

$$Y_n(v) = \frac{2}{\pi}J_n(v)\ln(\alpha v) - \sum_{r=0}^{n-1}\frac{(\frac{1}{2}v)^{2r-n}(n-r-1)!}{\pi r!}$$
$$- \sum_{r=0}^{\infty}\frac{(-1)^r(\frac{1}{2}v)^{n+2r}}{\pi r!(n+r)!}\left(\sum_{m=1}^{r}\frac{1}{m} + \sum_{m=1}^{n+r}\frac{1}{m}\right) \quad (5)$$

A complete solution of Bessel's equation when n is an integer is

$$R_n(v) = A'J_n(v) + B'Y_n(v) \quad (6)$$

We note that $Y_n(0)$ is infinite and shall see in 5.303 that $Y_n(\infty)$ is zero. There are many notations for the function defined by (5). Watson and the British Association Tables use $Y_n(v)$, Jahnke and Emde, Schelkunoff, and Stratton use $N_n(v)$, and Gray, Mathews, and MacRobert use $\bar{Y}_n(v)$.

Setting $M = 1/v$ and $N = 1 - (n/v)^2$ in 5.111 (6) reduces it to 5.291 (3) so that, if $J_n(v)$ is the known solution, 5.111 (7) becomes

$$Y_n(v) = J_n(v)\{A + B\int[vJ_n^2(v)]^{-1}\,dv\} \quad (7)$$

Differentiating this and omitting the argument give

$$\frac{d}{dv}\left(\frac{Y_n}{J_n}\right) = \frac{Y'_n}{J_n} - \frac{J'_nY_n}{J_n^2} = \frac{B}{vJ_n^2} \quad (8)$$

From the recurrence formulas, B is independent of both n and v and so may be evaluated for integral n's by taking the simple case where n is zero and v is very small. Then only the ln v term in (5) is important, so it and its derivative may be used for $Y_n(v)$ and $Y'_n(v)$ in (8). The logarithm terms cancel and B comes out $2/\pi$ so (8) may be rearranged in the form

$$Y'_n(v)J_n(v) - J'_n(v)Y_n(v) = \frac{2}{\pi v} \quad (9)$$

For cylindrical electromagnetic waves we need the "Hankel" functions which combine with $e^{j\omega t}$ to give traveling waves and are defined by

$$H_n^{(1)}(v) = J_n(v) + jY_n(v), \qquad H_n^{(2)}(v) = J_n(v) - jY_n(v) \quad (10)$$

5.294. Recurrence Formulas for Bessel Functions.—If we multiply 5.293 (3) by v^n and differentiate, we obtain

$$\frac{d[v^nJ_n(v)]}{dv} = \frac{v^{2n-1}}{2^{n-1}\Gamma(n)}\left[1 - \frac{v^2}{2^2n} + \frac{v^4}{2^42!n(n+1)} - \cdots\right]$$

Factoring out v^n on the right side and comparing with 5.293 (3), we see that

$$\frac{d[v^nJ_n(v)]}{dv} = v^nJ_{n-1}(v)$$

Differentiate left side, rearrange, divide out v^n, and indicate differentiation with respect to v by a prime, and we have

$$J'_n = J_{n-1} - \frac{n}{v}J_n \tag{1}$$

If we use the same procedure multiplying by v^{-n} instead of by v^n, we obtain

$$J'_n = -J_{n+1} + \frac{n}{v}J_n = \frac{1}{2}(J_{n-1} - J_{n+1}) \tag{2}$$

Subtracting (1) from (2), we have

$$\frac{2n}{v}J_n = J_{n-1} + J_{n+1} \tag{3}$$

Writing $(-1)^n J_{-n}(v)$ for $J_n(v)$ in (1), (2), and (3) gives the recurrence formulas for $J_{-n}(v)$. Differentiation of 5.293 (4) and substitution for $J'_\nu(v)$ from (1) and for $J'_{-\nu}(v)$ from a similar formula give, as $\nu \to n$, so that $\cos \nu\pi$ may be replaced by $-\cos(\nu - 1)\pi$ and $\sin \nu\pi$ by $-\sin(\nu - 1)\pi$,

$$\frac{J'_\nu \cos \nu\pi - J'_{-\nu}}{\sin \nu\pi} \underset{\nu \to n}{\longrightarrow} \frac{J_{\nu-1}\cos(\nu-1)\pi - J_{-\nu+1}}{\sin(\nu-1)\pi} - \frac{\nu(J_\nu \cos \nu\pi - J_{-\nu})}{v \sin \nu\pi}$$

A similar process can be applied to (2) so that

$$Y'_n = Y_{n-1} - \frac{n}{v}Y_n \tag{4}$$

$$Y'_n = -Y_{n+1} + \frac{n}{v}Y_n \tag{5}$$

Subtraction of these equations gives

$$\frac{2n}{v}Y_n = Y_{n-1} + Y_{n+1} \tag{6}$$

Two useful integral formulas are got by integrating the equation from which (1) was derived and the analogous equation associated with (4).

$$\int v^n J_{n-1}(v)\,dv = v^n J_n(v) \tag{7}$$
$$\int v^n Y_{n-1}(v)\,dv = v^n Y_n(v) \tag{8}$$

A similar integration of (2) and (5) gives

$$\int v^{-n} J_{n+1}(v)\,dv = -v^{-n} J_n(v) \tag{9}$$
$$\int v^{-n} Y_{n+1}(v)\,dv = -v^{-n} Y_n(v) \tag{10}$$

5.295. Values of Bessel Functions at Infinity.—In potential problems involving $\rho = \infty$, one must know how $J_n(k\rho)$ and $Y_n(k\rho)$ behave there. To find this limiting value, we use a trick often employed by Sommerfeld. In 5.291 (3), as $v \to \infty$ let us, as a first approximation, drop the terms containing v^{-1} and v^{-2}. This gives the approximate differential equation

$$\frac{d^2R}{dv^2} + R = 0 \tag{1}$$

The solution of this equation is

$$R = R'e^{\pm jv} \tag{2}$$

We now insert this trial solution in Bessel's equation 5.291 (3) and consider R' to vary so slowly with v that d^2R'/dv^2, $v^{-1}\,dR'/dv$, and $v^{-2}R'$ can be neglected compared with dR'/dv and $v^{-1}R'$, and we obtain

$$2\frac{dR'}{dv} + \frac{R'}{v} = 0 \qquad \text{or} \qquad R' = Cv^{-\frac{1}{2}}$$

so that, from (2) our asymptotic solution becomes

$$R = Cv^{-\frac{1}{2}}e^{\pm jv} \tag{3}$$

We now see that the largest term neglected was of the order $v^{-\frac{3}{2}}$. $J_n(v)$ and $Y_n(v)$ must be real linear combinations of the two solutions given by taking the plus or minus sign so that they must be of the form

$$J_n(v) \xrightarrow[v\to\infty]{} Av^{-\frac{1}{2}} \cos (v + \alpha) \tag{4}$$

$$Y_n(v) \xrightarrow[v\to\infty]{} Bv^{-\frac{1}{2}} \cos (v + \beta) \tag{5}$$

To find how A and α depend on n, put (4) into 5.294 (1) and 5.294 (2) which give, as $v \to \infty$, $J'_n = J_{n-1}$ and $J'_n = -J_{n+1}$, respectively. This gives the relation $\alpha_{n\pm1} = \alpha_n \mp \frac{1}{2}\pi$ which is satisfied by $\alpha_n = -\frac{1}{2}n\pi + \gamma$ and shows that A does not depend on n. Because n need not be an integer, we may write $n = \frac{1}{2}$ in (4) and compare it with 5.31 (2) which shows that $\gamma = -\frac{1}{4}\pi$ and gives

$$J_n(v) \xrightarrow[v\to\infty]{} \left(\frac{2}{\pi v}\right)^{\frac{1}{2}} \cos\left(v - \frac{1}{2}n\pi - \frac{1}{4}\pi\right) \tag{6}$$

where terms in $v^{-\frac{1}{2}m}$ have been neglected if $m > 3$ and n is real.

To get $Y_n(v)$ substitute (6), with ν and $-\nu$ for n, in 5.293 (4). The result gives 0/0 when ν is integral, but replacement of numerator and denominator by their derivatives with respect to ν gives, when $\nu = n$,

$$Y_n(v) \xrightarrow[v\to\infty]{} \left(\frac{2}{\pi v}\right)^{\frac{1}{2}} \sin\left(v - \frac{1}{2}n\pi - \frac{1}{4}\pi\right) \tag{7}$$

Thus both Bessel functions vanish at infinity. From (6), (7), and 5.293 (10) we find that for the Hankel functions

$$H_n^{(1)}(v) \xrightarrow[v\to\infty]{} \left(\frac{2}{\pi v}\right)^{\frac{1}{2}} e^{j(v-\frac{1}{2}n\pi-\frac{1}{4}\pi)}, \qquad H_n^{(2)}(v) \xrightarrow[v\to\infty]{} \left(\frac{2}{\pi v}\right)^{\frac{1}{2}} e^{-j(v-\frac{1}{2}n\pi-\frac{1}{4}\pi)} \tag{8}$$

5.296. Integrals of Bessel Functions.—In 5.261, we made an expansion in spherical harmonics satisfy the condition $V = 0$ on the cone $\theta = \alpha$ by choosing only such orders n of the harmonics $\Theta_n^m(\cos\theta)$ as made $\Theta_n^m(\cos\alpha) = 0$. To determine the coefficients in this expansion, it was first necessary, in 5.26, to evaluate the integral of the product of two such harmonics over the range of θ from 0 to α. In the same way, if we are to get an expansion in Bessel functions that meets the condition $V = 0$ or $E = 0$ on the cylinder $\rho = a$, we must evaluate the integral of the product of $R_n(k_p\rho)$ and $R_n(k_q\rho)$ over this range, where k_p and k_q are chosen to meet the boundary conditions.

Let $u = R_n(k_p\rho)$ and $v = R_n(k_q\rho)$ be two solutions of Bessel's equation. Then from 5.291

$$\frac{1}{\rho}\frac{d}{d\rho}\left(\rho\frac{du}{d\rho}\right) + \left(k_p^2 - \frac{n^2}{\rho^2}\right)u = 0$$
$$\frac{1}{\rho}\frac{d}{d\rho}\left(\rho\frac{dv}{d\rho}\right) + \left(k_q^2 - \frac{n^2}{\rho^2}\right)v = 0$$

Multiply the first by ρv and the second by ρu subtract, and integrate, and we have

$$(k_p^2 - k_q^2)\int_0^a \rho uv\, d\rho = -\int_0^a\left[v\frac{d}{d\rho}\left(\rho\frac{du}{d\rho}\right) - u\frac{d}{d\rho}\left(\rho\frac{dv}{d\rho}\right)\right] d\rho$$

Integrate each term on the right side by parts, and the integrals cancel, leaving

$$(k_p^2 - k_q^2)\int_0^a \rho uv\, d\rho = -\left(\rho v\frac{du}{d\rho} - \rho u\frac{dv}{d\rho}\right)_0^a$$
$$= -a[k_pR_n(k_qa)R_n'(k_pa) - k_qR_n(k_pa)R_n'(k_qa)]$$

This is zero if

$$R_n(k_pa) = R_n(k_qa) = 0 \tag{1}$$

or if

$$R_n'(k_pa) = R_n'(k_qa) = 0 \tag{2}$$

or if

$$k_paR_n'(k_pa) + BR_n(k_pa) = k_qaR_n'(k_qa) + BR_n(k_qa) = 0 \tag{3}$$

Thus if $k_p \neq k_q$, we have the result

$$\int_0^a \rho R_n(k_p\rho)R_n(k_q\rho)\, d\rho = 0 \tag{4}$$

If $R_n(k_pa) = R_n(k_qa) = R_n(k_pb) = R_n(k_qb) = 0$ we have, since

$$\int_a^b f(x)\, dx = \int_0^b f(x)\, dx - \int_0^a f(x)\, dx$$

the result

$$\int_a^b \rho R_n(k_p\rho)R_n(k_q\rho)\, d\rho = 0 \tag{5}$$

To evaluate the integral when $k_p = k_q$, multiply Bessel's equation [5.291 (3)] through by $v^2(dR_n/dv)\, dv$ giving

$$v^2\frac{dR_n}{dv}d\left(\frac{dR_n}{dv}\right) + v\left(\frac{dR_n}{dv}\right)^2 dv + v^2R_n\frac{dR_n}{dv}\, dv - n^2R_n\frac{dR_n}{dv}\, dv = 0$$

Integrate from 0 to a using integration by parts on the first and third terms. We thus find the following expression to be zero;

$$\left|\frac{v^2}{2}\left(\frac{dR_n}{dv}\right)^2\right|_0^a - \int_0^a v\left(\frac{dR_n}{dv}\right)^2 dv + \int_0^a v\left(\frac{dR_n}{dv}\right)^2 dv + \left|\frac{v^2}{2}R_n^2\right|_0^a - \int_0^a vR_n^2\,dv - \left|\frac{n^2R_n^2}{2}\right|_0^a$$

Canceling the second and third terms and solving for the fifth give

$$\int_0^a v[R_n(v)]^2\,dv = \frac{1}{2}\left|v^2\left[\frac{dR_n(v)}{dv}\right]^2 + (v^2 - n^2)[R_n(v)]^2\right|_0^a \tag{6}$$

Substituting for the derivative from 5.294 (2) gives

$$\begin{aligned}\int_0^a v[R_n(v)]^2\,dv &= \tfrac{1}{2}a^2\{[R_n(a)]^2 + [R_{n+1}(a)]^2\} - naR_n(a)R_{n+1}(a)\\ &= \tfrac{1}{2}a^2\{[R_n(a)]^2 + [R_{n-1}(a)]^2\} - naR_n(a)R_{n-1}(a)\end{aligned} \tag{7}$$

In dealing with the vector potential we shall have occasion to make use of the orthogonal properties of the vector function defined by

$$\mathbf{R}_n(k_p\rho) = {}^{\boldsymbol{\rho}}_{\boldsymbol{\phi}}R_n'(k_p\rho) \pm {}^{\boldsymbol{\phi}}_{\boldsymbol{\rho}}\frac{n}{k_p\rho}R_n(k_p\rho) \tag{8}$$

The integral of the scalar product of two such functions from 0 to a is

$$\int_0^a \mathbf{R}_n(k_p\rho)\cdot\mathbf{R}_n(k_q\rho)\rho\,d\rho$$
$$= \int_0^a\left[R_n'(k_p\rho)R_n'(k_q\rho) + \frac{n^2}{k_pk_q\rho^2}R_n(k_p\rho)R_n(k_q\rho)\right]\rho\,d\rho \tag{9}$$

With the aid of (1), (2), (3), (4), (5), (6) of Art. 5.294 we may write this as the sum of two integrals of the form of (4). Thus

$$\tfrac{1}{2}\int_0^a R_{n+1}(k_p\rho)R_{n+1}(k_q\rho)\rho\,d\rho + \tfrac{1}{2}\int_0^a R_{n-1}(k_p\rho)R_{n-1}(k_q\rho)\rho\,d\rho \tag{10}$$

Evaluate each integral by the formula for $\int\rho uw\,d\rho$ already given and add the results, then eliminate the derivatives by 5.294 (1) and 5.294 (2), cancel terms not involving the nth order, and combine the resultant $n + 1$ and $n - 1$ orders by the same formulas. These operations give, if $v = k_pa$ and $v' = k_qa$,

$$\int_0^a \mathbf{R}_n(k_p\rho)\cdot\mathbf{R}_n(k_q\rho)\rho\,d\rho = (k_p^2 - k_q^2)^{-1}[vR_n(v')R_n'(v) - v'R_n(v)R_n'(v')] \tag{11}$$

Thus, if $k_p \neq k_q$, the integral vanishes under conditions (1), (2), or (3). But, if $k_p = k_q$, evaluation of each integral in (10) by (7) gives for their sum

$$\tfrac{1}{2}[a^2 - (n/k_p)^2][R_n(k_pa)]^2 + \tfrac{1}{2}[aR_n'(k_pa)]^2 + (a/k_p)R_n'(k_pa)R_n(k_pa) \tag{12}$$

Thus a surface vector function of ρ and ϕ, one component of which vanishes at $\rho = a$, may be written as a sum of terms of the form

$$\mathbf{R}_n(k_p\rho)\sin(n\phi + \delta_n)$$

5.297. Expansion in Series of Bessel Functions.—Consider a function $f(v)$ which satisfies the conditions for expansion into a Fourier series in the range from $v = 0$ to $v = a$ and which fulfills one of the following boundary conditions:

(a) $f(a) = 0$. This case arises, if $f(a)$ is a potential function, when the boundary is at zero potential.

(b) $f'(a) = 0$. This case occurs when the boundary is a line of force.

(c) $af'(a) + Bf(a) = 0$. This case reduces to (a) if $B = \infty$ and to (b) if $B = 0$. An example of its use is given in 11.08.

The function $f(v)$ may be expanded in the form

$$f(v) = \sum_{r=1}^{\infty} A_r J_n(\mu_r v) \tag{1}$$

where the values of μ_r are chosen so that in case (a) $J_n(\mu_r a) = 0$, in case (b) $J'_n(\mu_r a) = 0$, and in case (c) $\mu_r a J'_n(\mu_r a) + BJ_n(\mu_r a) = 0$. To determine A_r, multiply both sides of (1) by $vJ_n(\mu_s v)$, and integrate from $v = 0$ to $v = a$. By 5.296 (4), all terms on the right vanish except $A_s \int v[J_n(\mu_s v)]^2\,dv$ so that

$$A_s = \frac{\int_0^a vf(v)J_n(\mu_s v)\,dv}{\int_0^a v[J_n(\mu_s v)]^2\,dv} \tag{2}$$

We can evaluate the lower integral by 5.296 (7) giving

$$\int_0^a v[J_n(\mu_s v)]^2\,dv = \mu_s^{-2}\int_0^{\mu_s a} x[J_n(x)]^2\,dx$$
$$= \frac{1}{2}a^2\{[J_n(\mu_s a)]^2 + [J_{n\pm1}(\mu_s a)]^2\} - \frac{na}{\mu_s}J_n(\mu_s a)J_{n\pm1}(\mu_s a) \tag{3}$$

In case (a), substituting (3) in (2) gives

$$A_s = \frac{2}{[aJ_{n\pm1}(\mu_s a)]^2}\int_0^a vf(v)J_n(\mu_s v)\,dv \tag{4}$$

In case (b), substituting (3) in (2) gives

$$A_s = \frac{2}{(a^2 - n^2\mu_s^{-2})[J_n(\mu_s a)]^2}\int_0^a vf(v)J_n(\mu_s v)\,dv \tag{5}$$

In case (c), substituting (3) in (2) gives

$$A_s = \frac{2}{[a^2 + (B^2 - n^2)/\mu_s^2][J_n(\mu_s a)]^2}\int_0^a vf(v)J_n(\mu_s v)\,dv \tag{6}$$

5.298. Green's Function for Cylinder. Inverse Distance.—We shall calculate by the principles of the last few articles the potential when a point charge q is placed at the point $z = 0$, $\rho = b$, $\phi = \phi_0$, inside an earthed conducting cylinder. By a point charge, we mean one whose dimensions, although too small to measure physically, are different from zero, so that the field-intensity function and the potential function are everywhere bounded. From 5.291 (6), a solution that vanishes when $z = \infty$, is symmetrical about the $\phi = \phi_0$ plane, gives $V = 0$ when $\rho = a$, and is valid for positive values of z, is

$$V = \sum_{r=1}^{\infty} \sum_{s=0}^{\infty} A_{rs} e^{-\mu_r z} J_s(\mu_r \rho) \cos s(\phi - \phi_0) \tag{1}$$

where μ_r is chosen so that $J_s(\mu_r a) = 0$. $Y_s(\mu_r \rho)$ is excluded because it is infinite on the axis.

From symmetry, the whole plane $z = 0$ is formed by lines of force except at the point charge itself. To complete the boundary condition in this plane, we shall consider $(\partial V/\partial z)_0$ to be zero except in a small area ΔS at $\phi = \phi_0$, $\rho = b$. Differentiating (1) and setting $z = 0$ give

$$\left(\frac{\partial V}{\partial z}\right)_0 = -\sum_{r=1}^{\infty} \sum_{s=0}^{\infty} \mu_r A_{rs} J_s(\mu_r \rho) \cos s(\phi - \phi_0)$$

We determine A_{rs} in this expansion as in 5.296 and 5.297 by multiplying through by $\rho J_p(\mu_q \rho) \cos p(\phi - \phi_0)$ and integrating from $\rho = 0$ to $\rho = a$ and from $\phi = 0$ to $\phi = 2\pi$. By *Dw* 858.2 or *Pc* 488 and 5.296 (4), all terms on the right disappear unless $p = s$ and $q = r$. In the latter case, we see by *Dw* 858.4 or *Pc* 489 that the ϕ integration introduces a factor π on the right if $s > 0$ and 2π if $s = 0$, so that we have, by 5.297 (4),

$$A_{rs} = \frac{-(2 - \delta_s^0)}{\pi \mu_r [a J_{s+1}(\mu_r a)]^2} \int \int \left(\frac{\partial V}{\partial z}\right)_0 \rho J_s(\mu_r \rho) \cos s(\phi - \phi_0)\, d\rho\, d\phi \tag{2}$$

where $\delta_s^0 = 1$ if $s = 0$ and $\delta_s^0 = 0$ if $s \neq 0$. In the $z = 0$ plane, the area ΔS in which $(\partial V/\partial z)_0 \neq 0$ is taken so small that in it $J_s(\mu_r \rho)$ has the constant value $J_s(\mu_r b)$ and $\cos s(\phi - \phi_0) = 1$. The integral then becomes

$$J_s(\mu_r b) \int \int \left(\frac{\partial V}{\partial z}\right)_0 \rho\, d\phi\, d\rho = J_s(\mu_r b) \int_{\Delta S} \frac{\partial V}{\partial n}\, dS = -\frac{q}{2\epsilon} J_s(\mu_r b)$$

from Gauss's electric flux theorem [1.10 (1)] it being remembered that only half the flux passes upward. Substituting for the integral in (2) gives

$$A_{rs} = \frac{q(2 - \delta_s^0) J_s(\mu_r b)}{2\pi \epsilon \mu_r a^2 [J_{s+1}(\mu_r a)]^2} \tag{3}$$

Thus we obtain, for the potential,

$$V = \frac{q}{2\pi \epsilon a^2} \sum_{r=1}^{\infty} \sum_{s=0}^{\infty} (2 - \delta_s^0) e^{-\mu_r |z|} \frac{J_s(\mu_r b) J_s(\mu_r \rho)}{\mu_r [J_{s+1}(\mu_r a)]^2} \cos s(\phi - \phi_0) \tag{4}$$

This is Green's function (see 3.08) for a circular cylinder. If the coordinates of q are $\rho = b$, $z = z_0$, and $\phi = \phi_0$, we should substitute $|z - z_0|$ for $|z|$ in the above formula. If the charge is on the axis, all but the first term of the s-summation drops out and $J_0(\mu_r b) = 1$.

When $a \to \infty$ in (4), it gives the potential of a charge q in unbounded space. From 5.295 (6) $J_n(\mu_r a)$ oscillates sinusoidally as $\mu_r a \to \infty$ so that its zeros are uniformly spaced at intervals $\pi = \mu_{r+1}a - \mu_r a = a\,\Delta\mu$ and so that, when $J_n(\mu_r a)$ is zero, $J_{n+1}(\mu_r a)$ is $(\frac{1}{2}\pi\mu_r a)^{-\frac{1}{2}}$. Thus (4) becomes

$$V = \frac{q}{2\pi\epsilon a^2}\sum_{s=0}^{\infty}(2 - \delta_s^0)\cos s(\phi - \phi_0)\sum_{r=1}^{\infty} e^{-r\Delta\mu|z-z_0|}J_s(r\,\Delta\mu b)J_s(r\,\Delta\mu\rho)\frac{1}{2}a^2\,\Delta\mu$$

If $k = r\,\Delta\mu$, then as $\Delta\mu \to 0$ and $a \to \infty$ this takes the integral form

$$V = \frac{q}{4\pi\epsilon R} = \frac{q}{4\pi\epsilon}\sum_{s=0}^{\infty}(2 - \delta_s^0)\cos s(\phi - \phi_0)\int_0^\infty e^{-k|z-z_0|}J_s(kb)J_s(k\rho)\,dk \quad (5)$$

where $R^2 = (z - z_0)^2 + \rho^2 + b^2 - 2\rho b\cos(\phi - \phi_0)$. If $b = z_0 = 0$, this gives

$$(\rho^2 + z^2)^{-\frac{1}{2}} = \int_0^\infty e^{-k|z|}J_0(k\rho)\,dk \quad (6)$$

With z, z_0, and ϕ_0 zero, write R for ρ in (6) and compare with (5). Thus

$$J_0[(\rho^2 + b^2 - 2\rho b\cos\phi)^{\frac{1}{2}}] = \sum_{s=0}^{\infty}(2 - \delta_s^0)J_s(\rho)J_s(b)\cos s\phi \quad (7)$$

5.299. Green's Function for Cylindrical Box.—Superimposing potentials of the form of 5.298 (4) according to the principle of images 5.07 and 5.101, we may make the planes $z = 0$ and $z = L$ at zero potential as well as the cylinder $\rho = a$. If the coordinates of the positive charge q are $z = c$, $\rho = b$, and $\phi = \phi_0$, there are positive images at $z = 2nL + c$ and negative images at $z = 2nL - c$ where n has all integral values from $-\infty$ to $+\infty$. If $z < c$, the factor involving z in the resultant potential is

$$\sum_{n=0}^{\infty} e^{-\mu_r(2nL+c-z)} + \sum_{n=1}^{\infty} e^{-\mu_r(2nL-c+z)} - \sum_{n=1}^{\infty} e^{-\mu_r(2nL-c-z)} - \sum_{n=0}^{\infty} e^{-\mu_r(2nL+c+z)}$$

$$= 2\left(e^{-\mu_r c}\sum_{n=0}^{\infty} e^{-2n\mu_r L} - e^{\mu_r c}\sum_{n=1}^{\infty} e^{-2n\mu_r L}\right)\sinh\mu_r z$$

$$= 2\left[(e^{-\mu_r c} - e^{+\mu_r c})\sum_{n=0}^{\infty} e^{-2n\mu_r L} + e^{\mu_r c}\right]\sinh\mu_r z$$

Summing the series, by *Dw* 9.04 or *Pc* 755, multiplying numerator and denominator of the sum by $e^{\mu_r L}$, and putting over a common denominator give

$$\frac{2 \sinh \mu_r(L - c) \sinh \mu_r z}{\sinh \mu_r L}$$

Substituting this value in 5.298 (4) gives, when $z < c$, the potential

$$V = \frac{q}{\pi \epsilon a^2} \sum_{r=1}^{\infty} \sum_{s=0}^{\infty} (2 - \delta_s^0) \frac{\sinh \mu_r(L - c) \sinh \mu_r z}{\sinh \mu_r L} \frac{J_s(\mu_r b) J_s(\mu_r \rho)}{\mu_r [J_{s+1}(\mu_r a)]^2} \cos s(\phi - \phi_0) \quad (1)$$

When $z > c$, substitute $L - z$ for z and $L - c$ for c. If the charge is on the axis of the cylindrical box, drop the summation with respect to s in (1), retaining only the $s = 0$ term.

If, in addition, the planes $\phi = 0$ and $\phi = \phi_1$, where $0 < \phi_0 < \phi_1$, were at zero potential, and if $\phi_1 = \pi/n$, where n is an integer, we could get the Green's function by superimposing, according to the principle of images (see 5.07), $2n$ solutions of the type of (1).

5.30. Bessel Functions of Zero Order.—In the important case where we are dealing with fields symmetrical about the z-axis, the potential is independent of ϕ so that Bessel's equation becomes, from 5.291 (3),

$$\frac{d^2R}{dv^2} + \frac{1}{v}\frac{dR}{dv} + R = 0 \quad (1)$$

and the solution [5.293 (3)] becomes

$$J_0(v) = 1 - \frac{v^2}{2^2} + \frac{v^4}{2^4(2!)^2} - \frac{v^6}{2^6(3!)^2} + \cdots \quad (2)$$

This series is evidently convergent for all values of v. As with $J_n(v)$, $J_0(\infty) = 0$ but $J_0(0) = 1$. Equation 5.293 (5) becomes, when $n = 0$,

$$Y_0(v) = \frac{2}{\pi}\left[J_0(v) \ln \alpha v + \frac{v^2}{2^2} - \frac{v^4(1 + \frac{1}{2})}{2^4(2!)^2} + \frac{v^6(1 + \frac{1}{2} + \frac{1}{3})}{2^6(3!)^2} - \cdots \right] \quad (3)$$

where $\ln \alpha$ is -0.11593.

5.301. Roots and Numerical Values of Bessel Functions of Zero Order.—If we plot the values of $J_0(v)$ and $Y_0(v)$ given by 5.30 (2) and (3), we get the curves shown in Figs. 5.301a and 5.301b. We see that they oscillate up and down across the v-axis. It can be shown that both $J_0(v)$ and $Y_0(v)$ have an infinite number of real positive roots. The same is true of $J_n(v)$ and $Y_n(v)$. As we have seen in finding the Green's function for a cylinder, the existence of these roots is very useful as it makes it possible to choose an infinite number of values of k which will make $J_n(k\rho)$ or $Y_n(k\rho)$ zero for any specified value of ρ. Many excellent tables exist which give numerical values, graphs, roots, etc., of Bessel functions. Care should be taken in observing the notation used as this varies widely with different authors. Asymptotic expansions provide an easy method of evaluating Bessel functions of large argument.

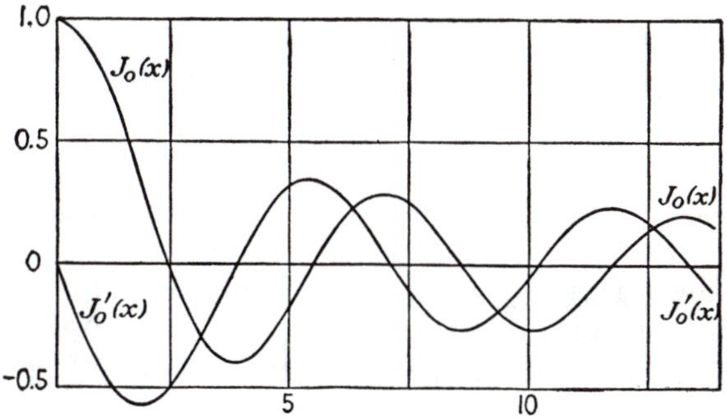

FIG. 5.301*a*.

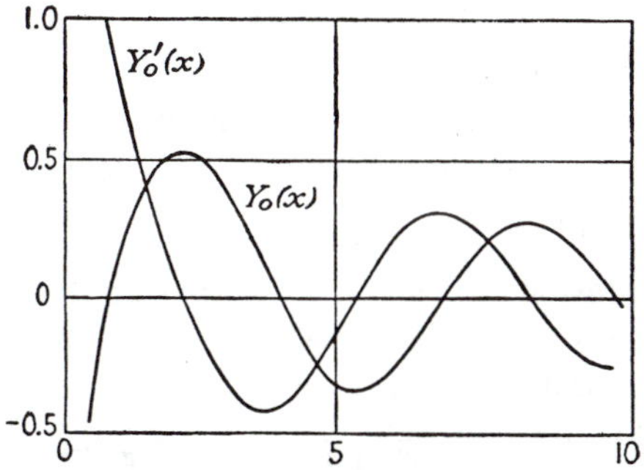

FIG. 5.301*b*.

5.302. Derivatives and Integrals of Bessel Functions of Zero Order. Putting $n = 0$ in 5.294 (2) and (5), we have

$$J_0'(v) = -J_1(v) \qquad \text{and} \qquad Y_0'(v) = -Y_1(v) \tag{1}$$

From 5.294 (7) and (8), we have

$$\int_0^v vJ_0(v)\,dv = vJ_1(v) \qquad \text{and} \qquad \int_0^v vY_0(v)\,dv = vY_1(v) + 2\pi^{-1} \tag{2}$$

There are several definite integrals involving $J_0(v)$ which will be useful. From 5.30 (2), using *Dw* 854.1 or *Pc* 483, we have

$$J_0(v) = \sum_{n=0}^{\infty} \frac{(-1)^n v^{2n}}{2^{2n}(n!)^2} = \sum_{n=0}^{\infty} (-1)^n \frac{v^{2n}}{(2n)!} \frac{(2n)!}{2^{2n}(n!)^2}$$

$$= \frac{1}{\pi} \sum_{n=0}^{\infty} \frac{(-1)^n 2v^{2n}}{(2n)!} \frac{\pi}{2} \frac{1 \cdot 3 \cdots (2n-1)}{2 \cdot 4 \cdots 2n}$$

$$= \frac{1}{\pi} \sum_{n=0}^{\infty} (-1)^n \frac{v^{2n}}{(2n)!} \int_0^{\pi} \cos^{2n} t\,dt$$

Interchanging integration and summation symbols and using *Dw* 415.02 or *Pc* 773, we obtain

$$J_0(v) = \frac{1}{\pi}\int_0^\pi \sum_{n=0}^{\infty}(-1)^n \frac{(v\cos t)^{2n}}{(2n)!}\,dt = \frac{1}{\pi}\int_0^\pi \cos(v\cos t)\,dt$$

so that

$$J_0(k\rho) = \frac{1}{\pi}\int_0^\pi \cos(k\rho\cos t)\,dt = \frac{1}{\pi}\int_0^\pi \cos(k\rho\sin t)\,dt \tag{3}$$

We can show easily, by using the trigonometric formulas for sums and differences of angles and integrating by parts, that an expression that satisfies the recurrence formulas 5.294 (1), (2), and (3), is

$$J_n(v) = \frac{1}{\pi}\int_0^\pi \cos(nt - v\sin t)\,dt \tag{4}$$

This evidently reduces to (3) when $n = 0$.

Direct substitution in 5.291 (3) and integration by parts show that

$$J_n(v) = \frac{(\frac{1}{2}v)^n}{\Gamma(\frac{1}{2})\Gamma(n+\frac{1}{2})}\int_0^\pi \cos(v\cos\theta)\sin^{2n}\theta\,d\theta \tag{5}$$

satisfies Bessel's equation if $n > -\frac{1}{2}$. Comparing the value of $(\frac{1}{2}v)^{-n}J_n(v)$ given by (5) as $v \to 0$ with that given by 5.293 (3) fixes the constant.

It can be shown that 5.298 (6) still holds if we write jz for z, thus

$$\int_0^\infty e^{-jkz}J_0(k\rho)\,dk = (\rho^2 - z^2)^{-\frac{1}{2}}$$

whence, equating real and imaginary parts, we have, when $\rho^2 > z^2$,

$$\int_0^\infty \cos kzJ_0(k\rho)\,dk = (\rho^2 - z^2)^{-\frac{1}{2}} \tag{6}$$

$$\int_0^\infty \sin kzJ_0(k\rho)\,dk = 0 \tag{7}$$

When $\rho^2 < z^2$, we have

$$\int_0^\infty \cos kzJ_0(k\rho)\,dk = 0 \tag{8}$$

$$\int_0^\infty \sin kzJ_0(k\rho)\,dk = (z^2 - \rho^2)^{-\frac{1}{2}} \tag{9}$$

5.303. Point Charge and Dielectric Plate.—As a first application of the last two articles, let us find the potential on one side of an infinite plate of material of thickness c and relative capacitivity K, due to a point charge q on the opposite side. We shall take the point charge at the origin and the z-axis perpendicular to the plate. The equation of the surfaces of the plate will be $z = a$ and $z = b$ where $b - a = c$. From 5.298 (6), the potential due to the point charge alone is

$$V = \frac{q}{4\pi\epsilon_v r} = \frac{q}{4\pi\epsilon_v}\int_0^\infty J_0(k\rho)e^{-k|z|}\,dk \tag{1}$$

Since this is a solution of Laplace's equation, which involves z and ρ only, it is evident that we shall still have a solution if we insert any function of k under the integral sign. Let V_1 be the potential when $-\infty < z < a$, and write

$$V_1 = (4\pi\epsilon_v)^{-1}q\left[\int_0^\infty J_0(k\rho)e^{-k|z|}\,dk + \int_0^\infty \Phi(k)J_0(k\rho)e^{+kz}\,dk\right] \tag{2}$$

The second term represents the potential in the region below the plate due to the polarization of the plate and is finite everywhere in this region. The potential in the plate may be written

$$V_2 = (4\pi\epsilon_v)^{-1}q\left[\int_0^\infty \Psi(k)J_0(k\rho)e^{-kz}\,dk + \int_0^\infty \Theta(k)J_0(k\rho)e^{+kz}\,dk\right] \tag{3}$$

This evidently is finite if $a < z < b$. The potential above the plate, where $b < z < +\infty$, must vanish at $z = \infty$ and so may be written

$$V_3 = (4\pi\epsilon_v)^{-1}q\int_0^\infty \Omega(k)J_0(k\rho)e^{-kz}\,dk \tag{4}$$

We must now determine $\Phi(k)$, $\Psi(k)$, $\Theta(k)$, and $\Omega(k)$, so the boundary conditions are satisfied for all values of ρ between 0 and ∞. This requires that the integrands alone satisfy these conditions. To prove this, we make use of the Fourier Bessel integral which states that, with suitable restrictions on n and the form of $f(x)$,

$$f(x) = \int_0^\infty tJ_n(tx)\left[\int_0^\infty uf(u)J_n(tu)\,du\right]dt$$

Thus, if we have

$$\int_0^\infty f_1(k)J_0(k\rho)\,dk = \int_0^\infty f_2(k)J_0(k\rho)\,dk$$

we may multiply both sides by $\rho J_0(m\rho)\,d\rho$ and integrate from 0 to ∞ and obtain, after multiplying through by m,

$$f_1(m) = f_2(m)$$

Applying this to the present case, at $z = a$ we have, after canceling out $J_0(k\rho)$, $V_1 = V_2$, giving

$$e^{-ka} + \Phi(k)e^{ka} - \Psi(k)e^{-ka} - \Theta(k)e^{ka} = 0 \tag{5}$$

and $\partial V_1/\partial z = K\partial V_2/\partial z$ which gives, canceling $J_0(k\rho)$,

$$-e^{-ka} + \Phi(k)e^{ka} + K\Psi(k)e^{-ka} - K\Theta(k)e^{ka} = 0 \tag{6}$$

At $z = b$, we have in the same way $V_2 = V_3$, giving

$$\Psi(k)e^{-kb} + \Theta(k)e^{kb} - \Omega(k)e^{-kb} = 0 \tag{7}$$

and $K\partial V_2/\partial z = \partial V_3/\partial z$ which gives

$$-K\Psi(k)e^{-kb} + K\Theta(k)e^{kb} + \Omega(k)e^{-kb} = 0 \tag{8}$$

Solving (5), (6), (7), and (8) for $\Omega(k)$ gives

$$\Omega(k) = \frac{4K}{(K+1)^2 - (K-1)^2 e^{2k(a-b)}} \tag{9}$$

To simplify this we write $b - a = c$ and $\beta = (K-1)/(K+1)$ so that $1 - \beta^2 = 4K(K+1)^{-2}$ which gives

$$\Omega(k) = \frac{1-\beta^2}{1-\beta^2 e^{-2kc}}$$

Substituting in (4), we have, for V_3, the result

$$V_3 = \frac{q(1-\beta^2)}{4\pi\epsilon_v}\int_0^\infty \frac{J_0(k\rho)e^{-kz}\,dk}{1-\beta^2 e^{-2kc}} \tag{10}$$

If we wish to expand this as a series, we may expand the denominator by *Dw* 9.04 or *Pc* 755.

$$4\pi\epsilon_v V_3 = q(1-\beta^2)\left[\int_0^\infty J_0(k\rho)e^{-kz}\,dk + \beta^2\int_0^\infty J_0(k\rho)e^{-k(z+2c)}\,dk + \cdots\right]$$

Substituting from 5.298 (6) for each integral gives

$$V_3 = \frac{q(1-\beta^2)}{4\pi\epsilon_v}\left\{\frac{1}{(z^2+\rho^2)^{\frac{1}{2}}} + \frac{\beta^2}{[(z+2c)^2+\rho^2]^{\frac{1}{2}}} + \frac{\beta^4}{[(z+4c)^2+\rho^2]^{\frac{1}{2}}} + \cdots\right\}$$

which may be written

$$V_3 = \frac{q(1-\beta^2)}{4\pi\epsilon_v}\sum_{n=0}^{\infty}\frac{\beta^{2n}}{[(z+2nc)^2+\rho^2]^{\frac{1}{2}}} \tag{11}$$

Another, but more tedious, method of obtaining this result would have been to use images. The potential in the other regions can be obtained in the same way by solving (5), (6), (7), and (8) for $\Phi(k)$, $\Psi(k)$, and $\Theta(k)$ and substituting in (2) and (3). Clearly this method could be applied to any number of plates.

5.304. Potential inside Hollow Cylindrical Ring.—As another example, let us find the potential at any point in the region bounded by the two cylinders $\rho = a$ and $\rho = b$, both of which are at zero potential, and the two planes $z = 0$ at potential zero and $z = c$ at potential $V = f(\rho)$. Since both $\rho = 0$ and $\rho = \infty$ are excluded, we may have both $J_0(k\rho)$ and $Y_0(k\rho)$. From 5.291 (5) and 5.293 (6), it is evident that a solution of Laplace's equation which satisfies the boundary conditions at $z = 0$ and $\rho = b$ is given by

$$V_k = A_k \sinh(\mu_k z)\left[J_0(\mu_k\rho) - \frac{J_0(\mu_k b)}{Y_0(\mu_k b)}Y_0(\mu_k\rho)\right] \tag{1}$$

We can satisfy the conditions $V_k = 0$ at $\rho = a$ by choosing values of μ_k so that

$$J_0(\mu_k a) - \frac{J_0(\mu_k b)}{Y_0(\mu_k b)} Y_0(\mu_k a) = 0$$

Hence, all the boundary conditions except that at $z = c$ will be satisfied by a sum of such solutions

$$V = \sum_{k=1}^{\infty} V_k \tag{2}$$

The final boundary condition $V = f(\rho)$ at $z = c$ will be satisfied if we choose A_k so that

$$f(\rho) = \sum_{k=1}^{\infty} A_k \sinh \mu_k c \left[J_0(\mu_k \rho) - \frac{J_0(\mu_k b)}{Y_0(\mu_k b)} Y_0(\mu_k \rho) \right]$$

Represent the term in brackets by $R_0(\mu_k\rho)$, since it satisfies Bessel's equation, multiply through by $\rho R_0(\mu_s\rho)$ where $R_0(\mu_s a) = R_0(\mu_s b) = 0$, and integrate from a to b. Since $R_0(\mu_k a) = R_0(\mu_k b) = 0$, we see from 5.296 (5) that all terms on the right drop out except the one for which $k = s$. For this term, we have, by 5.296 (6),

$$\int_a^b \rho f(\rho) R_0(\mu_k\rho)\, d\rho = \left[b^2 \left(\frac{dR_0}{d(\mu_k\rho)} \right)^2_{\rho=b} - a^2 \left(\frac{dR_0}{d(\mu_k\rho)} \right)^2_{\rho=a} \right] \frac{A_k \sinh \mu_k c}{2}$$

Differentiating $R_0(\mu_k\rho)$ by 5.302 (1), we have

$$R_0'(\mu_k\rho) = -\left\{ J_1(\mu_k\rho) - \frac{J_0(\mu_k b)}{Y_0(\mu_k b)} Y_1(\mu_k\rho) \right\}$$

Solving for A_k gives

$$A_k = \frac{2\int_a^b \rho f(\rho) R_0(\mu_k\rho)\, d\rho}{\{b^2[R_0'(\mu_k b)]^2 - a^2[R_0'(\mu_k a)]^2\} \sinh \mu_k c} \tag{3}$$

Substituting (3) in (1) and then (1) in (2) gives the required solution.

If we wish the potential inside the earthed cylinder $\rho = a$ with $V = 0$ when $z = 0$ and $V = f(\rho)$ when $z = c$, we drop the Y_0 term in (1), and (3) becomes

$$A_k = \frac{2\int_0^a \rho f(\rho) J_0(\mu_k\rho)\, d\rho}{a^2[J_1(\mu_k a)]^2 \sinh \mu_k c} \tag{4}$$

5.31. Nonintegral Order and Spherical Bessel Functions.—When n is not an integer, a complete solution of Bessel's equation is

$$R_n(v) = AJ_n(v) + BJ_{-n}(v) \tag{1}$$

We no longer need $Y_n(v)$ because it must be merely a linear combination of $J_n(v)$ and $J_{-n}(v)$. The formulas of 5.294 are equally valid for these values. The function $J_{\pm(n+\frac{1}{2})}(v)$ where n is an integer are especially simple, for if we substitute $n = \frac{1}{2}$ and $n = -\frac{1}{2}$ in 5.293 (3) and compare with the sine and cosine expansion *Pc* 772 and 773 or *Dw* 415.01 and 415.02, we have

$$J_{\frac{1}{2}}(v) = \left(\frac{2}{\pi v}\right)^{\frac{1}{2}} \sin v \qquad \text{and} \qquad J_{-\frac{1}{2}}(v) = \left(\frac{2}{\pi v}\right)^{\frac{1}{2}} \cos v \tag{2}$$

If we then put $n = \frac{1}{2}$ and $n = -\frac{1}{2}$ in 5.294 (3), we get

$$J_{\frac{3}{2}}(v) = \left(\frac{2}{\pi v}\right)^{\frac{1}{2}}\left(\frac{1}{v}\sin v - \cos v\right) \tag{3}$$

$$J_{-\frac{3}{2}}(v) = \left(\frac{2}{\pi v}\right)^{\frac{1}{2}}\left(-\sin v - \frac{1}{v}\cos v\right) \tag{4}$$

Putting $n = \frac{3}{2}$ and $n = -\frac{3}{2}$ gives

$$J_{\frac{5}{2}}(v) = \left(\frac{2}{\pi v}\right)^{\frac{1}{2}}\left[\left(\frac{3}{v^2} - 1\right)\sin v - \frac{3}{v}\cos v\right] \tag{5}$$

$$J_{-\frac{5}{2}}(v) = \left(\frac{2}{\pi v}\right)^{\frac{1}{2}}\left[\frac{3}{v}\sin v + \left(\frac{3}{v^2} - 1\right)\cos v\right] \tag{6}$$

In this fashion we could eventually obtain the formula

$$J_{\pm(n+\frac{1}{2})}(v) = \left(\frac{2}{\pi v}\right)^{\frac{1}{2}}\left\{\sin\left[v \mp \tfrac{1}{4}(2n+1\mp 1)\pi\right]\sum_{s=0}^{\leqslant\frac{1}{2}n}\frac{(-1)^s(n+2s)!}{(2s)!(n-2s)!(2v)^{2s}}\right.$$
$$\left.+\cos\left[v \mp \tfrac{1}{4}(2n+1\mp 1)\pi\right]\sum_{s=0}^{\leqslant\frac{1}{2}(n-1)}\frac{(-1)^s(n+2s+1)!}{(2s+1)!(n-2s-1)!(2v)^{2s+1}}\right\} \tag{7}$$

where the upper limit of each summation is the nearest integer to the value given and the upper or the lower sign is to be used throughout. We observe that, because $\cos(n+\frac{1}{2})\pi$ is zero, 5.293 (4) becomes

$$Y_{n+\frac{1}{2}}(v) = (-1)^{n+1}J_{-(n+\frac{1}{2})}(v) \xrightarrow[v\to 0]{} \infty \tag{8}$$

When $n + \frac{1}{2}$ is written for n in 5.293 (3) and $v \to 0$, there results

$$J_{n+\frac{1}{2}}(v) \xrightarrow[v\to 0]{} \frac{v^{n+\frac{1}{2}}}{2^{n+\frac{1}{2}}\Gamma(n+\frac{3}{2})} = \left(\frac{2v}{\pi}\right)^{\frac{1}{2}}\frac{v^n}{(2n+1)!!} \tag{9}$$

The spherical Bessel functions $j_n(v)$ and $n_n(v)$ are defined as

$$j_n(v) = \left(\frac{\pi}{2v}\right)^{\frac{1}{2}}J_{n+\frac{1}{2}}(v) = \frac{1}{v}\hat{J}_n(v), \qquad n_n(v) = \left(\frac{\pi}{2v}\right)^{\frac{1}{2}}Y_{n+\frac{1}{2}}(v) = \frac{1}{v}\hat{N}_n(v) \tag{10}$$

where $\hat{J}_n(v)$ and $\hat{N}_n(v)$ are the functions used by Schelkunoff. They arise in connection with spherical electromagnetic waves and have recently been tabulated. From (10), 5.293 (9), 5.294 (2), and 5.294 (3) it follows that

$$n'_n(v)j_n(v) - n_n(v)j'_n(v) = v^{-2} \tag{11}$$

$$(2n+1)v^{-1}j_n = j_{n-1} + j_{n+1}, \qquad j'_n = \pm j_{n\mp 1} \mp (n + \tfrac{1}{2} \pm \tfrac{1}{2})v^{-1}j_n \tag{12}$$

Similar formulas hold for $n_n(v)$. Differentiation of (12) and elimination of $j_{n-1}(v)$ give the differential equation for $j_n(v)$ or $n_n(v)$.

$$\frac{d}{dv}\left[v^2\frac{d}{dv}j_n(v)\right] + [v^2 - n(n+1)]j_n(v) = 0 \tag{13}$$

5.32. Modified Bessel Functions.—We can obtain a solution of 5.292 (1), the modified Bessel equation

$$\frac{d^2R^0}{dv^2} + \frac{1}{v}\frac{dR^0}{dv} - \left(1 + \frac{n^2}{v^2}\right)R^0 = 0 \tag{1}$$

by substituting jv for v in 5.293 (3). We see that the bracket is real but the coefficient j^n occurs. Since we wish a real solution, we define it to be

$$I_\nu(v) = j^{-\nu}J_\nu(jv) = \sum_{r=0}^{\infty}\frac{(\frac{1}{2}v)^{\nu+2r}}{r!\Gamma(\nu+r+1)} \tag{2}$$

When ν is not an integer, a second solution is

$$I_{-\nu}(v) = \sum_{r=0}^{\infty}\frac{(\frac{1}{2}v)^{-\nu+2r}}{r!\Gamma(-\nu+r+1)} \tag{3}$$

When ν is an integer, we may write $(n+r)!$ for $\Gamma(\nu+r+1)$ and find that $I_n(v)$ equals $I_{-n}(v)$ so that we need a solution which we may define by

$$K_n(v) = \tfrac{1}{2}\pi j^{n+1}[J_n(jv) + jY_n(jv)] \tag{4}$$

This with the aid of 5.293 (3), 5.293 (5), and (2) becomes

$$K_n(v) = (-1)^{n+1}I_n(v)\ln \alpha v + \sum_{r=0}^{n-1}\frac{(-1)^r(\frac{1}{2}v)^{-n+2r}(n-r-1)!}{2r!}$$
$$+ (-1)^n\sum_{r=0}^{\infty}\frac{(\frac{1}{2}v)^{n+2r}}{2r!(n+r)!}\left(\sum_{m=1}^{r}\frac{1}{m} + \sum_{m=1}^{n+r}\frac{1}{m}\right) \tag{5}$$

where $\ln \alpha = -0.11593$. A complete solution of (1) when n is integral is

$$R_n^0(v) = AI_n(v) + BK_n(v) \tag{6}$$

From (2), (4), and 5.293 (9), we have

$$I'_n(v)K_n(v) - K'_n(v)I_n(v) = \tfrac{1}{2}j\pi[Y'_n(jv)J_n(jv) - J'_n(jv)Y_n(jv)] = v^{-1} \tag{7}$$

We observe from (4) and 5.293 (10) and from (2) that

$$(2/\pi)K_n(v) = j^{n+1}H_n^{(1)}(jv) = (-j)^{n+1}H_n^{(2)}(-jv) \qquad I_n(v) = j^{-n}J_n(jv) \quad (8)$$

These are the functions tabulated in Jahnke and Emde.

5.321. Recurrence Formulas for Modified Bessel Functions.—Equation 5.294 (1) may be written

$$\frac{dJ_n(jv)}{j\,dv} = J_{n-1}(jv) - \frac{n}{jv}J_n(jv)$$

Substituting $j^nI_n(v)$ for $J_n(jv)$ and dividing out j^{n-1} give

$$I'_n = I_{n-1} - \frac{n}{v}I_n \quad (1)$$

and similarly from 5.294 (2) we get

$$I'_n = I_{n+1} + \frac{n}{v}I_n \quad (2)$$

Subtracting (1) from (2) gives

$$\frac{2n}{v}I_n = I_{n-1} - I_{n+1} \quad (3)$$

Substitution of the recurrence formulas of 5.294 in 5.32 (4) gives

$$-K'_n = K_{n-1} + \frac{n}{v}K_n \quad (4)$$

$$-K'_n = K_{n+1} - \frac{n}{v}K_n \quad (5)$$

$$-\frac{2n}{v}K_n = K_{n-1} - K_{n+1} \quad (6)$$

We use the notation of Watson and of Gray, Mathews, and MacRobert. Some authors omit the $(-1)^n$ in 5.32 (5) in which case the recurrence formulas for I_n and K_n are identical.

Two integral formulas that are sometimes useful can be obtained by integrating (1) and (4), as in 5.294. These integrals are

$$\int v^nI_{n-1}(v)\,dv = v^nI_n(v) \quad (7)$$

$$\int v^nK_{n-1}(v)\,dv = -v^nK_n(v) \quad (8)$$

A similar integration of (2) and (5) gives

$$\int v^{-n}I_{n+1}(v)\,dv = v^{-n}I_n(v) \quad (9)$$

$$\int v^{-n}K_{n+1}(v)\,dv = -v^{-n}K_n(v) \quad (10)$$

5.322. Values of Modified Bessel Functions at Infinity.—We can obtain the values of $I_n(v)$ and $K_n(v)$ as $v \to \infty$ from those for $J_n(v)$ in 5.295 by substituting jv for v. The results are good to the same order, *i.e.*, neglecting $v^{-\frac{3}{2}}$ compared with $v^{-\frac{1}{2}}$. Putting jv for v in 5.295 (6), writing $e^{-\frac{1}{2}j\pi}$ for $j^{-\frac{1}{2}}$, using the exponential form *Dw* 408.02 or *Pc* 613, and neglecting the e^{-v} term, we have

$$I_n(v) = j^{-n}J_n(jv) \xrightarrow[v\to\infty]{} \left(\frac{1}{2\pi v}\right)^{\frac{1}{2}} e^v j^{-n} e^{\frac{1}{2}jn\pi}$$

Evaluating the last term, n being an integer, by *Dw* 409.04 and 409.05 or *Pc* 609 we have

$$I_n(v) \xrightarrow[v\to\infty]{} \left(\frac{1}{2\pi v}\right)^{\frac{1}{2}} e^v \tag{1}$$

From 5.32 (8) and 5.295 (8), we have

$$K_n(v) \xrightarrow[v\to\infty]{} \left(\frac{\pi}{2v}\right)^{\frac{1}{2}} e^{-v} \tag{2}$$

Although we have derived these equations assuming n to be an integer, we see, by substitution in 5.32 (1), that they hold for any value of n.

5.323. Integral of a Product of Complex Modified Bessel Functions. In Chap. X, when we calculate the power dissipated by eddy currents, we frequently find it necessary to evaluate the integral of the product of a modified Bessel function $R_n^0[(jp)^{\frac{1}{2}}x]$ by its conjugate complex $R_n^0[(-jp)^{\frac{1}{2}}x]$, where $(j)^{\frac{1}{2}} = 2^{-\frac{1}{2}}(1+j)$ and $(-j)^{\frac{1}{2}} = 2^{-\frac{1}{2}}(1-j)$. In cylindrical problems, n is usually an integer, and in spherical problems it is usually half an odd integer. We may evaluate this integral, for any value of n, by using the equations of 5.296. Let us take

$$k_p = -j(jp)^{\frac{1}{2}} = (-jp)^{\frac{1}{2}} \qquad \text{and} \qquad k_q = j(-jp)^{\frac{1}{2}} = (jp)^{\frac{1}{2}}$$
$$u = R_n[-j(jp)^{\frac{1}{2}}x] = (-j)^n R_n^0[(jp)^{\frac{1}{2}}x]$$
$$v = R_n[j(-jp)^{\frac{1}{2}}x] = j^n R_n^0[(-jp)^{\frac{1}{2}}x]$$
$$R_n'[-j(jp)^{\frac{1}{2}}x] = (-j)^{n-1} R_n^{0\prime}[(jp)^{\frac{1}{2}}x]$$
$$R_n'[j(-jp)^{\frac{1}{2}}x] = j^{n-1} R_n^{0\prime}[(-jp)^{\frac{1}{2}}x]$$

Substituting these values in the equation preceding 5.296 (1), we obtain, writing x for ρ,

$$\begin{aligned}\int_0^a xR_n^0[(jp)^{\frac{1}{2}}x]R_n^0[(-jp)^{\frac{1}{2}}x]\,dx &= \int_0^a xR_n^0(k_qx)R_n^0(k_px)\,dx \\ &= \tfrac{1}{2}aj^{-1}p^{-\frac{1}{2}}[j^{\frac{1}{2}}R_n^0(k_pa)R_n^{0\prime}(k_qa) - (-j)^{\frac{1}{2}}R_n^0(k_qa)R_n^{0\prime}(k_pa)] \\ &= (2p)^{-1}[k_paR_n^0(k_pa)R_{n-1}^0(k_qa) + k_qaR_n^0(k_qa)R_{n-1}^0(k_pa)] \qquad (1) \\ &= (2p)^{-1}[k_paR_{n-2}^0(k_pa)R_{n-1}^0(k_qa) + k_qaR_{n-2}^0(k_qa)R_{n-1}^0(k_pa) \\ &\qquad - 4(n-1)R_{n-1}^0(k_pa)R_{n-1}^0(k_qa)]\end{aligned}$$

5.324. Green's Function for a Hollow Cylindrical Ring.—As an example of the use of modified Bessel functions, we shall find the potential due to a small charge q at $z = c$, $\rho = b$, and $\phi = \phi_0$ inside the hollow conducting ring whose walls are given by $z = 0$, $z = L$, $\rho = d$, and $\rho = a$ where $a > d$. The special case, when $d = 0$, gives us the cylindrical box which we have already solved in terms of Bessel functions in 5.299.

Since neither $I_m(v)$ nor $K_m(v)$ has any real roots, it is evident that we shall need a combination of the two to get a function that is zero for a given value of ρ. Clearly, such a function is

$$R_m^0(k, s, t) = K_m(ks)I_m(kt) - I_m(ks)K_m(kt) \tag{1}$$

Since, in general, this function is zero for only one value $t = s$ of t, we must use different functions for the region near the inner face and that near the outer, but they must have the same value at $\rho = b$, where the regions meet. We can now write down, by inspection, two functions that are zero over the conducting surfaces and identical when $\rho = b$. These are

$d \leqslant \rho < b$ or $\rho = b$, $\phi \neq \phi_0$

$$V' = \sum_{n=1}^{\infty}\sum_{m=0}^{\infty} C_{mn} R_m^0\left(\frac{n\pi}{L}, a, b\right) R_m^0\left(\frac{n\pi}{L}, d, \rho\right) \sin\frac{n\pi z}{L} \cos m(\phi - \phi_0) \tag{2}$$

$b < \rho \leqslant a$ or $\rho = b$, $\phi \neq \phi_0$

$$V'' = \sum_{n=1}^{\infty}\sum_{m=0}^{\infty} C_{mn} R_m^0\left(\frac{n\pi}{L}, d, b\right) R_m^0\left(\frac{n\pi}{L}, a, \rho\right) \sin\frac{n\pi z}{L} \cos m(\phi - \phi_0) \tag{3}$$

These solutions have the proper symmetry about $\phi = \phi_0$. To determine C_{mn}, we shall apply Gauss's electric flux theorem to the region surrounding the charge.

Consider the charge q concentrated in a small area of the cylinder $\rho = b$ at $z = c$, $\phi = \phi_0$. This area is taken too small to measure physically but not a mathematical point, so that the potential and intensity functions are everywhere bounded. By 1.10 (1), the integral over the two faces S_2 of this cylinder is

$$-\int_{S_2} \frac{\partial V}{\partial \rho} dS_2 = \int_S \left(\frac{\partial V'}{\partial \rho} - \frac{\partial V''}{\partial \rho}\right) dS = \frac{q}{\epsilon} \tag{4}$$

By (2) and (3), we have

$$\frac{\partial V'}{\partial \rho} - \frac{\partial V''}{\partial \rho} = -\sum_{n=1}^{\infty}\sum_{m=0}^{\infty} C_{mn}\frac{n\pi}{L}\left[R_m^0\left(\frac{n\pi}{L}, d, b\right) R_m^{0\prime}\left(\frac{n\pi}{L}, a, b\right)\right.$$
$$\left. - R_m^0\left(\frac{n\pi}{L}, a, b\right) R_m^{0\prime}\left(\frac{n\pi}{L}, d, b\right)\right] \sin\frac{n\pi z}{L} \cos m(\phi - \phi_0)$$

With the aid of 5.32 (7), this may be written

$$\frac{\partial V'}{\partial \rho} - \frac{\partial V''}{\partial \rho} = -\sum_{n=1}^{\infty}\sum_{m=0}^{\infty} C_{mn}\frac{1}{b} R_m^0\left(\frac{n\pi}{L}, d, a\right) \sin\frac{n\pi z}{L} \cos m(\phi - \phi_0) \tag{5}$$

Now let $\phi - \phi_0 = \phi'$, multiply both sides by $\sin (p\pi z/L) \cos q\phi' \, b \, d\phi' \, dz$, and integrate over the cylinder $\rho = b$. On the right, all terms go out except those for which $p = n$ and $q = m$, by *Pc* 488 or *Dw* 858.1 and 858.2. For these terms, the integrals are evaluated by *Pc* 489 or *Dw* 858.4, except when $q = m = 0$. On the left, we notice that, since $dS = b \, d\phi' \, dz$, we have exactly the integral of (4) if we take the charge so small that $\cos m\phi'$ and $\sin (n\pi z/L)$ have the constant values one and $\sin (n\pi c/L)$, respectively, and may be taken out of the integral. We then obtain, by (4), if $m \neq 0$,

$$\frac{q}{\epsilon} \sin \frac{n\pi c}{L} = -\frac{1}{2} C_{mn} L\pi R_m^0\left(\frac{n\pi}{L}, d, a\right) \tag{6}$$

If $m = 0$, the $\frac{1}{2}$ is omitted on the right since $\int_0^{2\pi} d\phi = 2\pi$. Solving for C_{mn} and substituting in (2) and (3), we have

$$V' = \frac{-q}{\pi\epsilon L} \sum_{n=1}^{\infty} \sum_{m=0}^{\infty} \frac{(2 - \delta_m^0) R_m^0\left(\frac{n\pi}{L}, a, b\right) R_m^0\left(\frac{n\pi}{L}, d, \rho\right)}{R_m^0\left(\frac{n\pi}{L}, d, a\right)} \sin \frac{n\pi c}{L} \sin \frac{n\pi z}{L} \cos m(\phi - \phi_0) \tag{7}$$

$$V'' = \frac{-q}{\pi\epsilon L} \sum_{n=1}^{\infty} \sum_{m=0}^{\infty} \frac{(2 - \delta_m^0) R_m^0\left(\frac{n\pi}{L}, d, b\right) R_m^0\left(\frac{n\pi}{L}, a, \rho\right)}{R_m^0\left(\frac{n\pi}{L}, d, a\right)} \sin \frac{n\pi c}{L} \sin \frac{n\pi z}{L} \cos m(\phi - \phi_0) \tag{8}$$

where $\delta_m^0 = 0$ if $m \neq 0$ and $\delta_m^0 = 1$ if $m = 0$.

For the cylindrical box, solved in terms of Bessel functions in 5.299, we have $d = 0$ and the above potentials become

$$V' = \frac{-q}{\pi\epsilon L} \sum_{n=1}^{\infty} \sum_{m=0}^{\infty} \frac{(2 - \delta_m^0) R_m^0\left(\frac{n\pi}{L}, a, b\right) I_m\left(\frac{n\pi\rho}{L}\right)}{I_m\left(\frac{n\pi a}{L}\right)} \sin \frac{n\pi c}{L} \sin \frac{n\pi z}{L} \cos m(\phi - \phi_0) \tag{9}$$

$$V'' = \frac{-q}{\pi\epsilon L} \sum_{n=1}^{\infty} \sum_{m=0}^{\infty} \frac{(2 - \delta_m^0) I_m\left(\frac{n\pi b}{L}\right) R_m^0\left(\frac{n\pi}{L}, a, \rho\right)}{I_m\left(\frac{n\pi a}{L}\right)} \sin \frac{n\pi c}{L} \sin \frac{n\pi z}{L} \cos m(\phi - \phi_0) \tag{10}$$

Inspection of V' shows that it satisfies all the boundary conditions, being finite, but not zero, on the axis. If the charge is on the axis, we use (10), dropping the summation with respect to m and setting $m = 0$.

5.33. Modified Bessel Functions of Zero Order.—When $n = 0$, 5.32 (2) becomes

$$I_0(v) = 1 + \frac{v^2}{2^2} + \frac{v^4}{2^4(2!)^2} + \frac{v^6}{2^6(3!)^2} + \cdots \tag{1}$$

Thus $I_0(v)$ is real but has no real roots and

$$I_0(0) = 1 \qquad \text{and} \qquad I_0(\infty) = \infty \tag{2}$$

When $n = 0$, 5.32 (5) becomes

$$K_0(v) = -I_0(v) \ln \alpha v + \frac{v^2}{2^2} + \frac{v^4(1 + \frac{1}{2})}{2^4(2!)^2} + \frac{v^6(1 + \frac{1}{2} + \frac{1}{3})}{2^6(3!)^2} + \cdots \tag{3}$$

The derivatives of $I_0(v)$ and $K_0(v)$ are given by 5.321 (2) and (5) to be

$$I_0' = I_1 \qquad \text{and} \qquad K_0' = -K_1 \tag{4}$$

From 5.321 (7) and (8), we have

$$\int_0^v vI_0(v)\, dv = vI_1(v) \tag{5}$$

$$\int_0^v vK_0(v)\, dv = -vK_1(v) + 1 \tag{6}$$

5.331. Definite Integrals for the Modified Bessel Function of the Second Kind. Value at Infinity.—Substitute $R^0 = \int_0^\infty e^{-v \cosh \phi}\, d\phi$ in the left side of 5.32 (1), setting $n = 0$, and it becomes

$$\int_0^\infty \cosh^2 \phi e^{-v \cosh \phi}\, d\phi - \frac{1}{v}\int_0^\infty \cosh \phi e^{-v \cosh \phi}\, d\phi - \int_0^\infty e^{-v \cosh \phi}\, d\phi$$

Combining first and third terms by *Dw* 650.01 or *Pc* 657 and integrating the result by parts give

$$\int_0^\infty (\sinh \phi)[e^{-v \cosh \phi}\, d(\cosh \phi)] = \int u\, dv$$

$$= \left| \frac{-\sinh \phi}{v} e^{-v \cosh \phi} \right|_0^\infty + \frac{1}{v}\int_0^\infty \cosh \phi e^{-v \cosh \phi}\, d\phi$$

which cancels the second term so that the integral satisfies 5.32 (1). Since any solution of 5.32 (1) must be of the form $R^0 = AI_0(v) + BK_0(v)$ and since the integral is zero when $v = \infty$, it cannot contain $I_0(v)$ and must be of the form $BK_0(v)$. In 5.35, we shall show that to agree with 5.33 (3) we must take $B = 1$, giving

$$K_0(v) = \int_0^\infty e^{-v \cosh \phi}\, d\phi = \tfrac{1}{2} j\pi H_0^{(1)}(jv) = -\tfrac{1}{2} j\pi H_0^{(2)}(-jv) \tag{1}$$

We see that

$$v^m \frac{d^n K_0(v)}{dv^n} = (-1)^n \int_0^\infty v^m \cosh^n \phi e^{-v \cosh \phi}\, d\phi$$

is also zero when $v = \infty$ because v appears in the integrand in the form $v^m e^{-av}$ and $a > 1$. From the recurrence formulas 5.321 (5) and (6), we see that

$$K_1 = -K_0' \qquad \text{so that} \qquad K_1(\infty) = 0$$

Thus we get from 5.321 (6)

$$K_n(\infty) = K_{n-2}(\infty) = \cdots = K_p(\infty) = 0$$

where $p = 1$ if n is odd and $p = 0$ if n is even. It follows now from 5.321 (4) that $K_n'(\infty) = 0$, etc.

Another definite integral for $K_0(v)$ is

$$K_0(v) = \int_0^\infty \cos(v \sinh \phi)\, d\phi \tag{2}$$

Prof. H. Bateman proves this relation as follows: Consider the function

$$W = \int_0^\infty e^{-x\cosh\phi} \cos(y \sinh \phi)\, d\phi$$

where $x = r\cos\theta$ and $y = r\sin\theta$. Differentiate,

$$\begin{aligned}\frac{\partial W}{\partial \theta} &= x\frac{\partial W}{\partial y} - y\frac{\partial W}{\partial x}\\ &= -\int_0^\infty e^{-x\cosh\phi}[x\sinh\phi \sin(y\sinh\phi) - y\cosh\phi\cos(y\sinh\phi)]\, d\phi\\ &= \int_0^\infty \frac{d}{d\phi}[e^{-x\cosh\phi}\sin(y\sinh\phi)]\, d\phi = -\Big| e^{-x\cosh\phi}\sin(y\sinh\phi)\Big|_0^\infty = 0\end{aligned}$$

Therefore W is independent of θ. Putting $\theta = 0$ gives $x = r$, $y = 0$ so that $W = K_0(r)$ by (1), and putting $\theta = \frac{1}{2}\pi$ gives $x = 0$, $y = r$ so that W is the integral of (2). Thus (2) is proved.

To get an integral useful in slit diffraction theory, we insert (1) in 5.33 (6), replace the 0 to ∞ integral by half the $-\infty$ to ∞ integral, and interchange the integration order. This gives, by *Pc* 402 or *Dw* 567.1,

$$2vK_1(v) = \pi v H_1^{(2)}(-jv) = \int_{-\infty}^\infty e^{-v\cosh\phi}(1 + v\cosh\phi)\operatorname{sech}^2\phi\, d\phi \tag{3}$$

5.34. Definite Integrals for Bessel Functions of Zero Order.—Substituting v/j for v in 5.331 (1), we have

$$K_0\left(\frac{v}{j}\right) = \int_0^\infty e^{jv\cosh\phi}\, d\phi = \int_0^\infty \cos(v\cosh\phi)\, d\phi + j\int_0^\infty \sin(v\cosh\phi)\, d\phi$$

Making the same substitution in 5.32 (4) and equating real and imaginary parts gives for $J_0(v)$ and $Y_0(v)$, if we use Watson's notation, see 5.293,

$$J_0(v) = \frac{2}{\pi}\int_0^\infty \sin(v\cosh\phi)\, d\phi \tag{1}$$

$$Y_0(v) = -\frac{2}{\pi}\int_0^\infty \cos(v\cosh\phi)\, d\phi \tag{2}$$

5.35. Inverse Distance in Terms of Modified Bessel Functions.—We now derive an expression for the inverse distance between two points of which the cylindrical coordinates are ρ_0, ϕ_0, z_0 and ρ, ϕ, z. This could be done by the method of 5.304, but it is shorter to start with the expression for the potential of a point charge midway between two plates at a distance L apart. To obtain this potential, let $a = \infty$ in 5.324 (9) and shift the origin to $L/2$. Only odd values of n survive so that, when $\rho < \rho_0$,

$$V = \frac{q}{\pi\epsilon L}\sum_{n=0}^{\infty}\sum_{m=0}^{\infty}(2-\delta_m^0)K_m(N\rho_0)I_m(N\rho)\cos Nz\cos Nz_0\cos m(\phi-\phi_0) \qquad (1)$$

where $N = (2n+1)\pi/L$. Now let $2n\pi/L = u_n$ and $2\pi/L = \Delta u$, and this becomes

$$\frac{\frac{1}{2}q}{\pi^2\epsilon}\sum_{m=0}^{\infty}(2-\delta_m^0)\cos m(\phi-\phi_0)\sum_{n=0}^{\infty}K_m\left[\left(u_n+\frac{\Delta u}{2}\right)\rho_0\right]$$
$$I_m\left[\left(u_n+\frac{\Delta u}{2}\right)\rho\right]\cos\left[\left(u_n+\frac{\Delta u}{2}\right)z\right]\Delta u$$

Next let $L \to \infty$ so that $\Delta u \to 0$, and from the basic definition of an integral the n-summation, $0 \leqslant n < \infty$, becomes the u-integral, $0 \leqslant u < \infty$. Thus we have

$$V = \frac{q}{4\pi\epsilon R} = \frac{q}{4\pi\epsilon[z^2+\rho^2+\rho_0^2-2\rho\rho_0\cos(\phi-\phi_0)]^{\frac{1}{2}}}$$
$$= \frac{q}{2\pi^2\epsilon}\sum_{m=0}^{\infty}(2-\delta_m^0)\left[\int_0^\infty K_m(u\rho_0)I_m(u\rho)\cos uz\,du\right]\cos m(\phi-\phi_0) \qquad (2)$$

If $\rho > \rho_0$ interchange ρ and ρ_0. At $\rho_0 = 0$ the summation disappears so that

$$r^{-1} = (\rho^2+z^2)^{-\frac{1}{2}} = 2\pi^{-1}\int_0^\infty \cos kzK_0(k\rho)\,dk \qquad (3)$$

With z and ϕ_0 zero, write R for ρ in (3) and compare with (2). Thus we get when $\rho < \rho_0$

$$K_0[(\rho^2+\rho_0^2-2\rho\rho_0\cos\phi)^{\frac{1}{2}}] = \sum_{m=0}^{\infty}(2-\delta_m^0)K_m(\rho_0)I_m(\rho)\cos m\phi \qquad (4)$$

To verify the choice of constant in 5.331 (1) substitute it for $K_0(k\rho)$ in (3), integrate first with respect to k using *Dw* 577.2 or *Pc* 415, and then with respect to ϕ using *Dw* 120.01 or *Pc* 99, and so get the left side.

5.351. Cylindrical Dielectric Boundaries.—The integral for the reciprocal distance just derived may be used to solve problems involving cylindrical boundaries when the fields extend to infinity. In such cases, expansion in a series of Bessel functions is not feasible since the integrals of 5.297 are infinite at $a = \infty$ as can be seen by inspection of the asymptotic values in 5.295. In such cases, we can usually obtain the solution as a definite integral which can be evaluated numerically by graphical integration.

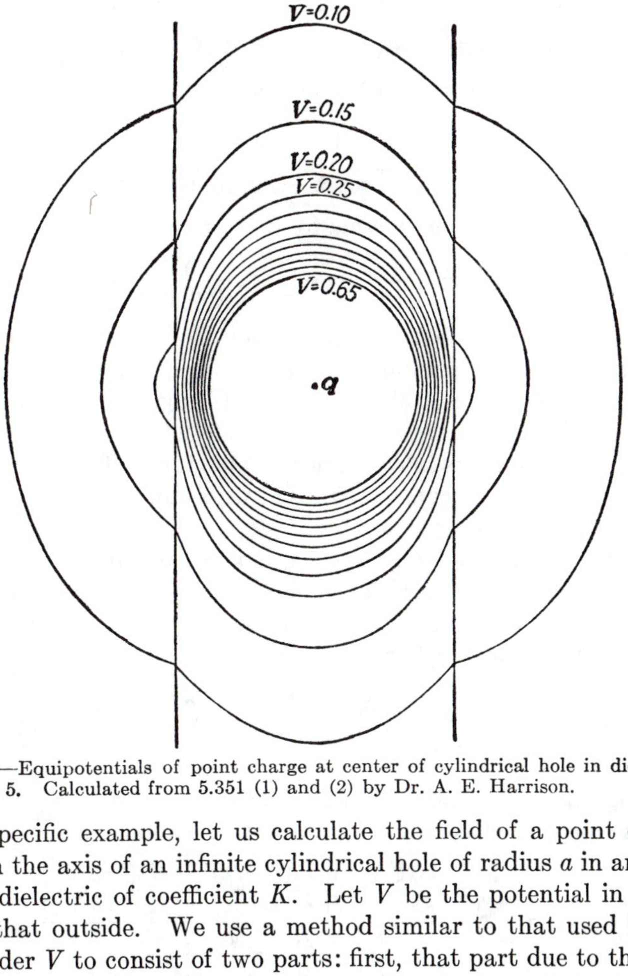

FIG. 5.351.—Equipotentials of point charge at center of cylindrical hole in dielectric for which $K = 5$. Calculated from 5.351 (1) and (2) by Dr. A. E. Harrison.

As a specific example, let us calculate the field of a point charge q placed on the axis of an infinite cylindrical hole of radius a in an infinite block of dielectric of coefficient K. Let V be the potential in the hole and V_K that outside. We use a method similar to that used in 5.303. We consider V to consist of two parts: first, that part due to the charge alone, given by 5.35 (3) and second, that due to the polarization of the dielectric, which must be finite on the axis. Thus V is of the form

$$V = \frac{q}{2\pi^2\epsilon_v}\int_0^\infty [K_0(k\rho) + \Psi(k)I_0(k\rho)] \cos kz \, dk \tag{1}$$

In the dielectric the potential must be finite at infinity and can have only the form

$$V_K = \frac{q}{2\pi^2\epsilon_v}\int_0^\infty \Phi(k)K_0(k\rho)\cos kz\,dk \tag{2}$$

The boundary condition $V = V_K$ when $\rho = a$ requires that

$$K_0(ka) + \Psi(k)I_0(ka) = \Phi(k)K_0(ka) \tag{3}$$

The boundary condition $\partial V/\partial\rho = K\,\partial V_K/\partial\rho$ at $\rho = a$ gives

$$K_0'(ka) + \Psi(k)I_0'(ka) = K\Phi(k)K_0'(ka) \tag{4}$$

Eliminating $\Psi(k)$ from (3) and (4) and simplifying by the use of 5.32 (7), we obtain

$$\Phi(k) = \frac{1}{1 - ka(K-1)I_0(ka)K_0'(ka)} \tag{5}$$

Substituting (5) in (3) and solving for $\Psi(k)$ give

$$\Psi(k) = \frac{ka(K-1)K_0(ka)K_0'(ka)}{1 - ka(K-1)I_0(ka)K_0'(ka)} \tag{6}$$

The field when $K = 5$, obtained by plotting the integrands in (1) and (2) and integrating with a planimeter, is shown in Fig. 5.351.

5.36. Potential inside Hollow Cylindrical Ring.—The most important electrical applications of the modified Bessel functions occur in connection with alternating currents in cylindrical conductors. There are, however, a few applications in electrostatics. Let us, as an example, consider the potential in the region bounded by $\rho = a$, $\rho = b$, $z = 0$ and $z = c$. Let the potential be zero on all these boundaries except the first. When $\rho = a$, let $V = f(z)$. Let us expand $f(z)$ as a Fourier's series in the interval $0 < z < c$. Since $f(z) = 0$ when $z = 0$ or $z = c$, we have the well-known expansion, *Pc* 815,

$$f(z) = \sum_{k=1}^{\infty} A_k \begin{matrix}\cos\\ \sin\end{matrix}\left(\frac{k\pi z}{c}\right), \qquad A_k = \frac{2}{c}\int_0^c f(z) \begin{matrix}\cos\\ \sin\end{matrix}\left(\frac{k\pi z}{c}\right) dz \tag{1}$$

Consulting 5.292 (2), 5.292 (3), and 5.32 (6), we see that a solution of Laplace's equation satisfying all boundary conditions is

$$V = \sum_{k=1}^{\infty} \frac{A_k \begin{matrix}\cos\\ \sin\end{matrix}\left(\frac{k\pi z}{c}\right)\left[\frac{I_0(k\pi\rho/c)}{I_0(k\pi b/c)} - \frac{K_0(k\pi\rho/c)}{K_0(k\pi b/c)}\right]}{\frac{I_0(k\pi a/c)}{I_0(k\pi b/c)} - \frac{K_0(k\pi a/c)}{K_0(k\pi b/c)}} \tag{2}$$

If we wish the potential inside the cylinder $\rho = a$, on which the potential is $V = f(z)$, and between the two earthed planes $z = 0$ and $z = c$, this expression becomes

$$V = \sum_{k=1}^{\infty} A_k \sin\frac{k\pi z}{c}\frac{I_0(k\pi\rho/c)}{I_0(k\pi a/c)} \tag{3}$$

If we wish the potential outside the cylinder $\rho = a$, on which the potential is $V = f(z)$, and between the two earthed planes $z = 0$ and $z = c$, we have

$$V = \sum_{k=1}^{\infty} A_k \sin \frac{k\pi z}{c} \frac{K_0(k\pi\rho/c)}{K_0(k\pi a/c)} \tag{4}$$

5.37. Modified Bessel Functions of Nonintegral Order.—From 5.31 (1), putting $I_n(v) = j^{-n}J_n(jv)$, we see that a general solution of Bessel's modified equation, when n is not an integer, is given by

$$R_n^0(v) = AI_n(v) + BI_{-n}(v) \tag{1}$$

$K_n(v)$ is a linear combination of $I_n(v)$ and $I_{-n}(v)$. The recurrence formulas are equally valid for these values. The Eqs. 5.31 (2), (3), (4), (5), and (6) lead to

$$I_{\frac{1}{2}}(v) = \left(\frac{2}{\pi v}\right)^{\frac{1}{2}} \sinh v \tag{2}$$

$$I_{-\frac{1}{2}}(v) = \left(\frac{2}{\pi v}\right)^{\frac{1}{2}} \cosh v \tag{3}$$

$$I_{\frac{3}{2}}(v) = \left(\frac{2}{\pi v}\right)^{\frac{1}{2}}\left(\cosh v - \frac{1}{v} \sinh v\right) \tag{4}$$

$$I_{-\frac{3}{2}}(v) = \left(\frac{2}{\pi v}\right)^{\frac{1}{2}}\left(\sinh v - \frac{1}{v} \cosh v\right) \tag{5}$$

$$I_{\frac{5}{2}}(v) = \left(\frac{2}{\pi v}\right)^{\frac{1}{2}}\left[\left(1 + \frac{3}{v^2}\right) \sinh v - \frac{3}{v} \cosh v\right] \tag{6}$$

$$I_{-\frac{5}{2}}(v) = \left(\frac{2}{\pi v}\right)^{\frac{1}{2}}\left[\left(1 + \frac{3}{v^2}\right) \cosh v - \frac{3}{v} \sinh v\right] \tag{7}$$

The general formula is simplest in exponential form

$$I_{\pm(n+\frac{1}{2})}(v) = \frac{1}{(2\pi v)^{\frac{1}{2}}} \sum_{s=0}^{n} \frac{[(-1)^s e^v \mp (-1)^n e^{-v}](n+s)!}{s!(n-s)!(2v)^s} \tag{8}$$

where the upper or lower sign is used throughout.

Both $I_{n+\frac{1}{2}}(v)$ and $I_{-n-\frac{1}{2}}(v)$ are infinite when $v = \infty$. A solution that is zero at ∞ is $K_{n+\frac{1}{2}}(v)$ as defined in 5.32 (4) when $n + \frac{1}{2}$ replaces n. By 5.31 (8), eliminate $Y_{n+\frac{1}{2}}(jv)$ then write in terms of $I_{\pm(n+\frac{1}{2})}(v)$ by 5.32 (2) and substitute (8). The positive exponential disappears and leaves

$$K_{n+\frac{1}{2}}(v) = \tfrac{1}{2}\pi(-1)^n[I_{-n-\frac{1}{2}}(v) - I_{n+\frac{1}{2}}(v)] = \left(\frac{\pi}{2v}\right)^{\frac{1}{2}} \sum_{s=0}^{n} \frac{(n+s)!e^{-v}}{s!(n-s)!(2v)^s} \tag{9}$$

The simplest cases are

$$K_{\frac{1}{2}}(v) = \left(\frac{\pi}{2v}\right)^{\frac{1}{2}} e^{-v} \qquad K_{\frac{3}{2}}(v) = \left(\frac{\pi}{2v}\right)^{\frac{1}{2}}\left(1 + \frac{1}{v}\right)e^{-v} \tag{10}$$

If $v = 0$ in (9), the $s = n$ term of the summation dominates so that

$$K_{n+\frac{1}{2}}(v) \underset{v\to 0}{\longrightarrow} \left(\frac{\pi}{2v}\right)^{\frac{1}{2}} \frac{(2n-1)!!}{v^n} \tag{11}$$

The spherical Bessel function $k_n(v)$ may be defined by

$$k_n(v) = \left(\frac{2}{\pi v}\right)^{\frac{1}{2}} K_{n+\frac{1}{2}}(v) = \frac{1}{v}\sum_{s=0}^{n} \frac{(n+s)!e^{-v}}{s!(n-s)!(2v)^s} \tag{12}$$

With the aid of 5.32 (4) and 5.31 (8), we obtain

$$k_n(jv) = -j^{-n}[j_n(v) - jn_n(v)] \tag{13}$$

From this equation and 5.31 (11)

$$k_n(jv)j_n'(v) - jk_n'(jv)j_n(v) = -j^{-n+1}v^{-2} \tag{14}$$

The relations with Schelkunoff's $\hat{K}_n(v)$ and Stratton's $h_n^{(1)}(v)$ are

$$k_n(v) = v^{-1}\hat{K}_n(v) = -j^n h_n^{(1)}(jv) = -j^n[j_n(jv) + jn_n(jv)] \tag{15}$$

From (12) and 5.321, the recurrence relations are

$$-(2n+1)v^{-1}k_n = k_{n-1} - k_{n+1}, \quad -k_n' = k_{n\mp 1} \pm (n + \tfrac{1}{2} \pm \tfrac{1}{2})v^{-1}k_n \tag{16}$$

5.38. Wedge Functions.—The functions of 5.291 and 5.292 are inadequate when the potential vanishes on the cylinders $\rho = \rho_1$, $\rho = \rho_2$ and on the planes $z = z_1$, $z = z_2$ and has arbitrary values over the planes $\phi = \phi_1$, $\phi = \phi_2$. This problem requires functions orthogonal in ρ and z obtained by substitution of $j\nu$ for n and jk for k in 5.291 (1), (2), and (3). Thus,

$$\Phi(\nu\phi) = C\cosh\nu\phi + D\sinh\nu\phi = C'e^{\nu\phi} + D'e^{-\nu\phi} \tag{1}$$

$$Z(kz) = E\sin kz + F\cos kz \tag{2}$$

$$R_\nu(k\rho) = AF_\nu(k\rho) + BG_\nu(k\rho) \tag{3}$$

where $R_\nu(v)$ satisfies the differential equation

$$\frac{d^2R}{dz^2} + \frac{1}{v}\frac{dR}{dv} - \left(1 - \frac{\nu^2}{v^2}\right)R = 0 \tag{4}$$

The function $F_\nu(v) \to \infty$ as $v \to \infty$ and is defined by

$$F_\nu(v) = \operatorname{csch}\nu\pi \int_0^\pi e^{v\cos\theta}\cosh\nu\theta\,d\theta - \int_0^\infty e^{-v\cosh t}\sin\nu t\,dt \tag{5}$$

The function $G_\nu(v) \to 0$ as $v \to \infty$ and is defined by

$$G_\nu(v) = \int_0^\infty e^{-v\cosh t}\cos\nu t\,dt = K_{j\nu}(v) \tag{6}$$

Below a certain value of v, which increases with ν, both functions are periodic and oscillate as $\sin[\nu\ln(\frac{1}{2}v)]$ and $\cos[\nu\ln(\frac{1}{2}v)]$ as v approaches zero. The integral with respect to v of v^{-1} times a product vanishes unless the factors have the same ν so that an arbitrary function of v suitable for expansion in Fourier series can be expanded in terms of $R_\nu(k\rho)$. The three-dimensional dielectric wedge problem is solved by methods similar to those used for the two-dimensional wedge in Art. 4.07.

5.39. Mixed Boundaries. Charged Right Circular Cylinder.—Many three-dimensional boundaries met in practice do not fit any separable coordinate system; thus the methods so far used are not applicable. For example, if a charged circular disk lies inside a coaxial circular cylinder, a system of coordinates fitting the boundaries can be generated by rotating the last part of Fig. 4.22*b* about the y-axis. Laplace's equation, 4.11 (6), in this system cannot be separated, and the surfaces, from 5.00 (2), are not equipotentials. Clearly the boundaries fit into the cylindrical system, but at $y = 0$, V is a constant only over the disk while between the disk and the cylinder $\partial V/\partial y$ is zero. Thus this becomes a mixed boundary value problem. It can be solved (see Art. 5.42). There are many cases where V or $\partial V/\partial n$ is known on part of a coordinate surface and nothing over the remainder. One of these is the charged solid right circular cylinder. A possible approach to this is to express the charge density as a series having arbitrary constants adjustable by 3.15 (3) to give a uniform potential inside the cylinder. When magnified the edge looks like a right-angle wedge for which 4.18 (1) gives $dz/dW = W^{\frac{1}{2}}$ so that $W = z^{\frac{2}{3}}$ and $dW/dz = z^{-\frac{1}{3}}$. Thus the series will diverge at the edge and converge slowly elsewhere unless the infinity appears explicitly in σ. A series of Gegenbauer or Jacobi polynomials with the proper weighting factor has some advantages but complicates integrations and numerical work.

If the cylinder is bounded by $z = \pm b$ and $\rho = a$, then simple forms for the charge densities on the end σ_e and on the side σ_s are

$$\sigma_s = \sum_{n=0}^{N} A_n(1 - b^{-2}z^2)^{n-\frac{1}{3}} \qquad \sigma_e = \sum_{m=0}^{M} B_m(1 - a^{-2}\rho^2)^{m-\frac{1}{3}} \tag{1}$$

Note that at the edge where $z = b$ and $\rho = a$ only the A_0 and B_0 terms survive. Thus the matching of σ_e and σ_s there requires that

$$b^{\frac{1}{3}}A_0 = a^{\frac{1}{3}}B_0 \tag{2}$$

There are two ways in which 3.15 (3) may be used to get A_n and B_m. One is to calculate $\partial^{2p}V(z,0)/\partial z^{2p}$ on the axis, set $z = 0$, and require that

$$\left[\frac{\partial^{2p}V(z,0)}{\partial^{2p}z}\right]_{z=0} = \delta_{0p}V_0 \tag{3}$$

where $\delta_{0p} = 0$ if $p \neq 0$ and $\delta_{0p} = 1$ if $p = 0$. This and (2) yield $p + 2$ equations which may be solved for $N + M + 2$ of the low-order terms in (1). Alternatively the potential in the plane $z = 0$ due to (1) may be expanded in even powers $2p$ of ρ and coefficients of all but the $p = 0$ term equated to zero. For a given b/a ratio the number of significant digits carried fixes the best values of M and N, which are those that give potentials nearest V_0 at pole and equator where maximum deviations occur. Fewer σ coefficients can be found for the more remote point than for the nearer one. If too big values are chosen for M and N, large terms alternating in sign appear, which reduce the number of significant digits in the sum. All the values of A_n and B_m found from (1) must be used in subsequent calculations as omission of one may change its neighbors so they are no longer correct. If the polynomials already mentioned replace (1) (see Taylor, *J. Research, Natl. Bur. Std.*, Vol. 64B, pages 142 and 143), A_n and B_m decrease rapidly as n and m increase, so that dropping the last terms has little effect on the rest.

The chief labor is finding the derivatives in (3). The axial values of the expressions for the potential of a charged ring in the text and problems of this chapter must be integrated over the charge distribution of (1). It is usually necessary to expand some of the functions in series in order to get fractional integrals of the form tabulated in Vol. II, Chap. XIII, of "Integral Transforms" (see bibliography). A hypergeometric series usually results. By summing these and using recursion formulas, the simultaneous equations of (3) are set up and then solved for A_n and B_m. Numerical values of b/a have been calculated (*J. Appl. Phys.*, Vol. 33, page 2966) for 0.125, 0.25, 0.50, 1, 2, 4, 8, which show that for the extreme values 0.125 and 8 only one coefficient can be found for the more remote surface. The coefficients for $b = a$ appear in the table below. Clearly for practical purposes the solution is exact when $b = a$, as the potentials V_p and V_e and the surface deviations Δb and Δa at pole and equator show.

Charge density coefficients for unit potential when $b = a$.

$A_0 =$	0.55941519	$B_3 =$	−3.9209295
$A_1 =$	0.24032463	$B_4 =$	6.2012014
$A_2 =$	−0.46123818	$B_5 =$	−5.6962826
$A_3 =$	0.71795706	$B_6 =$	2.8184609
$A_4 =$	−0.67534061	$B_7 =$	−0.5816914
$A_5 =$	0.34357563	$V_p =$	1.0000002
$A_6 =$	−0.07271528	$V_e =$	1.0000001
$B_0 =$	0.55941519	Surface deviations	
$B_1 =$	−0.35462716	$\Delta b =$	$-0.0000004b$
$B_2 =$	1.4624910	$\Delta a =$	$-0.0000002a$

The potential outside the sphere passing through the cylinder edges is found by the second integral in 1.06 (6) to be

$$V = \frac{aV_0}{2\epsilon}\sum_{s=0}^{\infty}\sum_{p=0}^{\infty}\frac{(-1)^p(C_e + C_s)(2p+2s)!a^{2p}b^{2s}}{4^p(p!)^2(2s)!r^{2s+2p+1}}P_{2s+2p}(\cos\theta) \tag{4}$$

$$C_e = a\sum_{m=0}^{M}B(p+1, m+\tfrac{2}{3})B_m, \qquad C_s = b\sum_{n=0}^{N}B(s+\tfrac{1}{2}, n+\tfrac{2}{3})A_n \tag{5}$$

where $B(x,y)$ is a beta function. Note that at a great distance (4) takes the form $V = Q/(4\pi\epsilon r)$. The capacitance is then Q/V_0. We get an empirical interpolation formula for C by adding a correction term containing two adjustable constants to the exact known capacitance of a circular disk given in 5.03 (2). This curve, which passes through the seven calculated points with an error of 0.2 per cent or less, is

$$C = \left[0.708 + 0.615\left(\frac{b}{a}\right)^{0.76}\right] \times 10^{-10}a \qquad \text{farad} \tag{6}$$

It is clear that this method applies to other geometrical forms such as the spindle formed by two finite circular cones fitted base to base. It also works when such objects are placed in uniform fields as seen in the foregoing Taylor reference.

5.40. Mixed Spherical, Spheroidal, and Cylindrical Boundaries.—Often different boundaries of an electric field belong to different coordinate systems. No method given so far handles this case. The formulas needed to relate the four systems, oblate, prolate, spherical, and cylindrical, naturally split into two groups: those relating the exterior harmonics, which satisfy Laplace's equation and vanish at infinity, and those relating the interior harmonics, solutions that are finite at the origin. The former depend on a set of Fourier transforms due to J. C. Cooke (see *Monatshefte fur Math.* Vol. 60, page 322, 1956; and *Zeit. fur Math. und Mech.* Vol. 42, page 305, 1962). These transforms have been modified to fit the notation of this chapter. Note that Hobson, HTF, HMF, and others insert $(-1)^m$ in the right side of 5.23 (6) and (7), so in their notation the sign of m in the exponent of j in the following formulas should be reversed. To shorten the formulas R_{mn} is written for $(n-m)!/(n+m)!$.

$$\int_0^\infty t^{-\frac{1}{2}}K_m(t\rho)I_{n+\frac{1}{2}}(tc)\cos(tz)\,dt = j^{1-m}(\tfrac{1}{2}\pi c^{-1})^{\frac{1}{2}}R_{mn}Q_n^m(j\zeta)P_n^m(\xi) \tag{1}$$

$$\int_0^\infty t^{-\frac{1}{2}}K_m(t\rho)I_{n+\frac{1}{2}}(tc)\sin(tz)\,dt = j^{2-m}(\tfrac{1}{2}\pi c^{-1})^{\frac{1}{2}}R_{mn}Q_n^m(j\zeta)P_n^m(\xi) \tag{2}$$

$$\int_0^\infty t^nK_m(t\rho)\cos(tz)\,dt = j^{n-m}(\tfrac{1}{2}\pi)(n-m)!r^{-n-1}P_n^m(\cos\theta) \tag{3}$$

$$\int_0^\infty t^nK_m(t\rho)\sin(tz)\,dt = j^{n-m-1}(\tfrac{1}{2}\pi)(n-m)!r^{-n-1}P_n^m(\cos\theta) \tag{4}$$

$$\int_0^\infty t^{-\frac{1}{2}}K_m(t\rho)J_{n+\frac{1}{2}}(tc)\cos(tz)\,dt = j^{n+m}(\tfrac{1}{2}\pi c^{-1})^{\frac{1}{2}}R_{mn}Q_n^m(\eta)P_n^m(\xi) \tag{5}$$

$$\int_0^\infty t^{-\frac{1}{2}}K_m(t\rho)J_{n+\frac{1}{2}}(tc)\sin(tz)\,dt = j^{n+m+1}(\tfrac{1}{2}\pi c^{-1})^{\frac{1}{2}}R_{mn}Q_n^m(\eta)P_n^m(\xi) \tag{6}$$

In (1), (3), and (5) $n + m$ is even and in (2), (4), and (6) $n + m$ is odd, if $-1 < \xi < +1$ and $0 < \zeta < \infty$ in (1) and (2), but the reverse is true when $0 < \xi < 1$ and $-\infty < \zeta < +\infty$. The interfocal distance in the spheroids is $2c$ as in Arts. 5.27 to 5.281.

Additional formulas needed for combining spherical and spheroidal forms are found by substitution of the power series 5.32 (1) and 5.293 (2) for $I_{n+\frac{1}{2}}(t)$ and $J_{n+\frac{1}{2}}(t)$ in (1) and (5) or (2) and (6), giving the integrals the form of (3) or (4). Thus

$$Q_n^m(j\zeta)P_n^m(\xi) = \sum_{s=\frac{1}{2}n}^{\infty}(-1)^{m+s}j^{-1}B_{mns}\left(\frac{c}{r}\right)^{2s+1}P_{2s}^m(\cos\theta) \tag{7}$$

$$Q_n^m(\eta)P_n^m(\xi) = \sum_{s=\frac{1}{2}n}^{\infty}B_{mns}\left(\frac{c}{r}\right)^{2s+1}P_{2s}^m(\cos\theta) \tag{8}$$

$$B_{mns} = \frac{(-1)^n(2s-m)!(n+m)!}{(n+2s+1)!!(n-m)!(2s-n)!!} \tag{9}$$

In (3) or (4), t^n may be written as a Neumann series involving $I_{n+\frac{1}{2}}(t)$ and $J_{n+\frac{1}{2}}(t)$ by Watson (1), page 138, and HMF 9.1.87, page 364. The resultant integrals are those in (1) and (5) or (2) and (6). Thus

$$\left(\frac{c}{r}\right)^{n+1}P_n^m(\cos\theta) = \sum_{s=\frac{1}{2}n}^{\infty}(-1)^{m+s}jC_{mns}Q_{2s}^m(j\zeta)P_{2s}^m(\xi) \tag{10}$$

$$= \sum_{s=\frac{1}{2}n}^{\infty}(-1)^nC_{mns}Q_{2s}^m(\eta)P_{2s}^m(\xi) \tag{11}$$

$$C_{mns} = \frac{(2s+1)(n+2s-1)!!(2s-m)!}{(n-m)!(2s-n)!!(2s+m)!} \tag{12}$$

From these a set of formulas for combining oblate and prolate boundaries may be derived. Let the interfocal distances and the ξ coordinate in the oblate case be $2c_1$ and ξ_1 and in the prolate case $2c_2$ and ξ_2. In (7) write c_1 for c and ξ_1 for ξ, and in (11), $(c_2/c_1)^{n+1}(c_1/r)^{n+1}$ for $(c/r)^{n+1}$, p for s, and then $2s$ for n. Elimination of spherical harmonics gives

$$Q_n^m(j\zeta)P_n^m(\xi_1) = \sum_{s=\frac{1}{2}n}^{\infty}\sum_{p=s}^{\infty}(-1)^{m+p+n}j^{-1}C_{mnsp}\left(\frac{c_2}{c_1}\right)^{2s+1}Q_{2p}^m(\eta)P_{2p}^m(\xi_2) \tag{13}$$

$$C_{mnsp} = \frac{(2p+1)(n+m)!(2s+2p-1)!!(2p-m)!}{(n+2s+1)!!(n-m)!(2s-n)!!(2p-2s)!!(2p+m)!} \tag{14}$$

A similar operation on (8) and (10) gives

$$Q_n^m(\eta)P_n^m(\xi_2) = \sum_{s=\frac{1}{2}n}^{\infty}\sum_{p=s}^{\infty}(-1)^{m+p+n}jC_{mnsp}\left(\frac{c_1}{c_2}\right)^{2s+1} Q_{2p}^m(j\zeta)P_{2p}^m(\xi_1) \quad (15)$$

In the same way for nonconfocal coaxial spheroids of the same type

$$Q_n^m(j\zeta_1)P_n^m(\xi_1) = \sum_{s=\frac{1}{2}n}^{\infty}\sum_{p=s}^{\infty}(-1)^{n+s+p}C_{mnsp}\left(\frac{c_1}{c_2}\right)^{2s+1} Q_{2p}^m(j\zeta_2)P_{2p}^m(\xi_2) \quad (16)$$

$$Q_n^m(\eta_1)P_n^m(\xi_1) = \sum_{s=\frac{1}{2}n}^{\infty}\sum_{p=s}^{\infty}C_{mnsp}\left(\frac{c_1}{c_2}\right)^{2s+1} Q_{2p}^m(\eta_2)P_{2p}^m(\xi_2) \quad (17)$$

The interior harmonics, like (1) to (6), depend on Bessel function formulas due to Cooke. In Hobson's notation the right sides of (18) to (23) should be multiplied by $(-1)^m$. The relations are

$$\cos(tz)I_m(t\rho) = \left(\frac{\pi}{2tc}\right)^{\frac{1}{2}}\sum_{n=m}^{\infty}\frac{j^{-m}(2n+1)(n-m)!}{(n+m)!}I_{n+\frac{1}{2}}(ct)P_n^m(\xi)P_n^m(j\zeta) \quad (18)$$

$$= \sum_{n=m}^{\infty}j^{n+m}(tr)^n[(n+m)!]^{-1}P_n^m(\cos\theta) \quad (19)$$

$$= \left(\frac{\pi}{2tc}\right)^{\frac{1}{2}}\sum_{n=m}^{\infty}\frac{(-1)^m(2n+1)(n-m)!}{(n+m)!}J_{n+\frac{1}{2}}(ct)P_n^m(\xi)P_n^m(\eta) \quad (20)$$

$$\sin(tz)I_m(t\rho) = \left(\frac{\pi}{2tc}\right)^{\frac{1}{2}}\sum_{n=m+1}^{\infty}\frac{(-j)^{m+1}(2n+1)(n-m)!}{(n+m)!} I_{n+\frac{1}{2}}(ct)P_n^m(\xi)P_n^m(j\zeta) \quad (21)$$

$$= \sum_{n=m+1}^{\infty}j^{n+m-1}(tr)^n[(n+m)!]^{-1}P_n^m(\cos\theta) \quad (22)$$

$$= \left(\frac{\pi}{2tc}\right)^{\frac{1}{2}}\sum_{n=m+1}^{\infty}\frac{(-1)^m(2n+1)(n-m)!}{(n+m)!}J_{n+\frac{1}{2}}(ct)P_n^m(\xi)P_n^m(\eta) \quad (23)$$

In (18), (19), and (20) $m+n$ is even, and in (21), (22), and (23) $m+n$ is odd. The interior equations analogous to (7) to (12) are now polynomials instead of infinite series. They may be derived by writing out the left sides in products of powers of its coordinates, then changing over to the coordinates of the right side by 5.27 (4) and (5) or 5.28 (4) and (5) and regrouping to form harmonics. Alternatively write $I_{n+\frac{1}{2}}(ct)$ and $J_{n+\frac{1}{2}}(ct)$ in (18) and (20) or in (21) and (23) as power series in ct by 5.32 (2) and 5.293 (3), replace $(tr)^n$ by $(r/c)^n(tc)^n$ in (19), and equate coefficients of $(tc)^n$. Thus we get

$$P_n^m(\xi)P_n^m(j\zeta) = \sum_{s=S}^{\frac{1}{2}n} j^n D_{mns}\left(\frac{r}{c}\right)^{2s} P_{2s}^m(\cos\theta) \tag{24}$$

$$P_n^m(\xi)P_n^m(\eta) = \sum_{s=S}^{\frac{1}{2}n} (-1)^{\frac{1}{2}n-s} D_{mns}\left(\frac{r}{c}\right)^{2s} P_{2s}^m(\cos\theta) \tag{25}$$

$$D_{mns} = \frac{(n+2s-1)!!(n+m)!}{(2s+m)!(n-m)!(n-2s)!!} \tag{26}$$

$$\left(\frac{r}{c}\right)^n P_n^m(\cos\theta) = \sum_{s=S}^{\frac{1}{2}n} j^{-n} G_{mns} P_{2s}^m(\xi)P_{2s}^m(j\zeta) \tag{27}$$

$$= \sum_{s=S}^{\frac{1}{2}n} (-1)^{\frac{1}{2}n-s} G_{mns} P_{2s}^m(\xi)P_{2s}^m(\eta) \tag{28}$$

$$G_{mns} = \frac{(4s+1)(2s-m)!(n+m)!}{(2s+m)!(n+2s+1)!!(n-2s)!!} \tag{29}$$

Here S equals $\frac{1}{2}m$ or $\frac{1}{2}(m+1)$ and is an integer if n is even and an integer plus $\frac{1}{2}$ if n is odd, and s increases by unit steps. Equations (30) to (33) derive from (24) to (29) in the same way that (13) to (15) derive from (7) to (12). Thus

$$P_n^m(j\zeta)P_n^m(\xi_1) = \sum_{s=S}^{\frac{1}{2}n} \sum_{p=P}^{s} j^n(-1)^{s-p} F_{mnsp}\left(\frac{c_2}{c_1}\right)^{2s} P_{2p}^m(\eta)P_{2p}^m(\xi_2) \tag{30}$$

$$P_n^m(\eta)P_n^m(\xi_1) = \sum_{s=S}^{\frac{1}{2}n} \sum_{p=P}^{s} (-1)^{\frac{1}{2}n} F_{mnsp}\left(\frac{c_2}{c_1}\right)^{2s} P_{2p}^m(j\zeta)P_{2p}^m(\xi_2) \tag{31}$$

$$P_n^m(j\zeta_1)P_n^m(\xi_1) = \sum_{s=S}^{\frac{1}{2}n} \sum_{p=P}^{s} j^n(-1)^{s} F_{mnsp}\left(\frac{c_2}{c_1}\right)^{2s} P_{2p}^m(j\zeta_2)P_{2p}^m(\xi_2) \tag{32}$$

$$P_n^m(\eta_1)P_n^m(\xi_1) = \sum_{s=S}^{\frac{1}{2}n} \sum_{p=P}^{s} (-1)^{\frac{1}{2}n-p} F_{mnsp}\left(\frac{c_2}{c_1}\right)^{2s} P_{2p}^m(\eta_2)P_{2p}^m(\xi_2) \tag{33}$$

$$F_{mnsp} = \frac{(4p+1)(n+2s-1)!!(2p-m)!(n+m)!}{(2s+2p+1)!!(2p+m)!(n-m)!(n-2s)!!(2s-2p)!!} \tag{34}$$

Here S and P equal $\frac{1}{2}m$ or $\frac{1}{2}(m+1)$ and are integer if n is even and integer plus $\frac{1}{2}$ if n is odd and s and p increase by unit steps.

There are two more formulas that will prove useful in working problems. The first is IT II (6), page 29, and the second is derivable from HTF (25), page 159, by expansion of the integrand binomial and integration of the real part. In both $P_n^m(\cos\theta)$ has been multiplied by $(-1)^m$ to change it from Hobson's notation to that of 5.23 (6) and (10). Thus

$$\int_0^\infty e^{-tz}J_m(t\rho)t^n\,dt = (n-m)!r^{-n-1}P_n^m(\cos\theta) \tag{35}$$

$$\left(\frac{r}{c}\right)^n P_n^m(\cos\theta) = (n+m)!\left(\frac{z}{c}\right)^n \sum_{s=m}^{S} \frac{(-1)^{\frac{1}{2}(s-m)}}{(n-s)!(s+m)!!(s-m)!!}\left(\frac{\rho}{z}\right)^s \tag{36}$$

where S is n or $n-1$ and $s+m$ is even so s jumps by two in successive terms of the summation.

5.41. Sphere in Concentric, Coaxial, Earthed Spheroid.—Suppose that the potential inside and on the surface of a sphere of radius a is

$$V = \sum_{n=0}^{N} B_n\left(\frac{r}{a}\right)^{2n} P_{2n}(\cos\theta) \tag{1}$$

and that outside it there is a concentric coaxial spheroid ζ_0 (or η_0) with interfocal distance $2c$, at potential zero. It is desired to find the potential between them and the charge densities. Let the interior and exterior potentials of the charge distribution on the sphere in the presence of the spheroid be, using 5.40 (27) and 5.40 (10) for the oblate case,

$$\begin{aligned} V_{in} &= \sum_{n=0}^{N} A_n\left(\frac{r}{a}\right)^{2n} P_{2n}(\cos\theta) \\ &= \sum_{n=0}^{N}\sum_{s=0}^{n} A_n \frac{(-1)^n(4s+1)(2n)!(a/c)^{2n}}{(2n+2s+1)!!(2n-2s)!!} P_{2s}(j\zeta)P_{2s}(\xi) \end{aligned} \tag{2}$$

$$\begin{aligned} V_{ex} &= \sum_{n=0}^{N} A_n\left(\frac{a}{r}\right)^{2n+1} P_{2n}(\cos\theta) \\ &= \sum_{n=0}^{N}\sum_{s=n}^{\infty} A_n \frac{j(2s+1)(2n+2s-1)!!}{(-1)^s(2n)!(2s-2n)!!}\left(\frac{c}{a}\right)^{2n+1} Q_{2s}(j\zeta)P_{2s}(\xi) \end{aligned} \tag{3}$$

In the prolate case 5.40 (28) and 5.40 (11) are used. The potential V'_{in} of the charge density on the spheroid $\zeta=\zeta_0$ (or $\eta=\eta_0$) must exactly cancel V_{ex} there and thus is given by (3) when $-Q_{2s}(j\zeta_0)P_{2s}(j\zeta)/P_{2s}(j\zeta_0)$

replaces $Q_{2s}(j\zeta)$. Thus

$$V'_{in} = \sum_{n=0}^{N}\sum_{s=n}^{\infty} -A_n \frac{j(2s+1)(2n+2s-1)!!}{(-1)^s(2n)!(2s-2n)!!}\left(\frac{c}{a}\right)^{2n+1}\frac{Q_{2s}(j\zeta_0)}{P_{2s}(j\zeta_0)}P_{2s}(j\zeta)P_{2s}(\xi) \quad (4)$$

The sum of the coefficients of $P_{2s}(j\zeta)P_{2s}(\xi)$ due to $V_{in} + V'_{in}$ must equal its original coefficient as given by (1), which is (2) with B_n written for A_n. Thus we have $N + 1$ simultaneous equations, one for each value of s from 0 to N inclusive, to be solved for $A_0, \ldots, A_N$, which are

$$\begin{array}{ll} 0 & A_0C_{00} + A_1C_{01} + A_2C_{02} + \cdots + A_NC'_{0N} = \sum_{n=0}^{N} B_nC_{0n} \\ 1 & A_0C_{10} + A_1C_{11} + A_2C_{12} + \cdots + A_NC'_{0N} = \sum_{n=1}^{N} B_nC_{1n} \\ \cdots & \cdots \\ N & A_0C_{N0} + A_1C_{N1} + A_2C_{N2} + \cdots + A_NC'_{NN} = B_NC_{NN} \end{array} \quad (5)$$

It is evident from (2) and (4) that when $s \leqslant n$ only the sphere's charge, and when $s \geqslant n$ only the spheroidal charge, contributes to the coefficients of $P_{2s}(j\zeta)P_{2s}(\xi)$ so that

$$s < n \qquad C_{sn} = \frac{(-1)^n(4s+1)(2n)!}{(2n+2s+1)!!(2n-2s)!!}\left(\frac{a}{c}\right)^{2n} \quad (6)$$

$$C_{nn} = (-1)^n\left[\frac{(2n)!}{(2n-\frac{1}{2})!}\left(\frac{a}{2c}\right)^{2n} + \frac{(2n+1)(4n-1)!!}{j(2n)!!}\left(\frac{c}{a}\right)^{2n+1}\frac{Q_{2n}(j\zeta_0)}{P_{2n}(j\zeta_0)}\right] \quad (7)$$

$$s > n \qquad C'_{sn} = \frac{j(2s+1)(2n+2s-1)!!Q_{2s}(j\zeta_0)}{(-1)^{s+1}(2n)!(2s-2n)!!P_{2s}(j\zeta_0)}\left(\frac{c}{a}\right)^{2n+1} \quad (8)$$

For the prolate case write $(-1)^s$ for $(-1)^n$ in (6) and -1 for j in (7), omit j and write $(-1)^n$ for $(-1)^{s+1}$ in (8), and replace $j\zeta_0$ by η_0 in (7) and (8). The potential between sphere and spheroid, $V_{ex} + V'_{in}$, is

$$\sum_{n=0}^{N}\sum_{s=n}^{\infty} A_nC_{sn}\left(\frac{c}{a}\right)^{2n+1}\left[\frac{P_{2s}(j\zeta_0)}{Q_{2s}(j\zeta_0)}Q_{2s}(j\zeta) - P_{2s}(j\zeta)\right]P_{2s}(\xi) \quad (9)$$

Use of the Wronskian, 5.214 (5), will simplify the charge density formula for the spheroid. For the sphere (9) should be written in spherical coordinates by 5.40 (7) and (24) before the differentiation.

Where the sphere is an equipotential, $B_0 = V_0$ and $B_1, B_2, \ldots, B_N$ are zero; therefore the right sides of all but the first equation in (5) are zero. The charge on the sphere is the coefficient of $1/r$ in (3) multiplied by $4\pi\epsilon$, so the capacitance between sphere and spheroid is

$$C = \frac{Q}{V_0} = \frac{4\pi\epsilon A_0 a}{V_0} \tag{10}$$

The C_{pn} factors are very easy to calculate and arrange by a digital computer if recursion formulas, 5.233, are used for the Legendre functions. If the capacitance is wanted to five significant figures, then N should be taken just large enough so that using $N + 1$ instead of N equations in (5) leaves the fifth digit in C unchanged.

If the sphere with the fixed potential distribution is outside the spheroid, exactly similar steps are taken starting with

$$V = \sum_{n=0}^{N} B_n\left(\frac{a}{r}\right)^{2n+1} P_{2n}(\cos\theta) \tag{11}$$

valid on and outside the sphere but with the roles of internal and external potentials reversed. In the more general case V will depend on r, θ, and ϕ, so double summations involving A_{mn} coefficients will occur. If the spheroids are concentric but the axis of the fixed potential on one spheroid does not coincide with that of the second spheroid, it may be made to do so by writing it in spherical coordinates and using 5.24.

5.42. Charged Sphere in Cylinder.—As another application of the formulas of 5.40, consider a sphere of radius c charged to potential V_0, which lies on the axis of an infinite cylinder of radius a at zero potential. The internal and external potentials of the charge on the sphere are, using 5.40 (3),

$$V_{in} = \sum_{n=0}^{N} A_n\left(\frac{r}{c}\right)^{2n} P_{2n}(\cos\theta) \tag{1}$$

$$V_{ex} = \sum_{n=0}^{N} A_n\left(\frac{c}{r}\right)^{2n+1} P_{2n}(\cos\theta)$$

$$= \frac{2}{\pi}\sum_{n=0}^{N}(-1)^n A_n c^{2n+1}\int_0^\infty t^{2n}K_0(t\rho)\cos(tz)\,dt \tag{2}$$

The induced charges on the cylinder must exactly cancel V_{ex} at $\rho = a$, so that its internal potential V'_{in} is formed by replacing $K_0(t\rho)$ by $-I_0(t\rho)K_0(ta)/I_0(ta)$ in (2). Elimination of $I_0(t\rho)\cos(tz)$ by 5.40 (19), where n is replaced by $2s$ and t by u/a, gives

$$V'_{in} = \frac{2}{\pi}\sum_{n=0}^{N}\sum_{s=0}^{\infty}\frac{(-1)^{n+s+1}I(2n+2s)A_n}{(2n+2s+1)(2s)!}\left(\frac{c}{a}\right)^{2n+2s+1}\left(\frac{r}{c}\right)^{2s}P_{2s}(\cos\theta) \tag{3}$$

$$I(2n+2s) = \int_0^\infty \frac{u^{2n+2s}\,du}{I_0^2(u)} = (2n+2s+1)\int_0^\infty \frac{u^{2n+2s}K_0(u)}{I_0(u)}\,du \tag{4}$$

The third form in (4) was integrated by parts to give the second. A table of numerical values of $I(2n+2s)$ is given in *J. Math. Phys.*, Vol. 4, page 834, 1963. These are needed in all cylindrical applications of the formulas of 5.40. Inside the sphere $V_{in} + V'_{in}$ must equal V_0, so that the coefficient of $(r/c)^{2n}P_{2n}(\cos\theta)$ appearing in (1) and (3) must vanish if $n \neq 0$ and be V_0 when $n = 0$. Thus there is a set of simultaneous equations like 5.41 (5) to be solved where $B_0 = V_0$ and $B_n = 0$ if $n \neq 0$. From (1) and (3)

$$n \neq p \qquad C_{np} = \frac{2(-1)^{n+p+1}I(2n+2p)}{\pi(2n+2p+1)(2n)!}\left(\frac{c}{a}\right)^{2n+2p+1} \tag{5}$$

$$n = p \qquad C_{nn} = 1 - \frac{2I(4n)}{\pi(4n+1)(2n)!}\left(\frac{c}{a}\right)^{4n+1} \tag{6}$$

The potential between sphere and cylinder is $V_{ex} + V'_{in}$. An idea of the number of coefficients N needed for different values of the radii of the sphere (c) and of the cylinder (a) can be found by looking at the data taken from the reference cited, where a comparison of (1) and (2) of this section with (3) and (4) of the reference shows that $A_n = (A_n)_{ref}/(4n+1)$. The potential is taken to be 1.

c/a	A_0	A_1	A_2	A_3	A_4	A_5
0.1	1.095373	−0.000226	0.000001			
0.3	1.353616	−0.007532	0.000181	−0.000004		
0.5	1.773104	−0.046118	0.002161	−0.000216	0.000014	−0.000001

In the reference cited numerical values are given for a wide range of c/a and for prolate and oblate spheroids including the disk. The reference formulas could be somewhat simplified by using more relations from 5.40, some of which were not available when it appeared. As in the last article the capacitance is

$$C = \frac{Q}{V} = \frac{4\pi\epsilon A_0 c}{V_0} \tag{7}$$

5.43. Cylinder in Hole in Sheet.—The formulas of 5.40 also apply to the case of an infinite circular conducting cylinder of radius a at potential

V_0 which passes through the center of a circular hole of radius c in an infinite thin conducting sheet normal to its axis as shown in Fig. 5.43*a*. With no cylinder the problem is solved in 5.272 using oblate spheroidal harmonics where $-\infty < \zeta < \infty$ and $0 \leqslant \xi \leqslant 1$. In the present case with a cylindrical boundary, 5.40 (1) applies, but now $P_n(\xi)$ is always even as regards z and $Q_n(j\zeta)$ is even only when n is odd from 5.213 (3). Thus a solution of Laplace's equation giving $V = 0$ on the plane outside the hole is

$$V = \sum_{n=0}^{N} A_n Q_{2n+1}(j\zeta) P_{2n+1}(\xi) \tag{1}$$

The problem is to choose A_n so that the cylinder $\rho = a$ is at potential V_0. An equation derivable from 5.40 (1) facilitates the analysis. Let (1*a*)

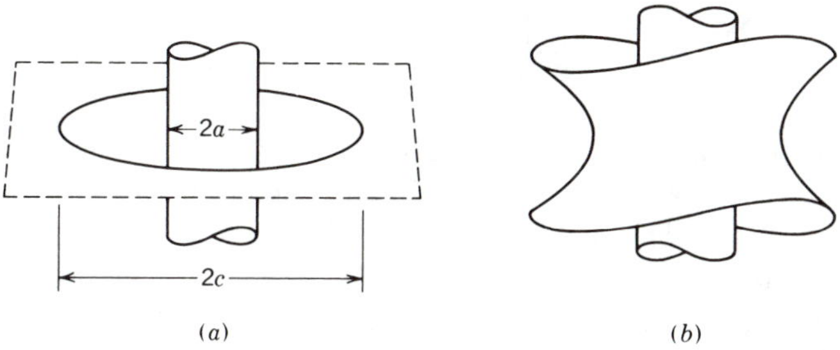

FIG. 5.43*a*, *b*.

designate 5.40 (1) or 5.40 (2) where $n + 2$ replaces n. Differentiate (1*a*) with respect to c, and eliminate $I'_{n+\frac{5}{2}}$ by 5.321 (1) and then one integral by (1*a*) to get Eq. (1*b*). Differentiate this with respect to c, and eliminate $I'_{n+\frac{3}{2}}$ by 5.321 (1) and then one integral by (1*b*). Note that the negative of the final integral is that produced by differentiation of 5.40 (1) or (2) twice with respect to z. Thus

$$\frac{\partial^2[Q_n(j\zeta)P_n(\xi)]}{\partial z^2} = -c^{-2}n(n+2)Q_{n+2}(j\zeta)P_{n+2}(\xi) \tag{2}$$

Evidently repetitions of this operation will yield

$$\frac{\partial^{2p}[Q_n(j\zeta)P_n(\xi)]}{\partial z^{2p}} = (-1)^{p+1}\frac{(n+2p-2)!!(n+2p)!!}{(n-2)!!n!!\,c^{2p}}Q_{n+2p}(j\zeta)P_{n+2p}(\xi) \tag{3}$$

The boundary condition applied to (1) is that $V = V_0$ and $\partial^{2p}V/\partial z^{2p} = 0$ when $z = 0$ and $\rho = a$ or when $\zeta = 0$ and $c\xi = (c^2 - a^2)^{\frac{1}{2}}$. Equation 5.214 (4) gives $Q_{2n+1}(j0)$, so that the $N + 1$ simultaneous equations, one for each value of p from 0 to N inclusive, to be solved for A_n are

$$V_0\delta_{0p} = \sum_{n=0}^{N} A_n \frac{(-1)^n(2n+2p)!!(2n+2p-1)!!}{(2n+1)!!(2n-1)!!} P_{2n+2p+1}[(c^2-a^2)^{\frac{1}{2}}c^{-1}] \tag{4}$$

where δ_{0p} is zero unless $p = 0$. The larger the value of c/a the smaller the N needed in (1) for a given accuracy.

The equipotentials of (1) will somewhat resemble hyperboloids of one sheet. If the solution of the problem of a cylinder of radius a passing axially through the hole in the hyperboloid $\xi = \xi_0$ is desired as in Fig. 5.43b, then (1) is replaced by

$$V = \sum_{n=0}^{N} A_n Q_{2n+1}(j\zeta)\left[P_{2n+1}(\xi) - \frac{Q_{2n+1}(\xi)P_{2n+1}(\xi_0)}{Q_{2n+1}(\xi_0)}\right] \tag{5}$$

The manipulation is similar to that given except that another formula, which appears on page 326 of the first Cooke reference of 5.40, is used for the new term. It is like 5.40 (1) but with $P_n^m(\xi)$ replaced by $-(2/\pi)Q_n^m(\xi)$.

5.44. Off-axis Charged Sphere in Cylinder.—In Art. 5.40 formulas for the solution of mixed coordinate problems where the two boundaries are concentric but not necessarily coaxial were given. These are inadequate if the surfaces are not concentric, so a different approach is needed. One is to express the external fields in terms of their multipole sources. A good example of this method is the solution for the potential between a cylinder at potential zero and an off-axis sphere at potential V_0 inside it. Two coordinate systems are needed. In the first ρ, ψ, z system the cylinder wall lies at $\rho = b$. The center of the sphere at $z = 0$, $\rho = a$, $\psi = 0$ is the origin of the second system r, θ, ϕ whose $\theta = 0$ axis parallels the z-axis and whose $\phi = 0$ plane lies in the $\psi = 0$ plane. The potential of the sphere of radius c due to its own charge density in the presence of the cylinder is

$$V_s = \sum_{n=0}^{\infty}\sum_{m=0}^{n} A_{nm} r^{-n-1} P_n^m(\cos\theta)\cos m\phi \qquad r > c$$
$$= \sum_{n=0}^{\infty}\sum_{m=0}^{n} A_{nm} r^n c^{-2n-1} P_n^m(\cos\theta)\cos m\phi \qquad r < c \tag{1}$$

This must have even symmetry in both z and ϕ. The $\cos m\phi$ term takes care of ϕ and the choice of even $m + n$ takes care of z. The potential of the induced charge on the cylinder must be added to this and if the sphere is conducting the sum must have the constant value V_0 when $r \lessgtr c$. The steps in the solution are:

1. Write the external harmonics $r^{-n-1}P_n^m(\cos\theta)\cos m\phi$ in terms of x-, y-, z-derivatives of r^{-1}.
2. Express the reciprocal distance r^{-1} in terms of ρ, ψ, z, and obtain by differentiation the potentials of the multipoles of the system. This gives spherical harmonics $r^{-n-1}P_n^m(\cos\theta)\cos m\phi$ in terms of ρ, ψ, z.
3. Write down the interior potential V_c of the induced charge density on the cylinder $\rho = b$ by noting that V_c must cancel V_s there.
4. Express this V_c in terms of r, θ, ϕ.
5. Add V_s when $r < c$ to V_c, and equate the coefficients of the sum of the harmonics containing r^k to V_0 when $k = 0$ and to zero when $k \neq 0$.
6. Solve the resultant linear simultaneous algebraic equations for A_{mn}.
7. Put these values into V_s with $r > c$ plus V_c. The sum is V.

The multipole expansion for $r^{-n-1}P_n^m(\cos\theta)\cos m\phi$ is given in complex form in HTF II (26), page 251, or Magnus 10.3.45. In the notation of 5.23 (6) the real part becomes, when $\partial^2(r^{-1})/\partial y^2$ is replaced by $-\partial^2(r^{-1})/\partial x^2 - \partial^2(r^{-1})/\partial z^1$, for field point insert $(-1)^n$ on right side of (2) and (3).

$$r^{-n-1}P_n^m(\cos\theta)\cos m\phi = \sum_{s=0}^{S}\frac{(2-\delta_m^0)^{m-1}m(m-s-1)!}{(n-m)!4^s(m-2s)!s!}\frac{\partial^n(r^{-1})}{\partial x_0^{m-2s}\,\partial z_0^{n-m+2s}} \tag{2}$$

Similarly the imaginary part becomes

$$r^{-n-1}P_n^m(\cos\theta)\sin m\phi = \sum_{s=0}^{S}\frac{2^{m-1}(m-s-1)!}{(n-m)!4^s(m-2s-1)!s!}\frac{\partial^n(r^{-1})}{\partial y_0\,\partial x_0^{m-2s-1}\,\partial z_0^{n-m+2s}} \tag{3}$$

The limit S is $\frac{1}{2}m$ or $\frac{1}{2}(m-1)$, whichever is an integer, and the source coordinates are x_0, y_0, z_0. To avoid ambiguity in some cases, put in the s value and simplify before insertion of m.

For step 2 use 5.35 (2), which gives the reciprocal distance from the source point at $\rho = a$, $z = z_0$, $\psi = 0$ to the field point ρ, z, ψ to be

$$r^{-1} = \pi^{-1}\sum_{p=0}^{\infty}\epsilon_p\int_0^\infty K_p(t\rho)I_p(ta)\cos t(z-z_0)\,dt\cos p\psi \tag{4}$$

where $2-\delta_p^0$ has been replaced by ϵ_p. Substitute (4) in (2) writing $\lambda_s(m,n)$ for the product of $(-1)^{\frac{1}{2}n-\frac{1}{2}m+s}$ by the factors preceding the derivative in (2) and $I_p^{(m-2s)}(u)$ for $\partial^{m-2s}I_p(u)/\partial u^{m-2s}$, and obtain for

$r^{-n-1}P_n^m(\cos\theta)\cos m\phi$ the expression

$$\sum_{p=0}^{\infty}\frac{\epsilon_p\cos p\psi}{\frac{1}{2}\pi}\sum_{s=0}^{S}\lambda_s(m,n)\int_0^\infty t^nK_p(t\rho)I_p{}^{(m-2s)}(ta)\cos tz\,dt \tag{5}$$

Step 3 consists of replacement of $K_p(t\rho)$ by $-A_{mn}K_p(tb)I_p(t\rho)/I_p(tb)$, which yields a solution of Laplace's equation that cancels

$$A_{nm}r^{-n-1}P_n^m(\cos\theta)\cos m\phi$$

on the cylinder $\rho = b$ and is valid inside it. For step 4 note that ρ^2 is $r^2\sin^2\theta + a^2 - 2ar\sin\theta\cos\phi$ and that $r\sin\theta$ is the ρ of 5.40 (19), so that from HTF II (36), page 102, or HMF 9.1.79 there results

$$\begin{aligned} I_p(t\rho)\cos p\psi &= \sum_{q=-\infty}^{+\infty}(-1)^qI_{p+q}(ta)I_q(tr\sin\theta)\cos q\phi \\ &= \sum_{q=0}^{\infty}\{(-1)^q[I_{p+q}(ta) + (1-\delta_{p-q}^0)I_{p-q}(ta)]\}I_q(tr\sin\theta)\cos q\phi \end{aligned} \tag{6}$$

The second form follows from the first because $I_n(v) = I_{-n}(v)$ by Art. 5.32. Multiplication of (6) by $\cos tz$, application of 5.40 (19) to the terms with r and z, replacement of the brace in (6) by $f_{pq}(ta)$ and of m and n by q and w, where $w + q$ is even by symmetry, give

$$I_p(t\rho)\cos tz\cos p\psi = \sum_{q=0}^{\infty}\sum_{w=q}^{\infty}\frac{f_{pq}(ta)t^w}{(-1)^w(w+q)!}r^wP_w^q(\cos\theta)\cos q\phi \tag{7}$$

This equation and (5) as modified by step 3 show that the coefficient C_{wq}^{nm} of $r^wP_w^q(\cos\theta)\cos q\phi$ in V_c due to $A_{nm}r^{-n-1}P_n^m(\cos\theta)\cos m\phi$ in V_s is

$$C_{wq}^{nm} = -\sum_{p=0}^{\infty}\sum_{s=0}^{S}\frac{2\epsilon_p\lambda_s(m,n)A_{nm}}{\pi(-1)^w(w+q)!}\int_0^\infty\frac{f_{pq}(ta)t^{w+n}K_p(tb)I_p{}^{(m-2s)}(t)}{I_p(tb)}\,dt \tag{8}$$

By 5.32 (2) $f_{pq}(ta)$ is written in powers of ta, and differentiation of 5.32 (2) gives $I_p{}^{(m-2s)}(ta)$ in powers of ta. Thus the integrals needed are

$$\int_0^\infty\frac{t^nK_p(t)}{I_p(t)}\,dt = \frac{1}{n+1}\int_0^\infty\frac{t^n\,dt}{[I_p(t)]^2} \tag{9}$$

This integral is tabulated only for $p = 0$ [see 5.42 (4)] and for $p = 1$ (see

6.18). Step 5 requires that

$$\begin{aligned} V_0 &= (c^{-1} + c^{00}_{00})A_{00} + c^{20}_{00}A_{20} + \cdots \\ &\qquad + c^{11}_{00}A_{11} + c^{31}_{00}A_{31} + \\ 0 &= c^{00}_{11}\,A_{00} + c^{20}_{11}A_{20} + \cdots \\ &\qquad + (c^{-3} + c^{11}_{11})A_{11} + c^{31}_{11}A_{31} + \\ 0 &= c^{00}_{20}\,A_{00} + (c^{-5} + c^{20}_{20})A_{20} + \cdots \\ &\qquad + c^{11}_{20}A_{11} + c^{31}_{20}A_{31} + \\ &\cdots\cdots\cdots\cdots\cdots\cdots\cdots\cdots \end{aligned} \tag{10}$$

These equations are solved for A_{nm}, which is then inserted in (1) with $r > c$ to get V_s and in (4) as modified by step 3 to get V_c. Then

$$V = V_s + V_c \tag{11}$$

Another form for the potential may be more convenient. Once the A_{pq}'s are known, the sphere can be replaced by a set of multipoles of known magnitude by (2). In the Green's function 5.298 (4), $z - z_0$ is written for z, b for a, a for b, and 0 for ϕ_0. Then, by differentiation with respect to z_0 and b, the potential in an earthed cylinder of each multipole may be found, and the sum of these, weighted according to their magnitude, gives V for a sphere centered at $z = 0$, $\rho = b$, $\phi = 0$.

If the sphere is replaced by a spheroid, the problem can be set up in the same way. By 5.40 (7) or (8) the spheroidal harmonics are expressed as a sum of spherical harmonics, which are then treated as in this article up to (7), at which point the spherical are converted back into spheroidal by 5.40 (10) or (11), and so forth. Before the advent of digital computers the numerical work involved in such methods would have been too laborious to be practical, but with these devices, solutions of any desired accuracy can be found.

Problems

Problems marked C are taken from the Cambridge examination questions as reprinted by Jeans by permission of the Cambridge University Press.

1. Two similar charges are placed at a distance $2b$ apart. Show that a grounded conducting sphere placed midway between them must have a radius of approximately $b/8$ to neutralize their mutual repulsion.

2. The radius vector from the center of a conducting sphere of radius a, carrying a charge Q, to a point charge q is $\mathbf{c}$. When the system is placed in a uniform field $\boldsymbol{E}$ along the direction of $\mathbf{c}$ show that, in order that no force act on q, Q must have the value

$$-c^2\left[4\pi\epsilon E\left(1 + 2\left(\frac{a}{c}\right)^3\right) - q\left(\frac{a}{c}\right)^3 \frac{2c^2 - a^2}{(c^2 - a^2)^2}\right]$$

3. An infinite conducting plane forms one boundary of a uniform field of unknown strength X. A known charge q, when placed at a distance r from the plane, experiences an unknown force F. It is found that if a hemispherical piece of conducting material, of radius a, is placed flat against the plane opposite the charge the force is unchanged. Show that

$$X = \frac{qr^6}{2\pi\epsilon(r^4 - a^4)^2} \qquad F = \frac{q^2}{4\pi\epsilon}\left[\frac{2r^6}{(r^4 - a^4)^2} - \frac{1}{4r^2}\right]$$

4. A hollow conducting sphere of radius a is half filled with a dielectric. On the axis of symmetry at a distance $a/3$ from the plane dielectric surface, a charge experiences no image force. Show that the capacitivity is $1.541\epsilon_v$.

5*C*. An infinite conducting plane at zero potential is under the influence of a charge of electricity at a point O. Show that the charge on any area of the plane is proportional to the angle it subtends at O.

6*C*. If two infinite plane uninsulated conductors meet at an angle of 60° and there is a charge e at a point equidistant from each and distant r from the line of intersection, find the electrification at any point of the planes. Show that at a point in a principal plane through the charged point at a distance $r3^{\frac{1}{2}}$ from the line of intersection, the surface density is

$$\frac{-e}{4\pi r^2}\left(\frac{3}{4} + \frac{1}{7^{\frac{3}{2}}}\right)$$

7*C*. What is the least positive charge that must be given to a spherical conductor, insulated and influenced by an external point charge e at distance r from its center, in order that the surface density may be everywhere positive?

8*C*. An uninsulated conducting sphere is under the influence of an external electric charge. Find the ratio in which the induced charge is divided between the part of its surface in direct view of the external charge and the remaining part.

9*C*. A point charge e is brought near to a spherical conductor of radius a having a charge E. Show that the particle will be repelled by the sphere, unless its distance from the nearest point of its surface is less than $\frac{1}{2}a(e/E)^{\frac{1}{2}}$, approximately (take $e \ll E$).

10*C*. A hollow conductor has the form of a quarter of a sphere bounded by two perpendicular diametral planes. Find the images of a charge placed at any point inside.

11*C*. A conducting surface consists of two infinite planes which meet at right angles and a quarter of a sphere of radius a fitted into the right angle. If the conductor is at zero potential and a point charge e is symmetrically placed with regard to the planes and the spherical surface at a great distance f from the center, show that the charge induced on the spherical portion is approximately $-5ea^3/(\pi f^3)$.

12*C*. A thin plane conducting lamina of any shape and size is under the influence of a fixed electrical distribution on one side of it. If σ_1 is the density of the induced charge at a point P on the side of the lamina facing the fixed distribution and σ_2 that at the corresponding point on the other side, prove that $\sigma_1 - \sigma_2 = \sigma_0$ where σ_0 is the density at P of the distribution induced on an infinite plane conductor coinciding with the lamina.

13*C*. A conducting plane has a hemispherical boss of radius a, and at a distance f from the center of the boss and along its axis there is a point charge e. If the plane and the boss are kept at zero potential, prove that the charge induced on the boss is

$$-e\left(1 - \frac{f^2 - a^2}{f(f^2 + a^2)^{\frac{1}{2}}}\right)$$

14*C*. Electricity is induced on an uninsulated spherical conductor of radius a by a uniform surface distribution, density σ, over an external concentric nonconducting spherical segment of radius c. Prove that the surface density at the point A of the conductor at the nearer end of the axis of the segment is

$$\frac{-\sigma c}{2a^2}(c + a)\left(1 - \frac{AB}{AD}\right)$$

where B is the point of the segment on its axis and D is any point on its edge.

15*C*. Two conducting disks of radii a, a' are fixed at right angles to the line that joins their centers, the length of this line being r large compared with a. If the first has potential V and the second is uninsulated, prove that the charge on the first is

$$\frac{8\pi^2\epsilon a r^2 V}{(\pi^2 r^2 - 4aa')}$$

16*C*. A spherical conductor of diameter a is kept at zero potential in the presence of a fine uniform wire in the form of a circle of radius c in a tangent plane to the sphere with its center at the point of contact, which has a charge Q of electricity. Prove that the electrical density induced on the sphere at a point whose direction from the center of the ring makes an angle ψ with the normal to the plane is

$$\frac{-c^2 Q \sec^3 \psi}{4\pi^2 a}\int_0^{2\pi}(a^2 + c^2 \sec^2 \psi - 2ac \tan \psi \cos \theta)^{-\frac{3}{2}}\, d\theta$$

17. A stationary ion with charge q and mass m is formed in a highly evacuated conducting bulb of radius a at a distance c from its center. Show that the time for this ion to reach the wall is

$$t = q^{-1}k(4\pi m\epsilon a^3)^{\frac{1}{2}}(K - E) \qquad \text{where } k^2 = (a^2 - c^2)a^{-2}$$

K and E are complete elliptic integrals of modulus k.

18. An earthed conducting sphere of radius a has its center on the axis of a charged circular ring, any radius vector c from its center to the ring making an angle α with the axis. Show that the force sucking the sphere into the ring is

$$\frac{Q^2 E(c^2 - a^2)k^3 \cos \alpha}{16\pi^2\epsilon c^2 a^2 \sin^3 \alpha(1 - k^2)} \qquad \text{where } k^2 = \frac{4a^2c^2 \sin^2 \alpha}{a^4 + c^4 - 2a^2c^2 \cos 2\alpha}$$

E is the complete elliptic integral of modulus k.

19*C*. If a particle charged with a quantity e of electricity is placed at the middle point of the line joining the centers of two equal spherical conductors kept at zero potential, show that the charge induced on each sphere is

$$-2em(1 - m + m^2 - 3m^3 + 4m^4)$$

neglecting higher powers of m, which is the ratio of the radius to the distance between the centers of the spheres.

20*C*. Two insulated conducting spheres of radii a, b, the distance c of whose centers is large compared with a and b, have charges Q_1, Q_2, respectively. Show that the potential energy is approximately

$$(8\pi\epsilon)^{-1}[(a^{-1} - b^3c^{-4})Q_1^2 + 2c^{-1}Q_1Q_2 + (b^{-1} - a^3c^{-4})Q_2^2]$$

21*C*. Show that the force between two insulated spherical conductors of radius a placed in an electric field of uniform intensity E perpendicular to their line of centers is $12\pi\epsilon E^2a^6c^{-4}(1 - 2a^3c^{-3} - 8a^5c^{-5} + \cdots)$, c being the distance between their centers.

22*C*. Two uncharged insulated spheres, radii a, b, are placed in a uniform field of force so that their line of centers is parallel to the lines of force, the distance c between their centers being great compared with a and b. Prove that the surface density at the point at which the line of centers cuts the first sphere a is approximately

$$\epsilon E(3 + 6b^3c^{-3} + 15ab^3c^{-4} + 28a^2b^3c^{-5} + 57a^3b^3c^{-6} + \cdots)$$

23*C*. Two equal spheres each of radius a are in contact. Show that the capacity of the conductor so formed is $8\pi\epsilon a \ln 2$.

24. Two spheres of radii a and b are in contact. Show that the capacitance of the conductor so formed is

$$C = -\frac{4\pi\epsilon ab}{a+b}\left[2C' + \psi\left(\frac{a}{a+b}\right) + \psi\left(\frac{b}{a+b}\right)\right]$$

where C' is Euler's constant 0.5772 and $\psi(z) = \Gamma'(z)/\Gamma(z)$.

25*C*. A conducting sphere of radius a is in contact with an infinite conducting plane. Show that if a unit point charge is placed beyond the sphere and on the diameter through the point of contact at distance c from that point, the charges induced on the plane and sphere are

$$-\frac{\pi a}{c}\cot\left(\frac{\pi a}{c}\right) \qquad \text{and} \qquad \frac{\pi a}{c}\cot\left(\frac{\pi a}{c}\right) - 1$$

26*C*. Prove that if the centers of two equal uninsulated spherical conductors of radius a are at a distance $2c$ apart, the charge induced on each by a unit charge at a point midway between them is

$$\sum_{1}^{\infty}(-1)^n \operatorname{sech} n\alpha \qquad \text{where } c = a\cosh\alpha$$

27. Two equal spheres, of radius a, with their centers a distance c apart are connected by a thin wire. Show that the capacitance of the system is

$$8\pi\epsilon a \sinh\beta \sum_{n=1}^{\infty}(-1)^{n+1}\operatorname{csch} n\beta$$

where $\cosh\beta = \frac{1}{2}c/a$.

28. Two equal spheres, of radius a, with their centers a distance c apart, are charged to the same potential V. Show that the repulsion between them is

$$2\pi\epsilon V^2 \sum_{n=1}^{\infty}(-1)^{n+1}(\coth\beta - n\coth n\beta)\operatorname{csch} n\beta$$

where $\cosh\beta = \frac{1}{2}c/a$.

29*C*. An insulated conducting sphere of radius a is placed midway between two parallel infinite uninsulated planes at a great distance $2c$ apart. If $(a/c)^2$ is neglected, show that the capacity of the sphere is approximately $4\pi\epsilon a[1 + (a/c)\log 2]$.

30*C*. Two spheres of radii r_1, r_2 touch each other, and their capacities in this position are c_1, c_2. Show that

$$c_1 = 4\pi\epsilon r_2\left(f^2\sum_{1}^{\infty}n^{-2} + f^3\sum_{1}^{\infty}n^{-3} + f^4\sum_{1}^{\infty}n^{-4} + \cdots\right)$$

where $f = r_1(r_1 + r_2)^{-1}$.

31*C*. A point charge e is placed between two parallel uninsulated infinite conducting planes, at distances a and b from them, respectively. Show that the potential at a point between the planes which is at a distance z from the charge and is on the line through the charge perpendicular to the planes is

$$\frac{e}{8\pi\epsilon(a+b)}\left[-\Psi\left(\frac{z}{2a+2b}\right)+\Psi\left(\frac{2a-z}{2a+2b}\right)+\Psi\left(\frac{2b+z}{2a+2b}\right)-\Psi\left(\frac{2a+2b-z}{2a+2b}\right)\right]$$

where $\Psi(z) = [\Gamma'(z)]/[\Gamma(z)]$.

32. An electron, charge e, mass m, traveling horizontally in a high vacuum with a velocity v, must pass over an uncharged horizontal dielectric plate of length d. Show that, if it comes within a distance a of the front edge of the plate, it will be drawn into it by image forces before clearing the back edge, where, if edge effects are neglected,

$$a^3 = \frac{(K-1)e^2d^2}{2m\pi^3\epsilon_v(K+1)v^2}$$

33*C*. A point charge is placed in front of an infinite slab of dielectric, bounded by a plane face. The angle between a line of force in the dielectric and the normal to the face of the slab is α; the angle between the same two lines in the immediate neighbourhood of the charge is β. Prove that α, β are connected by the relation

$$\sin \tfrac{1}{2}\beta = \left(\frac{2K}{1+K}\right)^{\frac{1}{2}} \sin \tfrac{1}{2}\alpha$$

34*C*. Two dielectrics of inductive capacities K_1 and K_2 are separated by an infinite plane face. Charges e_1, e_2 are placed at points on a line at right angles to the plane, each at a distance a from the plane. Find the forces on the two charges, and explain why they are unequal.

35*C*. Two conductors of capacities c_1, c_2 in air are on the same normal to the plane boundary between two dielectrics with coefficients k_1, k_2, at great distances a, b from the boundary. They are connected by a thin wire and charged. Prove that the charge is distributed between them approximately in the ratio

$$k_1\left[\frac{4\pi\epsilon_v}{c_2}-\frac{k_1-k_2}{2b(k_1+k_2)}-\frac{2k_2}{(k_1+k_2)(a+b)}\right]:k_2\left[\frac{4\pi\epsilon_v}{c_1}+\frac{k_1-k_2}{2a(k_1+k_2)}-\frac{2k_1}{(k_1+k_2)(a+b)}\right]$$

36*C*. A conducting sphere of radius a is placed in air, with its center at a distance c from the plane face of an infinite dielectric. Show that its capacity is

$$4\pi\epsilon_v a \sinh\alpha \sum_1^\infty \left(\frac{K-1}{K+1}\right)^{n-1} \operatorname{cosech} n\alpha$$

where $\cosh\alpha = c/a$.

37. Show that the charge induced on an earthed conductor consisting of two similar spheres of radius a, intersecting orthogonally, by a charge q in their plane of intersection, at a distance $3a2^{-\frac{1}{2}}$ from the axis of symmetry is $(1/3 - 2/5^{\frac{1}{2}})q$.

38. A charge q is at a distance b from the center of a circular hole of radius a in an infinite earthed flat conducting sheet and lies in the same plane. Show that the charge density induced on the sheet at a distance r from the charge and c from the center of the hole is $-q(a^2-b^2)^{\frac{1}{2}}/[2\pi^2r^2(c^2-a^2)^{\frac{1}{2}}]$.

39. Show that, if in the preceding problem the charge is distributed around a ring of radius b concentric with the hole and coplanar with the sheet, the induced charge density at P is $-q(a^2-b^2)^{\frac{1}{2}}/[2\pi^2(c^2-b^2)(c^2-a^2)^{\frac{1}{2}}]$.

40. Invert the preceding problem, apply Green's reciprocation theorem, and show that the potential at a point P on a sphere, part of whose surface is occupied by a thin conducting bowl at potential V_0, is $(2V_0/\pi)\sin^{-1}(\cos\alpha/\cos\theta)$ where θ is the angle, measured at the bowl end of the rotational symmetry axis, between this axis and P and α that between this axis and the edge of the bowl.

41. Starting with the preceding result find the first term in the spherical harmonic expansion for the potential of a freely charged spherical bowl all elements of which are at a distance a from the origin where its radius subtends an angle β. From this show that its capacitance is $4\epsilon a(\beta + \sin\beta)$.

42. Invert the result of problem 39 and find the charge density induced on a spherical earthed conducting bowl by a uniform charge distribution $-\sigma_1$ over the remainder of the sphere. By superimposing a uniform spherical surface charge of density σ_1 show that the charge density at P on the inside or outside, respectively, of a freely charged conducting spherical bowl at potential V_0 is

$$\sigma_i = \frac{\epsilon V_0}{\pi a}\left[\frac{\sin\alpha}{(\sin^2\theta - \sin^2\alpha)^{\frac{1}{2}}} - \sin^{-1}\frac{\sin\alpha}{\sin\theta}\right], \qquad \sigma_o = \frac{\epsilon V_0}{a} + \sigma_i$$

where the angles are defined as in problem 40.

43*C*. A conductor is formed by the outer surfaces of two equal spheres, the angle between their radii at a point of intersection being $2\pi/3$. Show that the capacity of the conductor so formed is $2\pi a\epsilon(5 - 3^{-\frac{1}{2}}4)$, where a is the radius of either sphere.

44*C*. An uncharged insulated conductor formed of two equal spheres of radius a, cutting one another at right angles, is placed in a uniform field of force of intensity E, with the line joining the centers parallel to the lines of force. Prove that the charges induced on the two spheres are $\pm 11\pi\epsilon a^2E/2$.

45*C*. A conductor is bounded by the larger portions of two equal spheres of radius a cutting at an angle $\pi/3$ and of a third sphere of radius c cutting the two former orthogonally. Show that the capacity of the conductor is

$$4\pi\epsilon\{c + a(\tfrac{5}{2} - \tfrac{2}{3}3^{\frac{1}{2}}) - ac[2(a^2 + c^2)^{-\frac{1}{2}} - 2(a^2 + 3c^2)^{-\frac{1}{2}} + (a^2 + 4c^2)^{-\frac{1}{2}}]\}$$

46*C*. A spherical shell of radius a with a little hole in it is freely electrified to potential V. Prove that the charge on its inner surface is less than $\frac{1}{2}\epsilon VS/a$, where S is the area of the hole.

47*C*. A thin spherical conducting shell from which any portions have been removed is freely electrified. Prove that the difference of densities inside and outside at any point is constant.

48*C*. Prove that the capacity of an elliptic plate of small eccentricity e and area A is approximately

$$8\epsilon\left(\frac{A}{\pi}\right)^{\frac{1}{2}}\left(1 + \frac{e^4}{64} + \frac{e^6}{64}\right)$$

49*C*. An ellipsoidal conductor differs but little from a sphere. Its volume is equal to that of a sphere of radius r, its axes are $2r(1 + \alpha)$, $2r(1 + \beta)$, $2r(1 + \gamma)$. Show that neglecting cubes of α, β, γ, its capacity is $4\pi\epsilon r[1 + (\frac{2}{15})(\alpha^2 + \beta^2 + \gamma^2)]$.

50. An oblate conducting spheroid with semiaxes a and b has a charge Q. Show that the repulsion between the two halves into which it is divided by its diametral plane is $Q^2[16\pi\epsilon(b^2 - a^2)]^{-1}\ln(b/a)$, where $b > a$.

51*C*. A prolate conducting spheroid, semiaxes a, b, has a charge Q of electricity. Show that repulsion between the two halves into which it is divided by its diametral plane is $Q^2[16\pi\epsilon(a^2 - b^2)]^{-1}\ln(a/b)$ where $a > b$. Determine the value of the force in the case of a sphere.

52C. A thin circular disk of radius a is electrified with charge Q and surrounded by a spheroidal conductor with charge Q_1, placed so that the edge of the disk is the locus of the focus S of the generating ellipse. Show that the energy of the system is $[Q^2 \angle BSC + (Q + Q_1)^2 \angle SBC]/(8\pi\epsilon a)$, B being an extremity of the polar axis of the spheroid and C the center.

53C. If the two surfaces of a condenser are concentric and coaxial oblate spheroids of small ellipticities δ and δ' and polar axes $2c$ and $2c'$, prove that the capacity is $4\pi\epsilon cc'(c' - c)^{-2}[c' - c + \frac{2}{3}(\delta c' - \delta' c)]$, neglecting squares of the ellipticities. Find the distribution of electricity on each surface to the same order of approximation.

54C. An accumulator is formed of two confocal prolate spheroids, and the specific inductive capacity of the dielectric is $Kl/\tilde{\omega}$, where $\tilde{\omega}$ is the distance of any point from the axis. Prove that the capacity of the accumulator is

$$\frac{2\pi^2\epsilon_v Kl}{\ln\left(\dfrac{a_1 + b_1}{a + b}\right)}$$

where a, b and a_1, b_1 are the semiaxes of the generating ellipses.

55C. A thin spherical bowl is made by the part of the sphere $x^2 + y^2 + z^2 = cz$ bounded by and lying within the cone $(x/a)^2 + (y/b)^2 = (z/c)^2$ and is put in connection with the earth by a fine wire. O is the origin, and C, diametrically opposite to O, is the vertex of the bowl; Q is any point on the rim, and P is any point on the great circle arc CQ. Show that the surface density induced at P by a charge E placed at O is

$$-\frac{Ec}{4\pi abI}\,\frac{CQ}{OP^2(OP^2 - OQ^2)^{\frac{1}{2}}}, \qquad \text{where } I = \int_0^{\frac{1}{2}\pi} \frac{d\theta}{(a^2 \sin^2 \theta + b^2 \cos^2 \theta)^{\frac{1}{2}}}$$

56. A charge q is placed at a distance c from the center of a spherical hollow of radius a in an infinite dielectric of relative capacitivity K. Show that the force acting on the charge is

$$\frac{(K - 1)q^2}{4\pi\epsilon_v c^2} \sum_{n=0}^{\infty} \frac{n(n + 1)}{n + K(n + 1)} \left(\frac{c}{a}\right)^{2n+1}$$

57. An earthed conducting sphere of radius a has its center on the axis of a charged circular ring, any radius vector c from this center to the ring making an angle α with the axis. Show that the force sucking the sphere into the ring is

$$-\frac{Q^2}{4\pi\epsilon c^2} \sum_0^{\infty} (n + 1) P_{n+1}(\cos \alpha) P_n(\cos \alpha) \left(\frac{a}{c}\right)^{2n+1}$$

58. The spherical coordinates of a circular ring are a, α. A sphere of relative capacitivity K and radius b has its center at the origin. If the ring carries a charge of linear density τ, show that the potential between the sphere and the ring is

$$\frac{\tau}{2\epsilon_v} \sum_{n=0}^{\infty} P_n(\cos \alpha) \sin \alpha \left[\left(\frac{r}{a}\right)^n - \frac{n(K - 1)b^{2n+1}}{a^n(n(K + 1) + 1)r^{n+1}}\right] P_n(\cos \theta)$$

59. That portion of a sphere of radius a, lying between $\theta = \alpha$ and $\theta = \pi - \alpha$ is uniformly electrified with a surface density σ. Show that the potential at an external point is

$$\frac{a\sigma}{\epsilon}\left\{\frac{a}{r}\cos\alpha + \sum_{1}^{\infty}\frac{1}{4n+1}[P_{2n+1}(\cos\alpha) - P_{2n-1}(\cos\alpha)]\left(\frac{a}{r}\right)^{2n+1}P_{2n}(\mu)\right\}$$

60. A circular ring of radius a and charge Q is used as a collector at the center of an earthed spherical photoelectric cell of radius b. Show that the field strength at the spherical surface is

$$\frac{Q}{4\pi\epsilon}\sum_{n=0}^{\infty}(-1)^n\frac{1\cdot 3\cdots(2n-1)}{2\cdot 4\cdots 2n}\frac{4n+1}{b^2}\left(\frac{a}{b}\right)^{2n}P_{2n}(\cos\theta)$$

61. A point charge q is placed at a distance b from the center of two concentric earthed conducting spheres of radii a and c, where $a < b < c$. Show that when $a < r < b$, the potential is

$$V = \frac{q}{4\pi\epsilon}\sum_{n=0}^{\infty}\frac{b^{2n+1} - c^{2n+1}}{b^{n+1}(a^{2n+1} - c^{2n+1})}\left(r^n - \frac{a^{2n+1}}{r^{n+1}}\right)P_n(\cos\theta)$$

62. An infinite flat conducting plate is influenced by a charge q at a distance b from it, and there is a solid dielectric hemisphere of radius a flat against the plate opposite the charge. Show that the surface density induced on that part of the plate lying outside the dielectric is

$$\frac{-q}{4\pi\epsilon}\left[\frac{b}{(b^2+r^2)^{\frac{3}{2}}} + \sum_{n=0}^{\infty}(-1)^{n+1}\frac{(1-K)(2n+1)(2n+2)}{1+(2n+1)(K+1)}\frac{1\cdot 3\cdots(2n+1)}{2\cdot 4\cdots(2n+2)}\frac{a^{4n+3}}{b^{2n+2}r^{2n+3}}\right]$$

63. Show that the potential at any point due to a circular disk of radius c raised to a potential V_1 is $(2V_1/\pi)\Sigma(-1)^n(2n+1)^{-1}(c/r)^{2n+1}P_{2n}(\cos\theta)$ where $r > c$. Find the value for $r < c$.

64. Two similar rings of radius a lie opposite each other in parallel planes, so spaced that the radius of one ring subtends an angle α at the center of the other. Show that when they carry charges Q and Q', the repulsion between them is

$$\frac{QQ'}{4\pi\epsilon a^2}\sum_{n=0}^{\infty}(-1)^n\frac{1\cdot 3\cdot 5\cdots(2n+1)}{2\cdot 4\cdot 6\cdots 2n}(\sin\alpha)^{2n+2}P_{2n+1}(\cos\alpha)$$

65. The inside of a conducting sphere of inner radius a is coated with a uniform layer of dielectric of inner radius b. Show that the force on a point charge q, in the cavity at a distance c from the center, is

$$\frac{q^2}{4\pi\epsilon_v}\sum_{n=0}^{\infty}\frac{nc^{2n-1}\{(K-1)(n+1)a^{2n+1} + [(K+1)n+1]b^{2n+1}\}}{b^{2n+1}\{[(K+1)n+K]a^{2n+1} + (K-1)nb^{2n+1}\}}$$

66. A point charge q is placed at a distance c from the center of an earthed conducting sphere of radius a on which is a dielectric layer of outer radius b and relative capacitivity K. Show that the potential in this layer is

$$\frac{q}{4\pi\epsilon_v c}\sum_{n=0}^{\infty}\frac{(2n+1)b^{2n+1}(r^n - a^{2n+1}r^{-n-1})}{c^n\{[(K+1)n+1]b^{2n+1} + (n+1)(K-1)a^{2n+1}\}}P_n(\cos\theta)$$

67*C*. A conducting spherical shell of radius a is placed, insulated and without charge, in a uniform field of electric force of intensity F. Show that, if the sphere is cut into two hemispheres by a plane perpendicular to the field, these hemispheres tend to separate and require forces equal to $\frac{9}{4}\pi\epsilon a^2F^2$ to keep them together.

68*C*. A spherical conductor of internal radius b, which is uncharged and insulated, surrounds a spherical conductor of radius a, the distance between their centers being c, which is small. The charge on the inner conductor is Q. Find the potential function for points between the conductors, and show that the surface density at a point P on the inner conductor is

$$\frac{Q}{4\pi}\left(\frac{1}{a^2} - \frac{3c\cos\theta}{b^3 - a^3}\right)$$

where θ is the angle that the radius through P makes with the line of centers and terms in c^2 are neglected.

69*C*. The equation of the surface of a conductor is $r = a(1 + \delta P_n)$, where δ is very small, and the conductor is placed in a uniform field of force F parallel to the axis of harmonics. Show that the surface density of the induced charge at any point is greater than it would be if the surface were perfectly spherical by the amount

$$\left[\frac{3n\epsilon F\delta}{2n+1}\right][(n+1)P_{n+1} + (n-2)P_{n-1}]$$

70*C*. A conductor at potential V whose surface is of the form $r = a(1 + \delta P_n)$ is surrounded by a dielectric (K) whose boundary is the surface $r = b(1 + \eta P_n)$, and outside this the dielectric is air. Show that the potential in the air at a distance r from the origin is

$$\frac{KabV}{(K-1)a+b}\left\{\frac{1}{r} + \frac{(2n+1)\delta a^n b^{2n+1} + (K-1)\eta b^n[nb^{2n+1} + (n+1)a^{2n+1}]}{(1+n+nK)b^{2n+1} + (K-1)(n+1)a^{2n+1}}\,\frac{P_n}{r^{n+1}}\right\}$$

where squares and higher powers of δ and η are neglected.

71*C*. The surface of a conductor is nearly spherical, its equation being

$$r = a(1 + \eta S_n)$$

where η is small. Show that if the conductor is uninsulated, the charge induced on it by a unit charge at a distance f from the origin and of angular coordinates θ, ϕ is approximately

$$-\frac{a}{f}\left[1 + \left(\frac{a}{f}\right)^n \eta S_n(\theta, \phi)\right]$$

72*C*. A uniform circular wire of radius a charged with electricity of line density e surrounds an uninsulated concentric spherical conductor of radius c. Prove that the electrical density at any point of the surface of the conductor is

$$-\frac{e}{2c}\left[1 - 5\frac{1}{2}\left(\frac{c}{a}\right)^2 P_2 + 9\frac{1\cdot 3}{2\cdot 4}\left(\frac{c}{a}\right)^4 P_4 - 13\frac{1\cdot 3\cdot 5}{2\cdot 4\cdot 6}\left(\frac{c}{a}\right)^6 P_6 + \cdots\right]$$

73*C*. A dielectric sphere is surrounded by a thin circular wire of larger radius b carrying a charge Q. Prove that the potential within the sphere is

$$\frac{Q}{4\pi\epsilon_v b}\left[1 + \sum_1^\infty (-1)^n \frac{1+4n}{1+2n(1+K)}\,\frac{1\cdot 3\cdot 5\cdots(2n-1)}{2\cdot 4\cdot 6\cdots 2n}\left(\frac{r}{b}\right)^{2n} P_{2n}\right]$$

74*C*. A conducting sphere of radius a is embedded in a dielectric (K) whose outer boundary is a concentric sphere of radius $2a$. Show that, if the system is placed in a uniform field of force F, equal quantities of positive and negative electricity are separated of amount $36\pi a^2\epsilon_v FK/(5K + 7)$.

75*C*. A spherical conductor of radius a is surrounded by a uniform dielectric K, which is bounded by a sphere of radius b having its center at a small distance γ from the center of the conductor. Prove that, if the potential of the conductor is V and there are no other conductors in the field, the surface density at a point where the radius makes an angle θ with the line of centers is, approximately,

$$\frac{\epsilon_v KVb}{a[(K-1)a+b]}\left[1 + \frac{6(K-1)\gamma a^2 \cos\theta}{2(K-1)a^3 + (K+2)b^3}\right]$$

76*C*. A shell of glass of inductive capacity K, which is bounded by concentric spherical surfaces of radii a, b, ($a < b$), surrounds an electrified particle with charge q which is at a point Q at a small distance c from O, the center of the spheres. Show that the potential at a point P outside the shell at a distance r from Q is approximately

$$\frac{q}{4\pi\epsilon_v}\left[\frac{1}{r} + \frac{2c(b^3 - a^3)(K-1)^2}{2a^3(K-1)^2 - b^3(K+2)(2K+1)}\,\frac{\cos\theta}{r^2}\right]$$

where θ is the angle which QP makes with OQ produced.

77. Show that the attraction between a sphere of radius a and relative capacitivity K, and a point charge q at a distance c from its center is

$$\frac{(1-K)q^2}{4\pi\epsilon_v c^2}\sum_{n=1}^{\infty}\frac{n(n+1)}{1+(K+1)n}\left(\frac{a}{c}\right)^{2n+1}$$

78. A conducting sphere of radius a is supported by an orthogonal conducting cone whose exterior half angle is α. Show that, if this system is charged, the potential is $A(r^n - a^{2n+1}r^{-n-1})P_n(\cos\theta)$ where $0 < n < 1$ if $\frac{1}{2}\pi < \alpha$ and $P_n(\cos\alpha) = 0$.

79. Two conducting coaxial cones whose half angles, measured from the positive axis, are α and β intersect a conducting sphere of radius a orthogonally. If the system is charged, show that the potential outside the sphere between the cones is

$$A(r^n - a^{2n+1}r^{-n-1})[Q_n(\cos\beta)P_n(\cos\theta) - P_n(\cos\beta)Q_n(\cos\theta)]$$

where n is the smallest number for which

$$Q_n(\cos\beta)P_n(\cos\alpha) - P_n(\cos\beta)Q_n(\cos\alpha) = 0$$

80. An oblate dielectric spheroid whose surface is ζ_0 is placed in a uniform electric field E parallel to the axis $\xi = 1$. Show that the resultant field inside is uniform and that its strength is $-E/\{(K-1)\zeta_0[(1+\zeta_0^2)\cot^{-1}\zeta_0 - \zeta_0] - K\}$.

81. The spheroid of the last problem is placed in a uniform field E normal to the axis $\xi = 1$. Show that the electrical intensity inside is uniform and of magnitude

$$\frac{2E}{2 + (K-1)\zeta_0[(1+\zeta_0^2)\cot^{-1}\zeta_0 - \zeta_0]}$$

82. A prolate dielectric spheroid whose surface is η_0 is placed in a uniform electric field E parallel to the axis $\xi = 1$. Show that the resultant field inside is uniform and that its strength is $-E/\{(K-1)\eta_0[(1-\eta_0^2)\coth^{-1}\eta_0 + \eta_0] - K\}$.

83. The spheroid of the last problem is placed in a uniform field E normal to the axis $\xi = 1$. Show that the electrical intensity inside is uniform and of magnitude

$$\frac{2E}{2 + (K - 1)\eta_0[(1 - \eta_0^2)\coth^{-1}\eta_0 + \eta_0]}$$

84. A spheroid of relative capacitivity K is placed in an electric field E, its axis of revolution making an angle α with the field. Show that the torque acting on it is

$$\tfrac{2}{3}\pi\epsilon_v(K - 1)m^2nE(E_1 - E_2)\sin 2\alpha$$

where n is the semiaxis in the direction of the axis of revolution and m is the semiaxis normal to it. For an oblate spheroid, E_1 and E_2 are the results of problems 80 and 81, respectively, where $\zeta_0 = n(m^2 - n^2)^{-\frac{1}{2}}$. For the prolate spheroid, E_1 and E_2 are the results of problems 82 and 83, respectively, where $\eta_0 = n(n^2 - m^2)^{-\frac{1}{2}}$.

85. An earthed conducting disk of radius a whose equation is $\zeta = 0$ is influenced by a point charge q at $1, \zeta_0, 0$. Show that the potential due to the induced charge on the disk is

$$V_i = \frac{q}{2\pi^2\epsilon a}\sum_{n=0}^{\infty}(4n + 1)Q_{2n}(j\zeta_0)Q_{2n}(j\zeta)P_{2n}(\xi)$$

86. Show, using 5.275 and Green's reciprocation theorem, that the potential due to a charged ring whose equations are $\zeta = \zeta_0$, $\xi = \xi_0$ is

$$V_i = \frac{jQ}{4\pi\epsilon a}\sum_{n=0}^{\infty}(2n + 1)P_n(j\zeta_0)P_n(\xi_0)Q_n(j\zeta)P_n(\xi)$$

when $\zeta > \zeta_0$ and the charge on the ring is Q. Write the expression valid when $\zeta < \zeta_0$.

87. An earthed conducting disk of radius a is influenced by a ring carrying a charge Q whose equation is $\zeta = \zeta_0$, $\xi = \xi_0$. Show that the potential due to the induced charge is

$$V_i = \frac{Q}{2\pi^2\epsilon a}\sum_{n=0}^{\infty}(4n + 1)Q_{2n}(j\zeta_0)P_{2n}(\xi_0)Q_{2n}(j\zeta)P_{2n}(\xi)$$

88. The upper and lower halves of an oblate spheroidal shell $\zeta = \zeta_0$ are insulated from each other and charged to potentials $+V_0$ and $-V_0$, respectively. Show that the potential at any external point is

$$V = V_0\sum_{n=0}^{\infty}(-1)^n\frac{1\cdot 3\cdots(2n + 1)}{2\cdot 4\cdots[2(n + 1)]}\,\frac{(4n + 3)Q_{2n+1}(j\zeta)}{(2n + 1)Q_{2n+1}(j\zeta_0)}P_{2n+1}(\xi)$$

Write down the potential for the region inside.

89. Show, by the method used in 5.274, that if the density of electric charge on a prolate spheroid $\eta = \eta_0$ is $\sigma(\xi, \phi)$ then the potential inside is

$$V = \sum_{n=0}^{\infty}\sum_{m=0}^{n}M_{mn}P_n^m(\eta)P_n^m(\xi)\cos m(\phi - \phi_m)$$

where M_{mn} is

$$\frac{(-1)^m(2 - \delta_m^0)(2n + 1)}{4\pi\epsilon c_2}\left[\frac{(n - m)!}{(n + m)!}\right]^2 Q_n^m(\eta_0)\int_{-1}^{+1}\int_0^{2\pi}\sigma P_n^m(\xi)\cos m(\phi - \phi_m)h_1h_3\,d\xi\,d\phi$$

where h_1h_3 are given by 5.28 (6) and (7). Write the analogous formula for the potential outside.

90. Using the results of the last problem, show that the potential due to a point charge q at ξ_0, η_0, ϕ_0 is given by

$$\frac{q}{4\pi\epsilon c_2}\sum_{n=0}^{\infty}\sum_{m=0}^{n}(2-\delta_m^0)(-1)^m(2n+1)\left[\frac{(n-m)!}{(n+m)!}\right]^2 P_n^m(\xi)Q_n^m(\eta)P_n^m(\eta_0)P_n^m(\xi_0)\cos m(\phi-\phi_0)$$

when $\eta > \eta_0$. Write out the result when $\eta < \eta_0$.

91. Starting with the potential inside an earthed cylindrical box of radius a due to a point charge q on the axis and using Green's reciprocation theorem, find the potential on the axis of a ring of radius b coaxial with and inside the box. Hence, show that the potential anywhere inside the box due to this ring is

$$\frac{q}{\pi\epsilon a^2}\sum_{k=1}^{\infty}\frac{\sinh\mu_k z\sinh\mu_k(L-c)}{\sinh\mu_k L}\,\frac{J_0(\mu_k b)J_0(\mu_k\rho)}{\mu_k[J_1(\mu_k a)]^2}$$

where $z < c$ and the coordinates of the bottom and top of the box and the plane of the ring are, respectively, $z = 0$, $z = L$, and $z = c$. Take μ_k so that $J_0(\mu_k a) = 0$.

Show by 3.08 (2) that the force toward the bottom of the box is

$$\frac{q^2}{2\pi\epsilon a^2}\sum_{k=1}^{\infty}\frac{\sinh\mu_k(L-2c)}{\sinh\mu_k L}\left[\frac{J_0(\mu_k b)}{J_1(\mu_k a)}\right]^2$$

92. The walls of a conducting box are given by $z = \pm c$, $\rho = a$. The halves above and below the plane $z = 0$ are insulated from each other and charged to potentials $+V_0$ and $-V_0$, respectively. Show that the potential anywhere inside is given by

$$V = \pm V_0\left\{1-\frac{2}{a}\sum_{k=1}^{\infty}\frac{\sinh[\mu_k(c-|z|)]J_0(\mu_k\rho)}{\mu_k\sinh\mu_k c J_1(\mu_k a)}\right\}$$

where $J_0(\mu_k a) = 0$ and the sign is that of z, and also by

$$V = V_0\left[\frac{z}{c}+\frac{2}{\pi}\sum_{n=1}^{\infty}\frac{I_0(n\pi\rho/c)}{nI_0(n\pi a/c)}\sin\left(\frac{n\pi}{c}z\right)\right]$$

93. The walls of a conducting box, which are given by $z = \pm c$, $\rho = a$, are earthed with the exception of two disk-shaped areas at the top and bottom bounded by $\rho = b$, which are charged to potentials $+V_0$ and $-V_0$, respectively. Show that the potential inside is given by

$$\frac{2bV_0}{a^2}\sum_{k=1}^{\infty}\frac{\sinh\mu_k z J_1(\mu_k b)J_0(\mu_k\rho)}{\mu_k\sinh\mu_k c[J_1(\mu_k a)]^2}$$

where $J_0(\mu_k a) = 0$, and also by

$$\frac{2bV_0}{c}\sum_{n=1}^{\infty}(-1)^n\frac{K_0(n\pi a/c)I_0(n\pi\rho/c)-I_0(n\pi a/c)K_0(n\pi\rho/c)}{I_0(n\pi a/c)}I_1\left(\frac{n\pi}{c}b\right)\sin\frac{n\pi z}{c}$$

when $\rho > b$ and when $\rho < b$ by

$$\frac{V_0}{c}\left[z + 2b\sum_{n=1}^{\infty}(-1)^n \frac{K_0(n\pi a/c)I_1(n\pi b/c) + I_0(n\pi a/c)K_1(n\pi b/c)}{I_0(n\pi a/c)} I_0\left(\frac{n\pi}{c}\rho\right)\sin\frac{n\pi z}{c}\right]$$

94. A semi-infinite conducting cylindrical shell of radius a is closed at one end by a plane conducting plate normal to its axis and at the same potential. Show that the image force on a charge q placed on the axis at a distance b from the plate is

$$\frac{q^2}{2\pi\epsilon a^2}\sum_{k=1}^{\infty}\left[\frac{e^{-\mu_k b}}{J_1(\mu_k a)}\right]^2 \qquad \text{where } J_0(\mu_k a) = 0$$

95. The infinite conducting cylinder $\rho = a$ is filled below the $z = 0$ plane with a dielectric of relative capacitivity K. There is a charge q on the axis at $z = b$. Show that the potential above the dielectric is

$$\frac{q}{2\pi\epsilon_v a^2}\sum_{k=1}^{\infty}\left[e^{-\mu_k|z-b|} - \frac{K-1}{K+1}e^{-\mu_k(z+b)}\right]\frac{J_0(\mu_k\rho)}{\mu_k[J_1(\mu_k a)]^2}$$

where $J_0(\mu_k a) = 0$. Find the potential in the dielectric.

96. The surface of a conductor is generated by rotating a circle of diameter a about one of its tangents. Show that the capacitance of the conductor is

$$8\pi\epsilon a\sum_{k=1}^{\infty}[J_1(\mu_k a)]^{-1}\int_0^{\infty} e^{-\mu_k a \sinh\phi}\,d\phi = 8\pi\epsilon a\sum_{k=1}^{\infty}S_{0,0}(\mu_k a)[J_1(\mu_k a)]^{-1}$$

where $J_0(\mu_k a) = 0$ and $S_{0,0}(\mu_k a)$ is a Lommel function.

97. The conductor in the last problem is placed, uncharged, in a uniform field E parallel to its axis of rotational symmetry. Show that, in polar coordinates, the potential at any point is

$$\frac{2Ea^2}{r}\sum_{1=k}^{\infty} e^{-\mu_k a^2 r^{-1}\cos\theta}\frac{J_0(\mu_k a^2 r^{-1}\sin\theta)}{[J_1(\mu_k a)]^2}$$

where $J_0(\mu_k a) = 0$.

98. The conductor in the last problem is rotated to make E normal to its axis. Show that, if $J_0(\mu_k a) = 0$, the potential at any point is

$$\frac{2Ea^2}{r}\sum_{k=1}^{\infty} e^{-\mu_k a^2 r^{-1}\cos\theta}\frac{J_1(\mu_k a^2 r^{-1}\sin\theta)}{[J_2(\mu_k a)]^2}\cos\phi$$

99. Show that the potential due to a ring of radius a carrying a charge Q is

$$\frac{Q}{2\pi^2\epsilon}\int_0^{\infty} K_0(ka)I_0(k\rho)\cos kz\,dk \qquad \text{if } \rho < a$$

or

$$\frac{Q}{2\pi^2\epsilon}\int_0^{\infty} I_0(ka)K_0(k\rho)\cos kz\,dk \qquad \text{if } \rho > a$$

100. A ring of radius a carrying a charge Q is coaxial with an infinite dielectric cylinder of radius b. Show that the potential in the dielectric is

$$\frac{Q}{2\pi^2\epsilon}\int_0^\infty \Psi(k)I_0(k\rho)K_0(ka)\cos kz\,dk$$

where $\Psi(k) = [1 + kb(K - 1)I_0'(kb)K_0(kb)]^{-1}$.

101. The earthed toroidal ring generated by rotating a circle of radius a centered at $\rho = b$ about the coplanar z-axis lies halfway between the plane $z = c$ at potential Ec and the plane $z = -c$ at $-Ec$. Observe that ring multipoles at $\rho = b$, $z = 0$, which give fields normal to the planes and zero potential halfway between them, have potentials $\partial^{2r}V_1/\partial z^{2r}$ where V_1 is the potential of a ring dipole of strength M given by

$$\frac{M}{2\epsilon c^2}\sum_{n=1}^{\infty} nK_0\left(\frac{n\pi b}{c}\right)I_0\left(\frac{n\pi\rho}{c}\right)\sin\frac{n\pi z}{c} \qquad \text{or} \qquad \frac{M}{2\epsilon c^2}\sum_{n=1}^{\infty} nI_0\left(\frac{n\pi b}{c}\right)K_0\left(\frac{n\pi\rho}{c}\right)\sin\frac{n\pi z}{c}$$

according as $\rho < b$ or $\rho > b$. Hence show that a potential function that gives $V = 0$ on the ring at $2m$ lines $\rho = \rho_s$, $z = \pm z_s$ where $0 < s \leqslant m$ and $(\rho_s - b)^2 + z_s^2 = a^2$ is

$$V = Ez - \sum_{n=1}^{\infty}(A_1 n + A_2 n^3 + \cdots + A_m n^{2m+1})K_0\left(\frac{n\pi b}{c}\right)I_0\left(\frac{n\pi\rho}{c}\right)\sin\frac{n\pi z}{c}$$

This holds if $\rho < b$. If $\rho > b$, interchange ρ and b. The coefficient A_r is given by

$$A_r = E\begin{vmatrix} V_{11} & \cdots & V_{r-1,1}z_1V_{r+1,1} & \cdots & V_{m1} \\ V_{12} & \cdots & V_{r-1,2}z_2V_{r+1,2} & \cdots & V_{m2} \\ \cdots & \cdots & \cdots & \cdots & \cdots \\ V_{1m} & \cdots & V_{r-1,m}z_mV_{r+1,m} & \cdots & V_{mm} \end{vmatrix} \div \begin{vmatrix} V_{11}V_{21} & \cdots & V_{m1} \\ V_{12}V_{22} & \cdots & V_{m2} \\ \cdots & \cdots & \cdots \\ V_{1m}V_{2m} & \cdots & V_{mm} \end{vmatrix}$$

$$\rho_s < b \qquad V_{rs} = \sum_{n=1}^{\infty} n^{2r+1}K_0\left(\frac{n\pi b}{c}\right)I_0\left(\frac{n\pi\rho_s}{c}\right)\sin\frac{n\pi z_s}{c}$$

If $\rho_s > b$, interchange ρ_s and b. Using only A_1 fitted at $\rho = b$, $z = \pm a$ gives excellent results when $a \ll c$ and $a \ll b$. This is a typical approximate method.

102. Referring to 5.03 (3), 5.302 (7), and 5.302 (9), show that the potential due to a disk of radius a, carrying a charge Q, is

$$\frac{Q}{4\pi\epsilon a}\int_0^\infty e^{-k|z|}J_0(k\rho)\frac{\sin ka}{k}\,dk$$

103. If V_0 is the potential due to an infinite plane conducting sheet having a uniform charge density σ_0 on each side, show, by referring to 5.272 and 5.302, that the potential due to the same sheet when it has a hole of radius a in it is

$$V_0 + \frac{2\sigma_0}{\pi\epsilon}\int_0^\infty e^{-k|z|}J_0(k\rho)\left(\frac{a\cos ka}{k} - \frac{\sin ka}{k^2}\right)dk$$

104. A cylinder $\rho = a$ makes contact with the earthed plane $z = 0$. There is a uniform potential gradient along the cylinder which has the potential V_0 where it passes through a second earthed plane $z = c$, from which it is insulated. Show that the potential between the planes outside the cylinder is

$$V = \frac{2V_0}{\pi}\sum_{n=1}^{\infty}(-1)^{n+1}\frac{\sin(n\pi z/c)}{n}\frac{K_0(n\pi\rho/c)}{K_0(n\pi a/c)}$$

105. The potential on the walls of a hollow ring bounded by $\rho = a$, $\rho = b$, $z = 0$, $z = c$ is given by $f_1(z)$, $f_2(z)$, $f_3(\rho)$, $f_4(\rho)$, respectively. Show that the potential at any point inside is given by the superposition of four potentials, two of the type of 5.304 and two of the type of 5.36.

106. A point charge q is placed at $z = z_0$, $\rho = b$, $\phi = \beta$. The planes $\phi = 0$ and $\phi = \alpha$ and the cylinder $\rho = a$ are at zero potential where $0 < \beta < \alpha$ and $0 < b < a$. Show that the potential is given by

$$V = \frac{2q}{\epsilon\alpha a^2}\sum_{s=1}^{\infty}\sum_{r=1}^{\infty}\frac{J_{\frac{s\pi}{\alpha}}(\mu_r b)\sin\frac{s\pi\beta}{\alpha}}{\mu_r[J_{\frac{s\pi}{\alpha}+1}(\mu_r a)]^2}e^{-\mu_r|z-z_0|}J_{\frac{s\pi}{\alpha}}(\mu_r\rho)\sin\frac{s\pi\phi}{\alpha}$$

where $J_{\frac{s\pi}{\alpha}}(\mu_r a) = 0$.

107. A charged semi-infinite conducting sheet has an oblate spheroidal boss whose axis coincides with its edge and whose equation is $\zeta = \zeta_0$. Show that the potential is

$$V = C\left\{(1+\zeta^2)^{\frac{1}{4}} - \frac{(1+\zeta_0^2)^{\frac{1}{4}}[(1+\zeta^2)^{\frac{1}{2}}-\zeta]}{(1+\zeta^2)^{\frac{1}{4}}[(1+\zeta_0^2)^{\frac{1}{2}}-\zeta_0]}\right\}(1-\xi^2)^{\frac{1}{4}}\cos\frac{1}{2}\phi$$

108. A solid dielectric sphere of radius a has a sector removed so that it fits the edge of an infinite conducting wedge of external angle α, its surface meeting the wedge faces orthogonally. If the wedge is charged, show that the potentials inside and outside the sphere are

$$V_i = C\frac{\alpha+2\pi}{\alpha+\pi(1+K)}(r\sin\theta)^{\frac{\pi}{\alpha}}\cos\frac{\pi\phi}{\alpha}$$

$$V_o = C\left[r^{\frac{\pi}{\alpha}} - \frac{\pi(K-1)a^{\frac{\pi}{\alpha}}}{\alpha+\pi(1+K)}\left(\frac{a}{r}\right)^{\frac{\pi}{\alpha}+1}\right](\sin\theta)^{\frac{\pi}{\alpha}}\cos\frac{\pi\phi}{\alpha}$$

109. The conducting wedge whose faces are $\phi = +\alpha$ and $\phi = -\alpha$ and the conducting plane, $z = 0$, which intersects it orthogonally, are charged. Show that the potential in the space $-\alpha < \phi < +\alpha$ and $0 < z$ is given by $Cz\rho^m\cos m\phi$, where m is the smallest number for which $\cos m\alpha = 0$.

110. Three orthogonal conducting surfaces, the wedge $\phi = +\alpha$ and $\phi = -\alpha$, the plane $z = 0$, and the cylinder $\rho = a$, are charged. Show that in the space $-\alpha < \phi < +\alpha$, $z > 0$ and $\rho > a$ the potential is given by $Cz(\rho^m - a^{2m}\rho^{-m})\cos m\phi$, where m is the smallest number for which $\cos m\alpha = 0$.

111. Three orthogonal conducting surfaces, the wedge $\phi = +\alpha$ and $\phi = -\alpha$, the plane $\theta = \frac{1}{2}\pi$, and the sphere $r = a$, are charged. Show that when $-\alpha < \phi < +\alpha$, $z > 0$ and $r > a$ the potential is given by

$$C(r^{m+1} - a^{2m+3}r^{-m-2})\cos\theta\sin^m\theta\cos m\phi$$

where m is the smallest number for which $\cos m\alpha = 0$.

112. A charged conducting body has a deep rectangular hole whose boundaries are given by $x = 0, x = a, y = 0, y = b$, and $z = 0$. Show that, far from the opening, the potential is given by

$$V = C \sin \frac{\pi x}{a} \sin \frac{\pi y}{b} \sinh \left[(a^2 + b^2)^{\frac{1}{2}} \frac{\pi z}{ab} \right]$$

113. The walls of an earthed rectangular conducting tube of infinite length are given by $x = 0, x = a$ and $y = 0, y = b$. A point charge is placed at $x = x_0$, $y = y_0$, and $z = z_0$ inside it. Show that the potential is given by

$$V = \frac{2q}{\pi\epsilon} \sum_{n=1}^{\infty} \sum_{m=1}^{\infty} (m^2a^2 + n^2b^2)^{-\frac{1}{2}} e^{-\frac{(m^2a^2+n^2b^2)^{\frac{1}{2}}\pi|z-z_0|}{ab}} \sin \frac{n\pi x_0}{a} \sin \frac{n\pi x}{a} \sin \frac{m\pi y_0}{b} \sin \frac{m\pi y}{b}$$

114. The walls of an earthed conducting box are given by $x = 0, x = a, y = 0, y = b$, and $z = 0, z = c$. A point charge q is placed at x_0, y_0, z_0. Show that the potential inside the box is given by

$$V = \frac{4q}{\epsilon ab} \sum_{n=1}^{\infty} \sum_{m=1}^{\infty} \frac{\sinh A_{mn}(c - z_0) \sinh A_{mn}z}{A_{mn} \sinh A_{mn}c} \sin \frac{n\pi x_0}{a} \sin \frac{n\pi x}{a} \sin \frac{m\pi y_0}{b} \sin \frac{m\pi y}{b}$$

where $A_{mn} = \dfrac{(m^2a^2 + n^2b^2)^{\frac{1}{2}}\pi}{ab}$ and $z < z_0$. If $z > z_0$, we interchange z and z_0. Show that the z-component of the force on the charge is

$$F_z = -\frac{2q^2}{\epsilon ab} \sum_{n=1}^{\infty} \sum_{m=1}^{\infty} \operatorname{csch} A_{mn}c \sinh A_{mn}(c - 2z_0) \sin^2 \frac{n\pi x_0}{a} \sin^2 \frac{m\pi y_0}{b}$$

For F_x, substitute c for a, a for c, x_0 for z_0 and z_0 for x_0 in F_z and A_{mn}. For F_y, substitute c for b, b for c, y_0 for z_0 and z_0 for y_0 in F_z and A_{mn}.

115. Show that the potential of a dipole $\boldsymbol{M}$ parallel to the θ-axis at $b, \alpha, 0$ is

$$\frac{M}{4\pi\epsilon b^2} \sum_{n=0}^{\infty} \sum_{m=0}^{n} (2 - \delta_m^0)(m - n - 1) \frac{(n-m)!}{(n+m)!} \left(\frac{r}{b}\right)^n P_{n+1}^m(\cos \alpha) P_n^m(\cos \theta) \cos m\phi$$

when $r < b$. If $r > b$, write $(m + n)(b/r)^{n+1} P_{n-1}^m(\cos \alpha)$ for

$$(m - n - 1)\left(\frac{r}{b}\right)^n P_{n+1}^m(\cos \alpha)$$

116. Show that the torque exerted by a sphere of radius a and relative capacitivity K at the origin on a dipole $\boldsymbol{M}$ at $\boldsymbol{r}$ is, if the angle between $\boldsymbol{M}$ and $\boldsymbol{r}$ is α,

$$T = M(r \sin \alpha)^{-1} C_{mn}[(n + 1)P_n^m(\cos \alpha) + (m - n - 1) \cos \alpha P_{n+1}^m(\cos \alpha)]$$

where C_{mn} is a summation operator given by

$$C_{mn} = \frac{(K - 1)M}{4\pi\epsilon_v r^2} \sum_{n=0}^{\infty} \sum_{m=0}^{n+1} \frac{(2 - \delta_m^0)a^{2n+1}(m - n - 1)(n - m)!n}{(nK + n + 1)r^{2n+1}(n + m)!} P_{n+1}^m(\cos \alpha)$$

Show that the radial force is

$$F = -Mr^{-2}(n+1)C_{mn}[\cos\alpha\, P_n^m(\cos\alpha) + (n-m+1)P_{n+1}^m(\cos\alpha)]$$

117. Rotate the transformation of Art. 4.13 about the x-axis letting $u_1 = U$ and $u_2 = V$ and set up 5.11 (6) for this case. Substitute $(\cosh u_1 - \cos u_2)^{\frac{1}{2}}U_1(u_1)U_2(u_2)$ for U in 5.11 (6) and show that the differential equations for U_2 and U_1 are

$$\frac{d^2U_2}{du_2^2} = -n^2U_2, \qquad \frac{d^2U_1}{du_1^2} + \coth u_1\frac{dU_1}{du_1} - \left(n^2 - \frac{1}{4} + \frac{m^2}{\sinh^2 u_1}\right)U_1 = 0$$

The latter is identical with 5.23 (3) if $n - \frac{1}{2}$ is written for n and $\cosh u_1$ for μ.

118. The torus generated by the circle of radius b given by $u_1 = u_0$ in the last problem is charged to a potential V_0. Show that the potential outside it is

$$2V_0(\cosh u_1 - \cos u_2)^{\frac{1}{2}}\sum_{n=0}^{\infty}\frac{(-2)^n(2n+1)P^n_{-\frac{1}{2}}(\coth u_0)P_{n-\frac{1}{2}}(\cosh u_1)}{(2n+1)!!(\sinh u_0)^{\frac{1}{2}}P_{n-\frac{1}{2}}(\cosh u_0)}\cos nu_2$$

where $\cosh u_0$ is c/b, c being the distance from the center of u_0 to the line $u_1 = 0$.

119. Find the total charge on the torus of the last problem by comparing the potential at $u_1 = 0$, $u_2 = 0$ with $q/(4\pi\epsilon r)$ and thus show that its capacitance is

$$8\pi\epsilon 2^{\frac{1}{2}}b(\sinh u_0)^{\frac{1}{2}}\sum_{n=0}^{\infty}\frac{(-2)^n(2n+1)P^n_{-\frac{1}{2}}(\coth u_0)}{(2n+1)!!P_{n-\frac{1}{2}}(\cosh u_0)}$$

120. The torus of the last problem lies in a uniform field E parallel to its axis of rotational symmetry. Show that the potential of the field is

$$V = Ex + (\cosh u_1 - \cos u_2)^{\frac{1}{2}}\sum_{n=1}^{\infty}A_nP_{n-\frac{1}{2}}(\cosh u_1)\sin nu_2$$

where, if a is defined by 4.13 (3), $U = u_0$,

$$A_n = \frac{4(2n+1)(-2)^n naE_0P^n_{-\frac{1}{2}}(\coth u_0)}{(2n+1)!!(\sinh u_0)^{\frac{1}{2}}P_{n-\frac{1}{2}}(\cosh u_0)}$$

121. The torus of problem 118 is placed in a uniform electric field E normal to its axis of rotational symmetry. Show that the potential outside it is

$$E\cos\phi[\rho + (\cosh u_1 - \cos u_2)^{\frac{1}{2}}\sum_{n=0}^{\infty}A_nP^1_{n-\frac{1}{2}}(\cosh u_1)\cos nu_2]$$

where

$$A_n = -\frac{(2n-1)(2n+1)(-2)^{n+1}b(\sinh u_0)^{\frac{1}{2}}P^n_{\frac{1}{2}}(\coth u_0)}{(2n+1)!!P^1_{n-\frac{1}{2}}(\cosh u_0)}$$

122. Rotate the transformation of 4.13 about the y-axis letting $u_1 = U$ and $u_2 = V$, and set up 5.11 (6) for this case. Substitute $(\cosh u_1 - \cos u_2)^{\frac{1}{2}}U_1(u_1)U_2(u_2)$ for U in 5.11 (6) and show that the differential equations for U_2 and U_1 are

$$\frac{d^2U_1}{du_1} = \left(n + \frac{1}{2}\right)^2 U_1, \qquad \frac{d^2U_2}{du_2^2} + \cot u_2\frac{dU_2}{du_2} + \left[n(n+1) - \frac{m^2}{\sin^2 u_2}\right]U_2 = 0$$

The latter is identical with 5.14 (2) with u_2 in place of θ.

123. The two spheres $u_1 = u_0$ and $u_1 = -u_0$ of radius b generated in the last problem are aligned with a uniform field of potential Ez so that their potentials are V_0 and $-V_0$, respectively. Show that the potential outside them is

$$Ez + (\cosh u_1 - \cos u_2)^{\frac{1}{2}} \sum_{m=0}^{\infty} A_n \sinh (n + \tfrac{1}{2}) u_1 P_n(\cos u_2)$$

$$A_n = 2^{\frac{3}{2}}[V_0 - (2n+1)Eb \sinh u_0](e^{(2n+1)u_0} - 1)^{-1}$$

where $\cosh u_0$ is c/b, c being the distance between the centers of the spheres.

124. If the spheres in the last problem are uncharged, show that

$$V_0 = Eb \sinh u_0 \sum_{n=0}^{\infty} \frac{2n+1}{e^{(2n+1)u_0} - 1} \left[\sum_{n=0}^{\infty} \frac{1}{e^{(2n+1)u_0} - 1} \right]^{-1}$$

125. Show that the force between the spheres in the last problem is

$$F = \pi\epsilon \sum_{n=0}^{\infty} A_n[(n + \tfrac{1}{2})A_n - (n+1)A_{n+1} - 2^{\frac{3}{2}} aE]$$

where A_n is found by substitution of V_0 from problem 124 into A_n in problem 123.

126. The potential on the surface of a sphere of radius a due to external sources is $f(\theta)$, where θ is the polar angle. Show by expanding as in 5.156 and summing the harmonic series under the integral by comparison with 5.16 that, when $r < a$,

$$V(r, \theta) = \frac{ar^{\frac{1}{2}}}{\pi} \frac{\partial}{\partial r} \int_{-\pi}^{+\pi} \int_0^{\pi} \frac{r^{\frac{1}{2}} f(\alpha) \sin \alpha \, d\alpha \, d\phi}{[a^2 + r^2 - 2ar(\cos \alpha \cos \theta + \sin \alpha \sin \theta \cos \phi)]^{\frac{1}{2}}}$$

127. Solve the problem of a dipole normal to and halfway between two parallel earthed conducting planes. Now by inversion show that if two conducting spheres of radius a are in contact in a uniform electric field E parallel to their line of centers, the charge density is

$$\sigma = 2\pi^2 \epsilon E \sec^3 \alpha \sum_{n=1}^{\infty} (-1)^n n^2 K_0(n\pi \tan \alpha)$$

where α is measured about the point of contact from a diameter passing through this point.

128. The surface of an oblate spheroid ζ_0 is maintained at a potential

$$V = \sum_{n=0}^{N} B_n P_{2n}(\xi)\, Q_{2N}(j\zeta_0)$$

Inside it a concentric sphere of radius a is at zero potential. Show that in the region between the two the potential is

$$V = \sum_{n=0}^{N} \sum_{s=0}^{n} (-1)^n A_n C_{sn} \left(\frac{c}{a}\right)^{2s+1} \left[\left(\frac{r}{a}\right)^{2s} - \left(\frac{a}{r}\right)^{2s+1} \right] P_{2s}(\cos \theta)$$

where A_n is given by 5.41 (5) with the following values of C_{sn}

$$s < n \qquad C_{sn} = \frac{(-1)^n(2n+2s-1)!!}{(2s)!(2n-2s)!}\left(\frac{a}{c}\right)^{4s+1}\frac{Q_{2n}(j\zeta_0)}{P_{2n}(j\zeta_0)}$$

$$C_{nn} = (-1)^n\left[\frac{(2n)!}{j(4n+1)!!} + \frac{(4n-1)!!}{(2n)!}\left(\frac{a}{c}\right)^{4n+1}\frac{Q_{2n}(j\zeta_0)}{P_{2n}(j\zeta_0)}\right]$$

$$s > n \qquad C_{sn} = \frac{(-1)^s(2s)!}{j(2n+2s+1)!!(2s-2n)!!}$$

and with the right side given by $\sum_{n=0}^{s} B_n C_{sn}$.

129. The potential V of a concentric sphere of radius r is maintained at the value

$$\sum_{n=0}^{N} B_n P_{2n}(\cos\theta)$$

Inside it is an oblate spheroid $\zeta = \zeta_0$ at potential zero with an interfocal distance $2c$. Show that the potential between spheroid and concentric sphere is

$$V = \sum_{n=0}^{N}\sum_{s=0}^{n} A_n C_{sn}\left[\frac{P_{2s}(j\zeta_0)}{Q_{2s}(j\zeta_0)}Q_{2s}(j\zeta) - P_{2s}(j\zeta)\right]P_{2s}(\xi)$$

where A_n is found from 5.41 (5) with

$$s < n \qquad C_{sn} = -\left(\frac{c}{a}\right)^{2n}\frac{(-1)^n(4s+1)(2n)!}{(2n+2s+1)!!(2n-2s)!!}\frac{P_{2s}(j\zeta_0)}{Q_{2s}(j\zeta_0)}$$

$$C_{nn} = (-1)^n\left[\left(\frac{a}{c}\right)^{2n+1}\frac{j(2n+1)(4n-1)!!}{(2n)!} - \left(\frac{c}{a}\right)^{2n}\frac{(2n)!}{(4n-1)!!}\frac{P_{2n}(j\zeta_0)}{Q_{2n}(j\zeta_0)}\right]$$

$$n > s \qquad C_{sn} = \left(\frac{a}{c}\right)^{2n+1}\frac{j(-1)^s(2s+1)(2n+2s-1)!!}{(2n)!(2s-2n)!!}$$

and the right side of 5.41 (5) is $\sum_{n=0}^{s} C_{sn}B_n$.

130. In the last problem if the sphere is at potential V_0, then $B_0 = V_0$ and $B_n = 0$ if $n \neq 0$. If the concentric spheroid inside it is a disk of radius c, then $\zeta_0 = 0$, so $P_{2s}(j\zeta_0)/Q_{2s}(j\zeta_0)$ is $2j/\pi$ by 5.214 (3). Show with the aid of 5.27 (6), 5.214 (5), and 5.214 (3) that the charge density on the disk is

$$\sigma = \frac{2\epsilon}{j\pi c\xi}\sum_{n=0}^{N}\sum_{s=0}^{n}\frac{(-1)^s(2s)!!}{(2s-1)!!}A_n C_{sn} P_{2s}(\xi)$$

131. The surface of an oblate spheroid is maintained at a potential

$$V = \sum_{n=0}^{N} B_n P_{2n}(\xi)$$

Outside it a concentric sphere of radius a is at zero potential. Show that in the region between the two the potential is

$$V = \sum_{n=0}^{N} \sum_{s=n}^{\infty} A_n C_{sn} \left[\left(\frac{r}{a}\right)^{2s} - \left(\frac{a}{r}\right)^{2s+1} \right] P_{2s}(\cos\theta)$$

where A_n is found by solving 5.41 (5) with the following values of C_{sn}:

$$s < n, \qquad C_{sn} = \frac{(-1)^n(2n+2s-1)!!}{(2s)!(2n-2s)!!}$$

$$C_{nn} = (-1)^n \left[\frac{(4n-1)!!}{(2n)!} - \frac{(2n)!(c/a)^{2n+1}}{j(4n+1)!!} \frac{P_{2n}(j\zeta_0)}{Q_{2n}(j\zeta_0)} \right]$$

$$s > n, \qquad C_{sn} = \frac{(-1)^s(2s)!j(c/a)^{2s+1}}{(2n+2s+1)!!(2s-2n)!!} \frac{P_{2n}(j\zeta_0)}{Q_{2n}(j\zeta_0)}$$

and with the right side given by $\sum_{n=s}^{N} B_n C_{sn}$.

132. The upper and lower halves of a conducting sphere of radius a are insulated and charged to potentials $+V_0$ and $-V_0$. It is surrounded by a concentric coaxial oblate spheroid of interfocal distance $2c$ at potential zero. Without the spheroid the external potential of the sphere is

$$V_1 = \sum_{n=0}^{\infty} B_n \left(\frac{a}{r}\right)^{2n+2} P_{2n+1}(\cos\theta), \qquad B_n = (-1)^n \frac{(4n+3)(2n+1)!!}{(2n+1)(2n+2)!!}$$

Show that with the spheroid in place the potential between them is

$$V_2 = \sum_{n=0}^{N} \sum_{s=n+\frac{1}{2}}^{\infty} A_n C_{sn} \left[\frac{P_{2s}(j\zeta_0)}{Q_{2s}(j\zeta_0)} P_{2s}(j\zeta) - Q_{2s}(j\zeta) \right] P_{2s}(\xi)$$

where A_n is found by solving 5.41 (5) and C_{sn} is given by

$$s < n + \tfrac{1}{2}, \qquad C_{sn} = \frac{(-1)^n j(4s+1)(2n+1)!}{(2n+2s+2)!!(2n-2s+1)!!} \left(\frac{a}{c}\right)^{2n+1}$$

$$C_{nn} = (-1)^n(4n+1)!! \left[j\left(\frac{a}{c}\right)^{2n+1} - \frac{2n+2}{(2n+1)!} \left(\frac{c}{a}\right)^{2n+2} \frac{Q_{2s}(j\zeta_0)}{P_{2s}(j\zeta_0)} \right]$$

$$s > n + \tfrac{1}{2}, \qquad C_{sn} = \frac{(-1)^s(2s+1)(2n+2s)!!}{j(2n+1)!(2s-2n-1)!!} \left(\frac{c}{a}\right)^{2n+2} \frac{Q_{2s}(j\zeta_0)}{P_{2s}(j\zeta_0)}$$

Note that s is an integer plus $\frac{1}{2}$ so right side of 5.41 (5) involves $C_{0.1n}$, $C_{1.5n}$, etc.

133. An uncharged conducting sphere of radius a lies at a distance b from an infinite line charge of uniform density q where $b > a$. Take the origin at the center of the sphere, and write the potential of the line charge in circular harmonics by 4.02 (2) using ϕ for θ, ρ for r, $\theta_0 = 0$, and $V = 0$ when $\rho = 0$. Express this in spherical coordinates by writing $r \sin\theta$ for ρ, so it has the form $\Sigma r^n S_n(\theta,\phi)$ where $n \neq 0$. Add to this the induced charge potential $-\Sigma(a^{2n+1}/r^{n+1})S_n(\theta,\phi)$, which cancels it at $r = a$. Sum by 4.02 and show that the total resultant potential is

$$V = \frac{q}{4\pi\epsilon_v} \ln \frac{b^2(b^2r^2 + a^4 \sin^2\theta - 2a^2br \sin\theta \cos\phi)^{a/r}}{(br)^{2a/r}(b^2 + r^2 \sin^2\theta - 2br \sin\theta \cos\phi)}$$

Show that the attractive force on the sphere is

$$F = \frac{q^2a^2}{\pi\epsilon_v b(b^2 - a^2)^{\frac{1}{2}}} \sin^{-1}\frac{a}{b}$$

134. If the conducting sphere in the preceding problem is replaced by a dielectric one ϵ, show that the total potential outside it is

$$V_0 = \frac{q}{2\pi\epsilon_v} \sum_{n=1}^{\infty} \left\{ \frac{r^n}{n} - \frac{(\epsilon - \epsilon_v)a^{2n+1}}{[n\epsilon + (n+1)\epsilon_v]r^{n+1}} \right\} \frac{\sin^n\theta}{b^n} \cos n\phi$$

and write down the potential inside. Show that the force on the sphere is

$$F = \frac{-q^2}{\pi\epsilon_v} \sum_{n=1}^{\infty} \frac{(2n-2)!!n(\epsilon - \epsilon_v)}{(2n-1)!![n\epsilon + (n+1)\epsilon_v]} \left(\frac{a}{b}\right)^{2n+1}$$

135. A sphere of radius a at potential V_0 lies at the origin between two infinite parallel earthed planes spaced a distance b apart, and its center is a distance c from the nearest plane. This is one section of the field produced by an infinite set of pairs of V_0 and $-V_0$ spheres at a distance $2c$ apart similarly aligned along the z-axis with pair centers spaced at intervals $2b$. The potential between sphere and planes is

$$V = \sum_{p=0}^{\infty} C_p \sum_{n=0}^{\infty} (ar_n^{-1})^{p+1} P_p(\cos\theta_n)$$

where r_n is the distance from the field point to the center of the nth sphere and θ_n is the angle at the center of the nth sphere between r_n and the z-axis. With the aid of 5.44(2) and 5.298 (6) write down $\partial^s V/\partial z^s$ on the axis for all pairs except the pair that includes the origin. Sum this for all but the center pair. Add to the sum the $\partial^s V/\partial z^s$ term inside the sphere at $z = 0$ due to it and its nearest image. Thus as in Art. 5.41 show that the P equations, one for each value of s from 0 to P, to be solved for C_p are, after dividing out $s!a^{-s}$,

$$\delta_{0s} V_0 = \sum_{p=0}^{\infty} C_p \left[\delta_{sp} - \left(\frac{-a}{2c}\right)^{p+s+1} + \left(\frac{a}{4b}\right)^{p+s+1} f\left(\frac{c}{2b}\right) \right]$$

$$f(v) = \frac{1}{2} \frac{\partial^{p+s+1}}{\partial v^{p+s+1}} \left(\frac{1}{v^2} - \frac{\pi \cot \eta v}{v} \right) + [1 - (-1)^{p+s}]\zeta(p+s+1)$$

where $\zeta(n)$ is the Riemann zeta function tabulated in HMF, page 811. Note that as $b \to \infty$ only the first two terms in the bracket survive, so the result applies to a single sphere at a distance c from a plane. The capacitance is $4\pi\epsilon a C_0/V_0$.

136. A sphere of radius a at potential V_0 lies halfway between two earthed parallel planes at a distance $2b$ apart. Observe that the potential in the region $0 < r < b$

which is V_0 on the sphere is

$$V = V_0 + \sum_{n=0}^{N} C_n \left[\left(\frac{a}{r}\right)^{2n+1} - \left(\frac{r}{a}\right)^{2n} \right] P_{2n}(\cos\theta)$$

where the $(r/a)^{2n}$ term arises from the induced charge on the planes. Expand V at $z = b$ when ρ is small in the form $\Sigma D_s \rho^{2s}$, so that there are N simultaneous equations $D_s = -V_0\delta_{0s}$, one for each value of s from 0 to $N - 1$, to be solved for C_n. With the aid of MO, page 50, 5.298 (6), and HMF 22.10.12 show that

$$D_s = \frac{(-1)^s}{[(2s)!!]^2} \left[\sum_{n=0}^{N} \frac{(2n+2s)!a^{2n+1}}{(2n)!b^{2n+2s+1}} C_n - \sum_{n=s}^{N} \frac{(2n)!b^{2n+2s}}{(2n-2s)!a^{2n}} C_n \right]$$

Show that the capacitance is $4\pi\epsilon C_0 a/V_0$.

137. Obtain an expression for the potential in the last problem when $r > b$ by writing down the Green's function $V_1(\rho,z)$ for a charge between two planes by 5.324 (10), shifting the origin to halfway between the planes, taking $\partial V_1^{2n}/\partial z^{2n}$ as $r \to 0$, and comparing this with $\partial^{2n}(1/r)/\partial z^{2n}$ in MO, page 50, as $r \to 0$. Thus

$$V = V_0 - \sum_{n=0}^{N} \frac{2\pi^{2n}}{(2n)!} \left(\frac{a}{b}\right)^{2n+1} C_n \sum_{s=0}^{\infty} (s + \tfrac{1}{2})^{2n} K_0 \left[\frac{(2s+1)\pi\rho}{2b} \right] \cos \frac{(2s+1)\pi z}{2b}$$

Note that this expression takes the indeterminate form $0 \cdot \ln \rho$ as $\rho \to 0$ except at $z = 0$ where it is infinite as it should be.

138. The potential at a spherical surface of radius c due to a concentric cylinder inside it having the charge density of 5.39 (1) is given by 5.39 (4) when $r = c$. If the spherical surface conducts, its charge density must cancel this potential and give a potential V_1 when $r < c$ like 5.39 (4) when $r^{-(2n+2p+1)}$ is replaced by $-r^{2n+2p}/c^{4n+4p+1}$. Show that the coefficients a_n and b_n can be found by solving equations 5.39 (2) and 5.39 (3) provided $V + V_1$ is used in place of V. Thus show that the capacitance is still found from 5.39 (4) with $s = p = 0$ but with the new C_e and C_s values.

References

The following references will be found useful in connection with this chapter.

A. Electrical

Buchholz, H.: "Electrische und Magnetische Potentialfelder," Springer, 1957. Elaborate discussions of spherical caps, toroids, and many other cases.

Durand, E.: "Electrostatique et Magnetostatique," Vol. I, 1953; Vol. II, 1966, Masson et Cie. Well-illustrated treatment of solutions in separable coordinate systems and of graphical and numerical methods.

Flugge, S.: "Handbuch der Physik," Vol. XVI, Springer, 1958. G. Wendt gives a clear discussion of three-dimensional potential problems, including numerical and graphical methods.

Geiger-Scheel.: "Handbuch der Physik," Vol. XII, Berlin, 1927.

Gray, A.: "Absolute Measurements in Electricity and Magnetism," Vol. I, Macmillan, 1888. Extended treatment of images and inversion.

Jeans, J. H.: "The Mathematical Theory of Electricity and Magnetism," Cambridge, 1925. Gives extended treatment but omits Bessel functions.

MAXWELL, J. C.: "Electricity and Magnetism," 3d ed. (1891), Dover, 1954. Treats images, inversion, confocal surfaces, and spherical harmonics.

MORSE, P. M., and H. FESHBACH: "Methods of Theoretical Physics," 2 vols., McGraw-Hill, 1953. Discusses the functions and methods of this chapter and tabulates solutions of Laplace's equation in separable coordinate systems.

PLANCK, M. K. E. L.: "Theory of Electricity and Magnetism," Macmillan, 1932. Gives excellent treatment of confocal coordinates.

STRATTON, J. A.: "Electromagnetic Theory," McGraw-Hill, 1941. Gives considerable material on potential theory.

THOMSON, W.: "Papers on Electrostatics and Magnetism," Macmillan, 1884. Solves many problems and gives numerical tables on spheres and spherical segments.

THOMSON, W., and P. G. TAIT: "Treatise on Natural Philosophy," Cambridge, 1912. Treats spherical harmonics.

VAN BLADEL, J.: "Electromagnetic Fields," McGraw-Hill, 1964. Treats approximately dielectric cubes and similar problems not in other books.

WEBER, E.: "Electromagnetic Fields," Vol. I, Wiley, 1950. Gives many examples and literature references.

WIEN-HARMS: "Handbuch der Experimentalphysik," Vol. X, Leipzig, 1930.

B. MATHEMATICAL

BATEMAN, H.: "Electrical and Optical Wave Motion," Cambridge, 1915. Applies the functions of this chapter.

BATEMAN, H.: "Partial Differential Equations," Dover, 1944. Important applications of toroidal and other special coordinate systems.

BRITISH ASSOCIATION: "Bessel Functions, Part I," Cambridge, 1937. Extensive numerical tables of J_0, J_1, Y_0, Y_1, I_0, I_1, K_0, and K_1.

BRITISH ASSOCIATION: "Bessel Functions, Part II," Cambridge, 1952. Numerical tables of J_n, Y_n, I_n, K_n for $2 < n < 20$.

CARSLAW, H. S.: "Mathematical Theory of the Conduction of Heat in Solids," Macmillan, 1921. Bessel functions and spherical harmonies are applied to problems in heat conduction having boundary conditions identical with electrical problems.

COPSON, E. F.: "Theory of Functions," Oxford, 1935.

ERDELYI, A., W. MAGNUS, F. OBERHETTINGER, and F. G. TRICOMI: "Higher Transcendental Functions," 3 vols., McGraw-Hill, 1953. Very complete and concise descriptions of functions with formulas. (HTF)

ERDELYI, A., et al.: "Integral Transforms," 2 vols. McGraw-Hill, 1953. A large collection of all types of integral transforms. (IT)

FLETCHER, A., J. C. P. MILLER, L. ROSENHEAD, and L. J. COMRIE: "An Index of Mathematical Tables," 2 vols., Blackwell, Oxford, 1962.

GEIGER-SCHEEL: "Handbuch der Physik," Vol. III, Berlin, 1928.

GRAY, A., G. B. MATHEWS, and F. M. MACROBERT: "Bessel Functions," Macmillan, 1922. Includes physical applications, numerical tables, and problems, as well as mathematical developments.

HARVARD UNIVERSITY COMPUTATION LABORATORY: "Tables of Bessel Functions of the First Kind of Orders 0 to 78," 11 vols., Harvard, 1946 to 1950.

HOBSON, E. W.: "Spherical and Ellipsoidal Harmonics," Chelsea, 1955. Includes almost everything known about the subject.

JAHNKE, E., F. EMDE, and F. LOSCH: "Tables of Higher Functions," McGraw-Hill, 1960. Formulas and numerical tables with outstanding illustrations.

MacRobert, F. M.: "Spherical Harmonics," Dutton, 1927. Includes physical applications and a chapter on Bessel functions.

Magnus, W., and F. Oberhettinger: "Formulas and Theorems for the Special Functions of Mathematical Physics," Chelsea, 1949. A fine collection of formulas for the functions of this chapter. (MO)

McLachlan, N. W.: "Bessel Functions for Engineers," Oxford, 1934. Many applications.

Morgan, S. P.: "Table of Bessel Functions of Imaginary Order and Imaginary Argument," California Institute of Technology, Pasadena, 1941.

National Bureau of Standards: "Handbook of Mathematical Functions," U.S. Bureau of Commerce, 1964. The best collection of formulas and numerical tables for the functions of this chapter. (HMF)

Riemann-Weber: "Differentialgleichungen der Physik," Vieweg, Braunsweig, 1925.

Ryshik, I. M., and I. S. Gradstein: "Tafeln," Deutscher Verlag der Wissenschaften, 1957. A large collection of useful formulas.

Watson, G. N.: "Theory of Bessel Functions," Cambridge, 1922. The standard treatise on the subject. It includes some numerical tables.

Webster, A. G.: "Partial Differential Equations of Mathematical Physics," Hafner, 1950.

Whittaker, E. T., and G. N. Watson: "Modern Analysis," Cambridge, 1935.

CHAPTER VI

ELECTRIC CURRENT

6.00. Electric Current Density. Equation of Continuity.—If we touch two conductors A and B, charged to potentials V_A and V_B, respectively, to two points in a third conducting body, we have seen in 1.00 that electric charge flows from one to the other until the potential of A equals that of B. We observe two phenomena associated with this transfer; one is the heating of the conductor and the other is the existence of a magnetic field in the neighborhood, while the flow is taking place. The second phenomenon will be considered in the next chapter. The rate of transfer of electric charge between A and B at any instant is called the electric current, which is therefore defined, in any system of units, by

$$I = \frac{dQ}{dt} \tag{1}$$

When Q is measured in coulombs and t in seconds, I is measured in amperes. If, by some mechanical electrostatic means, such as a moving insulated belt, we transfer, continuously, a charge from the point of contact of B to that of A at such a rate that the difference of potential $V_A - V_B$ is kept constant and at the same time cool the conductor so that its temperature has a fixed value, we find that the current and the magnetic field remain constant. The latter, therefore, need not enter into our considerations of steady currents. Electric current, at any point, is evidently a directed quantity. If, at the point P in a conducting medium, we take an element of area dS, which is normal to the direction of the electric current at P, and if the current flowing through this element of area is $d\boldsymbol{I}$, then we define the current density at P to be

$$\boldsymbol{i} = \frac{d\boldsymbol{I}}{dS} \tag{2}$$

When a steady current flows, the amount of electricity entering any element of volume must equal that leaving it. Thus, the surface integral of the normal component of the current density over this elementary volume is zero. This gives, by Gauss's theorem, 3.00 (2),

$$\int_S \boldsymbol{i} \cdot \boldsymbol{n}\, dS = \int_V \boldsymbol{\nabla} \cdot \boldsymbol{i}\, dv = 0$$

Since this holds for all volume elements

$$\boldsymbol{\nabla} \cdot \boldsymbol{i} = \operatorname{div} \boldsymbol{i} = 0 \tag{3}$$

This is known as the equation of continuity, and a vector which satisfies this equation everywhere is said to be solenoidal.

6.01. Electromotance.—In the experiment considered, the belt is an agent, known as an electromotance, which exerts a mechanical force on the electric charges carried by it just sufficient to overcome the electrostatic forces between A and B. The work in joules which the belt does in moving 1 coulomb of positive charge from B to A, after eliminating friction losses in the driving mechanism, etc., measures the magnitude, $\mathcal{E}$, of the electromotance in volts. If the force driving the belt is increased, it will accelerate until the additional charge conveyed increases $V_A - V_B$ so that the electrostatic forces on the charged belt just balance the driving force. If the contact between A and B is broken, the belt stops, since it cannot drive against increased electrostatic forces, and work against belt friction ceases, so that the electromotance $\mathcal{E}$ exactly equals $V_A - V_B$.

In order to maintain an electric current, at ordinary temperatures, it is necessary to have some source of electromotance in the circuit. In the present case, the location of the electromotance is perfectly clear. It is distributed along the belt between A and B. In chemical sources of electromotance, such as storage batteries and primary cells, it is located at the surfaces of the electrodes. In thermocouples, it lies in the interface between the members. When a whole metallic circuit is placed in a varying magnetic field, the induced electromotance, discussed in Chap. VIII, may be distributed over all the elements of the circuit. When a dynamo supplies power to a circuit, the electromotance is spread over its coils. In all these cases, if the circuit is opened and the difference of potential across the opening is measured by some device drawing no current, such as an electrostatic instrument or a potentiometer which is screened from induced effects, the result equals the sum of the electromotances around the circuit.

We usually define the electromotance around any given path in a different way. We defined the electrostatic field intensity at a point as the force acting on a stationary unit positive charge placed there. The nature of the force was not specified, except that it was taken proportional to the charge. As we shall see in subsequent sections, this measurement is not affected by the electrical resistance of the medium, as the latter, like viscous friction, acts on moving bodies only. Thus, if we examine the force acting on a charge on the belt, we find it to be zero, since the mechanical and electrostatic forces balance. In the part of the circuit where the current is flowing, only the electrostatic forces are present. Thus the line integral of $\boldsymbol{E}$ around the circuit just equals the potential difference between A and B, which, as we have seen, equals $\mathcal{E}$. Thus we find

$$\mathcal{E} = \oint \boldsymbol{E} \cdot d\boldsymbol{s} \tag{1}$$

which defines the electromotance around the path of integration. It is sometimes instructive to consider the electric field intensity $\boldsymbol{E}$, as made up of two parts, *viz.*, the electrostatic part $\boldsymbol{E}'$, which can be derived from a scalar potential, and for which the line integral around any closed path vanishes; and a part $\boldsymbol{E}''$ due to the electromotance, which is solenoidal in nature and does not necessarily vanish when integrated around a closed path. The intensity $\boldsymbol{E}''$ helps to visualize the case of a distributed electromotance but is not so satisfactory when the electromotance is located in a surface layer, since, in the latter case, it is infinite in these layers and zero everywhere else. The value of the integral of $\boldsymbol{E}''$ will be different around different paths. This is illustrated, in the example given, by the fact that if the path is chosen entirely on the belt or entirely off it the integral is zero.

In many cases, it is possible to place barriers in the path of integration so as to exclude the sources of electromotance, such as the belt in the example given. The line integral of $\boldsymbol{E}$ over all permitted closed paths is then zero so that we can use a scalar potential throughout this region. Even when the electromotance is distributed, a barrier may exist such that, for permitted paths, the integral of (1) is zero. In this case, we are unable to distinguish between $\boldsymbol{E}'$ and $\boldsymbol{E}''$ and can use the terms electromotance and potential interchangeably. This is the sense in which $\mathcal{E}$ appears in the next article except in Eq. (4). We may write, in this case,

$$\boldsymbol{E} = -\nabla V = \nabla \mathcal{E} \tag{2}$$

6.02. Ohm's Law. Resistivity.—If, in the experiment of 6.00, all physical conditions, such as temperature and working efficiency, are kept constant and the electromotance is increased we find that when equilibrium is reached the current has increased in the same ratio. This is Ohm's law. For a perfectly efficient machine the ratio of the electromotance between A and B to the current flowing is called the electrical resistance R_{AB} between these two points, and so we have

$$R_{AB} = \frac{V_A - V_B}{I_{AB}} = \frac{\mathcal{E}_{AB}}{I_{AB}} \tag{1}$$

Let us consider an elementary cylinder at the point P in a conducting medium whose ends, of area dS, are normal to the direction of the current at P and whose walls, of length ds, are parallel to it. The electromotance between the ends is then $(\partial\mathcal{E}/\partial s)\,ds = \nabla\mathcal{E} \cdot \boldsymbol{ds}$, and the current flowing through them is $i\,dS$. Assuming that Ohm's law applies to this cylinder, we may write (1) in the form

$$R_P = \frac{\nabla\mathcal{E} \cdot \boldsymbol{ds}}{i\,dS} = \frac{|\nabla\mathcal{E}| \cos\alpha\,ds}{i\,dS}$$

where α is the angle between $\nabla\mathcal{E}$ and $\boldsymbol{ds}$. If we take ds numerically equal to dS, we write τ for R_P and call it the resistivity or specific resistance of the medium at the point P. Thus

$$\tau = \frac{|\nabla\mathcal{E}|}{i} \cos\alpha \tag{2}$$

If τ is to be independent of the direction of the current, then $\alpha = 0$ so that the potential gradient and the current must be in the same direction. In this case, we say the conductor is isotropic. The conductivity γ is defined as the reciprocal of the resistivity τ, so that for an isotropic medium (2) becomes

$$\boldsymbol{i} = \frac{\nabla\mathcal{E}}{\tau} = \gamma\nabla\mathcal{E} \tag{3}$$

For a closed path in a conducting medium, we have from (3) and 6.01 (1)

$$\oint \boldsymbol{i}\cdot d\boldsymbol{s} = \gamma\mathcal{E} \tag{4}$$

6.03. Heating Effect of Electric Current.—We mentioned in 6.00 that the passage of an electric current heats a conductor. To keep the temperature of the conductor constant, we must remove this heat by some cooling device. If no changes are taking place in the conductor, then the energy which we thus obtain must enter the system in some way. Evidently, we do mechanical work against electrostatic forces in carrying the charges from B to A on our insulating belt. From the definition of potential, this work is $W = Q(V_A - V_B)$, so that the power supplied is

$$P = \frac{dW}{dt} = (V_A - V_B)\left(\frac{dQ}{dt}\right) = (V_A - V_B)I_{AB} \tag{1}$$

Substituting from 6.02 (1), we have

$$P = I_{AB}^2 R_{AB} \tag{2}$$

If I, V, and R are all in esu or if they are all in absolute emu, then P is in ergs per second. If I is in amperes, V in volts, and R in ohms, which is the practical mks system of units, then P is in watts. A complete table of these units appears in the Appendix. The heat that is generated in this manner is called Joule heat.

6.04. Steady Currents in Extended Mediums.—In 6.00 (3), we showed that if, in the steady state, no charge is to accumulate at any point in a conducting medium, the divergence of the current density $\boldsymbol{i}$ must be zero at all points. In 6.02 (3), we found that Ohm's law, in an isotropic medium, requires that $\boldsymbol{i}$ be proportional to the potential gradient and inversely proportional to the resistivity. Combining these two equations, we have

$$\nabla \cdot i = -\nabla \cdot \left(\frac{1}{\tau}\nabla V\right) = 0 \tag{1}$$

If the medium is homogeneous, τ is constant throughout it, so that this becomes

$$\nabla^2 V = 0 \tag{2}$$

Comparing (1) and (2) with 3.02 (2) and (3), we see that these equations are identical with Laplace's equation and that the reciprocal of τ, called the conductivity, plays exactly the same role as the capacitivity does in electrostatics. It follows that all mathematical technique used in electrostatics also applies here. The tubes of flow here bear the same relation to the equipotential surfaces as the tubes of force do in electrostatics. At the boundary between two conducting mediums, both the potential and the normal component of the current density must be continuous so that, from 1.17 (5) and (6), the conditions are

$$\frac{1}{\tau'}\frac{\partial V'}{\partial n} = \frac{1}{\tau''}\frac{\partial V''}{\partial n} \tag{3}$$

$$V' = V'' \tag{4}$$

These two equations determine completely the relations between the potential gradients on the two sides of the boundary, so that if the relative capacitivities of the two mediums also differ and are not proportional to the conductivities, another variable is required to satisfy this additional condition. By taking a charge density σ on the boundary, this can be done, giving, by (1),

$$\frac{\sigma}{\epsilon_v} = K_1\frac{\partial V_1}{\partial n} - K_2\frac{\partial V_2}{\partial n} = -(K_1\tau_1 - K_2\tau_2)i_n \tag{5}$$

Anyone observing the magnetic forces between conductors carrying currents, which is the subject of the next chapter, might be led to believe that the current distributions computed by solving (1) or (2) would be erroneous owing to the mutual displacement of current elements by magnetic interaction. Such is not the case. It is true that, in isotropic nonferromagnetic conductors, there is always an increase of resistance in a magnetic field, sometimes known as the "longitudinal Hall effect." The resistance of those portions of an extended conductor where the magnetic field is most intense is therefore increased relative to that of other portions, producing a decrease in the relative current density. Thus when a steady current flows in a cylindrical conductor, the current density is slightly larger near the axis. If this conductor is placed in an external uniform transverse magnetic field, which adds to the field of the current on one side of the cylinder and subtracts from it on the other, we obtain a lateral displacement of the current in the same direction as the force on the conductor. This does not change the orientation of the equipotential surfaces and does not imply any transverse force on the current carriers. Although the change of resistance can be measured, the resultant current displacement is too small to observe at ordinary temperatures. The usual transverse Hall effect changes the orientation of the equipotential surfaces in either direction, depending on the material, without, in this example, affecting the current distribution. It is possible, however, to arrange a circuit so that the latter is changed. This effect can be observed only by using very strong magnetic fields and sensitive instruments. Thus, we may conclude that the distortion of the current distribution in extended conductors caused by magnetic interactions is negligible, and the results obtained by fitting solutions of (1) and (2) to the given boundary conditions may be considered rigorous, as far as such effects are concerned.

Since, as we have seen, heat is generated by the passage of electric current and since, in general, the resistivity depends on the temperature, there will actually be deviations from the solutions obtained where heavy currents heat those portions of the conductor where the current density is highest. The solutions are therefore strictly true only when the temperature coefficient of resistance of the medium is small, when the current density is small, or when the current has been circulating only a short time.

6.05. General Theorems.—The general theorems derived from Green's theorem in Chap. III may now be restated in a form suitable for the present subject.

1. If the value of the potential V is given over all boundaries of a conductor as well as the size and location of all sources or sinks of current inside, then the value of V is uniquely determined at all points in the conductor.

2. If the normal component of the current density is given over all boundaries of a conductor as well as the size and location of all sources or sinks of current inside, then the value of the potential difference between any two points in the conductor is known.

3. If the resistivity of any element in a conductor is increased, then the resistance of the whole conductor will be increased or remain unaltered.

4. If the resistivity of any element in a conductor is decreased, then the resistance of the whole conductor will be decreased or remain unaltered.

To these we may add another theorem which we shall then prove. This is

5. The current density in a conductor distributes itself in such a way that the generation of heat is a minimum.

To prove this, let us suppose that there is a deviation from the distribution given by Ohm's law [6.02 (3)], the additional current density being j. In order that there may be no accumulation of electricity, j has to satisfy the equation of continuity $\nabla \cdot j = 0$. Applying 6.03 to an element of a tube of flow in the conductor, we have, for the heat generated in it, from 6.02 (3) and 6.03 (2),

$$dP = \left[\left(-\frac{1}{\tau}\nabla V + j\right) dS\right]^2 \tau\left(\frac{ds}{dS}\right) = \tau\left(-\frac{1}{\tau}\nabla V + j\right)^2 dv$$

where dS is the cross section of the elementary tube and ds is its length. Integrating gives

$$P = \int_V \left[\frac{1}{\tau}(\nabla V)^2 - 2j \cdot \nabla V + \tau j^2\right] dv \qquad (1)$$

Applying Green's theorem [3.06 (3)] to the second term, where $j = \nabla\Phi$ and $\Psi = V$, we have

$$\int_V j \cdot \nabla V\, dv = -\int_V V\nabla \cdot j\, dv + \int_S V\boldsymbol{n} \cdot j\, dS$$

The first term on the right is zero since $\nabla \cdot j = 0$. The second term is zero since the total electrode current is fixed. Thus, the first and third terms in (1) remain. The first gives the rate of heat generation when Ohm's law is followed, and the third, being positive, shows that any deviation from this law increases the rate of generation of heat.

6.06. Current Flow in Two Dimensions.—As pointed out in 4.00, there are, strictly speaking, no two-dimensional problems in electrostatics, since all cylindrical conductors are finite in length and there is no way to terminate the electric field sharply at the end, there being no known medium whose capacitivity is zero. In the case of current flow in conductors, we have, however, many rigorously two-dimensional problems, since the current may be confined to a finite region by an insulating boundary. All cases of flow in thin plane conducting sheets are of this type. We thus have available all the methods of Chap. IV for treating these problems. The most powerful of these is that of conjugate functions.

We saw in 4.09 that a solution of 6.04 (2), when the potential is a function of x and y only, is

$$U(x, y) \qquad \text{or} \qquad V(x, y) \qquad (1)$$

where

$$W = U + jV = f(z) = f(x + jy) \qquad (2)$$

The current density i at any point is, from 4.11 (2) and 6.02 (2), if V is the potential function,

$$i = \frac{1}{\tau}\left|\frac{dW}{dz}\right| = \frac{1}{\tau}\frac{\partial V}{\partial n} = -\frac{1}{\tau}\frac{\partial U}{\partial s} \tag{3}$$

If U is the potential function, this becomes

$$i = \frac{1}{\tau}\left|\frac{dW}{dz}\right| = \frac{1}{\tau}\frac{\partial U}{\partial n} = +\frac{1}{\tau}\frac{\partial V}{\partial s} \tag{4}$$

the same sign convention being used as in 4.11 (2).

Thus, if the conductor is bounded by the equipotentials U_1 and U_2 and by the lines of force V_1 and V_2, the current flowing through it will be

$$I = \int_{V_1}^{V_2} i\,ds = \frac{1}{\tau}\int_{V_1}^{V_2}\frac{\partial U}{\partial n}\,ds = \frac{1}{\tau}\int_{V_1}^{V_2}\frac{\partial V}{\partial s}\,ds = \frac{V_2 - V_1}{\tau}$$

By Ohm's law, the resistance of the conductor is

$$R = \frac{|U_2 - U_1|}{|I|} = \tau\frac{|U_2 - U_1|}{|V_2 - V_1|} \tag{5}$$

If the equipotentials U_1 and U_2 in (5) are closed curves, then the electrostatic capacitance between the electrodes, *in vacuo*, is, from 4.11,

$$C = \frac{\epsilon_v[V]}{|U_2 - U_1|} \tag{6}$$

where $[V]$ is the integral of V around U_1 or U_2. So that if C is known, we may find the resistance between U_1 and U_2, if the resisting medium fills the same space as the electrostatic field would, by combining (5) and (6), giving

$$R = \frac{\tau\epsilon_v}{C} \tag{7}$$

Thus, from 4.13 (5), we see that the resistance per unit length between two parallel cylindrical electrodes of radii R_1 and R_2 with a distance D between their centers is

$$R = \frac{\tau}{2\pi}\cosh^{-1}\left(\pm\frac{D^2 - R_1^2 - R_2^2}{2R_1R_2}\right) \tag{8}$$

where τ is the resistivity of the medium between them, and we use the positive sign if they are outside each other and the negative sign if one is inside the other.

6.07. Long Strip with Abrupt Change in Width.—Let us apply the theory just developed by computing the current distribution in a long straight strip of conducting material of uniform thickness whose half-width changes abruptly from h to k. The region near the junction is shown in Fig. 6.07*b*. Such a boundary can evidently be obtained by bending the real axis in the z_1-plane, the angles being $\pi/2$, $3\pi/2$, 0, $3\pi/2$, and $\pi/2$ at -1, $-a$, 0, $+a$, and $+1$, respectively. To have a finite distance between the points $-a$ and $+a$ after the bending, it will be

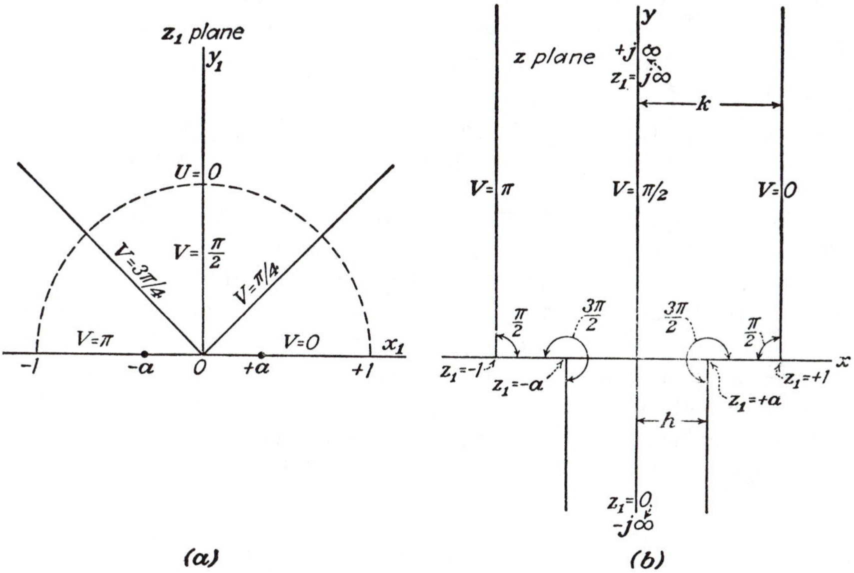

Fig. 6.07.

necessary to put the origin in the z_1-plane at $z = -j\infty$ in the z-plane. Putting these values for the α's and the u's in 4.18 (7), we have

$$\frac{dz}{dz_1} = c\frac{(z_1^2 - a^2)^{\frac{1}{2}}}{z_1(z_1^2 - 1)^{\frac{1}{2}}} = \frac{cz_1}{[(z_1^2 - 1)(z_1^2 - a^2)]^{\frac{1}{2}}} - \frac{ca^2}{z_1[(z_1^2 - 1)(z_1^2 - a^2)]^{\frac{1}{2}}} \qquad (1)$$

We shall evaluate the constants c and a by the method of 4.28 before performing the main integration. As in 4.28, when r_1 is constant, we have $dz_1 = jr_1e^{j\theta_1}\,d\theta_1 = jz_1\,d\theta_1$. When r_1 is a very small constant and θ_1 goes from 0 to π in the z_1 plane, then y is a large negative constant and x goes from $+h$ to $-h$ in the z-plane. Thus, substituting in (1), we have

$$\int_{+h}^{-h} dz = jc\left[\int_0^{\pi}\left(\frac{r_1^2e^{2j\theta_1} - a^2}{r_1^2e^{2j\theta_1} - 1}\right)^{\frac{1}{2}} d\theta_1\right]_{r_1\to 0} = \pm jc\int_0^{\pi} a\,d\theta_1$$

giving $h = \mp\frac{1}{2}jc\pi a$. Similarly, when $r_1 \to \infty$, y is a large positive constant so that

$$\int_{+k}^{-k} dz = jc\left[\int_0^{\pi}\left(\frac{r_1^2e^{2j\theta_1} - a^2}{r_1^2e^{2j\theta_1} - 1}\right)^{\frac{1}{2}} d\theta_1\right]_{r_1\to\infty} = \pm jc\int_0^{\pi} d\theta_1$$

giving $k = \mp\frac{1}{2}jc\pi$. Solving for c and a gives

$$a = \frac{h}{k} \qquad \text{and} \qquad c = \pm\frac{2jk}{\pi} \tag{2}$$

Putting these values into (1) and substituting $(u^2 + a^2)/(1 + u^2)$ for z_1^2 and a^2/z_1^2 in the first and second terms of (1), respectively, we may integrate by *Pc* 49 or *Dw* 120, giving

$$z = \frac{2}{\pi}\left\{k \tan^{-1}\left[\frac{(z_1^2 - a^2)}{(1 - z_1^2)}\right]^{\frac{1}{2}} + h \tan^{-1}\left[a\frac{(1 - z_1^2)^{\frac{1}{2}}}{(z_1^2 - a^2)^{\frac{1}{2}}}\right]\right\}$$

No integration constants are added since this gives, when $y_1 = 0$ and $x_1 = \pm a$, $z = \pm h$, and when $y_1 = 0$ and $x_1 = \pm 1$, $z = \pm k$.

The electrical conditions require that we have a different line of flow on each boundary. Clearly, if we take $W = \ln z_1$ or $z_1 = e^W$, where U is the potential function, then the line of flow $V = 0$ passes from $x_1 = +\infty$ to $x_1 = 0$ in Fig. 6.07*a* and the line of flow $V = \pi$ passes from $x_1 = -\infty$ to $x_1 = 0$. In Fig. 6.07*b*, therefore, these lines follow the right and left boundaries of the strip, respectively, from top to bottom, giving a total current $I = \pi/s$ in the strip. Substituting for z_1, we have

$$z = \frac{2}{\pi}\left[k \tan^{-1}\left(\frac{e^{2W} - a^2}{1 - e^{2W}}\right)^{\frac{1}{2}} + h \tan^{-1} a\left(\frac{1 - e^{2W}}{e^{2W} - a^2}\right)^{\frac{1}{2}}\right] \tag{3}$$

where the real part of z is positive when the real part of z_1 is positive.

If s is the resistance between opposite sides of a one meter square cut from the same piece as the strip, then the resistance of the length y_k of the wide part of the strip is $R_k = s|y_k|/2k$ and that of the narrow part alone is $R_h = s|y_h|/2h$. When they are joined as in Fig. 6.07*b*, the resistance between the ends is not $R_k + R_h$ but there is an additional resistance ΔR due to the distortion of the lines of flow near the junction. When $y_k \gg k$ and $y_h \gg h$, equipotentials are parallel to the axis in Fig. 6.07*b*. Let us compute ΔR in this case. On the y-axis, $V = \frac{1}{2}\pi$ so that $e^{2W} = -e^{2U}$ and (3) becomes, writing $j \tanh^{-1} u$ for $\tan^{-1} ju$, $x = 0$ and

$$y = \frac{2}{\pi}\left[k \tanh^{-1}\left(\frac{e^{2U} + a^2}{1 + e^{2U}}\right)^{\frac{1}{2}} - h \tanh^{-1} a\left(\frac{1 + e^{2U}}{e^{2U} + a^2}\right)^{\frac{1}{2}}\right] \tag{4}$$

Let $U = U_1 \to +\infty$, then e^{2U} is very large and from *Pc* 753 or *Dw* 4 and 5.3

$$y_k = \frac{2}{\pi}\left(k \tanh^{-1}\frac{2e^{2U_1} + a^2}{2e^{2U_1} + 1} - h \tanh^{-1} a\frac{2e^{2U_1} + 1}{2e^{2U_1} + a^2}\right)$$

Using *Pc* 681 or *Dw* 702 and neglecting 1, a, and a^2 compared with e^{2U} give

$$y_k = \frac{1}{\pi}\left(k \ln \frac{4e^{2U_1}}{1-a^2} - h \ln \frac{1+a}{1-a}\right)$$
$$= \frac{2kU_1}{\pi} + \frac{1}{\pi}\left(k \ln \frac{4k^2}{k^2-h^2} - h \ln \frac{k+h}{k-h}\right) = \frac{2kU_1}{\pi} + A$$

Let $U = U_2 \to -\infty$, then e^{2U} is very small and, as before,

$$y_h = \frac{2}{\pi}\left[k \tanh^{-1} \frac{2a^2 + e^{2U_2}}{a(2+e^{2U_2})} - h \tanh^{-1} \frac{a^2(2+e^{2U_2})}{2a^2+e^{2U_2}}\right]$$
$$= \frac{1}{\pi}\left[k \ln \frac{1+a}{1-a} - h \ln \frac{4a^2}{(1-a^2)e^{2U_2}}\right]$$
$$= \frac{2hU_2}{\pi} + \frac{1}{\pi}\left(k \ln \frac{k+h}{k-h} - h \ln \frac{4h^2}{k^2-h^2}\right) = \frac{2hU_2}{\pi} + B$$

Subtracting and solving for $U_1 - U_2$ give

$$U_1 - U_2 = \frac{\pi y_k}{2k} - \frac{\pi y_h}{2h} - \pi\left(\frac{A}{2k} - \frac{B}{2h}\right)$$

By 6.06 (5),

$$R = s\frac{U_1 - U_2}{\pi} = -\frac{s}{2}\left(\frac{A}{k} - \frac{B}{h}\right) + \frac{sy_k}{2k} + \frac{s|y_h|}{2h} = \Delta R + R_k + R_h$$

Substituting for A and B, solving for ΔR, and simplifying give

$$\Delta R = \frac{s}{2\pi}\left[\frac{h^2+k^2}{hk} \ln \frac{k+h}{k-h} + 2 \ln \frac{k^2-h^2}{4hk}\right] \tag{5}$$

Other examples of this method will be found in the problems at the end of this chapter.

6.08. Current Flow in Three Dimensions.—If the entire volume between two electrodes is filled with a uniform isotropic conducting medium, then we may obtain the current distribution and the resistance between the electrodes from the solution of the electrostatic problem for the capacitance between the same electrodes when the medium is insulating. In both cases, the equation to be solved is

$$\nabla^2 V = 0 \tag{1}$$

In the electrostatic case, the boundary conditions on the electrode a *in vacuo* are, from 1.15 (1),

$$V = V_a \qquad Q_a = -\int_S \epsilon_v \frac{\partial V}{\partial n}\, dS_a \tag{2}$$

In the current case, they are, from 6.01 (2), and 6.02 (3),

$$V = V_a \qquad I_a = -\int_S \frac{1}{\tau}\frac{\partial V}{\partial n}\,dS_a \tag{3}$$

The equipotential surfaces correspond exactly in the two cases, since the boundary conditions are identical. From Ohm's law, the resistance is

$$R = \frac{|V_b - V_a|}{|I_a|} = \tau\epsilon_v\frac{|V_b - V_a|}{|Q_a|} = \frac{\tau\epsilon_v}{C} \tag{4}$$

where C is the capacitance *in vacuo* in the electrostatic case. If an electrostatic case can be found such that the walls of a tube of force are identical in shape with the insulating boundary of a conductor of resistivity τ and the equipotential ends of the tube have the same shape as the perfectly conducting terminals of this conductor, then the resistance between these terminals can be computed by (4) from the "capacitance" of the tube of force. By the capacitance, we mean the ratio of the charge on an end to the difference of potential between the ends.

6.09. Systems of Electrodes. Two Spheres. Distant Electrodes.— When n perfectly conducting electrodes are immersed in a homogeneous isotropic conducting medium of resistivity τ, we may find all the relations between the currents entering and leaving the electrodes and their potentials by the methods of Arts. 2.13 to 2.18. It is necessary only to write the current I_s from the sth electrode in place of Q_s, in the electrostatic case and to multiply the capacitances by $(\tau\epsilon_v)^{-1}$ and the elastances by $\tau\epsilon_v$.

Thus, to find the resistance between two spheres of radii a and b, internal or external to each other, having a distance c between their centers, we have, from 2.15 (1), since $I_1 = -I_2$,

$$V_1 - V_2 = \tau\epsilon_v(s_{11} - 2s_{12} + s_{22})I_1$$

so that the resistance is

$$R = \left|\frac{V_1 - V_2}{I_1}\right| = \tau\epsilon_v(s_{11} - 2s_{12} + s_{22}) \tag{1}$$

In 5.08, we computed by images the values of c_{11}, c_{12}, and c_{22} in this case. In terms of these, by 2.16 (2), we have

$$R = \tau\epsilon_v\frac{c_{11} + 2c_{12} + c_{22}}{c_{11}c_{22} - c_{12}^2} \tag{2}$$

If one sphere is inside the other, this formula is of little use and the method of 5.19 can be used.

If two electrodes are at a great distance from each other in an infinite conducting medium, we have, from (1) and 2.18 (1), the result

$$R = \tau\left(\frac{\epsilon_v}{C_a} - \frac{1}{2\pi r} + \frac{\epsilon_v}{C_b}\right) \tag{3}$$

where r is the distance between them and C_a and C_b are the electrostatic capacitances of a and b alone.

6.10. Source and Sink in Solid Sphere. Spherical Bubble.—The problem of the calculation of the potential when a current I flows from a source to a sink in an insulated sphere of resistivity τ indicates how spherical harmonics apply to insulating boundaries. Superposition of the potential of a source I at P_1 and a sink $-I$ at the center on that of a sink $-I$ at P_2 and a source I at the center gives the potential with a source at P_1 and a sink at P_2 and simplifies calculation. Let the source of the first pair be at $\theta = 0$, $r = b$, which is a distance r_1 from the field point. Add to the potential of this pair a solution V_s of Laplace's equation which makes $\partial V/\partial r$ zero at $r = a$ and has no poles if $r < a$. From 5.153, if $b < r < a$,

$$V_1 = \frac{\tau I}{4\pi}\left(\frac{1}{r_1} - \frac{1}{r}\right) + V_s = \frac{\tau I}{4\pi}\sum_{n=1}^{\infty}\left[\frac{b^n}{r^{n+1}} + \frac{(n+1)r^n b^n}{na^{2n+1}}\right]P_n(\cos\theta) \tag{1}$$

The V_s term may be written, with the aid of 5.153 (1),

$$V_s = \frac{\tau I}{4\pi}\sum_{n=1}^{\infty}\left(\frac{r^n b^n}{a^{2n+1}} + \int_0^r \frac{r^n b^n}{a^{2n+1}}\frac{dr}{r}\right) = \frac{\tau I}{4\pi}\left[\frac{a}{br_1'} - \frac{1}{a} + \int_0^r\left(\frac{a}{br_1'} - \frac{1}{a}\right)\frac{dr}{r}\right] \tag{2}$$

where the distance from P to the source image at $\theta = 0$, $r = a^2b^{-1}$ is

$$r_1' = (r^2 + a^4b^{-2} - 2a^2rb^{-1}\cos\theta)^{\frac{1}{2}} \tag{3}$$

Insertion of (3) in (2), integration by *Dw* 380.111, and substitution of the resultant expression for V_s in (1) give

$$V_1 = \frac{\tau I}{4\pi}\left[\frac{1}{r_1} - \frac{1}{r} + \frac{a}{br_1'} - \frac{1}{a}\ln\left(\frac{br_1' + a^2 - br\cos\theta}{2a^2}\right) - \frac{1}{a}\right] \tag{4}$$

Equation (3) gives $\cos\theta$ in terms of a, b, r, and r_1'. In the general case where the radius vectors to the source, sink, and field point are a_1, a_2, and r, two potentials, one with a source at a_1, sink at center, and one with sink at a_2 and source at center, are superimposed. Equation (4) then gives

$$V = \frac{\tau I}{4\pi}\left[\frac{1}{r_1} - \frac{1}{r_2} + \frac{a}{a_1r_1'} - \frac{a}{a_2r_2'} - \frac{1}{a}\ln\frac{(a^2 + a_1r_1')^2 - a_1^2r^2}{(a^2 + a_2r_2')^2 - a_2^2r^2}\right] \tag{5}$$

Here r_1, r_2, r_1', and r_2' are the distances from the field point to the source, sink, image source, and image sink, respectively.

When source and sink are at opposite ends of a diameter, $r_1 = r_1'$, $r_2 = r_2'$, and $a_1 = a_2 = a$. Thus Eq. (5) becomes

$$V = \frac{\tau I}{2\pi}\left[\frac{1}{r_1} - \frac{1}{r_2} - \frac{1}{2a}\ln\frac{r_1(1 + \cos\alpha_1)}{r_2(1 + \cos\alpha_2)}\right] \tag{6}$$

where α_1 and α_2 are the angles at which r_1 and r_2 intersect the diameter. Any tube of flow in this case is a surface of revolution whose equation can be found by equating to the current I' in the tube the integral of the normal current density over any cap bounded by it. It is simpler to use a spherical cap centered at the source for the r_1 terms in (6) and one centered at the sink for the r_2 terms. Thus

$$I_1' = -\frac{1}{\tau}\int\frac{\partial V_1}{\partial r_1}\,dS = I\left(1 + \frac{r_1}{2a}\right)\int_0^\theta \sin\theta_1\,d\theta_1$$

Since the fluxes from source and sink add, integration and the addition of a similar expression for I_2' give

$$I' = I[(1 + \tfrac{1}{2}r_1a^{-1})(1 - \cos\alpha_1) + (1 + \tfrac{1}{2}r_2a^{-1})(1 - \cos\alpha_2)] \tag{7}$$

If β is the obtuse angle between r_1 and r_2, this simplifies to

$$I' = I[1 + \tfrac{1}{2}(r_1 + r_2)a^{-1}\cos\beta] \tag{8}$$

A solution similar to those given applies to the case of flow from a point source to a point sink in an infinite homogeneous conductor which has a spherical hollow of radius a in it. The equation analogous to (1) for a source at b and a sink at infinity has direct powers of r in the first term since $a < b$. The perturbation potential of the hole vanishes at infinity and so contains reciprocal powers of r. In the equation analogous to (2) the second and fourth terms cancel, so that

$$V_s = \frac{\tau I a}{4\pi b}\left[\frac{1}{r_1} - \int_r^\infty \frac{dr}{rr_1'}\right] \tag{9}$$

where r_1' is still given by (3). Integration and addition of a similar term for the sink yield a result like (5) except that the last term is replaced by

$$-\frac{1}{a}\ln\frac{a_2(1 - \cos\theta_2)(a_1r_1' + a^2 - a_1r\cos\theta_1)}{a_1(1 - \cos\theta_1)(a_2r_2' + a^2 - a_2r\cos\theta_2)} \tag{10}$$

6.11. Solid Conducting Cylinder.—To illustrate the application of Bessel functions to insulating cylindrical boundaries, we shall compute the potential anywhere inside a solid circular conducting cylinder of length $2c$, radius a, and resistivity τ when the current I enters and leaves by thin band electrodes at a distance b on either side of the equator. We take the width of the band too small to measure physically but mathematically not zero so that the current density and potential functions are everywhere bounded. Taking the equatorial plane at potential zero, we see that a solution of the equation of continuity which is finite on the axis is, from 5.292 (3) and 5.32,

$$V = \sum_n A_n I_0(k_n\rho) \sin k_n z \tag{1}$$

Since $z = c$ is to be an insulating boundary, we must make $\partial V/\partial z = 0$ there. Since $\cos\,(2n + 1)\frac{1}{2}\pi = 0$, we satisfy this boundary condition by taking

$$k_n = \frac{(2n+1)\pi}{2c} \tag{2}$$

To determine A_n, we differentiate (1) with respect to ρ by 5.33 (4), put $r = a$, multiply by $\sin k_p z$, and integrate from 0 to c. By *Dw* 435 or *Pc* 359, the only term left on the right is that for which $k_p = k_n$ so we have, writing n for p,

$$\int_0^c \frac{\partial V}{\partial \rho} \sin k_n z\, dz = A_n I_1(k_n a) \int_0^{k_n c} \sin^2 k_n z\, d(k_n z)$$

We use *Dw* 430.20 or *Pc* 362 for the right-hand integral. On the boundary, $\rho = a$, $\partial V/\partial\rho$ is zero except in the area ΔS, at $z = b$, covered by the electrode, which is so small that we may give $\sin k_n z$ the constant value $\sin k_n b$ therein. From 6.02 (3), setting $2\pi a\, dz = dS$, the left-hand integral becomes

$$\frac{\sin k_n b}{2\pi a}\tau \int_{\Delta S} i\, dS = \frac{\tau I}{2\pi a} \sin k_n b$$

Solving for A_n gives

$$A_n = \frac{2\tau I}{(2n+1)\pi^2 a} \frac{\sin k_n b}{I_1(k_n a)} \tag{3}$$

The desired solution is given by (1), (2), and (3).

The boundaries of the tubes of flow, in this case, are surfaces of revolution. The equation of the tube through which a current I' flows can be obtained by integrating the current density $-(1/\tau)(\partial V/\partial z)$ over a disk bounded by the tube and equating it to $-I'$, giving

$$I' = +\frac{1}{\tau}\sum_{n=0}^{\infty} 2\pi k_n A_n \cos k_n z \int_0^{\rho} I_0(k_n\rho)\rho\, d\rho$$

Integrating by 5.33 (5), we have, substituting for A_n from (3),

$$I' = \frac{4I\rho}{\pi a}\sum_{n=0}^{\infty}\frac{\sin k_n b \cos k_n z I_1(k_n\rho)}{(2n+1)I_1(k_n a)} \tag{4}$$

where k_n is given by (2).

6.12. Earth Resistance.—Geophysicists sometimes investigate the structure below the earth's surface by observing the distribution of potential on the surface when current is passed through the soil between two or more surface electrodes. Let us investigate the simplest case, in which to a depth a the resistivity is τ_1 and below this depth it is τ_2. We shall use the method of 5.303 to find the distribution about a single-point electrode. The case of two or more electrodes may then be found by superposition. The potential due to the electrode alone may be written down from 5.303 (1), by substituting for q, from 6.08, the value $2\tau\epsilon I$. It should be remembered that $2I$, in this case, corresponds to I in 6.08, since all the current flows in half the field. Then, we have

$$V = \frac{\tau_1 I}{2\pi r} = \frac{\tau_1 I}{2\pi}\int_0^\infty J_0(k\rho)e^{-k|z|}\,dk \tag{1}$$

owing to the electrode alone. As in 5.303 in the region of τ_1, we shall superimpose a second potential due to the discontinuity at $z = a$ that may contain both the e^{-kz} and e^{kz} terms since z is finite in this region, thus we have

$$V_1 = \frac{\tau_1 I}{2\pi}\left[\int_0^\infty \Phi(k)J_0(k\rho)e^{-kz}\,dk + \int_0^\infty \Psi(k)J_0(k\rho)e^{+kz}\,dk + \int_0^\infty J_0(k\rho)e^{-kz}\,dk\right] \tag{2}$$

Since the earth's surface $z = 0$ must be a line of flow and since the last term already meets this condition, the remaining terms must satisfy it independently so that, if $\partial V_1/\partial z = 0$ when $z = 0$,

$$-\Phi(k) + \Psi(k) = 0 \tag{3}$$

The potential in the region τ_2 must vanish at infinity and so can have only the form

$$V_2 = \frac{\tau_1 I}{2\pi}\int_0^\infty \Theta(k)J_0(k\rho)e^{-kz}\,dk \tag{4}$$

At $z = a$, the boundary conditions are, from 6.04 (3) and (4),

$$V_1 = V_2 \qquad \text{and} \qquad \frac{1}{\tau_1}\frac{\partial V_1}{\partial z} = \frac{1}{\tau_2}\frac{\partial V_2}{\partial z}$$

Substituting from (2), (3), and (4) and canceling out $J_0(k\rho)$ give

$$e^{-ka} + \Phi(k)(e^{ka} + e^{-ka}) - \Theta(k)e^{-ka} = 0 \tag{5}$$
$$-\tau_2 e^{-ka} + \tau_2\Phi(k)(e^{ka} - e^{-ka}) + \tau_1\Theta(k)e^{-ka} = 0 \tag{6}$$

Eliminating $\Theta(k)$ and substituting in (2), we have, on the earth's surface where $z = 0$ and setting $(\tau_1 - \tau_2)/(\tau_1 + \tau_2) = \beta$, the result

$$V_s = \frac{\tau_1 I}{2\pi}\int_0^\infty \frac{1 - \beta e^{-2ka}}{1 + \beta e^{-2ka}} J_0(k\rho)\, dk \tag{7}$$

Expanding the denominator by *Pc* 755 or *Dw* 9.04 and interchanging the integral and summation, we have

$$V_s = \frac{\tau_1 I}{2\pi}\left[\int_0^\infty J_0(k\rho)\, dk + 2\sum_{n=1}^{\infty}(-1)^n\beta^n \int_0^\infty e^{-2nka}J_0(k\rho)\, dk\right]$$

Substituting for the integral, from 5.298 (6), gives

$$V_s = \frac{\tau_1 I}{2\pi}\left[\frac{1}{\rho} + 2\sum_{n=1}^{\infty}\frac{(-1)^n\beta^n}{(4n^2a^2 + \rho^2)^{\frac{1}{2}}}\right] \tag{8}$$

Now if we take the earth's surface as the xy-plane and the current I passes from an electrode at $x = +b$ to one at $x = -b$, the potential at the surface is, since $\rho_+ = [(x - b)^2 + y^2]^{\frac{1}{2}}$ and $\rho_- = [(x + b)^2 + y^2]^{+\frac{1}{2}}$,

$$V_s = \frac{\tau_1 I}{2\pi}\left\{\frac{1}{\rho_+} - \frac{1}{\rho_-} + 2\sum_{n=1}^{\infty}(-\beta)^n[(4n^2a^2 + \rho_+^2)^{-\frac{1}{2}} - (4n^2a^2 + \rho_-^2)^{-\frac{1}{2}}]\right\} \tag{9}$$

The case for any number of layers has been worked out by Stefanesco and Schlumberger, *Journal de Physique*, Vol. I, p. 132, 1930.

6.13. Currents in Thin Curved Sheets.—If a thin curved sheet of uniform thickness can be developed into a plane, then, by so doing, the current distribution in such a sheet can be reduced to two dimensions and 6.06 applied. If, however, the surface cannot be so developed, other treatment is necessary. From 3.03 (4) and 6.04 (2), the equation of continuity in orthogonal curvilinear coordinates is

$$\frac{\partial}{\partial u_1}\left(\frac{h_2h_3}{h_1}\frac{\partial V}{\partial u_1}\right) + \frac{\partial}{\partial u_2}\left(\frac{h_3h_1}{h_2}\frac{\partial V}{\partial u_2}\right) + \frac{\partial}{\partial u_3}\left(\frac{h_1h_2}{h_3}\frac{\partial V}{\partial u_3}\right) = 0 \tag{1}$$

Let us suppose that u_1 and u_2 are a set of orthogonal curvilinear coordinates drawn on the surface and let u_3 be a distance measured normal to it. The surface is so thin that the current distribution on each side is the same, making $\partial V/\partial u_3 = 0$. The thickness of the surface is uniform so that h_3 is independent of u_1 and u_2. Under these conditions, (1) takes the form

$$\frac{\partial}{\partial u_1}\left(\frac{h_2}{h_1}\frac{\partial V}{\partial u_1}\right) + \frac{\partial}{\partial u_2}\left(\frac{h_1}{h_2}\frac{\partial V}{\partial u_2}\right) = 0 \qquad (2)$$

This is the equation that we must solve to get the current distribution.

6.14. Current Distribution on Spherical Shell.—From 3.05 (1), letting V be independent of r, we get 6.13 (2) for a spherical shell to be

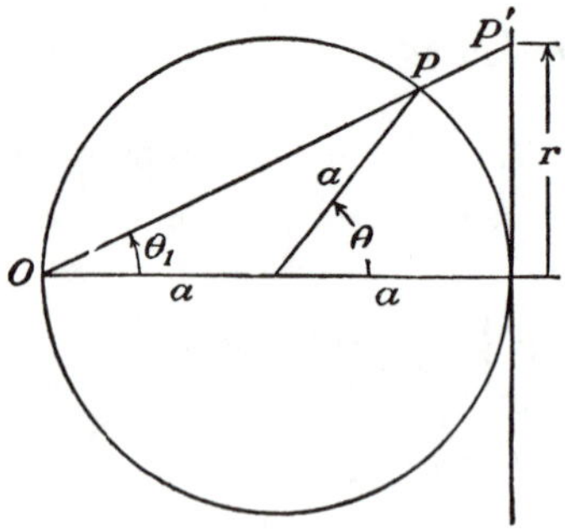

Fig. 6.14.

$$\sin\theta\frac{\partial}{\partial\theta}\left(\sin\theta\frac{\partial V}{\partial\theta}\right) + \frac{\partial^2 V}{\partial\phi^2} = 0 \qquad (1)$$

where θ is the colatitude angle and ϕ is the longitude angle. Although a spherical shell cannot be developed in a plane, it is possible to represent every point of a spherical surface on an infinite plane in such a way that angles are preserved. This process is known as stereographic projection and closely resembles inversion. A plane is taken tangent to the sphere at one end of a diameter. A line passing through the other end O of this diameter and the point P (Fig. 6.14) intersects the plane at P' which is called the projection of P. Let Ψ be a solution of the equation of continuity in the plane, where, from 4.01 (2), it takes the form

$$r\frac{\partial}{\partial r}\left(r\frac{\partial\Psi}{\partial r}\right) + \frac{\partial^2\Psi}{\partial\phi^2} = 0 \qquad (2)$$

In projecting the system of curves, which Ψ represents, onto the sphere, the longitude angle will be unchanged. From Fig. 6.14, we see that r and θ are connected by the relation

$$r = 2a\tan\theta_1 = 2a\tan\tfrac{1}{2}\theta$$

The equation that the projected curves satisfy is obtained by substituting θ and ϕ for r and ϕ in (2).

$$r\frac{\partial}{\partial r} = r\frac{d\theta}{dr}\frac{\partial}{\partial\theta} = \sin\theta\frac{\partial}{\partial\theta}$$

so that, from (2),

$$\sin\theta\frac{\partial}{\partial\theta}\left(\sin\theta\frac{\partial\Psi}{\partial\theta}\right) + \frac{\partial^2\Psi}{\partial\phi^2} = 0$$

This is identical with (1) so that the stereographic projection of a solution of the equation of continuity in a plane gives the solution of this equation for a thin uniform spherical shell.

We found, in 4.09, that both U and V are solutions of $\nabla^2 V = 0$ where

$$U + jV = f(x + jy) = f(r \cos \phi + jr \sin \phi)$$

if $f(z)$ is an analytic function. It follows from the preceding that if

$$U + jV = f[2a \tan \tfrac{1}{2}\theta(\cos \phi + j \sin \phi)] \qquad (3)$$

then both U and V are solutions of the equation of continuity on the surface of a sphere of radius a and that if U is the potential function then V is the stream function, and vice versa. From the laws of inversion, it follows that lines in the plane project into circles through O on the sphere and circles project into circles.

As an example, let us find the potential at any point on a spherical shell of radius a and surface resistivity s when a current I enters at $\theta = \alpha$, $\phi = \frac{1}{2}\pi$ and leaves at $\theta = \alpha$, $\phi = -\frac{1}{2}\pi$. From 4.13 (2.1), writing from 6.08, $\epsilon_v sI$ for the charge which, in 4.13, is $2\pi\epsilon_v$, we have for the potential U at any point on a plane sheet

$$\coth \frac{2\pi U}{sI} = \frac{x^2 + y^2 + b^2}{2by} = \frac{r^2 + b^2}{2br \sin \phi}$$

But from Fig. 6.14, we have $r = 2a \tan \frac{1}{2}\theta$ so that

$$\coth \frac{2\pi U}{sI} = \frac{\tan^2 \frac{1}{2}\theta + \tan^2 \frac{1}{2}\alpha}{2 \tan \frac{1}{2}\theta \tan \frac{1}{2}\alpha \sin \phi} = \frac{1 - \cos \alpha \cos \theta}{\sin \alpha \sin \theta \sin \phi} \qquad (4)$$

In a similar fashion, the equation of the lines of flow is, from 4.13 (4),

$$\cot \frac{2\pi V}{sI} = \frac{r^2 - b^2}{2br \cos \phi} = \frac{\cos \alpha - \cos \theta}{\sin \alpha \sin \theta \cos \phi} \qquad (5)$$

6.15. Surface of Revolution.—As another example, let us solve 6.13 (2) for the surface of revolution formed by rotating the curve $y = f(z)$, where $f(z)$ is a single-valued function, about the z-axis. Clearly, on such a surface, any point P can be located by the orthogonal coordinates z and ϕ, where ϕ is the longitude angle about z. In 6.13, let u_1, which lies along the surface, equal z numerically and let $u_2 = \phi$. From symmetry, h_1 and h_2 are independent of ϕ; so 6.13 (2) becomes

$$\frac{h_2}{h_1}\frac{\partial}{\partial z}\left(\frac{h_2}{h_1}\frac{\partial V}{\partial z}\right) + \frac{\partial^2 V}{\partial \phi^2} = 0 \qquad (1)$$

To solve this equation, let us take a new variable u that is zero when z is z_0 and satisfies the relation

$$\frac{\partial}{\partial u} = \frac{h_2}{h_1}\frac{\partial}{\partial z} \qquad \text{or} \qquad u = \int_{z_0}^{z} \frac{h_1}{h_2}\, dz \qquad (2)$$

Then (1) becomes

$$\frac{\partial^2 V}{\partial u^2} + \frac{\partial^2 V}{\partial \phi^2} = 0 \tag{3}$$

From 4.09, a solution of this equation is known to be either $U(u, \phi)$ or $V(u, \phi)$ where

$$U + jV = F(u + j\phi) \tag{4}$$

To evaluate u in terms of given quantities, we must calculate h_1 and h_2. The equation of the surface is

$$\rho = f(z) \qquad \text{so that} \qquad d\rho = f'(z)\,dz \tag{5}$$

and we have

$$ds^2 = d\rho^2 + dz^2 + \rho^2\,d\phi^2 = \{[f'(z)]^2 + 1\}\,dz^2 + [f(z)]^2\,d\phi^2$$

giving

$$h_1 = \{[f'(z)]^2 + 1\}^{\frac{1}{2}}, \qquad h_2 = f(z) \tag{6}$$

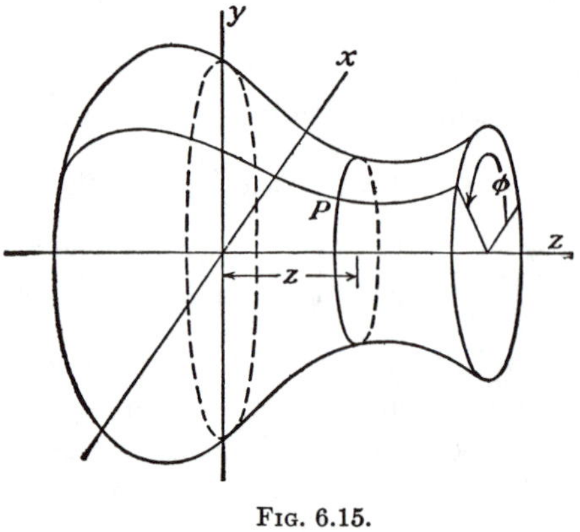

FIG. 6.15.

Putting these values in (2), we obtain

$$u = \int_{z_0}^{z} \frac{\{[f'(z)]^2 + 1\}^{\frac{1}{2}}}{f(z)}\,dz \tag{7}$$

Since an increase of $2n\pi$ in ϕ returns us to the same line on the surface, it is necessary, in order to have single-valued solutions for the potential, that $F(u + j\phi)$ be periodic in ϕ with a period 2π. The solution is identical in form with the solution for a cylindrical surface which can be unrolled into a flat strip. If one end of the surface is closed, as in Fig. 6.15, then $f(z)$ is zero at this point and, from (7), the equivalent cylinder will extend to $u = -\infty$. If one end is open, as in Fig. 6.15, then the equivalent cylinder will terminate at some positive value of u and the boundary conditions on the edge of the cylinder will be the same function of ϕ as on the edge of the surface illustrated in the figure.

6.16. Limits of Resistance.—The general theorems of 6.05 frequently enable us to compute limits between which the resistance of a conductor must lie, even when we cannot compute the rigorous value. To get a lower limit, we endeavor to insert into the conductor thin sheets of perfectly conducting material in such a way that they coincide as nearly as possible with the actual equipotentials but, at the same time, enable us to compute the resistance. In this case, we know, from 6.05, that the

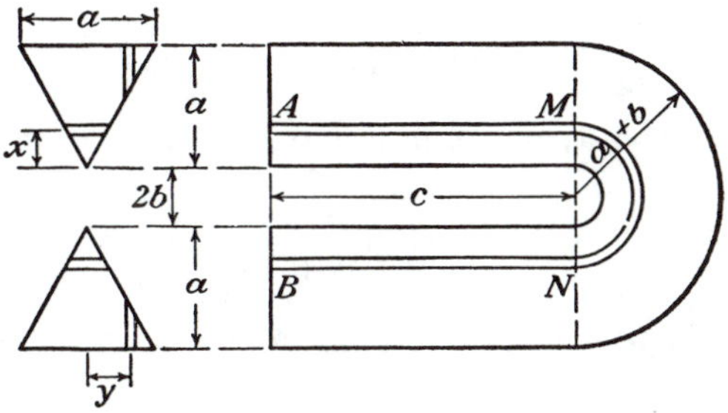

FIG. 6.16.

result is equal to or less than the actual resistance. To compute the upper limit, we insert thin insulating sheets as nearly as possible along the actual lines of flow, in such a way as to enable us to carry out the resistance computation. We know, from 6.05, that this resistance equals or exceeds the actual resistance.

For example, the resistance between two electrodes of a given shape lies between that of two circumscribed electrodes and that of two inscribed electrodes. As a specific example, let us compute the resistance between perfectly conducting electrodes applied at the ends A and B of the horseshoe-shaped conductor of triangular section, shown in Fig. 6.16. To get an upper limit for the resistance, we insert infinitely thin insulating layers very close together which require the current to flow only in straight lines and a semicircle. The length of such a layer, from the figure, is $2c + \pi(b + x)$. The area is $x\,dx$ so that the resistance is

$$dR_u = \frac{\tau[2c + \pi(b + x)]}{x\,dx}$$

All these layers are in parallel so that the upper limit of the resistance is, if *Pc* 29 or *Dw* 91.1 is used, given by (1),

$$R_u = \left(\int \frac{1}{dR_u}\right)^{-1} = \left(\frac{1}{\tau}\int_0^a \frac{x\,dx}{2c+\pi b+\pi x}\right)^{-1}$$
$$= \pi^2\tau\left[\pi a - (2c+\pi b)\ln\frac{2c+\pi(a+b)}{2c+\pi b}\right]^{-1} \tag{1}$$

To get a lower limit of resistance, we insert perfectly conducting sheets across each leg of the horseshoe at M and N. The current in each leg then is uniformly distributed, and so the resistance of the legs is $4c\tau/a^2$. In the semicircular part, the conductor has the shape of a triangular tube of force in the two-dimensional electrostatic field given by $W = \ln z$. This gives, for the stream function and the potential function the values, respectively,

$$U = \ln r \qquad \text{and} \qquad V = \theta \tag{2}$$

The resistance of the strip at y in Fig. 6.16 is then, from 6.06 (5),

$$dR_l'' = \tau\pi\left(\ln\frac{a+b}{b+2y}\,dy\right)^{-1}$$

The strips are in parallel so that, when *Pc* 426 or *Dw* 620 is used,

$$R_l'' = \left(\int\frac{1}{dR_l''}\right)^{-1} = \pi\tau\left(2\int_0^{\frac{1}{2}a}\ln\frac{a+b}{b+2y}\,dy\right)^{-1} = \frac{-\pi\tau}{b\ln(a+b) - b\ln b - a}$$

The legs and the curves are in series, so that

$$R_l = \tau\left[\frac{4c}{a^2} - \frac{\pi}{b\ln(a+b) - b\ln b - a}\right] \tag{3}$$

6.17. Currents in Nonisotropic Mediums. Earth Strata.—Taking the divergence of 6.02 (3), we have

$$\nabla\cdot(\gamma\nabla V) = -\nabla\cdot\boldsymbol{i} = 0 \tag{1}$$

If the medium is not isotropic, the conductivity in different directions varies so that γ cannot be factored out. If, however, the medium is homogeneous, the conductivity in any direction remains unchanged throughout it. In this case, we may choose a set of rectangular axes as in 1.19 (7) and 3.02 (5) so that (1) becomes

$$\gamma_x\frac{\partial^2 V}{\partial x^2} + \gamma_y\frac{\partial^2 V}{\partial y^2} + \gamma_z\frac{\partial^2 V}{\partial z^2} = 0 \tag{2}$$

Owing to the method of formation of the soil it frequently happens that the conductivity in the horizontal direction is greater than that in the vertical direction. If we take the z-axis vertical, (2) becomes

$$\gamma_h\left(\frac{\partial^2 V}{\partial x^2} + \frac{\partial^2 V}{\partial y^2}\right) + \gamma_v \frac{\partial^2 V}{\partial z^2} = 0 \tag{3}$$

If we choose a new variable defined by

$$u = \left(\frac{\gamma_h}{\gamma_v}\right)^{\frac{1}{2}} z = \alpha z \tag{4}$$

this becomes

$$\frac{\partial^2 V}{\partial x^2} + \frac{\partial^2 V}{\partial y^2} + \frac{\partial^2 V}{\partial u^2} = 0 \tag{5}$$

We have all the methods of Chap. V available for solving this equation but we must translate the boundary conditions into the x, y, u system and fit the solution to them, or vice versa. Suppose we have a spherical electrode of radius R half-buried in the earth. The boundary condition is then that $V = V_0$ when $x^2 + y^2 + z^2 = x^2 + y^2 + (u/\alpha)^2 = R^2$. Thus, in the x, y, u system, our boundary condition is that for a spheroid whose equation is

$$\frac{x^2 + y^2}{R^2} + \frac{u^2}{\alpha^2 R^2} = 1$$

From 5.02 (1) and (2), taking $a^2 = b^2 = R^2$ and $c^2 = \alpha^2 R^2$, the solution is

$$V = V_0\left[\int_\theta^\infty \frac{d\theta}{(a^2 + \theta)(c^2 + \theta)^{\frac{1}{2}}}\right] \bigg/ \left[\int_0^\infty \frac{d\theta}{(a^2 + \theta)(c^2 + \theta)^{\frac{1}{2}}}\right]$$

Using *Pc* 114 or letting $x = a^2 + \theta$ and using *Dw* 192.11, we have, since $\alpha > 1$,

$$V = V_0 \frac{\tanh^{-1}[(c^2 - a^2)/(c^2 + \theta)]^{\frac{1}{2}}}{\tanh^{-1}[(c^2 - a^2)/c^2]^{\frac{1}{2}}}$$

But from 5.28 (1) and (2)

$$c^2 + \theta = (c^2 - a^2)\eta^2$$

so, on substituting for a and c, this becomes

$$\begin{aligned} V &= V_0\left[\tanh^{-1}\left(\frac{\gamma_h - \gamma_v}{\gamma_h}\right)^{\frac{1}{2}}\right]^{-1} \tanh^{-1}\frac{1}{\eta} \\ &= V_0\left[\cosh^{-1}\left(\frac{\gamma_h}{\gamma_v}\right)^{\frac{1}{2}}\right]^{-1} \sinh^{-1}(\eta^2 - 1)^{-\frac{1}{2}} \end{aligned} \tag{6}$$

and from 5.28 (4) and (5)

$$u = \alpha z = R(\alpha^2 - 1)^{\frac{1}{2}}\eta\xi$$
$$x^2 + y^2 = R^2(\alpha^2 - 1)(1 - \xi^2)(\eta^2 - 1)$$

Eliminating ξ, we have

$$\frac{x^2 + y^2}{R^2(\alpha^2 - 1)(\eta^2 - 1)} + \frac{\alpha^2 z^2}{R^2(\alpha^2 - 1)\eta^2} = 1 \tag{7}$$

The smallest value of η is on the electrode where $\eta = \alpha/(\alpha^2 - 1)^{\frac{1}{2}}$. Hence, the denominator of the first term is always larger than that of the second so that the equipotential surfaces are nonconfocal oblate spheroids. On the earth's surface, where $z = 0$, the equipotentials become

$$V = V_0\left[\cosh^{-1}\left(\frac{\gamma_h}{\gamma_v}\right)^{\frac{1}{2}}\right]^{-1} \sinh^{-1} R\left[\frac{\gamma_h - \gamma_v}{\gamma_v(x^2 + y^2)}\right]^{\frac{1}{2}} \tag{8}$$

If R is very small, we have a point electrode. Writing the angle for the arc sinh and putting a constant times the current I, for remainder, (8) becomes

$$V = CI(x^2 + y^2)^{-\frac{1}{2}} \tag{9}$$

These curves thus have the same form as for an isotropic medium.

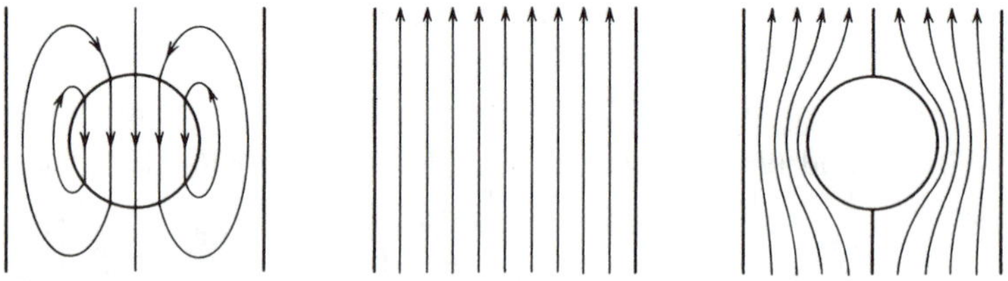

(a) (b) (c)

FIG. 6.18*a*, *b*, *c*.

6.18. Vector Potential. Flow around Sphere in Tube.—So far only scalar potentials have been used for current flow. With axial symmetry the vector potential whose curl is $\boldsymbol{i}_0$ is easier to use for it simplifies the boundary conditions and gives $2\pi\rho A_\phi = I$ for the equation of a tube of flow carrying a current I where ρ is the radius of a circular tube section. Formulas needed appear in Chap. VII if $\boldsymbol{B}$ is replaced by the current density $\boldsymbol{i}_0$. An instructive example is the calculation of the current flow in a solid circular conducting cylinder of radius a when a concentric spherical nonconducting obstacle or cavity partially obstructs the flow. All the formulas needed for solution are in Art. 5.40 except the expression for A_ϕ given in 7.06 (3). The approach is illustrated by Fig. 6.18*a*, *b*, and *c*. The flow (*a*) plus the uniform flow (*b*) yields the flow (*c*) around the sphere in the tube. If the current density far from the sphere is i_0, then the vector potential of this flow is $\frac{1}{2}i_0\rho$ from 7.02 (8). The sources on the sphere surface required to maintain the circulation of Fig. 6.18*a* produce the internal and external vector potentials

$$(A_\phi)_{in} = \sum_{n=0}^{N} D_n\left(\frac{r}{c}\right)^{2n+1} P^1_{2n+1}(\cos\theta), \qquad (A_\phi)_{ex} = \sum_{n=0}^{N} D_n\left(\frac{c}{r}\right)^{2n+2} P^1_{2n+1}(\cos\theta) \tag{1}$$

The physical nature of the sources needed to maintain the circulation in (*a*) is more evident if $\boldsymbol{i}_0$ is replaced by $\boldsymbol{B}$, in which case they are obviously coaxial current loops that lie in the cylinder and spherical shell. The vector potentials of (1) are single-valued even functions of z. Values of D_n must be found such that, in the presence of the cylinder, the internal current density in (*a*) cancels that of the uniform flow in (*b*). In cylindrical coordinates $(A_\phi)_{ex}$ may be written, by 5.40 (3),

$$(A_\phi)_{ex} = \sum_{n=0}^{N} \frac{(-1)^n 2D_n a^{2n+2}}{\pi(2n)!} \int_0^\infty t^{2n+1} K_1(t\rho)\cos(tz)\,dt \tag{2}$$

Evidently from Fig. 6.18*a*, since no net current traverses the cylinder, A_ϕ is zero at both $\rho = 0$ and $\rho = b$, so the internal vector potential of the sources in the wall that cancel (2) must be, using 5.40 (19),

$$\begin{aligned}(A'_\phi)_{in} &= -\sum_{n=0}^{N} \frac{(-1)^n 2D_n}{\pi(2n)!} \int_0^\infty \frac{(ta)^{2n+1} K_1(ta) I_1(t\rho)}{I_1(ta)} \cos(tz)\,d(ta) \\ &= \sum_{n=0}^{N}\sum_{s=0}^{\infty} \frac{(-1)^{n+s} 2D_n r^{2s+1}}{\pi(2n)!(2s+2)!a^{s+1}} P^1_{2s+1}(\cos\theta)\,\frac{I'(2n+2s+2)}{2n+2s+3}\end{aligned} \tag{3}$$

The integral $I'(2n + 2s + 2)$ is that of 5.42 (4) but with first- instead of zero-order Bessel functions and is tabulated (*Phys. Fluids*, Vol. 7, page 635, 1964) and will be written I_{ns}. Inside the sphere only the harmonic $-\frac{1}{2}i_0 r P^1_1(\cos\theta)$ or $-\frac{1}{2}i_0\rho$, which must cancel the uniform flow, can appear. So if $s \neq 0$, the coefficients of $r^{2s+1}P^1_{2s+1}(\cos\theta)$ must vanish. Thus each value of s gives a linear relation between the coefficients D_n. So as in 5.41 a set of $N + 1$ simultaneous equations must be solved for D_0, D_1, . . . , D_N. The value of the right side is zero for all but the first equation, for which it is $-\frac{1}{2}i_0$. The values of C_{sn} in 5.41 (5) are, from (3) and (1),

$$s \neq n \qquad C_{sn} = \frac{(-1)^{n+s} 2I_{ns}}{\pi(2n)!(2s+1)!a^{2s+1}}, \qquad C_{nn} = \frac{1}{c^{2n+1}} + \frac{2I_{nn}}{\pi(2n)!(2n+2)!a^{2n+1}} \tag{4}$$

The above procedure works well for oblate and prolate spheroids including the disk. Extensive numerical values are given in the reference cited. Some are given below to show the interrelation of various factors. The D_n of this section equals $i_0 c C_n/(4n + 3)$ in the reference.

c/a	D_0	D_1	D_2	D_3
0.1	0.0500399	−0.0000001		
0.3	0.153298	−0.000037	0.000001	
0.5	0.277673	−0.000870	0.000045	−0.000003

6.19. Space-charge Current. Child's Equation.—So far, we have considered currents in conductors where the net charge is zero. Let us now consider currents in which charges of only one sign are present so that the net charge is not zero. We must assume the motion of the charges sufficiently slow so that the formulas of electrostatics are valid. In this case, Poisson's equation [3.02 (3)] must be satisfied so that

$$\nabla^2 V = -\frac{\rho}{\epsilon_v} \tag{1}$$

where V is the potential and ρ the charge density. The charges are supposed to be similar and associated with a mass m and to acquire their energy entirely from a superimposed field. Thus, if their velocity is v and charge q, their energy at a point where the potential is V will be

$$mv^2 = 2q(V_0 - V) \tag{2}$$

where V_0 is the potential of their point of origin. The current density at any point is given by

$$\boldsymbol{i} = \rho \boldsymbol{v} \tag{3}$$

The simplest case is the one in which the charges are freed in unlimited quantity at the plane $x = 0$ and accelerated with a total potential V_0 to a plane $x = b$. At the surface $x = 0$, charges will be freed until there is no longer an electric field to move them away so that the boundary condition there is

$$\left(\frac{\partial V}{\partial x}\right)_{x=0} = 0 \tag{4}$$

Here, all the velocities are in the x-direction so that, eliminating ρ and v from (1) by means of (2) and (3), we have

$$\frac{\partial^2 V}{\partial x^2} = -\frac{i}{\epsilon_v}\left[\frac{m}{2q(V_0 - V)}\right]^{\frac{1}{2}}$$

Multiplying through by dV/dx and integrating from $V = V_0$ and $dV/dx = 0$ to V and dV/dx, we have

$$\left(\frac{dV}{dx}\right)^2 = \frac{4i}{\epsilon_v}\left[\frac{m(V_0 - V)}{2q}\right]^{\frac{1}{2}} \tag{5}$$

Taking the square root of both sides, integrating from $V = V_0$, $x = 0$ to $V = 0$ and $x = b$, and solving for i, we have

$$i = \frac{4\epsilon_v}{9}\left(\frac{2q}{m}\right)^{\frac{1}{2}}\frac{V_0^{\frac{3}{2}}}{b^2} \tag{6}$$

This is known as Child's equation. It shows that, with an unlimited supply of charges at one plate, the current between the plates varies as the three halves power of the potential. Such a current is said to be "space-charge limited." We see from (6) that the space-charge limitation is much more serious for charged atoms than for electrons, because of their greater mass.

In practice, we frequently have the emitter in the form of a small circular cylinder and accelerate the charges to a larger concentric cylinder. In this case, we use cylindrical coordinates, and if I is the total current per unit length of the cylinders, (3) becomes

$$I = 2\pi r\rho v \tag{7}$$

Writing (1) in cylindrical coordinates by 3.05 (2), eliminating ρ and v by (2) and (7), and writing V for $V_0 - V$, we have

$$r\frac{d^2V}{dr^2} + \frac{dV}{dr} = \frac{-I}{2\pi\epsilon_v}\left(\frac{m}{2qV}\right)^{\frac{1}{2}} \tag{8}$$

The direct solution of this equation in finite terms is difficult, if not impossible. We may, however, obtain a solution in series as follows: Let us see if the assumption that q, m, and V enter into this solution in the same way as into (6) leads to a solution of (8). Try the solution

$$-I = \frac{8\pi\epsilon_v}{9}\left(\frac{2q}{m}\right)^{\frac{1}{2}}\frac{V^{\frac{3}{2}}}{r\beta^2} \tag{9}$$

and see if β^2 can be determined to satisfy (8). Substituting (9) in (8) gives the equation

$$3\beta\frac{d^2\beta}{d\gamma^2} + \left(\frac{d\beta}{d\gamma}\right)^2 + 4\beta\frac{d\beta}{d\gamma} + \beta^2 - 1 = 0 \tag{10}$$

where

$$\gamma = \ln\left(\frac{r}{a}\right) \tag{11}$$

This equation can be solved in the regular way in series which gives

$$\beta = \gamma - \tfrac{2}{5}\gamma^2 + \tfrac{11}{120}\gamma^3 - \tfrac{47}{3300}\gamma^4 + \cdots \tag{12}$$

Here a is the radius of the inner cylinder. Tables of β as a function of r/a have been published by Langmuir.

Problems

The problems marked C in the following list are taken from the Cambridge examinations as reprinted by Jeans, with the permission of the Cambridge University Press.

1. A cylindrical column of length l and radius a of material of resistivity τ connects normally the parallel plane faces of two semi-infinite masses of the same material. Show that, if R is the resistance between the masses,

$$\frac{\tau l}{\pi a^2} + \frac{\tau}{2a} < R < \frac{\pi\tau}{2[\pi a - l \ln (1 + \pi a/l)]}$$

Observe that, when $l = 0$, both limits give the exact value.

2*C*. A cylindrical cable consists of a conducting core of copper surrounded by a thin insulating sheath of material of given specific resistance. Show that, if the sectional areas of the core and sheath are given, the resistance to lateral leakage is greatest when the surfaces of the two materials are coaxial right circular cylinders.

3*C* (Modified). Current enters and leaves a uniform circular disk through two circular perfectly conducting wires, of radius a, whose centers are a distance d apart and which intersect the edge of the disk orthogonally. Show that the resistance between the wires is $2(s/\pi) \cosh^{-1} (d/2a)$, where s is the resistivity.

4. Two electrodes each of small radius δ are situated at a distance $2a$ apart on a line lying midway between the edges of an infinite strip of width π and resistivity s with insulated boundaries. Show that the resistance between them is, approximately,

$$\frac{s}{\pi} \ln \frac{\sinh 2a}{\delta}$$

5. Instead of being situated as in the last problem, the two electrodes lie symmetrically on a line normal to the edges of the strip. Show that the resistance between them is, approximately,

$$\frac{s}{\pi} \ln \left(\frac{2 \tan a}{\delta}\right)$$

6. The electrodes are situated as in problem 4 but the edges of the strip are perfectly conducting. Show that the resistance is, approximately,

$$\frac{s}{\pi} \ln \left(\frac{2 \tanh a}{\delta}\right)$$

7*C*. A circular sheet of copper, of specific resistance s_1 per unit area, is inserted in a very large sheet of tinfoil (s_0), and currents flow in the composite sheet, entering and leaving at electrodes. Prove that the current function in the tinfoil, corresponding to an electrode at which a current e enters the tinfoil, is the coefficient of i in the imaginary part of

$$-\frac{s_0 e}{2\pi}\left[\ln (z - c) + \frac{s_0 - s_1}{s_0 + s_1} \ln \frac{cz}{cz - a^2}\right]$$

where a is the radius of the copper sheet, z is a complex variable with its origin at the center of the sheet, and c is the distance of the electrode from the origin, the real axis passing through the electrode.

8*C*. A uniform conducting sheet has the form of the catenary of revolution $y^2 + z^2 = c^2 \cosh^2 (x/c)$. Prove that the potential at any point due to an electrode at x_0, y_0, z_0, introducing a current C, is

$$\text{constant} - \frac{Cs}{4\pi} \ln \left\{\cosh \frac{x - x_0}{c} - \frac{yy_0 + zz_0}{[(y^2 + z^2)(y_0^2 + z_0^2)]^{\frac{1}{2}}}\right\}$$

9. The edges of two thin sheet electrodes are held in a circular cylinder of resisting material of length L. Both electrodes and the axis of the cylinder lie in the same plane and their edges are parallel to, and at a distance c from, this axis. Take c small compared with a, the radius of the cylinder. If τ is the resistivity and R is the resistance between electrodes, show that

$$\frac{\tau\pi}{2L \cosh^{-1} (a/c)} < R < \frac{\tau\pi}{2L \sinh^{-1} (a/c)}$$

10. Show from Art. 4.29 that the resistance in the last problem is given rigorously by

$$R = \frac{2\tau K(c^2 a^{-2})}{LK[(1 - c^4 a^{-4})^{\frac{1}{2}}]}$$

11. Show from Art. 4.30 that if, in the foregoing problems, the edges penetrate to different depths, so that the distance of the cylinder axis to one edge is c_1 and to the other c_2, then the resistance is

$$R = \frac{\tau K(k)}{LK[(1 - k^2)^{\frac{1}{2}}]} \qquad \text{where } k = \frac{a(c_1 + c_2)}{a^2 + c_1 c_2}$$

Here $K(k)$ is a complete elliptic integral of modulus k.

12. A semi-infinite plate of resistivity τ is bounded by $x = 0$, $x = l$, and $y = 0$. Two perfectly conducting electrodes are applied to the edge, so that their contact areas are bounded by $x = 0$, $x = l$, $z = b$, $z = a$ and $x = 0$, $x = l$, $z = -b$, $z = -a$ where $b < a$. Show from Art. 4.29 that the resistance between these strips is

$$R = \frac{2\tau K(b/a)}{lK\{[1 - (b/a)^2]^{\frac{1}{2}}\}}$$

13. The contact areas in the preceding problem are of unequal width. The distance between their near edges is g, far edges h, and the width of one is w. Show from 4.30 that the resistance between them is

$$R = \frac{\tau K(k)}{lK[(1 - k^2)^{\frac{1}{2}}]} \qquad \text{where } k = \left[\frac{hg}{(w + g)(h - w)}\right]^{\frac{1}{2}}$$

14. A long slab of resistivity τ, thickness $2b$, and width l has two perfectly conducting strip electrodes of width $2a$ opposite each other on opposite faces, at right angles to the sides. Show from problem 61, Chap. IV, that the resistance between them is

$$R = \frac{2\tau K(e^{-\pi a/b})}{lK[(1 - e^{-2\pi a/b})^{\frac{1}{2}}]}$$

15. A long strip of width $2b$ has a hole of radius c bored on its center line far from the ends. Using the results of 4.28, show that the hole increases the resistance by an amount equivalent to adding ΔL to the length of the strip where

$$\Delta L = (1 + \lambda)^{-1}2b\pi^{-1} \ln \cos [\tfrac{1}{2}\pi cb^{-1}(1 + \lambda)]$$

and the parameter λ must be found from 4.28 (6).

16. The curved surface of a right circular cylinder of radius r, length d, and resistivity τ is given a perfectly conducting coat. Two thin coplanar saw cuts are made, so that the two halves are connected only by a strip of width $2b$ whose center line is the cylinder axis. Show that the resistance between the two halves of the conducting coat is

$$R = \frac{\tau K[(1 - b^4 r^{-4})^{\frac{1}{2}}]}{2dK(b^2 r^{-2})}$$

where $K(k)$ is a complete elliptic integral of modulus k.

17. There is a dipole source of current of moment M in a sphere of radius a and resistivity τ with an insulated surface. Take the sphere center as the origin and the $\theta = 0$ axis parallel to the dipole which is located at $r = b$, $\theta = \theta_0$, and $\phi = 0$. Show from problem 115, Chap. V, that the potential when $b < r < a$ is

$$V = \frac{\tau M}{4\pi b^2} \sum_{n=0}^{\infty} \sum_{m=0}^{n} \frac{(2 - \delta_m^0)(n - m)!}{(n + m - 1)!} \left[\left(\frac{b}{r}\right)^{n+1} + \frac{(n + 1)b^{n+1}r^n}{na^{2n+1}} \right] P_{n-1}^m(\mu_0) P_n^m(\mu) \cos m\phi$$

where $\mu_0 = \cos \theta_0$, $\mu = \cos \theta$, and $r > b$. If $r < b$, replace $(b/r)^{n+1}P_{n-1}^m(\mu_0)$ by $-(r/b)^n[(n - m + 1)/(n + m)]P_{n+1}^m(\mu_0)$.

18. A solid conducting cylinder of resistivity τ is bounded by the planes $z = \pm c$ and the cylinder $\rho = a$. A current I enters at $\rho = a$, $\phi = 0$, $z = 0$ and leaves at $\rho = a$, $\phi = \pi$, $z = 0$. Find the potential anywhere inside the cylinder and show that, if the electrodes are small spheres of radius r, half embedded in the surface, the resistance between them is

$$\frac{2\tau}{\pi^2 a} \sum_{n=0}^{\infty} \sum_{m=0}^{\infty} (2 - \delta_n^0) \frac{I_{2m+1}[n\pi(a - r)/c]}{nI'_{2m+1}(n\pi a/c)}$$

19. A perfectly conducting rod of radius b is partially insulated by a semiconducting bushing of radius a and length $2c$ where it passes through a perfectly conducting thin sheet at $z = 0$. If the total leakage current is I, show that the current density on the rod when $-c < z < +c$ is

$$\frac{I}{4\pi^2 ab} \sum_{n=0}^{\infty} \frac{(2 - \delta_n^0) \cos (n\pi z/c)}{n[I_1(n\pi a/c)K_0(n\pi b/c) + I_0(n\pi b/c)K_1(n\pi a/c)]}$$

20. Current enters an infinite plane conducting sheet at some point P and leaves at infinity. A circular hole, which does not include P, is cut anywhere in the sheet. Show that the potential difference between any two points on the edge of the hole is twice what it was between the same two points before the hole was cut.

21. An insulator has the form of a truncated cone of height h, base radius a_1, and top radius a_2. The base rests on a metal plate, and the top supports a concentric metal post of radius a_3. If the specific surface resistance is s, show that the surface resistance between the plate and the post is

$$\frac{s}{2\pi}\left\{\frac{[h^2 + (a_1 - a_2)^2]^{\frac{1}{2}}}{a_1 - a_2}\ln\left(\frac{a_1}{a_2}\right) + \ln\left(\frac{a_2}{a_3}\right)\right\}$$

22. On a spherical conducting shell, current is introduced at the point $\theta = \alpha$. $\phi = 0$ and taken out at the point $\theta = \alpha$, $\phi = \pi$, where θ is the colatitude angle and ϕ the longitude angle. Show that the potential on the surface is of the form

$$A \ln\left(\frac{1 - \cos\alpha\cos\theta - \sin\alpha\sin\theta\cos\phi}{1 - \cos\alpha\cos\theta + \sin\alpha\sin\theta\cos\phi}\right) + C$$

23. The ends of two metal rods of radius b are embedded in a sphere of insulating material of radius a whose body resistance is very high but whose surface resistance is s. If the axes of the rods were extended, they would intersect at the center of the sphere at an angle α. Show that the surface resistance between the rods is

$$\frac{s}{\pi}\cosh^{-1}\left(\frac{a}{b}\sin\frac{\alpha}{2}\right)$$

24. A current flows in a thin shell whose equation is $(z/a)^2 + (\rho/b)^2 = 1$, where $\rho^2 = x^2 + y^2$ and $a > b$. Show that the potential V on the shell is either U_1 or U_2 where $U_1 + jU_2 = f(\alpha + j\phi)$, $\tan\phi = y/x$ and

$$\alpha = \tanh^{-1}\left\{\frac{bz}{[a^4 - (a^2 - b^2)z^2]^{\frac{1}{2}}}\right\} + \frac{(a^2 - b^2)^{\frac{1}{2}}}{b}\sin^{-1}\left[\frac{z(a^2 - b^2)^{\frac{1}{2}}}{a^2}\right]$$

The function $f(\alpha + j\phi)$ is periodic in ϕ with a period of 2π, and we must also have $\partial V/\partial\alpha = 0$ at $z = \pm a$ or $\alpha = \pm\infty$ if there are no sources or sinks of current there, otherwise $\partial V/\partial\phi = 0$ there.

25. If we have an oblate spheroidal shell carrying current instead of the prolate one treated above, so that $a < b$, show that the preceding considerations apply except that

$$\alpha = \tanh^{-1}\left(\frac{bz}{[a^4 + (b^2 - a^2)z^2]^{\frac{1}{2}}}\right) - \frac{(b^2 - a^2)^{\frac{1}{2}}}{b}\sinh^{-1}\left[\frac{z(b^2 - a^2)^{\frac{1}{2}}}{a^2}\right]$$

26. Consider the surface of the earth as plane and its resistivity to be τ_0 except for a region in the shape of a semi-infinite vertical cone with its apex at the surface which has the resistivity τ_1. A current I is introduced at the surface at $\phi = 0$, at a distance $r = a$ from the apex. Show that the potential at any point $r < a$ outside the cone is given by

$$V_0 = I(ar)^{-\frac{1}{2}}\pi^{-2}\sum_{m=0}^{\infty}\cos m\phi\int_0^{\infty}[A_m(p)P^m_{jp-\frac{1}{2}}(\mu) + B_m(p)P^m_{jp-\frac{1}{2}}(-\mu)]\cos(p\ln ra^{-1})\,dp$$

$$A_m(p) = \frac{(\tau_1 - \tau_2)B_m(p)}{\tau_2 S' - \tau_1 S} = \frac{\tau_1(\tau_1 - \tau_2)(2 - \delta^0_m)}{P^{m'}_{jp-\frac{1}{2}}(0)[\tau_1 - \tau_2 + (-1)^m(\tau_1 S - \tau_2 S')]}$$

$$S' = \frac{\partial[P^m_{jp-\frac{1}{2}}(\cos\alpha)]/\partial(\cos\alpha)}{\partial[P^m_{jp-\frac{1}{2}}(-\cos\alpha)]/\partial(\cos\alpha)}, \qquad S = \frac{P^m_{jp-\frac{1}{2}}(\cos\alpha)}{P^m_{jp-\frac{1}{2}}(-\cos\alpha)}$$

27. A thin circular loop current source I is coaxial with, and immersed in, a solid cylinder of resisting material of radius a and length L with perfectly conducting earthed ends. Show that the potential distribution in the cylinder is, if $z < c$,

$$\frac{\tau I}{\pi a^2}\sum_{k=1}^{\infty}\frac{\sinh \mu_k z \sinh \mu_k(L-c)}{\sinh \mu_k L}\,\frac{J_0(\mu_k b)J_0(\mu_k \rho)}{\mu_k[J_0(\mu_k a)]^2}$$

where μ_k is so chosen that $J_1(\mu_k a) = 0$ and the loop is at $\rho = b$, $z = c$.

28. Two rings made of wire of small radius d are coaxial with an infinite circular cylinder of resisting material of radius a and embedded in its surface to a depth d. Show that, if μ_k is chosen so that $J_1(\mu_k a) = 0$, the resistance between them when their centers are at a distance c apart is, approximately,

$$R = \frac{\tau}{\pi a^2}\sum_{k=1}^{\infty}\frac{1}{\mu_k}[1 - e^{-\mu_k(c-2d)}]e^{-\mu_k d}$$

29. A conducting circular cylinder with inner and outer radii a and b is cut on one side longitudinally, the edges of the cut being maintained at potentials of $\frac{1}{2}V_0$ and $-\frac{1}{2}V_0$, so that current flows around the cylinder. Show that the electrostatic fields, when $r < a$ and $r > b$, are given, respectively, by the transformations

$$W = \frac{V_0}{\pi}\ln(a+z) \qquad \text{and} \qquad W = \frac{V_0}{\pi}\ln\frac{z}{z+b}$$

30. A current I is introduced at one pole of a thin conducting spherical shell of radius a and uniform area resistivity s and removed at the other. Show that the electrostatic potential at any point inside the shell is

$$V = \frac{sI}{2\pi}\sum_{n=0}^{\infty}\frac{4n+3}{(2n+1)(2n+2)}\left(\frac{r}{a}\right)^{2n+1}P_{2n+1}(\mu)$$

31. A wire of radius a and resistivity τ_1 is coaxial with the z-axis. The medium outside the wire is insulating except for that portion lying between $z = -c$ and $z = c$, which has a resistivity τ_2. If $\tau_2 \gg \tau_1$ show that the ratio of the resistance across the gap of a perfectly insulated wire to that of the actual wire is, approximately,

$$\frac{R_\infty}{R} = 1 + \frac{32c\tau_1}{\pi^3 a\tau_2}\sum_{n=0}^{\infty}\frac{(-1)^n}{(2n+1)^3}\,\frac{K_1[\frac{1}{2}(2n+1)\pi a/c]}{K_0[\frac{1}{2}(2n+1)\pi a/c]}$$

32. The corners of the wider strip in Fig. 6.07b are cut off at 45° so that the half width of the strip increases linearly from h to k. Show that the presence of the tapered section near the center of a long strip increases its resistance over the sum of the resistances of the straight sections by an amount

$$\frac{s}{\pi hk}\left[(h^2+k^2)\tanh^{-1}\frac{h}{k} + (k^2-h^2)\tan^{-1}\frac{h}{k} + hk\ln\frac{k^4-h^4}{8h^2k^2}\right]$$

33. An infinite slab of resistive material bounded by $z = \alpha$ and $z = -\alpha$ with perfectly conducting surfaces has a spherical cavity of radius 1 centered between its faces on the z-axis. From 6.18 the vector potential for unit current density far from the cavity is $-\frac{1}{2}\rho$. Assume as in 6.18 that inside the cavity it has the form

$$A_{\phi i} = \sum_{n=0}^{N} C_n \left[r_0^{2n+1} P_{2n+1}^1(\cos\theta_0) + \sum_{s=-\infty}^{\infty} (1 - \delta_s^0) r_s^{2n-2} P_{2n+1}^1(\cos\theta_s) \right]$$

where r_s is the distance from the sth image sphere to the field point and θ_s is the acute angle between r_s and the z-axis. Express $A_{\phi i}$ in cylindrical coordinates by 5.40 (35) and (36), expand in powers of ρ, and require that the coefficient of ρ^{2n+1} be zero if $n \neq 0$ and $\frac{1}{2}$ if $n = 0$. Thus show that the $N + 1$ equations to be solved for C_n are

$$\tfrac{1}{2}\delta_n^0 = C_n + \sum_{l=0}^{N} C_l \frac{\zeta(2l + 2n + 3)(2l + 2n + 2)!}{(2n + 2)!(2l)!(2\alpha)^{2l+2n+2}}$$

where $\zeta(x)$ is the Riemann zeta function. Show that the vector potential in the current region is given by

$$A_{\phi o} = \sum_{n=0}^{N} C_n \sum_{s=-\infty}^{\infty} r_s^{-2n-2} P_{2n+1}^1(\cos\theta_s)$$

For a complete discussion of this problem see Paul Michael, *Phys. Fluids*, Vol. 8, page 1263, 1965.

References

Some material on the subject matter of this chapter will be found in most of the electrical references of Chap. V.

CHAPTER VII

MAGNETIC INTERACTION OF CURRENTS

7.00. Definition of the Ampere in Terms of the Magnetic Moment.—As mentioned in the last chapter, electric currents flowing in neighboring conductors exert forces on each other known as magnetic forces. This magnetic interaction was carefully investigated by Ampère. One of his experiments showed that, when two wires carrying equal currents in opposite directions lie sufficiently close together, their power of interacting magnetically with other circuits is destroyed. Let us now construct several small plane loops of wire, twisting together the thin wires by which the current is to enter and leave the loops so that any magnetic forces observed will be due to the loop alone. Keeping them at distances great compared with their dimensions, we find that, when carrying steady currents, the forces and torques that they produce on each other are in every way like those between electric dipoles, provided that we orient the normal to the plane of the loop as we would the axis of an electric dipole. Thus, using three small loops carrying fixed currents, we can, by taking them two at a time and measuring the forces and distances between them, determine by 1.071 (6) the products $M_1M_2 = A$, $M_1M_3 = B$ and $M_2M_3 = C$ and so find $M_1^2 = AB/C$, $M_2^2 = AC/B$, and $M_3^2 = BC/A$. If the experiment is done *in vacuo* and $1/\epsilon$ is replaced by $\mu_v = 4\pi \times 10^{-7}$, then M for each loop equals the product of its area in square meters by its current in amperes and is the magnitude of its "magnetic moment." Thus an ampere may be defined as that current which, flowing in a small plane loop of wire, gives it a magnetic moment equal to its area. In a medium other than a vacuum the forces and torques are different and the factor μ which replaces μ_v is called the permeability of the medium.

7.01. Magnetic Induction and Permeability.—If a conductor, carrying a current, when placed in a certain region, experiences a magnetic force we say a magnetic field exists in that region. We shall map the magnetic field by means of a small exploring loop just as we could map the electric field by means of a small electric dipole. When free to move, the loop assumes a certain orientation. We shall define the positive direction of the magnetic field to be normal to the loop and directed as a right-handed screw moves when rotated in the sense in which the current traverses the loop. As in the electrostatic case, we can measure the strength of the magnetic field in terms of the torque acting on the loop.

Thus, we define the magnetic induction or flux density $\boldsymbol{B}$ to be a vector in the magnetic field direction whose magnitude, in webers per square meter, is the torque in newton meters on a loop of moment one whose axis is set normal to this direction. This torque, and hence $\boldsymbol{B}$, depends on the medium in which the experiment is performed. We define the relative magnetic permeability K_m of a substance to be the ratio of the magnetic induction when all space is occupied by the substance to its value at the same point *in vacuo*, the configuration of all circuits and the magnitude of all currents being the same. The permeability μ is the product $\mu_v K_m$.

The experiments on which the definition of μ and $\boldsymbol{B}$ depend involve no theoretical difficulty in the case of liquid and gaseous mediums. It is obviously impossible, however, to manipulate a loop of wire in a solid medium. If all space outside the small region in which our loop is situated is filled with one liquid and this region is filled with another, then we find that the measured value of $\boldsymbol{B}$ is independent of the shape and size of this region only when the liquids have the same permeability. Thus, to determine $\boldsymbol{B}$ and μ at a point in a homogeneous isotropic solid, we excavate a cavity at P and fill it with a liquid such that the measurement at P is independent of the shape and size of the cavity. The value of $\boldsymbol{B}$ and μ at P, so determined, is the same as before the cavity was created. In 7.24, by using the boundary conditions derived in 7.21, we shall describe an experimental method of defining $\boldsymbol{B}$ and μ which includes the case of magnetically anisotropic crystals.

Suppose that we take a great number of small loops, each carrying a current I and, without changing the area, fit them together into a mesh always keeping the direction of the induction due to adjacent loops parallel. From Ampère's experiment, the magnetic effect of all loop boundaries, except those forming the outer edge of the mesh, disappears, so that the resultant magnetic effect is exactly the same as if the current I circulated around the boundary alone and, hence, is independent of the shape of the surface of the mesh. But we know that the magnetic induction due to any elementary loop of area dS is identical with the electric field due to an electric dipole of moment $\mu\epsilon I\,dS$. The magnetic induction due to the whole circuit is therefore the same as the electric field due to a uniform electric double layer of strength $\mu\epsilon I$, whose boundary coincides with the circuit and which we discussed in 1.12. This magnetic equivalent of a double layer is called a magnetic shell. If we look for this shell by following a magnetic line of force produced by an electric circuit, we find that we pierce the same surfaces repeatedly, without arriving at any magnetic discontinuity corresponding to the electric double layer. Thus, the magnetic shell has no physical existence and is merely a convenient device of use in computing magnetic fields. Furthermore, this shows that there are neither sources nor sinks of a magnetic nature corresponding to the electrostatic charge so that

$$\text{div}\, \boldsymbol{B} = \boldsymbol{\nabla} \cdot \boldsymbol{B} = 0 \tag{1}$$

The work required to take a unit electric charge from one face of a uniform electric double layer of strength $\mu\epsilon I$ to the other, by a path passing around its edges is μI, by 1.12 (3), since the difference in the solid angle subtended by the layer at the two ends of the path is 4π. Thus, if the electric intensity is $\boldsymbol{E}$, we have

$$\oint \boldsymbol{E} \cdot \boldsymbol{ds} = \mu I$$

when $\boldsymbol{ds}$ is an element of the path. As we have seen, the magnetic induction due to a current I is everywhere the same as the electric intensity due to a uniform electric double layer of strength $\mu\epsilon I$, having the circuit as a boundary, so that, *in vacuo*, $\oint \boldsymbol{B} \cdot \boldsymbol{ds} = \mu_v I$, and in a medium of uniform permeability μ we would have

$$\oint \boldsymbol{B} \cdot \boldsymbol{ds} = \mu I \tag{2}$$

But from 6.00 (2), we have $I = \int \boldsymbol{i} \cdot \boldsymbol{n}\, dS$ where $\boldsymbol{i}$ is the current density and S is the area inside the path s, thus we have $\oint \boldsymbol{B} \cdot \boldsymbol{ds} = \mu \int \boldsymbol{i} \cdot \boldsymbol{n}\, dS$. We can transform the left side by Stokes's theorem [3.01 (1)], giving

$$\int_S \boldsymbol{\nabla} \times \boldsymbol{B} \cdot \boldsymbol{n}\, dS = \mu \int_S \boldsymbol{i} \cdot \boldsymbol{n}\, dS \tag{2.1}$$

It is easy to see that this holds for any path, even if it includes only part of the current; for in that case either that part of the shell due to currents not encircled will be crossed twice, once in the positive and once in the negative direction, or it will not be crossed at all. In either case, it contributes nothing to the integral. Thus, this relation holds for the integrands alone, giving in a region of uniform permeability,

$$\boldsymbol{\nabla} \times \boldsymbol{B} = \mu \boldsymbol{i} \tag{3}$$

7.02. Magnetic Vector Potential. Uniform Field.—It is well known that the divergence of a vector which is itself the curl of another vector is zero and so satisfies 7.01 (1). Thus, we may define a new vector $\boldsymbol{A}$, which we shall call the magnetostatic vector potential as a vector whose divergence is zero and whose curl is $\boldsymbol{B}$, thus

$$\boldsymbol{\nabla} \times \boldsymbol{A} = \boldsymbol{B} \qquad \boldsymbol{\nabla} \cdot \boldsymbol{A} = 0 \tag{1}$$

The standard formula for curl curl $\boldsymbol{A}$ now becomes

$$\boldsymbol{\nabla} \times [\boldsymbol{\nabla} \times \boldsymbol{A}] = \boldsymbol{\nabla}(\boldsymbol{\nabla} \cdot \boldsymbol{A}) - (\boldsymbol{\nabla} \cdot \boldsymbol{\nabla})\boldsymbol{A} = -\nabla^2 \boldsymbol{A}$$

Substituting in 7.01 (3), we have, in a region of uniform permeability,

$$\nabla^2 \boldsymbol{A} = -\mu \boldsymbol{i} \tag{2}$$

On writing out the components, this is

$$\mathbf{i}\nabla^2 A_x + \mathbf{j}\nabla^2 A_y + \mathbf{k}\nabla^2 A_z = -\mu(\mathbf{i}i_x + \mathbf{j}i_y + \mathbf{k}i_z) \tag{3}$$

Thus A_x, A_y, and A_z all satisfy Poisson's equation [3.02 (3)]. We have found a solution of this equation in 3.09 (1) to be

$$A_x = \frac{\mu}{4\pi}\int_V \frac{i_x\,dv}{r}, \qquad A_y = \frac{\mu}{4\pi}\int_V \frac{i_y\,dv}{r}, \qquad A_z = \frac{\mu}{4\pi}\int_V \frac{i_z\,dv}{r} \tag{4}$$

Adding components gives

$$\mathbf{A} = \frac{\mu}{4\pi}\int_V \frac{\mathbf{i}\,dv}{r} \tag{4.1}$$

In the region outside the wire, $\mathbf{i}$ is zero and since, in a thin wire of cross-sectional area dS, we have $i_x\,dv = i\,dS\,dx = I\,dx$, we may write, summing up the vector components,

$$\mathbf{A} = \frac{\mu}{4\pi}\oint \frac{I\,\mathbf{ds}}{r} \tag{5}$$

Physically, this is a more satisfactory concept than the magnetic shell since it depends only on the configuration of the circuit, the current strength, and the location of the point of measurement and involves no artificial discontinuities. It will give the correct value for the line integral of $\mathbf{B}$ for any path. Evidently, we can consider the vector contribution to $\mathbf{A}$ by any element of the closed circuit as parallel to that element.

The vector potential in rectangular coordinates associated with a uniform magnetic induction $\mathbf{B}$ in the x-direction has only two components A_y and A_z, and their values are

$$A_y = -\alpha z B \qquad A_z = (1 - \alpha)yB \tag{6}$$

where α is an arbitrary number. Obviously, the x-component of the curl is B and the other components are zero. In spherical coordinates, the vector potential of a uniform field parallel to the $\theta = 0$ axis is

$$A_\phi = \tfrac{1}{2}Br\sin\theta \tag{7}$$

In cylindrical coordinates, the vector potential of a uniform field parallel to the z-axis is

$$A_\phi = \tfrac{1}{2}B\rho \tag{8}$$

In oblate spheroidal coordinates, the vector potential of a uniform field parallel to the $\xi = 1$ axis is

$$A_\phi = \frac{c_1 B}{2j}P_1^1(\xi)P_1^1(j\zeta) \tag{9}$$

In prolate spheroidal harmonics, it is

$$A_\phi = \frac{c_2 B}{2}P_1^1(\xi)P_1^1(\eta) \tag{10}$$

The forms of A given in the last five formulas are not, of course, the most general forms that give the required B. We could add to each the gradient of a scalar without affecting B.

7.03. Uniqueness Theorems for Magnetostatics.—In 3.10 we found just what information concerning the interior and boundary surface of a region is necessary to determine uniquely the electrostatic potential inside it. This proof applies without change to the magnetomotance discussed in 7.28, provided there are no electric currents within the region which would make it multiple valued. The vector potential requires a different proof. We shall show that, if the location and magnitude of all fixed currents inside a closed surface are specified as well as the value of the tangential component of either the vector potential or the magnetic induction over this surface, then the value of the magnetic induction everywhere inside it is uniquely determined. The contribution of internal sources given by 7.02 (1) and (5) is unique. Suppose there were two internal values B and B' having identical external sources and boundary values and derived from A and A'. If $\nabla \cdot A = \nabla \cdot A' = 0$, then $\nabla^2 A' - \nabla^2 A = -\mu i$ and $\nabla \times \nabla \times (A - A')$ vanishes in v so that putting

$$\Psi = \Phi = A - A'$$

in 3.06 (5) gives

$$\int_{\mathrm{V}} (B - B')^2 \, d\mathrm{v} = \sum_{j=1}^{n} \int_{S_j} (\mathbf{A} - \mathbf{A}') \cdot [(B - B') \times n] \, dS_j \qquad (1)$$

If, on S, $n \times A = n \times A'$ or $B \times n = B' \times n$, then the surface integral is zero so that in either case $B = B'$ throughout v because $(B - B')^2$ is positive.

If $B - B'$ is zero throughout the volume, then A and A' can differ therein only by the gradient of a scalar. By 3.10, if the gradient of a scalar is zero over S, it vanishes in v so that if $A - A'$ vanishes over S it vanishes in v and A is uniquely determined therein by its value over S.

7.04. Orthogonal Expansions for Vector Potential.—In electrostatics, after finding the potential due to a fixed charge distribution, we superimposed suitable perturbing potentials to satisfy the conditions at dielectric or conducting boundaries. To use the same method with vector potentials, it is necessary that, having found by 7.02 (4.1) that part of the vector potential due to the given current distribution, we know suitable forms of perturbing potentials to superimpose to meet magnetic boundary conditions.

The direct method of solving Laplace's equation for the vector potential is not easy for, where Laplace's operator is applied to a vector, it operates not only upon the magnitudes of the components of the vector, but also upon the unit vectors themselves, as illustrated in 10.05 (2) and (3). Except in rectangular coordinates, this gives rise to a set of three simultaneous partial differential equations whose solution may be very complicated. We are led, therefore, to look for a simpler method.

We have seen [3.02 (1)] that in free space where there are no electric charges the divergence of the electric field is zero and, since from 1.06 (4) it is the gradient of a scalar, its curl is also zero. Similarly, from 7.01 (1) and (3), in free space, where there are no electric currents, the divergence and curl of the magnetic induction are zero. We would expect, in such regions, to write expansions for the two fields in terms of orthogonal functions, in the same form. In Chap. V, we obtained these expansions by solving Laplace's second-order partial differential equation. This equation was broken up into three total differential equations, each involving a single coordinate, connected by two indices. Thus, each term in the expansion involves two indices and six integration constants. As just pointed out, owing to the mathematical similarity of electric and magnetic fields, we should expect that part of the vector potential which contributes to the magnetic induction to have the same number of indices and constants.

This potential should be derivable from three scalar potential functions because, in rectangular coordinates, each component of the vector potential satisfies Laplace's equation. These components are not, however, independent, but are connected by the relation $\nabla \cdot \boldsymbol{A} = 0$, so that only two independent scalar functions, at most, can be used. Furthermore, as just pointed out, when properly chosen, only one of these will contribute to the magnetostatic field. The general expression for the vector potential, giving zero divergence, is

$$\boldsymbol{A} = \nabla \times \boldsymbol{W} \tag{1}$$

where $\boldsymbol{W}$ is a vector which should be derivable from two scalar potential functions. To fit boundary conditions, when dealing with eddy currents and electromagnetic radiation, we shall find it convenient to split $\boldsymbol{W}$ into two components, normal to each other, each of which is derivable from a different scalar potential function. Thus we shall write

$$\boldsymbol{W} = \boldsymbol{u}W_1 + \boldsymbol{u} \times \nabla W_2 \tag{2}$$

where $\nabla^2 W_1 = \nabla^2 W_2 = 0$ and $\boldsymbol{u}$ is an arbitrary vector to be chosen so that

$$\nabla^2 \boldsymbol{A} = \nabla^2(\nabla \times \boldsymbol{W}) = \nabla \times (\nabla^2 \boldsymbol{W}) = 0 \tag{3}$$

This holds when $\nabla^2 W$ is replaced by $\nabla^2 W + CW$ unless C depends on x, y, z. Now it is easily verified, by writing out in rectangular coordinates, that

$$\nabla^2 \boldsymbol{u} W_1 = \boldsymbol{u}\nabla^2 W_1 \text{ or } \boldsymbol{u}\nabla^2 W_1 + 2\nabla W_1 \text{ and } \nabla^2(\boldsymbol{u} \times \nabla W_2) = \boldsymbol{u} \times \nabla(\nabla^2 W_2) \tag{4}$$

where $\boldsymbol{u} = \boldsymbol{i}, \boldsymbol{j}, \boldsymbol{k}$, or $\boldsymbol{r}$, so that these are suitable choices of $\boldsymbol{u}$. From the similarity of $\boldsymbol{B}$ and $\boldsymbol{E}$, we suspect that $\boldsymbol{B}$, like $\boldsymbol{E}$, can be obtained from a single scalar potential function. This surmise is confirmed by examining the contribution of W_2 to A. Thus we have

$$u = i, j, \text{ or } k, \quad \nabla \times (u \times \nabla W_2) = u(\nabla^2 W_2) - \nabla(u \cdot \nabla W_2) \tag{5}$$
$$u = r, \quad \nabla \times (u \times \nabla W_2) = u(\nabla^2 W_2) - \nabla(u \cdot \nabla W_2) - \nabla W_2 \tag{6}$$

Since $\nabla^2 W_2 = 0$, the part of A derived from W_2 is the gradient of a scalar and contributes nothing to B in the magnetostatic case.

If u_1, u_2 and u_3 are orthogonal curvilinear coordinates, if the u specified above lies in the direction of u_1, and if W_1 is of the form $U(u_1)F(u_2, u_3)$, then $B \cdot A = 0$. This is proved in Lass, pp. 48 and 57.

In the following articles, we shall use the theory just developed to obtain solutions of the equation

$$\nabla^2 A = 0 \tag{7}$$

in the form

$$A = u_1' U_{11} U_{12} U_{13} + u_2' U_{21} U_{22} U_{23} + u_3' U_{31} U_{32} U_{33}$$

where u_1', u_2', and u_3' are unit vectors in the directions of the coordinates u_1, u_2, and u_3, and U_{rs} is a function of u_s only. The solution should be in such form that the vector potential anywhere inside a volume containing no sources and bounded by a set of surfaces, on each of which one of the coordinates is constant, can be calculated when the value of its tangential components on these surfaces is given.

7.05. Vector Potential in Cylindrical Coordinates.—In Chap. V we found that the general solution of Laplace's equation in cylindrical coordinates could be built up of a sum of terms involving, except in particular cases, Bessel functions. We now wish to find analogous solutions for the vector potential which possess orthogonal properties on the surfaces of a right circular cylinder so that we can express the tangential components of the vector potential thereon as a sum of such solutions and thus determine its value at any interior point. Choosing the Bessel function solution of $\nabla^2 W_1 = 0$ given in 5.291 and setting $u = k$, the W of 7.04 (1) becomes

$$W = kk^{-1}(Ae^{kz} + Be^{-kz})[CJ_n(k\rho) + DY_n(k\rho)] \sin (n\phi + \delta_n) \tag{1}$$

The vector potential derived from this by 7.04 (1) is

$$A_\rho = -(Ae^{kz} + Be^{-kz})[CJ_n(k\rho) + DY_n(k\rho)]n(k\rho)^{-1} \cos (n\phi + \delta_n)$$
$$A_\phi = (Ae^{kz} + Be^{-kz})[CJ_n'(k\rho) + DY_n'(k\rho)] \sin (n\phi + \delta_n) \tag{2}$$

This is the orthogonal surface vector function defined by 5.296 (8). If, for a given z-value, either component of $k \times A$ vanishes when $\rho = a$ then at this z-value $k \times A$ may be expressed as a sum of such functions. A rarely used form containing ϕ is obtained by setting $\delta_n = 0$, letting $n \to 0$ and keeping the products of nA and nB finite.

In order to obtain suitable solutions for expressing the tangential components of A over the curved surfaces, we use the forms of 5.292 (3) which are orthogonal in z and ϕ. This gives in place of (2)

$$A_\rho = -[CI_n(k\rho) + DK_n(k\rho)]n(k\rho)^{-1} \cos (kz + \gamma_k) \cos (n\phi + \delta_n)$$
$$A_\phi = [CI'_n(k\rho) + DK'_n(k\rho)] \cos (kz + \gamma_k) \sin (n\phi + \delta_n) \quad (3)$$

The z-component satisfies the scalar Laplace equation and is written

$$A_z = [C'I_n(k\rho) + D'K_n(k\rho)] \cos (kz + \gamma'_k) \cos (n\phi + \delta'_n) \quad (4)$$

The above solutions are inadequate when $k = 0$ and $n \neq 0$. For the curved surface, solutions corresponding to (3) and (4) are

$$A_\rho = (A\rho^{n-1} + B\rho^{-n-1})(Cz + D) \sin (n\phi + \delta_n)$$
$$A_\phi = (-A\rho^{n-1} + B\rho^{-n-1})(Cz + D) \cos (n\phi + \delta_n)$$
$$A_z = (A'\rho^n + B'\rho^{-n})(C'z + D') \cos (n\phi + \delta'_n) \quad (5)$$

When $k = 0$, there are no solutions orthogonal in both ρ and ϕ suitable for the end surfaces. When both k and n are zero, some forms of interest can be found from 5.291 (9) by taking

$$W = k[(Az\phi + B\phi + Cz + D) \ln \rho + (Ez + F)\phi] + rGz \quad (6)$$
$$A_\rho = A\rho^{-1}z \ln \rho + B\rho^{-1} \ln \rho + E\rho^{-1}z + F$$
$$A_\phi = -A\rho^{-1}z\phi - B\rho^{-1}\phi - C\rho^{-1}z - D\rho^{-1} + G\rho$$
$$A_z = (G\phi + H) \ln \rho + I\phi \quad (7)$$

Other solutions not often needed are found by using $\mathbf{i}$, $\mathbf{j}$, or $\mathbf{r}$ in W.

7.06. Vector Potential in Spherical Coordinates.—In Chap. V the general solution of Laplace's equation in polar spherical coordinates was built up of a sum of terms involving spherical harmonics. We need analogous solutions for the vector potential which possess orthogonal properties on the surface of a sphere so that we can express the tangential components of the vector potential thereon as a sum of such solutions and thus determine its value at any interior point. Choosing the spherical harmonic solution of $\nabla^2 W_1 = 0$ given in 5.23 and setting $\mathbf{u} = \mathbf{r}$, the W and A of 7.04 become

$$W = \mathbf{r}(Ar^n + Br^{-n-1})[CP_n^m(\cos\theta) + DQ_n^m(\cos\theta)] \sin (m\phi + \delta_m) \quad (1)$$
$$A_\theta = (Ar^n + Br^{-n-1})[CP_n^m(\cos\theta) + DQ_n^m(\cos\theta)]m \csc\theta \cos (m\phi + \delta_m)$$
$$A_\phi = (Ar^n + Br^{-n-1})[CP_n^{m\prime}(\cos\theta) + DQ_n^{m\prime}(\cos\theta)] \sin\theta \sin (m\phi + \delta_m) \quad (2)$$

This A is the orthogonal surface vector function mentioned in 5.231. When $m = 0$ and $\delta_m = \frac{1}{2}\pi$, this equation becomes

$$A_\phi = (Ar^n + Br^{-n-1})[CP_n^1(\cos\theta) + DQ_n^1(\cos\theta)] \quad (3)$$

It is often useful to have $\mathbf{B}$ in terms of $\mathbf{W}$. If $\mathbf{u} = \mathbf{r}$ we have

$$\mathbf{B} = \nabla \times (\nabla \times \mathbf{r}W) = -\nabla \times (\mathbf{r} \times \nabla W) = r\frac{\partial}{\partial r}(\nabla W) + 2\nabla W$$

Writing out the components of this equation gives

$$B_r = \frac{\partial^2(rW)}{\partial r^2}, \qquad B_\theta = \frac{\partial^2(rW)}{r\,\partial r\,\partial\theta}, \qquad B_\phi = \frac{1}{r\sin\theta}\frac{\partial^2(rW)}{\partial r\,\partial\phi} \quad (4)$$

7.07. Vector Potential in Terms of Magnetic Induction on Axis.—Magnetic lenses for focusing electron beams are usually constructed from coaxial current coils or equivalent magnets so that the vector potential has only a ϕ-component. To calculate the focusing properties of such lenses, it is convenient to have the vector potential expressed in terms of the magnetic induction and its derivatives along the axis. If $\mathbf{W} = \mathbf{k}W$, where W is the solution of Laplace's equation given in 3.15 (1), 7.04 (1) yields, if Ψ is written for Φ',

$$A_\phi = -\frac{\partial W}{\partial \rho} = -\frac{j}{2\pi}\int_0^{2\pi} \Psi(z + j\rho \sin\theta)\sin\theta\, d\theta \tag{1}$$

where $\Psi(z)$ is a real function of z. The Taylor expansion is, by *Dw* 39,

$$\Psi(z + j\rho\sin\theta) = \Psi(z) + \frac{\partial\Psi(z)}{\partial z}\frac{j\rho\sin\theta}{1!} + \frac{\partial^2\Psi(z)}{\partial z^2}\frac{(j\rho\sin\theta)^2}{2!} + \cdots \tag{2}$$

Substitution of (2) in (1) and integration from 0 to 2π give

$$A_\phi = \sum_{n=0}^{\infty}\frac{(-1)^n}{n!(n+1)!}\frac{\partial^{2n+1}\Psi(z)}{\partial z^{2n+1}}\left(\frac{\rho}{2}\right)^{2n+1} \tag{3}$$

The magnetic induction $\nabla \times \mathbf{A}$ on the axis where $\rho = 0$ is

$$B_0(z) = \left[\frac{1}{\rho}\frac{\partial(\rho A_\phi)}{\partial\rho}\right]_{\rho=0} = \frac{\partial\Psi(z)}{\partial z} \tag{4}$$

This, when put in (3), gives $A_\phi(\rho, z)$ in terms of $B_0(z)$ and its derivatives.

$$A_\phi = \sum_{n=0}^{\infty}\frac{(-1)^n}{n!(n+1)!}\frac{\partial^{2n}B_0(z)}{\partial z^{2n}}\left(\frac{\rho}{2}\right)^{2n+1} \tag{5}$$

The current elements that generate fields of this type are coaxial circular current loops so the range of ρ and z over which (5) holds are the same as for such loops. To find the range insert $B_0(z)$ from 7.10 (8) into (5) and compare the result with that obtained from problem 27, which holds for all values of ρ and z, when $J_1(k\rho)$ is expanded by 5.293 (3), the summation and integration are interchanged, and the integrals are replaced by derivatives of 5.298 (6). The identity of the results shows that (5) gives a unique vector potential at all values of ρ and z which can be reached from the axis without crossing a current sheet. The restriction is needed because the external fields of any currents inside a closed surface are unaffected if these currents are removed, provided suitable currents are set up in the surface.

7.08. Equation of Axially Symmetrical Tubes of Induction.—A magnetic field is most easily visualized by mapping the stream lines or lines of magnetic induction. When we have axial symmetry so that all sections of the field made by planes passing through the axis look the same and all currents are normal to these planes, we can obtain the equation of these lines very simply from the vector potential which, in this case, has only the component A_ϕ. When rotated about the axis, each stream line generates the surface of a tube of induction. Any particular tube may be specified by stating the flux N through it. This is obtained by integrating the normal component of the induction over any cross section S of the tube. Thus the equation of the surface of such a tube is given by

$$N = \int_S \mathbf{B} \cdot \mathbf{n}\, dS = f(\rho, z) \tag{1}$$

By means of Stokes's theorem [3.01 (1)], this becomes

$$N = \int_S \nabla \times \mathbf{A} \cdot \mathbf{n}\, dS = \oint \mathbf{A} \cdot d\mathbf{s} \tag{2}$$

If the section is taken by a plane normal to the z-axis, then the path s is a circle on which A_ϕ is constant and its length is $2\pi\rho$ so that the equation of the tube of force becomes

$$N = 2\pi\rho A_\phi(\rho, z) \tag{3}$$

In an axially symmetrical system of orthogonal curvilinear coordinates u_1, u_2, and ϕ, the distance ρ will be a function of u_1 and u_2 so that the equation of the tubes of force is

$$N = 2\pi\rho(u_1, u_2)A_\phi(u_1, u_2) \tag{4}$$

7.09. Vector Potential and Field of Bifilar Circuit.—We shall first apply 7.02 (5) to finding the magnetic field due to a long straight wire with

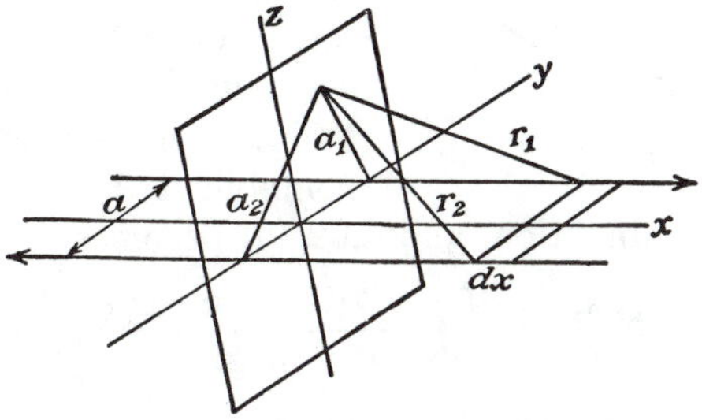

FIG. 7.09.

a parallel return at a distance a from it. Since all elements of the wire lie in the x-direction in Fig. 7.09, the vector A has only the component A_x Thus, we have, from 7.02 (5),

$$A_x = \frac{\mu I}{4\pi}\int_{-\infty}^{+\infty}\left(\frac{1}{r_1} - \frac{1}{r_2}\right)dx = \frac{\mu I}{2\pi}\int_0^\infty [(a_1^2 + x^2)^{-\frac{1}{2}} - (a_2^2 + x^2)^{-\frac{1}{2}}]\, dx$$
$$= \frac{\mu I}{2\pi}\left|\ln \frac{r_1 + x}{r_2 + x}\right|_0^\infty = \frac{\mu I}{2\pi} \ln \frac{a_2}{a_1} \tag{1}$$

This shows the surfaces of constant vector potential to be circular cylinders coinciding exactly with the equipotentials in the electrostatic case where $V = -q(2\pi\epsilon)^{-1} \ln (a_2/a_1)$ if the two wires carry charges $+\mu\epsilon I$ and $-\mu\epsilon I$. For $\mu I = 2\pi$, the equation of these cylinders is given by 4.13 (3), where A_x is substituted for U and z for x. The magnetic field which is given by $\nabla \times A$ is at right angles to the electric field given by ∇V but has the same numerical value. From 3.04 (1), (2), and (3), the components of the magnetic induction, are, *in vacuo*, since $h_1 = h_2 = h_3 = 1$,

$$B_x = \frac{\partial A_z}{\partial y} - \frac{\partial A_y}{\partial z} = 0, \qquad B_y = \frac{\partial A_x}{\partial z}, \qquad B_z = -\frac{\partial A_x}{\partial y} \tag{2}$$

Since $a_1 = [(y - \frac{1}{2}a)^2 + z^2]^{\frac{1}{2}}$ and $a_2 = [(y + \frac{1}{2}a)^2 + z^2]^{\frac{1}{2}}$, this gives

$$B_y = \frac{\mu I z}{2\pi}\left(\frac{1}{a_2^2} - \frac{1}{a_1^2}\right) \tag{3}$$

$$B_z = -\frac{\mu I}{2\pi}\left(\frac{y + \frac{1}{2}a}{a_2^2} - \frac{y - \frac{1}{2}a}{a_1^2}\right) \tag{4}$$

7.10. Vector Potential and Field of Circular Loop.—Let us compute the vector potential at the point P shown in Fig. 7.10. From symmetry, we know that in spherical polar coordinates the magnitude of A is independent of ϕ. Therefore, for simplicity, we choose the point P in the xz-plane where $\phi = 0$. We notice that when equidistant elements of length ds at $+\phi$ and $-\phi$ are paired, the resultant is normal to ρz. Thus, A has only the single component A_ϕ. Let ds_ϕ be the component of $\boldsymbol{ds}$ in this direction, and 7.02 (5) becomes

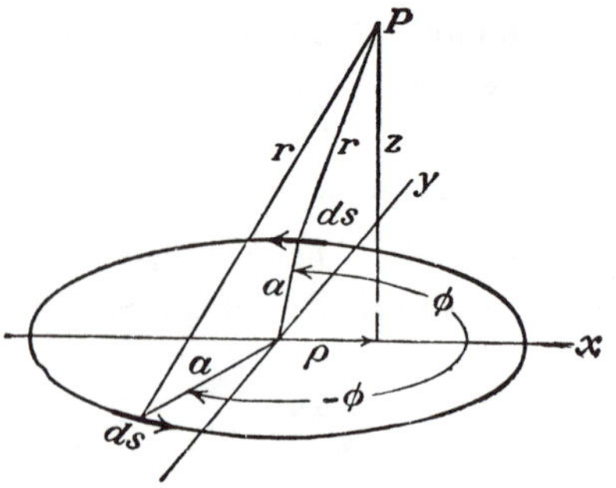

FIG. 7.10.

$$A_\phi = \frac{\mu I}{4\pi}\oint \frac{ds_\phi}{r} = \frac{\mu I}{2\pi}\int_0^\pi \frac{a \cos \phi \, d\phi}{(a^2 + \rho^2 + z^2 - 2a\rho \cos \phi)^{\frac{1}{2}}}$$

If our loop is very small, this becomes, since $r_o = (\rho^2 + z^2)^{\frac{1}{2}} \gg a$,

$$A_\phi = \frac{\mu I}{2\pi}\int_0^\pi \frac{a \cos \phi}{r_o}\left(1 + \frac{a\rho \cos \phi}{r_o^2}\right) d\phi \approx \frac{a^2 \mu I \rho}{4r^3} = \frac{a^2 \mu I \sin \theta}{4r^2} \tag{1}$$

By taking, from 7.00 the magnetic moment $\boldsymbol{M}$ of the loop equal to $\pi a^2 I$ and directed upward, this may be written

$$A = \mu(\boldsymbol{M} \times \boldsymbol{r})(4\pi r^3)^{-1} \tag{2}$$

If this approximation does not hold, let $\phi = \pi + 2\theta$ so that $d\phi = 2\, d\theta$

and $\cos\phi = 2\sin^2\theta - 1$, then this becomes

$$A_\phi = \frac{\mu a I}{\pi}\int_0^{\frac{1}{2}\pi} \frac{(2\sin^2\theta - 1)\,d\theta}{[(a+\rho)^2 + z^2 - 4a\rho\sin^2\theta]^{\frac{1}{2}}}$$

Rearrange and let

$$k^2 = 4a\rho[(a+\rho)^2 + z^2]^{-1}, \qquad m = [1-(1-k^2)^{\frac{1}{2}}][1+(1-k^2)^{\frac{1}{2}}]^{-1} \tag{3}$$

$$A_\phi = \frac{\mu I}{\pi k}\left(\frac{a}{\rho}\right)^{\frac{1}{2}}\left[\left(1 - \frac{1}{2}k^2\right)K - E\right] = \frac{\mu I}{32}\left(\frac{a}{\rho}\right)^{\frac{1}{2}} k^3\left(1 + \frac{3}{4}k^2 + \frac{75}{128}k^4 + \cdots\right)$$
$$= \frac{\mu I}{\pi}\left(\frac{a}{m\rho}\right)^{\frac{1}{2}}[K(m) - E(m)] = \frac{\mu I}{4}\left(\frac{a}{\rho}\right)^{\frac{1}{2}} m^{\frac{3}{2}}\left(1 + \frac{3}{8}m^2 + \frac{15}{64}m^4 + \cdots\right) \tag{4}$$

where K and E are complete elliptic integrals of the first and second kind. When $z = 0$, the m modulus is simply ρ/a or a/ρ.

To determine the magnetic induction, we must write, from 3.04 (1), (2), and (3), the components of the curl in cylindrical coordinates. From 3.05, this gives $h_1 = 1$, $h_2 = \rho$, and $h_3 = 1$ so that

$$\begin{aligned} B_\rho &= -\frac{1}{\rho}\frac{\partial}{\partial z}(\rho A_\phi) + \frac{1}{\rho}\frac{\partial}{\partial\phi}(A_z) = -\frac{\partial A_\phi}{\partial z} \\ B_\phi &= \frac{\partial}{\partial z}(A_\rho) - \frac{\partial}{\partial\rho}(A_z) = 0 \\ B_z &= -\frac{1}{\rho}\frac{\partial}{\partial\phi}(A_\rho) + \frac{1}{\rho}\frac{\partial}{\partial\rho}(\rho A_\phi) = \frac{1}{\rho}\frac{\partial}{\partial\rho}(\rho A_\phi) \end{aligned} \tag{5}$$

For the derivatives of K and we use the formulas *Dw* 789.1 and 789.2

$$\frac{\partial K}{\partial k} = \frac{E}{k(1-k^2)} - \frac{K}{k} \qquad \text{and} \qquad \frac{\partial E}{\partial k} = \frac{E}{k} - \frac{K}{k}$$

We also have, from (3), that

$$\frac{\partial k}{\partial z} = -\frac{zk^3}{4a\rho} \qquad \text{and} \qquad \frac{\partial k}{\partial \rho} = \frac{k}{2\rho} - \frac{k^3}{4\rho} - \frac{k^3}{4a}$$

Carrying out the differentiation, collecting terms, and substituting for k give

$$B_\rho = \frac{\mu I}{2\pi}\frac{z}{\rho[(a+\rho)^2 + z^2]^{\frac{1}{2}}}\left[-K + \frac{a^2+\rho^2+z^2}{(a-\rho)^2+z^2}E\right] \tag{6}$$

$$B_z = \frac{\mu I}{2\pi}\frac{1}{[(a+\rho)^2 + z^2]^{\frac{1}{2}}}\left[K + \frac{a^2-\rho^2-z^2}{(a-\rho)^2+z^2}E\right] \tag{7}$$

Numerical values of B_ρ and B_z can be computed for any values of ρ and z by finding k from (3) and looking up the corresponding values of K and E in a table (*Dw* pp. 208–210 or *Pc* p. 121). On the axis $\rho = 0$ we have

$$B_\rho = \frac{0}{0} \to 0 \qquad \text{and} \qquad B_z = \frac{\frac{1}{2}\mu a^2 I}{(a^2+z^2)^{\frac{3}{2}}} \tag{8}$$

7.11. Field of Currents in Spherical Shell.—We define the value of the stream function ψ, at any point P on a thin spherical shell of radius a, to be the total current flowing across any line drawn on the surface of the shell between P and a point where ψ is zero. The components of the current density are therefore related to ψ by the equation

$$i_\theta = \frac{-1}{a \sin\theta}\frac{\partial\psi}{\partial\phi}, \qquad i_\phi = \frac{1}{a}\frac{\partial\psi}{\partial\theta} \tag{1}$$

We wish to find the vector potential and magnetic field due to these currents. Since any possible ψ can be expressed as a sum of surface harmonics, it will suffice to calculate the field of the distribution $\psi_n^m = S_n^m(\theta, \phi)$, for we can then, by superimposing, obtain that due to any ψ. We shall designate the induction outside the shell by $\boldsymbol{B}_o$ and that inside by $\boldsymbol{B}_i$. Let us apply 7.01 (2) to a small circuit around an element of shell of length $\Delta\theta$, lying in a constant ϕ plane. We obtain

$$\mu i_\phi a\,\Delta\theta = [(B_\theta)_o - (B_\theta)_i]a\,\Delta\theta$$

Using (1) and introducing the scalar function W, of 7.06 (4), we have

$$\mu\frac{\partial\psi}{\partial\theta} = \frac{\partial^2(rW_o)}{\partial r\,\partial\theta} - \frac{\partial^2(rW_i)}{\partial r\,\partial\theta}$$

Similarly, taking a circuit in the ϕ direction gives

$$\frac{-\mu}{\sin\theta}\frac{\partial\psi}{\partial\phi} = \frac{-1}{\sin\theta}\frac{\partial^2(rW_o)}{\partial r\,\partial\phi} + \frac{1}{\sin\theta}\frac{\partial^2(rW_i)}{\partial r\,\partial\phi}$$

If we multiply the first equation through by $d\theta$ and the second by $\sin\theta\; d\phi$ and subtract, both sides are total differentials so that, if we integrate and remember that $W_o = W_i$ when $r = a$, we obtain

$$\mu\psi = \frac{\partial}{\partial r}(rW_o - rW_i) = a\left(\frac{\partial W_o}{\partial r} - \frac{\partial W_i}{\partial r}\right) \tag{2}$$

To give ψ the harmonic form $S_n^m(\theta, \phi)$, to make $W_o = W_i$ at $r = a$, and to have W_o finite at infinity and W_i finite at the origin, we must choose, from 7.06 (1), the forms

$$W_o = \frac{-\mu}{2n+1}\left(\frac{a}{r}\right)^{n+1} S_n^m(\theta, \phi), \qquad W_i = \frac{-\mu}{2n+1}\left(\frac{r}{a}\right)^n S_n^m(\theta, \phi) \tag{3}$$

Hence, from 7.04 (1), the vector potential has the form

$$(A_r)_o = 0, \qquad (A_\theta)_o = \frac{-\mu}{(2n+1)\sin\theta}\left(\frac{a}{r}\right)^{n+1}\frac{\partial}{\partial\phi}[S_n^m(\theta, \phi)]$$

$$(A_\phi)_o = \frac{\mu}{2n+1}\left(\frac{a}{r}\right)^{n+1}\frac{\partial}{\partial\theta}[S_n^m(\theta, \phi)]$$

$$(A_r)_i = 0, \qquad (A_\theta)_i = \frac{-\mu}{(2n+1)\sin\theta}\left(\frac{r}{a}\right)^n\frac{\partial}{\partial\phi}[S_n^m(\theta, \phi)]$$

$$(A_\phi)_i = \frac{\mu}{2n+1}\left(\frac{r}{a}\right)^n\frac{\partial}{\partial\theta}[S_n^m(\theta, \phi)] \tag{4}$$

The θ terms in A_ϕ and A_θ have exactly the form of the first and second terms of 5.231 (7), so A is an orthogonal surface vector function on the spherical surface.

7.12. Zonal Currents in Spherical Shell.—When the currents in a spherical shell flow along parallels of latitude, we have the most important practical case. In this case, axial symmetry makes all quantities independent of ϕ. The stream function may then be written

$$\psi = \sum_{n=1}^{\infty} C_n P_n(\cos\theta) \tag{1}$$

The current density having only a ϕ component, becomes

$$i_\phi = \frac{1}{a}\frac{\partial\psi}{\partial\theta} = \sum_{n=1}^{\infty}\frac{C_n}{a}\frac{\partial P_n(u)}{\partial\theta} = -\sum_{n=1}^{\infty}\frac{C_n\sin\theta}{a}\frac{\partial P_n(u)}{\partial u} = -\sum_{n=1}^{\infty}\frac{C_n}{a}P_n^1(u) \tag{2}$$

where $u = \cos\theta$. The function W, if all space has the permeability μ, takes, from 7.11 (3), the form

$$W_o = \sum_{n=1}^{\infty}\frac{-\mu C_n}{2n+1}\left(\frac{a}{r}\right)^{n+1}P_n(u), \qquad W_i = \sum_{n=1}^{\infty}\frac{-\mu C_n}{2n+1}\left(\frac{r}{a}\right)^{n}P_n(u) \tag{3}$$

The vector potential, from 7.11 (4), may be written

$$r < a \qquad A = \boldsymbol{\phi}\mu\sum_{n=1}^{\infty}\frac{-C_n}{2n+1}\left(\frac{r}{a}\right)^{n}P_n^1(u) \tag{4}$$

$$r > a \qquad A = \boldsymbol{\phi}\mu\sum_{n=1}^{\infty}\frac{-C_n}{2n+1}\left(\frac{a}{r}\right)^{n+1}P_n^1(u) \tag{5}$$

We may note here that $(\boldsymbol{\phi})_x = -\sin\phi$ and $(\boldsymbol{\phi})_y = \cos\phi$ so that A_x and A_y satisfy 7.02 (3) in free space where $\boldsymbol{i} = 0$. We also see that A has the form given in 7.06 (3). The components of the magnetic induction may be obtained from (3) or (4) and (5). Thus, by 7.06 (4),

$$r < a \qquad B_r = \frac{\partial^2(rW_i)}{\partial r^2} = -\frac{\mu}{a}\sum_{n=1}^{\infty}\frac{n(n+1)C_n}{2n+1}\left(\frac{r}{a}\right)^{n-1}P_n(u) \tag{6}$$

$$r > a \qquad B_r = \frac{\partial^2(rW_o)}{\partial r^2} = -\frac{\mu}{a}\sum_{n=1}^{\infty}\frac{n(n+1)C_n}{2n+1}\left(\frac{a}{r}\right)^{n+2}P_n(u) \tag{7}$$

From (4) and (5), we have

$$r < a \qquad B_\theta = -\frac{1}{r}\frac{\partial(rA_\phi)}{\partial r} = +\frac{\mu}{a}\sum_{n=1}^{\infty}\frac{(n+1)C_n}{2n+1}\left(\frac{r}{a}\right)^{n-1}P_n^1(u) \tag{8}$$

$$r > a \qquad B_\theta = -\frac{1}{r}\frac{\partial(rA_\phi)}{\partial r} = -\frac{\mu}{a}\sum_{n=1}^{\infty}\frac{nC_n}{2n+1}\left(\frac{a}{r}\right)^{n+2}P_n^1(u) \tag{9}$$

These equations show us not only how to find the induction due to wires wound with any density on the sphere, but also how to wind the wires to produce a given field. For a uniform field inside, we have $B_r = B\cos\theta$ and $B_\theta = -B\sin\theta$ so that, from (6) and (8), $C_n = 0$ except for $n = 1$. Thus, from (2), the density of winding is proportional to $d[P_1(\cos\theta)]/d\theta = -\sin\theta$. By 7.08 (3), the equations of the tubes of force, if we use (4) and (5), are

$$N = 2\pi r\sin\theta A_\phi \tag{10}$$

7.13. Field of Circular Loop in Spherical Harmonics.—Suppose that the current density is zero on the surface of the sphere except in a band at $\theta = \alpha$ whose width is too small to measure physically but is different from zero so that the current density function and vector potential are everywhere bounded. From 7.12 (2) and (3), i may be written

$$i = -\frac{1}{a}\sum_{n=1}^{\infty}C_nP_n^1(u)$$

We determine C_n as in Chap. V by multiplying by $P_m^1(u)$ and integrating from $u = -1$ to $u = 1$. By 5.13 (2), all terms on the right drop out except when $n = m$, and by 5.231 (3) this gives

$$C_n = -\frac{(n-1)!}{(n+1)!}\frac{2n+1}{2}\int_{-1}^{1}aiP_n^1(u)\,du$$

The current density i is zero except in a band of width Δs at $\theta = \alpha$, which is so narrow that $P_n^1(\cos\theta)$ has the constant value $P_n^1(\cos\alpha)$ on it. Then the integral becomes

$$P_n^1(\cos\alpha)\int ia\sin\theta\,d\theta = P_n^1(\cos\alpha)\sin\alpha\int_S i\,ds = I\sin\alpha P_n^1(\cos\alpha)$$

so that we have

$$C_n = -\frac{(2n+1)I\sin\alpha}{2n(n+1)}P_n^1(\cos\alpha) \tag{1}$$

The vector potential becomes, from 7.12 (4), if $r < a$,

$$\mathbf{A} = \boldsymbol{\phi}\frac{\mu I}{2}\sum_{n=1}^{\infty}\frac{\sin\alpha}{n(n+1)}\left(\frac{r}{a}\right)^n P_n^1(\cos\alpha)P_n^1(\cos\theta) \tag{2}$$

and the components of the induction are, from 7.12 (6) and (8),

$$B_r = \frac{\mu I \sin \alpha}{2a} \sum_{n=1}^{\infty} \left(\frac{r}{a}\right)^{n-1} P_n^1(\cos \alpha) P_n(\cos \theta) \quad (3)$$

$$B_\theta = -\frac{\mu I \sin \alpha}{2a} \sum_{n=1}^{\infty} \frac{1}{n} \left(\frac{r}{a}\right)^{n-1} P_n^1(\cos \alpha) P_n^1(\cos \theta) \quad (4)$$

where $u = \cos \theta$. The expression for $r > a$ may be obtained from this as before with 7.12 (7) and (9). When the origin is at the center of the loop, $\alpha = \frac{1}{2}\pi$, $\cos \alpha = 0$, and $P_n^1(0) = -(n+1)P_{n+1}(0)$ so that even values of n are out and by 5.157 we have, when $r < a$,

$$B_r = -\frac{\mu I}{2a} \sum_{n \text{ odd}}^{1 \text{ to } \infty} (-1)^{\frac{n+1}{2}} \frac{1 \cdot 3 \cdot 5 \cdots n \cdot (n+1)}{2 \cdot 4 \cdot 6 \cdots (n-1)(n+1)} \left(\frac{r}{a}\right)^{n-1} P_n(\cos \theta) \quad (5)$$

$$B_\theta = \frac{\mu I}{2a} \sum_{n \text{ odd}}^{1 \text{ to } \infty} (-1)^{\frac{n+1}{2}} \frac{1 \cdot 3 \cdot 5 \cdots n \cdot (n+1)}{n \cdot 2 \cdot 4 \cdot 6 \cdots (n-1)(n+1)} \left(\frac{r}{a}\right)^{n-1} P_n^1(\cos \theta) \quad (6)$$

When r is greater than a, we must substitute $(a/r)^{n+2}$ for $(r/a)^{n-1}$ in (5) and $-n(a/r)^{n+2}/(n+1)$ for $(r/a)^{n-1}$ in (6).

By 7.08 (3), the equations of the tubes of force, using (2), are

$$N = 2\pi r \sin \theta A_\phi \quad (7)$$

7.14. Biot and Savart's Law. Field of Straight Wire.—Taking the curl of both sides of 7.02 (5) we have

$$\boldsymbol{B} = \nabla \times \boldsymbol{A} = \frac{\mu I}{4\pi} \oint \nabla \times \frac{\boldsymbol{ds}}{r}$$

but

$$\nabla \times \frac{\boldsymbol{ds}}{r} = \frac{1}{r} \nabla \times \boldsymbol{ds} + \nabla\left(\frac{1}{r}\right) \times \boldsymbol{ds}$$

Since the coordinates of the point at which $\boldsymbol{B}$ is being computed do not appear in $\boldsymbol{ds}$, which depends only on the circuit configuration, $\boldsymbol{ds}$ is constant as far as ∇ is concerned and the first term on the right is zero. But $\nabla(1/r) = -(\boldsymbol{r}/r^3)$ so that

$$\boldsymbol{B} = \frac{\mu I}{4\pi} \oint \frac{\boldsymbol{ds} \times \boldsymbol{r}}{r^3} \quad (1)$$

If θ is the angle measured from $\boldsymbol{ds}$ to $\boldsymbol{r}$, this may be written

$$B = \frac{\mu I}{4\pi} \oint \frac{\sin \theta \, ds}{r^2} \quad (2)$$

$\boldsymbol{B}$ is normal to the plane containing ds and r. If θ is clockwise, $\boldsymbol{B}$ points away from the observer. Biot and Savart established this law only for straight currents.

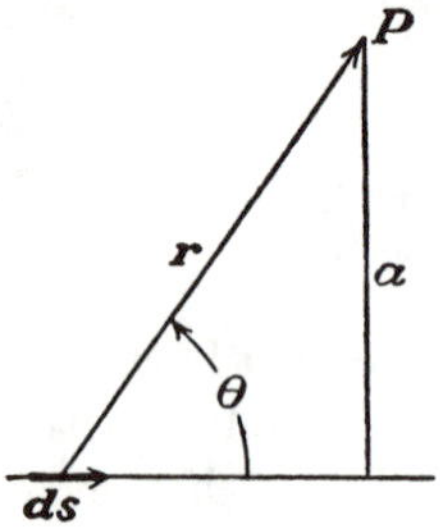

FIG. 7.14.

In this case, if a is the distance of P from the wire (Fig. 7.14), we have

$$r^2 = a^2 + s^2, \qquad \sin\theta = \frac{a}{r}$$

so that (2) becomes, by *Pc* 138 or *Dw* 200.03,

$$B = \frac{\mu I}{2\pi}\int_0^\infty \frac{a\,ds}{(a^2+s^2)^{\frac{3}{2}}} = \frac{\mu I}{2\pi a}\left|\frac{s}{(a^2+s^2)^{\frac{1}{2}}}\right|_0^\infty = \frac{\mu I}{2\pi a} \tag{3}$$

7.15. Field of Helical Solenoid.—A very frequently used form of electric circuit is the solenoid. This consists of wire wound in the form of a helix, shown in Fig. 7.15. Let the equation of this helix be

$$x = a\cos\phi, \qquad y = a\sin\phi, \qquad z = a\phi\tan\alpha \tag{1}$$

where α is the pitch of the winding so that z increases by $2\pi a \tan\alpha$ when ϕ increases by 2π. The z component of $\boldsymbol{B}$ [7.14 (1)] is

$$B_z = \frac{\mu I}{4\pi}\oint\frac{[\boldsymbol{ds}\times\boldsymbol{r}]_z}{r^3} = \frac{\mu I}{4\pi}\oint\frac{r_y\,ds_x - r_x\,ds_y}{r^3} \tag{2}$$

The components of $\boldsymbol{r}$ and $\boldsymbol{ds}$ are, for a point on the axis,

$$r_x = -a\cos\phi, \qquad ds_x = -a\sin\phi\,d\phi,$$
$$r_y = -a\sin\phi, \qquad ds_y = a\cos\phi\,d\phi,$$
$$r_z = -a\phi\tan\alpha + C \qquad ds_z = a\tan\alpha\,d\phi$$

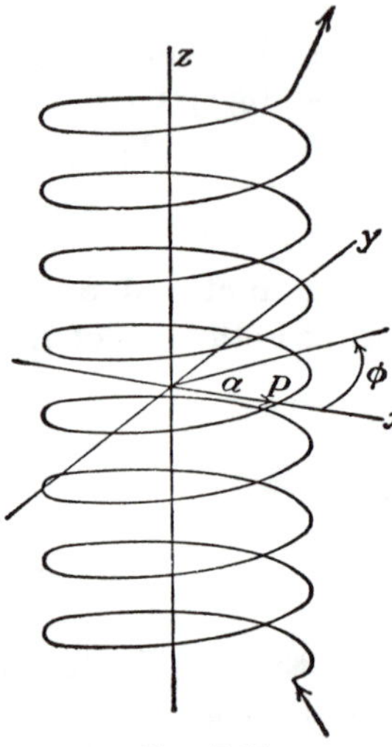

FIG. 7.15.

We shall find the field on the axis of a solenoid having an integral number N of turns, neglecting the field of the leads, which may be calculated separately. Taking the origin at the point at which B_z is to be found, setting $r = (x^2+y^2+z^2)^{\frac{1}{2}} = a(1+\phi^2\tan^2\alpha)^{\frac{1}{2}}$ in (2), and choosing the limits

$$\phi_1 = -N\pi + b/(a\tan\alpha)$$

and $\phi_2 = N\pi + b/(a\tan\alpha)$, we find, after integrating by *Pc* 138 or *Dw* 200.03, that B_z at a distance b from the center is

$$B_z = \frac{\mu I}{4\pi a}\int_{\phi_1}^{\phi_2}\frac{d\phi}{(1+\phi^2\tan^2\alpha)^{\frac{3}{2}}}$$
$$= \frac{\mu I}{4\pi a\tan\alpha}\left\{\frac{N\pi a\tan\alpha + b}{[a^2+(N\pi a\tan\alpha+b)^2]^{\frac{1}{2}}} + \frac{N\pi a\tan\alpha - b}{[a^2+(aN\pi\tan\alpha-b)^2]^{\frac{1}{2}}}\right\}$$

If β_1 and β_2 are the angles between the axis of the helix and the vectors drawn from b to the extreme ends of the wire, this may be written

$$B_z = \frac{\mu I}{4\pi a \tan \alpha}(\cos \beta_2 - \cos \beta_1) \tag{3}$$

It should be noted that either β_1 or β_2 must be greater than $\frac{1}{2}\pi$ if b lies inside the solenoid so that one cosine is negative. If the distance between the extreme ends of the wire in the solenoid is L, then

$$L = 2\pi N a \tan \alpha$$

and if there are n turns per unit length, $N = Ln$, so we can write

$$B_z = \tfrac{1}{2}\mu n I(\cos \beta_2 - \cos \beta_1) \tag{4}$$

Formulas (3) and (4) are rigorous expressions for the axial component of the field on the axis. Unless $\alpha = 0$, this is not, however, the only component. Proceeding as before gives

$$B_x = \frac{\mu I}{4\pi}\oint \frac{r_z\, ds_y - r_y\, ds_z}{r^3} = \frac{\mu I}{4\pi}\int_{\phi_1}^{\phi_2} \frac{a^2 \tan \alpha(-\phi \cos \phi + \sin \phi)\, d\phi}{r^3} \tag{5}$$

This is not, in general, zero.

$$B_y = \frac{\mu I}{4\pi}\oint \frac{r_x\, ds_z - r_z\, ds_x}{r^3} = \frac{\mu I}{4\pi}\int_{\phi_1}^{\phi_2} \frac{-a^2 \tan \alpha(\cos \phi + \phi \sin \phi)\, d\phi}{r^3} \tag{6}$$

This integral is also not, in general, zero. Thus the lines of force are not straight but spiral.

In the important case of an infinitely long solenoid, (3) and (4) become

$$B_z = \frac{\mu I}{2\pi a \tan \alpha} = n\mu I \tag{7}$$

The x-component vanishes since (5) is the integral of an odd function, but (6) becomes

$$B_y = -2\pi n^2 a \mu I \int_0^\infty \frac{\cos \phi + \phi \sin \phi}{[(2n\pi a)^2 + \phi^2]^{\frac{3}{2}}}\, d\phi$$

Integrate the second term under the integral by parts, letting $u = \sin \phi$ and $dv = \phi[(2n\pi a)^2 + \phi^2]^{-\frac{3}{2}}\, d\phi$, and we have

$$B_y = -2\pi n^2 a \mu I \left\{ \int_0^\infty \frac{\cos \phi\, d\phi}{[(2n\pi a)^2 + \phi^2]^{\frac{3}{2}}} + \int_0^\infty \frac{\cos \phi\, d\phi}{[(2n\pi a)^2 + \phi^2]^{\frac{1}{2}}} \right\}$$

If we let $\phi = 2n\pi a \sinh \psi$, the second integral becomes, by 5.331 (2), $K_0(2n\pi a)$ which is the modified Bessel function of the second kind treated in 5.331. Evidently, the first integral can be obtained by differentiating the second with respect to $2n\pi a$ and dividing by $-2n\pi a$ so that, by 5.331 (2),

$$B_y = -n\mu I[2n\pi a K_0(2n\pi a) + K_1(2n\pi a)] \tag{8}$$

When $\alpha = 0$ or $n = \infty$, $B_y = 0$ so that the field is axial. If $\alpha = \frac{1}{2}\pi$ or

$n = 0$, our solenoid becomes a straight wire parallel to and at a distance a from the z-axis. Then $B_z = 0$ and, from 5.33, $K_0 \to -\ln(2n\pi a)$ and $K_1 \to 1/(2n\pi a)$ so that

$$B_y \xrightarrow[n\to 0]{} -\frac{\mu I}{2\pi a}$$

which agrees with 7.14 (3).

7.16. Field in Cylindrical Hole in Conducting Rod.—The computation of the magnetic field due to a certain current density distribution can frequently be simply obtained by superimposing known solutions. As

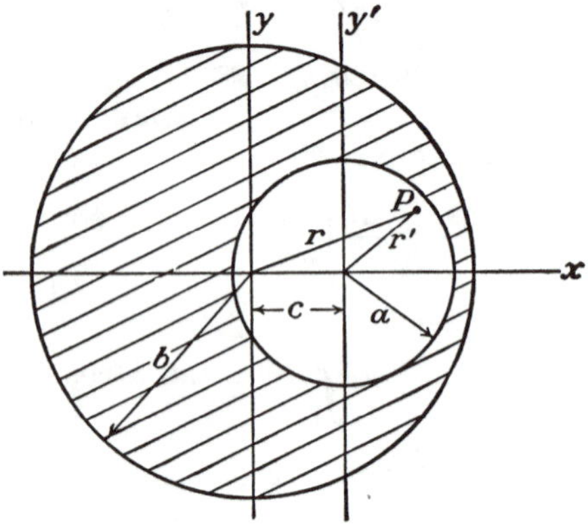

Fig. 7.16.

an example of this method, let us suppose conducting material with μ_v permeability occupies all the space between two infinite cylinders. We wish to find the induction inside the smaller cylinder of radius a, which lies entirely inside the larger cylinder, of radius b, the distance between axes being c (see Fig. 7.16). The current density i_z is taken as uniform and in the z-direction. If there were no hole, B_θ would be constant on the circle of radius r through P, and from 7.01 (2) its value would be

$$\mu\pi r^2 i = \oint \boldsymbol{B}' \cdot d\boldsymbol{s} = 2\pi r B_\theta' \qquad \text{or} \qquad \boldsymbol{B}' = \tfrac{1}{2}\mu_v \boldsymbol{i} \times \boldsymbol{r}$$

Addition of a similar expression for a current density $-i$ flowing only in the inner cylinder gives zero current density there so that

$$\boldsymbol{B} = \tfrac{1}{2}\mu_v \boldsymbol{i} \times (\boldsymbol{r} - \boldsymbol{r}') = \tfrac{1}{2}\mu_v \boldsymbol{i} \times \boldsymbol{c} = \tfrac{1}{2}\mu_v ic(\boldsymbol{k} \times \boldsymbol{i}_1) = \boldsymbol{j}\tfrac{1}{2}\mu_v ic \tag{1}$$

where $\boldsymbol{i}_1$, $\boldsymbol{j}$, and $\boldsymbol{k}$ are unit vectors in the x, y, and z directions. The total current I is $(\pi b^2 - \pi a^2)i$ so the uniform field in the hole is

$$B_y = \frac{\mu_v c I}{2\pi(b^2 - a^2)} \tag{2}$$

7.17. Field of Rectilinear Currents in Cylindrical Conducting Shell.—Suppose currents in an infinite circular cylindrical shell flow everywhere parallel to the axis and that the z-directed current density i can be expressed in circular surface harmonics; thus,

$$i_z = \sum_{n=0}^{\infty} (C_n \cos n\alpha + D_n \sin n\alpha) \qquad (1)$$

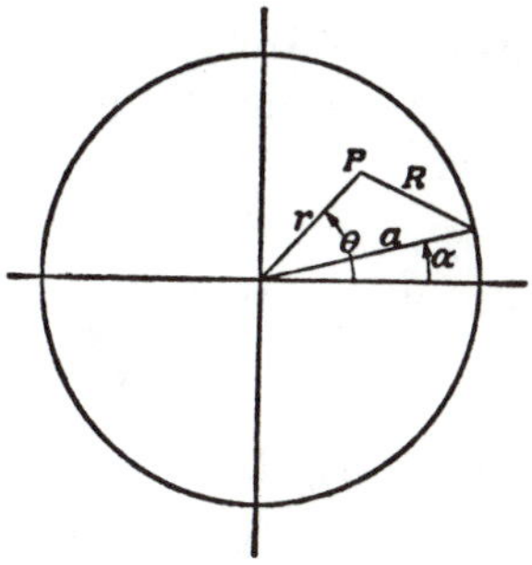

Fig. 7.17.

From 7.09, the vector potential at P due to a long straight wire carrying a current I is parallel to it and equals $(\frac{1}{2}\mu I/\pi) \ln R$ if the shortest distance from P to the wire is R. Thus, from 7.02 (4), the vector potential due to the cylinder is (see Fig. 7.17)

$$A_z = \tfrac{1}{2}\mu\pi^{-1}\int_0^{2\pi} ia \ln R \, d\alpha \qquad (2)$$

4.02 (2) $$\ln R = \ln a - \sum_{m=1}^{\infty} \frac{1}{m}\left(\frac{r}{a}\right)^m (\cos m\theta \cos m\alpha + \sin m\theta \sin m\alpha) \qquad (3)$$

Substitute (1) and (3) in (2), and we have integrals from 0 to 2π of the products $\cos mx \cos nx$, $\sin mx \sin nx$, and $\sin mx \cos nx$. From *Dw* 858.1 and 858.2 or *Pc* 488 and 360, these are all zero unless $m = n$. If $m = n$, the integral of $\sin 2m\alpha = 2 \sin m\alpha \cos m\alpha$ is still zero, but that of $\sin^2 m\alpha$ and $\cos^2 m\alpha$ is π by *Pc* 489 or *Dw* 858.4 so that (2) becomes

$$A_z = \mu a C_0 \ln a + \frac{\mu a}{2}\sum_{n=1}^{\infty} \frac{1}{n}\left(\frac{r}{a}\right)^n (C_n \cos n\theta + D_n \sin n\theta) \qquad (4)$$

7.18. Force on Electric Circuit in Magnetic Field.—We found experimentally in 7.00 that the forces acting on a small loop of wire in a magnetic field are exactly like those acting on an electric dipole in an electric field. We also saw that any electric circuit may be considered as a mesh of such loops. From 1.07 (3), we can therefore write the force on an element dS of the mesh of a circuit carrying a current I by substituting $\mathbf{n}I\,dS$ for $\boldsymbol{M}$ and $\mathbf{B}$ for $\mathbf{E}$; thus

$$dF = I(\mathbf{n} \cdot \nabla)\mathbf{B}\, dS \qquad (1)$$

We have the well-known relation

$$\nabla(\mathbf{n} \cdot \mathbf{B}) = (\mathbf{B} \cdot \nabla)\mathbf{n} + \mathbf{B} \times (\nabla \times \mathbf{n}) + (\mathbf{n} \cdot \nabla)\mathbf{B} + \mathbf{n} \times (\nabla \times \mathbf{B}) \qquad (2)$$

The first two terms on the right are zero since ∇ does not operate on $\mathbf{n}$. Since no current flows inside the loop, $\nabla \times \mathbf{B} = \mu i = 0$ from 7.01 (3). We may therefore write for the total force on the circuit, substituting $\nabla \times \mathbf{A}$ for $\mathbf{B}$ from 7.02 (1),

$$F = I\nabla\int_S n \cdot B\,dS = I\nabla\int_S n \cdot \nabla \times A\,dS$$

Applying Stokes's theorem [3.01 (1)], making use of (2) again where $\boldsymbol{ds}$ replaces $\boldsymbol{n}$ and $\boldsymbol{A}$ replaces $\boldsymbol{B}$, and noting that ∇ does not operate on $\boldsymbol{ds}$, we have

$$F = I\nabla\oint A \cdot ds = I\oint[(ds \cdot \nabla)A + ds \times (\nabla \times A)] \qquad (3)$$

But we see that

$$\oint (ds \cdot \nabla)A = \oint (i\,dA_x + j\,dA_y + k\,dA_z) = 0$$

since dA_x, dA_y, and dA_z are total differentials and A_x, A_y, and A_z are single-valued functions. From (1), $\nabla \times A = B$ so that

$$F = I\oint ds \times B \qquad \text{or} \qquad I\oint B \sin\theta\,ds = \oint dF \qquad (4)$$

where θ, which is less than π, is the angle from $\boldsymbol{ds}$ to $\boldsymbol{B}$. If, when the line of observation is normal to $\boldsymbol{ds}$ and $\boldsymbol{B}$, θ appears clockwise the force is away from the observer.

Ampère's experiments give no unique law of force for circuit *elements* because (4) and 7.14 (1) give the force between two circuits to be

$$F = I\oint ds \times B = \mu II'\oint\oint\frac{ds \times (ds' \times r)}{4\pi r^3} = \mu II'\oint\oint\frac{ds'(ds \cdot r) - r(ds \cdot ds')}{4\pi r^3} \qquad (5)$$

The line integral around the s-circuit of the first member of the last integral vanishes because r/r^3 is the gradient of the scalar $1/r$. Thus Ampère's *force* experiments give the force between circuit elements to be

$$C\frac{ds \times (ds' \times r)}{r^3} \qquad \text{or} \qquad -C\frac{r(ds \cdot ds')}{r^3} \qquad (6)$$

Comparing (4) with 7.14 (1), we see that if we have two circuits nearly in coincidence, carrying equal currents in opposite directions so that, from Ampère's experiments, the magnetic field is nearly zero, then only those contributions to the force from portions of the range of integration for which r is small are important. These are in the positive r direction. The circuits therefore tend to separate, thus creating a magnetic field. This is exactly the reverse of the electric case where the force on equal and opposite charges tends to bring them together and destroy the electric field. Thus, there is a fundamental difference in the nature of the energy in electric and magnetic fields. We shall consider this in the next chapter, in connection with Faraday's law of induction. All these formulas for forces on circuits placed in magnetic fields have been well verified by experiment.

7.19. Examples of Forces between Electric Circuits.—When we have two parallel infinite wires at a distance a apart, carrying currents I and I', we have, from 7.14 (3) and 7.18 (4), since $\sin\theta = 1$, the result

$$\boldsymbol{F}_a = \frac{\mu I I'}{2\pi a^2}\oint d\boldsymbol{s}' \times [\boldsymbol{s}_1 \times \boldsymbol{a}]$$

so that there is an apparent repulsive force per unit length of amount

$$F_1 = \pm\frac{\mu I I'}{2\pi a} \tag{1}$$

The positive sign is taken when I and I' are in opposite directions, and the negative sign when they are in the same direction.

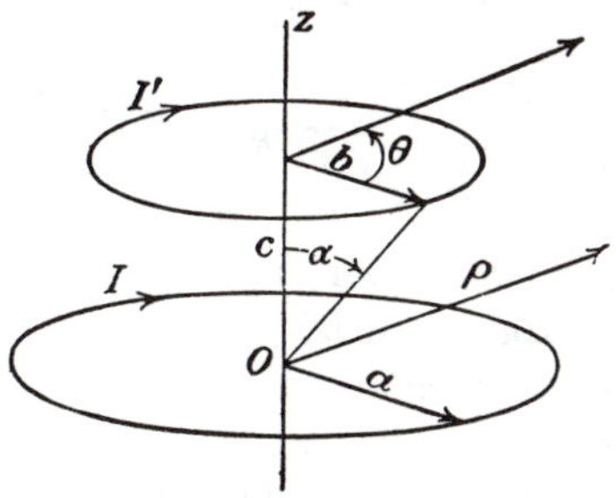

FIG. 7.19.

Next, let us compute the force between two parallel coaxial loops of wire of radii a and b, carrying currents I and I', as shown in Fig. 7.19. From symmetry, the force is one of attraction only so that, from 7.18 (4), the only effective component of the induction is B_ρ, which is the same for all elements of either circuit. Thus, since the coordinates of the I' circuit are $\rho = b$, $z = c$, we have for the force

$$F = I'B_\rho(a, b, c)\int_0^{2\pi} b\, d\theta = 2\pi b I' B_\rho(a, b, c)$$

Substituting from 7.10 (6), we have

$$F = \frac{\mu I I' c}{[(a+b)^2 + c^2]^{\frac{1}{2}}}\left[-K + \frac{(a^2 + b^2 + c^2)E}{(a-b)^2 + c^2}\right] \tag{2}$$

where, from 7.10 (3), the modulus k is

$$k^2 = \frac{4ab}{(a+b)^2 + c^2}$$

The force reverses if either current reverses.

We may easily get this force in series form by combining 7.18 (4) with 7.13 (3), giving

$$F = -\pi\mu I I' \sin\alpha \sum_{n=1}^{\infty}\left(\frac{a^2}{b^2 + c^2}\right)^{\frac{1}{2}n} P_n^1(\cos\alpha)P_n(0) \tag{3}$$

when $c^2 + b^2 > a^2$, since $u = 0$ in the latter, $r = a$ and $\oint ds = 2\pi a$.

Another, more rapidly converging expression for the force can be obtained, if the two loops have different radii, by choosing the origin at the apex of the circular cone, of half angle β, in which the two loops lie. If r is the radius vector to the smaller loop and s the radius vector to the larger, then the force between them is

$$F = 2\pi r I' \sin\beta(B_r \sin\beta + B_\theta \cos\beta)$$

where B_r and B_θ are given by 7.13 (3) and (4). Thus, we have

$$F = \pi\mu II' \sin^3 \beta \sum_{n=2}^{\infty} \left(\frac{r}{s}\right)^n \frac{1}{n} P_n^1(\cos\beta)[nP_n(\cos\beta) - \cos\beta P_n'(\cos\beta)]$$

We start the summation from $n = 2$ since the $n = 1$ term is zero. The last term may be simplified by 5.154 (2) and (4.1) so that, writing $n + 1$ for n, where $1 \leqslant n < \infty$, and noting that $r/s = a/b$ if $b > a$, we obtain

$$F = -\pi\mu II' \sin^2 \beta \sum_{n=1}^{\infty} \frac{1}{n+1} \left(\frac{a}{b}\right)^{n+1} P_{n+1}^1(\cos\beta) P_n^1(\cos\beta) \tag{4}$$

7.20. Vector Potential and Magnetization.—The definition of the vector potential given in 7.02 (4.1) has proved adequate for all the cases so far treated in which the whole region is occupied by a medium of a single constant permeability. If μ is variable or discontinuous, it is necessary, in order to define a magnetic vector potential uniquely, to consider the nature of magnetization. In 12.06, we shall give experimental evidence that magnetization is due to circulating currents or spinning electric charges within the body. We define a vector $\boldsymbol{M}$ to be the magnetic moment per unit volume of such currents or spins and call it the magnetization. With the aid of 7.10 (2), setting $\mu = \mu_v$ and measuring $\boldsymbol{r}$ from the field point the vector potential of $\boldsymbol{M}$ becomes

$$A_M = -\frac{\mu_v}{4\pi} \int_V \frac{\boldsymbol{M} \times \boldsymbol{r}}{r^3} \, dv = \frac{\mu_v}{4\pi} \int_V \boldsymbol{M} \times \nabla\left(\frac{1}{r}\right) dv \tag{1}$$

We can transform this equation by using the vector relations

$$\nabla \times p\boldsymbol{q} = \nabla p \times \boldsymbol{q} + p\nabla \times \boldsymbol{q} \tag{2}$$

and

$$\int_V \nabla \times \boldsymbol{M} \, dv = \int_S \boldsymbol{n} \times \boldsymbol{M} \, dS \tag{3}$$

where $\boldsymbol{n}$ is the unit outward normal vector to the surface S which encloses v. We prove (3) by putting $\boldsymbol{A} = \boldsymbol{M} \times \boldsymbol{a}$ in Gauss's theorem [3.00 (2)] where $\boldsymbol{a}$ is a constant vector so that $\nabla \cdot (\boldsymbol{M} \times \boldsymbol{a}) = \boldsymbol{a} \cdot \nabla \times \boldsymbol{M}$, giving

$$\boldsymbol{a} \cdot \int_V \nabla \times \boldsymbol{M} \, dv = \int_S \boldsymbol{M} \times \boldsymbol{a} \cdot \boldsymbol{n} \, dS = \boldsymbol{a} \cdot \int_S \boldsymbol{n} \times \boldsymbol{M} \, dS$$

Since this is true for all values of $\boldsymbol{a}$, Eq. (3) follows. Applying (2) and (3) to (1) gives

$$\frac{4\pi}{\mu_v} A_M = \int_V \frac{\nabla \times \boldsymbol{M}}{r} \, dv - \int_V \nabla \times \left(\frac{\boldsymbol{M}}{r}\right) dv = \int_V \frac{\nabla \times \boldsymbol{M}}{r} \, dv + \int_S \frac{\boldsymbol{M} \times \boldsymbol{n}}{r} \, dS \tag{4}$$

When the magnetization is uniform in the interior of a body, (4) is more useful than (1) since the volume integral is zero. If $\boldsymbol{i}$ is the actual current density, then the total vector potential is now given by

$$\boldsymbol{A} = \frac{\mu_v}{4\pi}\int_V \frac{\boldsymbol{i} + \nabla \times \boldsymbol{M}}{r}\,dv + \frac{\mu_v}{4\pi}\int_S \frac{\boldsymbol{M} \times \boldsymbol{n}}{r}\,dS \tag{5}$$

In order to apply this formula in most cases where the data include only the current distributions and the permeabilities, we must express $\boldsymbol{M}$ in terms of μ and $\boldsymbol{B}$. To do this, we note that, at great distances from closed circuits of finite extent, the induction and, hence, the magnetization are inversely proportional to the square or higher powers of r so that the surface integral in (4) vanishes. If μ is constant in this region, and $\boldsymbol{M}$ is due entirely to $\boldsymbol{i}$ the vector potential must also be given by 7.02 (4.1) so that, comparing it with (5), we have

$$\mu \boldsymbol{i} = \mu_v(\boldsymbol{i} + \nabla \times \boldsymbol{M}) \tag{6}$$

In a medium that is isotropic but not homogeneous we define μ so that this equation is still valid. With it we eliminate $\mu \boldsymbol{i}$ from 7.01 (3), and because the resultant equation must reduce to 7.01 (3) when μ is constant and it is known experimentally that $\boldsymbol{M}$ is proportional to $\boldsymbol{B}$ and in the same direction, we have

$$\nabla \times \left(\frac{\boldsymbol{B}}{\mu_v} - \boldsymbol{M}\right) = \boldsymbol{i} = \nabla \times \frac{\boldsymbol{B}}{\mu} \tag{7}$$

$$\boldsymbol{M} = \left(\frac{1}{\mu_v} - \frac{1}{\mu}\right)\boldsymbol{B} \tag{8}$$

Just as we can get 7.01 (2) from 7.01 (3), we get from (7) the relation

$$\oint \frac{\boldsymbol{B} \cdot d\boldsymbol{s}}{\mu} = I \tag{9}$$

7.21. Magnetic Boundary Conditions.—In the last article, we defined the vector potential $\boldsymbol{A}$ for regions where the magnetization is not uniform and on surfaces where it is discontinuous. To find the boundary conditions that $\boldsymbol{A}$ must satisfy, we observe that each of the three components of $\boldsymbol{A}$ is defined by a scalar equation analytically identical with 1.06 (6), which may be written *in vacuo*

$$4\pi\epsilon_v V = \int_V \frac{\rho\, dv}{r} + \int_S \frac{\sigma\, dS}{r} \qquad (1.06)\quad (6)$$

This defines the electrostatic potential in free space due to a volume distribution of electric charge of density ρ and a surface distribution of density σ. From electrostatic considerations, we know the value V_i of these integrals just inside the surface S is the same as the value V_o just outside. Furthermore, by applying Gauss's electric flux theorem to a small disk-shaped volume fitting closely an area dS of the surface and so thin that dv is negligible compared with dS, we have, after canceling out dS, the relation

$$\frac{\partial V_o}{\partial n} - \frac{\partial V_i}{\partial n} = -\frac{\sigma}{\epsilon_v}$$

Thus, we know the boundary conditions that apply to integrals of the form of 1.06 (6) and hence to each component of 7.20 (5). Adding the components and writing μ_v for $1/\epsilon_v$, we have

$$\boldsymbol{A}_o = \boldsymbol{A}_i \tag{1}$$

and with the aid of 7.20 (8),

$$\frac{\partial \boldsymbol{A}_o}{\partial n} - \frac{\partial \boldsymbol{A}_i}{\partial n} = -\mu_v(\boldsymbol{M} \times \boldsymbol{n}) = -\frac{\mu - \mu_v}{\mu}[(\nabla \times \boldsymbol{A}_i) \times \boldsymbol{n}] \tag{2}$$

If there is a magnetization $\boldsymbol{M}'$ on one side of the boundary and $\boldsymbol{M}''$ on the other, we may find the boundary conditions by imagining a thin layer of permeability μ_v between the boundaries, writing down (1) and (2) for each boundary, referring them to the same normal, and eliminating $\boldsymbol{A}_o$ and $\partial \boldsymbol{A}_o/\partial n$, giving

$$\boldsymbol{A}' = \boldsymbol{A}'' \tag{3}$$

and

$$\frac{\partial \boldsymbol{A}'}{\partial n} - \frac{\partial \boldsymbol{A}''}{\partial n} = \mu_v[(\boldsymbol{M}' - \boldsymbol{M}'') \times \boldsymbol{n}] \tag{4}$$

To get the second boundary condition in terms of the permeability, we write $\boldsymbol{n} \cdot \nabla$ for $\partial/\partial n$ and use the relation

$$\boldsymbol{n} \cdot \nabla(\boldsymbol{A}' - \boldsymbol{A}'') = \nabla[\boldsymbol{n} \cdot (\boldsymbol{A}' - \boldsymbol{A}'')] - \boldsymbol{n} \times [\nabla \times (\boldsymbol{A}' - \boldsymbol{A}'')] - (\boldsymbol{A}' - \boldsymbol{A}'') \cdot \nabla \boldsymbol{n} - (\boldsymbol{A}' - \boldsymbol{A}'') \times (\nabla \times \boldsymbol{n})$$

The last two terms drop out because of (3), so that, writing $\boldsymbol{B}' - \boldsymbol{B}''$ for $\nabla \times (\boldsymbol{A}' - \boldsymbol{A}'')$, by 7.02 (1), and rearranging, we have

$$\nabla[\boldsymbol{n} \cdot (\boldsymbol{A}' - \boldsymbol{A}'')] = \boldsymbol{n} \times (\boldsymbol{B}' - \mu_v \boldsymbol{M}' - \boldsymbol{B}'' + \mu_v \boldsymbol{M}'')$$

Substituting from 7.20 (8), we have

$$\nabla(A'_n - A''_n) = \mu_v \boldsymbol{n} \times \left(\frac{\boldsymbol{B}'}{\mu'} - \frac{\boldsymbol{B}''}{\mu''}\right)$$

From (3) $A'_n - A''_n$ is equal to zero over all the boundary so that its gradient along the boundary is zero. Thus $\nabla(A'_n - A''_n)$ is a vector normal to the boundary. But the right side is a vector directed along the boundary surface so that writing out the tangential components of this equation gives

$$\boldsymbol{n} \times \left(\frac{\boldsymbol{B}'}{\mu'} - \frac{\boldsymbol{B}''}{\mu''}\right) = 0 \tag{5}$$

We may now write down the boundary conditions on the components of A in the orthogonal curvilinear coordinate system u_r, u_s, and u_t, treated in 3.03, 3.04, and 3.05. Let us take u_r to be constant on the boundary; then, from (3), we have

$$A'_r = A''_r, \qquad A'_s = A''_s, \qquad \text{and} \qquad A'_t = A''_t \tag{6}$$

Writing $\nabla \times A$ for B in the left side of (5), we have, by 3.04 (1), 3.04 (2), and 3.04 (3), the two equations

$$\frac{1}{\mu'}\left[\frac{\partial(h_{r,s}A'_{r,s})}{\partial u_{t,r}} - \frac{\partial(h_{t,r}A'_{t,r})}{\partial u_{r,s}}\right] = \frac{1}{\mu''}\left[\frac{\partial(h_{r,s}A''_{r,s})}{\partial u_{t,r}} - \frac{\partial(h_{t,r}A''_{t,r})}{\partial u_{r,s}}\right] \tag{7}$$

These equations (6) and (7) give the required boundary conditions.

To obtain the boundary conditions on B, we note that (3) shows the differences in the vector potential between two points on the interface to be the same in either medium. Hence, the derivatives of the vector potential on the two sides of the interface, in the same direction parallel to it, are equal. The vector $n \times A$ lies in the interface, and so $\nabla \cdot n \times A$ involves only such derivatives. We have the well-known vector equation for the derivative of a cross product

$$\nabla \cdot n \times A = -n \cdot \nabla \times A + A \cdot \nabla \times n \tag{8}$$

When we substitute $A' - A''$ for A, the last term vanishes by (3) giving

$$n \cdot \nabla \times A' = n \cdot \nabla \times A'' \tag{9}$$

Writing B for $\nabla \times A$ gives

$$n \cdot B' = n \cdot B'' \tag{10}$$

Thus, the normal components of the induction are continuous. From (5), we have for the tangential components of the induction

$$\frac{1}{\mu'}(n \times B') = \frac{1}{\mu''}(n \times B'') \tag{11}$$

It is often possible to choose two vectors A' and A'', different from and simpler than A' and A'', whose curl gives the same induction at all points but which, as will be shown in 8.04, satisfy, instead of (3), the boundary conditions

$$\mathsf{A}'_r \neq \mathsf{A}''_r \qquad \text{and} \qquad \mathsf{A}'_{s,t} = \mathsf{A}''_{s,t} \tag{12}$$

It is evident in the example considered in the next article that these vectors may be more convenient for computation than A' and A''. They are not uniquely defined, however, as are A' and A'', by integrals of the type of 7.20 (5). We shall refer to them as quasi-vector potentials.

7.22. Example of the Use of A and A.—As an instructive example of the use of the boundary conditions in the last article, let us consider an infinite wire carrying a current I in the z-direction, positive and negative values of z having permeabilities μ' and μ'', respectively. Clearly, from 7.02 (5), the simplest vectors $\mathbf{A}'$ and $\mathbf{A}''$ that we can write which satisfy 7.21 (7) but not 7.21 (3) and whose curls give us the correct values of $\mathbf{B}'$ and $\mathbf{B}''$ are

$$\mathbf{A}' = -\mathbf{k}(\tfrac{1}{2}\pi^{-1}I\mu' \ln \rho) \qquad \text{and} \qquad \mathbf{A}'' = -\mathbf{k}(\tfrac{1}{2}\pi^{-1}I\mu'' \ln \rho) \tag{1}$$

The vector defined by 7.20 (5) requires that $\mathbf{A}' = \mathbf{A}''$ at the boundary. The part of the volume integral involving I in 7.20 (5) is given by 7.02 (5). We must superimpose on this a potential to meet the new boundary condition. Inspection of the solutions available in cylindrical coordinates in 7.05 (7) shows that the suitable form is cz/ρ. Putting this in and evaluating the constants, we obtain

$$\mathbf{A}' = \tfrac{1}{2}\pi^{-1}I\{-\mathbf{k}\mu_v(\ln \rho + C_1) + \boldsymbol{\rho}[(\mu' - \mu_v)z\rho^{-1} + C_2]\} \tag{2}$$

$$\mathbf{A}'' = \tfrac{1}{2}\pi^{-1}I\{-\mathbf{k}\mu_v(\ln \rho + C_1) + \boldsymbol{\rho}[(\mu'' - \mu_v)z\rho^{-1} + C_2]\} \tag{3}$$

The constants C_1 and C_2 are arbitrary. The second terms represent the contribution from the integral of 7.20 (1). We get the same value of $\mathbf{B}'$ and $\mathbf{B}''$ from (2) and (3) as from (1).

7.23. Current Images in Plane Face.—The similarity of the boundary conditions for $\mathbf{B}$ in 7.21 to those for $\mathbf{D}$ in 1.17 suggests that the image method of 5.05 might be used in getting the magnetic field when an electric circuit is near the plane face of a semi-infinite block of material of permeability μ. Let the plane face be at $z = 0$, the circuit lie in the region of positive z, and the material of permeability μ occupy the entire region of negative z. Let $\mathbf{A}$ be the vector potential due to the circuit alone. The image law of 5.05 suggests that the quasi-vector potential above the interface can be given by $\mathbf{A} + \mathbf{A}'$ and that below by $\mathbf{A}''$, where

$$\mathbf{A} = \mathbf{i}f_1(x, y, z) + \mathbf{j}f_2(x, y, z) + \mathbf{k}f_3(x, y, z) \tag{1}$$

$$\mathbf{A}' = \mathbf{i}C_1'f_1(x, y, -z) + \mathbf{j}C_2'f_2(x, y, -z) + \mathbf{k}C_3'f_3(x, y, -z) \tag{2}$$

$$\mathbf{A}'' = \mathbf{i}C_1''f_1(x, y, z) + \mathbf{j}C_2''f_2(x, y, z) + \mathbf{k}C_3''f_3(x, y, z) \tag{3}$$

From 7.21 (12) at $z = 0$ $A_z + A_z' \neq A_z''$ and $A_{x,y} + A_{x,y}'' = A_{x,y}''$, or

$$1 + C_1' = C_1'', \qquad 1 + C_2' = C_2'', \qquad 1 + C_3' \neq C_3'' \tag{4}$$

The boundary conditions at $z = 0$ on the derivatives given by 7.21 (7) are satisfied if we have

$$\mu(1 - C_1') = \mu_v C_1'', \qquad \mu(1 - C_2') = \mu_v C_2'', \qquad \mu(1 + C_3') = \mu_v C_3'' \tag{5}$$

Solving, we obtain, since $\mathbf{A}''$ and $\mathbf{A}$ have the same form,

$$C_1' = C_2' = -C_3' = \frac{\mu - \mu_v}{\mu + \mu_v}, \qquad C_1'' = C_2'' = C_3'' = \frac{2\mu}{\mu + \mu_v} \tag{6}$$

Thus, the magnetic induction outside the permeable medium appears to be caused by two circuits, the original carrying a current I, and an image circuit carrying a current $I' = \dfrac{\mu - \mu_v}{\mu + \mu_v}I$. The direction of I' is such that the projections of I and I' on the interface coincide in position and direction. In the permeable medium, the magnetic induction appears to be due to the original circuit alone but carrying a current $I'' = 2I\mu/(\mu + \mu_v)$ instead of I.

7.24. Magnetic Induction and Permeability in Crystals.—The similarity of the boundary conditions derived in 7.21 for $\boldsymbol{B}$ to those derived in 1.17 for $\boldsymbol{D}$ suggests that the method of measuring $\boldsymbol{D}$ and $\boldsymbol{E}$ in 1.18 may be applicable also to $\boldsymbol{B}$ and μ. Let us excavate a small disk-shaped cavity in the solid, whose thickness is very small compared with its radius so that the induction in the cavity, far from the edges, will be determined entirely by the boundary conditions over the flat surfaces. If we orient the cavity so that the induction inside is normal to the flat faces, then from 7.21 (10) this $\boldsymbol{B}$ equals that in the medium. Let us now excavate a long thin cylindrical cavity so that the induction inside, far from the ends, is determined entirely by the boundary conditions over the curved walls and orient it so that this induction is parallel to the axis. We now find, from 7.21 (11), that $\boldsymbol{B}'/\mu_v$ inside equals $\boldsymbol{B}/\mu$ in the solid. These two measurements give us both $\boldsymbol{B}$ and μ. If we perform this experiment in a magnetically anisotropic medium we shall find, in general, that $\boldsymbol{B}$ and $\boldsymbol{B}'$ are in different directions. There will be at least three orientations of the field with respect to the crystal for which $\boldsymbol{B}$ and $\boldsymbol{B}'$ have the same direction. These are the magnetic axes of the crystal, along each of which μ may have a different value. In the more common case, $\boldsymbol{B}$ and $\boldsymbol{B}'$ have the same direction for all orientations of the field in some plane, and for the field normal to it. This subject is further discussed in 9.02 and 9.03.

7.25. Two-dimensional Magnetic Fields.—In rectangular coordinates, the boundary conditions for the tangential component of the magnetostatic vector potential are identical with those for the electrostatic scalar potential when ϵ is replaced by $1/\mu$. In two-dimensional problems, involving only magnetic fields parallel to the xy-plane, the currents must flow in the z-direction and so the vector potential can have only one component A_z, which is, necessarily, tangential to all surfaces. When μ/μ_v becomes very large, we see, from 7.21 (7), that the surfaces on which A_z is constant, *i.e.*, the equivector potential surfaces, are normal to the boundary. Thus, at this point, A_z behaves as the electrostatic stream function does at an electric equipotential boundary. We also obtain $\boldsymbol{B}$ from A_z in the same way as we obtain the electric field from the stream function; for, from 3.04 (2) and (3), we see that, in rectangular coordinates, where $h_1 = h_2 = h_3 = 1$ we have

$$B_x = \frac{\partial A_z}{\partial y}, \qquad B_y = -\frac{\partial A_z}{\partial x} \tag{1}$$

These are the same as 4.11 (1).

Where there is no current, 7.02 (3) becomes

$$\nabla^2 A_z = 0 \tag{2}$$

so that all the methods of Chap. IV are available for its solution.

7.26. Magnetic Shielding of Bifilar Circuit.—To illustrate the application of the results of the last article, let us compute the magnetic induction outside a cylindrical shield of permeability μ, with internal and external radii a and b which surrounds, symmetrically, two parallel conducting wires carrying currents in opposite directions. Evidently, the problem calls for the use of the circular harmonics of 4.01. In 7.09, the vector potential due to two wires carrying currents in opposite directions was found to be

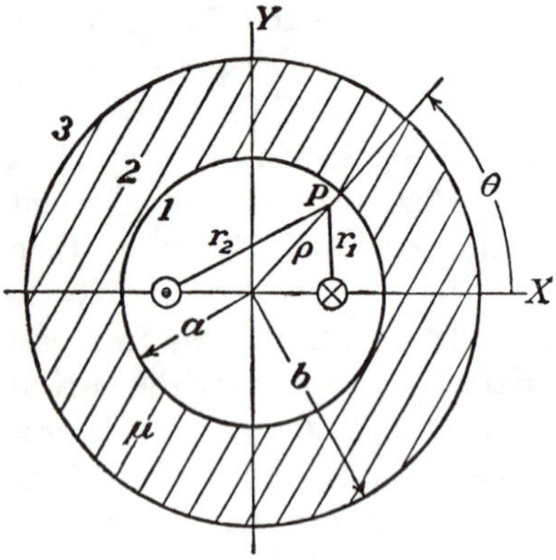

FIG. 7.26.

$$A_z = \frac{\mu_v I}{2\pi} \ln \frac{r_2}{r_1} \tag{1}$$

By 4.02 (1), setting $\theta_0 = 0$ and $\theta_0 = \pi$ and $\rho_0 = c$, we have the expansions, when $\rho > c$,

$$\ln r_1 = -\sum_{n=1}^{\infty} \frac{1}{n}\left(\frac{c}{\rho}\right)^n \cos n\theta + \ln \rho$$

$$\ln r_2 = -\sum_{n=1}^{\infty} \frac{1}{n}\left(\frac{c}{\rho}\right)^n (-1)^n \cos n\theta + \ln \rho$$

When we take the difference, even terms go out so that, writing $2n + 1$ for n, we get

$$\ln \frac{r_2}{r_1} = 2\sum_{n=0}^{\infty} \frac{1}{2n+1}\left(\frac{c}{\rho}\right)^{2n+1} \cos (2n+1)\theta \tag{2}$$

Let us now superimpose a potential inside the shell, due to its influence, which is finite at the origin, giving in this region

$$A_{1z} = \frac{\mu_v I}{\pi}\sum_{n=0}^{\infty}\left[A_{2n+1}\rho^{2n+1} + \frac{1}{2n+1}\left(\frac{c}{\rho}\right)^{2n+1}\right] \cos (2n+1)\theta \tag{3}$$

In the region 2, of permeability μ, ρ is neither zero nor infinity so we may have

$$A_{2z} = \frac{\mu_v I}{\pi}\sum_{n=0}^{\infty}(B_{2n+1}\rho^{2n+1} + C_{2n+1}\rho^{-2n-1}) \cos (2n+1)\theta \tag{4}$$

Outside the potential must vanish at infinity so that

$$A_{3z} = \frac{\mu_v I}{\pi}\sum_{n=0}^{\infty} D_{2n+1}\rho^{-2n-1}\cos(2n+1)\theta \tag{5}$$

These equations must be satisfied for all values of θ so that each term separately must meet the boundary conditions 7.21 (6) and (7), giving after multiplying through by a^{2n+1} or b^{2n+1},
when $\rho = a$

$$A_{2n+1}a^{4n+2} + \frac{1}{2n+1}c^{2n+1} = B_{2n+1}a^{4n+2} + C_{2n+1} \tag{6}$$

and

$$\mu A_{2n+1}a^{4n+2} - \frac{\mu}{2n+1}c^{2n+1} = \mu_v B_{2n+1}a^{4n+2} - \mu_v C_{2n+1} \tag{7}$$

when $\rho = b$

$$B_{2n+1}b^{4n+2} + C_{2n+1} = D_{2n+1} \tag{8}$$

and

$$\mu_v B_{2n+1}b^{4n+2} - \mu_v C_{2n+1} = -\mu D_{2n+1} \tag{9}$$

Solving for D_{2n+1}, we have

$$A_{3z} = \frac{4\mu I}{\pi}\sum_{n=0}^{\infty}\left[(K_m+1)^2-(K_m-1)^2\left(\frac{a}{b}\right)^{4n+2}\right]^{-1}\frac{1}{2n+1}\left(\frac{c}{\rho}\right)^{2n+1}\cos(2n+1)\theta \tag{10}$$

The field outside is given by 3.04 (2) and (3) where, from 3.05, $h_1 = 1$ and $h_2 = \rho$ to be

$$B_\rho = \frac{1}{\rho}\frac{\partial A_z}{\partial\theta}, \qquad B_\theta = -\frac{\partial A_z}{\partial\rho} \tag{11}$$

7.27. Current Images in Two Dimensions.—The vector potential due to a straight linear current I and the scalar potential due to a line charge q have exactly the same form. Furthermore, as we have seen, the magnetic vector potential and the electrostatic potential, in two dimensions, satisfy the same form of Laplace's equation and the boundary conditions, if $1/K_m$ is substituted for K, are the same. It follows that the results of Art. 4.04 apply to a linear current I parallel to and at a distance b from a circular cylinder of permeability μ and radius a. Thus, the vector potential in the region outside the cylinder due to its presence is the same as if the cylinder were replaced by a parallel image current I' located between I and the axis, at a distance a^2/b from the latter, plus a current of strength $-I'$ along the axis. The vector potential inside the cylinder is the same as if I were replaced by a linear current I''. I' and I'' are given by substituting $1/K_m$ for K in 4.04 (7). These images lead to the potentials

$$\rho < a \qquad A_{zi} = -\frac{\mu_v I}{2\pi}\sum_{n=1}^{\infty}\frac{2K_m}{K_m+1}\left(\frac{\rho}{b}\right)^n\frac{\cos n\phi}{n} \tag{1}$$

$$a < \rho < b \qquad A_{z0} = \frac{\mu_v I}{2\pi}\left\{\ln\frac{\rho}{a} - \sum_{n=1}^{\infty}\left[\left(\frac{\rho}{a}\right)^n + \frac{K_m-1}{K_m+1}\left(\frac{a^2}{b\rho}\right)^n\right]\frac{\cos n\phi}{n}\right\} \tag{2}$$

At $\rho = a$ when $K_m = \infty$, both B_ϕ and H_ϕ vanish outside the cylinder but only H_ϕ inside it.

If the current I lies at $\rho = b$, $\phi = 0$ inside a hole of radius a in an infinite medium of permeability μ, the potentials are

$$b < \rho < a \qquad A_{zi} = \frac{\mu_v I}{2\pi}\left\{\ln\rho - \sum_{n=1}^{\infty}\left[\left(\frac{b}{\rho}\right)^n + \frac{K_m-1}{K_m+1}\left(\frac{b\rho}{a^2}\right)^n\right]\frac{\cos n\phi}{n}\right\} \tag{3}$$

$$a < \rho \qquad A_{zo} = \frac{\mu_v I}{2\pi}\left\{-\frac{K_m(K_m-1)}{K_m+1}\ln\rho + \frac{2K_m}{K_m+1}\left[\ln\rho - \sum_{n=1}^{\infty}\left(\frac{b}{\rho}\right)^n\frac{\cos n\phi}{n}\right]\right\} \tag{4}$$

In A_{zo} an additional term I''', the first term in (4), appears at the origin. When $K_m = \infty$, only the ln ρ term in (3) contributes to B_ϕ at $\rho = a$ so its value $\mu_v I/(2\pi a)$ is independent of b. This is a fairly general situation and occurs again in Art. 7.31. Outside only the first term in (4) survives, so B_ϕ is infinite but H_ϕ has the value $I/(2\pi\rho)$ and H_ρ vanishes.

7.28. Magnetomotance and Magnetic Intensity.—In dealing with the magnetic field, we are immediately struck by the many properties that the magnetic induction and the electric current have in common. In the first place, we have

$$6.00\ (3)\ \nabla\cdot\boldsymbol{i} = 0 \qquad \text{and} \qquad 7.01\ (1)\ \nabla\cdot\boldsymbol{B} = 0 \tag{1}$$

Then, for a simple closed path, we have

$$6.02\ (4)\ \oint\frac{\boldsymbol{i}\cdot d\boldsymbol{s}}{\gamma} = \mathcal{E} \qquad \text{and} \qquad 7.20\ (9)\ \oint\frac{\boldsymbol{B}\cdot d\boldsymbol{s}}{\mu} = I \tag{2}$$

The magnetic vector potential was derived on the basis of 7.20 (5). It is clear that we can proceed equally well along the lines of Chap. VI and define a new scalar quantity, analogous to $\mathcal{E}$, which we shall call the magnetomotance Ω. Thus the magnetomotance in traversing once a path enclosing a current I is

$$\Omega = I \tag{3}$$

where I is in amperes and Ω in ampere-turns. As both the magnetomotance and the electromotance are multiple-valued functions, their value depends on the path chosen. Using the method applied to $\mathcal{E}$ in 6.01, we can often insert a suitable barrier so that the integral of (2) becomes zero for all permitted paths. Ω then takes on the character of a scalar potential. In the simplest cases, the barrier erected is known as a magnetic shell. In (2), we see that the permeability μ plays the role of the conductivity γ in Chap. VI, so that we may set up a relation equivalent to Ohm's law for magnetic circuits, which corresponds to 6.02 (3) and is

$$\boldsymbol{B} = -\mu \nabla\Omega \tag{4}$$

It is convenient to have a symbol for the gradient of the magnetomotance. We shall call the negative of this quantity the magnetic field intensity and designate it by $-\boldsymbol{H}$, so that

$$\boldsymbol{H} = -\nabla\Omega = \frac{\boldsymbol{B}}{\mu} \tag{5}$$

The mks unit for $\boldsymbol{H}$ is ampere-turns per meter. In terms of $\boldsymbol{H}$, (2) is

$$\oint \boldsymbol{H} \cdot d\boldsymbol{s} = I \tag{6}$$

This equation is independent of the permeability of the medium. From (1) and (4), we get the relation analogous to 6.04 (2) which is

$$\nabla^2\Omega = 0 \tag{7}$$

Starting with 6.00 (3), 6.02 (4), and 6.04 (2), we obtained the conditions that the electromotance must satisfy when at the boundary between two mediums of different conductivity. In the same way, we would get, from 7.28 (1), (2), and (7), the conditions Ω satisfies at the boundary between two mediums of different permeability. It is simpler, however, to start with 7.21 (10) and (11) and use (5). Thus from 7.21 (10)

$$\mu'\boldsymbol{H}' \cdot \boldsymbol{n} = \mu''\boldsymbol{H}'' \cdot \boldsymbol{n} \tag{8}$$

and from 7.21 (11)

$$\boldsymbol{H}' \times \boldsymbol{n} = \boldsymbol{H}'' \times \boldsymbol{n} \tag{9}$$

In terms of the magnetomotance, these are

$$\mu'\frac{\partial\Omega'}{\partial n} = \mu''\frac{\partial\Omega''}{\partial n} \tag{10}$$

and

$$\Omega' = \Omega'' \tag{11}$$

It is to be noted that the magnetomotance satisfies just the same equations (7), (10), and (11) as the electrostatic potential, 3.02 (6) and 1.17 (5) and (6). It differs from the latter, however, in being a multiple-valued function unless we erect a barrier in the field in such a way as to prevent our making a complete circuit of any currents in the field; then Ω is single-valued in the region outside the barrier and may be treated by the methods of electrostatics. When so restricted, Ω is called the scalar magnetic potential and we use it to obtain solutions of magnetic problems from electrostatic solutions.

For example, if we wish to find the shielding effect of a spherical shell of permeability μ and internal and external radii a and b when placed in a uniform magnetic field of induction $\boldsymbol{B}$, we can use 5.18 (5). This shows us that the field inside the shell is still uniform and that its induction is

$$\boldsymbol{B}_i = \frac{9K_m\boldsymbol{B}}{9K_m - 2(K_m - 1)^2(a^3b^{-3} - 1)} \tag{12}$$

Because (10) and (11) are identical in form with 1.17 (5) and (6) the law of refraction of magnetic lines of force at the boundary between two isotropic magnetic mediums, derived from the former, will be identical with that for electric lines of force derived from the latter and expressed by 1.17 (8). Thus, at such a boundary, the magnetic lines of force or induction make an abrupt change in direction given by

$$\mu' \cot \alpha' = \mu'' \cot \alpha'' \tag{13}$$

where α' is the angle that a line makes with the normal to the boundary in the medium of permeability μ' and α'' is the corresponding angle in the medium of permeability μ''.

7.29. The Magnetic Circuit. Anchor Ring.—In 7.28, we pointed out the close mathematical analogy between the magnetic induction and the electric current density. To use the technique of Chap. VI for the solution of magnetic problems analogous to electric current problems, it will be convenient to define several new magnetic quantities. The magnetic flux N, through a surface S, corresponds to the electric current I, the equations being

$$6.00\ (2)\ I = \int_S \boldsymbol{i}\cdot\boldsymbol{n}\,dS \qquad N = \int_S \boldsymbol{B}\cdot\boldsymbol{n}\,dS \tag{1}$$

The mks unit of flux is the weber. The magnetic reluctance R' between two points in a circuit corresponds to the electric resistance, and the analogous equations are

$$6.02\ (1)\ R_{AB} = \frac{\mathcal{E}_{AB}}{I_{AB}} \qquad R'_{AB} = \frac{\Omega_{AB}}{N_{AB}} \qquad (2)$$

To illustrate the computation of such circuits, let us find the flux in a torus of permeability μ wound with n turns of wire carrying a current I. Let the radius of a section of the torus be b and the distance from its center to the axis of the torus a. From 7.28 (3), the magnetomotance around the torus from 0 to 2π is nI. Since, by symmetry, H is independent of θ, the magnetomotance around that portion of the path between 0 and θ is proportional to θ, which gives

$$\Omega = \frac{nI\theta}{2\pi} \qquad (3)$$

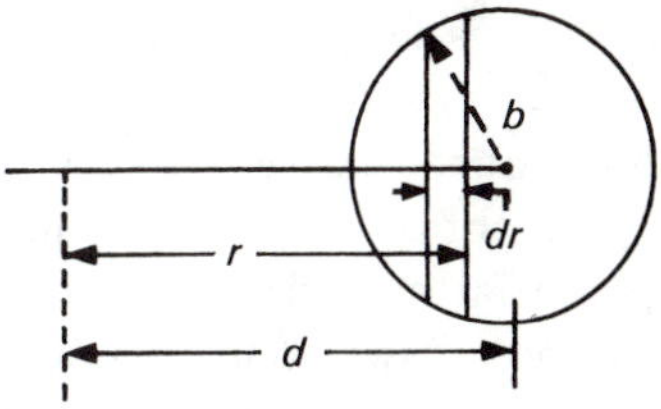

FIG. 7.29.

Thus, from 7.28 (5),

$$H_\theta = -\frac{\partial\Omega}{r\partial\theta} = -\frac{In}{2\pi r}$$

So, for a given value of r, H_θ is constant, and we may break up the torus into a set of cylindrical laminations, as shown in Fig. 7.29, without changing its reluctance. The reluctance of one of these laminas is

$$dR' = \frac{\text{length}}{\mu\ \text{area}} = \frac{2\pi r}{\mu\, dS}$$

But $dS = 2[b^2 - (a - r)^2]^{\frac{1}{2}}\, dr$, and since the laminas are in parallel we have

$$\frac{1}{R'} = \int \frac{1}{dR'} = \frac{\mu}{\pi}\int_{a-b}^{a+b} \frac{[b^2 - (a - r)^2]^{\frac{1}{2}}}{r}\, dr$$

Integrating by *Pc* 187 and putting in limits give

$$\frac{1}{R'} = \left(\frac{\mu}{\pi}\right)\pi[a - (a^2 - b^2)^{\frac{1}{2}}]$$

By (2), the flux through the ring is then

$$N = \frac{\Omega}{R'} = \mu n I[a - (a^2 - b^2)^{\frac{1}{2}}] \qquad (4)$$

7.30. Air Gaps in Magnetic Circuits.—It is evident that if there is a continuous path of high permeability and sufficient cross-sectional area at all points, the flux in the magnetic circuit will be confined almost entirely to the permeable medium. If, however, there is a break in the path, the flux must cross this gap and will distribute itself in such a way that the magnetomotance satisfies 7.28 (7) everywhere and 7.28 (10) and (11) at the gap boundaries. If the medium permeability μ'' is very high compared with the gap permeability μ', then $\alpha' \to 0$ from 7.28 (13), and the flux leaves it normally. If the product of μ' by the length of that part of the magnetic circuit lying in the permeable medium is small compared with the product of μ'' by the length of the air gap and if the cross section is nowhere too small, we may consider practically the entire reluctance of the circuit to lie in the gap. This is true in many electrical machines. When the faces of the gap are parallel and their area A is large compared with the distance apart d, the field in the gap may be taken as uniform and the effect of the edges neglected so that its reluctance is $d/(\mu' A)$.

To obtain rigorous solutions of other cases, we must use the vector potential or work with the induction, using Biot and Savart's law and the boundary conditions of 7.21. This frequently entails more mathematical labor than is justified by the accuracy required. In such cases, a sufficiently accurate result may be given by using the scalar magnetic potential which makes available the methods of Chaps. IV and V. As mentioned in 7.28, this necessitates the erection of an artificial barrier somewhere in the permeable part of the magnetic circuit. If the permeability is high, we then assign one value of the potential to that portion of the medium on one side of the barrier and another value to that portion on the other side. If the configuration is such that we can put the barrier far from the gap, the error introduced by its presence is negligible.

7.31. Field in Shell-type Transformer.—Calculation of the fields in a shell-type transformer with infinitely permeable walls when a one-way current flows in a winding as in Fig. 7.31*a* is difficult because the obvious simple assumption that the field is everywhere normal to the walls makes the line integral of $\boldsymbol{B}$ around the window zero. Thus by 7.01 (3) there is no net circulating current inside. For example, in Fig. 7.31*b*, *c*, and *d* of the second edition of this book there is a uniform reverse current sheet occupying the remainder of the cylinder on which the primary current sheet lies. A true one-way solution needs a different approach. A clue is found in Art. 7.27 where, as K_m approaches infinity, the I''' image dominates. Thus $\boldsymbol{B}$ is tangential to the wall in the exterior region regardless of the actual position of I inside the hole. This is logical because, viewed from the outside, the permeability of the hole is zero. Thus the $\boldsymbol{B}$ outside the transformer window encircles it without penetration, and the line integral of $\boldsymbol{H}$ around the window equals the current inside but is independent of its distribution. Finding the analytical solution of Laplace's equation that meets this boundary condition in Fig. 7.31*a* is a difficult, unsolved mixed boundary value problem, but approximate solutions by field plotting or electrolytic tank methods are possible. If solved, this gives the interior tangential $\boldsymbol{H}$ which, together with the known interior current distribution, is sufficient to specify the interior fields. The only restriction on the normal $\boldsymbol{B}$ is that the net flux through the wall is zero subject to the condition that Laplace's equation is satisfied outside the wires.

A more tractable problem is the calculation of the fields in the window due to two thin symmetrically placed layers of wire coaxial with the curved

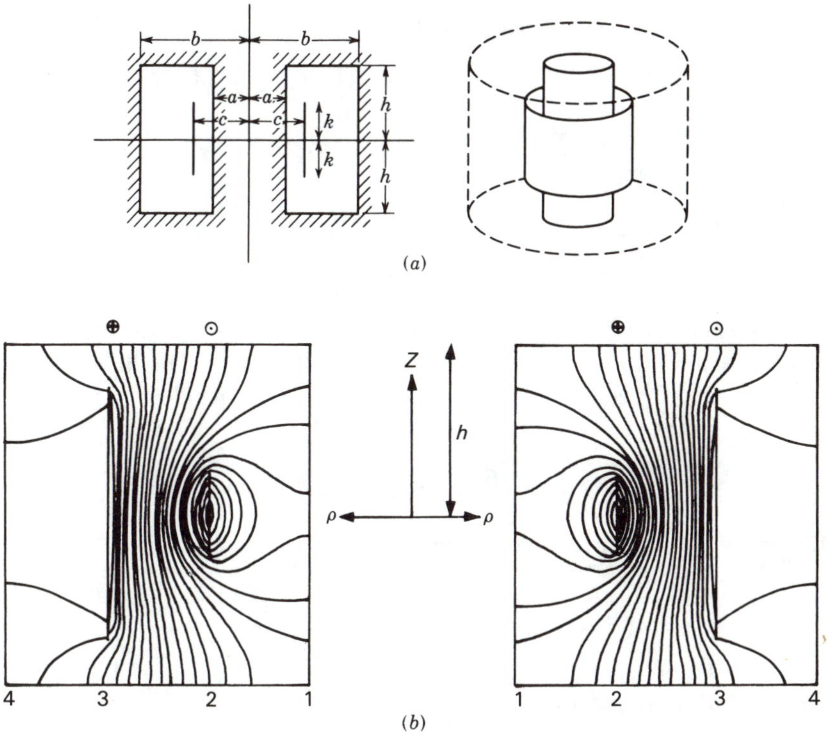

FIG. 7.31*a*, *b*.

walls having the same number of ampere-turns but oppositely directed currents. Let the window be bounded by $\rho = a$, $\rho = b$, $z = \pm h$ as in Fig. 7.31*a* and *b*. One layer, at $\rho = c$, extends from $z = -k$ to $z = +k$ and the other, at $\rho = c'$, extends from $z = -k'$ to $z = +k'$. Both are closely wound with n and n' turns per unit length, so that when carrying currents I and $-I'$ they act as uniform current sheets with total currents $2nIk = -2n'I'k' = N/\mu$. It is evident that the modified Bessel function solutions for A_ϕ of 7.05 (3) where $n = 0$ and the z factor is $\cos(m\pi z/h)$ will meet the boundary condition that $B_\rho = 0$ at $z = \pm h$. The $m = 0$

term comes entirely from $I_1(m\pi\rho/h)$ since by 5.32 (5) $K_1(0) = \infty$. From 5.32 (2) $I_1(m\pi\rho/h)$ becomes $\frac{1}{2}m\pi\rho/h$ as m approaches zero, and if there is an m in the denominator as in (11) and (12), this term has the form $B_0\rho$. From 7.02 (8) this gives a uniform z-directed field so that B_ρ is zero at $z = \pm h$. It cannot appear in regions adjacent to $\rho = a$ or $\rho = b$ because there is no z-component at these walls. Suitable solutions can now be written down. Formulas are shortened by the notation $M = m\pi/h$. Thus, taking zero order terms from 7.05 (7),

$$0 < \rho < c \qquad A_{1\phi} = \sum_{m=1}^{\infty} A_m[I_1(M\rho)K_0(Ma) + K_1(M\rho)I_0(Ma)] \cos Mz + B_0c^2/\rho \quad (1)$$

$$c < \rho < c' \qquad A_{2\phi} = \sum_{m=1}^{\infty} [B_mI_1(M\rho) + C_mK_1(M\rho)] \cos Mz + B_0\rho \quad (2)$$

$$c' < \rho < b \qquad A_{3\phi} = \sum_{m=1}^{\infty} D_m[I_1(M\rho)K_0(Mb) + K_1(M\rho)I_0(Mb)] \cos Mz + B_0c'^2/\rho \quad (3)$$

The brackets in (1) and (3) are chosen so that $B_z = 0 = [\partial(\rho A_\phi)/\partial\rho]/\rho$ when $\rho = a$ and $\rho = b$ from 5.321 (1) and 5.321 (4). The line integral fitting closely the coil between $z = 0$ and $z = z$ is independent of magnetization and other coils so that

$$\oint \boldsymbol{B} \cdot d\boldsymbol{s} = \int_0^z (B_{1z} - B_{2z})_{\rho=c}\, dz = \mu n I z$$

Differentiation with respect to z gives the boundary conditions

$$(B_{1z} - B_{2z})_{\rho=c} = \mu n I \qquad \text{when } -k < z < +k \quad (4)$$

$$(B_{1z} - B_{2z})_{\rho=c} = 0 \qquad \text{when } k < |z| < +h \quad (5)$$

A similar set holds at $\rho = c'$ for $B_{2z} - B_{3z}$. At $\rho = c$, $A_{1\phi} = A_{2\phi}$, and at $\rho = c'$, $A_{2\phi} = A_{3\phi}$, so that

$$A_m[I_1(Mc)K_0(Ma) + K_1(Mc)I_0(Ma)] = B_mI_1(Mc) + C_mK_1(Mc) \quad (6)$$

$$D_m[I_1(Mc')K_0(Mb) + K_1(Mc')I_0(Mb)] = B_mI_1(Mc') + C_mK_1(Mc') \quad (7)$$

From (1) and (2) $B_{1z} - B_{2z}$ is calculated and equated to (4) and (5). The summation as usual is eliminated by multiplying through by $\cos Mz$ and integrating from $-h$ to $+h$. A similar operation is done for $B_{2z} - B_{3z}$ at $\rho = c'$, and thus two additional equations are obtained:

$$N \sin Mk = Mhk\{A_m[I_0(Mc)K_0(Ma) - K_0(Mc)I_0(Ma)] - B_mI_0(Mc) + C_mK_0(Mc)\} \quad (8)$$

$$N \sin Mk' = Mhk'\{B_mI_0(Mc') - C_mK_0(Mc') - D_m[I_0(Mc')K_0(Mb) - K_0(Mc')I_0(Mb)]\} \quad (9)$$

Equations (6), (7), (8), and (9) are solved for A_m, B_m, C_m, and D_m, so that

$$F_m(a,c,k) = \frac{Nc}{k} \sin Mk[K_1(Mc)I_0(Ma) + I_1(Mc)K_0(Ma)] \tag{10}$$

$$-B_m = \frac{K_0(Mb)F_m(a,c,k) + K_0(Ma)F_m(b,c',k')}{Mh[I_0(Mb)K_0(Ma) - I_0(Ma)K_0(Mb)]} \tag{11}$$

$$-C_m = \frac{I_0(Mb)F_m(a,c,k) + I_0(Ma)F_m(b,c',k')}{Mh[I_0(Mb)K_0(Ma) - I_0(Ma)K_0(Mb)]} \tag{12}$$

$$A_m = \frac{B_mI_1(Mc) + C_mK_1(Mc)}{K_0(Ma)I_1(Mc) + I_0(Ma)K_1(Mc)} \tag{13}$$

$$D_m = \frac{B_mI_1(Mc') + C_mK_1(Mc')}{K_0(Ma)I_1(Mc') + I_0(Mb)K_1(Mc')} \tag{14}$$

The $m = 0$ term yields

$$B_0 = -\frac{\frac{1}{2}\mu nIk}{h} \tag{15}$$

From 7.08 (4) the equation of a tube of force or the stream function is

$$\Psi = 2\pi\rho A_\phi \tag{16}$$

Figure 7.31*b* shows the results of a calculation using the foregoing formulas in which $h = \frac{1}{2}\pi$, $k = \frac{1}{4}h$, $k' = \frac{3}{4}h$, $a = 1$, $b = 4$, $c = 2$, $c' = 3$, and $N = \frac{5}{8}\pi$. The values of ρA_ϕ are plotted at intervals of one-half. All the B-lines that reach the walls pass from top to bottom in the right-hand window with the current directions indicated.

7.32. Slotted Pole Piece. Effective Air Gap.—Suppose that one of the plane faces of an air gap of length B has parallel slots of width $2A$

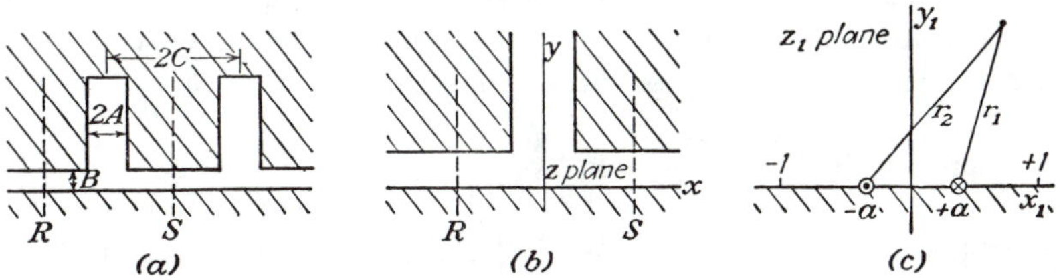

Fig. 7.32.

spaced at intervals $2C$ between centers. If $C - A$ is sufficiently large compared with B, the field near a slot is practically undisturbed by the adjacent slots and may be computed as if they were absent. If the depth of the slot is great compared with its width, the effect of its bottom is negligible and it can be treated as infinitely deep. Thus the section RS of Fig. 7.32*a* can be treated as section RS of Fig. 7.32*b* where the openings extend to infinity. We shall consider the permeability so high that the entire magnetomotance falls across the air gap as was discussed in 7.30. The opening has the shape of a polygon with interior angles zero at $x = \pm\infty$ and $3\pi/2$ at $x = \pm A, y = B$. Thus, an inverse Schwarz transformation will unfold this boundary into the real axis in the z_1-plane. Let us suppose that the $3\pi/2$ corners come to the points $x_1 = \pm 1$, then the 0 corners come to $x_1 = \pm a$ where $a^2 < 1$. By 4.18 (7), the z and z_1 planes are connected by the relation

$$\frac{dz}{dz_1} = C'\frac{(z_1^2 - 1)^{\frac{1}{2}}}{z_1^2 - a^2} \tag{1}$$

The magnetic circuit in the z_1-plane is shown in Fig. 7.32*c* which is a section of the bifilar circuit of 7.09. Since the reluctance of the part of the magnetic circuit below the x_1-axis is zero, the vector potential above the x_1-axis is twice that of 7.09 (1) and is

$$A_z = \frac{\mu_v I}{\pi}\ln\left(\frac{r_2}{r_1}\right) \tag{2}$$

From 4.12 (3), this is the real part of

$$W = \frac{\mu_v I}{\pi}\ln\frac{a + z_1}{a - z_1} \tag{3}$$

Before we determine W in the z-plane, we must integrate (1), which may be written

$$z = C'\int\frac{dz_1}{(z_1^2 - 1)^{\frac{1}{2}}} + C'(a^2 - 1)\int\frac{dz_1}{(z_1^2 - a^2)(z_1^2 - 1)^{\frac{1}{2}}}$$

For the first integral, use *Pc* 127 or *Dw* 320.01 after taking out j, and for the second, use *Pc* 229 or substitute $z_1 = au(u^2 - 1)^{-\frac{1}{2}}$ and use *Pc* 126*a* or *Dw* 200.01. After writing jC for C', the integrals may be written in several equivalent forms. In each, the signs must be checked by differentiation and comparison with (1). These forms are

$$z = C\sin^{-1} z_1 + C\frac{(1 - a^2)^{\frac{1}{2}}}{a}\sinh^{-1}\left\{\frac{1 - a^2}{(a/z_1)^2 - 1}\right\}^{\frac{1}{2}} \tag{4a}$$

$$= C\sin^{-1} z_1 + jC\frac{(1 - a^2)^{\frac{1}{2}}}{a}\sin^{-1}\left\{z_1\left[\frac{1 - a^2}{z_1^2 - a^2}\right]^{\frac{1}{2}}\right\} \tag{4b}$$

$$= \frac{1}{2}\pi C + jC\cosh^{-1} z_1 + jC\frac{(1 - a^2)^{\frac{1}{2}}}{a}\sin^{-1}\left\{\frac{1 - a^2}{1 - (a/z_1)^2}\right\}^{\frac{1}{2}} \tag{4c}$$

$$= C\sin^{-1} z_1 + jC\frac{\pi(1 - a^2)^{\frac{1}{2}}}{2a} + C\frac{(1 - a^2)^{\frac{1}{2}}}{a}\cosh^{-1}\left\{\frac{1 - a^2}{1 - (a/z_1)^2}\right\}^{\frac{1}{2}} \tag{4d}$$

when $x_1 > 1$, $x = +A = \frac{1}{2}C\pi$ from (4*c*) so that $C = 2A/\pi$, and when $1 > x_1 > a$, $y = B = Aa^{-1}(1 - a^2)^{\frac{1}{2}}$ from (4*d*) so that $a = A(A^2 + B^2)^{-\frac{1}{2}}$. From (3) by *Pc* 681 or *Dw* 702,

$$W = \frac{\mu_v I}{\pi}\ln\frac{1 + (z_1/a)}{1 - (z_1/a)} = \frac{2\mu_v I}{\pi}\tanh^{-1}\frac{z_1}{a}, \quad \text{or} \quad z_1 = a\tanh\frac{\pi W}{2\mu_v I} \tag{5}$$

Then, by *Pc* 659 or *Dw* 650.08,

$$\frac{z_1}{(a^2 - z_1^2)^{\frac{1}{2}}} = \sinh \frac{\pi W}{2\mu_v I}$$

This substituted in (4*a*) gives

$$z = \frac{2}{\pi}\left\{A \sin^{-1}\left[\frac{A}{(A^2 + B^2)^{\frac{1}{2}}} \tanh \frac{\pi W}{2\mu_v I}\right]\right. \\ \left. + B \sinh^{-1}\left[\frac{B}{(A^2 + B^2)^{\frac{1}{2}}} \sinh \frac{\pi W}{2\mu_v I}\right]\right\} \quad (6)$$

We have equivalent forms for (4*b*), (4*c*), and (4*d*). To find the effect of the slot on the reluctance of the air gap, we must find the flux between *R* and *S* with and without the slot. From (5), we are on the real axis when *W* is real. For large values of $U = A_z$ on the *x*-axis, which will occur when $C - A$ is considerably greater than *B*, we have

$$\tanh \frac{\pi A_z}{2\mu_v I} \to 1 \qquad \text{and} \qquad \sinh \frac{\pi A_z}{2\mu_v I} \to \frac{1}{2} e^{\frac{\pi A_z}{2\mu_v I}}$$

and, since $\sin^{-1} \dfrac{A}{(A^2 + B^2)^{\frac{1}{2}}} = \tan^{-1} \dfrac{A}{B} = \dfrac{1}{2}\pi - \tan^{-1} \dfrac{B}{A}$,

(6) is

$$x = A + \frac{2}{\pi}\left\{-A \tan^{-1} \frac{B}{A} + B \sinh^{-1}\left[\frac{B}{2(A^2 + B^2)^{\frac{1}{2}}} e^{\frac{\pi A_z}{2\mu_v I}}\right]\right\}$$

or

$$A_z = \frac{2\mu_v I}{\pi} \ln \left\{\frac{2}{B}(A^2 + B^2)^{\frac{1}{2}} \sinh \left[\frac{\pi(x - A)}{2B} + \frac{A}{B} \tan^{-1} \frac{B}{A}\right]\right\} \quad (7)$$

and, if $(x - A) \gg B$, this becomes

$$A_z = \frac{\mu_v I}{\pi} \ln \left(1 + \frac{A^2}{B^2}\right) + \frac{2\mu_v I}{\pi}\left[\frac{A}{B} \tan^{-1} \frac{B}{A} + \frac{(x - A)\pi}{2B}\right]$$

Since $A_z = 0$ at the origin from (3), its value at any point *x* is, from 7.25 (1),

$$A_z = \int_0^x \frac{\partial A_z}{\partial x}\, dx = -\int_0^x B_y\, dx$$

This is the flux leaving the lower face between $x = 0$ and $x = C$. If the same flux crossed the same interval with the same magnetomotance when there was no slot, *i.e.*, $A' = 0$, then the width B' of the air gap would be

$$B' = \frac{\mu_v I(A + C)}{A_z} = \frac{\pi(A + C)B}{\pi C + 2A \tan^{-1} (B/A) + B \ln (1 + (A/B)^2)} \quad (8)$$

Thus a gap of length B' without slots has the same reluctance as a gap of length B with slots of width $2A$ spaced at intervals $2C$.

Problems

The problems marked C in the following list are taken from the Cambridge examinations as reprinted by Jeans, with the permission of the Cambridge University Press.

1. Show that the magnetic induction at the center of a loop of wire carrying a current and shaped like a regular plane polygon of $2n$ sides, the distance between parallel sides being $2a$, is $[\mu nI/(\pi a)] \sin(\pi/2n)$, where I is the current.

2. A wire follows the circumference of a circle of radius a except for an arc of angular length 2ϕ across which it follows the chord. This loop is suspended from a point opposite the center of the chord so that its plane is perpendicular to a long straight wire that passes through its center. When the currents are i and i', show that the torque on the loop is $(\mu ii'a/\pi)(\sin\phi - \phi\cos\phi)$.

3. A filament is wound in a flat open spiral whose polar equations is $r = R\theta/(2\pi N)$, where N is the total number of turns and R is the radius vector from the center to the outer end of the spiral. If this spiral carries a current I, show rigorously that the axial component of the magnetic induction at a distance z along the axis from the coil is $(\frac{1}{2}\mu NI/R)(-R(R^2 + z^2)^{-\frac{1}{2}} + \ln\{[R + (R^2 + z^2)^{\frac{1}{2}}]z^{-1}\})$.

4. The plane of a circular loop of wire of radius b, carrying a current i, makes an angle β with a uniform magnetic field of induction B. Show that, if a sphere of radius a and permeability μ is placed inside the loop concentrically, the increase in torque is $[(K_m - 1)/(K_m + 2)]2\pi Bia^3b^{-1}\cos\beta$.

5. A certain lead consists of two parallel conducting strips of width A at a distance B apart, forming opposite sides of a rectangular prism. A current I goes out on one and returns on the other. If the current is uniformly distributed, show that the repulsion per unit length between strips is

$$\mu I^2(\pi A^2)^{-1}\{A\tan^{-1}(A/B) - \tfrac{1}{2}B\ln[(A^2 + B^2)/B^2]\}$$

6. An electrodynamometer consists of two square loops of wire, each side being of length a, one of which is free to turn about an axis which bisects the top and bottom sides of both loops. Assuming the loops intersect on this axis, show that the torque on one when it makes an angle $\frac{1}{2}\pi$ with the other and both carry a current I is

$$\frac{\mu I^2 a}{2\pi}\left\{2^{\frac{1}{2}}(3^{\frac{1}{2}} - 1) + 2\ln\left[\frac{5^{\frac{1}{2}}(1 + 2^{\frac{1}{2}})}{(1 + 6^{\frac{1}{2}})}\right] + 4\cot^{-1}(6^{\frac{1}{2}}\cdot 2) - 2\cos^{-1}\frac{1}{5}\right\}$$

7. Two circular cylinders lie with their axes at $x = +c$ and $x = -c$ and each carries a uniformly distributed current $\frac{1}{2}I$ in the z-direction. Surrounding these is a circular cylindrical shell with its axis at $x = 0$, carrying a uniformly distributed current I in the reverse direction. Show that the force toward the center acting on one of the inner cylinders is $\mu I^2/(16\pi c)$.

8*C*. Show that, at any point along a line of force, the vector potential due to a current in a circle is inversely proportional to the distance between the center of the circle and the foot of the perpendicular from the point on to the plane of the circle. Hence, trace the lines of constant vector potential.

9*C*. A current i flows in a circuit in the shape of an ellipse of area A and length l. Show that the field at the center is $\frac{1}{4}\mu il/A$.

10C. A current i flows around a circle of radius a, and a current i' flows in a very long straight wire in the same plane. Show that the mutual attraction is

$$\mu i i'(\sec \alpha - 1)$$

where 2α is the angle subtended by the circle at the nearest point of the straight wire.

11C. Two circular wires of radii a, b have a common center and are free to turn on an insulating axis which is a diameter of both. Show that when the wires carry currents i, i', a couple, of magnitude $\frac{1}{2}\pi\mu b^2 a^{-1}[1 - \frac{9}{16}a^{-2}b^2]ii'$, is required to hold them with their planes at right angles, it being assumed that b/a is so small that its fourth power can be neglected.

12C. Two currents i, i' flow round two squares each of side a, placed with their edges parallel to one another and at right angles to the distance c between their centers. Show that they attract with a force

$$\frac{2\mu i i'}{\pi}\left\{\frac{c(2a^2 + c^2)^{\frac{1}{2}}}{a^2 + c^2} + 1 - \frac{a^2 + 2c^2}{c(a^2 + c^2)^{\frac{1}{2}}}\right\}$$

13C. A current i flows in a rectangular circuit whose sides are of lengths $2a$, $2b$, and the circuit is free to rotate about an axis through its center parallel to the sides of length $2a$. Another current i' flows in a long straight wire parallel to the axis and at a distance d from it. Prove that the couple required to keep the plane of the rectangle inclined at an angle ϕ to the plane through its center and the straight current is $[2\mu i i' abd(b^2 + d^2) \sin \phi]/[\pi(b^4 + d^4 - 2b^2d^2 \cos 2\phi)]$.

14C. A circular wire of radius a is concentric with a spherical shell of soft iron of radii b and c. If a steady current I flows around the wire, show that the presence of the iron increases the number of lines of induction through the wire by

$$\frac{\pi I a^4 \mu_v (K_m - 1)(K_m + 2)(c^3 - b^3)}{2b^3[(K_m + 2)(2K_m + 1)c^3 - 2(K_m - 1)^2 b^3]}$$

approximately, where a is small compared with b and c.

15C. A right circular cylindrical cavity is made in an infinite mass of iron of permeability μ. In this cavity, a wire runs parallel to the axis of the cylinder, carrying a steady current of strength I. Prove that the wire is attracted toward the nearest part of the surface with a force $\mu_v I^2(K_m - 1)/[2\pi d(K_m + 1)]$ per unit length, where d is the distance of the wire from its electrostatic image in the cylinder.

16C. If the magnetic field within a body of permeability μ is uniform, show that any spherical portion can be removed and the cavity filled up with a concentric spherical nucleus of permeability μ_1 and a concentric shell of permeability μ_2 without affecting the external field, provided that μ lies between μ_1 and μ_2 and that the ratio of the volume of the nucleus to that of the shell is properly chosen. Prove also that the field inside the nucleus is uniform and that its intensity is greater or less than that outside according as μ is greater or less than μ_1.

17C. A sphere of soft iron of radius a is placed in a field of uniform magnetic force parallel to the axis of z. Show that the lines of force external to the sphere lie on surfaces of revolution, the equation of which is of the form

$$\left\{1 + \frac{2(K_m - 1)}{K_m + 2}\left(\frac{a}{r}\right)^3\right\}(x^2 + y^2) = \text{constant}$$

r being the distance from the center of the sphere.

18*C*. A sphere of soft iron of permeability μ is introduced into a field of force in which the scalar potential is a homogeneous polynomial of degree n in x, y, z. Show that the scalar potential inside the sphere is changed to its original value multiplied by $(2n + 1)/(nK_m + n + 1)$.

19*C*. If a shell of radii a, b is introduced in place of the sphere in the last question, show that the force inside the cavity is altered in the ratio

$$(2n + 1)^2K_m : (nK_m + n + 1)(nK_m + n + K_m) - n(n + 1)(K_m - 1)^2(a/b)^{2n+1}$$

20*C*. An infinitely long hollow iron cylinder of permeability μ, the cross section being concentric circles of radii a, b, is placed in a uniform field of magnetic force the direction of which is perpendicular to the generators of the cylinder. Show that the number of lines of induction through the space occupied by the cylinder is changed by inserting the cylinder in the field, in the ratio

$$b^2(K_m + 1)^2 - a^2(K_m - 1)^2 : 2K_m[b^2(K_m + 1) - a^2(K_m - 1)]$$

21. There is a zonal distribution of current flowing in an oblate spheroidal shell $\zeta = \zeta_0$ such that current between $\xi = 1$ and $\xi = \xi_0$ is given by $\psi(\xi_0) = \Sigma C_nP_n(\xi_0)$. Evaluating the integral of 7.10 by the result in 5.274, show that the vector potential due to this current, when $\zeta < \zeta_0$, is

$$-\Phi\mu j(1 + \zeta_0^2)^{\frac{1}{2}}\sum_1^\infty \frac{C_n}{n(n + 1)}Q_n^1(j\zeta_0)P_n^1(j\zeta)P_n^1(\xi)$$

22. Using the last problem, show that the vector potential due to a loop of wire at ζ_0, ξ_0, carrying a current I, $\zeta < \zeta_0$, is

$$-\Phi\tfrac{1}{2}\mu jI(1 + \zeta_0^2)^{\frac{1}{2}}(1 - \xi_0^2)^{\frac{1}{2}}\Sigma n^{-2}(n + 1)^{-2}(2n + 1)Q_n^1(j\zeta_0)P_n^1(\xi_0)P_n^1(j\zeta)P_n^1(\xi)$$

Write, by inspection, the result when $\zeta > \zeta_0$.

23. If the spheroid in 21 is prolate instead of oblate, show that, when $\eta < \eta_0$,

$$A = -\Phi\mu(\eta_0^2 - 1)^{\frac{1}{2}}\sum_1^\infty \frac{C_n}{n(n + 1)}Q_n^1(\eta_0)P_n^1(\eta)P_n^1(\xi)$$

24. In prolate spheroidal harmonics, the coordinates of a loop are ξ_0, η_0. Show that the vector potential due to such a loop carrying a current I is

$$-\Phi\tfrac{1}{2}\mu I[(\eta_0^2 - 1)(1 - \xi_0^2)]^{\frac{1}{2}}\Sigma n^{-2}(n + 1)^{-2}(2n + 1)P_n^1(\xi_0)Q_n^1(\eta_0)P_n^1(\eta)P_n^1(\xi)$$

25. In the last problem, a prolate spheroid of permeability μ whose surface is given by $\eta = \eta_1$ is inserted. Show that the additional vector potential due to its presence is given by

$$-\Phi\frac{\mu_v I}{2}(K_m - 1)(\eta_0^2 - 1)^{\frac{1}{2}}(1 - \xi_0^2)^{\frac{1}{2}}\sum_1^\infty \frac{2n+1}{n^2(n+1)^2}\,\frac{P_n^1(\xi_0)Q_n^1(\eta_0)P_n^1(\eta_1)P_n(\eta_1)}{Q_n^1(\eta_1)P_n(\eta_1) - K_mQ_n(\eta_1)P_n^1(\eta_1)}Q_n^1(\eta)P_n^1(\xi)$$

where $\eta > \eta_1$.

26. Show that the vector potential due to a plane loop of wire of radius a at $z = 0$ and carrying a current I is, if $\rho > a$,

$$A_\phi = \mu a I \pi^{-1} \int_0^\infty I_1(ka) K_1(k\rho) \cos kz \, dk$$

If $\rho < a$, interchange ρ and a under the integral.

27. Show that the vector potential due to a plane loop of wire of radius a, carrying a current I, is

$$A_\phi = \tfrac{1}{2} \mu a I \int_0^\infty J_1(ka) J_1(k\rho) e^{-k|z|} \, dk$$

28. A loop of wire of radius a is coaxial with an infinite cylinder of radius b and permeability μ and carries a current I. Show that that part of the vector potential outside the cylinder due to its presence is

$$\mu_v a I \pi^{-1} \int_0^\infty \Phi(k) K_1(ka) K_1(k\rho) \cos kz \, dk$$

where $$\Phi(k) = \frac{(K_m - 1) kb I_0(kb) I_1(kb)}{(K_m - 1) kb K_0(kb) I_1(kb) + 1}$$

Write out the vector potential for the region inside.

29. Show that the vector potential, due to a section of length $2c$ of a circular cylinder of radius a around which circulates a uniformly distributed current I, is, when $\rho > a$

$$A_\phi = \frac{\mu a I}{\pi c} \int_0^\infty \frac{1}{k} I_1(ka) K_1(k\rho) \cos kz \sin kc \, dk$$

30. The plane of a circular loop of wire of radius a, carrying a current I, is parallel to and at a distance b from the face of an infinite plate of material of permeability μ and thickness t. Show that the vector potentials on the loop side of the slab, in the slab, and on the other side of the slab are given, respectively, by

$$A_1 = \tfrac{1}{2} \mu_v a I \int_0^\infty J_1(k\rho) J_1(ka) [e^{-k|z|} + C(K_m^2 - 1)(1 - e^{-2kt}) e^{k(z-2b)}] \, dk$$

$$A_2 = \mu_v K_m a I \int_0^\infty J_1(k\rho) J_1(ka) C \{(K_m + 1) e^{-kz} - (K_m - 1) e^{k[z-2(b+t)]}\} \, dk$$

$$A_3 = 2\mu_v K_m a I \int_0^\infty J_1(k\rho) J_1(ka) C e^{-kz} \, dk$$

where ρ is measured from the axis of the loop, z from the plane of the loop, and

$$C = [(K_m + 1)^2 - (K_m - 1)^2 e^{-2kt}]^{-1}$$

31. An infinite tube of external radius a_1 and infinite permeability is coaxial with an infinite tube of the same material and internal radius a_2. In the $z = 0$ plane between a_1 and a_2 there is a current sheet. The current density i_ϕ in the ranges $a_1 < \rho < b_1$ and $b_2 < \rho < a_2$ is $-I[\rho \ln (a_2/a_1)]^{-1}$, and in the range $b_1 < \rho < b_2$ it is $I[\rho \ln (b_2/b_1)]^{-1} - I[\rho \ln (a_2/a_1)]^{-1}$, so that the total circulating current is zero and $A_\phi = 0$ on both walls. Show that the vector potential when $a_1 < \rho < a_2$ and $z > 0$ is

$$A_\phi = \sum_{n=1}^{\infty} C_n e^{-k_n z} R_1(k_n \rho)$$

where $R_m(k_n\rho) = Y_0(k_na_1)J_m(k_n\rho) - J_0(k_na_1)Y_m(k_n\rho)$ and k_n is chosen so that

$$R_0(k_na_2) = 0$$

(see HMF, Table 9.7), and

$$C_n = \frac{\mu\pi^2 I[R_0(k_nb_2) - R_0(k_nb_1)][J_1(k_na_2)]^2}{4\ln(b_2/b_1)\{[J_0(k_na_1)]^2 - [J_0(k_na_2)]^2\}}$$

32. The potential $A'_\phi = -B_0z/\rho$ of 7.05 (6) gives only a radial magnetic field B_0/ρ. Thus by superimposing the potentials $-\frac{1}{2}\mu Iz[\rho \ln(a_2/a_1-]^{-1}$ when $z > 0$ and $\frac{1}{2}\mu Iz[\rho \ln(a_2/a_1)]^{-1}$ when $z < 0$ on the result of the last problem, wipe out the circulating current except in the strip $b_1 < \rho < b_2$ of the $z = 0$ plane where i_ϕ becomes $I[\rho \ln(b_2/b_1)]^{-1}$. From 6.16 (2) this is the current distribution for a steady current in a conducting ring.

33. The permeability of the walls of the region bounded by $\rho = a_1$, $\rho = a_2$, and $z = 0$ is infinite. At $z = c$ there is a current sheet bounded by $\rho = b_2$ and $\rho = b_1$ in which the current density i_ϕ is $I[\rho \ln(b_2/b_1)]^{-1}$. Show from the last two problems that when $a_1 < \rho < a_2$ the values of the vector potential, A'_ϕ when $z > c$ and A''_ϕ when $c > z > 0$, are

$$A'_\phi = -Iz\left(\rho \ln \frac{a_2}{a_1}\right)^{-1} + 2\sum_{n=1}^{\infty} C_n e^{-k_nz} \cosh k_ncR_1(k_n\rho)$$

$$A''_\phi = -Ic\left(\rho \ln \frac{a_2}{a_1}\right)^{-1} + 2\sum_{n=1}^{\infty} C_n e^{-k_nc} \cosh k_nzR_1(k_n\rho)$$

34. An infinitely permeable rectangular conduit has its corners at $x = \pm a$, $y = 0$, and $y = b$. Wires parallel to its axis carry currents I at $c + jd$ and $-I$ at $f + jg$. Show with the aid of 4.29 that the vector potential in the conduit is

$$A_z = \frac{\mu I}{4\pi} \ln \frac{[(x_1 - c_1)^2 + y_1^2]^2 + 2[(x_1 - c_1)^2 - y_1^2]d_1^2 + d_1^4}{[(x_1 - f_1)^2 + y_1^2]^2 + 2[(x_1 - f_1)^2 - y_1^2]g_1^2 + g_1^4}$$

$$x_1 = \frac{\operatorname{sn} mx \operatorname{dn} my}{\operatorname{cn}^2 my + k \operatorname{sn}^2 mx \operatorname{sn}^2 my}, \qquad y_1 = \frac{\operatorname{sn} mx \operatorname{dn} mx \operatorname{sn} my \operatorname{dn} my}{\operatorname{cn}^2 my + k \operatorname{sn}^2 mx \operatorname{sn}^2 my}$$

Similar formulas hold for $x_1 = c_1$ or f_1 and $x = c$ or f and for $y_1 = d_1$ or g_1 and $y = d$ or g. The modulus k is found from $a/b = K(k)/K[(1 - k^2)^{\frac{1}{2}}]$ and $ma = K$.

35. An infinite wire carrying a current I is situated at $x = a$, $y = b$, between two faces of infinite permeability $y = 0$ and $y = c$. Show that the vector potential between the faces is

$$U = -\frac{\mu I}{4\pi} \ln \left[\left(\cos \frac{\pi b}{c} - \cosh \frac{\pi(x-a)}{c} \cos \frac{\pi y}{c}\right)^2 + \sinh^2 \frac{\pi(x-a)}{c} \sin^2 \frac{\pi y}{c}\right]$$

which is the real part of

$$W = -\frac{\mu I}{2\pi} \ln \left[\cos \frac{\pi b}{c} - \cosh \frac{\pi(z-a)}{c}\right]$$

Note that the field is uniform and oppositely directed when $x \gg a$ and when $x \ll a$.

36. An infinite wire carrying a current I lies at z_0 to the right of a semi-infinite block of material of infinite permeability whose face is the plane $x = 0$ except for a circular ridge or groove of radius b whose center is at $x = c$. Show with the aid of problem 49, Chap. IV, that the vector potential is the real part of

$$W = -\frac{\mu I}{2\pi} \ln \{[f(z) - f(z_0)][f(z) + f(z_0^*)]\}$$

where

$$f(z) = -\frac{\pi(b^2 - c^2)^{\frac{1}{2}}\{[z + j(b^2 - c^2)^{\frac{1}{2}}]^{\pi/\alpha} + [z - j(b^2 - c^2)^{\frac{1}{2}}]^{\pi/\alpha}\}}{\alpha\{[z + j(b^2 - c^2)^{\frac{1}{2}}]^{\pi/\alpha} - [z - j(b^2 - c^2)^{\frac{1}{2}}]^{\pi/\alpha}\}}$$

and $\cos\alpha = c/b$. For a ridge $0 < \alpha < \pi$ and for a groove $\pi < \alpha < 2\pi$.

37. Using the relation $-\ln(z - a) = \int_0^\infty k^{-1}[e^{-k(z-a)} - e^{-k}]dk$, where the real part of $z - a$ is positive, show that the transformations giving the field of a wire at $x = a$, carrying a current I between two semi-infinite blocks of material of permeability μ whose faces are at $x = 0$ and $x = b$, where $0 < a < b$, are, when $0 < x < b$,

$$W_1 = -\frac{\mu_v I}{2\pi}\left\{\ln(z - a) - 2\beta\int_0^\infty \frac{1}{k}\left\{\frac{\cosh[k(z+a-b)] + \beta e^{-kb}\cosh[k(z-a)]}{e^{kb} - \beta^2 e^{-kb}} - \frac{e^{-k}}{1-\beta}\right\}dk\right\}$$

when $b < x$, $$W_2 = \frac{\mu_v I}{\pi(K_m + 1)}\int_0^\infty \frac{K_m}{k}\left(e^{-kz}\frac{e^{ka} + \beta e^{-ka}}{1 - \beta^2 e^{-2kb}} - \frac{e^{-k}}{1 - \beta}\right)dk$$

when $x < 0$, $$W_3 = \frac{\mu_v I}{\pi(K_m + 1)}\int_0^\infty \frac{K_m}{k}\left[e^{kz}\frac{e^{-ka} + \beta e^{k(a-2b)}}{1 - \beta^2 e^{-2kb}} - \frac{e^{-k}}{1 - \beta}\right]dk$$

where U is the vector potential, $\mu^{-1}V$ the scalar potential, $W = U + jV$ and we have $\beta = (K_m - 1)(K_m + 1)^{-1}$.

38. Show that the transformations giving the field of a wire at $x = d$, carrying a current I between two semi-infinite blocks of material of permeability μ, whose faces are at $x = -c$ and $x = +c$, where $-c < d < +c$, are

$-c < x < c$, $$W_1 = -\frac{\mu_v I}{2\pi}\left(\ln\frac{z - d}{2c} + \sum_{n=1}^{\infty}\beta^n \ln\left\{1 - \frac{4nc^2 + [z - (-1)^n d]^2}{4n(n + 1)c^2}\right\}\right)$$

$c < x$, $$W_2 = -\frac{\mu_v K_m I}{\pi(K_m + 1)}\left(\ln\frac{z - d}{2c} + \sum_{n=1}^{\infty}\beta^n \ln\left[1 + \frac{-2c + z - (-1)^n d}{2(n + 1)c}\right]\right)$$

$x < -c$, $$W_3 = -\frac{\mu_v K_m I}{\pi(K_m + 1)}\left(\ln\frac{z - d}{2c} + \sum_{n=1}^{\infty}\beta^n \ln\left[-1 + \frac{2c + z - (-1)^n d}{2(n + 1)c}\right]\right)$$

where $\beta = (K_m - 1)(K_m + 1)^{-1}$, U is the vector potential, $\mu^{-1}V$ the scalar potential, and $W = U + jV$. This can be done directly by images or by expanding the denominator in the previous problem, integrating and writing $z + c$ for z, $2c$ for b, and d for $a - c$. To make U finite near the origin, add the infinite constant

$$\mu_v K_m I[\pi(K_m + 1)]^{-1}\sum_{n=1}^{\infty}\beta^n \ln[2(n + 1)c]$$

39. Compare the first A_ϕ integral of 7.10 with 5.23 (15) and thus show that the vector potential of a circular wire loop of radius a carrying a current I is

$$A_\phi = -2^{-\frac{1}{2}}\mu I(\cosh u_1 - \cos u_2)^{\frac{1}{2}} P^1_{-\frac{1}{2}}(\cosh u_1)$$

where, from 4.13, $2a\rho \coth u_1 = r^2 + a^2$ and $2az \cot u_2 = r^2 - a^2$.

40. By reference to 4.13, 7.06 (2), 5.233 (5), problem 117 of Chap. V, etc., add terms to the A_ϕ of the last problem such that the magnetic field is tangent to the torus, $\cosh u_1 = c/b$, obtained by rotation of the circle of radius b about a line in its plane at a distance c from its center. Show that the sum is

$$A_\phi = \mu I\left(\frac{\cosh u_1 - \cos u_2}{2}\right)^{\frac{1}{2}} \frac{P^1_{-\frac{1}{2}}(c/b)}{Q^1_{-\frac{1}{2}}(c/b)} \sum_{n=0}^{\infty} \frac{(2-\delta^0_n) Q^1_{n-\frac{1}{2}}(c/b) P^1_{n-\frac{1}{2}}(\cosh u_1)}{-(2n-1)(2n+1)P^1_{n-\frac{1}{2}}(c/b)} \cos nu_1$$

41. The current density i_ϕ at $\rho = a$ of period c in z is given by

$$i_\phi = \sum_{n=0}^{\infty} C_n \cos\frac{2n\pi z}{c} = \sum_{n=0}^{\infty} C_n \cos Nz$$

Show that the vector potentials, A'_ϕ inside and A''_ϕ outside, are

$$A'_\phi = \mu_v a \sum_{n=0}^{\infty} C_n I_1(N\rho) K_1(Na) \cos Nz$$

$$A''_\phi = \mu_v a \sum_{n=0}^{\infty} C_n I_1(Na) K_1(N\rho) \cos Nz$$

42. A semi-infinite conducting sheet occupies the region $0 < \phi < \pi$ of the $z = 0$ plane. A current I enters the edge at the origin and leaves at $\rho = \infty$. Show that at a point P, whose coordinates are ρ, ϕ, z, the magnetic-induction components are

$$B_\phi = \frac{\mu I}{4\pi\rho}\left(1 + \frac{2}{\pi}\tan^{-1}\frac{\rho\sin\phi}{z} - \frac{z}{r}\right), \qquad \frac{\rho}{z}B_\rho = B_z = \frac{\mu I}{2\pi^2 r}\tanh^{-1}\frac{\rho\cos\phi}{r}$$

where r, the distance from the origin to P, is $(\rho^2 + z^2)^{\frac{1}{2}}$.

43. A solenoidal current sheet has a density i_ϕ amperes per meter. A vector of length b from its center to any point on its edge makes an angle β with its axis. Show that the vector potential inside the current sheet is

$$A_\phi = \mu_v i_\phi b \sin\beta \sum_{n=0}^{\infty} C_n \left(\frac{r}{b}\right)^{2n+1} P^1_{2n+1}(\cos\theta)$$

where θ is measured from the axis about its center and

$$C_0 = -\tfrac{1}{2}\cot\beta, \qquad C_n = [2n(2n+1)(2n+2)]^{-1} P^1_{2n}(\cos\beta)$$

See 10.17 and problem 34, Chap. X. Outside the current sheet the same expression holds when $r < b$ except that $C_0 = \frac{1}{2}(\csc\beta - \cot\beta)$.

44. A plane loop of wire of radius a lies parallel to the plane faces of a gap of width c between the halves of an infinite block of infinite permeability at a distance b from one face. If it carries a current I, show that the vector potential is

$$A_\phi = \frac{2\mu a I}{c}\sum_{n=0}^{\infty} I_1\frac{n\pi\rho}{c}K_1\frac{n\pi a}{c}\cos\frac{n\pi b}{c}\cos\frac{n\pi z}{c}$$

when $\rho < a$, and write down A_ϕ when $\rho > a$. Show that the force pulling the loop toward the $z = 0$ plane is

$$\mu\left(\frac{\pi a I}{c}\right)^2\sum_{n=1}^{\infty} nI_1\frac{n\pi a}{c}K_1\frac{n\pi a}{c}\sin\frac{2n\pi b}{c}$$

45. Show that the magnetic induction in terms of the flux function N of 7.08 (3) and the differential equation satisfied by N where no current is present are

$$\frac{\partial N}{\partial\rho} = 2\pi\rho B_z, \qquad \frac{\partial N}{\partial z} = -2\pi\rho B_\rho, \qquad \frac{\partial^2 N}{\partial z^2} + \frac{\partial^2 N}{\partial\rho^2} - \frac{\partial N}{\rho\,\partial\rho} = 0$$

Note that the sign difference in the last term rules out the solutions of 5.29 (1) with $n = 0$ but yields instead $2\pi\rho A_\phi$ as given by 7.05 (2) and 7.05 (3). Show that the boundary conditions on N at the interface between μ_1 and μ_2 are

$$N_1 = N_2, \qquad \mu_2\left(\boldsymbol{\rho}_1\frac{\partial N_1}{\partial\rho} + \mathbf{k}\frac{\partial N_1}{\partial z}\right) = \mu_1\left(\boldsymbol{\rho}_1\frac{\partial N_2}{\partial\rho} + \mathbf{k}\frac{\partial N_2}{\partial z}\right)$$

References

A. Magnetic

Abraham, M., and R. Becker: "Classical Electricity and Magnetism," Hafner, 1950. Gives simple vector treatment.

Attwood, S. S.: "Electric and Magnetic Fields," Wiley, 1941. Gives graphical method for plotting fields.

Buchholz, H.: "Electrische und Magnetische Potentialfelder," Springer, 1957. Gives theory and many solutions of special problems.

Corson, D. R., and P. Lorrain: "Introduction to Electromagnetic Fields and Waves," Freeman, 1962. Well-illustrated exposition of basic ideas.

Durand, E.: "Electrostatique et Magnetostatique," Masson et Cie, 1953. Gives theory with many examples and field maps.

Geiger-Scheel: "Handbuch der Physik," Vol. XV, Berlin, 1927.

Gray, A.: "Absolute Measurements in Electricity and Magnetism," Vol. II, Macmillan, 1888. Extensive calculations of fields of circuits and shells.

Jeans, J. H.: "The Mathematical Theory of Electricity and Magnetism," Cambridge, 1925. Gives extensive treatment without vectors.

Mason, M., and W. Weaver,: "The Electromagnetic Field," University of Chicago Press, 1929, and Dover. Fine treatment of vector potential and boundary conditions.

Maxwell, J. C.: "Electricity and Magnetism," 3d ed. (1891), Dover, 1954. Gives extensive treatment with good field maps and no vector analysis.

Owen, G. E.: "Electromagnetic Theory," Allyn and Bacon, 1963. Well-illustrated exposition of basic ideas with simple examples.

PANOFSKY, W. K. H., and MELBA PHILLIPS: "Classical Electricity and Magnetism," Addison-Wesley, 1962. Excellent treatment of vector potential in Chap. VIII.

STRATTON, J. A.: "Electromagnetic Theory," McGraw-Hill, 1941. Gives complete theory of the material of this chapter.

WALKER, M.: "Schwarz-Christoffel Transformation and Its Applications," Dover, 1964. Works out important applications in detail.

WEBER, E.: "Electromagnetic Fields," Vol. I, Wiley, 1950. Gives many examples and literature references.

WIEN-HARMS: "Handbuch der Experimentalphysik," Vol. XI, Leipzig, 1932.

B. MATHEMATICAL

See references of Chap. V.

CHAPTER VIII

ELECTROMAGNETIC INDUCTION

8.00. Faraday's Law of Induction.—A century ago Faraday and Henry discovered, independently, that when the magnetic flux N, defined by 7.29 (1), through a closed conducting circuit changes, a current is generated in the circuit. The direction of this induced current is such as to set up a magnetic flux opposing the change. Thus if the flux through a circuit in a certain direction is increasing, the induced current sets up a flux in the opposite direction, and if the flux is decreasing, this current sets up a flux in the same direction. The induced currents always seek to maintain the *status quo* of the magnetic field. The induced electromotance in volts equals the negative rate of the change of flux in webers per second so that

$$\mathcal{E} = -\frac{dN}{dt} \tag{1}$$

with possible limitations mentioned later.

It is immaterial what means we use to change the flux. We may move the source of the flux with respect to our circuit, we may change its strength, or we may move or change the shape of our circuit. Faraday proved (1) for a closed metallic circuit, but since we know that the tangential component of $\nabla\mathcal{E} = E$ is the same on either side of the wire surface, (1) must apply equally well to the electromotance around a path just outside the wire. We may therefore assert that (1) holds for any path whatever and write, from 7.29 (1), for a two-sided surface S

$$\oint \boldsymbol{E} \cdot d\boldsymbol{s} = -\frac{d}{dt}\int_S \boldsymbol{B} \cdot \boldsymbol{n}\, dS \tag{2}$$

Applying Stokes's theorem [3.01 (1)] to the left side, we have

$$\int_S \boldsymbol{n} \cdot \nabla \times \boldsymbol{E}\, dS = -\int_S \frac{d}{dt}\boldsymbol{B} \cdot \boldsymbol{n}\, dS$$

Since this must hold for any arbitrary surface, we have

$$\nabla \times \boldsymbol{E} = -\frac{d\boldsymbol{B}}{dt} \tag{3}$$

in webers per square meter-second. If $\boldsymbol{E}$ is due entirely to electromagnetic induction, it is a special case of the $\boldsymbol{E}''$ of 6.01 and possesses no sources or sinks so that its divergence is zero. If A is obtained from 7.02 (5), its divergence is also zero. Let us now substitute $\nabla \times A$ for B in (3) and interchange d/dt and the curl, and we see that the curls of the vectors E and $-dA/dt$ are equal. If two vectors have everywhere the same curl and divergence, they differ at most by a constant, so that

$$E = -\frac{dA}{dt} \tag{4}$$

which connects the electric intensity or electromotance per meter and the vector potential whose change induces it.

Equation (1) seems adequate to deal with all cases involving a change of flux in rigid circuits and also with those cases in which the flux is changed by distorting the shape of the circuit, provided that, during the process, all elements of the circuit initially adjoining, remain adjoining. Experiments can be devised, using sliding contacts, to which the application of (1) is obscure or ambiguous. These cases can be treated by focusing our attention, not upon the area surrounded by the circuit, but upon the elements of the circuit themselves, and then applying (4).

Another formulation of the law of induction is worked out in Chap. XIV, where it is shown, from the principle of special relativity, that when an observer moves relative to the fixed circuits producing a magnetic induction B he will observe, in general, an electromotance. If v is the velocity of the motion and E that part of the observed electric intensity which is due to the motion, we have, from 14.13 (12), if E is volts per meter, v meters per second, and B webers per square meter,

$$E = [v \times B] \tag{5}$$

8.01. Mutual Energy of Two Circuits.—Let us now consider the work required to bring near each other two circuits carrying steady currents I and I'. Each circuit contains a source of electromotance that supplies or absorbs energy at the proper rate to maintain the currents constant at all times. In circuit 1 we have, from Ohm's law, 6.02 (1), and 8.00 (1),

$$IR = \mathcal{E} - \frac{dN}{dt}$$

where $\mathcal{E}$ is the electromotance just described. The energy consumed in circuit 1 while bringing it up from infinity in time t is then, from 6.03 (2),

$$I^2R\int_0^t dt = I\int_0^t \mathcal{E}\,dt \pm I\int_0^N dN = I\int_0^t \mathcal{E}\,dt \pm NI$$

where N is the flux through the first circuit due to the second circuit in the final position. Evidently, the last term represents the work done in the first circuit due to the magnetic field of the second. At the same time, work of amount $\pm N'I'$ has been done in the second circuit to maintain its current constant. Thus, the total energy expended by the electromotance sources in the two circuits to maintain the currents constant while they are brought together is

$$\pm(IN + I'N')$$

Substituting the vector potential in 7.29 (1) and applying Stokes's theorem and 7.02 (5), we have

$$N = \int_S \mathbf{B}' \cdot \mathbf{n}\, dS = \int_S \nabla \times \mathbf{A}' \cdot \mathbf{n}\, dS = \oint \mathbf{A}' \cdot d\mathbf{s} = \frac{\mu I'}{4\pi}\oint\oint \frac{d\mathbf{s} \cdot d\mathbf{s}'}{r}$$

and similarly

$$N' = \frac{\mu I}{4\pi}\oint\oint \frac{d\mathbf{s} \cdot d\mathbf{s}'}{r} \tag{1}$$

so that the energy expended becomes

$$W_{\mathcal{E}} = \pm(IN + I'N') = \pm\frac{\mu I I'}{2\pi}\oint\oint \frac{d\mathbf{s} \cdot d\mathbf{s}'}{r} \tag{2}$$

Faraday's law shows that if the two circuits attract each other the induced electromotance sets up an opposing field which tries to reverse the current in each circuit so that the sources of electromotance must supply energy to maintain I and I'. If they repel each other, the reverse is true.

Let us now consider the mechanical work done against magnetic forces in moving the two circuits. From 7.18 (5), the force on circuit 1 in newtons is

$$\mathbf{F} = I\oint d\mathbf{s} \times \mathbf{B}' = \frac{\mu I I'}{4\pi}\oint\oint \frac{d\mathbf{s} \times (d\mathbf{s}' \times \mathbf{r})}{r^3} = \frac{\mu I I'}{4\pi}\oint\oint \frac{d\mathbf{s}'(d\mathbf{s} \cdot \mathbf{r}) - \mathbf{r}(d\mathbf{s} \cdot d\mathbf{s}')}{r^3}$$

The first term vanishes when integrated around circuit 1 since the integrand $\mathbf{r}/r^3$ can be written as the gradient of the scalar $1/r$. The mechanical work done is then, for pure translation from $r = \infty$ to $r = r$,

$$W_{me} = \int_\infty^r \mathbf{F} \cdot d\mathbf{r} = -\frac{\mu I I'}{4\pi}\oint\oint\int_\infty^r \frac{d\mathbf{s} \cdot d\mathbf{s}'\mathbf{r} \cdot d\mathbf{r}}{r^3} = \frac{\mu I I'}{4\pi}\oint\oint \frac{d\mathbf{s} \cdot d\mathbf{s}'}{r} \tag{3}$$

Comparing (2) and (3), we see that half the energy supplied by the battery in the circuit is used in doing mechanical work. Since the only difference between the initial and final states is in the magnetic field surrounding the circuits, the remainder of the energy must be in the magnetic field. Thus, when two constant-current circuits are moved with respect to each other, the mechanical work done by the circuit and the energy in the magnetic field increase or decrease together and at the same rate. This explains the apparent anomaly mentioned in 7.18. It follows that if we know the energy W_B in the magnetic fields of two circuits we can get the mechanical force or torque trying to increase any coordinate θ by taking the positive derivative of this energy with respect to θ, so that

$$F = +\frac{\partial W_B}{\partial \theta} \tag{4}$$

8.02. Energy in a Magnetic Field.—We shall now find the energy required to establish the magnetic field of a single circuit, utilizing the results obtained in the last article and assuming all space filled with a homogeneous, isotropic medium of permeability μ. Let us build up the field step by step by bringing together infinitesimal current filaments. If the final current density is nowhere infinite, the denominator in 8.01 (3) causes no difficulty because finite current filaments are a finite distance apart. When 8.01 (3) is applied to a single circuit, we must include the factor $\frac{1}{2}$ to get the correct energy since our integration includes not only the work done in bringing filament a up to b, but also that in bringing b to a. When we write in current densities in place of currents by 6.00 (2), and make this correction, 8.01 (3) becomes

$$W_B = \frac{\mu}{8\pi}\int_{\mathrm{V}}\int_{\mathrm{V}'}\frac{\boldsymbol{i}\,d\mathrm{v}\cdot\boldsymbol{i}'\,d\mathrm{v}'}{r} \tag{1}$$

where r is the distance between the volume elements $d\mathrm{v}$ and $d\mathrm{v}'$, $\boldsymbol{i}$ and $\boldsymbol{i}'$ are the current densities in these elements, and the integration is performed twice throughout space, where, as in 8.01, we have assumed that μ is constant in the region of integration. Substituting for $\boldsymbol{i}$ from 7.01 (3) and for $\int_{\mathrm{V}'}(\boldsymbol{i}'/r)\,d\mathrm{v}'$ from 7.02 (4), we have

$$W_B = \frac{1}{2\mu}\int_{\mathrm{V}}\nabla\times\boldsymbol{B}\cdot\boldsymbol{A}\,d\mathrm{v} \tag{2}$$

From the formula for the divergence of a crossproduct, we have the relation

$$\boldsymbol{A}\cdot\nabla\times\boldsymbol{B} = \boldsymbol{B}\cdot[\nabla\times\boldsymbol{A}] - \nabla\cdot[\boldsymbol{A}\times\boldsymbol{B}] = B^2 - \nabla\cdot[\boldsymbol{A}\times\boldsymbol{B}]$$

which gives

$$W_B = \frac{1}{2\mu}\int_{\mathrm{V}}B^2\,d\mathrm{v} - \frac{1}{2\mu}\int_{\mathrm{V}}\nabla\cdot[\boldsymbol{A}\times\boldsymbol{B}]\,d\mathrm{v}$$

where the integrals are over all space. By Gauss's theorem [3.00 (2)], we may transform the second integral into a surface integral over the sphere at infinity. This integral vanishes, since, from 7.02 (4), $\boldsymbol{A}$ goes to zero as $1/r$ whereas $\boldsymbol{B} = \nabla\times\boldsymbol{A}$ goes to zero as $1/r^2$, and the surface area increases only by r^2, so that

$$\int_{\mathrm{V}}\nabla\cdot[\boldsymbol{A}\times\boldsymbol{B}]\,d\mathrm{v} = \int_S[\boldsymbol{A}\times\boldsymbol{B}]\cdot\boldsymbol{n}\,dS \xrightarrow[r\to\infty]{} 0$$

Thus, the final expression for the energy becomes

$$W_B = \frac{1}{2\mu}\int_V B^2\,dv \tag{3}$$

This energy can be looked upon as residing in the space occupied by the magnetic field surrounding a current, the energy density being $\frac{1}{2}B^2/\mu$. This may be compared with 2.08 (2) which gives the energy in an electrostatic field. We see that the magnetic field, just as well as the electrostatic field, can be visualized in terms of a system of stresses.

8.03. Mutual Inductance.—The coefficient of mutual inductance M_{12} between two circuits is defined as the flux N_{12} through circuit 1 produced by unit current in circuit 2. The mks mutual inductance unit is the henry. This is expressed mathematically by writing

$$M_{12} = \int_{S_1} \boldsymbol{B}_2 \cdot \boldsymbol{n}\,dS_1 = \oint \boldsymbol{A}_2 \cdot d\boldsymbol{s}_1 \tag{1),(2}$$

where $\boldsymbol{A}_2$ is the entire vector potential due to unit current in circuit 2. From 8.01 (1), we may write

$$M_{12} = \frac{\mu}{4\pi}\oint\oint\frac{d\boldsymbol{s}_1 \cdot d\boldsymbol{s}_2}{r} = M_{21} \tag{3}$$

The flux through circuit 1 due to a current I_2 in circuit 2 is, from (1),

$$N_{12} = M_{12}I_2 \tag{4}$$

From 8.00 (1), the electromotance in circuit 1 produced by a fluctuating current in 2 is

$$\mathcal{E}_1 = -M_{12}\frac{dI_2}{dt} \tag{5}$$

From 8.01 (3), the mutual energy of the two circuits is given by

$$W_{12} = M_{12}I_1I_2 \tag{6}$$

From 8.02 (3), the total energy of the two circuits is

$$\begin{aligned} W_t &= \frac{1}{2\mu}\int_V (\boldsymbol{B}_1 + \boldsymbol{B}_2) \cdot (\boldsymbol{B}_1 + \boldsymbol{B}_2)\,dv \\ &= \frac{1}{2\mu}\left(\int_V B_1^2\,dv + 2\int_V \boldsymbol{B}_1 \cdot \boldsymbol{B}_2\,dv + \int_V B_2^2\,dv\right) \end{aligned}$$

The first term is the energy required to establish I_1 alone and the last term that for I_2 alone so that the remaining term gives the energy used in bringing the two into interaction. Hence, from (6), we have

$$M_{12}I_1I_2 = \frac{1}{\mu}\int_V \boldsymbol{B}_1 \cdot \boldsymbol{B}_2\,dv \tag{7}$$

From 8.01 (4), the force or torque tending to increase any coordinate of position θ of one circuit relative to the other is

$$F = I_1 I_2 \frac{\partial M_{12}}{\partial \theta} \tag{8}$$

8.04. Boundary Conditions on $\boldsymbol{A}$.—In 7.21 appear the conditions that derivatives of the quasi-vector potential $\boldsymbol{A}$ satisfy at the boundary between two regions of different permeability. In 8.03 (2), we have, for the first time, a relation involving an integral of $\boldsymbol{A}$, and we must therefore consider what boundary conditions to impose on $\boldsymbol{A}$ itself in order that this equation will be valid where such discontinuities occur. Let us take a small rectangular circuit, the long sides of which are very close together but on opposite sides of the boundary between regions of different permeability. As these sides approach each other, the area of the rectangle approaches zero, so that the flux through this circuit, due to a current I_2 flowing in a second circuit, must become zero. But by 8.03 (2) and (4), this flux is $N = \oint \boldsymbol{A}_2 \cdot d\boldsymbol{s}$. The integral is taken around the rectangle, the ends of which are vanishingly small, so that the whole contribution must come from the sides, which are of equal length L and which are short enough so that the vector $\boldsymbol{A}$ is constant along a side.

From 7.21 (10) and (11), the tangential component of $\boldsymbol{A}$ has the same direction on both sides of the boundary, so orienting our rectangle along $\boldsymbol{u}_s$ and $\boldsymbol{u}_t$ we obtain for the corresponding components of $\boldsymbol{A}$, respectively,

$$\oint \boldsymbol{A}_{s,t} \cdot d\boldsymbol{s}_{s,t} = L(A'_{s,t} - A''_{s,t}) = 0$$

Thus, we have the two equations for the tangential components of $\boldsymbol{A}$.

$$A'_{s,t} = A''_{s,t} \tag{1}$$

8.05. Mutual Inductance of Simple Circuits.—As a simple example of the computation of a mutual inductance coefficient, consider the two closely wound coils A, B, shown in Fig. 8.05. Coil B contains n turns and is wound on a ring of permeability μ. Coil A has m turns. All the flux goes through the ring so that, from 7.29 (4) and 8.03 (4), we have

$$M_{12} = \mu n m[a - (a^2 - b^2)^{\frac{1}{2}}] \tag{1}$$

If $a \gg b$, we may factor a out of the radical and expand the remainder by *Pc* 753 or *Dw* 5.3, keeping only the square terms. This then becomes

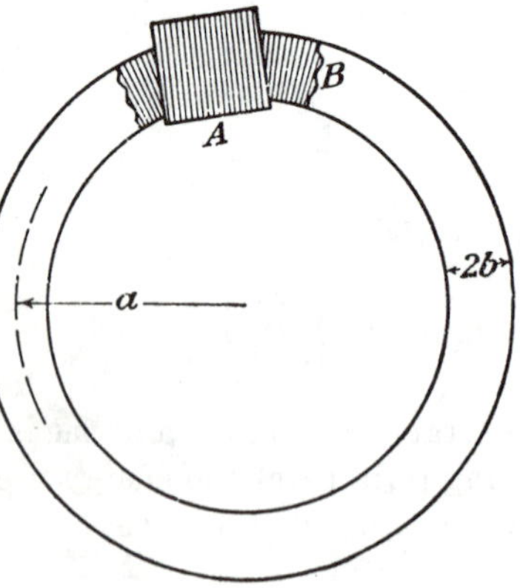

FIG. 8.05.

$$M_{12} = \frac{\mu m n b^2}{2a} \tag{2}$$

If $A = \pi b^2$ and $n_1 = n/(2\pi a)$ is the number of turns per unit length, then this gives the mutual inductance between an infinite solenoid wound on a cylinder of area A and permeability μ and a coil of m turns encircling it to be

$$M_{12} = \mu n_1 m A \tag{3}$$

This could have been written down directly from 7.15 (7).

8.06. Mutual Inductance of Circular Loops.—The coefficient of mutual inductance for two coaxial loops of wire can be written down by

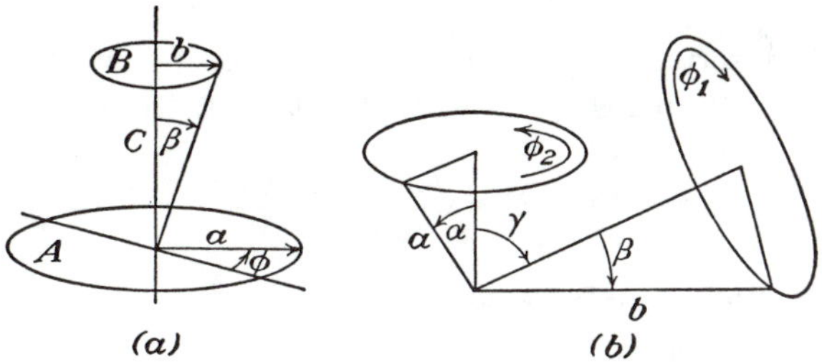

Fig. 8.06.

the use of 8.03 (2). The vector potential of A is entirely in the ϕ-direction and so has the same value for all elements of loop B and is parallel to each element so that, from 8.03 (2),

$$M_{BA} = \oint \boldsymbol{A}_A \cdot \boldsymbol{ds}_B = 2\pi b \Big| A_A \Big|_{\rho=b}^{z=c}$$

Thus, from 7.10 (4), we have

$$M_{12} = 2\mu k^{-1}(ab)^{\frac{1}{2}}[(1 - \tfrac{1}{2}k^2)K - E] \tag{1}$$

where, from 7.10 (3),

$$k^2 = 4ab[(a + b)^2 + c^2]^{-1} \tag{2}$$

Numerical values of K and E may be found in *Pc* 121 or *Dw* 1040 and 1041. Another expression for M_{12} may be written down in the same way from 7.13 (2), letting $\alpha = \frac{1}{2}\pi$ and $\theta = \beta$, giving, by 5.157, if $b^2 + c^2 < a^2$,

$$M_{12} = \pi\mu b \sum_{n\ \text{odd}}^{1\to\infty} (-1)^{\frac{n+1}{2}} \frac{1 \cdot 3 \cdot 5 \cdots n}{n \cdot 2 \cdot 4 \cdot 6 \cdots (n+1)} \left(\frac{b^2 + c^2}{a^2}\right)^{\frac{1}{2}n} P_n^1(\cos\beta) \tag{3}$$

When $b^2 + c^2 > a^2$, write $[a^2/(b^2 + c^2)]^{\frac{n+1}{2}}$ for $[(b^2 + c^2)/a^2]^{\frac{1}{2}n}$.

We may compute the mutual inductance when the two loops are in any position, provided that their axes intersect, by means of 8.03 (2), 7.13 (3), and 5.24 (5). Thus if, in Fig. 8.06*b*, $b > a$, we have, taking the origin at the point of intersection and measuring θ from the a-loop axis,

$$M_{12} = a^2 \int_0^{2\pi} \int_0^{\alpha} [B_r]_b \sin\theta \, d\theta \, d\phi \tag{4}$$

If θ' is the angle measured from the b-loop axis, then substitution of 5.24 (5) for $P_n(\cos\theta')$ in 7.13 (3), where $\gamma \equiv \Theta$, $a \equiv r$, and $I = 1$, give

$$[B_r]_b = \frac{\mu \sin\beta}{2a}\sum_{n=1}^{\infty}\left(\frac{a}{b}\right)^n P_n^1(\cos\beta)$$

$$\sum_{m=0}^{n}(2-\delta_m^0)\frac{(n-m)!}{(n+m)!}P_n^m(\cos\gamma)P_n^m(\cos\theta)\cos m\phi$$

When this is inserted in (4), the ϕ-integration eliminates all but the $m = 0$ terms. By 5.154 (5) the integration with respect to θ gives

$$-\int_1^{\cos\alpha} P_n(\cos\theta)d(\cos\theta) = \frac{\sin^2\alpha P'_n(\cos\alpha)}{n(n+1)} = \frac{\sin\alpha P_n^1(\cos\alpha)}{n(n+1)}$$

so that the mutual inductance is

$$M_{12} = \pi\mu a \sin\alpha \sin\beta \sum_{n=1}^{\infty}\frac{1}{n(n+1)}\left(\frac{a}{b}\right)^n P_n^1(\cos\alpha)P_n^1(\cos\beta)P_n(\cos\gamma) \quad (5)$$

8.07. Variable Mutual Inductance.—Variable mutual inductors are frequently built in the form shown in Fig. 8.07. Each coil consists of a single layer of wire wound on a bobbin with a spherical surface. One coil rotates, relative to the other, about a common axis. Let the angle between the axes of the coils be γ, let the coils subtend angles $2A$ and $2B$ at the center, and let the radii of the two surfaces be a and b. Consider each coil as being made up of plane loops of wire, the planes being uniformly spaced with a number n per unit length. The number of turns n' per unit angle in each of the coils will then be $n'_a = n_a a \sin\alpha$ and $n'_b = n_b b \sin\beta$. The mutual inductance between elements of the two coils will then be, from 8.06 (5),

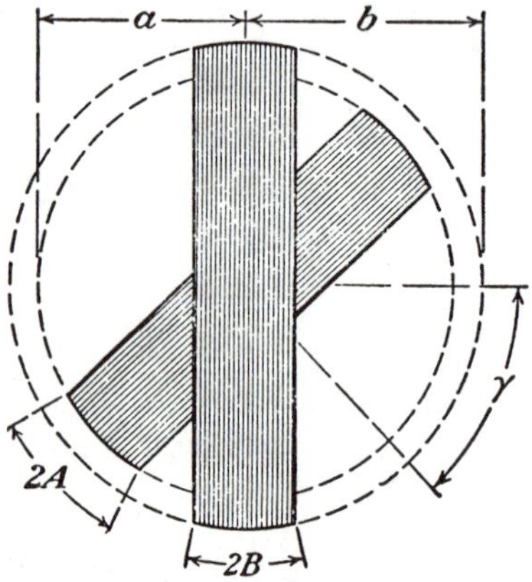

Fig. 8.07.

$$dM_{ab} = \pi a^2 b\mu n_a n_b \sin^3\alpha \sin^3\beta \sum \frac{1}{n(n+1)}\frac{a^n}{b^n}P'_n(\mu_a)P'_n(\mu_b)P_n(\cos\gamma)\,d\alpha\,d\beta$$

The total mutual inductance will be obtained by integrating on both coils; thus,

$$M_{ab} = \int_{\alpha=\frac{1}{2}\pi-A}^{\alpha=\frac{1}{2}\pi+A}\int_{\beta=\frac{1}{2}\pi-B}^{\beta=\frac{1}{2}\pi+B} dM_{ab}$$

Writing u for $\cos\alpha$ or $\cos\beta$, each integral takes the form, when 5.154 (5) is used,

$$\int (1-u^2)\frac{dP_n(u)}{du}\,du = \frac{n(n+1)}{2n+1}\int [P_{n-1}(u) - P_{n+1}(u)]\,du \qquad (1)$$

Integrating by 5.154 (3) and simplifying by repeated application of 5.154 (1), this becomes

$$f_n(A) = \frac{n}{(n-1)(n+2)}\Big\{[(n+1)-(n-1)u^2]P_n(u) - 2uP_{n-1}(u)\Big\}_{\sin A}^{-\sin A} \qquad (2)$$

When n is even, the value at both limits is the same by 5.157 so these values drop out and only odd values of n appear. Special treatment is required when $n = 1$. In this case, from (1), we have

$$f_1(A) = \left[u\left(1-\frac{u^2}{3}\right)\right]_{\sin A}^{-\sin A} = -\frac{2}{3}\sin A(3-\sin^2 A) \qquad (3)$$

Similar expressions hold for B so that

$$M_{ab} = \pi\mu a^2 b n_a n_b \sum_{n\ \text{odd}}^{1\to\infty} \frac{1}{n(n+1)}\left(\frac{a}{b}\right)^n f_n(A) f_n(B) P_n(\cos\gamma) \qquad (4)$$

From (2) and (4), the coefficients depending on the fixed dimensions of the coil can be computed and the formula then takes the form

$$M_{ab} = \sum_{n\ \text{odd}}^{1\to\infty} M_n P_n(\cos\gamma) \qquad (5)$$

where γ is the angle between the axes.

8.08. Self-inductance.—From 8.02 (3), we see that the magnetic energy density of a single circuit, carrying a current I_1, is proportional to B^2, and since B is proportional to I_1, we must have

$$W_B = \frac{1}{2\mu}\int_V B^2\,dv = \frac{1}{2}L_{11}I_1^2 \qquad (1)$$

We define the constant of proportionality L_{11} to be the self-inductance of the circuit. Another useful formula for self-inductance may be obtained by substituting 7.01 (3) in 8.02 (2), giving

$$W_B = \tfrac{1}{2}\int \boldsymbol{i}\cdot\boldsymbol{A}\,dv = \tfrac{1}{2}L_{11}I^2 \qquad (1.1)$$

where the integration must be taken throughout the region in which the current flows.

In the mks system the inductance unit is the henry. If the circuit consisted of infinitely thin wire carrying a finite current, its self-inductance would be infinite. This is evident because $\boldsymbol{B}$ is inversely proportional to the distance from the wire when we are close to it. Thus, for a finite length of wire, the integral in (1) is logarithmically infinite since dv may be written $r\,dr\,d\theta$. This means that in computing self-inductance we must consider the actual dimensions of the wire and take the current density i nowhere infinite. We may therefore treat the wire as a bundle of current filaments of infinitesimal cross-sectional area dS each carrying a current $i\,dS$. When the current is established in such a wire, the magnetic flux about each current filament cuts across neighboring filaments and, by Faraday's law [8.00 (1)], sets up in each an electromotance opposing the current, of amount $\mathcal{E} = -dN/dt$ where dN/dt is the flux cutting the circuit in one second. To maintain the current I_1, the source of power must do, in each second, an amount of work equal to

$$\mathcal{E}I_1 = I_1\frac{dN_1}{dt} \tag{2}$$

in addition to any work against resistance. This energy, stored in the magnetic field each second, equals dW/dt so that we have, from (1) and (2),

$$I_1\frac{dN_1}{dt} = L_{11}I_1\frac{dI_1}{dt}$$

Canceling out the I_1 and integrating from no current to final current, we have

$$N_{11} = L_{11}I_1 \tag{3}$$

Thus we may also define the self-inductance of a circuit as the change in the flux linking the circuit when the current changes by one unit. The electromotance in the circuit caused by the changing current is

$$\mathcal{E}_1 = -\frac{dN_{11}}{dt} = -L_{11}\frac{dI_1}{dt} \tag{4}$$

8.09. Computation of Self-inductance. Thin Wire.—From what has been said, it appears natural, when computing self-inductance, to treat separately the region lying outside the wire and that lying inside. The contribution of the latter is frequently negligible, and when it is not, but when the radius of the wire is small compared with other dimensions of the circuit, we may assume that the field outside is the same as if the current were concentrated at the axis of the wire and that inside the field is identical with that inside a long straight wire of the same size, carrying the same current. The lines of force near the surface are approximately circles about the axis of the wire so that all the flux outside it links the axial filament with any line drawn parallel to it on the surface of the wire. To get this part of the self-inductance we need, therefore, to find the mutual inductance between two parallel curvilinear circuits spaced a distance apart equal to the radius of the wire.

Inside the wire, the magnetomotance around any circle of radius r

about the axis depends, from 7.28 (3), only on the current inside the circle. Since, from symmetry, B depends only on r, we have, from 7.14,

$$B = \frac{\mu I_i}{2\pi r} = \frac{\mu r I}{2\pi a^2} \tag{1}$$

where I is the total current and a is the radius of the wire. The energy inside the wire is, from 8.08 (1),

$$W_i = \frac{1}{2\mu}\int_V B^2\,dv = \frac{\mu I^2 L}{8\pi^2 a^4}\int_0^a r^2 2\pi r\,dr = \frac{\mu I^2 L}{16\pi} = \frac{1}{2}L'_{11}I^2$$

where L is the length of the wire. The internal self-inductance, per unit length, becomes then

$$L'_{11} = \frac{\mu'}{8\pi} \tag{2}$$

where μ' is the permeability of the wire.

8.10. Self-inductance of Circular Loop.—Using the method of the last article and the result of a previous example, we shall now compute the self-inductance of a loop of radius b of wire of permeability μ' and radius a. The length of the wire is $2\pi b$ so that that part of the self-inductance due to the interior of the wire is, from 8.09 (2),

$$L'_{11} = \tfrac{1}{4}\mu' b \tag{1}$$

As shown in 8.09, we can get the rest of the self-inductance by applying the result of 8.06 (1) to two circuits, one coinciding with the axis of the wire and one with its inner edge. In this case, from 8.06 (2), setting $c = 0$, and taking $a \ll b$, we have

$$1 - k^2 = 1 - \frac{4b(b-a)}{(b+b-a)^2} = \frac{a^2}{(2b-a)^2} \approx \left(\frac{a}{2b}\right)^2 = k'^2$$

So that k is approximately unity.
When $k \approx 1$, the E of 8.06 (1) becomes

$$E = \int_0^{\frac{\pi}{2}}(1 - \sin^2\theta)^{\frac{1}{2}}\,d\theta = \int_0^{\frac{\pi}{2}}\cos\theta\,d\theta = 1$$

To get K, let $\phi = \frac{1}{2}\pi - \theta$, and split the integral into two parts

$$K = \int_0^{\frac{\pi}{2}}\frac{d\theta}{(1 - k^2\sin^2\theta)^{\frac{1}{2}}} = \int_0^{\phi_0}\frac{d\phi}{(1 - k^2\cos^2\phi)^{\frac{1}{2}}} + \int_{\phi_0}^{\frac{\pi}{2}}\frac{d\phi}{(1 - k^2\cos^2\phi)^{\frac{1}{2}}}$$

where $(1 - k^2) \ll \phi_0 \ll 1$. Now put $\sin\phi = \phi$ in the first integral where ϕ_0 is small, and then put $k = 1$ in both integrals, and we have

$$K = \int_0^{\phi_0}\frac{d\phi}{(k'^2 + \phi^2)^{\frac{1}{2}}} + \int_{\phi_0}^{\frac{\pi}{2}}\frac{d\phi}{\sin\phi}$$

Integrate by *Pc* 126*a* and 291 or by *Dw* 200.01 and 432.10; then put $\phi_0 = \tan\phi_0$, and neglect k'^2 compared with ϕ_0^2, and we have

$$K = \ln\left\{\frac{[\phi_0 + (\phi_0^2 + k'^2)^{\frac{1}{2}}]}{k'}\right\} - \ln\tan\frac{1}{2}\phi_0 = \ln\frac{2\phi_0}{k'}\frac{2}{\phi_0} = \ln\frac{4}{k'} = \ln\frac{8b}{a}$$

Substituting in 8.06 (1), and adding the internal inductance term (1), we get

$$L_{11} = b\left[\mu\left(\ln\frac{8b}{a} - 2\right) + \frac{1}{4}\mu'\right] \tag{2}$$

where μ' is the permeability inside the wire and μ is that outside.

8.11. Self-inductance of Solenoid.—If we treat a solenoid as an infinitely thin cylindrical current sheet of uniform strength, all lines of flow lying in planes normal to its axis, we may compute the self-inductance rigorously by using 8.06 (1) to find the mutual inductance between elements of the shell and integrating over the shell. The result is complicated, so that where only rough values are necessary and the solenoid is long compared with its diameter, the following approximation is useful. We treat the field inside the solenoid as uniform across any section taken normal to the axis. To compute the self-inductance by 8.08 (3), we must know the flux linking any element of the shell and then integrate over the shell. From 7.15 (4), the flux through the $n\,dx$ turns in dx is, per unit current,

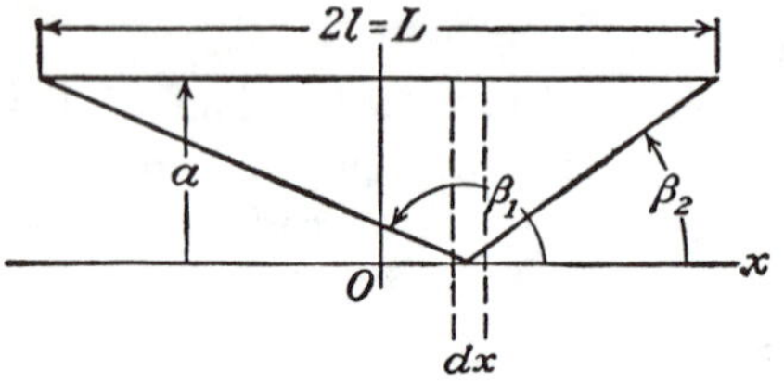

FIG. 8.11.

$$dN_1 = \frac{1}{2}\pi\mu a^2 n^2\left\{\frac{l-x}{[(l-x)^2+a^2]^{\frac{1}{2}}} + \frac{l+x}{[(l+x)^2+a^2]^{\frac{1}{2}}}\right\}dx$$

Integrating by *Pc* 132 or *Dw* 201.01 from $-l$ to $+l$, we have

$$L_{11} = \pi\mu a^2 n^2[(4l^2+a^2)^{\frac{1}{2}} - a] = \pi\mu a^2 n^2[(L^2+a^2)^{\frac{1}{2}} - a] \tag{1}$$

8.12. Self-inductance of Bifilar Lead.—We shall now calculate rigorously, by the method of 8.08 (1.1), the self-inductance per unit length of a bifilar lead in which the current in one direction is uniformly distributed in a wire of radius a and the return current in a wire of radius c, parallel to the first, the axes of the wires being a distance b apart. The permeability of the wire and the region surrounding them will be taken as unity. Clearly, the vector potential and the current density have z components only. If I is the total current, then the current densities are

$$i_1 = \frac{I}{\pi a^2} \qquad i_2 = \frac{-I}{\pi c^2} \tag{1}$$

To obtain the vector potential A' inside the wire due to its own current, we write 7.02 (2) in cylindrical coordinates with the aid of 3.05 (2) and

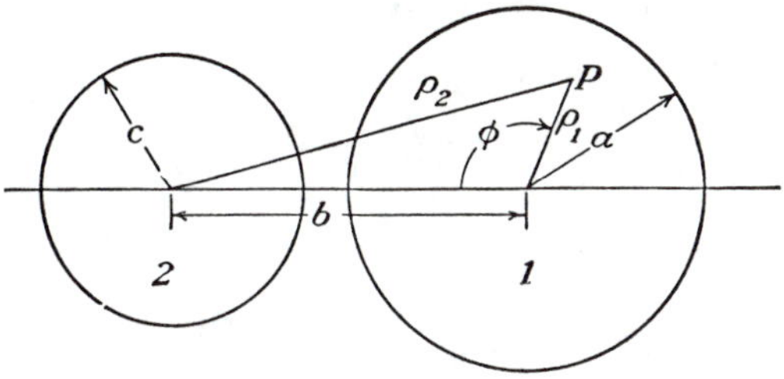

FIG. 8.12.

obtain, since A' is a function of ρ only,

$$\frac{\partial}{\rho\,\partial\rho}\left(\rho\frac{\partial A'}{\partial\rho}\right) = -\mu i = -\frac{\mu I}{\pi a^2} \qquad \text{or} \qquad +\frac{\mu I}{\pi c^2} \tag{2}$$

Integrating from 0 to ρ_1 twice, we have

$$\rho_1\frac{\partial A_1'}{\partial\rho_1} = -\frac{\mu I\rho_1^2}{2\pi a^2} \qquad A_1' = C_1 - \frac{\mu I\rho_1^2}{4\pi a^2} \tag{3}$$

Outside the wire, putting $i = 0$ in (2), integrating twice from the outer boundary of the wire to ρ_1, and observing that the lower limit in each integration is obtained by putting $\rho_1 = a$ in (3), we have

$$\rho_1\frac{\partial A_1''}{\partial\rho_1} + \frac{\mu I}{2\pi} = 0 \qquad A_1'' = C_1 - \frac{\mu I}{4\pi} - \frac{\mu I}{2\pi}\ln\frac{\rho_1}{a} \tag{4}$$

Similarly,

$$A_2' = C_2 + \frac{\mu I\rho_2^2}{4\pi c^2} \qquad A_2'' = C_2 + \frac{\mu I}{4\pi} + \frac{\mu I}{2\pi}\ln\frac{\rho_2}{c} \tag{5}$$

To make (4) and (5) agree with 7.09 (1), at infinity we must take

$$2\pi(C_1 + C_2) + \mu I\ln\frac{a}{c} = 0 \tag{6}$$

$$L_{11}I^2 = \int_{S_1} i_1(A_1' + A_2'')\,dS_1 + \int_{S_2} i_2(A_2' + A_1'')\,dS_2 \tag{7}$$

We shall calculate the first integral only and evaluate the second from symmetry. Thus we have, using *PC* 523 or *Dw* 868.4 and canceling out C_1,

$$(L_{11}I^2)_1 = \frac{\mu I^2}{4\pi^2a^2}\int_0^a\int_0^{2\pi}\left[-\frac{\rho_1^2}{a^2} + 1 + \ln\left(\frac{b^2+\rho_1^2-2b\rho_1\cos\phi}{a^2}\right)\right]\rho_1\,d\rho_1\,d\phi$$

$$= \frac{\mu I^2}{2\pi a^2}\left(-\int_0^a\frac{\rho_1^3}{a^2}\,d\rho_1 + \int_0^a\rho_1\,d\rho_1 + \ln\frac{b^2}{a^2}\int_0^a\rho_1\,d\rho_1\right)$$

$$= \frac{\mu I^2}{4\pi}\left(\frac{1}{2} + 2\ln\frac{b}{a}\right) \tag{8}$$

Similarly, for the second wire,

$$(L_{11}I^2)_2 = \frac{\mu I^2}{4\pi}\left(\frac{1}{2} + 2\ln\frac{b}{c}\right) \tag{9}$$

Adding (8) and (9), we obtain

$$L_{11} = \frac{\mu}{4\pi}\left(1 + 2\ln\frac{b^2}{ac}\right) \tag{10}$$

If the wires are alike, $a = c$, so that

$$L_{11} = \frac{\mu}{4\pi}\left(1 + 4\ln\frac{b}{a}\right) \tag{11}$$

8.13. Energy of *n* Circuits.—If we have circuits carrying currents I_1, I_2, . . . , I_n and the induction at any point due to each circuit alone is $\boldsymbol{B}_1$, $\boldsymbol{B}_2$, . . . , $\boldsymbol{B}_n$, then the total resultant induction is the vector sum of these and the total energy in the field from 8.02 (3) is

$$W = \frac{1}{2\mu}\int_V B^2\,dv = \frac{1}{2\mu}\int_V\left(\sum_{i=1}^{n}\boldsymbol{B}_i\right)\cdot\left(\sum_{j=1}^{n}\boldsymbol{B}_j\right)dv \tag{1}$$

$$= \frac{1}{2\mu}\int_V\left(\sum_{i=1}^{n}B_i^2 + \sum_{i=1}^{n}\sum_{j=1}^{n}\boldsymbol{B}_i\cdot\boldsymbol{B}_j\right)_{i\neq j}dv$$

in the double summation, both $\boldsymbol{B}_i\cdot\boldsymbol{B}_j$ and $\boldsymbol{B}_j\cdot\boldsymbol{B}_i$ occur so that, from 8.08 (1) and 8.03 (7), this may be written

$$W = \tfrac{1}{2}\left(\sum_{i=1}^{n}L_iI_i^2 + \sum_{i=1}^{n}\sum_{j=1}^{n}M_{ij}I_iI_j\right)_{i\neq j}$$

$$= \tfrac{1}{2}(L_1I_1^2 + 2M_{12}I_1I_2 + L_2I_2^2 + \cdots) \tag{2}$$

As we saw in 8.01 (4), the force or torque tending to increase any coordinate of position θ of the *i*th circuit is obtained by taking the positive derivative of the field energy with respect to this coordinate; thus,

$$F_i = \frac{1}{2}I_i\left(I_i\frac{\partial L_i}{\partial\theta} + 2\sum_{j=1}^{n}I_j\frac{\partial M_{ij}}{\partial\theta}\right)_{j\neq i} \tag{3}$$

8.14. Stresses in a Magnetic Field.—We saw in 8.02 (3) and 2.08 (2) that the energy densities dW/dv in magnetic and electric fields are given by

$$\left(\frac{dW}{dv}\right)_m = \frac{B^2}{2\mu} \quad\text{and}\quad \left(\frac{dW}{dv}\right)_e = \frac{D^2}{2\epsilon} \tag{1}$$

From 7.01 (1), we know that $\nabla\cdot\boldsymbol{B} = 0$ is true everywhere, and from 3.02 (1) that $\nabla\cdot\boldsymbol{D} = 0$ is true in the absence of electric charges. It follows that a system of stresses that will give equilibrium in an electric field will do the same in a magnetic field if we replace $\boldsymbol{D}$ by $\boldsymbol{B}$ and ϵ by μ. Thus we see, from 1.14, that we may visualize magnetic forces as transmitted by a tension T along the lines of force and a pressure P at right angles to them. Both T and P, per unit area, are given numerically by

$$|T| = |P| = \frac{B^2}{2\mu} \tag{2}$$

If the permeability is a function of the density τ, we proceed exactly as in 2.09 except that we deal with a section of permeable material in a magnetic circuit instead of with a dielectric slab in a capacitor. This leads to a conclusion, analogous to 2.09 (2), that there is required, in addition to the above stresses, a hydrostatic pressure of amount

$$P_h = -\frac{B^2}{2\mu^2}\tau\frac{\partial\mu}{\partial\tau} \tag{3}$$

To find the force on the plane boundary between two materials of different permeabilities μ' and μ'' in a magnetic field, we add up the normal components of the stresses given by (2) and (3) and in the same way in which we obtained 1.17 (7) and 2.09 (3) we get

$$F_n = \frac{1}{2}\left[\frac{\mu' - \mu''}{\mu'}\left(\frac{B_t'^2}{\mu'} + \frac{B_n'^2}{\mu''}\right) - \frac{B'^2\tau'}{\mu'^2}\frac{\partial\mu'}{\partial\tau'} + \frac{B''^2\tau''}{\mu''^2}\frac{\partial\mu''}{\partial\tau''}\right] \tag{4}$$

This is the normal stress directed from μ' to μ''. The normal and tangential components of the inductions $\boldsymbol{B}'$ and $\boldsymbol{B}''$ in the two mediums are indicated by the subscripts n and t, respectively. For a boundary between the medium μ and a vacuum, we set $\mu'' = \mu_v$, $\mu' = \mu_v K_m$, and $\partial\mu''/\partial\tau'' = 0$ and obtain

$$F_n = \frac{B'^2}{2\mu_v K_m^2}\left[(K_m - 1) - \tau\frac{\partial K_m}{\partial\tau}\right] + \frac{(K_m - 1)^2 B_n'^2}{2\mu_v K_m^2} \tag{5}$$

where

$$B'^2 = B_n'^2 + B_t'^2 \tag{6}$$

The formulas derived in this article hold fairly well for ordinary substances but break down for the ferromagnetic materials discussed in Chap. IX. The latter give many complicated effects, such as changes of shape without change of volume, which cannot all be included in any simple formula.

8.15. Energy of a Permeable Body in a Magnetostatic Field.—The last article shows that the force acting on a body of permeability μ placed in a medium of permeability μ_v in the presence of a magnetostatic field of induction $\boldsymbol{B}$ produced by currents fixed in magnitude and position is the same numerically as that calculated in 3.13 (6) for a dielectric body $\epsilon_v = \mu_v$ placed in an electric field of displacement $\boldsymbol{D} = \boldsymbol{B}$ produced by fixed electric charges. It appears in 8.01 (4) that, in the magnetic case, the force or torque acting in the direction θ is obtained from the magnetic field energy by taking its positive derivative. Thus

$$F = +\frac{\partial W_B}{\partial \theta} \tag{1}$$

A comparison with 3.13 (5) shows that if the magnetic induction produced in a region of permeability μ_v by a fixed set of currents is $\boldsymbol{B}$ inside a volume v before it is occupied by a body of permeability μ and $\boldsymbol{B}_i$ after it is occupied then the energy required to place this body in position is given by the following integral over its volume

$$W_B = \frac{1}{2}\int_{\mathrm{v}}\left(\frac{1}{\mu_v} - \frac{1}{\mu}\right)\boldsymbol{B}\cdot\boldsymbol{B}_i\, d\mathrm{v} \tag{2}$$

It is assumed that the body is isotropic, that μ is independent of B_i, that only induced magnetism is present, and that μ and B_i are the final values in $d\mathrm{v}$. This becomes, in terms of magnetization, by 7.20 (9),

$$W_B = \tfrac{1}{2}\int_{\mathrm{v}} \boldsymbol{M}\cdot\boldsymbol{B}\, d\mathrm{v} \tag{3}$$

Problems

Problems marked *C* in the following list are taken from the Cambridge examination questions as reprinted by Jeans, with permission of the Cambridge University Press.

1. Show that the coefficient of mutual inductance between a long straight wire and a coplanar equilateral triangular loop of wire is $[\mu_v/(3^{\frac{1}{2}}\pi)][(a+b)\ln(1+a/b) - a]$, where a is the altitude of the triangle and b is the distance from the straight wire to the side of the triangle parallel to and nearest it.

2. A circular loop of wire of radius a lies with its plane parallel to, and at a distance d from, the plane face of a semi-infinite block of material of relative permeability K_m. Show that the increase in the self-inductance of the loop due to the presence of the block is, if c is written for $a(a^2+d^2)^{-\frac{1}{2}}$,

$$\mu_v a\frac{K_m-1}{K_m+1}\left[\left(\frac{2}{c}-c\right)K(c) - \frac{2}{c}E(c)\right]$$

3. A sphere of radius b and very large permeability is concentric with a wire loop of radius a. Show that the additional self-inductance due to its presence is

$$\mu_v\pi b\sum_{n=0}^{\infty}\left[\frac{1\cdot3\cdot5\cdots(2n+1)}{2\cdot4\cdot6\cdots(2n+2)}\right]^2\left(\frac{2n+2}{2n+1}\right)^2\left(\frac{b}{a}\right)^{4n+2} = \frac{2\mu b^3}{a^2}K\left(\frac{b^2}{a^2}\right)$$

4. Show that the coefficient of mutual inductance between two coplanar concentric loops of wire whose radii are a and b is

$$\mu_v\pi a\sum_{n=0}^{\infty}\left[\frac{1\cdot3\cdot5\cdots(2n+1)}{2\cdot4\cdot6\cdots(2n+2)}\right]^2\frac{2n+2}{2n+1}\left(\frac{b}{a}\right)^{2(n+1)} = 2\mu a\left[K\left(\frac{b}{a}\right) - E\left(\frac{b}{a}\right)\right]$$

5. A circuit consists of a thin conducting cylindrical shell of radius a and a return wire of radius b inside. The axis of shell and wire are parallel but not coincident. Show that the self-inductance per unit length is $\frac{1}{8}(\mu_v/\pi)[1+4\ln(a/b)]$.

6. A plane circuit of any form is pressed into the face of a semi-infinite block of permeability μ so that it is coplanar with the face. Neglecting the field inside the wire, show that the block increases the self-inductance by a factor $2\mu/(\mu + \mu_v)$.

7*C*. A long straight current intersects at right angles a diameter of a circular current, and the plane of the circle makes an acute angle α with the plane through this diameter and the straight current. Show that the coefficient of mutual induction is $\mu_v[c \sec \alpha - (c^2 \sec^2 \alpha - a^2)^{\frac{1}{2}}]$ or $\mu_v c \tan (\frac{1}{4}\pi - \frac{1}{2}\alpha)$, according as the straight current passes without or within the circle, a being the radius of the circle and c the distance of the straight current from its center.

8. A cylindrical wire of permeability μ_1 is concentric with a thick layer of insulation of permeability μ_1. This wire is placed in a liquid of permeability μ_2 normal to a field of induction B. Show that the transverse force per unit length on the surface of the insulation is $[(\mu_2 - \mu_1)/(\mu_2 + \mu_1)]IB$ and that it is $2\mu_1 IB/(\mu_1 + \mu_2)$ on the metallic surface giving a resultant total force of IB.

9. A small sphere of radius a and permeability μ is placed near a circuit which, when carrying unit current, would produce a magnetic induction B at the point where the center of the sphere is placed. Show that the presence of the sphere increases the self-inductance of the circuit by approximately $4\pi a^3 B^2(K_m - 1)/[\mu_v(K_m + 2)]$.

10. A cylindrical shell of permeability μ and internal and external radii a and b surrounds two parallel wires, carrying currents in opposite directions, which are coplanar with, parallel to, and each at a distance c from its axis. Show that the additional self-inductance per unit length of the circuit due to the shell is, if $\mu = \mu_v K_m$,

$$\frac{2\mu_v(K_m^2 - 1)}{\pi} \sum_{n=0}^{\infty} \frac{(a^{4n+2} - b^{4n+2})(c/a)^{4n+2}}{(2n+1)[(K_m - 1)^2 a^{4n+2} - (K_m + 1)^2 b^{4n+2}]}$$

11*C* (Modified). A circular loop of radius a is concentric with a spherical shell of permeability μ and radii b and c. Show that the additional self-inductance of the loop due to the presence of the shell is approximately, if $\mu = \mu_v K_m$,

$$\frac{\mu_v \pi a^4 (K_m - 1)(K_m + 2)(c^3 - b^3)}{2b^3[(K_m + 2)(2K_m + 1)c^3 - (K_m - 1)^2 2b^3]}$$

where a is very small compared with b and c.

12. The coordinates of a loop of wire of radius a in the prolate spheroidal system are $\xi = \xi_1$ and $\eta = \eta_1$. A prolate spheroid of permeability μ is inserted so that its surface is $\eta = \eta_0$. Show that the additional self-inductance of the loop is, if $\mu = \mu_v K_m$,

$$\frac{\pi\mu_v a^2 (K_m - 1)}{c_2} \sum_{n=1}^{\infty} \frac{2n+1}{n^2(n+1)^2} \frac{[P_n^1(\xi_1)]^2 [Q_n^1(\eta_1)]^2 P_n^1(\eta_0) P_n(\eta_0)}{Q_n^1(\eta_0)P_n(\eta_0) - K_m Q_n(\eta_0) P_n^1(\eta_0)}$$

13. The coordinates of two loops of wire of radii a_1 and a_2 in the prolate spheroidal system are η_1, ξ_1, and η_2, ξ_2. A prolate spheroid of permeability μ is inserted in the loops so that its surface is $\eta = \eta_0$. Show that the additional mutual inductance due to the spheroid is, if $\mu = \mu_v K_m$,

$$\frac{\pi\mu_v a_1 a_2 (K_m - 1)}{c_2} \sum_{n=1}^{\infty} \frac{2n+1}{n^2(n+1)^2} \frac{P_n^1(\xi_1) Q_n^1(\eta_1) P_n^1(\xi_2) Q_n^1(\eta_2) P_n^1(\eta_0) P_n(\eta_0)}{Q_n^1(\eta_0)P_n(\eta_0) - K_m Q_n(\eta_0) P_n^1(\eta_0)}$$

14. Show that the additional mutual inductance between two parallel coaxial loops of wire of radii a and c with a distance s between centers, due to the insertion, coaxially, of an infinite cylinder of radius b and permeability μ, is, if $\mu = \mu_v K_m$,

$$2\mu_v ac\int_0^\infty \Phi(k)K_1(ka)K_1(kc)\cos ks\,dk$$

where $\Phi(k)$ is given in problem 28, Chap. VII.

15. Show that the additional self-inductance of a loop of wire of radius a, due to the insertion, coaxially, of an infinite cylinder of radius b and permeability μ, is

$$2\mu_v a^2\int_0^\infty \Phi(k)[K_1(ka)]^2\,dk$$

where $\Phi(k)$ is given in problem 28, Chap. VII.

16. Two circuits are formed by that part of the cylinder $\rho = a_1$ lying between $z = +c_1$ and $z = -c_1$ and that part of the cylinder $\rho = a_2$ lying between $z = +c_2$ and $z = -c_2$, where $a_2 > a_1$. Assuming that the currents will distribute themselves uniformly on the bands, show that the mutual inductance between them is

$$\frac{2\mu a_1 a_2}{c_1 c_2}\int_0^\infty k^{-2}I_1(ka_1)K_1(ka_2)\sin kc_1\sin kc_2\,dk$$

17. Using problem 30 of Chap. VII, show that if the plane of a circular loop of wire of radius a is parallel to, and at a distance b from, the face of an infinite plate of material of permeability μ and thickness t the self-inductance of the loop is increased by the amount

$$\Delta L_{11} = \beta\left[M_0 - (1-\beta^2)\sum_{n=1}^{\infty}\beta^{2(n-1)}M_n\right]$$

where $M_n = \dfrac{2\mu_v a}{k_n}\left[\left(1-\dfrac{k_n^2}{2}\right)K_n - E_n\right]$, $\beta = \dfrac{\mu-\mu_v}{\mu+\mu_v}$, $k_n^2 = \dfrac{a^2}{a^2+(nt+b)^2}$, and K_n and E_n are the complete elliptic integrals of the first and second kinds of modulus k_n.

18. An infinite plate of thickness t and permeability μ is inserted between the loops of Fig. 8.06a, with its faces parallel to them. Show that the mutual inductance between the loops is now given by

$$M = 2\mu_v(ab)^{\frac{1}{2}}(1-\beta^2)\sum_{n=0}^{\infty}\frac{\beta^{2n}}{k_n}\left[\left(1-\frac{k_n^2}{2}\right)K_n - E_n\right]$$

K_n and E_n are complete elliptic integrals of modulus k_n, where

$$k_n^2 = \frac{4ab}{(a+b)^2+(c+2nt)^2} \qquad \text{and} \qquad \beta = \frac{\mu-\mu_v}{\mu+\mu_v}$$

19. Show that the self-inductance of a solenoid of diameter d, length c, N turns, and small pitch is $\frac{1}{3}\mu N^2 d\{\csc\alpha[(\tan^2\alpha - 1)E + K] - \tan^2\alpha\}$ where $d/c = \tan\alpha$ and E and K are complete elliptic integrals of modulus $\sin\alpha$.

20. A wire loop of radius a is coaxial with a solenoid of radius b and n turns per unit length. If the greatest distance from a point on the loop to the near end of the solenoid is c and to the far end it is d, show from 7.10 (4) without further integration that the force between loop and solenoid when they carry currents I and I' is

$$2\mu nII'(ab)^{\frac{1}{2}}\{k_1^{-1}[(1-\tfrac{1}{2}k_1^2)K_1 - E_1] - k_2^{-1}[(1-\tfrac{1}{2}k_2^2)K_2 - E_2]\}$$

where the modulus of the complete elliptic integrals is $k_1 = 2(ab)^{\frac{1}{2}}/c$ and $k_2 = 2(ab)^{\frac{1}{2}}/d$.

21. Two coaxial solenoids of diameter d and small pitch, with n and m turns per unit length, are placed so that the distance between near ends is b and between far ends is c. If one is of length a, show that their mutual inductance is

$$\tfrac{1}{6}\mu nmd^3 \sum_{n=1}^{4} (-1)^n \sec \alpha_n[(1 - \tan^2 \alpha_n)E(k_n) + \tan^2 \alpha_n K(k_n)]$$

where $\tan \alpha_1 = c/d$, $\tan \alpha_2 = (c - a)/d$, $\tan \alpha_3 = b/d$, $\tan \alpha_4 = (a + b)/d$, and $k_n = \cos \alpha_n$.

22. Show that, if the currents in problem 21 are I and I', the force between the coils is

$$\tfrac{1}{2}\mu d^2 nmII' \sum_{n=1}^{4} (-1)^n \sin \alpha_n \sec^2 \alpha_n[E(k_n) - K(k_n)]$$

23. Show by 5.35 (4) that the mutual inductance between two loops of radii a and b with parallel axes a distance c apart and planes a distance d apart is, if $c > a + b$,

$$2\mu ab \int_0^\infty I_1(ka)I_1(kb)K_0(kc) \cos kd \, dk$$

24. Show that, if $a > b + c$, the mutual inductance in the last problem is

$$2\mu ab \int_0^\infty K_1(ka)I_1(kb)I_0(kc) \cos kd \, dk$$

25. Show with the aid of 5.298 (7) that the results of problems 23 and 24 may be written

$$\pi\mu ab \int_0^\infty J_1(ka)J_1(kb)J_0(kc)e^{-kd} \, dk$$

26. Show from problem 23 that the mutual inductance between two solenoids of radii a and b, length $2A$ and $2B$, and m and n turns per unit length, with parallel axes a distance c apart and central loop planes a distance d apart is, when $c > a + b$,

$$\frac{2}{AB} \mu abmn \int_0^\infty k^{-2}I_1(ka)I_1(kb)K_0(kc) \sin kA \sin kB \cos kd \, dk$$

27. If coil b of the last problem lies inside coil a, show from problem 24 that

$$M = \frac{2}{AB} \mu abmn \int_0^\infty k^{-2}K_1(ka)I_1(kb)I_0(kc) \sin kA \sin kB \cos kd \, dk$$

28. From problem 25 write the mutual inductances of problems 26 and 27 in the form

$$\frac{1}{AB}\pi\mu abmn \int_0^\infty k^{-2}J_1(ka)J_1(kb)I_0(kc) \sinh kA \sinh kB \, e^{-kd} \, dk$$

29. The most general position of a loop of radius a relative to one of radius b may be stated in terms of the shortest distance c between their axes, the distances A and B of the centers of each from c, and the angle β between their axes measured from each center toward c. Show that the mutual inductance is

$$\frac{\mu ab}{4\pi} \int_0^{2\pi} \int_0^{2\pi} \frac{(\cos \phi \cos \phi' \cos \beta + \sin \phi \sin \phi') \, d\phi \, d\phi'}{[D^2 + a^2 + b^2 + 2ab(\cos \phi \cos \phi' + \sin \phi \sin \phi' \cos \beta) - 2f(\phi, \phi')]^{\frac{1}{2}}}$$

where $D = (c^2 + A^2 + B^2 - 2AB \cos \beta)^{\frac{1}{2}}$ is the distance between their centers and

$$f(\phi, \phi') = a(c \cos \phi + B \sin \phi \sin \beta) + b(c \cos \phi' + A \sin \phi' \sin \beta)$$

30. A bifilar lead of two parallel wires with a spacing c is placed symmetrically in the gap of width a between two parallel sheets of infinite permeability with its plane normal to them. Show that the additional self-inductance per unit length due to the presence of the sheets is $(\mu/\pi) \ln \{[2a \tan (\frac{1}{2}\pi c/a)]/(\pi c)\}$.

31. Obtain a series solution of problem 25 by first establishing, from Watson, page 148, and HMF 15.4.14 and 8.2.5, the formula

$$J_1(ak)J_1(bk) = \sum_{m=0}^{\infty} \frac{(-1)^m(a^2-b^2)^{m+1}}{(m+1)!(m+2)!} P^1_{m+1}\left(\frac{a^2+b^2}{a^2-b^2}\right)\left(\frac{k}{2}\right)^{2m+2}$$

Substitute this in the integral of problem 25 and integrate the result by IT I (9), page 182. Thus show that the mutual inductance is

$$2\pi\mu ab \sum_{m=0}^{\infty} \frac{(-1)^m(2m+1)!!}{(2m+4)!!} \frac{(a^2-b^2)^{m+1}}{r^{2m+3}} P^1_{m+1}\left(\frac{a^2+b^2}{a^2-b^2}\right) P_{2m+2}(\cos\alpha)$$

where the angle between r, the line joining loop centers, and either axis is α.

32. If $a = b$ in the preceding problem, show that the mutual inductance becomes

$$2\pi\mu a \sum_{m=0}^{\infty} \frac{(2m+2)!(-1)^m(2m+1)!!}{m!(m+1)!(2m+4)!!}\left(\frac{a}{r}\right)^{2m+3} P_{2m+2}(\cos\alpha)$$

33. Apply the method of problem 31 to problem 28 and thus show that the mutual inductance is

$$\frac{\pi}{8AB}\mu abmn \sum_{s=0}^{3}\sum_{m=0}^{\infty} \frac{(-1)^{m+s}(2m-1)!!}{(2m+2)(2m+4)!!} \frac{(a^2-b^2)^{m+1}}{r_s^{2m+1}} P^1_{m+1}\left(\frac{a^2+b^2}{a^2-b^2}\right) P_{2m}(\cos\alpha_s)$$

where if $p_0 = d + A + B$, $p_1 = d + A - B$, $p_2 = d - A - B$, and $p_3 = d - A + B$, then $r_s^2 = c^2 + p_s^2$, α_s is the angle between r_s and either coil axis, and $d > A + B$.

34. If $a = b$ in the preceding problem, show that the mutual inductance becomes

$$\frac{\pi}{8AB}\pi\mu a^3 mn \sum_{s=0}^{3}\sum_{m=0}^{\infty} \frac{(2m+2)!(-1)^{m+s}(2m-1)!!}{(2m+2)m!(m+1)!(2m+4)!!}\left(\frac{a}{r_s}\right)^{2m+1} P_{2m}(\cos\alpha_s)$$

References

All the references of Chap. VII contain some of the material of this chapter. In addition there is one very important reference.

GROVER, F. W.: "Inductance Calculations," Dover, 1962. This contains the most extensive tables and instructions for making all kinds of inductance calculations.

CHAPTER IX

MAGNETISM

9.00. Paramagnetism and Diamagnetism.—In most substances, with some notable exceptions to be considered in Art. 9.04, the magnetic permeability depends very little on the field strength so that taking it constant, as we have done, introduces no appreciable error. Unlike the relative capacitivity, the relative permeability may be either greater or less than 1. Substances in which it is greater are called paramagnetic, and those in which it is less diamagnetic. Let us consider the forces acting on such bodies when placed in the field of a fixed circuit carrying a constant current. From 8.01 (4) and 8.03 (4) or 8.08 (3), we see that, if any infinitesimal displacement or rotation of the circuit will increase the flux through it, then there is a force or torque trying to produce this motion. If there are any bodies in the field of this circuit whose displacement or rotation will increase the flux through it then, by Newton's law of reaction, there will be corresponding forces or torques acting on these bodies.

In 7.29, we saw that the magnetic induction or flux density and the permeability are related in magnetic circuits in exactly the same way as the current density and the electric conductivity in electric circuits. Thus we may state theorems 3 and 4 of 6.05 in a form applicable to magnetic circuits. If the permeability of any element in the magnetic field of an electric current is increased or decreased, the reluctance of the magnetic circuit is decreased or increased, respectively. As we have seen, these are forces acting to increase the flux and hence to decrease the reluctance. Thus, in an inhomogeneous field, there is a tendency for bodies that are more paramagnetic or less diamagnetic than the ambient medium to move toward the more intense parts of the field, and vice versa. If a paramagnetic or diamagnetic body of elongated shape is placed in a uniform field, there is a torque tending to set its axis parallel to the field. This can be seen in the case of spheroids by substituting μ for ϵ, μ_v for ϵ_v, and H or B/μ for E in problem 84, Chap. V, and observing that the torque formula contains $H_1 - H_2$ which has the factor $\mu - \mu_v$ in it so that the torque depends on $(\mu - \mu_v)^2$ and has the same sign for $\mu > \mu_v$ and $\mu < \mu_v$.

Quantum mechanics gives a theoretical basis for the empirical fact that the permeability of diamagnetic bodies is usually independent of temperature. For weakly paramagnetic substances, the permeability is often independent of temperature. In strongly paramagnetic, but not ferromagnetic, substances, the permeability usually depends on temperature, obeying the equation

$$\mu = \mu_v + \frac{\mu_v C}{T + \theta} \tag{1}$$

where C and θ are constants and T is the absolute temperature. This empirical relation is called Curie's law. A theoretical basis for it is found in quantum mechanics. For gases, θ is usually zero.

9.01. Magnetic Susceptibility.—It is often convenient to use a new quantity magnetic susceptibility defined, in an isotropic medium, in terms of the quantities already treated in 7.20 and 7.28 by the equations

$$\kappa H = M = B\left(\frac{1}{\mu_v} - \frac{1}{\mu}\right) \tag{1}$$

Thus susceptibility and permeability are related by the equation

$$\kappa = \mu_v^{-1}(\mu - \mu_v) = K_m - 1 \tag{2}$$

When a substance is placed in a magnetic field, its energy is decreased if it is paramagnetic and increased if it is diamagnetic. From 8.02 (3) and (2), the change is given by

$$\Delta W = \frac{\mu_v H^2}{2} - \frac{B^2}{2\mu} = \frac{(\mu_v - \mu)H^2}{2} = -\frac{\mu_v}{2}\kappa H^2 \tag{3}$$

We notice that, for paramagnetic bodies, κ is positive; for diamagnetic bodies, it is negative. Curie's law [9.00 (1)] may be expressed more simply in terms of the susceptibility by the empirical equation

$$\kappa = \frac{C}{T + \theta} \tag{4}$$

A theoretical basis for this equation has also been worked out.

9.02. Magnetic Properties of Crystals.—Many substances, especially crystals, possess different magnetic properties in different directions. It is even possible with some materials, such as graphite, to prepare specimens that are paramagnetic in one direction and diamagnetic in another. In such cases, it is found however that for any given orientation the magnetic induction $\boldsymbol{B}$ is proportional to the field intensity $\boldsymbol{H}$ and makes a constant angle α with it. This relation is similar therefore to that connecting the electric displacement $\boldsymbol{D}$ and the electric field intensity $\boldsymbol{E}$ in a crystal and can be formulated in the same way as in 1.19 by writing a set of equations analogous to 1.19 (1) and (2). We find, therefore, from 1.19 (3) that the components of $\boldsymbol{B}$ and $\boldsymbol{H}$ are connected by the relations

$$\begin{aligned} B_x &= \mu_{11}H_x + \mu_{21}H_y + \mu_{31}H_z \\ B_y &= \mu_{12}H_x + \mu_{22}H_y + \mu_{32}H_z \\ B_z &= \mu_{13}H_x + \mu_{23}H_y + \mu_{33}H_z \end{aligned} \tag{1}$$

where

$$\mu_{12} = \mu_{21}, \qquad \mu_{13} = \mu_{31}, \qquad \mu_{23} = \mu_{32} \tag{2}$$

Thus **B** and **H** are now connected by a quantity having nine components of which six are different. The permeability, formerly a simple ratio, has become a symmetrical tensor. By a suitable orientation of axes, (1) may be written in a form analogous to 1.19 (7), giving

$$B_x = \mu_1 H_x, \qquad B_y = \mu_2 H_y, \qquad B_z = \mu_3 H_z \tag{3}$$

When (3) holds, the coordinate axes are said to lie along the magnetic axes of the crystal.

By means of 9.01 (1), we see that the corresponding equation connecting the magnetic susceptibility and intensity of magnetization are

$$M_x = \kappa_1 H_x, \qquad M_y = \kappa_2 H_y, \qquad M_z = \kappa_3 H_z \tag{4}$$

where $\mu_v \kappa_1 = \mu_1 - \mu_v$, etc. By rotation of coordinates, we can get a set of equations, connecting **M** and **H** in a crystal, analogous to (1) and involving a tensor magnetic susceptibility.

9.03. Crystalline Sphere in Uniform Magnetic Field.—As an illustration of the manipulation of the formulas of the last article, let us find the torque acting on a crystalline sphere when placed in a uniform magnetic field of induction **B**. Let the angle between **B** and the x-axis in 9.02 (3) be α, and let $\mu_2 = \mu_3$. The boundary conditions will evidently be satisfied if we superimpose the case of a field of induction $B \cos \alpha$ in the x-direction, acting on an isotropic sphere of permeability μ_1, and the case of a field $B \sin \alpha$ in the y-direction, acting on an isotropic sphere of permeability μ_2.

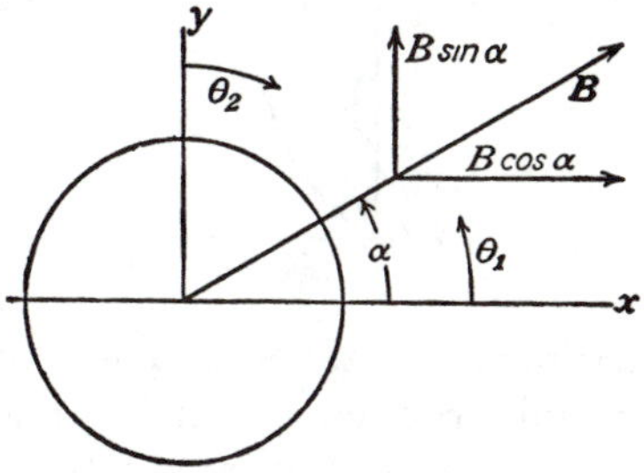

FIG. 9.03.

If we take **B** in the xy-plane (Fig. 9.03), then problem 18*C*, Chap. VII, and 7.10 (2) show that the field outside the sphere is exactly the same as if we superimposed on the original field a field due to a magnetic dipole of moment M_x in the x-direction and that of one of strength M_y in the y-direction where

$$M_x = \frac{\mu_1 - \mu_v}{\mu_1 + 2\mu_v} \frac{4\pi a^3 B \cos \alpha}{\mu_v} \qquad \text{and} \qquad M_y = \frac{\mu_2 - \mu_v}{\mu_2 + 2\mu_v} \frac{4\pi a^3 B \sin \alpha}{\mu_v}$$

The field due to these two dipoles equals that due to a single dipole of strength M making an angle β with the axis, where

$$\tan\beta = \frac{M_y}{M_x} = \frac{(\mu_2 - \mu_v)(\mu_1 + 2\mu_v)}{(\mu_1 - \mu_v)(\mu_2 + 2\mu_v)}\tan\alpha \tag{1}$$

$$M = (M_x^2 + M_y^2)^{\frac{1}{2}} = \frac{4\pi B a^3[(\mu_1 - \mu_v)^2(\mu_2 + 2\mu_v)^2(1 - \sin\alpha) + (\mu_2 - \mu_v)^2(\mu_1 + 2\mu_v)^2 \sin\alpha]^{\frac{1}{2}}}{\mu_v(\mu_1 - \mu_v)(\mu_2 + 2\mu_v)} \tag{2}$$

The angle that M makes with the field is, from *Dw* 405.02 or *Pc* 593,

$$\tan(\beta - \alpha) = \frac{3(\mu_2 - \mu_1)\mu_v \tan\alpha}{(\mu_1 - \mu_v)(\mu_2 + 2\mu_v) + (\mu_2 - \mu_v)(\mu_1 + 2\mu_v)\tan^2\alpha} \tag{3}$$

The torque acting is then, from 7.18,

$$T = MB\sin(\beta - \alpha) \tag{4}$$

This torque tends to rotate the sphere so that the axis of μ_1 lies in the direction of the field if $\mu_1 > \mu_2$ and so that it lies at right angles to the field if $\mu_1 < \mu_2$.

9.04. Ferromagnetism.—There is an important group of materials whose permeability varies with the magnetizing field, depends on the previous treatment of the specimen, and is much larger than that of

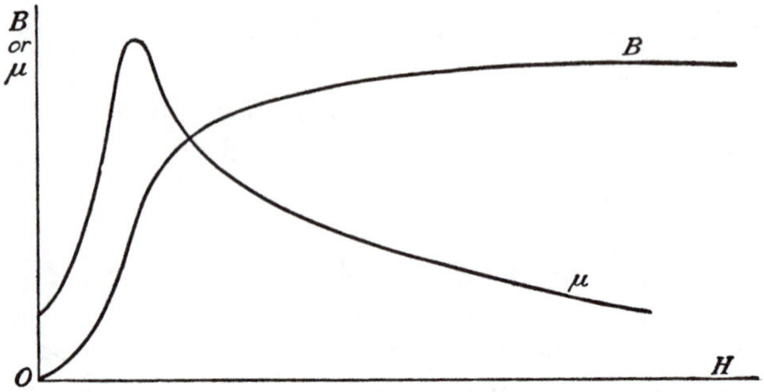

Fig. 9.04.

ordinary substances. These materials are said to be ferromagnetic and include iron, cobalt, nickel, and the Heusler alloys and, at low temperatures, some of the rare earth metals. Let us construct, of ferromagnetic material, a simple magnetic circuit such as the anchor ring considered in 7.29 in which the magnetizing force, or the magnetomotance per meter, can be computed easily. Starting with the zero values of both the magnetic induction $\boldsymbol{B}$ and the magnetizing force $\boldsymbol{H}$, we find that as $\boldsymbol{H}$ is increased $\boldsymbol{B}$ also increases but the ratio of $\boldsymbol{B}$ to $\boldsymbol{H}$ which is the permeability μ first increases and then decreases. Typical curves of $\boldsymbol{B}$ and μ as a function of $\boldsymbol{H}$, for a ferromagnetic material, are shown in Fig. 9.04.

In many ferromagnetic substances, if we measure the magnetization with sufficiently sensitive instruments, we find the steeper part of the magnetization curve to have a step structure. This is called the Barkhausen effect. It suggests that a large region, involving many similarly oriented atomic magnets, changes direction as a unit. Experiments on ferromagnetic single crystals indicate that certain preferred directions are easiest for magnetization. Since most ferromagnetic substances with which we work are polycrystalline, it is believed that the major part of the magnetization is due to the magnetic units in a microcrystal which are partially aligned with the magnetizing field growing at the expense of those which are not. When all units are so oriented, the Barkhausen effect ceases, and further increases in the magnetizing field force the magnetization of these units continuously from their preferred directions toward the direction of the field. This accounts for the absence of step structure in the flatter part of the curve where the magnetization approaches saturation. A theoretical basis for this picture is found in the electron configuration of the iron atom.

In the calculation of magnetic fields, inductances, magnetic forces, eddy currents, and such quantities, we have hitherto assumed that the permeability of a given substance is fixed. We see from these curves that if there is much variation in the magnetizing force in a region occupied by ferromagnetic material this assumption is not justified. In such cases, we take a mean value of μ, which is approximately correct for this substance, for the range of $\boldsymbol{H}$ used. Our knowledge of the magnetization curve of the particular specimen involved usually does not justify higher precision. In case, however, all ferromagnetic material used is very homogeneous and has been carefully annealed, we may be able to determine from this solution and the magnetization curve the value of μ which should be used in different regions. It is in geometrically simple cases only that this knowledge can be used to find a more rigorous analytical solution.

9.05. Hysteresis. Permanent Magnetism.—Suppose that, having increased $\boldsymbol{H}$ to $\boldsymbol{H}_1$, thereby increasing $\boldsymbol{B}$ to $\boldsymbol{B}_1$, as described in 9.04, we now decrease $\boldsymbol{H}$ to $-\boldsymbol{H}_1$. We find that the induction $\boldsymbol{B}$ does not retrace any portion of the path shown in Fig. 9.04 but decreases less rapidly, following the upper curve in Fig. 9.05*a*, and, for a normal specimen, reaches the value $-\boldsymbol{B}_1$ at $-\boldsymbol{H}_1$. If we now increase $\boldsymbol{H}$ again to $+\boldsymbol{H}_1$, we follow the lower curve in Fig. 9.05 and reach the original curve at $\boldsymbol{B}_1$, $\boldsymbol{H}_1$. This lag in the induction is called hysteresis, and the closed curve in Fig. 9.05*a* is called a hysteresis loop.

It is clear that with a given specimen we get a different hysteresis loop when we start with a different point on the magnetization curve, but if we vary $\boldsymbol{H}$ continuously from $\boldsymbol{H}_1$ to $-\boldsymbol{H}_1$ and back we repeatedly retrace the same loop.

The hysteresis loops for different ferromagnetic materials may differ greatly. For magnetically "soft" specimens, the area inside the loop is very small, giving the inner curve in Fig. 9.05*b*, whereas in magnetically "hard" substances it is very great as shown in curve *D*. In the latter case, we still have a large residual induction or retentivity B_m when the magnetizing force is zero and it actually takes a large reverse field H_c, called the coercive force, to destroy the induction in the specimen. Here, we observe for the first time the presence of a magnetic field when electric currents are apparently absent. Magnetic phenomena were

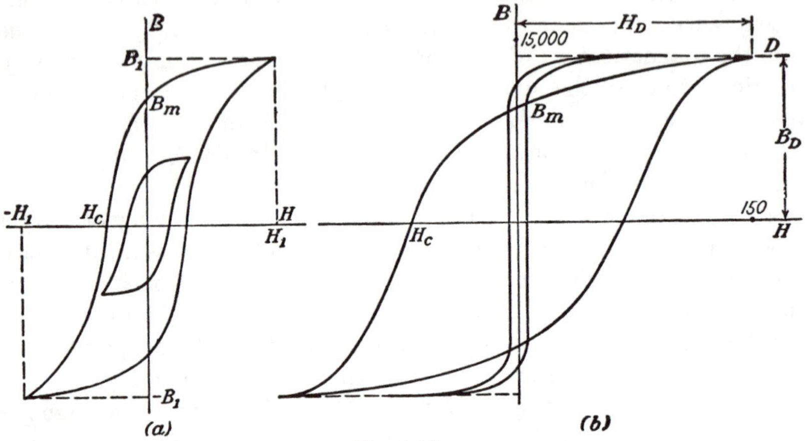

FIG. 9.05.

originally discovered associated with such "permanent" magnets, and the whole theory of magnetism was developed on the basis of experiments made with them.

9.06. The Nature of Permanent Magnetism.—Hitherto, we have considered the energy in a magnetic field as essentially kinetic, it being associated with the motion of electric charges. As the magnetic fields produced by permanent magnets appear in every respect to be identical with those produced by electric currents, it is natural to seek a similar origin for them. The nature of permanent magnetism is revealed by a group of phenomena known as the gyromagnetic effects. Since no electricity is entering or leaving a permanent magnet, any motion of electricity therein must be circulatory and this circulation or spin must be about axes that are oriented, on the average, in a definite direction to produce a definite external field. If, as we have seen in 1.04, the electrical carriers possess mechanical inertia, then when circulating in closed paths or spinning, they possess angular momentum and therefore are subject to gyroscopic forces. Such forces were predicted by Maxwell but could not be detected with the experimental technique of his day.

Two effects immediately suggest themselves. The first of these is magnetization by rotation. A well-known fact in mechanics is that when the supporting system of a gyroscope is rotated and its axis $\boldsymbol{a}$ is free to turn only in the plane common to it and the axis $\boldsymbol{b}$ of rotation of the system, then $\boldsymbol{a}$ tends to set itself parallel to $\boldsymbol{b}$.

Thus, if an unmagnetized body possessed circulating or spinning electricity with axes oriented at random, a rotation of this body should produce an alignment of these axes with the axis of rotation, and the body should become magnetized. Such effects have been detected and measured by Barnett in ferromagnetic substances. The second effect is the converse of the first, rotation by magnetization. From the law of conservation of angular momentum, if the random axes of rotation are aligned by a magnetic field, then the body as a whole must rotate in the reverse direction to keep the resultant angular momentum zero. This effect was first measured by Einstein and De Haas and has since been done with greater precision by other experimenters. Both effects show that there is a rotation of negative electricity in ferromagnetic bodies and that the average magnetic moment of the individual gyroscopes is slightly greater than that of a spinning electron. The excess is supposed to be due to an "orbital" motion of the electrons. Thus, we see that the magnetic fields of permanent magnets are not different from those already studied.

9.07. Uniform Magnetization. Equivalent Current Shell.—In permanent magnets, the magnetization $\boldsymbol{M}$ is, by definition, independent of applied fields. It can be defined, as in 7.20, as the magnetic moment per unit volume of the permanent circulating currents or spins. From 7.20 (1) and(4), the vector potential due to $\boldsymbol{M}$ is given by the equation

$$\boldsymbol{A}_M = -\frac{\mu_v}{4\pi}\int_V \frac{\boldsymbol{M}\times\boldsymbol{r}}{r^3}\,dv = \frac{\mu_v}{4\pi}\int_V \frac{\nabla\times\boldsymbol{M}}{r}\,dv + \frac{\mu_v}{4\pi}\int_S \frac{\boldsymbol{M}\times\boldsymbol{n}}{r}\,dS \tag{1}$$

where the volume integrals are throughout the volume of the magnet and the surface integral over its surface.

When $\boldsymbol{M}$ is the same in magnitude and direction in every element of a given region, we say this region is uniformly magnetized. For such a magnet, the second volume integral in (1) is zero. Let us examine the remaining surface integral. Suppose that $\boldsymbol{M}$ is in the x-direction so that $\boldsymbol{M} = \boldsymbol{i}M$. Let θ be the angle between $\boldsymbol{i}$ and $\boldsymbol{n}$, and let $\boldsymbol{ds}$ and $\boldsymbol{ds}_1$ be orthogonal vectors lying in the magnet surface, $\boldsymbol{ds}$ being normal to $\boldsymbol{i}$ and $\boldsymbol{n}$. Then, we have

$$\boldsymbol{M}\times\boldsymbol{n}\,dS = M\sin\theta\,ds_1\,\boldsymbol{ds} = M\,dx\,\boldsymbol{ds}$$

so that

$$\boldsymbol{A} = \frac{\mu_v}{4\pi}\int\int\frac{M\,dx}{r}\,\boldsymbol{ds} \tag{2}$$

We see that this is identical in form with 7.02 (5) so that the whole magnet can be replaced by a current shell, coinciding with its surface, around which the currents flow in planes normal to the direction of magnetization x. The current density, in terms of x, is uniform and equals the intensity of magnetization M. Such a shell is called an equivalent current shell and is frequently very useful in treating permanent magnets.

9.08. Magnetized Sphere and Cylinder. Magnetic Poles.—For a uniformly magnetized sphere, the current flowing in the equivalent current shell between θ and $\theta - d\theta$ is

$$i\,d\theta = M\,dx = Ma\sin\theta\,d\theta$$

where i is the angular current density. The stream function is then given by

$$\Psi = \int_0^\theta i\,d\theta = Ma(1 - \cos\theta) = Ma[P_0(u) - P_1(u)] \tag{1}$$

where $u = \cos\theta$. This is identical in form with 7.12 (1); and setting $n = 0$ and 1 in 7.12 (5) and (3), we see that the vector potential outside the sphere is

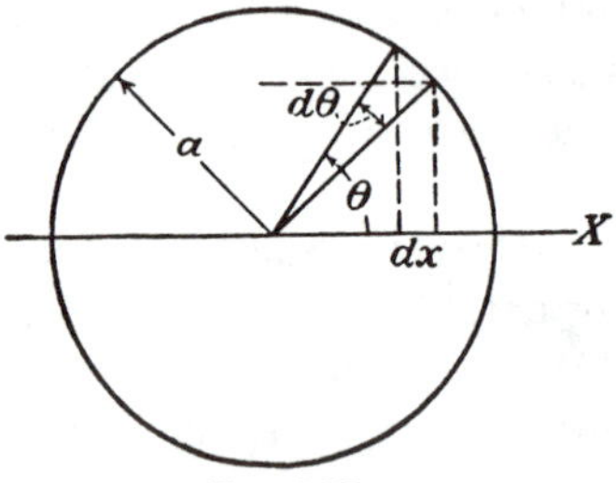

FIG. 9.08.

$$A_{o\phi} = \frac{\mu_v M a^3}{3r^2}\sin\theta \tag{2}$$

Comparing with 7.10 (1), we see that the field is identical with that of a small current loop of moment

$$M = \frac{4\pi M a^3}{3} \tag{3}$$

We saw in 7.00 that the magnitude and configuration of the magnetic field at a distance from such a loop are identical with that of the electric field about an electric dipole having the same moment.

For a right circular cylinder magnetized parallel to its axis, the equivalent current shell is seen to be a solenoid with zero pitch, the current per unit length being M. From 7.15 (4), the magnetic induction at any point P on the axis of such a solenoid is given by

$$B_x = \tfrac{1}{2}\mu_v M_x(\cos\beta_2 - \cos\beta_1) \tag{4}$$

where β_2 and β_1 are the angles subtended by radii of the two ends at P. The assumption that B can be found from the normal magnetization M_n on a magnet surface from the inverse-square law $\mu_v M(4\pi r^2)^{-1}$ also gives (4). Take M_x uniform over the flat ends of this magnet, and consider a ring of radius $\rho = c\tan\theta$ and width $d\rho = c\sec^2\theta\,d\theta$ which subtends an angle θ at an axial point a distance c from its center so $r = c\sec\theta$ and B_x at c is seen to be

$$B_x = \frac{\mu_v M}{4\pi} \int_0^{\beta_2} \frac{2\pi\rho\, d\rho \cos\theta}{r^2} = \frac{\mu_v M}{2} \int_0^{\beta_2} \sin\theta\, d\theta = \frac{\mu_v M}{2}(\cos\beta_2 - 1) \quad (5)$$

Subtraction of a similar term for the opposite end gives (4). For a magnet of length l, this is just the electric displacement given on the x-axis with electric charges $q_1 = \mu_v \pi a^2 M$ at $x = \frac{1}{2}l$ and $q_2 = -\mu_v \pi a^2 M$ at $x = -\frac{1}{2}l$. Thus, experiments with magnetic needles lead naturally to the hypothesis of magnetic charges or poles. The region from which the magnetic lines of force emerge is called the north pole, and the region in which they reenter the magnet is called the south pole. As we have seen, the actual existence of magnetic charges is impossible if the divergence of the magnetic induction is to be zero everywhere.

9.09. Boundary Conditions on Permanent Magnets.—As the name implies, we assume that the intensity of magnetization of a "permanent" magnet is unaffected by the fields in which it is placed. This means that if we replace the magnet with an equivalent current shell, the region inside is assumed to have a permeability μ_v. If the magnet is immersed in a region of permeability μ and an external field is superimposed, it will be distorted at the surface of the magnet to meet the boundary conditions between a medium of permeability μ_v and one of permeability μ. These boundary conditions have already been stated in 7.21, 7.22, and 7.28.

9.10. Spherical Permanent Magnet in Uniform Field.—As an example, let us compute the torque on a uniformly magnetized sphere immersed in a medium of permeability μ in which the magnetic induction $\boldsymbol{B}$, before the introduction of the sphere, was uniform. Let α be the angle between $\boldsymbol{M}$ and $\boldsymbol{B}$. Obviously, the current sheet can produce no torque on itself; so we need compute only the induction due to the external field. In spherical coordinates, the vector potential of $\boldsymbol{B}$ is

$$A_\phi = \tfrac{1}{2} Br \sin\theta \quad (1)$$

In 7.10 (1), we have a form of vector potential in which θ enters in the same way but which vanishes at infinity. This is therefore the logical form to choose for the additional term due to the presence of the sphere with permeability μ_v. Thus, outside the sphere, we have

$$A_{o\phi} = \frac{B}{2}\left(r + \frac{C}{r^2}\right) \sin\theta \quad (2)$$

Since A_i must be finite at the origin and must include θ in the same way, it must have the form

$$A_{i\phi} = Dr \sin\theta \quad (3)$$

To determine C and D, we apply the boundary conditions of 7.21 (6) and

(7) at $r = a$. These give the equations

$$\frac{B}{2}\left(1 - \frac{C}{2a^3}\right) = \frac{\mu D}{\mu_v} \quad \text{and} \quad \frac{B}{2}\left(a + \frac{C}{a^2}\right) = Da \tag{4}$$

Solving for C and substituting in (2) give

$$A_{o\phi} = \frac{1}{2}B\left(r - 2\frac{\mu - \mu_v}{2\mu + \mu_v}\frac{a^3}{r^2}\right)\sin\theta \tag{5}$$

When $r = a$, this has the same form as (1) and so, from (3), represents a uniform field in the direction of $\mathbf{B}$. The torque on the current ring element lying between θ_1 and $\theta_1 + d\theta_1$, where θ_1 is the colatitude angle measured from the axis of magnetization, equals the product of the current in the ring by its area, by the magnetic induction, and by $\sin\alpha$, giving

$$dT = \pi a^2 \sin^2\theta_1 B\left[1 - 2\frac{\mu - \mu_v}{2\mu + \mu_v}\right] i\, d\theta_1 \sin\alpha$$

Substituting for $i\, d\theta_1$ from 9.08 and integrating by *Dw* 854.1 or *Pc* 483 give

$$T = \frac{3\pi\mu_v a^3 MB \sin\alpha}{2\mu + \mu_v}\int_0^\pi \sin^3\theta_1\, d\theta_1 = \frac{4\pi\mu_v a^3 MB\sin\alpha}{2\mu + \mu_v} \tag{6}$$

If the field in which the sphere was placed were not homogeneous, we would have a force as well as a torque. This would consist of two parts, *viz.*, the force on a sphere of permeability μ_v and the force on the current shell in the field at the surface of this sphere. These forces could be calculated, using 10.06 (13) and 7.18.

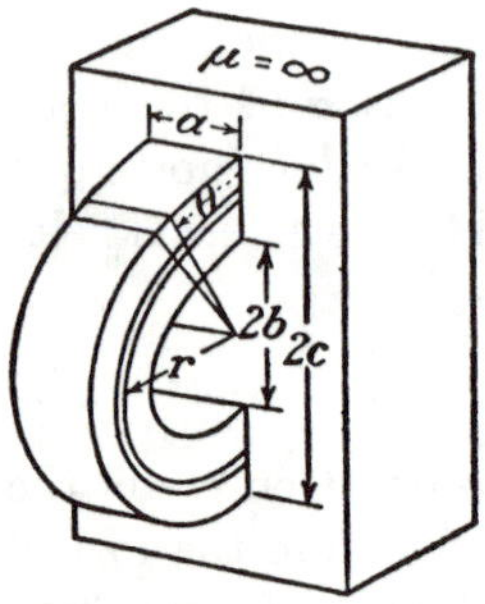

FIG. 9.11.

9.11. Lifting Power of Horseshoe Magnet.—For lifting ferromagnetic objects, one frequently uses a permanent magnet in the form of a horseshoe. In treating this, we shall assume for simplicity that the "legs" are eliminated, leaving only half a ring as shown in Fig. 9.11. If the legs were present, we could use an approximate method like that in 6.16. We shall assume that the half ring was magnetized by placing it in contact with another similar piece of steel to form a complete ring, winding uniformly with wire and passing a current. To a first approximation, the magnetization M will be the same function of r, the distance from the center of the ring, as the magnetomotance was, so that we have, from 7.29 (3),

$$M = \frac{C}{r} \tag{1}$$

To find the equivalent current shell, we take a thin layer of thickness dr and radius r in which M may be considered constant. Then, if i is the angular current density, we have for the current flowing between θ and $\theta - d\theta$, as in 9.08,

$$i\,d\theta = M\,dx = \left(\frac{C}{r}\right)r\,d\theta = C\,d\theta \tag{2}$$

When the layer current shells are superimposed, adjacent current layers cancel and we have a constant angular current density C around any section of the ring. Now let us suppose this magnet is placed in perfect contact with a block of infinite permeability. The lines of magnetic induction inside the current shell are then semicircles since they enter and leave the block normally by 7.21 (2). The magnetomotance around the circuit is, by 7.28 (3),

$$\Omega = \int_0^\pi C\,d\theta = \pi C \tag{3}$$

The reluctance of the layer dr is, as in 7.29,

$$dR = \frac{\text{length}}{\mu_v \times \text{area}} = \frac{\pi r}{\mu_v a\,dr}$$

since the permeability inside our current shell is μ_v. The flux in this layer is

$$Ba\,dr = \frac{\Omega}{dR} = \frac{\mu_v C}{r} a\,dr$$

The tension across a section of this layer is, from 8.14 (2),

$$dT = \frac{aB^2}{2\mu_v}\,dr = \frac{\mu_v a C^2}{2r^2}\,dr$$

The pull on the block at each contact is

$$T = \frac{\mu_v a C^2}{2}\int_b^c \frac{dr}{r^2} = \frac{\mu_v a C^2(c-b)}{2bc} \tag{4}$$

An upper limit for C can be computed roughly from the magnetization curve of the steel. If the total number of turns in the original magnetizing winding was n and the current used was i_m, then the magnetomotance per meter was $ni_m/(\pi r)$ ampere-turns per meter from 7.28 (3). Suppose that the wide loop in Fig. 9.05b corresponds to our specimen and that $H_D = ni_m/(\pi b)$. Then, for all values of r greater than b, the magnetizing force will be smaller than H_D and the hysteresis loops will lie inside of that shown. We have assumed that the ratio of B_m to B_D is the same for all these loops (let us call it P) and that the ratio of B_D to H_D is also a constant (let us call it μ'). Then, from (1),

$$M = \frac{C}{r} = \frac{B_m}{\mu_v} = \frac{ni_m\mu' P}{\mu_v \pi r}$$

so that we have

$$C_{max} = \frac{ni_m\mu' P}{\mu_v \pi} \tag{5}$$

This is an upper limit for C since any mechanical shock, heat-treatment, or "softness" of the steel will diminish C, especially after the ring has been separated into two magnets. The approximation made in assuming $\mu = \infty$ for the block is not serious, for in most soft iron specimens $\mu/\mu_v > 500$.

9.111. Field of Cylindrical Magnet.—In order to produce strong steady fields over small areas, permanent magnets are frequently made in the form of rings, with small gaps, which are magnetized, as described in the last article, by winding uniformly with wire and passing a current. To simplify the calculation of the field in this case, let us suppose that the ring is so wide in the direction parallel to its axis that we may take it as a long cylinder of internal and external radii a and b which has been magnetized, before the gap is made, by winding wire uniformly over the walls parallel to its axis. As explained in the last article, the resultant M is inversely proportional to r. After magnetizing, a gap with radial walls is created, without disturbing M, by removing part of the metal, leaving only the portions lying between $\theta_1 = +\alpha$ and $\theta_1 = -\alpha$. If b is sufficiently small compared with the length of the cylinder, the problem far from the ends becomes a two-dimensional one. In this case, the lines of magnetic induction coincide with the lines of constant vector potential as proved in 7.25 and may be most easily calculated by the method of circular harmonics considered in 4.01, 4.02, 4.03, and 7.17.

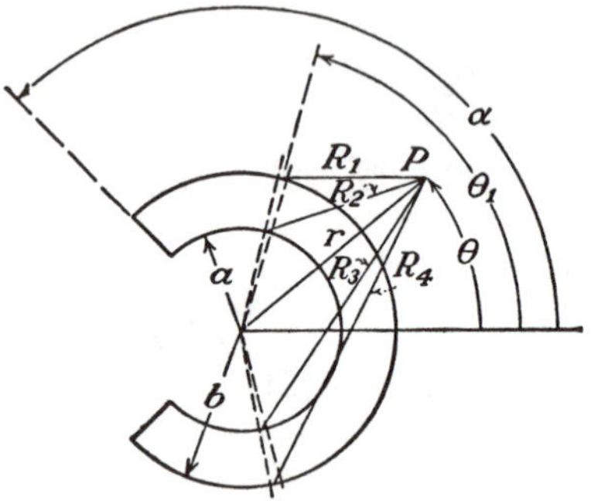

FIG. 9.111*a*.

As shown in the last article, the equivalent current shell consists of that portion of the original magnetizing winding not removed with the gap and carries the current $I\,d\theta$ between θ and $\theta + d\theta$. Since all elements of the winding are parallel to the axis and the ends are so distant that they do not affect the vector potential A, the latter is also everywhere parallel to the axis and the current elements may be considered infinite in length. Considering the contribution of the inner elements negative and that of the outer ones positive, we find that dA at the point P at r, θ due to the elements at $+\theta_1$ and $-\theta_1$ is given from 7.09 (1), by

$$dA = \tfrac{1}{2}\pi^{-1}\mu_v I(\ln R_1 - \ln R_2 - \ln R_3 + \ln R_4)\,d\theta_1$$

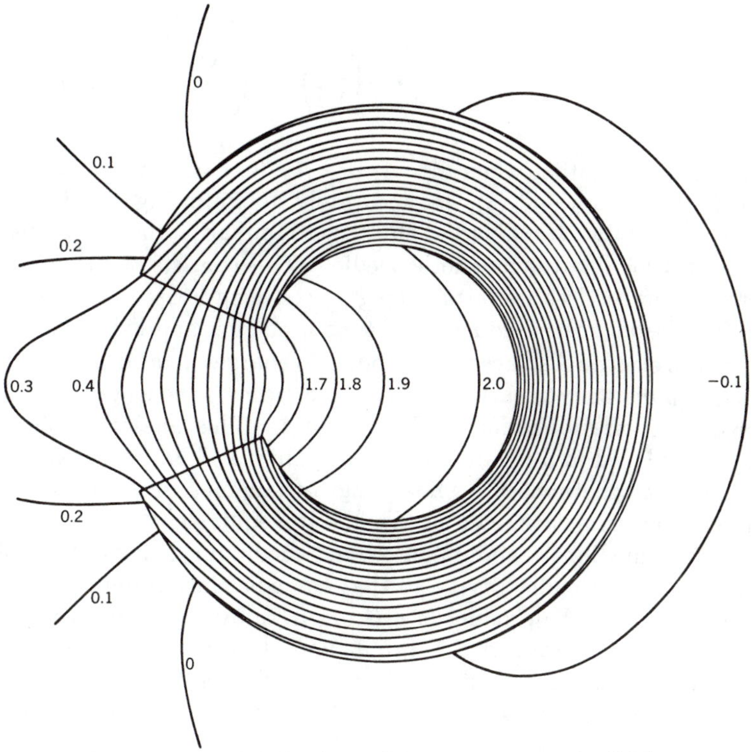

FIG. 9.111b.—Lines of magnetic induction and constant vector potential in a permanent magnet formed by magnetizing a long thick cylindrical shell and removing a sector to form the air gap. Calculated from Eqs. (1), (2), and (3) of Art. 9.111 in which $C = 1$, $\alpha = 7\pi/8$, $a = 1$, and $b = 2$.

Substituting for the logarithms from 4.02 (1) and writing C for $\mu_v I/\pi$ give

$$dA = C\sum_{n=1}^{\infty}\frac{1}{n}\left(\frac{a^n - b^n}{r^n}\right)\cos n\theta_1 \cos n\theta\, d\theta_1$$

Integrating this from $\theta_1 = 0$ to $\theta_1 = \alpha$ gives

$$r > b, \qquad A = C\sum_{n=1}^{\infty}\frac{1}{n^2}\left(\frac{a^n - b^n}{r^n}\right)\sin n\alpha \cos n\theta \tag{1}$$

Similarly,

$$b > r > a, \qquad A = C\left\{\alpha \ln\frac{b}{r} + \sum_{n=1}^{\infty}\frac{1}{n^2}\left[\left(\frac{a}{r}\right)^n - \left(\frac{r}{b}\right)^n\right]\sin n\alpha \cos n\theta\right\} \tag{2}$$

and

$$a > r, \qquad A = C\left\{\alpha \ln \frac{b}{a} + \sum_{n=1}^{\infty} \frac{1}{n^2}\left[\left(\frac{r}{a}\right)^n - \left(\frac{r}{b}\right)^n\right] \sin n\alpha \cos n\theta\right\} \tag{3}$$

The lines of induction when $C = 1$, $\alpha = 7\pi/8$, $a = 1$, and $b = 2$, calculated from (1), (2), and (3), are accurately drawn in Fig. 9.111*b*.

9.12. Magnetic Needles.—The most familiar form of permanent magnet is probably the magnetic needle. This consists of a thin cylinder of steel more or less uniformly magnetized lengthwise. As we saw in 9.08 (5), the magnetic induction of such a magnet approximates in form that of the electric displacement about two charges $\mu_v \pi a^2 M$ and $-\mu_v \pi a^2 M$ placed at a distance l apart, where a is the radius, l the length, and M the intensity of magnetization of the magnet. This analogy fails if the magnet is placed in a medium of permeability $\mu \neq \mu_v$. However, in the case of the magnetic needle whose length-to-diameter ratio is very great, another approximation can be used. Its equivalent current shell, if filled with a medium μ_v and immersed in a medium μ, behaves by 7.24 and 1.18 like a long slender cavity so that the axial flux through it from an external source is the product of $\mu_v H$ or $\mu_v B/\mu$ by $\pi a^2 \cos\theta$ where θ is the angle between its axis and $\mathbf{B}$. From 9.07, the current in a length dl of the equivalent current shell that links this flux is $M\,dl$ where M is the magnetization. From 8.03 (6) and 7.28 (5), the energy of the shell is

$$W = \mu_v M \pi a^2 \int_{l_1}^{l_2} H \cos\theta \, dl = \mu_v M \pi a^2 (\Omega_2 - \Omega_1) \tag{1}$$

This has the form of 1.071 (1) so that a thin needle uniformly magnetized lengthwise acts exactly like a pair of equal and opposite electric charges of size $\mu_v M \pi a^2$ placed in an electric field of potential $V = \Omega$, where Ω is the magnetomotance or scalar magnetic potential.

In the absence of external magnetic field sources, the flux threading the slender current shell is practically independent of the permeability of the medium surrounding the needle for the reluctance of the magnetic circuit involved lies almost entirely inside the shell where the medium does not penetrate and the permeability is μ_v. Thus the magnetic flux from a sufficiently thin magnetic needle, like the electric flux from an electric dipole, is independent of the medium surrounding it. The scalar magnetic potential or magnetomotance of such a needle is, therefore, inversely proportional to the permeability of the medium around it. These forces may be computed by the formulas of 1.071 as if the magnets possessed a mutual potential energy. Thus if $\mathbf{r}$ is the radius vector from $\mathbf{M}_1'$ to $\mathbf{M}_2'$ which make angles θ_1 and θ_2 with it and α with each other and, if ψ is the angle between the planes intersecting in $\mathbf{r}$ which contain $\mathbf{M}_1'$ and $\mathbf{M}_2'$ then, from (1), 1.071 (4), and 1.071 (5), this potential energy is

$$W = \frac{M_1'M_2'}{4\pi\mu r^3}(\cos\alpha - 3\cos\theta_1\cos\theta_2) \quad (2)$$

$$= \frac{M_1'M_2'}{4\pi\mu r^3}(\sin\theta_1\sin\theta_2\cos\psi - 2\cos\theta_1\cos\theta_2) \quad (3)$$

All forces and torques exerted by one dipole on the other can be found from these formulas by the usual method of differentiation with respect to the coordinate involved. In particular, the force of repulsion is

$$F = -\frac{\partial W}{\partial r} = \frac{3M_1'M_2'}{4\pi\mu r^3}(\sin\theta_1\sin\theta_2\cos\psi - 2\cos\theta_1\cos\theta_2) \quad (4)$$

The apparent potential energy of a needle M' in a field of induction $\boldsymbol{B}$, from which the forces are found by differentiation, is, from 1.071 (2),

$$-W = M' \cdot H = M' \cdot B\mu^{-1} \quad (5)$$

The vector potential at a distance r from a magnetic needle is given in spherical polar coordinates by 7.10 to be, when $r \gg l$.

$$A_\phi = \frac{M' \sin\theta}{4\pi r^2} \quad (6)$$

which, like its total flux, is independent of the medium's permeability. It should be noted that the classical magnetic moment M' used in this article differs dimensionally from the loop magnetic moment defined in 7.00 and 7.10 (2). The formulas of this article are rigorous only for infinitely thin magnets. Forces on or energies of permanent magnets of large cross section in mediums where $\mu \neq \mu_v$ cannot be found by integrating these formulas.

Problems

Problems in the following group marked C are taken, by permission of the Cambridge University Press, from the Cambridge examinations as reprinted by Jeans.

The classical magnetic moment is used throughout these problems.

1. An iron pipe runs east and west in a certain region. To locate the pipe, measurements of the dip at 5-ft intervals, starting from a fixed point, in a northerly direction are made. At distances x ft from the point, the dip deviates from the normal 60° dip by the following amounts:

x	185	190	195	200	205	210	215	220	225	230	235
$\Delta\phi$	−4.5°	−5.0°	−5.5°	−7.5°	−6.0°	+0.0°	+14.5°	+0.0°	−1.5°	−1.2°	−1.0°

Find x for the center of the pipe and its distance below the surface.

2. A steel pendulum bob of radius a is uniformly magnetized in a vertical direction with an intensity I. The strength of the vertical component of the earth's field at a certain location is V and the acceleration of gravity g. Find the ratio of the period to that when unmagnetized if it swings in an east and west plane, its mass being m and the distance from center of mass to pivot, l.

3. A sphere of permeability 1000 and radius 10 cm is placed 1 m northeast (magnetic) of a compass needle. Show that, within 1 per cent, the deviation of the needle from the magnetic meridian, neglecting the magnet's own image forces, is 5′.

4. Find the magnetic field outside and in the cavity of a thick spherical shell uniformly magnetized in the x-direction with an intensity M.

5C. Two small magnets float horizontally on the surface of water, one along the direction of the straight line joining their centers, and the other at right angles to it. Prove that the action of each magnet on the other reduces to a single force at right angles to the straight line joining the centers and meeting that line at one-third of its length from the longitudinal magnet.

6C. A small magnet ACB, free to turn about its center C, is acted on by a small fixed magnet PQ. Prove that in equilibrium the axis ACB lies in the plane PQC and that $\tan\theta = -\frac{1}{2}\tan\theta'$, where θ, θ' are the angles that the two magnets make with the line joining them.

7C. Three small magnets having their centers at the angular points of an equilateral triangle ABC, and being free to move about their centers, can rest in equilibrium with the magnet at A parallel to BC and those at B and C, respectively, at right angles to AB and AC. Prove that the magnetic moments are in the ratios $3^{\frac{1}{2}}:4:4$.

8C. The axis of a small magnet makes an angle ϕ with the normal to a plane. Prove that the line from the magnet to the point in the plane where the number of lines of force crossing it per unit area is a maximum makes an angle θ with the axis of the magnet, such that $2\tan\theta = 3\tan 2(\phi - \theta)$.

9C. Two small magnets lie in the same plane and make angles θ, θ' with the line joining their centers. Show that the line of act on of the resultant force between them divides the line of centers in the ratio $\tan\theta' + 2\tan\theta : \tan\theta + 2\tan\theta'$.

10C. Two small magnets having their centers at distances r apart make angles θ, θ' with the line joining them and an angle ϵ with each other. Show that the longitudinal force on the first magnet is $3MM'(4\pi\mu r^4)^{-1}(5\cos^2\theta\cos\theta' - \cos\theta' - 2\cos\epsilon\cos\theta)$. Show that the couple about the line r which the magnets exert on one another is $MM'(4\pi\mu r^4)^{-1}d\sin\epsilon$, where d is the shortest distance between their axes produced.

11C. Two magnetic needles of moments M, M' are soldered together so that their directions include an angle α. Show that when they are suspended so as to swing freely in a uniform horizontal magnetic field, their directions will make angles θ, θ' with the lines of force given by

$$\frac{\sin\theta}{M'} = \frac{\sin\theta'}{M} = (M^2 + M'^2 + 2MM'\cos\alpha)^{-\frac{1}{2}}\sin\alpha$$

12C. Prove that if there are two magnetic molecules, of moments M and M', with their centers fixed at A and B, where $AB = r$, and one of the molecules swings freely, whereas the other is acted on by a given couple, so that when the system is in equilibrium this molecule makes an angle θ with AB, the moment of the couple is

$$\frac{3MM'\sin 2\theta}{8\pi\mu r^3(3\cos^2\theta + 1)^{\frac{1}{2}}}$$

where there is no external field.

13C. Two small equal magnets have their centers fixed and can turn about them in a magnetic field of uniform intensity H, whose direction is perpendicular to the line r joining the centers. Show that the position in which the magnets both point in the direction of the lines of force of the uniform field is stable only if $H > 3M(4\pi\mu r^3)^{-1}$.

14C. Two magnetic particles of equal moment are fixed with their axes parallel to the axis of z, and in the same direction, and with their centers at the points $\pm a, 0, 0$. Show that if another magnetic molecule is free to turn about its center, which is fixed at the point $(0, y, z)$, its axis will rest in the plane $x = 0$ and will make with the axis of z the angle $\tan^{-1} [3yz/(2z^2 - a^2 - y^2)]$. Examine which of the two positions of equilibrium is stable.

15*C*. Prove that there are four positions in which a given bar magnet may be placed so as to destroy the earth's control of a compass needle so that the needle can point indifferently in any direction (and experiences no translational force). If the bar is short compared with its distance from the needle, show that one pair of these positions is about $1\frac{1}{4}$ times more distant than the other pair.

16*C*. Three small magnets, each of magnetic moment M, are fixed at the angular points of an equilateral triangle ABC, so that their north poles lie in the directions AC, AB, BC, respectively. Another small magnet, moment M', is placed at the center of the triangle and is free to move about its center. Prove that the period of a small oscillation is the same as that of a pendulum of length $4\pi\mu Ib^3g/(351)^{\frac{1}{2}}MM'$, where b is the length of a side of the triangle and I the moment of inertia of the movable magnet about its center.

17*C*. Three magnetic particles of equal moments are placed at the corners of an equilateral triangle and can turn about those points so as to point in any direction in the plane of the triangle. Prove that there are four and only four positions of equilibrium such that the angles, measured in the same sense of rotation, between the axes of the magnets and the bisectors of the corresponding angles of the triangle are equal. Also prove that the two symmetrical positions are unstable.

18*C*. Four small equal magnets are placed at the corners of a square and oscillate under the actions they exert on each other. Prove that the times of vibration of the principal oscillations are

$$2\pi\left[\frac{4\pi\mu mk^2\, d^3}{M^2 3(2 + \frac{1}{2}2^{-\frac{1}{2}})}\right]^{\frac{1}{2}}, \qquad 2\pi\left\{\frac{4\pi\mu mk^2\, d^3}{M^2[3 - \frac{1}{2}(2)^{-\frac{1}{2}}]}\right\}^{\frac{1}{2}}, \qquad 2\pi\left[\frac{4\pi\mu mk^2\, d^3 2(2)^{\frac{1}{2}}}{3M^2}\right]^{\frac{1}{2}}$$

where M is the magnetic moment, and mk^2 the moment of inertia, of a magnet, and d is a side of the square.

19*C*. A system of magnets lies entirely in one plane, and it is found that when the axis of a small needle travels round a contour in the plane that contains no magnetic poles, the needle turns completely round. Prove that the contour contains at least one equilibrium point.

20*C*. Prove that the scalar potential of a body uniformly magnetized with intensity I is, at any external point, the same as that due to a complex magnetic shell coinciding with the surface of the body and of strength Ix, where x is a coordinate measured parallel to the direction of magnetization.

21*C*. A sphere of hard steel is magnetized uniformly in a constant direction, and a magnetic particle is held at an external point with the axis of the particle parallel to the direction of magnetization of the sphere. Find the couples acting on the sphere and on the particle.

22*C*. A spherical magnetic shell of radius a is normally magnetized so that its strength at any point is S_i, where S_i is a spherical surface harmonic of positive order i. Show that the scalar potential at a distance r from the center is

$$\left[\frac{-4\pi(i+1)}{2i+1}\right]S_i\left(\frac{r}{a}\right)^i \qquad \text{when } r < a$$

$$\left[\frac{4\pi i}{2i+1}\right]S_i\left(\frac{a}{r}\right)^{i+1} \qquad \text{when } r > a$$

23*C*. If the earth was a uniformly magnetized sphere, show that the tangent of the dip at any point would be equal to twice the tangent of the magnetic latitude.

24*C*. Prove that if the horizontal component, in the direction of the meridian, of the earth's magnetic force was known all over its surface, all the other elements of its magnetic force might be theoretically deduced.

25*C*. From the principle that the line integral of the magnetic force round any circuit ordinarily vanishes, show that the horizontal components of the magnetic force at any station may be deduced approximately from the known values for three other stations which lie around it. Show that these six known elements are not independent but must satisfy one equation of condition.

26*C*. If the earth was a sphere and its magnetism due to two small straight bar magnets of the same strength situated at the poles, with their axes in the same direction along the earth's axis, prove that the dip δ in latitude λ would be given by

$$8 \cot (\delta + \tfrac{1}{2}\lambda) = \cot \tfrac{1}{2}\lambda - 6 \tan \tfrac{1}{2}\lambda - 3 \tan^3 \tfrac{1}{2}\lambda$$

27*C*. Assuming that the earth is a sphere of radius a and that the magnetic potential Ω is represented by $\Omega = S_1(r/a) + S_2(r/a)^2 + S_1'(a/r)^2 + S_2'(a/r)^3$, show that Ω is completely determined by observations of horizontal intensity, declination, and dip at four stations and of dip at four more.

28*C*. Assuming that in the expansion of the earth's magnetic potential the fifth and higher harmonics may be neglected, show that observations of the resultant magnetic force at eight points are sufficient to determine the potential everywhere.

29*C*. Assuming that the earth's magnetism is entirely due to internal causes and that in latitude λ the northerly component of the horizontal force is $A \cos \lambda + B \cos^3 \lambda$, prove that in this latitude the vertical component reckoned downward is

$$2\left(A + \frac{6B}{5}\right) \sin \lambda - \frac{4B}{3} \sin^3 \lambda$$

30*C*. A magnetic particle of moment M lies at a distance a in front of an infinite block of soft iron bounded by a plane face, to which the axis of the particle is perpendicular. Find the force acting on the magnet, and show that the potential energy of the system is

$$\frac{M^2(\mu - \mu_v)}{32\pi\mu_v a^3(\mu + \mu_v)}$$

31. A small magnet of moment M is held in the presence of a very large fixed mass of soft iron of permeability μ with a very large plane face, the magnet is at a distance a from the plane face and makes an angle θ with the shortest distance from it to the plane. Show that a certain force and a couple

$$\frac{(\mu - \mu_v)M^2 \sin \theta \cos \theta}{32\pi\mu_v(\mu + \mu_v)a^3}$$

are required to keep the magnet in position.

References

ATTWOOD, S. S.: "Electric and Magnetic Fields," Wiley, 1941. Clear practical treatment of ferromagnetism and permanent magnets with diagrams and tables.

BITTER, F. T.: "Introduction to Ferromagnetism," McGraw-Hill, 1937. Excellent modern treatment.

BOZORTH, R. M.: "Ferromagnetism," Van Nostrand, 1951. A classic in modern magnetic experiments.

EWING, J. A.: "Magnetic Induction in Iron and Other Substances," Van Nostrand, 1900. The standard work of its time.

FLUGGE, S.: "Handbuch der Physik," Vol. XVIII, Springer, 1966.

GEIGER-SCHEEL: "Handbuch der Physik," Vol. XII, Springer, 1927.

GRAY, D. E.: "American Institute of Physics Handbook," 2d ed., 5–164, McGraw-Hill, 1963. Valuable tables and bibliography.

MATTIS, D. C.: "The Theory of Magnetism," Harper and Row, 1965. A very modern and advanced treatment of magnetism and cooperative phenomena.

RADO, G. T., and N. SUHL: "Magnetism," Vol. III, Academic, 1963. First of three volumes on magnetic theory and materials by many contributors.

SPOONER, T.: "Properties and Testing of Magnetic Materials," McGraw-Hill, 1931. Gives detailed experimental information.

WIEN-HARMS: "Handbuch der Experimentalphysik," Vol. XI, Leipzig, 1932.

WILLIAMS, S. R.: "Magnetic Phenomena," McGraw-Hill, 1931. Treats experimental facts.

CHAPTER X

EDDY CURRENTS

10.00. Induced Currents in Extended Conductors.—This chapter covers the laws of magnetic interaction of currents and the induction laws of Faraday discussed in Chap. VIII as they apply to extended conductors, including skin effect, induction heating, drag on moving magnets, and the magnetic shielding by thin sheets. We should note that this treatment, and all that have preceded it, involves one approximation, *viz.*, the assumption that electric and magnetic fields are instantaneously propagated. This is equivalent to saying that Maxwell's "displacement current" is neglected. The error so introduced is perfectly negligible if the frequencies are such that the wave length is large compared with the dimensions of the apparatus. If this condition is not met, we must resort to the complete Maxwell equations treated in Chap. XI. .

Faraday's law of induction states that, if the magnetic induction $\boldsymbol{B}$ in a conductor, is changing, an electric field $\boldsymbol{E}$ is produced which is given, in magnitude and direction by 8.00 (3), to be

$$\nabla \times \boldsymbol{E} = -\frac{d\boldsymbol{B}}{dt} \tag{1}$$

In terms of the magnetic vector potential $\boldsymbol{A}$, 8.00 (4) gives

$$\boldsymbol{E} = -\frac{d\boldsymbol{A}}{dt} \tag{2}$$

Since this electric field is produced in a conductor, a current will flow according to Ohm's law. If τ is the resistivity and $\boldsymbol{i}$ the current density, we may, with the aid of 6.02 (3), write (1) and (2) in the form

$$\tau(\nabla \times \boldsymbol{i}) = -\frac{d\boldsymbol{B}}{dt} \tag{3}$$

$$\tau\boldsymbol{i} = -\frac{d\boldsymbol{A}}{dt} \tag{4}$$

These currents, flowing in the conductor of permeability μ, will produce magnetic fields which, by 7.01 (3) and 7.02 (2), are given by

$$\nabla \times \boldsymbol{B} = \mu\boldsymbol{i} \tag{5}$$

$$\nabla^2\boldsymbol{A} = -\mu\boldsymbol{i} \tag{6}$$

The equations that must be satisfied by $\boldsymbol{i}$, $\boldsymbol{B}$, and $\boldsymbol{A}$ in a conductor when a changing field is present derive from (3), (4), (5), and (6). Elimination of $\boldsymbol{B}$ by (3) from the time derivative of (5) gives, by 6.00 (3),

$$\frac{\mu}{\tau}\frac{d\boldsymbol{i}}{dt} = -[\nabla \times (\nabla \times \boldsymbol{i})] = \nabla^2\boldsymbol{i} - \nabla(\nabla \cdot \boldsymbol{i}) = \nabla^2\boldsymbol{i} \tag{7}$$

Elimination of $\boldsymbol{i}$ from (4) and (6) gives

$$\frac{\mu}{\tau}\frac{d\boldsymbol{A}}{dt} = \nabla^2\boldsymbol{A} \tag{8}$$

Elimination of $\boldsymbol{i}$ from (3) and (5) gives, with the aid of 7.01 (1),

$$\frac{\mu}{\tau}\frac{d\boldsymbol{B}}{dt} = -[\nabla \times (\nabla \times \boldsymbol{B})] = \nabla^2\boldsymbol{B} - \nabla(\nabla \cdot \boldsymbol{B}) = \nabla^2\boldsymbol{B} \tag{9}$$

Equations (7), (8), and (9) all have the form of the well-known equation of the conduction of heat, but the dependent variable is a vector instead of a scalar. In rectangular coordinates, each component, regarded as a scalar, satisfies the same equation. This is not true for the components in any other system of coordinates except in special cases.

Eddy current problems fall into two classes, transient and steady state. In the latter case the current density is often written as the real part of a complex Fourier series with terms of the form $\check{A}_n e^{jn\omega t}$ where n is an integer, ω is the angular frequency, and $\check{A}_n$ is a complex constant called a phasor. The latter is indicated by a small flat v or inverted circumflex over the symbol, and its conjugate complex by a circumflex over the symbol. Vectors, shown by boldface type, may also be phasors. If i, the current density with amplitude i_0, has its maximum value when t is zero, then the instantaneous rate of dissipation of energy in an element dv is, from 6.03 (2),

$$dP = \tau i^2\, dv = \tau i_0^2 \cos^2 \omega t\, dv = \tau\check{i}\hat{i} \cos^2 \omega t\, dv$$

The power averaged over a cycle is

$$d\bar{P} = \tfrac{1}{2}\tau\check{i}\hat{i}\, dv \tag{10}$$

10.01. Solution for Vector Potential of Eddy Currents.—We can obtain solutions of 10.00 (8) when $\tau \neq \infty$ in just the same way we did when $\tau = \infty$ in 7.04. As in 7.04 (1) and (2), let us write $\boldsymbol{A}$ in the form

$$\boldsymbol{A} = \nabla \times (\boldsymbol{u}W_1 + \boldsymbol{u} \times \nabla W_2) \tag{1}$$

where $\boldsymbol{u} = \boldsymbol{i}, \boldsymbol{j}, \boldsymbol{k}$, or $\boldsymbol{r}$. From 7.04 (3) and (4), we have

$$\nabla^2\boldsymbol{A} = \nabla \times [\boldsymbol{u}\nabla^2 W_1 + \boldsymbol{u} \times \nabla(\nabla^2 W_2)]$$

Substituting from these two equations in 10.00 (8) gives

$$\nabla \times \left[\boldsymbol{u}\left(\nabla^2 W_1 - \frac{\mu}{\tau}\frac{dW_1}{dt}\right) + \boldsymbol{u} \times \nabla\left(\nabla^2 W_2 - \frac{\mu}{\tau}\frac{dW_2}{dt}\right)\right] = 0$$

Thus, if W_1 and W_2 are solutions of the heat conduction equation, *viz.*,

$$\nabla^2 W = \frac{\mu}{\tau}\frac{dW}{dt} \tag{2}$$

then, with the aid of (1), we can obtain from them solutions for the vector potential of eddy currents.

We calculate B in terms of W_1 and W_2 by making use of 7.04 (5) and (6). Thus,

$$B = \nabla \times A = \nabla \times (\nabla \times \mathbf{u} W_1) + \nabla \times (\mathbf{u}\nabla^2 W_2) \tag{3}$$

$$= \nabla \times \left[\frac{\mu}{\tau}\frac{d}{dt}(\mathbf{u}W_2) - \mathbf{u} \times \nabla W_1\right] \tag{4}$$

where $\mathbf{u} = \mathbf{i}, \mathbf{j}, \mathbf{k}$, or $\mathbf{r}$. Since B and A satisfy identical equations, 10.00 (8) and (9), the similar form of (1) and (4) is to be expected. We notice that now both W_1 and W_2 contribute to B. We can simplify this expression somewhat further by using 7.04 (6), which gives

$$\mathbf{u} = \mathbf{i}, \mathbf{j}, \mathbf{k}, \qquad B = -\frac{\mu}{\tau}\frac{d}{dt}(\mathbf{u} \times \nabla W_2 + \mathbf{u}W_1) + \mathbf{u} \cdot \nabla(\nabla W_1) \tag{5}$$

$$\mathbf{u} = \mathbf{r}, \qquad B = -\frac{\mu}{\tau}\frac{d}{dt}(\mathbf{r} \times \nabla W_2 + \mathbf{r}W_1) + \mathbf{r} \cdot \nabla(\nabla W_1) + 2\nabla W_1 \tag{6}$$

When $\mathbf{u}$ lies along u_1, a solution of (2) of the form $U(u_1)F(u_2, u_3, t)$ makes B normal to A as in 7.04.

10.02. Steady-state Skin Effect.—A simple solution of 10.00 (7) applies to a medium of permeability μ and resistivity τ filling all positive z-space. If the phasor current density $\check{i}_x$ is uniform, x-directed, and of angular frequency ω on the surface, then in the interior 10.00 (7) becomes

$$\frac{j\omega\mu}{\tau}\check{i}_x = j\omega\mu\gamma\check{i}_x = \frac{\partial^2 \check{i}_x}{\partial z^2} = \frac{d^2\check{i}_x}{dz^2} \tag{1}$$

since $\check{i}_x$ is a function of z only. The solution of this equation is

$$\check{i}_x = \check{C}e^{-(j\omega\mu\gamma)^{\frac{1}{2}}z} + \check{D}e^{(j\omega\mu\gamma)^{\frac{1}{2}}z} \tag{2}$$

If $\check{i}_x$ is finite when $z = \infty$, then $\check{D} = 0$; so writing $1 + j$ for $(2j)^{\frac{1}{2}}$ gives

$$\check{i}_x e^{j\omega t} = i_0 e^{-(\frac{1}{2}\omega\mu\gamma)^{\frac{1}{2}}z} e^{j[\omega t - (\frac{1}{2}\omega\mu\gamma)^{\frac{1}{2}}z]} \tag{3}$$

where i_0 is the value of $\check{i}_x$ at the surface. Taking the real part gives

$$i_x = i_0 e^{-(\frac{1}{2}\omega\mu\gamma)^{\frac{1}{2}}z} \cos\left[\omega t - (\tfrac{1}{2}\omega\mu\gamma)^{\frac{1}{2}}z\right] \tag{4}$$

Thus the i_x amplitude decrease is exponential and the phase change uniform.

The net current $I \cos(\omega t + \psi)$ per meter width, given by expanding the cosine and using *Dw* 863.1 and 863.2 or *Pc* 506 and 507, is

$$\int_0^\infty i_x\, dz = i_0(2\omega\mu\gamma)^{-\frac{1}{2}}(\cos\omega t + \sin\omega t) = i_0(\omega\mu\gamma)^{-\frac{1}{2}}\cos(\omega t - \tfrac{1}{4}\pi) \tag{5}$$

This would be increased by removing all conducting matter below a certain depth since, for certain values of z, the current is reversed. The power absorbed as heat, per square meter of surface, is, from 10.00 (10), with the aid of *Dw* 440.20 and 565.1 or *Pc* 363 and 401,

$$\bar{P}_1 = \frac{\omega\tau}{2\pi}\int_0^\infty \int_0^{\frac{2\pi}{\omega}} i_x^2\,dt\,dz = \frac{\tau \check{i}_0^2}{2}\int_0^\infty e^{-2(\frac{1}{2}\omega\mu\gamma)^{\frac{1}{2}}z}\,dz$$
$$= \tfrac{1}{4}\tau\check{i}_0^2(\tfrac{1}{2}\omega\mu\gamma)^{-\frac{1}{2}} = \tau I_e^2(\tfrac{1}{2}\omega\mu\gamma)^{\frac{1}{2}} = I_e^2(\gamma\delta)^{-1} = \tfrac{1}{2}\omega\mu\delta I_e^2 \tag{6}$$

where $I_e = (2)^{-\frac{1}{2}}I$ is the effective value of I. Thus the resistance equals the d-c resistance of a skin of thickness $\delta = (\frac{1}{2}\omega\mu\gamma)^{-\frac{1}{2}}$.

A uniform y-directed magnetic induction $B_0 \cos(\omega t - \frac{1}{2}\pi)$ at the surface would induce the eddy currents just considered. From 10.00 (7) and (9) the same differential equation describes the behavior of $\boldsymbol{i}$ and $\boldsymbol{B}$ in the conductor so that both die out exponentially. To find the power dissipated, take the line integral of $\boldsymbol{B}$ around a rectangular path normal to x whose sides lie in the z-direction and whose meter long ends lie, one just outside the surface where the induction is B_0 and the other far inside the conductor where it is zero. Only the B_0 end contributes to the integral so that from 7.01 (2) (or from Maxwell's first equation)

$$B_0 = \mu_v I = 2^{\frac{1}{2}}\mu_v I_e \tag{7}$$

From (6), the power dissipated by eddy currents in a surface area S in terms of the magnetic induction amplitude B_0 just outside the surface is

$$\bar{P} = \int_S \bar{P}_1\,dS = (2\mu_v^2\gamma\delta)^{-1}\int_S B_0^2\,dS \tag{8}$$

The inductance per square L_i' contributed by the magnetic field inside the conductor is found from (4), (6), 8.08 (1.1), and 10.00 (3).

$$\frac{L_i' I_e^2}{2} = \frac{\omega}{\pi}\int_0^\infty \int_0^{\frac{\pi}{2\omega}} \boldsymbol{i}\cdot\boldsymbol{A}\,dt\,dz = \frac{\tau}{4\pi}\int_0^\infty \int_0^{\frac{2\pi}{\omega}} i_x^2\,dt\,dz = \frac{\bar{P}}{2\omega} = \frac{\tau I_e^2}{2\omega\delta} \tag{9}$$

Comparison with the resistance R_i' per square gives the relation

$$L_i' = \frac{\tau}{\omega\delta} = \frac{1}{\omega\delta\gamma} = \frac{R_i'}{\omega} \tag{10}$$

At high frequencies L_i' is completely negligible.

10.03. Skin Effect on Tubular Conductor.—If the frequency is so high that the depth of penetration is small compared with the curvature of the conducting surface, then the results of the last article can be applied to the usual cylindrical wire or tubular conductor. If this is not true, we must use cylindrical coordinates. In this case, if we have symmetry about the axis so that i_z is a function of ρ only, 10.00 (7) becomes

$$\frac{j\omega\mu}{\tau}\check{i}_z = j\omega\mu\gamma\check{i}_z = jp\check{i}_z = \frac{\partial^2\check{i}_z}{\partial\rho^2} + \frac{1}{\rho}\frac{\partial\check{i}_z}{\partial\rho} \tag{1}$$

Let $v = (jp)^{\frac{1}{2}}\rho$, and this becomes

$$\frac{\partial^2 \check{\imath}_z}{\partial v^2} + \frac{1}{v}\frac{\partial \check{\imath}_z}{\partial v} - \check{\imath}_z = 0 \tag{2}$$

This is exactly Bessel's modified Eq. 5.292 (1) of order zero, the solutions of which were found in 5.32 (6) to be

$$\check{\imath}_z = \check{C}I_0(v) + \check{D}K_0(v) = \check{C}I_0[(jp)^{\frac{1}{2}}\rho] + \check{D}K_0[(jp)^{\frac{1}{2}}\rho] \tag{3}$$

From 7.16 (2), we know that there can be no magnetic field in the cavity in this tube because of the symmetry. If the external and internal radii of the tube are a and b, respectively, we see that the boundary condition at $\rho = b$ is $\boldsymbol{B} = 0$. From 10.00 (3), this gives

$$(\nabla \times \check{\imath})_{\rho=b} = 0 \qquad \text{or} \qquad \left(\frac{\partial \check{\imath}_z}{\partial \rho}\right)_{\rho=b} = 0 \tag{4}$$

At $\rho = a$, the boundary condition is

$$(\check{\imath}_z)_{\rho=a} = i_0 \tag{5}$$

From (3) and (4) using 5.33 (4), we have

$$0 = \check{C}I_1[(jp)^{\frac{1}{2}}b] - \check{D}K_1[(jp)^{\frac{1}{2}}b]$$

and from (3) and (5),

$$i_0 = \check{C}I_0[(jp)^{\frac{1}{2}}a] + \check{D}K_0[(jp)^{\frac{1}{2}}a]$$

From these, we obtain

$$\check{C} = \frac{K_1[(jp)^{\frac{1}{2}}b]i_0}{I_0[(jp)^{\frac{1}{2}}a]K_1[(jp)^{\frac{1}{2}}b] + I_1[(jp)^{\frac{1}{2}}b]K_0[(jp)^{\frac{1}{2}}a]} \tag{6}$$

$$\check{D} = \frac{I_1[(jp)^{\frac{1}{2}}b]i_0}{I_0[(jp)^{\frac{1}{2}}a]K_1[(jp)^{\frac{1}{2}}b] + I_1[(jp)^{\frac{1}{2}}b]K_0[(jp)^{\frac{1}{2}}a]} \tag{7}$$

10.04. Skin Effect on Solid Cylindrical Conductor.—If we have a solid wire instead of a tube, we must set $\check{D} = 0$ in 10.03 (3) since $K_0(x)$ is infinite when $x = 0$ by 5.33. We would then have for $\check{\imath}_z$

$$\check{\imath}_z = \frac{I_0[(jp)^{\frac{1}{2}}\rho]}{I_0[(jp)^{\frac{1}{2}}a]}i_0 \tag{1}$$

instead of the result given by 10.03 (3), (6), and (7). For numerical computation, it is necessary to split $I_0[(j)^{\frac{1}{2}}x]$ and $K_0[(j)^{\frac{1}{2}}x]$ into real and imaginary parts. This is done by means of the *ber*, *bei*, *ker*, and *kei* functions of Lord Kelvin, for which numerous tables exist; see, for example, *Dw* 1050 or HMF, pages 430 to 433. Thus

$$I_0[(j)^{\frac{1}{2}}x] \equiv ber_0 x + j bei_0 x \tag{2}$$

$$K_0[(j)^{\frac{1}{2}}x] \equiv ker_0 x + j kei_0 x \tag{3}$$

Series for these functions are easily obtained by substituting in 5.33 (1) and (4). Using *ber* and *bei* functions in (1) multiplied by $e^{j\omega t}$ and taking the real part, we have for the current density in a solid wire of radius a

$$i_z = \left\{\frac{ber_0^2[(p)^{\frac{1}{2}}\rho] + bei_0^2[(p)^{\frac{1}{2}}\rho]}{ber_0^2[(p)^{\frac{1}{2}}a] + bei_0^2[(p)^{\frac{1}{2}}a]}\right\}^{\frac{1}{2}} i_0 \cos(\omega t + \alpha) \tag{4}$$

$$\alpha = \tan^{-1}\frac{ber_0[(p)^{\frac{1}{2}}a]bei_0[(p)^{\frac{1}{2}}\rho] - ber_0[(p)^{\frac{1}{2}}\rho]bei_0[(p)^{\frac{1}{2}}a]}{ber_0[(p)^{\frac{1}{2}}a]ber_0[(p)^{\frac{1}{2}}\rho] + bei_0[(p)^{\frac{1}{2}}\rho]bei_0[(p)^{\frac{1}{2}}a]} \tag{5}$$

We can obtain the total current in the wire at any instant from the magnetic field just inside the surface, since by 8.09 (1), they are connected by the relation $B = \mu I/(2\pi a)$. From (1), 10.00 (3), and 5.33 (4), we have, after dividing out $e^{j\omega t}$,

$$j\omega(\check{B})_a = -\tau(\nabla \times \check{\imath})_a = \phi\tau\left(\frac{\partial \check{\imath}_z}{\partial\rho}\right)_a = \phi\frac{(jp)^{\frac{1}{2}}I_0'[(jp)^{\frac{1}{2}}a]\tau i_0}{I_0[(jp)^{\frac{1}{2}}a]}$$

or

$$\check{I} = \frac{2\pi a\check{B}}{\mu} = \frac{2\pi\tau(jp)^{\frac{1}{2}}aI_0'[(jp)^{\frac{1}{2}}a]}{j\omega\mu I_0[(jp)^{\frac{1}{2}}a]}i_0 \tag{6}$$

The average power dissipated per unit length in a ring of radius ρ and thickness $d\rho$ is, from 10.00 (10),

$$d\bar{P}_{av} = \tfrac{1}{2}\tau|\check{\imath}_z|^2 2\pi\rho\, d\rho = \pi\tau\check{\imath}_z\hat{\imath}_z\rho\, d\rho$$

where $\hat{\imath}_z$ is the conjugate phasor of $\check{\imath}_z$. We should note here that for electrical purposes the conjugate complex of $(+j)^{\frac{1}{2}} = 2^{-\frac{1}{2}}(1 + j)$ is $(-j)^{\frac{1}{2}} = 2^{-\frac{1}{2}}(1 - j) = -j(j)^{\frac{1}{2}}$. We may write

$$\hat{I}_0[(jp)^{\frac{1}{2}}\rho] = \check{I}_0[(-jp)^{\frac{1}{2}}\rho] = \check{I}_0[-j(jp)^{\frac{1}{2}}\rho]$$

The total power consumed per unit length in the wire is then

$$\bar{P}_{av} = \int_0^a d\bar{P}_{av} = \frac{\pi\tau i_0^2}{I_0[(jp)^{\frac{1}{2}}a]I_0[-j(jp)^{\frac{1}{2}}a]}\int_0^a I_0[-j(jp)^{\frac{1}{2}}\rho]I_0[(jp)^{\frac{1}{2}}\rho]\rho\, d\rho$$

This integral is a special case of 5.323 (1) obtained by setting $n = 0$, and the result written in terms of $ber_0 x$ and $bei_0 x$ is

$$\bar{P}_{av} = \frac{\tau\pi a}{(p)^{\frac{1}{2}}}\frac{ber_0[(p)^{\frac{1}{2}}a]bei_0'[(p)^{\frac{1}{2}}a] - ber_0'[(p)^{\frac{1}{2}}a]bei_0[(p)^{\frac{1}{2}}a]}{ber_0^2[(p)^{\frac{1}{2}}a] + bei_0^2[(p)^{\frac{1}{2}}a]}i_0^2 \tag{7}$$

From (6), the rms current I_e is

$$I_e^2 = \frac{2\pi^2\tau^2pa^2}{\mu^2\omega^2}\frac{\check{I}_0'[(jp)^{\frac{1}{2}}a]\hat{I}_0'[(jp)^{\frac{1}{2}}a]}{\check{I}_0[(jp)^{\frac{1}{2}}a]\hat{I}_0[(jp)^{\frac{1}{2}}a]}i_0^2$$

$$= \frac{2\pi^2a^2}{p}\frac{\{ber_0'[(p)^{\frac{1}{2}}a]\}^2 + \{bei_0'[(p)^{\frac{1}{2}}a]\}^2}{\{ber_0[(p)^{\frac{1}{2}}a]\}^2 + \{bei_0[(p)^{\frac{1}{2}}a]\}^2}i_0^2 \tag{8}$$

If $R = \tau/(\pi a^2)$ is the resistance per unit length for direct currents, then the high-frequency resistance R' is

$$R' = \frac{\bar{P}_{av}\pi a^2}{\tau I_e^2}R = \frac{a(p)^{\frac{1}{2}}}{2}\frac{ber[(p)^{\frac{1}{2}}a]bei'[(p)^{\frac{1}{2}}a] - ber'[(p)^{\frac{1}{2}}a]bei[(p)]^{\frac{1}{2}}a]}{\{ber'[(p)^{\frac{1}{2}}a]\}^2 + \{bei'[(p)^{\frac{1}{2}}a]\}^2}R \tag{9}$$

where, as in many tables, the zero subscript is omitted and from 10.03 (1)

$$p = \tau^{-1}\mu\omega = \gamma\mu\omega \tag{10}$$

and a is the radius of the cylinder.

One gets the magnetic field energy inside the wire as in (6), thus

$$j\omega B = \boldsymbol{\phi}\tau\left(\frac{\partial \breve{\imath}_z}{\partial \rho}\right)_a = \frac{\boldsymbol{\phi}\tau i_0}{I_0[(jp)^{\frac{1}{2}}a]}\frac{\partial I_0[(jp)^{\frac{1}{2}}a]}{\partial a}$$

The average energy inside the wire is, from 8.02 (3),

$$\frac{\pi}{2\mu}\int_0^a \breve{B}\cdot\hat{B}\rho\,d\rho = \frac{\pi\tau^2 i_0^2 p}{2\mu\omega^2 I_0[(jp)^{\frac{1}{2}}a]I_0[(-jp)^{\frac{1}{2}}a]}\int_0^a \breve{I}_0'\hat{I}_0'\rho\,d\rho$$

where we have written I_0 for $I_0[(jp)^{\frac{1}{2}}\rho]$ and I_0' for $\partial I_0(x)/\partial x$. Since $I_0'(x)$ equals $I_1(x)$ by 5.33 (4), we see that this integral is identical with 5.323 (1) when $n = 1$, so that its value in terms of $ber_1 x$ and $bei_1 x$ is

$$ap^{-\frac{1}{2}}[ber_1(p^{\frac{1}{2}}a)bei_1'(p^{\frac{1}{2}}a) - ber_1'(p^{\frac{1}{2}}a)bei_1(p^{\frac{1}{2}}a)]$$

Using *Dw* 828.1, 828.2, 829.3, and 829.4 to reduce this to zero order and remembering that this average energy equals $\frac{1}{2}L_i I_e^2$ where L_i is the internal self-inductance per unit length and I_e^2 is given by (8), we obtain

$$L_i = \frac{\tau(p)^{\frac{1}{2}}}{2\pi\omega a}\frac{ber[(p)^{\frac{1}{2}}a]ber'[(p)^{\frac{1}{2}}a] + bei[(p)^{\frac{1}{2}}a]bei'[(p)^{\frac{1}{2}}a]}{\{ber'[(p)^{\frac{1}{2}}a]\}^2 + \{bei'[(p)^{\frac{1}{2}}a\}^2} \tag{11}$$

10.05. Solution in Spherical Coordinates for Axial Symmetry.—Let us assume that the magnetic field producing the eddy currents is independent of ϕ and has no ϕ-component. Then the vector potential has only a ϕ-component and may be written

$$A = \boldsymbol{\phi}A_\phi(r, \theta, t) \tag{1}$$

where $\boldsymbol{\phi}$ is a unit vector in the ϕ-direction given by

$$\boldsymbol{\phi} = -\boldsymbol{i}\sin\phi + \boldsymbol{j}\cos\phi \tag{2}$$

Apply Laplace's operator 3.05 (1) to $A_x = -A_\phi \sin\phi$ and $A_y = A_\phi \cos\phi$ separately. After recombining we see that 10.00 (8) becomes

$$\frac{\mu}{\tau}\boldsymbol{\phi}\frac{\partial A_\phi}{\partial t} = \nabla^2 A = \boldsymbol{\phi}\left[\nabla^2 A_\phi - \frac{A_\phi}{r^2\sin^2\theta}\right] = \boldsymbol{\phi}\nabla^2 A_\phi + A_\phi\nabla^2\boldsymbol{\phi} \tag{3}$$

Writing out ∇^2 in polar coordinates by 3.05 (1) and dividing out $\boldsymbol{\phi}$ give

$$\begin{aligned}\frac{\mu}{\tau}\frac{\partial A_\phi}{\partial t} &= \frac{1}{r^2}\frac{\partial}{\partial r}\left(r^2\frac{\partial A_\phi}{\partial r}\right) + \frac{1}{r^2\sin\theta}\frac{\partial}{\partial\theta}\left(\sin\theta\frac{\partial A_\phi}{\partial\theta}\right) - \frac{A_\phi}{r^2\sin^2\theta}\\ &= \frac{1}{r^2}\frac{\partial}{\partial r}\left(r^2\frac{\partial A_\phi}{\partial r}\right) + \frac{(1-u^2)^{\frac{1}{2}}}{r^2}\frac{\partial^2[(1-u^2)^{\frac{1}{2}}A_\phi]}{\partial u^2}\end{aligned} \tag{4}$$

where $u = \cos\theta$.

We shall now consider the steady-state eddy currents when the magnetic field oscillates with an angular frequency ω. As in 5.12, we shall seek a solution that is the product of a function of θ by a function of r. Then we write

$$A_\phi = \Theta r^{-\frac{1}{2}}\check{R}e^{j\omega t} \text{ (real part)} \tag{5}$$

Substituting in (4), multiplying through by r^2, and dividing through by $\Theta r^{-\frac{1}{2}}\check{R}e^{j\omega t}$ give

$$\frac{r^2}{\check{R}}\frac{d^2\check{R}}{dr^2} + \frac{r}{\check{R}}\frac{d\check{R}}{dr} - \frac{1}{4} - jpr^2 + \frac{(1-u^2)^{\frac{1}{2}}}{\Theta}\frac{d^2}{du^2}[(1-u^2)^{\frac{1}{2}}\Theta] = 0 \tag{6}$$

where, as in 10.04 (10), we have written

$$p = \tau^{-1}\mu\omega = \gamma\mu\omega \tag{7}$$

Proceeding as we did in solving 5.12 (1), we set the terms in (6) involving θ equal to $-n(n+1)$ and those involving r equal to $n(n+1)$, thus satisfying (6) and giving, after expanding the derivatives in (6),

$$(1-u^2)\frac{d^2\Theta_n}{du^2} - 2u\frac{d\Theta_n}{du} - \frac{\Theta_n}{1-u^2} + n(n+1)\Theta_n = 0 \tag{8}$$

$$\frac{d^2\check{R}_n}{dr^2} + \frac{1}{r}\frac{d\check{R}_n}{dr} - \left[jp + \frac{n(n+1)+\frac{1}{4}}{r^2}\right]\check{R}_n = 0 \tag{9}$$

The first of these is identical with the differential equation for Legendre's associated functions with $m = 1$ as written in 5.23 (4.1), and the second is the modified Bessel equation in x where $x = (jp)^{\frac{1}{2}}r$ as written in 5.32 (1). Thus, from 5.23 (5) and 5.32, A_ϕ is the real part of

$$r^{-\frac{1}{2}}[A_nP_n^1(u) + B_nQ_n^1(u)]\{\check{C}_nI_{n+\frac{1}{2}}[(jp)^{\frac{1}{2}}r] + \check{D}_nK_{n+\frac{1}{2}}[(jp)^{\frac{1}{2}}r]\}e^{j\omega t} \tag{10}$$

If n is an integer, as it must be for $P_n^1(u)$ and $Q_n^1(u)$ unless conical boundaries are involved, we can use $I_{-(n+\frac{1}{2})}$ instead of $K_{n+\frac{1}{2}}$ for the second solution, by 5.37.

In a region where the conductivity is zero, the left side of (4) is zero, and if we let $\check{A}_\phi e^{j\omega t} = \check{R}'\Theta e^{j\omega t}$, we obtain (8) as before but instead of (9) we get

$$\frac{d}{dr}\left(r^2\frac{d\check{R}'}{dr}\right) - n(n+1)\check{R}' = 0 \tag{11}$$

whose solution is given by 5.12 (3) to be

$$\check{R}' = \check{A}r^n + \check{B}r^{-n-1} \tag{12}$$

and in a nonconducting region this will replace terms involving r in (10).

10.06. Conducting Sphere in Alternating Field.—Consider now the specific example of a sphere of resistivity τ, permeability μ, and radius a placed in a uniform alternating z-directed magnetic field $\check{B}e^{j\omega t}$. The phasor vector potential of this field is, when $e^{j\omega t}$ is divided out,

$$\check{A} = \boldsymbol{\phi}\tfrac{1}{2}\check{B}r\sin\theta = \boldsymbol{\phi}\tfrac{1}{2}\check{B}rP_1^1(\cos\theta) \tag{1}$$

as can be verified easily by taking its curl by 3.04 (2) and (3). Thus, $n = 1$ in 10.05 (10) and (12); and, since the eddy current vector potential must vanish at infinity, we have, outside the sphere,

$$a < r < \infty, \qquad \breve{A}_o = \boldsymbol{\phi}\tfrac{1}{2}\breve{B}(r + \breve{D}r^{-2})\sin\theta \tag{2}$$

At $r = 0$, $\breve{A}_i$ is finite, so 5.37 (4) and (5) show that only $I_{\frac{3}{2}}[(jp)^{\frac{1}{2}}r]$ can occur inside the sphere. Thus, setting $n = 1$ in 10.05 (10), we have

$$0 < r < a, \qquad \breve{A}_i = \boldsymbol{\phi}\tfrac{1}{2}\breve{B}\breve{C}r^{-\frac{1}{2}}I_{\frac{3}{2}}[(jp)^{\frac{1}{2}}r]\sin\theta \tag{3}$$

From 7.21 (6) and (7), the boundary conditions when $r = a$ are

$$\breve{A}_o = \breve{A}_i \qquad \text{and} \qquad \mu_v\frac{\partial}{\partial r}(r\sin\theta\breve{A}_i) = \mu\frac{\partial}{\partial r}(r\sin\theta\breve{A}_o) \tag{4}$$

Putting $r = a$ in (2) and (3), we obtain with the aid of 5.321 (1), (2), and (3), after writing I_n for $I_n[(jp)^{\frac{1}{2}}a]$ and v for $(jp)^{\frac{1}{2}}a$,

$$a^3 + \breve{D} = a^{\frac{3}{2}}\breve{C}I_{\frac{3}{2}} = a^{\frac{3}{2}}\breve{C}[I_{-\frac{1}{2}} - v^{-1}I_{\frac{1}{2}}]$$
$$(2a^3 - \breve{D})\mu = \mu_v a^{\frac{3}{2}}[\tfrac{1}{2}I_{\frac{3}{2}} + vI_{\frac{3}{2}}']\breve{C} = \mu_v a^{\frac{3}{2}}[(v + v^{-1})I_{\frac{1}{2}} - I_{-\frac{1}{2}}]\breve{C}$$

Solving for $\breve{C}$ and $\breve{D}$ gives

$$\breve{C} = \frac{3\mu v a^{\frac{3}{2}}}{(\mu - \mu_v)vI_{-\frac{1}{2}} + [\mu_v(1 + v^2) - \mu]I_{\frac{1}{2}}} \tag{5}$$

$$\breve{D} = \frac{(2\mu + \mu_v)vI_{-\frac{1}{2}} - [\mu_v(1 + v^2) + 2\mu]I_{\frac{1}{2}}}{(\mu - \mu_v)vI_{-\frac{1}{2}} + [\mu_v(1 + v^2) - \mu]I_{\frac{1}{2}}}a^3 \tag{6}$$

This may be expressed in terms of hyperbolic functions by 5.37 or *Dw* 808.1 and 808.3. From 10.00 (4), the current density anywhere inside can be obtained from (3) by the equation

$$\breve{\mathbf{i}} = -j\omega\gamma\breve{A}_i = -jp\mu^{-1}\breve{A}_i \tag{7}$$

The magnetic field at any point outside is obtained from (2), 3.04 (3), and 3.04 (2) to be

$$\breve{B}_{o\theta} = -\frac{1}{r}\frac{\partial}{\partial r}(r\breve{A}_o) = -\breve{B}\left(1 - \frac{\breve{D}}{2r^3}\right)\sin\theta \tag{8}$$

$$\breve{B}_{or} = \frac{1}{r\sin\theta}\frac{\partial}{\partial\theta}(\sin\theta\breve{A}_o) = \breve{B}\left(1 + \frac{\breve{D}}{r^3}\right)\cos\theta \tag{9}$$

By a similar method (3) yields $\breve{B}_{i\theta}$ and $\breve{B}_{ir}$. Comparison of (8) or (9) with 7.10 (2) shows the eddy current field to be like that of a magnetic dipole loop of radius a carrying a current $\breve{I}e^{j\omega t}$ where $\mu_v a^2\breve{I} = 2\breve{B}\breve{D}$. If the magnetic field is not alternating, then $\omega = 0$ so that $p \to 0$ from 11.05 (7). From 5.37 and *Dw* 657.1 or 657.2, we see that

$$I_{\frac{1}{2}}(x) \underset{x\to 0}{\longrightarrow} \left(\frac{2}{\pi x}\right)^{\frac{1}{2}}\left(x + \frac{x^3}{6}\right) \qquad \text{and} \qquad I_{-\frac{1}{2}}(x) \underset{x\to 0}{\longrightarrow} \left(\frac{2}{\pi x}\right)^{\frac{1}{2}}\left(1 + \frac{x^2}{2}\right)$$

Thus (5) and (6) simplify so that (2) and (3) become

$$\breve{A}_o = \phi\frac{\breve{B}}{2}\left[r + \frac{2(K_m - 1)a^3}{(K_m + 2)r^2}\right]\sin\theta \tag{10}$$

$$\breve{A}_i = \phi\frac{3K_m\breve{B}}{2(K_m + 2)}r\sin\theta \tag{11}$$

These are the correct static fields. From (7), we obtain as a first approximation for slowly alternating fields

$$\breve{\imath}_\phi = -\frac{3j\omega K_m\gamma\breve{B}}{2(K_m + 2)}r\sin\theta \tag{12}$$

The same result is obtained if the resistivity is made infinite.

When the frequency is made very high, we see that

$$\breve{A}_i \to 0 \qquad \breve{A}_o \to \phi\tfrac{1}{2}\breve{B}(r - a^3r^{-2})\sin\theta \tag{13}$$

because, from 5.37,

$$I_{\frac{1}{2}}(x) \xrightarrow[x\to\infty]{} I_{\frac{3}{2}}(x) \xrightarrow[x\to\infty]{} I_{-\frac{1}{2}}(x) \xrightarrow[x\to\infty]{} \cdots \xrightarrow[x\to\infty]{} \left(\frac{2}{\pi x}\right)^{\frac{1}{2}}\frac{e^x}{2}$$

thus, there is no magnetic field inside, and the eddy currents are confined to the surface as we would expect.

To visualize the magnitudes of the numbers involved, we shall compute p by 10.05 (7) for a field alternating at 60 cycles per second (so that $\omega = 120\pi$) for several substances. If τ and μ are in mks units, we find for copper $\tau \approx 1.7 \times 10^{-8}$ ohm-meter, $\mu \approx \mu_v = 4\pi \times 10^{-7}$ henry per meter, and $p \approx 28{,}000$. For a typical specimen of iron, $\tau \approx 10^{-7}$, $\mu \approx 480\pi \times 10^{-7}$ for a magnetic field intensity of $1.5 \times 10^5/(4\pi)$ ampere-turns per meter, and $p = 570{,}000$. For graphite, $\tau \approx 8 \times 10^{-6}$, $\mu \approx 4\pi \times 10^{-7}$, and $p = 60$. Thus for this frequency, over distances of a few centimeters, (12) would apply to graphite but not to iron or copper. The initial assumption that $\boldsymbol{A}_i$ is zero and $\boldsymbol{B}_0$ tangential to the surface greatly simplifies the calculation.

All the results of this section, being written in phasors, give both the amplitude and phase of the quantities involved. The same forms for the electromotances and the same types of boundary conditions apply equally well to any number of thick concentric spherical shells and could be used, for example, to calculate their screening effect. The results would be much more complicated than those given here. The distribution of a given amount of material in several separated shells will increase the screening effect. There are optimum thicknesses and spacings.

10.07. Power Absorbed by Sphere in Alternating Magnetic Field.—We shall now compute the power absorbed by the sphere in the last article. From 10.00 (10), the power absorbed in the volume element dv is

$$d\bar{P} = \tfrac{1}{2}\tau\breve{\imath}\hat{\imath}\,dv = \pi\tau\breve{\imath}\hat{\imath}r^2\sin\theta\,dr\,d\theta$$

Substitution for $\check{\imath}$ and $\hat{\imath}$ from 10.06 (7) and (3) and integration with respect to θ from $\theta = 0$ to $\theta = \pi$ give

$$\bar{P} = \frac{\pi\omega^2 B^2}{3\tau}\check{C}\hat{C}\int_0^a I_{\frac{3}{2}}[(jp)^{\frac{1}{2}}r]I_{\frac{3}{2}}[(-jp)^{\frac{1}{2}}r]r\,dr \tag{1}$$

Integrate by the last form of 5.323 (1) and note from 5.37 that

$$I_{\frac{1}{2}}[(\pm jp)^{\frac{1}{2}}a] = [\tfrac{1}{2}\pi(\pm jp)^{\frac{1}{2}}a]^{-\frac{1}{2}} \sinh[\tfrac{1}{2}(2p)^{\frac{1}{2}}(1 \pm j)a]$$
$$I_{-\frac{1}{2}}[(\pm jp)^{\frac{1}{2}}a] = [\tfrac{1}{2}\pi(\pm jp)^{\frac{1}{2}}a]^{-\frac{1}{2}} \cosh[\tfrac{1}{2}(2p)^{\frac{1}{2}}(1 \pm j)a] \tag{2}$$

Application of *Dw* 651.06 to 651.09 or *Pc* 669 to 672 gives for the integral

$$(\pi p^{\frac{3}{2}}a)^{-1}\{\tfrac{1}{2}(2pa^2)^{\frac{1}{2}}[\sinh(2pa^2)^{\frac{1}{2}} + \sin(2pa^2)^{\frac{1}{2}}] - \cosh(2pa^2)^{\frac{1}{2}} + \cos(2pa^2)^{\frac{1}{2}}\}$$

Writing out $\check{C}\hat{C}$ by (2), 10.06 (5), and 11.05 (7) gives for $\bar{P}$, if $U = \mu - \mu_v$,

$$\frac{3\pi a^5\omega^2 B^2\mu^2\tau^{-1}[\frac{1}{2}u(S + s) - C + c]}{U^2[(pa^2+1)C + (pa^2-1)c - u(S+s)] + U\mu_v pa^2 u(S-s) + \mu_v^2 p^2 a^4(C-c)}$$

where $u = (2p)^{\frac{1}{2}}a$, $C = \cosh u$, $c = \cos u$, $S = \sinh u$, and $s = \sin u$.

10.08. Transients in Conducting Sphere.—In the last two articles we solved the steady-state problem of a sphere of permeability μ and resistivity τ in a uniform alternating magnetic field. Let us now solve a transient problem by calculating the effects when the same sphere is placed in a uniform magnetic field $\boldsymbol{B}$ which is suddenly removed at the time $t = 0$. Clearly, in this case, as in the last one, $\boldsymbol{A}$ will have a ϕ-component only. At $t = 0$, surface eddy currents prevent the interior field from changing, and since the vector potential must be continuous across the boundary, we have, from 10.06 (11), at $t = 0$,

$$A_i = \phi\frac{3K_m B}{2(K_m + 2)}rP_1^1(u), \qquad A_o = \phi\frac{3K_m B}{2(K_m + 2)}\frac{a^3}{r^2}P_1^1(u) \tag{1}$$

Since $\tau = \infty$ outside, the subsequent behavior of $\boldsymbol{A}$ is determined from 10.00 (8) by solving the equations

$$\nabla^2 A_i = \frac{\mu}{\tau}\frac{dA_i}{dt}, \qquad \nabla^2 A_o = 0 \tag{2}$$

In 10.05, we found a solution of the first of these equations when the time entered exponentially. Let us see if we can fit the boundary conditions in the present case by using a sum of such solutions. Clearly in the present case there will be no oscillations, so we shall need to substitute a negative exponent $-q_s t$ for $j\omega t$. Let us write

$$k_s^2 = \tau^{-1}\mu q_s = \gamma\mu q_s \tag{3}$$

We must then substitute $-k_s^2$ for jp throughout 10.05. This gives us jk_s instead of $(jp)^{\frac{1}{2}}$ in 10.05 (10) and leads to ordinary Bessel functions. From 10.05 (10), writing $\sin\theta$ for $P_1^1(u)$ and noting that A_i must be finite at $r = 0$ and contain θ in the same form as (1), we obtain

$$A_i = \phi \sum_s A_s r^{-\frac{1}{2}} J_{\frac{3}{2}}(k_s r) \sin\theta e^{-q_s t} \tag{4}$$

A_o is finite at infinity and equals A_i at $r = a$ for all values of t. From 10.05 (12), it therefore has the form

$$A_o = \phi \sum_s B_s r^{-2} \sin\theta\, e^{-q_s t} \tag{5}$$

In addition to satisfying (1) when $t = 0$, we must satisfy 10.06 (4). This gives at $r = a$, dividing out $\sin\theta$,

$$A_o = A_i, \qquad \mu_v \frac{\partial(rA_i)}{\partial r} = \mu \frac{\partial(rA_o)}{\partial r}$$

Therefore, at $r = a$ for all values of t, it is necessary that

$$A_s a^{\frac{3}{2}} J_{\frac{3}{2}}(k_s a) = B_s, \qquad \mu_v A_s a^2 \frac{d}{da}[a^{\frac{1}{2}} J_{\frac{3}{2}}(k_s a)] = -\mu B_s \tag{6}$$

Differentiating the product in the second equation, multiplying the first equation by μ, and adding and dividing out A_s, we obtain

$$\mu_v a \frac{d}{da}[J_{\frac{3}{2}}(k_s a)] + \left(\mu + \frac{1}{2}\mu_v\right) J_{\frac{3}{2}}(k_s a) = 0 \tag{7}$$

To meet the boundary conditions, we must choose values of k_s that satisfy this equation. This determines, by (3), the values of q_s that appear in (4) and (5). The values of k_s can be found with the aid of 5.31 (3) and a table of trigonometric functions.

From (1), (4), and (5), multiplying through by $r^{\frac{1}{2}}$ and setting $t = 0$, we have

$$\frac{3K_m B}{2(K_m + 2)} r^{\frac{3}{2}} = \sum_s A_s J_{\frac{3}{2}}(k_s r) \tag{8}$$

This is identical with 5.297 (1), last case, where $n = \frac{3}{2}$, and, from (7), the B of that article equals $(\mu/\mu_v) + \frac{1}{2}$. From 5.294 (7),

$$\int_0^a v f(v) J_{\frac{3}{2}}(k_s v)\, dv = \frac{3K_m B}{2(K_m + 2)} \int_0^a v^{\frac{5}{2}} J_{\frac{3}{2}}(k_s v)\, dv = \frac{3K_m B a^{\frac{5}{2}}}{2k_s(K_m + 2)} J_{\frac{5}{2}}(k_s a)$$

With the aid of (7) and 5.294 (2), we obtain

$$J_{\frac{5}{2}}(k_s a) = -J'_{\frac{3}{2}}(k_s a) + \frac{3}{2k_s a} J_{\frac{3}{2}}(k_s a) = \frac{K_m + 2}{k_s a} J_{\frac{3}{2}}(k_s a)$$

So that, from 5.297 (6) we have

$$A_s = \frac{3K_m B a^{\frac{3}{2}}}{[k_s^2 a^2 + (K_m - 1)(K_m + 2)] J_{\frac{3}{2}}(k_s a)} \tag{9}$$

Substituting this in (4) and (5) gives A_i and A_o. With the aid of 10.00 (4), we obtain for the current density inside

$$i = \phi\frac{3Ba^{\frac{1}{2}}\sin\theta}{\mu_v r^{\frac{1}{2}}}\sum_s \frac{k_s^2 J_{\frac{3}{2}}(k_s r)}{[k_s^2 a^2 + (K_m - 1)(K_m + 2)]J_{\frac{3}{2}}(k_s a)} e^{-k_s^2 \tau \mu^{-1} t} \quad (10)$$

10.09. Eddy Currents in Plane Sheets.—We shall now calculate the vector potential $A(x, y, z, t)$ of the eddy currents of area density $\boldsymbol{i}$ induced in a very thin plane sheet of area resistivity $\mathfrak{s}$ at $z = 0$ by a fluctuating magnetic field whose vector potential is $A'(x, y, z, t)$. In the interior of the sheet, the electric field $-\partial(A'_z + A_z)/\partial t$ drives to the surface charges whose electrostatic field exactly neutralizes it. The effect of the associated current is negligible, so that inside the sheet we need consider only the tangential components A_s and A'_s. Thus using 10.00 (4), we may write

$$-\frac{d(A'_s + A_s)}{dt} = \mathfrak{s}\boldsymbol{i} \quad (1)$$

Let the eddy currents be confined to a finite region of the sheet which may or may not extend to infinity, and let us define the stream function $\Phi(x, y)$ at any point P in the sheet to be the current flowing through any cross section of the sheet extending from P to its edge. From 7.01 (2), we have, for the closed path bounding this cross section, but not including the surface, since $\boldsymbol{B}$ is symmetrical with respect to the sheet,

$$\mu_v \Phi = \oint \boldsymbol{B} \cdot d\boldsymbol{s} = 2\int_x^\infty B_x\, dx = 2\int_y^\infty B_y\, dy \quad (2)$$

where the path of integration is in the positive x- or y-direction when z is positive. Differentiating this equation, we have

$$i_x = \frac{\partial \Phi}{\partial y} = -\frac{2B_y}{\mu_v} = -\frac{2}{\mu_v}\frac{\partial A_x}{\partial z} \qquad i_y = -\frac{\partial \Phi}{\partial x} = \frac{2B_x}{\mu_v} = -\frac{2}{\mu_v}\frac{\partial A_y}{\partial z}$$

Adding the components gives, inside the sheet

$$\boldsymbol{i} = -\frac{2}{\mu_v}\frac{\partial A_s}{\partial z} \quad (3)$$

Substituting this in (1) gives, inside the sheet,

$$\frac{d(A'_s + A_s)}{dt} = \frac{2\mathfrak{s}}{\mu_v}\frac{\partial A_s}{\partial z} \quad (4)$$

Outside the sheet, A consists of two parts, a magnetic part arising from the eddy currents and an electric part, the gradient of a scalar, arising from the electric double layer produced by A'_z and A_z. Since A'_s, the tangential component of the exciting potential, is known and since A_s is continuous across each face of the double layer, (4) gives the boundary condition on A_s just outside either surface of the sheet. This, combined with the equation $\nabla^2 A = 0$, determines A everywhere outside the sheet. When the sheet is finite, the boundary condition in its plane, but beyond its edge, becomes $i = 0$ or $da_s/dt = 0$. The right side of (4) is finite at all times, which means that if $\delta t \to 0$, then $\delta(A'_s + A_s) \to 0$. Thus an abrupt change in A' instantaneously induces eddy currents such as will maintain $A + A'$ and $B + B'$ unchanged in the sheet. Therefore, for a specified change in A', the initial value of A is known and, if no further changes occur, its subsequent values, as the eddy currents decay, can be determined by putting $dA'_s/dt = 0$ in (4) and solving. A second abrupt change in A' produces a second set of eddy currents, and so forth. At any instant, the actual field of the eddy currents is a superposition of these. As the magnitudes of the discontinuous changes in the external field become smaller and the intervals between them shorter, we approach, as a limit, a continuously changing magnetic field.

10.10. Eddy Currents in Infinite Plane Sheet by Image Method.—Suppose that a thin infinite plane conducting sheet at $z = 0$ lies in a magnetic field produced by sources in the region $z > 0$. At $t = 0$ the field is changed, the vector potential being given by $A'_1 = f_1(x, y, z)$ when $t < 0$ and by $A'_2 = f_2(x, y, z)$ when $0 < t$. From the last article, the eddy currents produced at the instant $t = 0$ keep the vector potential over the surface of the sheet the same so that, initially, the whole field on the negative side is unchanged. Thus we have, on the negative side of the sheet, from the eddy currents alone

$$(A)_{t=0} = A'_1 - A'_2 = f_1(x, y, z) - f_2(x, y, z) \tag{1}$$

This field could be produced by reversing the sign of the new source and replacing the old source. These hypothetical sources which can replace the actual eddy currents are images as in electrostatics.

Since A'_2 is not a function of t, 10.09 (4) reduces to

$$\frac{dA}{dt} = \frac{2s}{\mu_v}\frac{\partial A}{\partial z} \tag{2}$$

A general solution of this equation satisfying (1) at $t = 0$ is

$$A = f_1(x, y, -|z| - 2s\mu_v^{-1}t) - f_2(x, y, -|z| - 2s\mu_v^{-1}t) \tag{3}$$

The sign of z was chosen to make A identical at $\pm z$ as required by symmetry and to make it die out with time. Thus, the equation shows that, added to the field A'_2 which would exist if no sheet were present, there is a decaying field due to the eddy currents which appears, from either side of the sheet, to come from a pair of images on the opposite side which recedes with a uniform velocity $2s/\mu_v$.

Maxwell gives a formula for this law of images which applies to any type of field variation. Suppose the inducing field vector potential is

$$A' = f(t, x, y, z) \tag{4}$$

The change in this field in an infinitesimal interval of time $d\tau$ is

$$\frac{\partial A'}{\partial t}\,d\tau = \frac{\partial}{\partial t}f(t, x, y, z)\,d\tau$$

The initial field of the eddy currents formed in that interval must be equal and opposite to this. This field dies out, as we have seen, as if it were due to an image on the opposite side of the sheet moving away with a uniform velocity $2s/\mu_v$. Thus, the vector potential of the eddy currents at the present time t due to images formed in the interval $d\tau$ at a positive time τ before the present is given by

$$dA = -\frac{\partial}{\partial t}f\left(t - \tau, x, y, -|z| - \frac{2s}{\mu_v}\tau\right)d\tau$$

At the time t, the total vector potential due to eddy currents is

$$A = -\int_0^\infty \frac{\partial}{\partial t}f\left(t - \tau, x, y, -|z| - \frac{2s}{\mu_v}\tau\right)d\tau \qquad (5)$$

When $z < 0$, z replaces $-|z|$ and, by *Pc* 863, $\partial f/\partial\tau = -\partial f/\partial t - (2s/\mu_v)\partial f/\partial z$ so that substitution for $\partial f/\partial t$ in (5) and integration give, since f is zero at $\tau = \infty$ and equals A' at $\tau = 0$, for the total resultant field

$$z < 0, \qquad A + A' = +\frac{2s}{\mu_v}\frac{\partial}{\partial z}\int_0^\infty f\left(t - \tau, x, y, z - \frac{2s}{\mu_v}\tau\right)d\tau \qquad (6)$$

This is often simpler to integrate than (5).

10.11. Torque on Small Rotating Current Loop or Magnetic Dipole.—If a magnetic system is rotated about an axis normal to a conducting sheet, the field of induced eddy currents will, in general, set up a retarding torque on the system which is, as we shall see, proportional to its angular velocity if this is not too great. By measuring this torque, the angular velocity can be found, and this is the basic principle of some automobile speedometers. The simplest magnetic system is a dipole or small current loop, which rotates about its center keeping its axis parallel to the sheet. In order to visualize the images, we shall not use 10.10 (5) but shall suppose that at regular intervals of time τ the magnet is instantaneously rotated through an angle $\omega\tau$. From this result, by letting the intervals approach zero, we shall obtain the result for continuous motion. Figure 10.11 shows the images formed by the last four jumps at the instant the dipole, of moment M, arrives in the position indicated, as viewed from the upper side of the

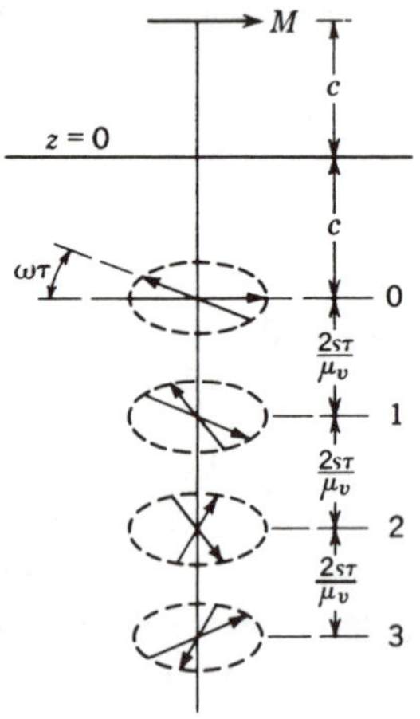

Fig. 10.11.

sheet. The actual torque on M will be the sum of torques due to all the images. Since all image dipoles are perpendicular to the axis of rotation, $\theta_1 = \theta_2' = \frac{1}{2}\pi$ in 1.071 (4) and, from 1.071 (7) and 7.00, the torque due to a single image making an angle ψ with M is

$$T = -\frac{\partial W}{\partial \psi} = \frac{\mu_v M^2}{4\pi r^3}\sin\psi \tag{1}$$

The torque due to all the images is then, if $p = \omega\mu_v/s$,

$$T = \frac{\mu_v M^2}{4\pi}\sum_{n=0}^{\infty}\frac{\sin n\omega\tau - \sin(n+1)\omega\tau}{[2c + (2ns/\mu_v)\tau]^3} \tag{2}$$

$$= \frac{\mu_v p^3 M^2}{32\pi}\sum_{n=0}^{\infty}\frac{\sin n\omega\tau - \sin n\omega\tau\cos\omega\tau - \cos n\omega\tau\sin\omega\tau}{(pc + n\omega\tau)^3}$$

As the motion becomes continuous, $\tau \to dt$, $n\tau \to t$, and $\sin\omega\tau \to \omega\tau \to \omega\, dt$ so that (2) passes over into the integral

$$T = \frac{\mu_v\omega p^3 M^2}{32\pi}\int_0^\infty \frac{-\cos\omega t}{(pc+\omega t)^3}\,dt \tag{3}$$

Let $x = pc + \omega t$ so that $\cos\omega t = \cos(x - pc)$, and this becomes

$$-T = \frac{\mu_v p^3 M^2}{32\pi}\left(\cos pc\int_{pc}^\infty \frac{\cos x}{x^3}\,dx + \sin pc\int_{pc}^\infty \frac{\sin x}{x^3}\,dx\right)$$

Applying *Dw* 441.13 and 431.13 or *Pc* 345 and 344 gives

$$-T = \frac{\mu_v p^3 M^2}{64\pi}\left(\frac{1}{p^2c^2} - \cos pc\int_{pc}^\infty \frac{\cos x}{x}\,dx - \sin pc\int_{pc}^\infty \frac{\sin x}{x}\,dx\right) \tag{4}$$

These integrals may be evaluated by the series *Pc* 346 and 347 or *Dw* 431.11 and 441.11, or they may be put in the form of the standard cosine integral *Ci* and sine integral *Si*. Thus

$$\int_{pc}^\infty \frac{\cos x}{x}\,dx = -\mathrm{Ci}(pc), \qquad \int_{pc}^\infty \frac{\sin x}{x}\,dx = \frac{\pi}{2} - \mathrm{Si}(pc) = -\mathrm{si}(pc) \tag{5}$$

Numerical tables and graphs of Ci and Si are available in Jahnke and Emde and other mathematical handbooks, especially HMF, Chap. V.

If pc is much greater than unity, we integrate (3) thrice by parts and obtain for $-T$,

$$\frac{\mu_v p^3 M^2}{32\pi}\left[\left|\frac{\sin\omega t}{(pc+\omega t)^3} - \frac{3\cos\omega t}{(pc+\omega t)^4} + \frac{12\sin\omega t}{(pc+\omega t)^5}\right|_0^\infty - 60\int_0^\infty \frac{\omega\sin\omega t}{(pc+\omega t)^6}\,dt\right]$$

Neglecting $20\omega/(pc)^2$ compared with unity, we have

$$pc \gg 1 \qquad T = -\frac{3\mu_v M^2}{32\pi pc^4} \tag{6}$$

Neglecting $\omega(pc)^2$ compared with unity (4) gives

$$pc \ll 1 \qquad T = \frac{\mu_v p M^2}{64\pi c^2} \tag{7}$$

For a copper sheet 0.1 mm thick $s = 0.00017$ so that for 100, 10,000, and 10,000,000 cycles per second we have for p, 4.67, 467, and 467,000, respectively. Thus, for these frequencies, we would use (7), (4), and (6), respectively, if $c \approx 0.01$ meter.

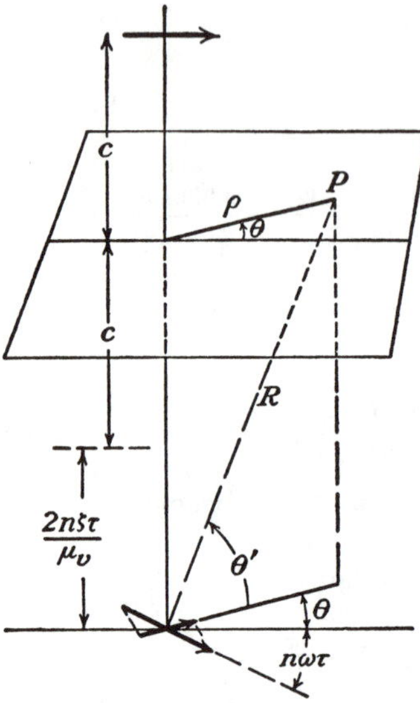

FIG. 10.12*a*.

10.12. Eddy Currents from Rotating Dipole.—It is instructive actually to calculate the equation for the eddy currents produced by the rotating dipole treated in the last article. In Fig. 10.12*a*, we draw one of the images shown in Fig. 10.11 and indicate at P the point on the sheet whose coordinates are ρ, θ at which we wish to calculate the stream function Φ. From 10.09 (2), we have

$$\Phi = 2\mu_v^{-1}\int_\rho^\infty B_\rho \, d\rho \tag{1}$$

by choosing a radial path for our integration and noting that the contribution from the lower half equals that from the upper. The figure shows the nth positive image. Only its ρ-component contributes to B_ρ. Using 7.10 (5) and (2) for B_ρ, adding the contribution from the nth negative image, and summing over all the images, we have

$$B_\rho = -\frac{\mu_v M}{4\pi}\sum_{n=0}^{\infty}\{\cos(n\omega\tau + \theta) - \cos[(n+1)\omega\tau + \theta]\}\frac{\partial}{\partial z}\left(\frac{\sin\theta'}{R^2}\right) \tag{2}$$

Substituting in (1), integrating, and breaking up the second cosine by *Dw* 401.03 or *Pc* 592, we have

$$\Phi = \frac{M\rho}{2\pi}\sum_{n=0}^{\infty}\frac{\cos(n\omega\tau + \theta) - \cos(n\omega\tau + \theta)\cos\omega\tau + \sin(n\omega\tau + \theta)\sin\omega\tau}{-\{\rho^2 + [c + (2ns/\mu_v)\tau]^2\}^{\frac{3}{2}}} \tag{3}$$

Passing over into the integral exactly as in the last article we have

$$\Phi = \frac{\omega M\rho}{2\pi}\int_0^\infty \frac{-\sin(\omega t + \theta)\,dt}{\{\rho^2 + [c + (2s/\mu_v)t]^2\}^{\frac{3}{2}}} \tag{4}$$

Choose a new variable u so that $(2s/\mu_v)t = u - c$ and the limits become c and ∞. Then let $\lambda = \frac{1}{2}\omega\mu_v/s$, substitute in (4), and expand the sine by *Dw* 401.01. Thus (4) becomes (see also problem 32)

$$\Phi = \frac{\omega\mu_v M\rho}{4\pi s}\left[\sin(\theta - \lambda c)\int_c^\infty \frac{\cos\lambda u\,du}{(\rho^2+u^2)^{\frac{3}{2}}} + \cos(\theta - \lambda c)\int_c^\infty \frac{\sin\lambda u\,du}{(\rho^2+u^2)^{\frac{3}{2}}}\right] \quad (5)$$

If $\frac{1}{2}\omega\mu c/s = \lambda c$ is very small, this reduces to

$$\Phi = \frac{\omega\mu_v M\rho}{4\pi s}\sin\theta\int_c^\infty \frac{du}{(\rho^2+u^2)^{\frac{3}{2}}} = -\frac{\omega\mu_v M}{4\pi s\rho}\left[1 - \frac{c}{(\rho^2+c^2)^{\frac{1}{2}}}\right]\sin\theta \quad (6)$$

The system of eddy currents given by this equation is shown in Fig. 10.12*b*, the value of $4\pi\Phi/\mu_v$ being indicated on the figure. This system of stream lines rotates with the same speed as the magnet, which lies at a

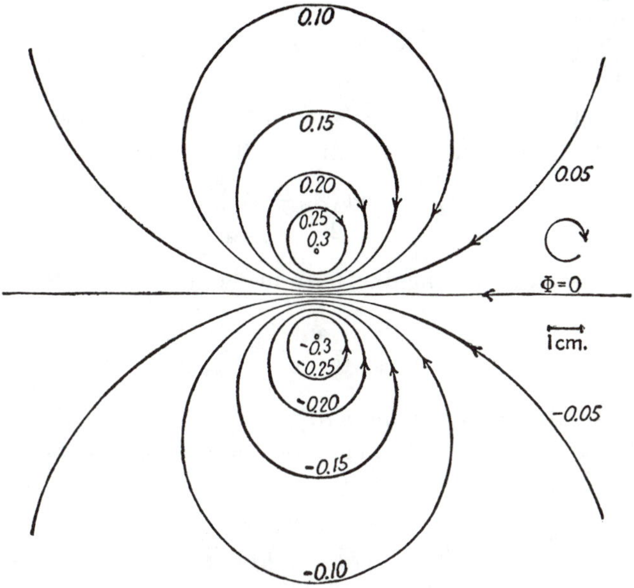

FIG. 10.12*b*.—Stream lines of the eddy currents induced in an infinite thin plane conducting sheet as seen by an observer on a magnetic dipole directed to the right which rotates with an angular velocity ω as shown in a plane 1 cm above the sheet. The values of the stream function Φ in Eq. 6, Art. 10.12, are shown when $\omega\mu_v M/(4\pi s) = 1$.

1 cm distance, on the scale shown, above the plane of the paper and is parallel to the $\Phi = 0$ line. We see at once that the currents shown will give a field at right angles to the magnet. This will produce the torque calculated in the last article.

10.13. Shielding of Circular Coil by Thin Conducting Sheet.—Let us use 10.10 (6) to find the effect of the insertion of an infinite thin conducting sheet on the flux linkage between two coaxial loops of radii a and b at a distance c apart. Without a sheet the vector potential at one loop due to the other is, by 7.10 and problem 27, Chap. VII, the real part of

$$\breve{A}'_\phi e^{j\omega t} = N\pi^{-1}e^{j\omega t}\int_0^\pi (u^2 + c^2)^{-\frac{1}{2}} \cos\phi\, d\phi = Ne^{j\omega t}\int_0^\infty J_1(ka)J_1(kb)e^{-kc}\, dk \tag{1}$$

where $u^2 = a^2 + b^2 - 2ab\cos\phi$ and $N = \frac{1}{2}\mu_v aI$. Then writing λ for $2s/\mu_v$ and $c + \lambda\tau$ for $-|z| - \lambda\tau$ in 10.10 (6) gives at the second loop

$$\begin{aligned} A_\phi + A'_\phi &= -N\lambda\int_0^\infty kJ_1(ka)J_1(kb)e^{-kc}\int_0^\infty e^{-(\lambda k + j\omega)\tau}\, d\tau\, dk \\ &= -N\int_0^\infty \lambda k(j\omega + \lambda k)^{-1}J_1(ka)J_1(kb)e^{-kc}\, dk \end{aligned} \tag{2}$$

The general solution is in problem 33, but if $\lambda k/\omega < 1$, this becomes

$$\frac{\mu_v aI}{2}\sum_{n=1}^{\infty}\left(\frac{2js}{\mu_v\omega}\right)^n\int_0^\infty k^nJ_1(ka)J_1(kb)e^{-kc}\, dk = \sum_{n=1}^{\infty}\left(-\frac{2js}{\mu_v\omega}\right)^n\frac{\partial^n A'_\phi}{\partial c^n} \tag{3}$$

If b and c replace ρ and z, the k series of 7.10 (4) is a good form for A'_ϕ. From 8.06 the ratio of the new to the old flux linkage is

$$R_0 = \frac{2\pi b(A_\phi + A'_\phi)_{\text{rms}}}{2\pi b(A'_\phi)_{\text{rms}}} \tag{4}$$

If $2s/(\omega\mu_v)$ is small, only the first term in the series is needed so that

$$R_0 = \frac{2s}{\omega\mu_v}\frac{1}{A'_\phi}\frac{\partial A'_\phi}{\partial c} = \frac{2s}{\omega\mu_v}\frac{B_\rho}{A_\phi} \tag{5}$$

where B_ρ and A_ϕ are given by 7.10 (6) and (4), respectively, and are evaluated at $\rho = b$, $z = c$. If, in addition, either a or b is small compared with c, then from 7.10 (1) we get

$$R_0 = \frac{6sc}{\omega\mu_v(a^2 + b^2 + c^2)} \tag{6}$$

As we saw in 10.12 for a sheet of copper 0.1 mm thick and a frequency of a million cycles per second, $\omega\mu_v/s$ is 46,700 which justifies the approximation in (5). Hence, if $a = 1$ cm, $b = 10$ cm, and $c = 10$ cm, then, from (6), $R_0 \approx 0.0006$ so that the linkage is greatly reduced by the shield.

10.14. Rotating Sheet in Magnetic Gap.—The reasoning of 10.10 applies to the scalar as well as to the vector magnetic potential and the magnetic induction. As an example consider an infinite thin horizontal sheet of area resistivity s parallel to the horizontal plane faces of a gap of height $2h$ in an infinite vertical circular cylinder of radius a having a uniform permanent magnetization M parallel to its axis. The sheet rotates with a constant angular velocity ω about an axis parallel to and at a distance c from the cylinder axis, or the magnet may rotate with the sheet stationary. In any case, the system of eddy currents is fixed relative to the magnet. The magnetization is assumed to be absolutely rigid, so that the magnet acts as a cylindrical current sheet and does not perturb the magnetic field of the eddy currents as a ferromagnetic magnet would. The integrations in this article will be carried out exactly, partly to show what can be done with the mathematical references now available, especially HTF, IT, and HMF, and partly because the resultant formulas which involve many repetitive operations are easily set up and evaluated on a digital computer.

From 5.16, 7.28, 9.08, and 5.298 (6) the potential on the axis of a ring of radius ρ' and magnetization M is, when $z < h$,

$$d\Omega_a = M2\pi\rho'\,d\rho'\,\{4\pi[\rho'^2 + (h^2 - z^2)]^{\frac{1}{2}}\}^{-1} = \tfrac{1}{2}M\rho'\,d\rho'\int_0^\infty J_0(k\rho')e^{-k(h-z)}\,dk \tag{1}$$

Here the loop definition for M is used so μ_v is omitted. Integration by 5.294 (7) from zero to a gives the potential on the axis of one pole face:

$$\Omega_a = \int d\Omega_a = \tfrac{1}{2}aM\int_0^\infty k^{-1}J_1(ka)e^{-k(h-z)}\,dk \tag{2}$$

Off the axis, in the plane $z = 0$ where the sheet lies, this becomes

$$\Omega' = \tfrac{1}{2}aM\int_0^\infty k^{-1}J_0(k\rho_1)J_1(ka)e^{-kh}\,dk \tag{3}$$

Expand this by 5.298 (7), write $\omega t + \phi$ for ϕ to express rotation, so that when $t = 0$, c lies on the $\phi = 0$ axis (see Fig. 10.14), and obtain

$$\Omega' = \tfrac{1}{2}aM\sum_{s=0}^{\infty}(2 - \delta_s^0)\int_0^\infty k^{-1}J_s(k\rho)J_s(kc)J_1(ka)e^{-kh}\,dk\cos[s(\omega t + \phi)] \tag{4}$$

This replaces 10.10 (4). Similar changes in 10.05 (5) give the eddy current Ω, so write $t - \tau$ for t and $h + (2s/\mu_v)\tau$ for h in (4) and put

$$p_s = \frac{1}{2}\frac{s\omega\mu_v h}{s}, \qquad \frac{2s}{\mu_v} = \frac{s\omega h}{p_s} \tag{5}$$

Now take $\partial/\partial t$, set $t = 0$, and integrate with respect to τ, so that

$$\Omega = aM\sum_{s=1}^{\infty}\frac{p_s}{h}\int_0^\infty \frac{J_0(k\rho)J_s(kc)J_1(ka)}{k[k^2 + h^{-2}p_s^2]}e^{-kh}(k\sin s\phi + p_s h^{-1}\cos s\phi)\,dk \tag{6}$$

The $s = 0$ term drops out since p_s contains an s factor. From 10.12 (1) the stream function Φ is, adding that of the lower pole,

$$\Phi = 4\int_\rho^\infty H_\rho\,d\rho = -4\int_\rho^\infty \frac{\partial\Omega}{\partial\rho}\,d\rho = 4\Omega \tag{7}$$

Now let $k = p_s t/h$, $2R = p_s\rho/h$, $2c_1 = p_s c/h$, and $2a_1 = p_s a/h$ so all symbols after the summation are dimensionless, then

$$\Phi = 4aM\sum_{s=1}^{\infty}\int_0^\infty \frac{J_s(2Rt)J_s(2c_1t)J_1(2a_1t)}{t(1+t^2)}e^{-p_st}(t\sin s\phi + \cos s\phi)\,dt \tag{8}$$

Watson gives on page 148 a power series in t for $J_s(2tc_1)J_1(2ta_1)$.

$$J_s(2tc_1)J_1(2ta_1) = a_1\sum_{m=0}^{\infty}(-1)^m c_1^{2m+s}C_{ms}(a_1,c_1)t^{2m+s+1} \tag{9}$$

$$\begin{aligned} C_{ms}(a_1,c_1) &= {}_2F_1(-m,-m-s;2;a_1^2c_1^{-2})[m!(m+s)!]^{-1} \\ &= (a_1^2c_1^{-2}-1)^m[(m+1)!(m+s)!]^{-1}P_m^{(1,s)}[(c_1^2+a_1^2)(c_1^2-a_1^2)^{-1}] \end{aligned} \tag{10}$$

by HMF 22.5.44, where $P_m^{(1,s)}(x)$ is a Jacobi polynomial whose recurrence relations and coefficients appear in HMF, pages 782 and 793. In the integral $J_s(2tc_1)J_1(2ta_1)$ is therefore replaced by t^{2m+s+1}, so the integrals in the sine or cosine terms are

$$\int_0^\infty \frac{t^{2m+s+1}J_s(2tR)}{1+t^2}e^{-p_st}\,dt \qquad \text{or} \qquad \int_0^\infty \frac{t^{2m+s}J_s(2tR)}{1+t^2}e^{-p_st}\,dt \tag{11}$$

Expand $J_s(2tR)$ in series by 5.293 (3) and write N for $m+s+r$, so (10) is

$$\sum_{r=0}^{\infty}\frac{(-1)^rR^{2r+s}}{r!(r+s)!}\left[\int_0^\infty \frac{t^{2N+1}e^{-p_st}}{1+t^2}dt \qquad \text{or} \qquad \int_0^\infty \frac{t^{2N}e^{-p_st}}{1+t^2}\,dt\right] \tag{12}$$

If N were zero, the integrals would be HMF 5.2.13 or 5.2.12. Since N is never zero because $s \geqslant 1$, divide the numerator by the denominator until the remainders have this form and integrate the quotient terms by *Dw* 860.07, so the integrals of (12) are

$$-F_N(p_s) = \sum_{i=1}^{N}\frac{(2N-2i+1)!}{(-1)^i p_s^{2N-2i+2}} + (-1)^N[\mathrm{Ci}(p_s)\cos p_s + \mathrm{si}(p_s)\sin p_s] \tag{13}$$

$$-G_N(p_s) = \sum_{i=1}^{N}\frac{(-1)^i(2N-2i)!}{p_s^{2N-2i+1}} - (-1)^N[\mathrm{Ci}(p_s)\sin p_s - \mathrm{si}(p_s)\cos p_s] \tag{14}$$

See 10.11 (5) for Ci and si and see HMF, pages 238 to 244, for tables. Thus the final value of the stream function Φ is

$$\Phi = -4aM \sum_{s=1}^{\infty} \sum_{m=0}^{\infty} (-1)^m a_1 c_1^{2m+s} C_{ms}(a_1,c_1) \sum_{r=0}^{\infty} (-1)^r R^{2r+s} \times [r!(r+s)!]^{-1}[F_n(p_s) \sin s\phi + G_n(p_s) \cos s\phi] \quad (15)$$

The torque about the axis of rotation will be calculated only in the case where h is so small compared with the radius a of the pole pieces that the fringing field is negligible. With this narrow gap the magnetic induction under the poles has the constant value $\mu_v M$, which is that in the equivalent current sheet solenoid, from 9.07. The torque equals the product of the normal induction B_z by the radial current density i_ρ and by the lever arm ρ integrated over the pole area.

$$T = \int_S \rho B_z i_\rho \, dS \qquad \text{where } i_\rho = -\frac{\partial \Phi}{\rho\, \partial \phi} \quad (16)$$

For the sine term in (8), i_ρ is even about $\phi = 0$, and for the cosine term it is odd. Thus for the latter the torque on one-half will cancel that

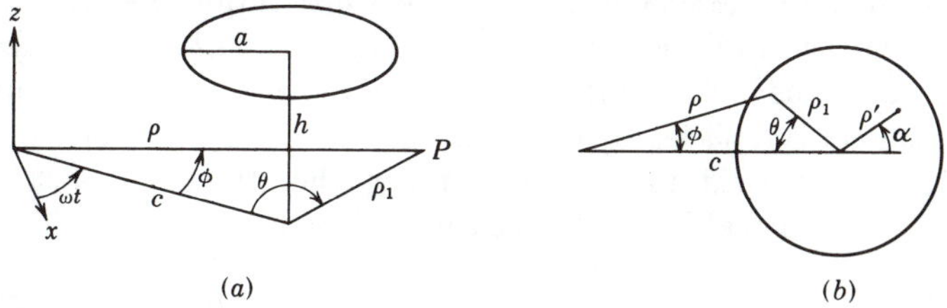

FIG. 10.14*a*, *b*.

on the other so it need not be considered. From Fig. 10.14, the torque will be given by (8) if the $J_s(2tR)t \sin s\phi$ factor is replaced by

$$-M\mu_v st \int_0^a \int_{-\pi}^{+\pi} \cos s\phi \, J_s(2tR) \rho_1 \, d\rho_1 \, d\theta \quad (17)$$

Let $2R_1 = p_s \rho_1 / h$. Then from HMF 9.1.79 we have

$$J_s(2tR) \cos s\phi = \sum_{n=-\infty}^{\infty} J_{n+s}(2tc_1) J_n(2tR_1) \cos n\theta \quad (18)$$

The limits $-\pi$ to $+\pi$ leave only $n = 0$. The R integral is 5.302 so (17) is

$$-2\pi M \mu_s s J_s(2tc_1) 4h^2 p_s^{-2} \int_0^a t J_0(2tR_1) R_1 \, dR_1 = -2\pi M \mu s a h p_s^{-1} J_s(2tc_1) J_1(2ta_1)$$

Insert the factor $2\pi M\mu_v sa$ in (8), omit the ϕ terms, and replace $J_s(2tR)$ in (11) by $hp_s^{-1}t^{-1}J_s(2tc_1)J_1(2ta_1)$. Then use of (9) gives

$$T = 2\pi\mu_v a^3 M^2 \sum_{s=1}^{\infty}\sum_{m=0}^{\infty}\sum_{r=0}^{\infty} sa_1(-1)^{m+s}c_1^{2N}C_{ms}(a_1,c_1)C_{rs}(a_1,c_1)F_N(p_s) \quad (19)$$

10.15. Rotating Disk in Magnetic Gap.—Suppose that the infinite rotating sheet of the last article is replaced by a disk of radius A rotating in a magnetic gap. The primary change is that the circulation of the eddy currents is restricted to the region inside $\rho = A$. A secondary result is that the magnetic fields of the eddy currents can now pass freely around the edge of the disk without penetrating the sheet. The main problem is therefore to find a means of confining the eddy currents to the region $\rho < A$. One method to look into is that of images. In 10.14 (3) if $\rho = a$, then ρ_1 becomes $(c^2 + A^2 - 2cA\cos\theta)^{\frac{1}{2}}$, and if A^2/c replaces c, this becomes $(A/c)(c^2 + A^2 - 2cA\cos\theta)^{\frac{1}{2}}$. Thus if a and h are replaced by Aa/c and Ah/c and the integration variable is changed from k to Ak/c, then the value of the integral in 10.14 (3) is unchanged and the new potential or magnetomotance Ω' is also the same at $\rho = A$ if aM is the same. In passing from 10.14 (3) to 10.14 (6), the substitution $h + (2s/\mu_v)\tau$ was made for h in 10.14 (3), so for the new magnet $(A/c)[h' + (2s/\mu_v)\tau]$ must replace h' and thus in 10.14 (5) $p_s' = (c/A)p_s$ replaces p_s. Therefore if the original magnet is replaced by one centered at c' with a pole face radius a', a gap $2h'$, a magnetization M', and a new p_s', then the value of the stream function at $\rho = A$ will be unchanged provided that

$$c' = \frac{A^2}{c}, \qquad a' = \frac{Aa}{c}, \qquad h' = \frac{Ah}{c}, \qquad M' = \frac{cM}{A}, \qquad p_s' = \frac{cp_s}{A} \quad (1)$$

Thus if both magnets are present, with M' reversed in the second, the circle $\rho = A$ becomes the stream line $\Phi = 0$. The stream function for a disk is then given by adding to the Φ of 10.14 (15) and (12) a function Φ' calculated by the formulas using c_1', a_1', h', $-M'$, and p' instead of c_1, a_1, h, M, and p_s in 10.14 (14).

If $s/(\mu_v\omega)$ is large compared with a^2 and c^2, a simpler procedure than that of 10.14 can be used. From Fig. 10.14, in the $z = 0$ plane (note that there is no μ_v because the loop definition is used for M)

$$\Omega' = \frac{M}{4\pi}\int_0^a\int_0^{2\pi}\frac{\rho'\,d\rho'\,d\alpha}{(h^2 + \rho'^2 + \rho_1^2 - 2\rho_1\rho'\cos\alpha)^{\frac{1}{2}}} \quad (2)$$

where ρ_1 is $[c^2 + \rho^2 - 2c\rho\cos(\phi + \omega t)]^{\frac{1}{2}}$ for a rotating field. The 10.10 (5) substitutions are now made, but because $\omega \ll s/\mu_v$, the $h + 2s\tau/\mu_v$, which replaces h, dominates the denominator as τ increases before the

$\omega(t - \tau)$ for t substitution can take effect in the sine and cosine terms so only ωt is retained. For a narrow gap h is small compared with a and may be omitted. The resultant integral is

$$\Omega = \frac{M}{4\pi}\int_0 \int_0^{2\pi} \int_0^{\infty} \frac{\omega c\rho \sin\phi(\rho_1 - \rho' \cos\alpha)\rho' \, d\rho' \, d\alpha \, d\tau}{[(2s/\mu_v)^2\tau^2 + (\rho')^2 + \rho_1^2 + 2\rho'\rho_1 \cos\alpha]} \tag{2.1}$$

where $\rho_1 = (c^2 + \rho^2 - 2c\rho \cos\phi)^{\frac{1}{2}}$. The τ integration by *Dw* 200.03 yields

$$\Omega = \frac{M\mu_v\omega c\rho \sin\phi}{8\pi s\rho_1}\int_0^a \int_0^{2\pi} \frac{(\rho_1 - \rho' \cos\alpha)\rho' \, d\rho' \, d\alpha}{(\rho')^2 + \rho_1^2 - 2\rho'\rho_1 \cos\alpha} \tag{3}$$

By *Dw* 859.124 the α integral is $2\pi/\rho_1$ if $\rho_1 > \rho'$ and 0 if $\rho_1 < \rho'$, so the upper limit for the ρ' integration is a when $\rho_1 > a$ and ρ_1 when $\rho_1 < a$. Multiplication by 4, because of 10.14 (9) and the lower pole, gives for the stream function for an infinite plane

$$\rho_1 > a \qquad \Phi = \frac{M\mu_v\omega c\rho a^2 \sin\phi}{2s\rho_1^2} \qquad \text{(outside pole piece)} \tag{4}$$

$$\rho_1 < a \qquad \Phi = \frac{M\mu_v\omega c\rho \sin\phi}{2s} \qquad \text{(under pole piece)} \tag{5}$$

These formulas replace 10.14 (11) and (12) when $\mu_v\omega/s$ is small.

For the disk of radius A, the stream function of the second pole found by making the substitutions for c, a, h, M, and p_s indicated in (1) must be subtracted from (4) and (5). Thus Φ_u under the pole piece and Φ_o between the pole piece and $\rho = A$ become

$$\Phi_u = \frac{M\mu_v\omega c\rho \sin\phi}{2s}\left[1 - \frac{A^2a^2}{c^2\rho^2 + A^4 - 2\rho c A^2 \cos\phi}\right] \tag{6}$$

$$\Phi_o = \frac{M\mu_v\omega c\rho a^2 \sin\phi}{2s}\left[\frac{1}{\rho^2 + c^2 - 2c\rho\cos\phi} - \frac{A^2}{c^2\rho^2 + A^4 - 2c\rho A^2 \cos\phi}\right] \tag{7}$$

Figure 10.15 shows Φ plotted from the above equations for a single pole (a) and, by superposition, for two poles of opposite polarity (b), so that the currents from both poles flow in the same direction along the line between centers. In Fig. 10.15a and b, $M\mu\omega s^{-1}$ is 10, c/a is 0.7, and A is 0.1 meter. The stream function in amperes has the value zero on the boundary and increases by steps of 0.1.

The torque about $\rho = 0$ is calculated by 10.14 (13). The use of polar coordinates r, θ, coaxial with the pole applying the torque, simplifies integration. The needed relations between r, θ and ρ, ϕ appear in Fig.

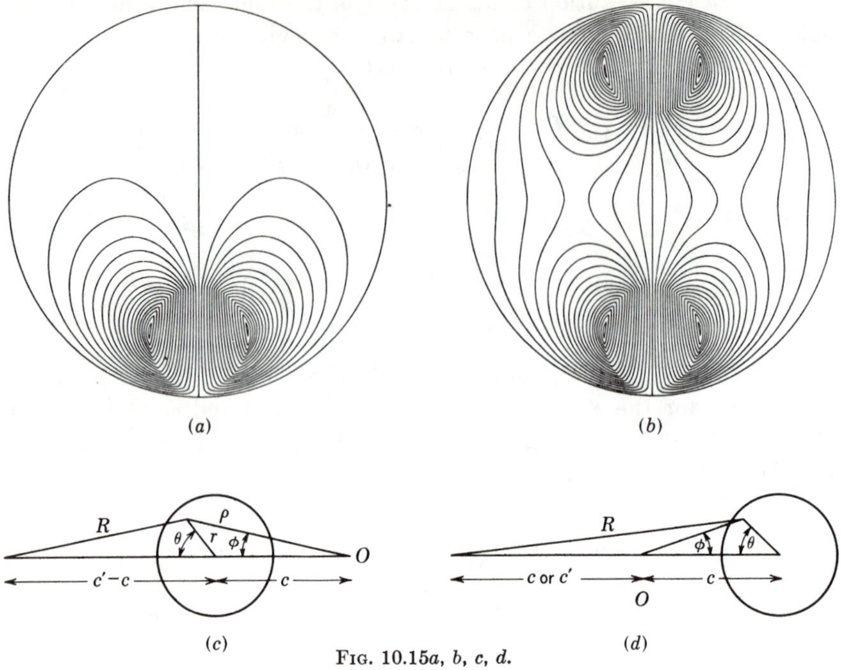

FIG. 10.15*a*, *b*, *c*, *d*.

10.15*c* and *d*. For the single magnet these are

$$\rho \cos \phi = c + r \cos \theta, \qquad \rho \sin \phi = r \sin \theta, \qquad c' = \frac{A^2}{c}$$

$$R^2 = \rho^2 + (c')^2 - 2\rho c' \cos \phi = (c' - c)^2 + r^2 - 2r(c' - c) \cos \theta \quad (8)$$

From (6) and 10.14 (16), the torque with one magnet is

$$T = -\tfrac{1}{2}M^2\mu^2\omega a^2 s^{-1}\int_S [(ca^{-2} - c'R^{-2})\rho \cos \phi + 2(c')^2R^{-4}\rho^2 \sin^2 \phi]\, dS$$

$$= -M^2\mu^2\omega a^2 s^{-1}\int_0^a\int_0^\pi [(ca^{-2} - c'R^{-2})(c + r \cos \theta) + 2(c')2R^{-4}r^2 \sin^2 \theta]r\, dr\, d\theta \quad (9)$$

Integration of the R^{-4} term by parts changes it to an R^{-2} term, so the θ integrals become *Dw* 859.121, 859.122. The simple r integration then gives the torque for a single magnet to be

$$T = \tfrac{1}{2}M^2\mu^2\omega c^2a^2\pi s^{-1}[1 - A^2a^2(A^2 - c^2)^{-2}] \quad (10)$$

The two magnets of Fig. 10.15*b* give approximately double the torque of one. The additional ΔT due to the eddy currents from one pole flowing under the other may be set up and calculated with the aid of Fig. 10.15*d*. In the i_ϕ given by (7) and 10.14 (16) for one gap, ϕ is replaced by $\pi - \phi$ to put it in the coordinates of the other gap. The integral is put in terms of r, θ by the geometrical relations of Fig. 10.15*d*. The resultant integrals, evaluated as before, give

$$\Delta T = \tfrac{1}{2}M^2\mu^2\omega c^2 a^2 \pi s^{-1}[\tfrac{1}{4}a^2c^{-2} - A^2a^2(A^2 + c^2)^{-2}] \tag{11}$$

The total torque for Fig. 10.15*b* where magnets have opposite polarity is

$$T' = 2(T + \Delta T) \tag{12}$$

When the poles have the same polarity it is

$$T'' = 2(T - \Delta T) \tag{13}$$

If the magnet gaps $2h$ and $2h'$ of the M and M' magnets are small compared with the distance of their edges from the boundary $\rho = A$, the primary eddy currents induced by each will be unaffected by the field of the other. There is, however, a slight distortion of the eddy currents on one side of the boundary by the magnetic fields of the eddy currents on the other side. This should have little effect on Φ. However, the magnetic field of the disk eddy currents, except near its surface, cannot be found from the scalar potentials 10.14 (6) and 10.15 (3) as modified to include the image magnets. They may be found by 7.02 (5) or 7.14 (1), where the integrals are taken over the regions $\rho < A$ and the i_ρ and i_ϕ used are derived from the stream function 10.14 (14) and its image or from 10.15 (6) and (7).

Note that Eqs. (2) to (13) apply to small rotation speeds or high-resistivity disks. This means one made of material of poor conductivity or one made very thin of good conductivity. Since the resistivity τ of copper is 1.75×10^{-8} ohm-meters and $\mu_v = 1.25 \times 10^{-6}$, for a sheet 0.1 mm in thickness, $2s/\mu_v$ is 280 and ω must be considerably less than this for (2.1) to hold accurately. The formulas of the last two articles should be applied with caution to electromagnets if the eddy currents are strong. These cannot be replaced by an equivalent current sheet as a permanent magnet can. The permeable core tends to short-circuit the magnetic fields of the eddy currents, which then produce a demagnetizing force that in extreme cases can greatly reduce the torque.

10.16. Zonal Eddy Currents in Spherical Shell.—Let us now consider the eddy currents in a thin spherical conducting shell when we have axial symmetry so that all the eddy currents flow in coaxial circles. Let the total vector potential be $\boldsymbol{A}' + \boldsymbol{A}$, that part due to the eddy currents being $\boldsymbol{A}$. The electromotance $\mathcal{E}$ producing the current in the ring of width $a\,d\theta$ at θ is induced by the change in the total flux through the ring. By 8.00 (1) and 7.08, we can write $\mathcal{E}$ in terms of $\boldsymbol{A}' + \boldsymbol{A}$ and then relate it to the current by Ohm's law 6.02 (1), giving

$$\mathcal{E} = -\frac{dN}{dt} = -\frac{d}{dt}[2\pi a \sin\theta(A'_\phi + A_\phi)] = \frac{2\pi a s \sin\theta}{a\, d\theta} i_\phi a\, d\theta$$

so that, if the vector potentials are zero at $\theta = 0$,

$$-\frac{d}{dt}(A'_\phi + A_\phi) = s i_\phi \tag{1}$$

where i_ϕ is the current density and s the area resistivity. Let the eddy currents in the shell be expressed as a series of zonal harmonics, the nth term being i_n. In 7.12 (2) and (5), we found a simple relation between i_n and the nth term in the expansion of the vector potential produced by these currents. The relation is

$$(A_{\phi n})_{r=a} = \frac{\mu_v a}{2n+1} i_n \tag{2}$$

If A'_ϕ is also expanded in spherical harmonics, we see by substituting (2) in (1) that the expansions of A'_ϕ and A_ϕ are related at the surface of the shell $r = a$ by the equation

$$-\sum_n \frac{d}{dt}(A'_n + A_n) = \sum_n \frac{2n+1}{\mu_v a} A_n s \tag{3}$$

If, after the time $t = 0$, the inducing field A'_ϕ is constant and if, at the time $t = 0$, the field of the eddy currents is known to be

$$A_\phi = \sum_n C_n A_n \tag{4}$$

then, clearly, a solution of (3), showing how the eddy currents decay is,

$$A_\phi = \sum_n C_n A_n e^{-\frac{2n+1}{\mu_v a} st} \tag{5}$$

Suppose that the vector potential of the inducing field can be written, on the surface of the sphere, in the form

$$A'_{\phi s} = \sum_n C_n A_{ns}(t) \tag{6}$$

The change in this field in the infinitesimal interval of time $d\tau$, at a time τ before the present time t, is given by

$$\frac{\partial A'_{\phi s}}{\partial t} d\tau = \sum_n C_n \frac{\partial A_{ns}(t-\tau)}{\partial t} d\tau$$

The eddy currents set up in that interval $d\tau$ exactly neutralized the change at the surface of the sphere at that time but have been dying out according to (5) since, so that the present effect is

$$dA_{\phi s} = -\sum_n C_n \frac{\partial A_{ns}(t-\tau)}{\partial t} e^{-\frac{2n+1}{\mu_v a}s\tau}\, d\tau$$

The total vector potential of the eddy currents at the surface of the sphere at the present time due to all these past changes is, then,

$$A_{\phi s} = -\sum_n C_n \int_0^\infty \frac{\partial A_{ns}(t-\tau)}{\partial t} e^{-\frac{2n+1}{\mu_v a}s\tau}\, d\tau \tag{7}$$

Replacement of $\partial A_{ns}/\partial t$ by $-\partial A_{ns}/\partial \tau$ and integration by parts with the aid of (6) give, for the total vector potential from all sources at $r = a$,

$$A_{\phi s} + A'_{\phi s} = +\frac{s}{\mu_v a}\sum_n (2n+1)C_n \int_0^\infty A_{ns}(t-\tau)e^{-\frac{2n+1}{\mu_v a}s\tau}\, d\tau \tag{8}$$

In the steady-state a-c case A'_ϕ, which may result from both external and internal sources, is given at $r = a$ by (6) where

$$A_{ns} = P_n^1(\cos\theta)\cos\omega t \tag{9}$$

Integration of (7), with N substituted for $(2n+1)s/(\mu_v a)$, by *Dw* 860.80 and 860.90, gives for the eddy currents alone at $r = a$

$$A_{\phi s} = \sum_{n=1}^\infty C_n P_n^1(\cos\theta)\omega(\omega^2 + N^2)^{-1}(N\sin\omega t - \omega\cos\omega t)$$

$$= -\sum_{n=1}^\infty C_n P_n^1(\cos\theta)\cos\psi_n\cos(\omega t + \psi_n) \tag{10}$$

where $\tan\psi_n$ is N/ω. Thus the present eddy currents are in phase with those formed at a time ψ_n/ω before the present. The phase lag ϵ_n is $-\psi_n$. The value of A_ϕ that vanishes at $r = \infty$ and gives (10) at $r = a$ is the vector potential outside the sphere due to eddy currents alone. Thus

$$A_o = \phi\sum_{n=1}^\infty -C_n\left(\frac{a}{r}\right)^{n+1} P_n^1(\cos\theta)\cos\epsilon_n\cos(\omega t - \epsilon_n) \tag{11}$$

$$\tan\epsilon_n = -\frac{(2n+1)s}{\mu_v a\omega} \tag{12}$$

Inside the shell, the eddy current vector potential A_i must be everywhere finite, so $(r/a)^n$ replaces $(a/r)^{n+1}$. For an external source, r enters A' as

$(r/a)^n$. From (6), (9), and (10), the total internal field is, by *Dw* 401.03,

$$A_i + A' = \phi \sum_{n=1}^{\infty} - C_n\left(\frac{r}{a}\right)^n P_n^1(\cos\theta)\sin\epsilon_n \sin(\omega t - \epsilon_n) \tag{13}$$

This could also have been found directly from (6), (8), and (9). The ratio of the nth term in the harmonic expansion of the new amplitude to the corresponding term in the expansion of the old amplitude is

$$R_{on} = \frac{|A_i + A'|}{|A'|} = \sin\epsilon_n \tag{14}$$

From (12), if s is very large, this is unity and the field is unchanged, but if the shield is a good conductor, *i.e.*, if s is small compared with $\mu_v\omega$, R_{on} is very small and the shielding is high.

The eddy current density in the shell as given by (2) and (10) is

$$i = \phi \sum_n - \frac{2n+1}{\mu_v a} C_n P_n^1(\cos\theta)\cos\epsilon_n \cos(\omega t - \epsilon_n) \tag{15}$$

The instantaneous energy dissipation rate in an element of area dS of the shell is $i^2 s\, dS$ from 6.03 (2). In the whole shell it is, writing u for $\cos\theta$,

$$s\int_S i^2\, dS = 2s\pi a^2 \int_0^{\pi} i^2 \sin\theta\, d\theta = -2s\pi a^2 \int_{-1}^{+1} i^2\, du$$

When (15) is squared and integrated from $u = -1$ to $u = 1$, all cross-product terms vanish by 5.13 (2) leaving only a sum of integrals of squares. Thus each harmonic component $i_n \cos(\omega t - \epsilon_n)$ of i behaves as an independent circuit so that the mean rate of energy dissipation from 6.03 (2) is

$$\tfrac{1}{2}s\int_S i^2\, dS = s\pi a^2 \int_{-1}^{+1}\left[\sum i_n\right]^2 du = s\pi a^2 \sum \int_{-1}^{+1} i_n^2\, du$$

Integration by 5.231 (3) gives for the energy dissipation rate in the sphere

$$\bar{P} = 2\pi s \mu_v^{-2} \sum_{n=1}^{\infty} n(n+1)(2n+1)C_n^2 \cos^2\epsilon_n \tag{16}$$

10.17. Spherical Shell in Alternating Field Solenoid Gap.—As an application of the formulas of the last article consider a spherical shell of radius a centered in a gap of width $2d$ in an infinite solenoid of radius c having a current density $i_\phi \cos\omega t$ amperes per meter, as shown in Fig. 10.17. The potential dA_ϕ due to a ring element of width dz_0 at $z = z_0$ may be written down by 7.13 (2) where I is replaced by $i_\phi\, dz_0$. Superimposing the results

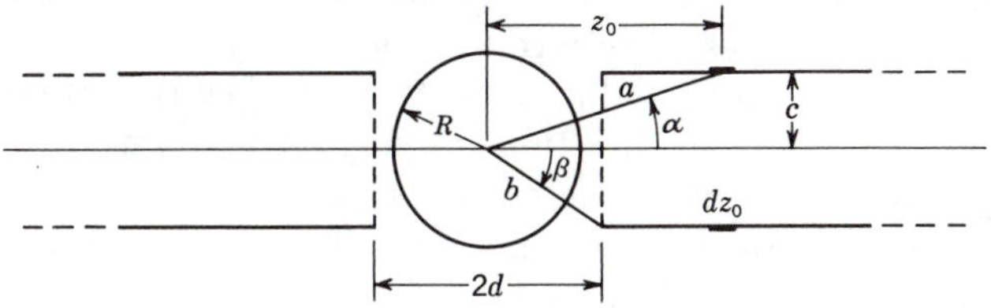

FIG. 10.17.

for a ring at $r = a$, $\theta = \alpha$ and one at $r = a$, $\theta = \pi - \alpha$ eliminates even values of n and doubles the odd values so that, writing $2n + 1$ for n,

$$dA_\phi = \mu i_\phi\, dz_0 \sum_{n=0}^{\infty} \frac{\sin\alpha}{(2n+1)(2n+2)}\left(\frac{r}{a}\right)^{2n+1} P^1_{2n+1}(\cos\alpha) P^1_{2n+1}(\cos\theta)\cos\omega t \tag{1}$$

The $n = 0$ integration gives

$$\int_d^\infty a^{-1}\sin^2\alpha\, dz_0 = \int_d^\infty c^2(c^2 + z_0^2)^{-\frac{3}{2}}\, dz_0 = 1 - d(c^2 + d^2)^{-\frac{1}{2}} = 1 - \cos\beta \tag{2}$$

For $s > 1$ a useful formula, verifiable by differentiation, is

$$a^{-s}\sin\alpha P^1_s(\cos\alpha) = [(s-1)!]^{-1}(-1)^{s+1}\frac{\partial^{s+1}a}{\partial z_0^{s+1}} \tag{3}$$

$$a^2 = c^2 + z_0^2, \qquad a\cos\alpha = z_0, \qquad a\sin\alpha = c, \qquad s > 1 \tag{4}$$

The z_0 integration of (3) merely reduces the derivative order by one, and (3) expresses the rêsult in terms of a, so that

$$A'_\phi = \sum_{n=0}^{\infty} C'_n r^{2n+1} P^1_{2n+1}(\cos\theta)\cos\omega t, \qquad C'_0 = \tfrac{1}{2}\mu i_\phi(1 - \cos\beta) \tag{5}$$

$$C'_n = \mu i_\phi b^{-2n}[2n(2n+1)(2n+2)]^{-1}\sin\beta P^1_{2n}(\cos\beta) \tag{6}$$

The applied vector potential is then 10.16 (6) where $r = R$, so that

$$A_{ns} = P^1_{2n+1}(\cos\theta)\cos\omega t, \qquad C_1 = \mu_v i_\phi R\sin^2\tfrac{1}{2}\beta \tag{7}$$

$$C_{2n+1} = \mu_v i_\phi R^{2n+1} b^{-2n}\sin\beta P^1_{2n}(\cos\beta)[2n(2n+1)(2n+2)]^{-1} \tag{8}$$

Thus the eddy current density i and the energy dissipation are given by 10.16 (15) and (16) provided n is replaced by $2n + 1$.

10.18. General Eddy Currents in Spherical Shell.—It was shown in 7.04 and 7.06 that the vector potential whose curl gives a magnetostatic $\boldsymbol{B}$ in spherical coordinates has only θ- and ϕ-components given by

$$A_{mn} = \nabla \times \boldsymbol{r}(Ar^n + Br^{-n-1})S^m_n(\theta,\phi) \tag{1}$$

In 7.11 the vector potential of an arbitrary current distribution in a spherical shell was expressed in terms of these components. These relations provide a basis for the solution of the general eddy current problem when a thin shell is placed in a time-varying field whose wave length is much longer than the shell diameter $2a$ and when the skin depth is much greater than the shell thickness. First the orthogonality relations of these solutions must be found. From 7.11 their form is

$$\mathbf{A}_{mn} = \boldsymbol{\theta}_1 \frac{-m}{\sin\theta} P_n^m(\cos\theta)\begin{bmatrix}\sin m\phi \\ \cos m\phi\end{bmatrix} + \boldsymbol{\phi}_1 \sin\theta P_n^{m\prime}(\cos\theta)\begin{bmatrix}\cos m\phi \\ \sin m\phi\end{bmatrix} \quad (2)$$

The ϕ integration of $\mathbf{A}_{mn} \cdot \mathbf{A}_{pq}$ over the unit sphere gives integrals of $\sin m\phi \sin p\phi$ or $\cos m\phi \cos p\phi$ from 0 to 2π. From *Dw* 858.516 and 858.517 these vanish unless $m = p$ in which case the result is π. Thus

$$\int_{-1}^{-1}\int_0^{2\pi} \mathbf{A}_{mn} \cdot \mathbf{A}_{mq} \sin\theta\, d\theta\, d\phi = \pi \int_{-1}^{+1} \mathbf{P}_n^m(\cos\theta) \cdot \mathbf{P}_q^m(\cos\theta) \sin\theta\, d\theta \quad (3)$$

From 5.232 (7) and 5.232 (8) this is zero if $n \neq q$, and if $n = q$, it is

$$\int_{-1}^{+1}\int_0^{2\pi} \mathbf{A}_{mn} \cdot \mathbf{A}_{mn} \sin\theta\, d\theta\, d\phi = \frac{2\pi n(n+1)(n+m)!}{(2n+1)(n-m)!} \quad (4)$$

Let the total vector potential be $\mathbf{A}' + \mathbf{A}$, the eddy current part being $\mathbf{A}$. Then from 10.00 (4), if $\mathbf{i}$ is surface current density and s is area resistivity,

$$\frac{-d(A' + A)}{dt} = s\mathbf{i} \quad (5)$$

The relation between the current $\mathbf{i}$ and the vector potential it produces at $r = a$ is found by a procedure like that in 7.11 and 7.12 to be

$$\mathbf{A}_{mn} = -\mu a \frac{\mathbf{i}_{mn}}{2n+1} \quad (6)$$

The reasoning of Art. 10.16 starting with 10.16 (1) and (2) applies without change to the present case, if C_{mn}, $\mathbf{A}_{mn}$, and $\mathbf{A}$ replace C_n, A_n, and A_ϕ, so that if the applied vector potential at the surface is

$$\mathbf{A}'_s = \sum_n \sum_m C_{mn} \mathbf{A}_{mns}(t) \quad (7)$$

then the total vector potential at the surface of the sphere at the present time due to all past fluctuations is

$$\mathbf{A}_s = -\sum_n \sum_m C_{mn} \int_0^\infty \frac{\partial \mathbf{A}_{mns}(t-\tau)}{\partial t} e^{-[(2n+1)/(\mu_v a)] s\tau}\, d\tau \quad (8)$$

10.19. Torque on Spinning Spherical Shell between Magnet Poles.— The theory of the last article yields the solution for the retarding torque on a spinning thin spherical conducting shell of radius a centered in a gap between the coaxial halves of an infinite cylindrical bar magnet having a uniform permanent magnetization M parallel to its axis. The edges of the plane magnet faces are at a distance b from the sphere center, where any face radius subtends an angle β as shown in Fig. 10.19. The vector potential for the equivalent current shell (see 9.07) is given by 10.17 (5) if M replaces i_ϕ. Operations are simplified by having the harmonic axis and the rotation axis coincide. To accomplish this, observe from 7.06 (1) and (2) that the A'_ϕ of 10.17 is given by

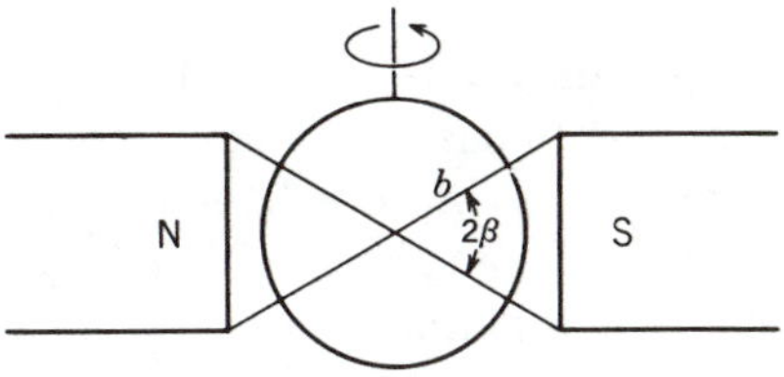

FIG. 10.19.

$$A' = \nabla \times \mathbf{r}W = \nabla \times \mathbf{r}\sum_{n=0}^{\infty} C'_n r^{2n+1} P_{2n+1}(\cos\theta') \tag{1}$$

where the prime indicates the 10.17 coordinate. By 5.24 (5), taking Θ to be $\frac{1}{2}\pi$, the expression for W referred to the new axis is

$$W = \sum_{n=0}^{\infty}\sum_{m=0}^{n} C'_n \frac{2(2n-2m-1)!!}{(2n+2m+2)!!} r^{2n+1} P_{2n+1}^{2m+1}(\cos\theta)\cos(2m+1)\phi \tag{2}$$

Now taking $\nabla \times \mathbf{r}W$ by 7.06 (1) and (2) gives Eqs. (4) and (5) for the vector potential referred to the rotation axis, in terms of

$$C_{mn} = 2(2n-2m-1)!![(2n+2m+2)!!]^{-1} C'_n \tag{3}$$

$$A'_\theta = \sum_{n=0}^{\infty}\sum_{m=0}^{n} (2m+1) C_{mn} r^{2n+1} \csc\theta P_{2n+1}^{2m+1}(\cos\theta)\sin(2m+1)\phi \tag{4}$$

$$A'_\phi = -\sum_{n=0}^{\infty}\sum_{m=0}^{n} C_{mn} r^{2n+1} \sin\theta (P_{2n+1}^{2m+1})'(\cos\theta)\cos(2m+1)\phi \tag{5}$$

Substitution of $\phi + \omega t$ for ϕ puts in rotation. The $A_{\phi s}$ of the eddy currents is now found from 10.18 (8). Substitute $t - \tau$ for τ, differentiate with respect to t, put $t = 0$, and perform the τ integration. Writing S for $\sin(2m+1)\phi$ and C for $\cos(2m+1)\phi$, the last factors in (4) and (5) are

$$(2m+1)\omega\int_0^\infty \left[\begin{matrix} C \\ S \end{matrix} \cos(2m+1)\omega\tau + \begin{matrix} S \\ -C \end{matrix} \sin(2m+1)\omega\tau \right] e^{-[(2n+1)/\mu_v a]\tau}\, d\tau \tag{6}$$

where the upper symbols refer to A_θ and the lower to A_ϕ. Integration by *Dw* 860.90 and *Dw* 860.80 gives the results for $(A_{mns})_\theta$ and $(A_{mns})_\phi$:

$$C_{mn} \cos \psi_{mn} r^{2n+1}(2m+1) \csc \theta P_{2n+1}^{2m+1}(\cos \theta) \sin [(2m+1)\phi + \psi_{mn}] \quad (7)$$
$$-C_{mn} \cos \psi_{mn} r^{2n+1} \sin \theta (P_{2n+1}^{2m+1})'(\cos \theta) \cos [(2m+1)\phi + \psi_{mn}] \quad (8)$$

where $\tan \psi_{mn}$ is $(2n+1)s/[(2m+1)\mu_v a\omega)]$. Clearly A_{mns} with these components has the same orthogonal properties as 10.18 (2) since from 0 to 2π the integration of the ϕ factor products is independent of ψ_{mn}. The torque on the sphere is found by using the $\boldsymbol{i}$ of 10.18 (6). The rate of energy dissipation is constant, as in 6.03 (2), so that

$$T = P\omega^{-1} = \omega^{-1} s \boldsymbol{i} \cdot \boldsymbol{i}\, dS = 2\pi\omega s \sum_{n=0}^{\infty} \sum_{m=0}^{n}$$
$$\frac{a^{4n+4}(2n+1)(2n+2)(4n+3)(2m+1)^2(2n+2m+2)!}{[(2n+1)^2 s^2 + (2m+1)^2 \mu_v^2 a^2 \omega^2](2n-2m)!} C_{mn}^2 \quad (9)$$

10.20. Eddy Currents in Thin Cylindrical Shell.—When all the eddy currents in a thin infinite conducting cylinder of radius a parallel the axis, the vector potential of the field producing them also parallels the axis and the magnetic field is a two-dimensional one. In this case, as we saw in 7.25, the vector potential behaves as an electrostatic stream function so its value at any point represents the flux between this point and some fixed point. Let the total vector potential be $A'_z + A_z$, that part due to the eddy currents being A_z. Then the electromotance per unit length induced in a strip of width $a\, d\theta$ at a, θ by the change in total flux N linking unit length of this strip gives, by Ohm's law, 6.02 (1),

$$\mathcal{E} = -\frac{dN}{dt} = -\frac{d(A'_z + A_z)}{dt} = \frac{s}{a\, d\theta} i_z a\, d\theta \quad (1)$$

where i_z is the current density and s the area resistivity. If i is written as a series of circular harmonics, then from 7.17 (1) and (4), there is a simple relation between the nth term of this expansion and the nth term of the vector potential produced by it. At the surface of the shell this relation is

$$(A_n)_{\rho=a} = \tfrac{1}{2}\mu_v a n^{-1} i_n \quad (2)$$

If both A'_z and A_z are expanded in circular harmonics, we see, by putting (2) in (1), that these expansions satisfy the relation

$$-\sum_n \frac{d}{dt}(A'_n + A_n) = \sum_n \frac{2ns}{\mu_v a} A_n \quad (3)$$

Except for a constant factor, this equation is identical with 10.16 (3), so that if the inducing field at $\rho = a$ is given, as in 10.16 (6), by

$$A'_{zs} = \sum_{n=1}^{\infty} C_n A_{ns}(t) \tag{4}$$

then the eddy current potential is given, as in 10.16 (7), by

$$A_{zs} = -\sum_{n=1}^{\infty} C_n \int_0^{\infty} \frac{\partial A_{ns}(t-\tau)}{\partial t} e^{-\frac{2ns}{\mu_v a}\tau}\, d\tau \tag{5}$$

Replacement of $\partial A_{ns}/\partial t$ by $-\partial A_{ns}/\partial \tau$ and integration by parts give

$$A_{zs} + A'_{zs} = -\frac{2s}{\mu_v a}\sum_{n=1}^{\infty} nC_n \int_0^{\infty} A_{ns}(t-\tau) e^{-\frac{2n}{\mu_v a}\tau}\, d\tau \tag{6}$$

For a sinusoidal inducing field we have, as in 10.16 (9),

$$A_{ns} = \cos(n\theta + \delta_n)\cos \omega t \tag{7}$$

As in 10.16 (11) and 10.16 (12), the eddy currents alone give, when $\rho > a$,

$$\boldsymbol{A}_0 = -\boldsymbol{k}\sum_{n=1}^{\infty} C_n a^n \rho^{-n} \cos(n\theta + \delta_n)\cos\epsilon_n \cos(\omega t - \epsilon_n) \tag{8}$$

$$\tan \epsilon_n = -2ns(\omega\mu_v a)^{-1} \tag{9}$$

When $\rho < a$, $(\rho/a)^n$ replaces $(a/\rho)^n$ and for external sources (see 10.14),

$$\boldsymbol{A}_i + \boldsymbol{A}' = -\boldsymbol{k}\sum_{n=1}^{\infty} C_n \rho^n a^{-n} \cos(n\theta + \delta_n)\sin\epsilon_n \sin(\omega t - \epsilon_n) \tag{10}$$

The ratio of the nth harmonic with the shell to that without the shell is

$$R_{on} = \frac{|A_i + A'|}{|A'|} = \sin \epsilon_n \tag{11}$$

The same remarks apply to this as to 10.16 (14). For internal excitation the new and old amplitude ratio outside the shell is still given by (11). The eddy current density in the shell is, from (2) and (8),

$$\boldsymbol{i} = -\boldsymbol{k}2(\mu_v a)^{-1}\sum_{n=1}^{\infty} nC_n \cos(n\theta + \delta_n)\cos\epsilon_n \cos(\omega t - \epsilon_n) \tag{12}$$

power

From 10.02 (6), the average power consumption per unit length is

$$\bar{P} = \tfrac{1}{2}a\int_0^{2\pi} si^2\,d\theta$$

When (12) is squared, the integration from 0 to 2π eliminates all the cross-product terms and leaves only the sum of integrals of $\cos^2(n\theta + \delta_n)$ which equal π by *Dw* 854.1. Thus each n term behaves as an independent circuit, and the rate of energy dissipation per unit length is

$$\bar{P} = 2\pi s(\mu_v^2 a)^{-1}\sum_{n=1}^{\infty} n^2 C_n^2 \cos^2 \epsilon_n \qquad (13)$$

10.21. Eddy Currents in Rotating Finite Cylindrical Shell.—A device often used for the damping of torsional oscillations consists of a thin

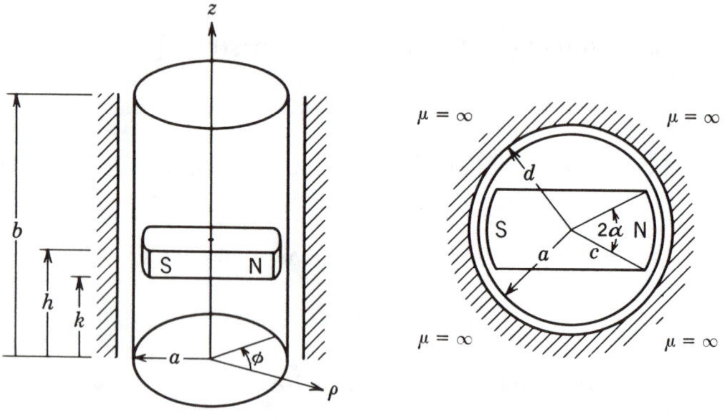

FIG. 10.21.

conducting cylinder of area resistivity s suspended in the gap between one or more transverse bar magnets and a soft iron external yoke which serves as a return path for the magnetic flux between poles as shown in Fig. 10.21. The treatment is rigorous, subject to the conditions of 10.00 and 10.09, for an infinite thin cylinder with a periodic arrangement of transverse uniformly magnetized bars so that the circles at $z = 0, \pm b, \pm 2b$, etc., are eddy current flow lines. It is nearly rigorous for a single section of length b as shown if the magnet poles are not too near the edge. It is pointed out in the next article that the damping can be worked out as soon as the solution for the torque on a uniformly rotating cylinder is known. It is convenient to use the scalar potential Ω throughout, and the procedure involves the same steps used in the preceding articles.

Suppose the stream function is expanded in the harmonic form

$$\Phi = \sum_n \sum_m C_{mn} \sin \frac{n\pi z}{b} \sin (m\phi + \delta_m) \tag{1}$$

so that $\Phi = 0$ at $z = 0$, $z = b$. From 7.01 (2), the line integral of $\boldsymbol{H}$ around a path from $(\Omega_o)_P$ on the outer face around the edge to the point $(\Omega_i)_P$ on the inner face is $\Omega_o - \Omega_i$ which equals Φ, the total current flowing between P and the edge. One can see by inspection that the potentials Ω_i and Ω_o are (see 5.292 and 5.324),

$$\Omega_i = \pi ab^{-1}\sum_n \sum_m nC_{mn}I_m(n\pi\rho b^{-1})K'_m(n\pi ab^{-1}) \sin (n\pi zb^{-1}) \sin (m\phi + \delta_m) \tag{2}$$

$$\Omega_o = \pi ab^{-1}\sum_n \sum_m nC_{mn}I'_m(n\pi ab^{-1})K_m(n\pi\rho b^{-1}) \sin (n\pi zb^{-1}) \sin (m\phi + \delta_m) \tag{3}$$

because if the difference $\Omega_i - \Omega_o$ at $\rho = a$ is taken and the modified functions eliminated by the Wronskian 5.32 (7), the result is (1). The rate of decay of these eddy currents is found from 10.00 (3), applied at $\rho = a$.

$$\boldsymbol{i} = -\boldsymbol{\phi}\frac{\partial\Phi}{\partial z} + \boldsymbol{k}\frac{\partial\Phi}{a\partial\phi}, \qquad \boldsymbol{\nabla} \times \boldsymbol{i} = \boldsymbol{\rho}_1\left(\frac{\partial^2\Phi}{a^2\,\partial\phi^2} + \frac{\partial^2\Phi}{\partial z^2}\right) - \boldsymbol{k}\frac{\partial\Phi}{a\,\partial z} \tag{4}$$

$$\frac{-\mu_v\,\partial\Omega}{\partial\rho} = -\mu_v n^2\pi^2 ab^{-2}I'_m(n\pi ab^{-1})K'_m(n\pi ab^{-1})C_{mn} \sin (n\pi zb^{-1}) \sin (m\phi + \delta_m) \tag{5}$$

As noted at the start of 10.09, B_z has negligible effect on a thin shell, so substitution of (5) for $\boldsymbol{B}$ and (4) for $\boldsymbol{\nabla} \times \boldsymbol{i}$ in 10.00 (3) gives

$$-\frac{\partial \boldsymbol{B}_{mn}}{\partial t} = \frac{s}{\mu_v a}\left[1 + \left(\frac{mb}{n\pi a}\right)^2\right]\left[I'_m\left(\frac{n\pi a}{b}\right)K'_m\left(\frac{n\pi a}{b}\right)\right]^{-1}\boldsymbol{B}_{mn} \tag{6}$$

The same law of decay applies to $\boldsymbol{B}$ and Ω, so that

$$[\Omega_{mn}]_t = [\Omega_{mn}]_{t=0}e^{-P_{mn}st} \tag{7}$$

$$P_{mn} = [1 + m^2b^2(n\pi a)^{-2}][\mu_v aI'_m(n\pi ab^{-1})K'_m(n\pi ab^{-1})]^{-1} \tag{8}$$

Suppose the inducing potential at the cylinder surface is

$$\Omega' = \sum_n \sum_m D_{mn}(\rho,\phi,z,t) \tag{9}$$

Then the change in this potential in an interval $d\tau$ at a time τ before the present was compensated by the eddy currents induced at that time. The present set of eddy currents is the result of all past changes, each change having died out according to (7) so that the present potential of the eddy currents is

$$\Omega = -\sum_n \sum_m \int_0^\infty \frac{\partial D_{mn}\,(\rho,\phi,z,t-\tau)}{\partial t} e^{-P_{mns}\tau}\,d\tau \tag{10}$$

The form of $D_{mn}(\rho,\phi,z,t)$ for Fig. 10.21 must now be found. The bar is uniformly magnetized with intensity M (loop definition), but the ends are rounded so the normal M is $M \cos \phi_0$ where $-\alpha < \phi_0 < +\alpha$. The pole strength in an area dS is then $cM \cos \phi_0\, dz_0\, d\phi_0$. The magnetic field is normal to the permeable boundary at $\rho = d$ and to the plane $z = b$ so Ω' vanishes there. Thus the problem is exactly that of a point charge in an earthed box which was solved in 5.324 (1) and (10). The following substitutions must be made: $cM \cos \phi_0$ for $-q/\epsilon$ and z_0, b, c, d for c, L, b , a. There is another positive element at $-\phi_0$, which by *Dw* 401.06 changes $\cos m(\phi - \phi_0)$ to $2 \cos m\phi \cos m\phi_0$. There are negative elements at $\pi - \phi_0$ and $\pi + \phi_0$, which cancel the positive element terms when m is even and double them when m is odd. For the four elements the last term is then replaced by $4 \cos (2m + 1)\phi \cos (2m + 1)\phi_0$ and the integration range is 0 to α. Thus from 5.324 (10), writing s for $2m + 1$,

$$d\Omega' = \frac{8cM \cos \phi_0\, dz_0\, d\phi_0}{\pi b} \sum_{n=1}^{\infty} \sum_{m=0}^{\infty} \frac{I_s(n\pi c b^{-1}) R_s^0(n\pi b^{-1},d,\rho)}{I_s(n\pi d b^{-1})} \sin (n\pi z_0 b^{-1}) \sin (n\pi z b^{-1}) \cos s\phi \cos s\phi_0 \tag{11}$$

$$R_s^0(n\pi b^{-1},d,\rho) = K_s(n\pi d b^{-1}) I_s(n\pi\rho b^{-1}) - I_s(n\pi d b^{-1}) K_s(n\pi \rho b^{-1}) \tag{12}$$

Integrations with respect to z_0 and ϕ_0 give

$$\int_0^\alpha \cos s\phi_0 \cos \phi_0\, d\phi_0 = (4m)^{-1} \sin 2m\alpha + (4m + 4)^{-1} \sin (2m + 2)\alpha \tag{13}$$

$$\int_k^h \sin (n\pi z_0 b^{-1})\, dz_0 = b(n\pi)^{-1}[\cos (n\pi k b^{-1}) - \cos (n\pi h b^{-1})] \tag{14}$$

Thus, if $s = 2m + 1$, (9) may be written

$$\Omega' = 2cM\pi^{-2} \sum_n \sum_m F_{mn} R_s^0(n\pi b^{-1},d,\rho) \sin (n\pi z b^{-1}) \cos s(\phi + \omega t) \tag{15}$$

$$F_{mn} = I_s\left(\frac{n\pi c}{b}\right)\left(\cos\frac{n\pi k}{b} - \cos\frac{n\pi h}{b}\right) \frac{(m + 1) \sin 2m\alpha + m \sin (2m + 2)\alpha}{nm(m + 1) I_s(n\pi d b^{-1})} \tag{16}$$

The potential of the eddy currents is now given by (10). Substitute $t - \tau$ for t, differentiate with respect to t, and put $t = 0$, so that the eddy currents are given at the instant that $\phi = 0$ splits the pole piece. The integral involved is then, if $\tan \psi_m$ is written for $P_{mn} \mathfrak{s} s^{-1} \omega^{-1}$,

$$s\omega \int_0^\infty \sin (s\phi - s\omega\tau) e^{-P_{ms}\mathfrak{s}\tau} d\tau = -\cos \psi_m \cos (s\phi + \psi_m) \tag{17}$$

This replaces $\cos s(\phi + \omega t)$ in (15) for $\rho > a$ to give Ω_0 at $t = 0$. Note that for a perfectly conducting cylinder or for a high frequency, ψ_m is zero, so that Ω completely neutralizes Ω' outside the cylinder. At $\rho = a$ on the inside surface $\Omega_i = -\Omega_o$, so Φ is $2\Omega_o$ at the surface and $\boldsymbol{i}$ is given by (4). The rate of energy dissipation is constant and equals

$$\bar{P} = \mathfrak{s} \int \boldsymbol{i} \cdot \boldsymbol{i}\, dS = 4\mathfrak{s} a \int_0^{2\pi} \int_0^b \left[\left(\frac{\partial \Omega}{\partial z} \right)^2_{\rho=a} + \left(\frac{\partial \Omega}{a \partial \phi} \right)^2_{\rho=a} \right] dz\, d\phi \tag{18}$$

The integrations are straightforward and the torque is given by P/ω. Thus writing out $\cos^2 \psi_m$ gives finally

$$T = \frac{8c^2 M^2 \omega \mathfrak{s}}{\pi^3 ab} \sum_{n=1}^{\infty} \sum_{m=0}^{\infty} \frac{F^2_{mn} [R^0_{2m+1}(n\pi b^{-1}, d, a)]^2 [n^2\pi^2 a^2 + (2m+1)^2 b^2]}{P^2_{mn} \mathfrak{s}^2 + (2m+1)^2 \omega^2} \tag{19}$$

where F_{mn} is given by (16), R^0_{2m+1} by (12), and P_{mn} by (8).

10.22. Eddy Current Damping.—Numerous calculations of eddy current torques and forces have been made in this chapter, especially in the case of conducting shells whose thickness is small compared with the skin depth at the frequencies involved. In every case where the linear velocity v or the angular velocity ω is small, the forces and torques are proportional to v and ω. There is nearly always, however, as in the last two articles, a term involving $\mu_v^2 \omega^2 a^2$ plus a $\mathfrak{s}^2$ member in the denominator somewhere which, if $\mu_v \omega a$ is large enough compared with $\mathfrak{s}$, will eventually take charge. Only the case where ω is small or $\mathfrak{s}$ large gives a simple damped solution. Then the differential equation is that of an R, L, C circuit where, in the angular case, R is the coefficient of ω in the torque expression such as 10.19 (10) or 10.20 (19), L is the moment of inertia I, and C^{-1} is the torque constant of the supporting wire or spring. Formulas for the various types of damping can be found in any circuit theory text.

10.23. Transient Shielding by a Thick Cylindrical Shell.—Sinusoidal longitudinal eddy currents in a thick cylindrical shell and arbitrarily varying longitudinal currents in a thin shell were treated in Arts. 10.03 and 10.20, respectively. As a relatively simple example involving both cross and longitudinal components, consider an infinite cylinder of permeability μ, conductivity γ, internal radius a, and external radius b which makes an angle $\frac{1}{2}\pi - \alpha$ with a uniform field of induction $\boldsymbol{B}$. Find the field inside at a time t after the external field is removed or established. Examination of 7.05 (5) shows that suitable solutions for the static vector potential when $\rho > b$, $b > \rho > a$, and $a > \rho$ are, respectively,

$$(A_1)_0 = B[\boldsymbol{\phi}\tfrac{1}{2} \sin \alpha(\rho - C_0\rho^{-1}) + \boldsymbol{k} \cos \alpha(\rho - C_0'\rho^{-1}) \sin \phi] \quad (1)$$
$$(A_2)_0 = B[\boldsymbol{\phi}\tfrac{1}{2} \sin \alpha(D_0\rho - E_0\rho^{-1}) + \boldsymbol{k} \cos \alpha(D_0'\rho - E_0'\rho^{-1}) \sin \phi] \quad (2)$$
$$(A_3)_0 = B[\boldsymbol{\phi}\tfrac{1}{2} \sin \alpha F_0\rho + \boldsymbol{k} \cos \alpha F_0'\rho \sin \phi] \quad (3)$$

These give the prescribed field at $\rho = \infty$, and the field is finite at $\rho = 0$. The values of the constants needed to match $\boldsymbol{A}_1$ and $(\boldsymbol{B} \times \boldsymbol{\rho})/\mu$ at $\rho = a$ and at $\rho = b$ are, if β is written for $(\mu - \mu_v)/(\mu + \mu_v)$ and G for $b^2/(b^2 - \beta^2a^2)$,

$$C_0 = \mu_v^{-1}(\mu - \mu_v)(b^2 - a^2),\ D_0 = \mu_v^{-1}\mu,\ E_0 = \mu_v^{-1}(\mu - \mu_v)b^2,\ F_0 = 1 \quad (4)$$
$$C_0' = \beta(a^2 - b^2)G,\ D_0' = (1 + \beta)G,\ E_0' = \beta(1 + \beta)a^2G,\ F_0' = (1 - \beta^2)G \quad (5)$$

After $t = 0$, the same solutions of 10.00 (8) are valid where $\gamma = 0$ except that in A_1 the ρ-terms disappear because $B = 0$ at $\rho = \infty$. For $a < \rho < b$, solutions are needed where A_z has the factor $\sin \phi$ and A_ϕ in independent of ϕ. As in 10.08 we try sums of exponentially damped solutions so that

$$A_2 = B\sum_s[\boldsymbol{\phi}R(\rho)e^{-p_st} + \boldsymbol{k}P(\rho) \sin \phi e^{-q_st}] \quad (6)$$

For A_z 10.00 (8) is a scalar equation from which $\sin \phi e^{-q_st}$ divides out and which, upon substitution of $v^2 = \mu\gamma q_s\rho^2$, yields 5.291 (3) with $n = 1$. For A_ϕ 10.00 (8) becomes 10.05 (3) which, upon inserting $-\boldsymbol{\phi}\rho^{-2}$ for $\nabla^2\boldsymbol{\phi}$ from 10.05 (5), dividing out e^{-p_st} and writing $v^2 = \mu\gamma p_s\rho^2$ gives again 5.291 (3) with $n = 1$. Thus, when $t > 0$, (1), (2) and (3) are replaced by

$$A_1 = B\sum_s[\boldsymbol{\phi} \sin \alpha C_s\rho^{-1}e^{-p_st} + \boldsymbol{k} \cos \alpha C_s'\rho^{-1} \sin \phi e^{-q_st}] \quad (7)$$

$$A_2 = B\sum_s\{\boldsymbol{\phi} \sin \alpha D_sR_1[(\mu\gamma p_s)^{\frac{1}{2}}\rho]e^{-p_st} + \boldsymbol{k} \cos \alpha D_s'P_1[(\mu\gamma q_s)^{\frac{1}{2}}\rho] \sin \phi e^{-q_st}\} \quad (8)$$

$$A_3 = B\sum_s[\boldsymbol{\phi} \sin \alpha F_s\rho e^{-p_st} + \boldsymbol{k} \cos \alpha F_s'\rho \sin \phi e^{-q_st}] \quad (9)$$

where R_1 and P_1 are Bessel functions. At $\rho = b$, equate $\boldsymbol{A}_1$ to $\boldsymbol{A}_2$, write k_s for $(\mu\gamma p_s)^{\frac{1}{2}}$, l_s for $(\mu\gamma q_s)^{\frac{1}{2}}$, and cancel $\sin \phi$ out of A_z. Thus we get

$$\boldsymbol{\phi}C_s + \boldsymbol{k}C_s' = \boldsymbol{\phi}bD_sR_1(k_sb) + \boldsymbol{k}bD_s'P_1(l_sb) \quad (10)$$

Equating tangential components of $\mu^{-1}\boldsymbol{B} = \mu^{-1}\nabla \times \boldsymbol{A}$ at $\rho = b$, canceling $\sin \phi$ out of the ϕ-component, and multiplying through by b^2 give

$$-\boldsymbol{\phi}\mu C_s' = \boldsymbol{k}\mu_v bD_s[k_sbR_1'(k_sb) + R_1(k_sb)] - \boldsymbol{\phi}\mu_v l_sb^2D_s'P_1'(l_sb) \quad (11)$$

The κ-term is zero so, from 5.294, $R_0(k_s b) = 0$ and, from 5.293 (9),

$$R_1(k_s b) = \tfrac{1}{2}\pi(k_s b)^{-1}, \qquad R_1(k_s\rho) = Y_0(k_s b)J_1(k_s\rho) - J_0(k_s b)Y_1(k_s\rho) \quad (12)$$

Similarly the boundary conditions at $\rho = a$ give, writing out $P_1(l_s a)$,

$$\phi F_s a + \kappa F'_s a = \phi D_s R_1(k_s a) + \kappa D'_s[E'_s J_1(l_s a) + Y_1(l_s a)] \quad (13)$$
$$\kappa 2\mu F_s - \phi\mu F'_s = \kappa\mu_v k_s D_s R_0(k_s a) - \phi\mu_v l_s D'_s P'_s(l_s a) \quad (14)$$

Eliminating F_s, D_s, F'_s, and D'_s from these four component equations gives

$$k_s a R_0(k_s a) - 2(\mu_v^{-1}\mu)R_1(k_s a) = k_s a R'_1(k_s a) + (1 - 2\mu_v^{-1}\mu)R_1(k_s a) = 0 \quad (15)$$
$$\mu_v l_s b P'_1(l_s b) + \mu P_1(l_s b) = 0 = \mu_v l_s a P'_1(l_s a) - \mu P_1(l_s a) \quad (16)$$

Only values of $p_s = k_s^2(\mu\gamma)$ which satisfy (15) can be used. Insertion of $Y_0(k_s b)$ from (15) into $R_1(k_s a)$ and application of 5.293 (9) give

$$R_1(k_s a) = 2\mu_v J_0(k_s b)\pi^{-1}[\mu_v k_s a J_0(k_s a) - 2\mu J_1(k_s a)]^{-1} \quad (17)$$

If P'_1 and P_1 are expressed in terms of P_0 and P_2 by 5.294 (2) and (3) and if β replaces $(\mu - \mu_v)/(\mu + \mu_v)$, (16) may be rearranged to give

$$-E'_s = \frac{\beta Y_0(l_s a) + Y_2(l_s a)}{\beta J_0(l_s a) + J_2(l_s a)} = \frac{Y_0(l_s b) + \beta Y_2(l_s b)}{J_0(l_s b) + \beta J_2(l_s b)} \quad (18)$$

Only values of $q_s = l_s^2/(\mu\gamma)$ satisfying the second equality can be used. Insertion of E'_s into $P_1(l_s a)$ and $P_1(l_s b)$ and application of the recurrence formulas of 5.294 and of 5.293 (9) give

$$P_1(l_s a) = \frac{2(1-\beta)}{\pi l_s a[\beta J_0(l_s a) + J_2(l_s a)]}, \quad P_1(l_s b) = \frac{-2(1-\beta)}{\pi l_s b[J_0(l_s b) + \beta J_2(l_s b)]} \quad (19)$$

It remains to determine D_s and D'_s so that, when $t = 0$, (8) agrees with (2) when $a < \rho < b$ or so that, dividing common factors out of each component,

$$\phi(D_0\rho - E_0\rho^{-1}) + \kappa(D'_0\rho - E'_0\rho^{-1}) = \sum_s[\phi D_s R_1(k_s\rho) + \kappa D'_s P_1(l_s\rho)] \quad (20)$$

Take the scalar product of both sides by $\rho[\phi R_1(k_n\rho) + \kappa P_1(l_n\rho)]\,d\rho$ and integrate from a to b or subtract the 0 to a from the 0 to b integral. By (11), (15), (16), and 5.296 (3) only the nth term on the right survives, and for A_ϕ its value is given by 5.297 (6) with $B = 1 - 2\mu_v^{-1}\mu$. Integrate the left side by 5.294 (7) and (8), solve for D_n, and from (13) F_n is

$$F_n = \frac{4\pi\mu\mu_v R_1(k_n a)}{k_n a\{4\mu_v^2 - (4\mu^2 - 4\mu\mu_v + k_n^2 a^2\mu_v^2)\pi^2[R_1(k_n a)]^2\}} \quad (21)$$

For A_z 5.297 (6) with $B = \mu_v^{-1}\mu$ gives the right side. Integrate the left side by 5.294 (7) and 5.294 (8), solve for D'_n, and from (5) F'_n becomes

$$F'_n = \frac{4\mu\mu_v b(1-\beta)P_1(l_n b)}{(\mu_v^2 l_n^2 b^2 + \mu^2 - \mu_v^2)[P_1(l_n b)]^2 - (\mu_v^2 l_n^2 a^2 + \mu^2 - \mu_v^2)[P_1(l_n a)]^2} \quad (22)$$

The total field inside at any time is given by (9), (21), (22), (17), and (19) where k_n and l_n are determined by (15) and (18). If there is no field when $t < 0$ but at $t = 0$ the field $\boldsymbol{B}$ is established, then it is clear from Eqs. (3) and (9) that the vector potential $\boldsymbol{A}'_3$ inside is

$$\rho \lessgtr a, \qquad \boldsymbol{A}'_3 = (\boldsymbol{A}_3)_0 - \boldsymbol{A}_3 \quad (23)$$

These solutions are obtained by a different method in *Phil. Mag.*, Vol. 29, page 18, 1940. The field, when $\boldsymbol{B}$ varies arbitrarily with time, can be found from these results by methods similar to those used in preceding articles.

Problems

Problems marked C in the following group are taken from the Cambridge examinations as reprinted by Jeans, with the permission of the Cambridge University Press.

1*C*. An infinite iron plate is bounded by the parallel planes $x = h$, $x = -h$; wire is wound uniformly round the plate, the layers of wire being parallel to the axis of y. If an alternating current is sent through the wire, producing outside the plate a magnetic force $H_0 \cos pt$ parallel to z, prove that H, the magnetic force in the plate at a distance x from the center, will be given by

$$H = H_0\left(\frac{\cosh 2mx + \cos 2mx}{\cosh 2mh + \cos 2mh}\right)^{\frac{1}{2}} \cos (pt + \beta)$$

$$\tan \beta = \frac{-\sinh m(h+x)\sin m(h-x) - \sinh m(h-x)\sin m(h+x)}{\cosh m(h+x)\cos m(h-x) + \cosh m(h-x)\cos m(h+x)}$$

where $m^2 = \frac{1}{2}\mu p/\tau$. Discuss the special cases of mh small and mh large.

2. All points of a circular loop of wire are at a distance c from the center of a ball of radius a, permeability μ, and resistivity τ, and any radius of the loop subtends an angle α at the center. A current $I \cos \omega t$ flows in the loop. Show that the vector potential inside the sphere is the real part of

$$\boldsymbol{\phi}\frac{\mu\mu_v a^{\frac{1}{2}} I \sin\alpha}{2r^{\frac{1}{2}}}\sum_{n=1}^{\infty}\frac{2n+1}{n(n+1)}\left(\frac{a}{c}\right)^n A_n I_{n+\frac{1}{2}}[(jp)^{\frac{1}{2}}r]P_n^1(\cos\theta)e^{j\omega t}$$

where

$$A_n = \{(\mu-\mu_v)nI_{n+\frac{1}{2}}[(jp)^{\frac{1}{2}}a] + \mu_v(jp)^{\frac{1}{2}}aI_{n-\frac{1}{2}}[(jp)^{\frac{1}{2}}a]\}^{-1}P_n^1(\cos\alpha) \text{ and } p = \mu\omega/\tau$$

3. A sphere of radius a, permeability μ, and resistivity τ lies in a region where, at $t = 0$, a field is established that, but for the sphere, would be of uniform induction $\boldsymbol{B}$.

Show in the notation of 10.08 that, when $t > 0$, the potentials are

$$A_o = \phi\left[\frac{1}{2}Br + \frac{(\mu - \mu_v)a^3}{(\mu + 2\mu_v)r^2}B - \sum_s B_s r^{-2} e^{-q_s t}\right]\sin\theta$$

$$A_i = \phi\left[\frac{3\mu Br}{2(\mu + 2\mu_v)} - \sum_s A_s r^{-\frac{1}{2}} J_{\frac{3}{2}}(k_s r) e^{-q_s t}\right]\sin\theta$$

4. An infinite circular cylinder of radius a, permeability μ, and resistivity τ is wound with wire carrying an alternating current. If H_z at the surface is uniform and of magnitude $H_0 \cos \omega t$, show that the magnetic field at any point inside is given by 10.04 (4) and (5) with B_z and B_0 substituted for i_z and i_0. Find the density and direction of the eddy currents by 10.00 (5).

5. A circular loop of wire is located at $r = a$, $\theta = \alpha$ and a spherical conducting shell of area resistivity s at $r = b$. A current is built up in the loop according to the equation $i = I(1 - e^{-Rt/L})$. If the sphere and loop do not move appreciably before the eddy currents have decayed, show that the momentum imparted is

$$\frac{\pi\mu_v^2 bI^2}{2s}\sum_{n=1}^{\infty}\frac{\sin^2\alpha}{(n+1)(2n+1)}\left(\frac{b}{a}\right)^{2n+1}\frac{(2n+1)sL + 2\mu_v bR}{(2n+1)sL + \mu_v bR}P_n^1(\cos\alpha)P_{n+1}^1(\cos\alpha)$$

6. A small loop of wire of moment $M \cos \omega t$ lies above an infinite thin plane sheet of area resistivity s, with its axis perpendicular to the sheet. Show that, if $1 \gg \omega\tau_0$ where $s\tau_0 \gg \mu_v r$, the magnetic induction on the far side of the sheet is

$$B_r = \frac{\mu_v M}{2\pi r^3}\left[\cos\theta\cos\omega t + \frac{\mu_v\omega r}{4s}\sin\omega t\right], \qquad B_\theta = \frac{\mu_v M}{4\pi r^3}\sin\theta\cos\omega t$$

7. Show, in the preceding case, that the current density i_ϕ at a distance ρ from the axis and the average rate of dissipation of power are given, approximately, by

$$\frac{\mu_v\omega\rho M\sin\omega t}{4\pi s(\rho^2 + c^2)^{\frac{3}{2}}} \quad \text{and} \quad \frac{\mu_v^2 M^2\omega^2}{64\pi sc^2}$$

8. The magnetic moment of a small loop carrying current falls linearly from the value M at $t = 0$ to 0 at $t = T$. If the loop is at a distance c above an infinite thin sheet of area resistivity s with its axis normal to it, show that the additional B on the axis at a distance z above the sheet due to its presence is

$$\Delta B = \frac{\mu_v^2 Mt[\mu_v(c + z) + st]}{2\pi T(c + z)^2[\mu_v(c + z) + 2st]^2}$$

when $0 < t < T$ and that, when $t > T$, it is

$$\Delta B = \frac{\mu_v^4 M[\mu_v(c + z) + s(2t - T)]}{2\pi[\mu_v(c + z) + 2s(t - T)]^2[\mu_v(c + z) + 2st]^2}$$

9. A magnetic loop dipole of moment M is parallel to and at a distance c from an infinite thin sheet of area resistivity s and is moved parallel to the sheet and normal to its axis with a uniform velocity v. Show that in a steady state the eddy current retarding force is $\frac{3}{2}\pi^{-1}\mu_v^2 M^2 vs(2c)^{-4}[(4s^2 + \mu_v^2 v^2)^{\frac{1}{2}} + 2s]^{-2}$.

10. A loop dipole of moment $\boldsymbol{M}$ moves with a uniform velocity $\boldsymbol{v}$ in a straight line at a distance c from an infinite plane sheet of area resistivity s. If $\boldsymbol{M}$ parallels the sheet and makes an angle ϕ with v, show that the retarding force is

$$F = \left[1 + \frac{4s\cos^2\phi}{(\mu_v^2 v^2 + 4s^2)^{\frac{1}{2}}}\right]F_0$$

where F_0 is the retarding force in the last problem.

11C. A slowly alternating current $I \cos \omega t$ is traversing a small circular coil whose magnetic moment for unit current is M. A thin spherical shell, of radius a and specific resistance s, has its center on the axis of the coil at a distance f from the center of the coil. Show that the currents in the shell form circles round the axis of the coil and that the strength of the current in any circle whose radius subtends an angle α at the center is, if $\tan \epsilon_n = -(2n+1)s/(\mu_v \omega a)$,

$$MI(4\pi f^2 a)^{-1}\Sigma(2n+1)(f^{-1}a)^n P_n^1(\cos\alpha)\cos\epsilon_n \cos(\omega t - \epsilon_n)$$

12. A thin spherical shell of area resistivity s and radius a rotates with a uniform angular velocity ω about an axis normal to a uniform field of induction $\boldsymbol{B}$. Show that the retarding torque is $6\pi B^2 s\omega a^4(9s^2 + \mu_v^2 a^2\omega^2)^{-1}$.

13. Show that the average power dissipation in the shell of the last problem when placed in the field $B \cos \omega t$ is $3\pi s B^2\omega^2 a^4(9s^2 + \mu_v^2 a^2\omega^2)^{-1}$.

14. Show that the field inside the shell in the last problem is, if $\tan \epsilon = -3s/(\mu_v \omega a)$,

$$3Bs(9s^2 + \mu_v^2 a^2\omega^2)^{-\frac{1}{2}}\sin(\omega t - \epsilon)$$

15. A thin circular cylinder of area resistivity s and radius a is placed in a field $B \cos \omega t$ normal to its axis. If $\tan \epsilon = -2s/(\mu_v \omega a)$, show that the field inside is

$$2sB(4s^2 + \omega^2\mu_v^2 a^2)^{-\frac{1}{2}}\sin(\omega t - \epsilon)$$

16. Show that the retarding torque per unit length, when the cylinder of the last problem is spinning on its axis with an angular velocity ω in a uniform magnetic field B normal to its axis, is $4\pi\omega s a^3 B^2(4s^2 + \omega^2\mu_v^2 a^2)^{-1}$.

17. The cylindrical coordinates of the wires of a bifilar circuit which carry the outgoing and incoming currents are, respectively, b, 0 and b, π. When the current in the circuit is $I \cos \omega t$, show that the vector potential at the point ρ, θ outside a thin tube $\rho = a$ of area resistivity s is

$$2\mu_v s I\pi^{-1}\Sigma[4(2n+1)^2 s^2 + \mu_v^2\omega^2 a^2]^{-\frac{1}{2}}(\rho^{-1}b)^{2n+1}\cos(2n+1)\theta\sin(\omega t - \epsilon_n)$$

where $\tan \epsilon_n = -2(2n+1)s(\mu_v \omega a)^{-1}$ and the summation is from 0 to ∞, $a > b$.

18. A thin oblate spheroidal shell $\zeta = \zeta_0$ has a variable thickness such that we may consider the area resistivity to be $s = h_2 s_0$. At the time $t = 0$, a uniform magnetic field of induction B is suddenly established parallel to the axis. Show that if $N = 2[(1+\zeta_0^2)^{\frac{3}{2}} Q_1^1(j\zeta_0)]^{-1}$ the eddy current density is given by

$$i_\phi = \frac{BN(1+\zeta_0^2)(1-\xi^2)^{\frac{1}{2}}}{2(\xi^2+\zeta_0^2)^{\frac{1}{2}}}e^{-Ns_0 t}$$

19. A long solid cylinder of radius $a = (x^2+y^2)^{\frac{1}{2}}$, permeability μ, and resistivity τ is placed in an x-directed alternating magnetic field $Be^{j\omega t}$ (real part). Show that the z-directed vector potentials A_i and A_0 inside and outside the cylinder are

$$A_i = \check{C}I_1[(jp)^{\frac{1}{2}}\rho]e^{j\omega t}\sin\phi \text{ (real part)}, \qquad A_o = B(\rho + \check{D}\rho^{-1})e^{j\omega t}\sin\phi \text{ (real part)}$$

where, writing I_0 and I_2 for $I_0[(jp)^{\frac{1}{2}}a]$ and $I_2[(jp)^{\frac{1}{2}}a]$,

$$\breve{C} = \frac{4\mu B}{(jp)^{\frac{1}{2}}[(\mu+\mu_v)I_0 - (\mu-\mu_v)I_2]} \quad \text{and} \quad \breve{D} = \frac{(\mu-\mu_v)I_0 - (\mu+\mu_v)I_2}{(\mu+\mu_v)I_0 - (\mu-\mu_v)I_2}a^2$$

20. Show that the mean power dissipated per unit length in the last problem is

$$\bar{P} = \frac{4\pi\omega\mu a^2 B^2[ber_0(p^{\frac{1}{2}}a)bei_2(p^{\frac{1}{2}}a) - bei_0(p^{\frac{1}{2}}a)ber_2(p^{\frac{1}{2}}a)]}{[(\mu+\mu_v)ber_0 + (\mu-\mu_v)ber_2]^2 + [(\mu+\mu_v)bei_0 + (\mu-\mu_v)bei_2]^2}$$

where $-(ber_2 x + jbei_2 x)$ is written for $I_2(j^{\frac{1}{2}}x)$. The arguments in numerator and denominator are the same. If p is small, show that this reduces to

$$\bar{P} = \tfrac{1}{2}\pi\mu^2\omega^2 a^4 B^2 \gamma(\mu+\mu_v)^{-2}$$

21. A long solid cylinder of radius a, permeability μ, and conductivity γ is rotated about its axis with a uniform angular velocity ω in a magnetic field B normal to this axis. Show that the retarding torque per unit length, due to eddy currents, is

$$T = \frac{8\pi\mu a^2 B^2[ber_0(p^{\frac{1}{2}}a)bei_2(p^{\frac{1}{2}}a) - bei_0(p^{\frac{1}{2}}a)ber_2(p^{\frac{1}{2}}a)]}{[(\mu+\mu_v)ber_0 + (\mu-\mu_v)ber_2]^2 + [(\mu+\mu_v)bei_0 + (\mu-\mu_v)bei_2]^2}$$

The arguments in numerator and denominator are the same, and $p = \gamma\mu\omega$.

22. A loop of radius a is coaxial with and at a distance d from the bottom of a cylindrical box of radius b and height c. If the skin effect is so large that $\boldsymbol{B}$ is nearly tangent to the walls, show that the decrease in self-inductance is

$$\Delta L = \left\{ \sum_{n=-\infty}^{\infty} M_{1n} + 4\pi\mu a^2 \sum_{n=0}^{\infty} [I_1(n\pi ac^{-1})]^2[cI_1(n\pi bc^{-1})]^{-1}K_1(n\pi bc^{-1}) \sin^2(n\pi dc^{-1}) \right\}$$

$$M_{1n} = 2\mu a\{k_1^{-1}[(1-\tfrac{1}{2}k_1^2)K_1 - E_1] - (1-\delta_n^0)k^{-1}[(1-\tfrac{1}{2}k^2)K - E]\}$$

where the modulus k_1 is $a[a^2 + (nc-d)^2]^{-\frac{1}{2}}$ and k is $a(a^2+n^2c^2)^{-\frac{1}{2}}$.

23. In problem 22 the skin depth is δ, the conductivity γ, and the current I. Show from 10.02 (8) that if k_m is chosen so that $J_1(k_m b)$ is zero, the power dissipation is

$$\bar{P} = \frac{2\pi I^2}{\gamma\delta}\left\{\frac{a^2}{bc}\sum_{n=1}^{\infty}\left[\frac{I_1(n\pi a/c)\sin(n\pi d/c)}{I_1(n\pi b/c)}\right]^2 + \frac{a^2}{b^2}\sum_{m=1}^{\infty}\frac{[J_1(k_m a)]^2[\sinh^2 k_m(c-d) + \sinh^2(k_m d)]}{[J_0(k_m b)\sinh(k_m c)]^2}\right\}$$

24. A thin disk of radius a and area resistivity s is placed in a region where, before its insertion, there was a uniform field of induction $\boldsymbol{B}_0 \cos\omega t$ normal to it. Find the secondary currents from the vector potential of the primary eddy currents induced by $\boldsymbol{B}_0 \cos\omega t$. Thus show that the total eddy current density is

$$\frac{\omega B_0}{2s}\left\{\rho\sin\omega t + \frac{\omega\mu}{6\pi\rho s}[(a-\rho)(2\rho^2+a^2)K + (a+\rho)(2\rho^2-a^2)E]\cos\omega t + \cdots\right\}$$

where the modulus of K and E is $2(a\rho)^{\frac{1}{2}}(a+\rho)^{-1}$ and terms in $(\omega\mu a/s)^2$ are neglected. Note that at 60 cycles this is good for 0.1 mm copper sheet.

25. If in the last problem the frequency is so high that the skin depth δ is much less than the disk thickness so that $\boldsymbol{B}$ can be taken tangent to its surface, show that the eddy current density, including both faces, at a distance ρ from the center is

$$i_\phi = \frac{\pi B \rho^2}{\mu a(a^2 - \rho^2)^{\frac{1}{2}}}$$

This i_ϕ is completely rigorous for a perfectly conducting disk.

26. The eddy current density in an infinitely long cylindrical shell of radius a, area resistivity $\mathfrak{s}$, and thickness much less than skin depth is $(i_\phi)_0$ at $t = 0$ and i_ϕ at $t = \tau$. If $(i_\phi)_0$ can be written as a Fourier integral, show that

$$(i_\phi)_0 = \int_0^\infty \Psi(k) \cos kz\, dk, \qquad i_\phi = \int_0^\infty \Psi(k) \cos kz e^{-[\mu a I_1(ka) K_1(ka)]^{-1} \mathfrak{s}\tau}\, dk$$

27. The vector potential of the inducing field at the shell's surface in problem 26 is

$$(A'_\phi)_{\rho=a} = -\int_0^\infty \Phi(k, a, t) \cos kz\, dk$$

Show by the reasoning of 10.16 and the last problem that, due to eddy currents,

$$(A_\phi)_{\rho=a} = -\int_0^\infty \int_0^\infty \frac{\partial}{\partial t}\Phi(k, a, t - \tau) e^{-[\mu a I_1(ka) K_1(ka)]^{-1} \mathfrak{s}\tau} \cos kz\, dk\, d\tau$$

28. A loop of radius b carries a current $I \cos \omega t$ and is coaxial with a very long shell at $\rho = a$ of area resistivity $\mathfrak{s}$ and thickness much less than the skin depth. Show that the shell increases the loop's resistance and inductance by amounts

$$\Delta R = 2\mu b^2 \omega \int_0^\infty \sin\theta \cos\theta [I_1(kb)]^2 K_1(ka) [I_1(ka)]^{-1}\, dk$$

$$\Delta L = -2\mu b^2 \int_0^\infty \sin^2\theta [I_1(kb)]^2 K_1(ka) [I_1(ka)]^{-1}\, dk$$

if $b < a$. If $b > a$, K and I are interchanged and $\tan\theta = \mathfrak{s}^{-1}\omega\mu a I_1(ka) K_1(ka)$.

29. From problem 40, Chap. VII, and 7.04 show that the external self-inductance of the torus formed by rotating a circle of radius b about a coplanar line c meters from its center at high frequencies where there is a strong skin effect is

$$L = 2\mu c \cos\alpha \cos\frac{1}{2}\alpha \left[\left(1 - \frac{1}{2}k^2\right)K - E\right]\left[1 - 2\sum_{n=1}^{\infty} \frac{(-1)^n Q^1_{2n-\frac{1}{2}}(\csc\alpha)}{(4n-1)(4n+1)Q^1_{-\frac{1}{2}}(\csc\alpha)}\right]$$

where $\sin\alpha = b/c$ and the modulus k of K and E is given by $k^2 = \cos\alpha \sec^2 \frac{1}{2}\alpha$.

30. A wire loop of radius a is coaxial with and outside an infinite conducting cylinder of radius b and carries an alternating current I of such high frequency that the eddy current density in the cylinder is confined to its skin. Show with the aid of problems 26 and 28 of Chap. VII that this density is

$$\check{i}_\phi = \frac{a\check{I}}{b\pi} \int_0^\infty \frac{K_1(ka)}{K_1(kb)} \cos kz\, dk$$

31. In the foregoing problem show with the aid of 8.06 (1) that the presence of the cylinder decreases the self-inductance of the loop by an amount

$$|\Delta L| = 2\mu a^2 \int_0^\infty I_1(kb) K_1^2(ka) K_1^{-1}(kb)\, dk$$

Show that the eddy current losses increase the resistance of the loop by an amount

$$R = \frac{4a^2}{\gamma\delta b}\int_0^\infty \left[\frac{K_1(ka)}{K_1(kb)}\right]^2 dk$$

where γ is the conductivity and δ the skin depth for the cylinder material.

32. Show from HMF 9.6.25, 12.2.3, and 12.2.4 that the integrals of 10.12 (5) are

$$\frac{\lambda}{\rho}K_1(\lambda\rho) - \int_0^c \frac{\cos\lambda u\,du}{(\rho^2+u^2)^{\frac{3}{2}}} \quad\text{and}\quad -\frac{\pi\lambda}{2\rho}[I_1(\lambda\rho) - \mathbf{L}_1(\lambda\rho)] + \frac{\lambda}{\rho} - \int_0^c \frac{\sin\lambda u\,du}{(\rho^2+u^2)^{\frac{3}{2}}}$$

Note that $K_1(z)$ and $I_1(z) - \mathbf{L}_1(z)$ are both tabulated in HMF. If $\lambda c \lessdot 1$, then $\cos\lambda u$ and $\sin\lambda u$ may be expanded in series and the integrals done. If $(\lambda c)^2$ is neglected compared with 1, show that Φ in 10.12 (5) becomes, if ψ is $\theta - \lambda c$.

$$\frac{\omega\mu_v M}{4\pi s}\left\{\left[\lambda K_1(\lambda\rho) - \frac{c}{\rho(\rho^2+c^2)^{\frac{1}{2}}}\right]\sin\psi - \left(\frac{\pi\lambda}{2}[I_1(\lambda\rho) - \mathbf{L}_1(\lambda\rho)] - \frac{\lambda\rho}{(\rho^2+c^2)^{\frac{1}{2}}}\right)\cos\psi\right\}$$

33. Solve the shielding problem of 10.13 exactly as follows: From Watson, page 148, and HMF 15.4.14 and 8.2.5 establish the formula

$$J_1(az)J_1(bz) = \sum_{m=0}^{\infty} \frac{(-1)^m(a^2-b^2)^{m+1}}{(m+1)!(m+2)!}P^1_{m+1}\left(\frac{a^2+b^2}{a^2-b^2}\right)\left(\frac{z}{2}\right)^{2m+2} = \sum_{m=0}^{\infty} C_m z^{2m+2}$$

Substitute for $J_1(az)J_1(bz)$ in the integral of 10.13 (2), integrate the result by IT I(11), page 135, and evaluate $\mathrm{Ei}(-j\omega c/\lambda)$ by HTF II (4), page 145. Replace $\omega c/\lambda = \frac{1}{2}\mu_v\omega c/s$ by p and thus show that

$$\breve{A}_\phi + \breve{A}'_\phi = -\frac{\mu_v a I}{2}\sum_{m=0}^{\infty} C_m\left(\frac{jp}{c}\right)^{2m+3}\left\{e^{jp}(\mathrm{Ci}\,p - j\,\mathrm{si}\,p) - \sum_{s=1}^{2m+3}\frac{(s-1)!}{(-jp)^s}\right\}$$

For Ci and si see 10.11 (5).

34. It is desired to calculate the induction heating of a solid sphere of radius a, permeability μ, and resistivity τ at the center of a cylindrical current sheet of radius c and length $2d$ whose current density is $i_\phi \cos\omega t$ amperes per meter. From 7.15 (7) and 7.02 (8) the vector potential inside an infinite solenoid with opposite current density is $-\frac{1}{2}\mu i_\phi r P_1^1(\cos\theta)\cos\omega t$. Superimpose on this the potential of 10.17 (6) so as to wipe out the current sheet except for a length $2d$. Thus show that the potential inside the sheet is the real part of

$$A'_\phi = \sum_{n=0}^{\infty} C_n r^{2n+1} P^1_{2n+1}(\cos\theta)e^{j\omega t}$$

$$C_0 = -\tfrac{1}{2}\mu_v i_\phi \cos\beta, \qquad C_n = \mu_v i_\phi d^{-2n}[2n(2n+1)(2n+2)]^{-1}\sin\beta P^1_{2n}(\cos\beta)$$

where $\tan\beta$ is c/d. Now apply the theory as in 10.06, and show that the vector potential inside the sphere is, writing $\breve{P}$ for $(jp)^{\frac{1}{2}}$,

$$\breve{A}_{\phi i} = r^{-\frac{1}{2}}\sum_{n=0}^{\infty} \breve{A}_n I_{2n+\frac{3}{2}}(\breve{P}r)P^1_{2n+1}(\cos\theta)$$

$$\breve{A}_n = K_m a^{\frac{1}{2}}(4n+3)a^{2n+1}C_n[(K_m-1)(2n+1)I_{2n+\frac{3}{2}}(\breve{P}a) + (\breve{P}aI_{2n+\frac{1}{2}}(\breve{P}a)]^{-1}$$

Now show from 10.07 and 5.231 (1) that the power absorbed by the sphere is

$$\bar{P} = \pi\tau pa\mu^{-2}\sum_{n=0}^{\infty}(2n+1)(2n+2)(4n+3)^{-1}\check{A}_n\hat{A}_n(\hat{P}\hat{I}_{2n+\frac{3}{2}}\check{I}_{2n+\frac{1}{2}} + \check{P}\check{I}_{2n+\frac{3}{2}}\hat{I}_{2n+\frac{1}{2}})$$

where $\check{I}_m$ and $\hat{I}_m$ are written for $I_m(\check{P}a)$ and $I_m(\hat{P}a)$ and K_m is μ/μ_v.

35. A solid sphere of radius a, permeability μ, and resistivity τ is spinning between the magnet poles as shown in Fig. 10.19. Show that the interior and exterior vector potentials are given by 10.19 (4) and (5) provided that

$$\check{A}_{mn}r^{-\frac{1}{2}}I_{2n+\frac{3}{2}}[(jp)^{\frac{1}{2}}r]e^{j[(2m+1)\phi+\omega t]} \quad \text{and} \quad (C_{mn}r^{2n+1} + \check{D}_{mn}r^{-2n-2})e^{j[(2m+1)\phi+\omega t]}$$

respectively, replace $C_{mn}r^{2r+1}\cos[(2m+1)\phi+\omega t]$. Show by using the boundary conditions of 10.06 and writing K_m for μ/μ_v that A_{mn} is given by

$$\check{A}_{mn} = \frac{(4n+3)K_m a^{2n+\frac{3}{2}}C_{mn}}{(2n+1)(K_m-1)\check{I}_{2n+\frac{3}{2}} + (jp)^{\frac{1}{2}}a\check{I}_{2n+\frac{1}{2}}}$$

where $\check{I}_{2n+\frac{3}{2}}$ is written for $I_{2n+\frac{3}{2}}[(jp)^{\frac{1}{2}}a]$ and C_{mn} is defined in 10.19 (3). Find the power dissipation and show that the retarding torque is

$$\frac{\pi\omega a}{\mu}\sum_{n=0}^{\infty}\sum_{m=0}^{n}\check{A}_{mn}\hat{A}_{mn}\frac{(2n+1)(2n+2)(2n+2m+2)!}{(4n+3)(2n-2m)!}[(-jp)^{\frac{1}{2}}\hat{I}_{2n+\frac{3}{2}}\check{I}_{2n+\frac{1}{2}} + (jp)^{\frac{1}{2}}\check{I}_{2n+\frac{3}{2}}\hat{I}_{2n+\frac{1}{2}}]$$

References

Bateman, H.: "Electrical and Optical Wave Motion," Cambridge, 1915. Discusses boundary conditions, gives solutions of wave equations and an extensive list of references.

Frenkel, J.: "Lehrbuch der Elektrodynamik," Vol II, Springer, Berlin, 1928. Gives vector potential treatment of eddy currents.

Geiger-Scheel: "Handbuch der Physik," Vol. XV, Berlin, 1927. Gives surprisingly meager treatment.

Jeans, J. H.: "The Mathematical Theory of Electricity and Magnetism," Cambridge, 1925. Follows Maxwell.

MacDonald, H. M.: "Electromagnetism," Bell, 1934. Uses long notation but gives rigorous treatment of eddy currents in solids.

Maxwell, J. C.: "Electricity and Magnetism," 3d ed. (1891), Dover, 1954. Treats current sheets.

Moullin, E. B.: "Principles of Electromagnetism," Oxford, 1932. Gives simple approximate treatment of eddy currents.

Ollendorf, F.: "Die Grundlagen der Hochfrequenztechnik," Springer, Berlin, 1926. Treats a-c resistance of wires and coils.

Russell, A.: "Alternating Currents," Cambridge, 1914. Treats manipulation of *ber* and *bei* functions.

Schelkunoff, S. A.: "Electromagnetic Waves," Van Nostrand, 1943. Treats chiefly very high frequency eddy currents.

Stratton, J. A.: "Electromagnetic Theory," McGraw-Hill, 1941. Treats eddy currents associated with electromagnetic waves.

CHAPTER XI

PLANE ELECTROMAGNETIC WAVES

11.00. Maxwell's Field Equations.—In a region where μ and ϵ are continuous and in which there may be an electric charge density together with electric convection or conduction currents, we shall assume that the electric and magnetic quantities are connected by the equations

$$\nabla \times H = \nabla \times \frac{B}{\mu} = i + \frac{\partial D}{\partial t} + \rho \mathbf{v} \tag{1}$$

$$\nabla \times E = -\frac{\partial B}{\partial t} \tag{2}$$

$$\nabla \cdot D = \rho \tag{3}$$

$$\nabla \cdot B = 0 \tag{4}$$

Without the last terms the first of these equations expresses Ampère's law 7.01 (3), the second generalizes Faraday's law of induction 8.00 (3), and the third and fourth are 3.02 (1) and 7.01 (1), respectively. The D term in the first equation, introduced by Maxwell, is new and, when currents and charges are absent, completes the symmetry of the group and provides a companion equation for (2). It asserts that a change in the electric flux through a closed curve in space produces a magnetomotance around the curve; just as Faraday's law of induction states that a change in the magnetic flux through a closed curve produces an electromotance around it. Maxwell apparently thought of the magnetic field as being caused by an actual displacement of electric charge, but such a picture is not needed to justify (1). These equations are justified by the far-reaching validity of the conclusions based on them. They may be only approximately true, but for the present precision of measurement we may consider them exact. If charge is conserved, the equation of continuity, 6.00 (3), holds so that

$$\nabla \cdot i + \frac{\partial \rho}{\partial t} = 0 \tag{5}$$

In isotropic bodies, i and E are connected by Ohm's law, 6.02 (3), which is

$$E = \tau i \qquad \text{or} \qquad i = \gamma E \tag{6}$$

In anisotropic bodies, with a proper choice of axes, 6.23 (2) gives 6.17

$$i = i\gamma_x E_x + j\gamma_y E_y + k\gamma_z E_z \tag{7}$$

The relations between $\boldsymbol{D}$ and $\boldsymbol{E}$ for isotropic bodies [1.13 (1)] or for anisotropic bodies with proper choice of axes [1.19 (7)] are

$$\boldsymbol{D} = \epsilon \boldsymbol{E} \qquad \text{or} \qquad \boldsymbol{D} = \boldsymbol{i}\epsilon_1 E_x + \boldsymbol{j}\epsilon_2 E_y + \boldsymbol{k}\epsilon_3 E_z \tag{8}$$

In (6), (7), and (8) μ was taken independent of field strength so these equations cannot be used for ferromagnetic materials at some frequencies.

11.01. Propagation Equation. Dynamic Potentials. Gauges. Hertz Vector.—If, in the vector formula $\nabla \times \nabla \times \boldsymbol{V} = \nabla(\nabla \cdot \boldsymbol{V}) - \nabla^2 \boldsymbol{V}$, $\boldsymbol{B}$ or $\boldsymbol{E}$ replaces $\boldsymbol{V}$, 11.00 (1) or (2) replaces $\nabla \times \boldsymbol{B}$ or $\nabla \times \boldsymbol{E}$, zero replaces $\nabla \cdot \boldsymbol{B}$ or $\nabla \cdot \boldsymbol{E}$, and then 11.00 (2) or (1) replaces the remaining $\nabla \times \boldsymbol{E}$ or $\nabla \times \boldsymbol{B}$ and $\gamma \boldsymbol{E}$ from 11.00 (6) replaces $\boldsymbol{i}$, the resultant propagation equations are

$$\nabla^2 \boldsymbol{B} = \mu\gamma \frac{\partial \boldsymbol{B}}{\partial t} + \mu\epsilon \frac{\partial^2 \boldsymbol{B}}{\partial t^2} \qquad \text{or} \qquad \nabla^2 \boldsymbol{E} = \mu\gamma \frac{\partial \boldsymbol{E}}{\partial t} + \mu\epsilon \frac{\partial^2 \boldsymbol{E}}{\partial t^2} \tag{1}$$

These hold in charge-free space. The first term on the right gives energy dissipation by heat and a wave damping that is absent when γ is zero. The static scalar and vector potentials may be redefined to include rapidly fluctuating fields. The vector potential whose curl is $\boldsymbol{B}$ must reduce for constant fields to the magnetostatic potential. Replacing $\boldsymbol{B}$ in 11.00 (2) by $\nabla \times \boldsymbol{A}$ and changing the differentiation order gives

$$\boldsymbol{B} = \nabla \times \boldsymbol{A} \qquad \nabla \times \boldsymbol{E} = -\nabla \times \frac{\partial \boldsymbol{A}}{\partial t} \tag{2}$$

Integration of $\nabla \times \boldsymbol{E}$ removes the curl but adds an integration constant with zero curl which must be the gradient of a scalar. Thus

$$\boldsymbol{E} = -\frac{\partial \boldsymbol{A}}{\partial t} - \nabla\psi \tag{3}$$

The potentials for a specified $\boldsymbol{E}$ and $\boldsymbol{B}$ are not unique. If $\boldsymbol{A}$ is solenoidal, then $\nabla \cdot \boldsymbol{E}$ along with 11.00 (3) and (5) yields Poisson's equation:

$$\nabla^2\psi = -\rho/\epsilon = \int \nabla \cdot \boldsymbol{i}\, dt/\epsilon \tag{4}$$

This depends only on the lamellar part of $\boldsymbol{i}$. Also, when $\nabla \cdot \boldsymbol{A}$ is zero, writing $\nabla \times \boldsymbol{A}$ for $\boldsymbol{B}$ in 11.00 (1) and using the above formula yield

$$\nabla \times \nabla \times \boldsymbol{A} = \mu \boldsymbol{i} + \mu\epsilon \frac{\partial \boldsymbol{E}}{\partial t} = \nabla^2 \boldsymbol{A}$$

Let $\boldsymbol{E}_s$ and $\boldsymbol{i}_s$ be the solenoidal parts of (3) and of the source $\boldsymbol{i}$, and let $\gamma \boldsymbol{E}_s$ be the current induced in the medium by $\boldsymbol{E}_s$. Then from (3)

$$\nabla^2 \boldsymbol{A} - \mu\gamma \frac{\partial \boldsymbol{A}}{\partial t} - \mu\epsilon \frac{\partial^2 \boldsymbol{A}}{\partial t^2} = -\mu \boldsymbol{i}_s \tag{5}$$

Clearly for static fields, this becomes 7.02 (2). The potential choice in which $\boldsymbol{A}$ derives from the solenoidal part of $\boldsymbol{i}$ and ψ from the lamellar part is called the Coulomb gauge. When time-dependent, only $\boldsymbol{A}$ propagates and ψ gives a localized field in the same time phase everywhere. This $\boldsymbol{A}$ comes from a scalar solution W of the propagation equation by using $\nabla \times \boldsymbol{u}W$ or $\nabla \times (\boldsymbol{u} \times \nabla W)$, a method that appeared in 10.01 and is applied later in Chaps. XII and XIII to wave guides and cavity calculations.

Another and more usual choice of potential is the Lorentz gauge, which requires that both $\boldsymbol{A}$ and ψ have the same propagation equations as $\boldsymbol{E}$ and $\boldsymbol{B}$. In a region of conductivity γ with no sources, these equations are

$$\nabla^2 \boldsymbol{A} = \mu\gamma \frac{\partial \boldsymbol{A}}{\partial t} + \mu\epsilon \frac{\partial^2 \boldsymbol{A}}{\partial t^2} \qquad \nabla^2 \psi = \mu\gamma \frac{\partial \psi}{\partial t} + \mu\epsilon \frac{\partial^2 \psi}{\partial t^2} \tag{6}$$

This gives another relation between $\boldsymbol{A}$ and ψ for, if ρ is zero, taking the divergence of both sides of (3) gives zero for $\nabla \cdot \boldsymbol{E}$ by 11.00 (3), so that

$$\nabla^2 \psi = -\frac{\partial \nabla \cdot \boldsymbol{A}}{\partial t}$$

Comparison of this with (6) shows that for $\nabla^2 \psi$ to agree requires that

$$\nabla \cdot \boldsymbol{A} = -\mu\gamma\psi - \mu\epsilon \frac{\partial \psi}{\partial t} \tag{7}$$

To prove this also gives $\nabla^2 \boldsymbol{A}$ in (6), take the gradient of both sides of (7), expand $\nabla\nabla \cdot \boldsymbol{A}$ by the initial vector relation of this article, write $\nabla \times \boldsymbol{B}$ in terms of $\boldsymbol{E}$ by 11.00 (1), (6), and (8), and eliminate $\nabla\psi$ by (3). Cancellation of the $\boldsymbol{E}$ terms gives $\nabla^2 \boldsymbol{A}$ as in (6).

The whole electromagnetic field may be represented by a single vector $\boldsymbol{Z}$, the Hertz vector, from which $\boldsymbol{A}$ and ψ can be obtained by the equations

$$\boldsymbol{A} = \mu\gamma \boldsymbol{Z} + \mu\epsilon \frac{\partial \boldsymbol{Z}}{\partial t}, \qquad \psi = -\nabla \cdot \boldsymbol{Z} \tag{8}$$

These equations satisfy (7) and also (3) if we take

$$\boldsymbol{E} = \nabla(\nabla \cdot \boldsymbol{Z}) - \nabla^2 \boldsymbol{Z} = \nabla \times (\nabla \times \boldsymbol{Z}) \tag{9}$$

$$\nabla^2 \boldsymbol{Z} = \mu\gamma \frac{\partial \boldsymbol{Z}}{\partial t} + \mu\epsilon \frac{\partial^2 \boldsymbol{Z}}{\partial t^2} \tag{10}$$

The magnetic induction is given by (2) and (8) in terms of $\boldsymbol{Z}$.

$$\boldsymbol{B} = \mu\gamma \nabla \times \boldsymbol{Z} + \mu\epsilon \frac{\partial (\nabla \times \boldsymbol{Z})}{\partial t} \tag{11}$$

Equation (10) contains all the properties of the electromagnetic wave. In a lossless medium the fields of (9) and (11) may be written

$$E = \nabla \times \nabla \times Z_e \qquad \text{and} \qquad H = \frac{\epsilon\partial(\nabla \times Z_e)}{\partial t} \tag{12}$$

This Z_e is the electric Hertz vector. If 11.00 (1) and (2), in a medium without charges or currents, are put in terms of H and E, then the same two equations result if E and H as well as $-\mu$ and ϵ are interchanged. Making the same interchanges in (12) gives a new Hertz vector Z_m, thus

$$H = \nabla \times \nabla \times Z_m \qquad \text{and} \qquad E = -\frac{\mu\partial(\nabla \times Z_m)}{\partial t} \tag{13}$$

This is often called the magnetic Hertz vector and is sometimes more convenient than the Coulomb gauge potentials in solving wave-guide problems.

11.02. Poynting's Vector.—Multiplication of 11.00 (1) by $-E$ and of 11.00 (2) by B/μ and addition of the results give

$$\frac{B}{\mu}\cdot(\nabla \times E) - E\cdot\left(\nabla \times \frac{B}{\mu}\right) = -i\cdot E - E\cdot\frac{\partial D}{\partial t} - \frac{B}{\mu}\cdot\frac{\partial B}{\partial t} \tag{1}$$

Integration of this over any volume v gives by Gauss's theorem, 3.00 (2),

$$\int_V\left[\frac{B}{\mu}\cdot(\nabla \times E) - E\cdot\left(\nabla \times \frac{B}{\mu}\right)\right]dv = \int_V \nabla\cdot\left(E \times \frac{B}{\mu}\right)dv = \int_S n\cdot\left(E \times \frac{B}{\mu}\right)dS$$

for the left side. The right side remains a volume integral so that

$$-\int_S n\cdot\left(E \times \frac{B}{\mu}\right)dS = \int_V\left[i\cdot E + \frac{\partial}{\partial t}\left(\frac{D\cdot E}{2} + \frac{B^2}{2\mu}\right)\right]dv \tag{2}$$

By 6.03 (2) and Ohm's law, the first term on the right represents the electric power absorbed as heat in v. By 2.08 (2) and 8.02 (3), the second and third term represent the rate of change of the electric and magnetic field energies in v. If all these terms are positive and if the law of the conservation of energy holds, then v absorbs power from outside, and the rate of flow of energy through its surface must be given by the left side of (2). Thus the outward normal component of the vector

$$\Pi = \frac{E \times B}{\mu} \tag{3}$$

when integrated over a closed surface, represents the rate of flow of energy outward through that surface. The vector Π is called Poynting's vector We should note that a meaning has been given only to the surface integral of this vector over a closed surface. We have not shown that Π can represent the rate of flow of energy through any element of surface.

11.03. Plane Waves in Homogeneous Uncharged Dielectric Insulator. An electromagnetic disturbance is a plane wave when the instantaneous values of $\boldsymbol{B}$, $\boldsymbol{E}$, ψ, $\boldsymbol{A}$, and $\boldsymbol{Z}$ are constant in phase over any plane parallel to a fixed plane. These planes are called the wave fronts and their normal, given by the unit vector $\boldsymbol{n}$, is the wave normal. In the dielectric where γ is zero so that the first term on the right of 11.01 (11) is missing, differentiation shows that, if $v = (\mu\epsilon)^{-\frac{1}{2}}$, a general solution is

$$\begin{aligned} \boldsymbol{Z} &= f_1(\boldsymbol{n}\cdot\boldsymbol{r} - vt) + f_2(\boldsymbol{n}\cdot\boldsymbol{r} + vt) \\ &= f_1[(\boldsymbol{n}\cdot\boldsymbol{r} + vt') - v(t + t')] + f_2[(\boldsymbol{n}\cdot\boldsymbol{r} - vt') + v(t + t')] \end{aligned} \tag{1}$$

so that f_1 has the same value at the point $\boldsymbol{n}\cdot\boldsymbol{r} + vt'$ at the time $t + t'$ as at the point $\boldsymbol{n}\cdot\boldsymbol{r}$ at the time t. Thus it represents a wave moving in the $\boldsymbol{n}$-direction with a velocity v. Similarly, f_2 represents a wave of the same velocity in the $-\boldsymbol{n}$-direction. The vacuum velocity of an electromagnetic wave is $c = (\mu_v\epsilon_v)^{-\frac{1}{2}}$ and equals 3×10^8 meters per second. The ratio of c to the velocity v in a medium is its index of refraction n.

$$n = \frac{c}{v} \tag{2}$$

Since Eqs. (1), (2), (6), (7), and (10) of 11.01 are identical, the simplest solutions for a wave in the positive $\boldsymbol{n}$-direction have the form

$$\boldsymbol{D} = \boldsymbol{D}_0 f(\boldsymbol{n}\cdot\boldsymbol{r} - vt) \tag{3}$$

$$\boldsymbol{E} = \boldsymbol{E}_0 f(\boldsymbol{n}\cdot\boldsymbol{r} - vt) \tag{4}$$

$$\boldsymbol{B} = \boldsymbol{B}_0 f_1(\boldsymbol{n}\cdot\boldsymbol{r} - vt) \tag{5}$$

where $\boldsymbol{D}_0$, $\boldsymbol{E}_0$, and $\boldsymbol{B}_0$ are the vector amplitudes of $\boldsymbol{D}$, $\boldsymbol{E}$, and $\boldsymbol{B}$, respectively, and $f(\boldsymbol{n}\cdot\boldsymbol{r} - vt)$ is a scalar. When ρ is zero, (3) and 11.00 (3) give

$$\nabla\cdot\boldsymbol{D} = \boldsymbol{n}\cdot\boldsymbol{D}_0 f'(\boldsymbol{n}\cdot\boldsymbol{r} - vt) = 0$$

Thus, either f' is zero, giving a static field of no interest here, or

$$\boldsymbol{n}\cdot\boldsymbol{D}_0 = 0 \tag{6}$$

A similar substitution of (5) in 11.00 (4) gives the similar result

$$\boldsymbol{n}\cdot\boldsymbol{B}_0 = 0 \tag{7}$$

Thus $\boldsymbol{D}$ and $\boldsymbol{B}$ lie in the wave front. From (4), (5), and 11.00 (2) we have

$$\nabla\times\boldsymbol{E} = \boldsymbol{n}\times\boldsymbol{E}_0 f'(\boldsymbol{n}\cdot\boldsymbol{r} - vt) = -\frac{\partial\boldsymbol{B}}{\partial t} = v\boldsymbol{B}_0 f_1'(\boldsymbol{n}\cdot\boldsymbol{r} - vt)$$

This holds at all times so that f_1 is proportional to f, and we take

$$\boldsymbol{n}\times\boldsymbol{E}_0 = v\boldsymbol{B}_0 \qquad \text{or} \qquad \boldsymbol{n}\times\boldsymbol{E} = v\boldsymbol{B} \tag{8}$$

Similarly, substitution of (3) and (5) in 11.00 (1) where $\boldsymbol{i}$ is zero gives

$$\boldsymbol{n} \times \boldsymbol{B}_0 = -\mu v \boldsymbol{D}_0 \qquad \text{or} \qquad \boldsymbol{n} \times \boldsymbol{B} = -\mu v \boldsymbol{D} \tag{9}$$

The scalar product of (8) by (9) gives, since $\boldsymbol{n} \cdot \boldsymbol{B}_0$ is zero by (7),

$$(\boldsymbol{n} \times \boldsymbol{E}_0) \cdot (\boldsymbol{n} \times \boldsymbol{B}_0) = -\mu v^2 \boldsymbol{D}_0 \cdot \boldsymbol{B}_0 = (\boldsymbol{n} \cdot \boldsymbol{n})(\boldsymbol{E}_0 \cdot \boldsymbol{B}_0) - (\boldsymbol{n} \cdot \boldsymbol{B}_0)(\boldsymbol{n} \cdot \boldsymbol{E}_0) = \boldsymbol{E}_0 \cdot \boldsymbol{B}_0$$

By (8) the right side of this equation is zero, so that

$$\boldsymbol{E}_0 \cdot \boldsymbol{B}_0 = 0 \qquad \text{and} \qquad \boldsymbol{D}_0 \cdot \boldsymbol{B}_0 = 0 \tag{10}$$

Thus both $\boldsymbol{E}_0$ and $\boldsymbol{D}_0$ are normal to $\boldsymbol{B}_0$. From 11.02 (3), Poynting's vector is

$$\boldsymbol{\Pi} = \mu^{-1} \boldsymbol{E}_0 \times \boldsymbol{B}_0 [f(\boldsymbol{n} \cdot \boldsymbol{r} - vt)]^2 = \boldsymbol{\Pi}_0 [f(\boldsymbol{n} \cdot \boldsymbol{r} - vt)]^2 \tag{11}$$

whereas from (6) and (7) the normal to the wave front has the direction of $\boldsymbol{D}_0 \times \boldsymbol{B}_0$. Thus the energy propagation direction $\boldsymbol{\Pi}$ makes the same angle with the wave normal $\boldsymbol{n}$ that $\boldsymbol{E}$ makes with $\boldsymbol{D}$. When all $\boldsymbol{B}$ vectors parallel a fixed line, then from (10) all $\boldsymbol{E}$ vectors parallel a line normal to it and the wave is plane polarized. The polarization plane is defined as that parallel to $\boldsymbol{\Pi}$ and $\boldsymbol{B}$ in optics and as that parallel to $\boldsymbol{\Pi}$ and $\boldsymbol{E}$ in radio.

11.04. Plane Wave Velocity in Anisotropic Mediums.—We now consider the propagation of electromagnetic effects in a homogeneous anisotropic medium of zero conductivity and permeability μ_v. By 1.19 (7), in such a medium there are axes for which $\boldsymbol{D}$ and $\boldsymbol{E}$ are connected by the equations

$$D_x = \epsilon_1 E_x, \qquad D_y = \epsilon_2 E_y, \qquad D_z = \epsilon_3 E_z \tag{1}$$

These coordinate axes are chosen to coincide with the electric axes of the medium. By 2.08, the electrical energy density is given by

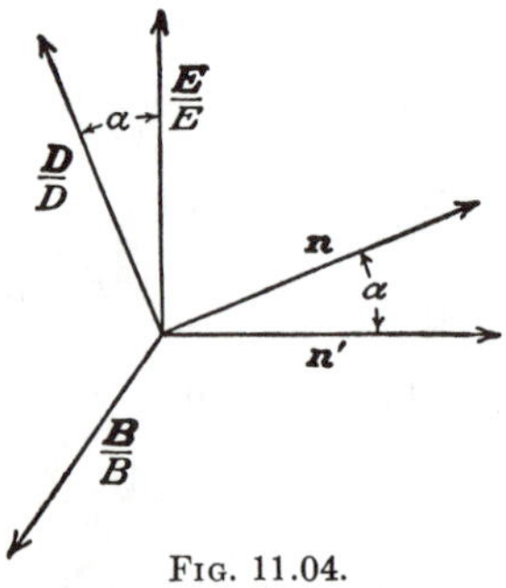

FIG. 11.04.

$$\frac{\partial W}{\partial v} = \frac{1}{2} \boldsymbol{D} \cdot \boldsymbol{E} = \frac{1}{2}(\epsilon_1 E_x^2 + \epsilon_2 E_y^2 + \epsilon_3 E_z^2) = \frac{B^2}{2\mu_v} \tag{2}$$

The last relation is obtained by using 11.03 (9) for $\boldsymbol{D}$, interchanging dot and cross, and substituting 11.03 (8) for $\boldsymbol{n} \times \boldsymbol{E}$. It shows the equality of electric and magnetic energy densities whose sum is twice (2). Let the direction of ray or energy propagation be $\boldsymbol{n}'$, then $\boldsymbol{B}$, $\boldsymbol{D}$, $\boldsymbol{E}$, $\boldsymbol{n}$, and $\boldsymbol{n}'$ are related as shown in Fig. 11.04. Evidently the vectors $\boldsymbol{D}$, $\boldsymbol{E}$, $\boldsymbol{n}$, and $\boldsymbol{n}'$ lie in the same plane since all are normal to $\boldsymbol{B}$. We also notice that

$$\sin \alpha = \frac{\boldsymbol{n} \cdot \boldsymbol{E}}{E} = -\frac{\boldsymbol{n}' \cdot \boldsymbol{D}}{D} \tag{3}$$

To find the wave front velocity v along $\boldsymbol{n}$, we start with 11.03 (8) and (9).

$$\mu_v v^2 \boldsymbol{D} = -\boldsymbol{n} \times (\boldsymbol{n} \times \boldsymbol{E}) = \boldsymbol{E} - \boldsymbol{n}(\boldsymbol{n} \cdot \boldsymbol{E}) \tag{4}$$

We now introduce the crystal constants v_1, v_2, and v_3 defined by

$$\mu_v \epsilon_1 v_1^2 = 1, \qquad \mu_v \epsilon_2 v_2^2 = 1, \qquad \mu_v \epsilon_3 v_3^2 = 1 \tag{5}$$

Elimination of $\boldsymbol{E}$ from the components of (4) by (1) and (5) gives

$$\mu_v(v_1^2 - v^2)D_x = l(\boldsymbol{n} \cdot \boldsymbol{E}), \qquad \mu_v(v_2^2 - v^2)D_y = m(\boldsymbol{n} \cdot \boldsymbol{E}),$$
$$\mu_v(v_3^2 - v^2)D_z = n(\boldsymbol{n} \cdot \boldsymbol{E}) \tag{6}$$

where l, m, and n are the components or direction cosines of $\boldsymbol{n}$. From 11.03 (6) $\boldsymbol{n} \cdot \boldsymbol{D} = lD_x + mD_y + nD_z = 0$. Insertion of D_x, D_y, and D_z from (6) gives

$$\frac{l^2}{v^2 - v_1^2} + \frac{m^2}{v^2 - v_2^2} + \frac{n^2}{v^2 - v_3^2} = 0 \tag{7}$$

This is Fresnel's equation and gives the velocity of the wave front propagation in terms of the direction cosines of its normal. In general, for every particular direction of the wave normal, there are two different values of this velocity. We can easily verify by substitution in (7) that, if $v_1 > v_2 > v_3$, the only two directions in which v has a single value are

$$l_0 = \pm(v_1^2 - v_2^2)^{\frac{1}{2}}(v_1^2 - v_3^2)^{-\frac{1}{2}}, \quad m_0 = 0, \quad n_0 = \pm(v_2^2 - v_3^2)^{\frac{1}{2}}(v_1^2 - v_3^2)^{-\frac{1}{2}} \tag{8}$$

and that the velocity in these directions is v_2. The directions defined by l_0, m_0, and n_0 are called the optic axes of this biaxial crystal. When $v_1 = v_2 \neq v_3$, v is single-valued in the z-direction only, for which $n_0 = 1$ and $l_0 = m_0 = 0$. If $v_1 \neq v_2 = v_3$, v is single-valued in the x-direction only, for which $l_0 = 1$ and $m_0 = n_0 = 0$. In such cases there is only one optic axis and the crystal is said to be uniaxial.

11.05. Ray Surface and Polarization in Anisotropic Mediums.—The energy propagation velocity v' in any direction in a crystal is found by dividing 11.02 (3) by $E^2/(\mu v)$ and multiplying through by $\boldsymbol{n} \cdot \boldsymbol{E}$ giving by 11.03 (8)

$$\mu v \frac{\boldsymbol{n} \cdot \boldsymbol{E}}{E^2} \boldsymbol{\Pi} = v\boldsymbol{n} \cdot \boldsymbol{E} \frac{\boldsymbol{E} \times \boldsymbol{B}}{E^2} = \boldsymbol{n}(\boldsymbol{n} \cdot \boldsymbol{E}) - \frac{(\boldsymbol{n} \cdot \boldsymbol{E})^2}{E^2}\boldsymbol{E} \tag{1}$$

We can then solve 11.04 (4) for $\boldsymbol{n}(\boldsymbol{n} \cdot \boldsymbol{E})$, write out components, eliminate D_x, D_y, and D_z by 11.04 (6), and insert in the right side of (1), giving

$$\mu v \frac{\boldsymbol{n} \cdot \boldsymbol{E}}{E^2} \Pi_{x,y,z} = \left[1 - \frac{v^2}{v_{1,2,3}^2} - \frac{(\boldsymbol{n} \cdot \boldsymbol{E})^2}{E^2}\right] E_{x,y,z} \tag{2}$$

Figure 11.04 and 11.04 (3) show that the relation between v and v' is

$$\left(\frac{v}{v'}\right)^2 = \cos^2\alpha = 1 - \sin^2\alpha = 1 - \frac{(\boldsymbol{n}\cdot\boldsymbol{E})^2}{E^2} \tag{3}$$

Also $\Pi_{x,y,z} = \Pi_0 n'_{x,y,z}$ so that we can rearrange (2) in the form

$$-\frac{n'_{x,y,z}}{(v/v_{1,2,3})^2 - (v/v')^2} = \frac{E^2}{\mu v \Pi_0(\boldsymbol{n}\cdot\boldsymbol{E})} E_{x,y,z}$$

Let l', m', and n' be the direction cosines or components of $\boldsymbol{n}'$ and notice that, from 11.03 (11), $\boldsymbol{n}'\cdot\boldsymbol{E} = l'E_x + m'E_y + n'E_z$. Then multiplying these equations through by l', m', and n' and adding, we obtain

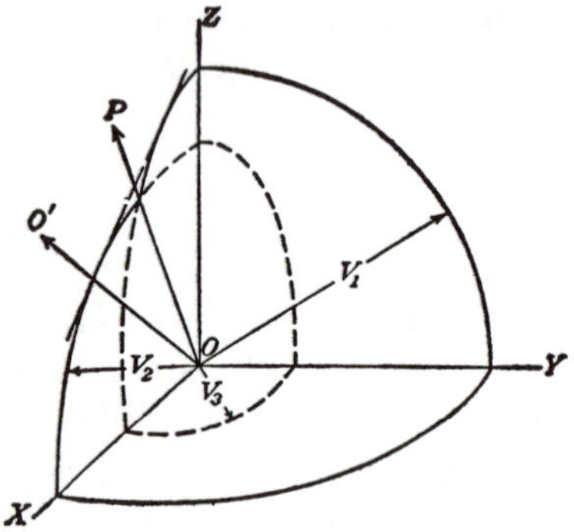

FIG. 11.05.

$$\frac{l'^2v_1^2}{v'^2 - v_1^2} + \frac{m'^2v_2^2}{v'^2 - v_2^2} + \frac{n'^2v_3^2}{v'^2 - v_3^2} = 0 \tag{4}$$

This equation gives the ray velocity v' in the direction l', m', and n' as a function of the crystal constants v_1, v_2, and v_3. In general for every particular direction there are two ray velocities. If, at some instant, an electromagnetic disturbance starts from the origin, then in 1 sec the wave will be given by $r = v'$. When $v_1 > v_2 > v_3$, the form of this wave in one octant is shown in Fig. 11.05. In the direction of OP and of its image in the yz-plane, both rays have the same velocity. Substitution in (4) shows the directions of the ray and optic axes to be related by

$$l_r = \frac{l_0 v_3}{v_2}, \qquad m_r = m_0 = 0, \qquad n_r = \frac{n_0 v_1}{v_2} \tag{5}$$

The optic axis is shown by OO' in Fig. 11.05. In each of the coordinate planes, the section of one of the sheets of the double surface is circular and of radius v_1, v_2, or v_3 as shown. If two of the quantities v_1, v_2, and v_3 are equal, one sheet of the wave surface is a prolate or oblate spheroid whose axis is a diameter of the other sheet which is spherical. The rays that produce the spherical surface are said to be ordinary rays, and those producing the spheroidal surface are called extraordinary rays.

Let us designate the two solutions of 11.04 (7) by v_a and v_b and the components of the electric displacement corresponding to these two normal velocities by $D^{(b)}_{x,y,z}$ and $D^{(a)}_{x,y,z}$. Multiply each equation of 11.04

(6) with $v = v_a$ by the same equation with $v = v_b$ and rearrange. This gives

$$\frac{\mu^2 D^{(a)}_{x,y,z} D^{(b)}_{x,y,z}}{(\boldsymbol{n}\cdot\boldsymbol{E})^2} = \frac{n^2_{x,y,z}}{(v^2_{1,2,3} - v^2_a)(v^2_{1,2,3} - v^2_b)} \tag{6}$$

Split the right side by partial fractions and add the three equations:

$$\frac{\mu^2 \boldsymbol{D}^{(a)}\cdot\boldsymbol{D}^{(b)}}{(\boldsymbol{n}\cdot\boldsymbol{E})^2} = \frac{1}{v_a^2 - v_b^2}\sum_{x,y,z}\left(\frac{n^2_{x,y,z}}{v^2_{1,2,3} - v_a^2} - \frac{n^2_{x,y,z}}{v^2_{1,2,3} - v_b^2}\right)$$

By 11.04 (7), each term in the sum on the right is zero so that we have

$$\boldsymbol{D}^{(a)}\cdot\boldsymbol{D}^{(b)} = 0 \tag{7}$$

Thus the v_a and v_b rays are plane polarized at right angles to each other.

11.06. Energy, Pressure, and Momentum of a Plane Wave.—Imagine a plane wave in an isotropic medium striking a thin infinite plane absorbing sheet normally, and consider a right prism which encloses a unit area of the sheet and whose axis is normal to it. An integration of Poynting's vector $\boldsymbol{\Pi}$ over its surface gives the radiant energy entering it. But we have already seen, from 11.03 (6) and (7), that on the sides of the prism $vB_n = E_n = 0$ so that Π_n is zero. It is also zero on that end which is on the opposite side of the absorbing sheet from the impinging wave. Thus, the only contribution to the integral is from the base on which the wave impinges. Therefore, from 11.03 (11) and (8), the instantaneous rate of energy absorption in the sheet, in watts per square meter, is

$$\boldsymbol{\Pi} = \frac{\boldsymbol{E}\times\boldsymbol{B}}{\mu} = \frac{\boldsymbol{n}E^2 - (\boldsymbol{n}\cdot\boldsymbol{E})\boldsymbol{E}}{\mu v}$$

In isotropic mediums $\boldsymbol{n}\cdot\boldsymbol{E} = 0$ and $\bar{B} = (\mu\epsilon)^{\frac{1}{2}}\bar{E}$ from 11.03 (8), so that

$$\boldsymbol{\Pi}_i = \boldsymbol{n}\frac{\epsilon E^2}{(\mu\epsilon)^{\frac{1}{2}}} = \boldsymbol{n}\frac{B^2}{\mu(\mu\epsilon)^{\frac{1}{2}}} = \boldsymbol{n}\frac{1}{(\mu\epsilon)^{\frac{1}{2}}}\left(\frac{B^2}{2\mu} + \frac{\epsilon E^2}{2}\right) \tag{1}$$

But, by 2.08 (2) and 8.02 (3), the energy density in the wave field is

$$\frac{\partial W}{\partial v} = \frac{\epsilon E^2}{2} + \frac{B^2}{2\mu} \tag{2}$$

The propagation velocity is $(\mu\epsilon)^{-\frac{1}{2}}$ so that the energy per square meter of wave front equals that in a cylinder of unit area and length $(\mu\epsilon)^{-\frac{1}{2}}$.

From 1.14 (6) and 8.14 (2) there appear to be in electric and magnetic fields pressures normal to the lines of force of $\frac{1}{2}\epsilon E^2$ and $\frac{1}{2}B^2/\mu$, respectively. Since in the plane wave case, such fields parallel the face of the absorbing sheet on one side but not on the other, we should expect a total pressure, in newtons per square meter, against the sheet, of amount

$$P = \tfrac{1}{2}\epsilon E^2 + \tfrac{1}{2}\mu^{-1}B^2 \tag{3}$$

From (2) this equals the energy density at the surface of the sheet.

Consider the absorbing sheet free to move but so heavy that P gives it a negligible velocity. Then the relation between P and the momentum p is

$$P = \frac{dp}{dt} \tag{4}$$

The mechanical law of conservation of momentum, if applied here, requires that the momentum acquired by the sheet must have been carried by the electromagnetic wave which impinged on it. A comparison of (4), (3), and (1) shows that this implies that the wave possesses a momentum g_n per unit volume, in the direction of propagation $\boldsymbol{n}$, equal to

$$g_n = \mu\epsilon\Pi_n \tag{5}$$

11.07. Refraction and Reflection of a Plane Wave.—Let us now consider a plane wave in a medium of permeability μ and capacitivity ϵ, impinging on the infinite plane boundary of a second medium whose constants are μ'' and ϵ''. This wave may give rise to two waves: one, known as the reflected wave, returning into the first medium, and the other, known as the refracted wave, entering the second medium. Let the unit vectors in the direction of propagation of the incident, reflected, and refracted waves be $\boldsymbol{\sigma}$, $\boldsymbol{\sigma}'$, and $\boldsymbol{\sigma}''$, respectively. Let $\boldsymbol{\sigma}$ and $\boldsymbol{\sigma}''$ make angles θ and θ'' with the normal to the plane surface drawn into the second medium, and $\boldsymbol{\sigma}'$ make an angle θ' with the opposite normal. The plane of incidence is perpendicular to the interface and contains $\boldsymbol{\sigma}$.

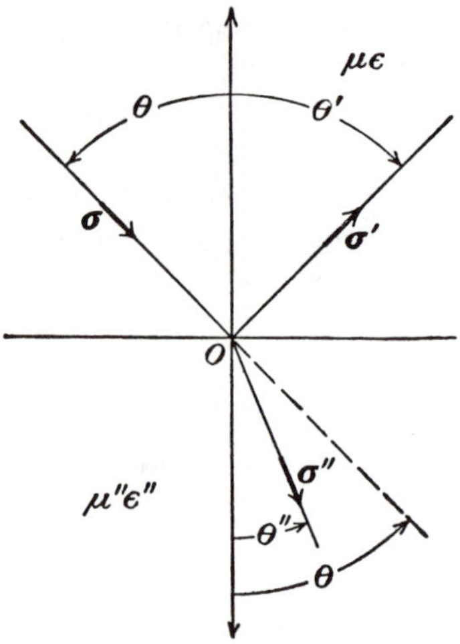

FIG. 11.07.

Let M and N be any two points in space, the vector from M to N being $\boldsymbol{p}$. If, as before, we take $\boldsymbol{n}$ to be a unit vector normal to the wave front, then $\boldsymbol{n}\cdot\boldsymbol{p}/v$ seconds after passing through M this wave, which has a velocity v, will pass through N. In the case being considered, the same wave front passes through a given point P in the first medium twice, once before and once after reflection. A wave front which passes through a point O on the reflecting surface at the time $t = 0$ will pass through P before reflection at $t = \boldsymbol{\sigma}\cdot\boldsymbol{r}/v$ and after reflection at $t = \boldsymbol{\sigma}'\cdot\boldsymbol{r}/v$, where

$\mathbf{r}$ is the radius vector from O to P. If the same law of reflection holds for all parts of the surface, the interval of time between these two passages must be the same for any other point Q at the same distance from the surface as P. The vector from O to Q will be $\mathbf{r} + \mathbf{s}$ where $\mathbf{s}$ is a vector lying in the surface. Equating these time intervals and multiplying through by the velocity give

$$\mathbf{\sigma} \cdot \mathbf{r} - \mathbf{\sigma}' \cdot \mathbf{r} = \mathbf{\sigma} \cdot (\mathbf{r} + \mathbf{s}) - \mathbf{\sigma}' \cdot (\mathbf{r} + \mathbf{s}) \qquad \text{or} \qquad \mathbf{\sigma} \cdot \mathbf{s} = \mathbf{\sigma}' \cdot \mathbf{s} \quad (1)$$

Therefore $\mathbf{\sigma}$ and $\mathbf{\sigma}'$ make the same angle with any vector drawn in the interface which is possible only if $\mathbf{\sigma}$ is the mirror image of $\mathbf{\sigma}'$ in the surface. Thus both reflected and incident wave normals lie in the plane of incidence on opposite sides of the normal to the interface and make the same (acute) angle with it.

To locate the refracted beam, suppose a wave front requires a certain time to pass from P in the first medium to its image point P'' in the second. Let the radius vector from O to P be $\mathbf{r}$ and from O to P'' be $\mathbf{r}''$. If two other points Q and its image point Q'' are at the same distances from the interface as P and P'', the time for the wave front to pass from Q to Q'' will be the same. The radius vectors to Q and Q'' are $\mathbf{r} + \mathbf{s}$ and $\mathbf{r}'' + \mathbf{s}$, where $\mathbf{s}$ lies in the interface. Equating the time intervals in the same way as for the reflected wave normal, we have

$$(\mu''\epsilon'')^{\frac{1}{2}}(\mathbf{\sigma}'' \cdot \mathbf{r}'') - (\mu\epsilon)^{\frac{1}{2}}(\mathbf{\sigma} \cdot \mathbf{r}) = (\mu''\epsilon'')^{\frac{1}{2}}[\mathbf{\sigma}'' \cdot (\mathbf{r}'' + \mathbf{s})] - (\mu\epsilon)^{\frac{1}{2}}[\mathbf{\sigma} \cdot (\mathbf{r} + \mathbf{s})]$$
$$(\mu''\epsilon'')^{\frac{1}{2}}(\mathbf{\sigma}'' \cdot \mathbf{s}) = (\mu\epsilon)^{\frac{1}{2}}(\mathbf{\sigma} \cdot \mathbf{s}) \quad (2)$$

If $\mathbf{s}$ is taken perpendicular to $\mathbf{\sigma}$, which means normal to the plane of incidence, then the right side of (2) is zero, so that $\mathbf{\sigma}''$ is also perpendicular to $\mathbf{s}$. Thus both reflected and refracted wave normals lie in the plane of incidence. If $\mathbf{s}$ is taken in the plane of incidence then, if v and v'' are the wave velocities in the first and second mediums, respectively,

$$\left(\frac{\mu''\epsilon''}{\mu\epsilon}\right)^{\frac{1}{2}} = \frac{v}{v''} = \frac{\cos \mathbf{\sigma}\mathbf{s}}{\cos \mathbf{\sigma}''\mathbf{s}} = \frac{\sin\theta}{\sin\theta''} = n \quad (3)$$

This is Snell's law of refraction. The ratio n of $\sin\theta$ to $\sin\theta''$ is called the index of refraction. The preceding reasoning is equally valid for an anisotropic medium, although v and v'' may be different for each angle of incidence so that n depends on θ.

11.08. Intensity of Reflected and Refracted Waves.—The law of conservation of energy requires that the energy passing into a medium through a square meter of surface must equal the difference between the incident and the reflected energies. From Fig. 11.08 and 11.03 (11), this gives

$$(\Pi - \Pi')\cos\theta = \Pi''\cos\theta'' \quad (1)$$

We shall identify all vectors associated with a beam whose magnetic vector lies in the plane of incidence by the subscript 1, and those associated with a beam whose electric vector lies in this plane with the subscript 2. In finding the intensity of reflected and refracted beams, it is necessary to treat these cases separately.

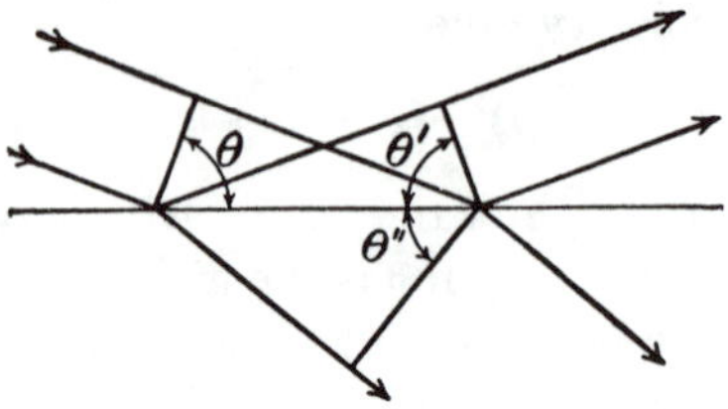

FIG. 11.08.

For the beam whose magnetic vector lies in the plane of incidence, E_1 is parallel to the surface; so, from 1.17 (3), we have

$$E_1 + E_1' = E_1'' \tag{2}$$

In this case, from 11.06 (1), setting $\mu = \mu'' = \mu_v$, (1) becomes

$$\epsilon^{\frac{1}{2}}(E_1^2 - E_1'^2)\cos\theta = \epsilon''^{\frac{1}{2}}E_1''^2\cos\theta''$$

Dividing this by (2) and using 11.07 (3), we get

$$E_1 - E_1' = \frac{\sin\theta\cos\theta''}{\sin\theta''\cos\theta}E_1'' \tag{3}$$

Solving (2) and (3) for E_1' and E_1'' gives

$$E_1' = -\frac{\sin(\theta - \theta'')}{\sin(\theta + \theta'')}E_1 \tag{4}$$

$$E_1'' = \frac{2\sin\theta''\cos\theta}{\sin(\theta + \theta'')}E_1 \tag{5}$$

It follows from 11.03 (8) and 11.07 (3) that

$$B_1' = -\frac{\sin(\theta - \theta'')}{\sin(\theta + \theta'')}B_1 \tag{6}$$

$$B_1'' = \frac{\sin 2\theta}{\sin(\theta + \theta'')}B_1 \tag{7}$$

Using 11.06 (1), we get for the reflected intensity

$$\Pi_1' = \frac{\sin^2(\theta - \theta'')}{\sin^2(\theta + \theta'')}\Pi_1 \tag{8}$$

Substituting this in (1), we get for the refracted intensity

$$\Pi_1'' = \frac{2\sin\theta''\cos\theta\sin 2\theta}{\sin^2(\theta + \theta'')}\Pi_1 \tag{9}$$

For the beam whose electric vector lies in the plane of incidence, B_2 is parallel to the surface so, setting $\mu = \mu'' = \mu_v$, we have

$$B_2 + B_2' = B_2'' \tag{10}$$

From 11.06 (1), setting $\mu = \mu'' = \mu_v$, (1) becomes

$$\epsilon^{-\frac{1}{2}}(B_2^2 - B_2'^2) \cos \theta = B_2''^2 \epsilon''^{-\frac{1}{2}} \cos \theta''$$

Dividing this by (10) and using 11.07 (3), we get

$$B_2 - B_2' = \frac{\sin 2\theta''}{\sin 2\theta} B_2'' \tag{11}$$

Solving (10) and (11) for B_2' and B_2'' gives

$$B_2' = \frac{\tan (\theta - \theta'')}{\tan (\theta + \theta'')} B_2 \tag{12}$$

$$B_2'' = \frac{\sin 2\theta}{\sin (\theta + \theta'') \cos (\theta - \theta'')} B_2 \tag{13}$$

It follows, from 11.03 (8) and 11.07 (3), that

$$E_2' = \frac{\tan (\theta - \theta'')}{\tan (\theta + \theta'')} E_2 \tag{14}$$

$$E_2'' = \frac{2 \sin \theta'' \cos \theta}{\sin (\theta + \theta'') \cos (\theta - \theta'')} E_2 \tag{15}$$

Using 11.06 (1), we get for the reflected intensity

$$\Pi_2' = \frac{\tan^2 (\theta - \theta'')}{\tan^2 (\theta + \theta'')} \Pi_2 \tag{16}$$

Substituting this in (1), we get for the refracted intensity

$$\Pi_2'' = \frac{2 \sin \theta'' \cos \theta \sin 2\theta}{\sin^2 (\theta + \theta'') \cos^2 (\theta - \theta'')} \Pi_2 \tag{17}$$

At normal incidence, the cosines in (1) are unity, so that in place of (3)

$$E - E' = \left(\frac{\epsilon''}{\epsilon}\right)^{\frac{1}{2}} E'' = \frac{v}{v''} E''$$

Combining this with (2) gives

$$E' = -\frac{v - v''}{v + v''} E \tag{18}$$

$$E'' = \frac{2v''}{v + v''} E \tag{19}$$

From 11.06 (1), the reflected intensity is

$$\Pi' = \left(\frac{v - v''}{v + v''}\right)^2 \Pi \tag{20}$$

Combining this with (1) gives

$$\Pi'' = \frac{4vv''}{(v + v'')^2}\Pi \tag{21}$$

It should be noted that the intensity of radiation given in (8), (9), (16), (17), (20), and (21) is the energy passing in 1 sec through 1 sq m of area parallel to the wave front. To get the actual radiation energy in 1 cu m divide Π by the wave velocity in the medium.

Evidently from (12), if $\theta + \theta'' = \frac{1}{2}\pi$, $B_2' = 0$, which means that at this incident angle radiation whose electric vector lies in the plane of incidence is not reflected. This is called the polarizing angle because unpolarized radiation incident at this angle is reflected with its magnetic field in the plane of incidence. From 11.07 (3), this "Brewster's angle" is

$$\frac{\sin \theta_p}{\sin (\frac{1}{2}\pi - \theta_p)} = \tan \theta_p = \frac{v}{v''} = n \tag{22}$$

11.09. Frequency, Wave Length, Elliptic Polarization.—So far we have used the general solution for a plane wave, $f(\boldsymbol{n} \cdot \boldsymbol{r} - vt)$, in deriving the laws of reflection and refraction. For discussing circular and elliptic polarization it is convenient to take a regularly periodic function which can be built up by Fourier's series from simple sinusoidal terms. Thus

$$f(\boldsymbol{n} \cdot \boldsymbol{r} - vt) = D \cos [\omega(t - v^{-1}\boldsymbol{n} \cdot \boldsymbol{r}) + \phi] \tag{1}$$

where $\boldsymbol{n}$ is a unit vector in the propagation direction. The mathematical manipulation is considerably simplified by writing

$$D \cos [\omega(t - v^{-1}\boldsymbol{n} \cdot \boldsymbol{r}) + \phi] = De^{j\phi}e^{j\omega(t - \boldsymbol{n}\cdot\boldsymbol{r}/v)} \text{ (real part)} \tag{2}$$

The term $De^{j\phi}$ is usually written as a phasor $\check{D}$. The angular frequency ω and the cyclical frequency ν are related by

$$\omega = 2\pi\nu \tag{3}$$

The wave length λ is the shortest distance in the direction of propagation along a "frozen" wave in which the electrical conditions repeat. It is related to the frequencies and the wave velocity v by the equations

$$\lambda = \frac{v}{\nu} = \frac{2\pi v}{\omega} = \frac{1}{\nu(\mu\epsilon)^{\frac{1}{2}}} = \frac{2\pi}{\omega(\mu\epsilon)^{\frac{1}{2}}} = \frac{2\pi}{\beta} \tag{4}$$

Now consider the superposition of two plane electromagnetic waves of the same frequencies traveling in the z-direction. Let one be plane polarized with a y-directed magnetic vector and the other with an x-directed one so that the equations for the electrical intensities are

$$E_x = E_1 \cos [\omega(t - v^{-1}z)] \tag{5}$$
$$E_y = E_2 \cos [\omega(t - v^{-1}z) + \delta] \tag{6}$$

E_x and E_y are shown as a function of z in Fig. 11.09. To find the locus

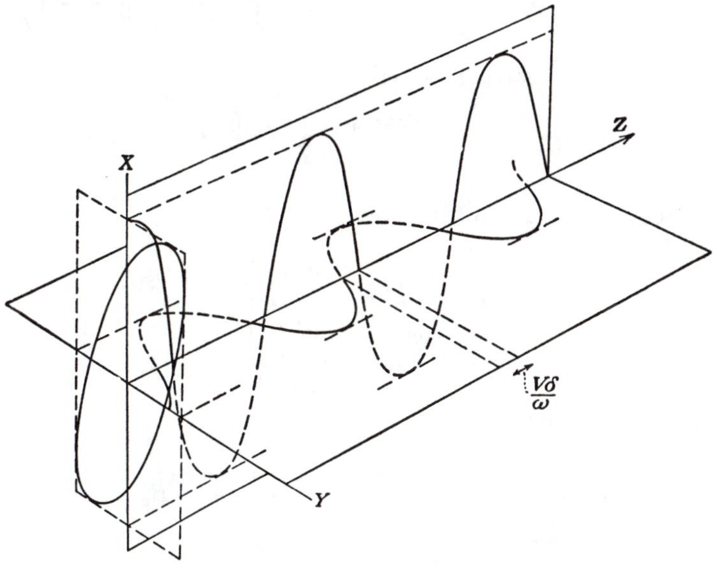

FIG. 11.09.

in the xy-plane of the end of the vector $\boldsymbol{E}$ whose components are E_x and E_y, set $z = 0$ and eliminate t from (5) and (6). This gives

$$\frac{E_x^2}{E_1^2} + \frac{E_y^2}{E_2^2} - \frac{2E_xE_y}{E_2E_1} \cos \delta = \sin^2 \delta \tag{7}$$

This is the equation of an ellipse, shown at the left in Fig. 11.09. Such waves are called elliptically polarized. If $\delta = n\pi$, (7) becomes

$$E_2E_x \pm E_1E_y = 0 \tag{8}$$

which represents a straight line, so the resultant wave is plane polarized. If $\delta = \frac{1}{2}(2n + 1)\pi$ and, in addition, $E_1 = E_2 = E$, then (7) becomes

$$E_x^2 + E_y^2 = E^2 \tag{9}$$

This locus is a circle, and the resultant wave is circularly polarized.

11.10. Total Reflection.—Our first example of elliptically polarized radiation occurs in connection with the phenomenon of total reflection. If $\epsilon'' < \epsilon$ in 11.07 (3), then $\theta'' > \theta$ so that θ'' may equal $\frac{1}{2}\pi$ when $\theta < \frac{1}{2}\pi$. The value of θ which gives $\theta'' = \frac{1}{2}\pi$ is called the critical angle θ_c. The agreement between experimental facts and the results of the following interpretation of this phenomenon justifies its use. When $\theta > \theta_c$, let us write 11.07 (3) in the following form, when $\mu = \mu''$,

$$\cos\theta'' = (1 - \sin^2\theta'')^{\frac{1}{2}} = \left(1 - \frac{\epsilon}{\epsilon''}\sin^2\theta\right)^{\frac{1}{2}} = j\left(\frac{\epsilon}{\epsilon''}\sin^2\theta - 1\right)^{\frac{1}{2}}$$

Equation 11.08 (3) must then be written

$$\breve{E}_1 - \breve{E}_1' = j\frac{[\sin^2\theta - (\epsilon''/\epsilon)]^{\frac{1}{2}}}{\cos\theta}\breve{E}_1'' \tag{1}$$

which, combined with 11.08 (2), gives

$$\breve{E}_1' = \frac{\cos\theta - j[\sin^2\theta - (\epsilon''/\epsilon)]^{\frac{1}{2}}}{\cos\theta + j[\sin^2\theta - (\epsilon''/\epsilon)]^{\frac{1}{2}}}E_1 = e^{j\psi_1}E_1 \tag{2}$$

$$\tan\psi_1 = -\frac{2\cos\theta[\sin^2\theta - (\epsilon''/\epsilon)]^{\frac{1}{2}}}{\cos^2\theta - \sin^2\theta + (\epsilon''/\epsilon)} \tag{3}$$

Corresponding to 11.08 (11), we have

$$B_2 - \breve{B}_2' = \frac{j\{(\epsilon/\epsilon'')[(\epsilon/\epsilon'')\sin^2\theta - 1]\}^{\frac{1}{2}}}{\cos\theta}\breve{B}_2''$$

which, combined with 11.08 (10), gives

$$\breve{B}_2' = \frac{\cos\theta - j\{(\epsilon/\epsilon'')[(\epsilon/\epsilon'')\sin^2\theta - 1]\}^{\frac{1}{2}}}{\cos\theta + j\{(\epsilon/\epsilon'')[(\epsilon/\epsilon'')\sin^2\theta - 1]\}^{\frac{1}{2}}}B_2 = e^{j\psi_2}B_2 \tag{4}$$

$$\tan\psi_2 = -\frac{2\cos\theta\{(\epsilon/\epsilon'')[(\epsilon/\epsilon'')\sin^2\theta - 1]\}^{\frac{1}{2}}}{\cos^2\theta - (\epsilon/\epsilon'')^2\sin^2\theta + (\epsilon/\epsilon'')} \tag{5}$$

It is evident from (2) and (4) that $|E_1'| = |E_1|$ and $|B_2'| = |B_2|$ so that the reflected and incident rays have the same intensity. Thus the law of the conservation of energy permits no energy in the refracted beam. This can be verified by computing E_1'' and B_2'' and from this Π''. A comparison of (2) and (4) with 11.09 (2) shows that there is a phase change ψ_1 for the beam whose magnetic vector lies in the plane of incidence and ψ_2 for the beam whose magnetic vector lies normal to it. Thus, for plane polarized incident radiation, whose two electric vector components are E_1 and E_2, the reflected radiation is elliptically polarized. The phase difference δ can be obtained from the ratio of (2) to (4), since $B_2' = (\mu\epsilon)^{\frac{1}{2}}E_2'$ and

$B_2 = (\mu\epsilon)^{\frac{1}{2}}E_2$ by 11.03 (8), giving

$$e^{j\delta} = e^{j(\psi_1-\psi_2)} = \frac{\sin^2\theta + j\cos\theta[\sin^2\theta - (\epsilon''/\epsilon)]^{\frac{1}{2}}}{\sin^2\theta - j\cos\theta[\sin^2\theta - (\epsilon''/\epsilon)]^{\frac{1}{2}}}$$

Taking the ratio of the imaginary to the real part of this gives us $\tan\delta$, which, with the aid of *Pc* 573 or *Dw* 406.02, may be reduced to

$$\tan\frac{\delta}{2} = \frac{\cos\theta[\sin^2\theta - (\epsilon''/\epsilon)]^{\frac{1}{2}}}{\sin^2\theta} \tag{6}$$

11.11. Electromagnetic Waves in Homogeneous Conductors.—In a medium whose conductivity is not zero, we must use all the terms in the propagation equations 11.01 (1) and (2). In deriving 11.01 (2), we set $\rho = 0$. To justify this, take the divergence of 11.00 (1) which gives

$$\mu\epsilon\frac{\partial}{\partial t}(\nabla\cdot E) + \mu\gamma\nabla\cdot E = 0$$

Write ρ for $\epsilon\nabla\cdot E$ by 11.00 (3) and integrate from 0 and ρ_0' to t' and ρ'.

$$\rho' = \rho_0' e^{-\frac{\gamma t'}{\epsilon}} \tag{1}$$

Thus, an electrical distribution ρ_0' is dissipated independently of electromagnetic disturbances so that, if ρ_0' is zero initially, ρ' is always zero, which justifies setting $\rho = 0$. We define the relaxation time T to be

$$T = \tau\epsilon = \gamma^{-1}\epsilon \tag{2}$$

Take a simple periodic disturbance so that both B and E have the form

$$\boldsymbol{B} = \breve{\boldsymbol{B}}e^{j\omega t} \text{ (real part)} = \boldsymbol{B}_0 e^{j(\omega t+\phi)} \text{ (real part)}$$

Now $\partial/\partial t$ may be replaced by $j\omega$, and 11.01 (1) and (2) become

$$\nabla^2\boldsymbol{B} = \mu(j\omega\gamma - \omega^2\epsilon)\boldsymbol{B} \tag{3}$$

$$\nabla^2\boldsymbol{E} = \mu(j\omega\gamma - \omega^2\epsilon)\boldsymbol{E} \tag{4}$$

These are identical in form with 11.01 (1) for a periodic disturbance in an insulator. The equations will have identical solutions if we write $\epsilon - j\gamma/\omega$ for ϵ. The complex quantity $(\mu\epsilon)^{\frac{1}{2}}$ which occurs frequently may be separated into real and imaginary parts by *Dw* 58.2.

$$[\mu(\epsilon - j\omega^{-1}\gamma)]^{\frac{1}{2}} = n + jk \tag{5}$$

$$n = \left\{\frac{\mu\epsilon}{2}\left[\left(1 + \frac{\gamma^2}{\omega^2\epsilon^2}\right)^{\frac{1}{2}} + 1\right]\right\}^{\frac{1}{2}}, \qquad k = -\left\{\frac{\mu\epsilon}{2}\left[\left(1 + \frac{\gamma^2}{\omega^2\epsilon^2}\right)^{\frac{1}{2}} - 1\right]\right\}^{\frac{1}{2}} \tag{6}$$

11.12. Plane Waves in Homogeneous Isotropic Conductors.—Certain properties of plane waves in conductors can now be obtained, by the substitution suggested above, from the corresponding properties in a dielectric derived in 11.03. In isotropic mediums the nature of ϵ does not affect (6), (7), or (8) of 11.03 so that in a homogeneous conductor $\breve{\boldsymbol{B}}$ is normal to $\breve{\boldsymbol{E}}$ and both lie in the wave front. From 11.03 (9)

$$\breve{B}_0 = (n + jk)E_0 = (n^2 + k^2)^{\frac{1}{2}} E_0 e^{j\tan^{-1}(k/n)} \tag{1}$$

Thus, the electric and magnetic vectors differ in phase by $\tan^{-1}(k/n)$. Writing $\breve{\boldsymbol{E}}$ in the form of 11.09 (2) and substituting in (1) give

$$\breve{\boldsymbol{E}}e^{j\omega t} = \boldsymbol{\varepsilon} A_1 e^{\omega k \sigma \cdot r} e^{j\omega(t - n\sigma\cdot r)} \tag{2}$$

$$\breve{\boldsymbol{B}}e^{j\omega t} = (\boldsymbol{\delta} \times \boldsymbol{\varepsilon})(n^2 + k^2)^{\frac{1}{2}} A_1 e^{\omega k\sigma\cdot r} e^{j[\omega(t - n\sigma\cdot r) + \tan^{-1}(k/n)]} \tag{3}$$

where $\boldsymbol{\varepsilon}$ is a unit vector along $\breve{\boldsymbol{E}}$, and $\boldsymbol{\delta}$ is a unit vector normal to the wave front. We note that both $\breve{\boldsymbol{E}}$ and $\breve{\boldsymbol{B}}$ decrease exponentially as the wave advances, indicating a rapid absorption if k is large. We call n the index of refraction of the conductor and k the coefficient of extinction.

It is useful to know the magnitude of these quantities in nonferromagnetic conductors whose permeability is nearly equal to μ_v. In conductors the capacitivity is not accurately known but seems to be of the same order of magnitude as in insulators. As a typical example take copper, whose conductivity is about 5.8×10^7 mhos/m. Designating wave length *in vacuo* by λ_0 and using 11.09 (4), we have, for copper, writing K for ϵ/ϵ_v,

$$\frac{\gamma}{\omega\epsilon} = \frac{\lambda_0\gamma}{2\pi c\epsilon} \approx \frac{\lambda_0}{K} \times 3.48 \times 10^9$$

Except for very short waves, this quantity is certainly large compared with unity so that, from 11.11 (6), n and k are well approximated by

$$n \approx -k \approx (\tfrac{1}{4}\mu\lambda_0\gamma)^{\frac{1}{2}}(\pi c)^{-\frac{1}{2}} \tag{4}$$

The distance an electromagnetic wave must travel in order that $\breve{\boldsymbol{E}}$ and $\breve{\boldsymbol{B}}$ fall to $e^{-2\pi} = 0.002$ times their initial values is, from (3),

$$d = -\frac{2\pi}{\omega k} = -\frac{\lambda_0}{kc} = \left(\frac{4\pi\lambda_0}{\mu\gamma c}\right)^{\frac{1}{2}} \text{ meters} \tag{5}$$

For air wave lengths of 1 cm, 1 m, 100 m, and 10 km, this gives d the values, in copper, of 0.0025, 0.024, 0.24, and 2.4 mm, respectively. It is also of interest to compare the average electric and magnetic energies in the wave. From (3) and (4) and 11.06 (2), we have

$$\frac{W_e}{W_m} = \mu\epsilon\frac{E^2}{B^2} = \frac{\mu\epsilon}{n^2 + k^2} \approx \frac{2\pi\epsilon c}{\lambda_0\gamma} \tag{6}$$

On the assumption that $\epsilon = 2\epsilon_v$, this gives, for the waves just considered, 5.7×10^{-8}, 5.7×10^{-10}, 5.7×10^{-12}, and 5.7×10^{-14}, respectively, so that nearly the entire energy is in the magnetic field.

11.13. Reflection from Conducting Surface.—We saw in 11.11 that in a conducting medium $(\mu''\epsilon'')^{\frac{1}{2}}$ is replaced by

$$[\mu''(\epsilon'' - j\omega^{-1}\gamma)]^{\frac{1}{2}} = n + jk$$

Thus, sin θ'' and cos θ'' become, by 11.07 (3) with the aid of 11.11 (5),

$$\sin\theta'' = \frac{(\mu\epsilon)^{\frac{1}{2}}}{n + jk}\sin\theta \qquad \cos\theta'' = \left[1 - \frac{\mu\epsilon\sin^2\theta}{(n + jk)^2}\right]^{\frac{1}{2}} \tag{1}$$

We saw in the last article that, for electromagnetic waves over 1 cm long.

$$-k \approx n > 1.4 \times 10^4$$

So that, from (1), θ'' is a complex and extremely small angle. Thus, from 11.08 (8), when the magnetic vector lies in the plane of incidence,

$$\Pi_1' = \frac{\sin^2(\theta - \theta'')}{\sin^2(\theta + \theta'')}\Pi_1 \xrightarrow[\theta''\to 0]{} \Pi_1 \tag{2}$$

When the electric vector lies in the plane of incidence, we get, from 11.08 (16), provided θ is not too close to $\frac{1}{2}\pi$,

$$\Pi_2' = \frac{\tan^2(\theta - \theta'')}{\tan^2(\theta + \theta'')}\Pi_2 \xrightarrow[\theta''\to 0]{} \Pi_2 \tag{3}$$

Thus, for all angles of incidence and for all states of polarization, a sufficiently long electromagnetic wave is completely reflected from a conducting surface. The same results apply to longer waves at lower conductivities. From the last article, the extremely small fraction not reflected is very rapidly absorbed so that quite thin sheets of metal can be very opaque for short waves. To study the state of polarization of the reflected beam, we divide 11.08 (14) by 11.08 (4) and obtain

$$\frac{E_2'}{E_1'} = -\frac{\cos(\theta + \theta'')E_2}{\cos(\theta - \theta'')E_1} = -\frac{(\cos\theta'' - \sin\theta''\tan\theta)E_2}{(\cos\theta'' + \sin\theta''\tan\theta)E_1} \tag{4}$$

For the wave lengths just considered, we have seen that

$$n + jk \approx n(1 - j)$$

If $\mu\epsilon/|n + jk|^2 \ll 1$, then $\cos\theta'' = 1$ from (1) so that

$$\frac{E_2'}{E_1'} = -\frac{[n - (\mu\epsilon)^{\frac{1}{2}}\sin\theta\tan\theta - jn]E_2}{[n + (\mu\epsilon)^{\frac{1}{2}}\sin\theta\tan\theta - jn]E_1}$$

$$= \left\{\frac{[n - (\mu\epsilon)^{\frac{1}{2}}\sin\theta\tan\theta]^2 + n^2}{[n + (\mu\epsilon)^{\frac{1}{2}}\sin\theta\tan\theta]^2 + n^2}\right\}^{\frac{1}{2}} e^{j\delta}\frac{E_2}{E_1} \tag{5}$$

$$\tan\delta = -\frac{2n(\mu\epsilon)^{\frac{1}{2}}\sin\theta\tan\theta}{2n^2 - \mu\epsilon\sin^2\theta\tan^2\theta} \tag{6}$$

Because n is a large quantity, we note that, except when $\theta \rightarrow \frac{1}{2}\pi$,

$$\frac{E_2'}{E_1'} = -\frac{E_2}{E_1}$$

When $\tan\theta \rightarrow \infty$, there is not only a change in the magnitudes of E_2' and E_1' but a phase shift δ as well. Thus, for wave lengths of a centimeter or more, we obtain elliptically polarized radiation only when plane polarized radiation is reflected near grazing incidence from the conducting surface.

11.14. Plane Waves on Cylindrical Perfect Conductors.—Plane waves can be propagated not only in free space and infinite dielectric mediums with plane faces but also in the z-direction along a set of perfect conductors, provided all sections of the set taken normal to the z-axis are identical. From 10.02, we know that on perfect conductors the currents flow in an infinitely thin surface layer with no fields inside and no energy dissipation. This problem can be set up in several ways. The results of 7.25 suggest taking the vector potential and hence, from 11.01, the Hertz vector, in the z-direction parallel to the currents. The solution of the propagation equation for the Hertz vector yields both a scalar and a vector potential. Elimination of the scalar potential by 11.01 (12) gives a vector potential normal to z. The same result can be obtained more directly by solving the scalar propagation equation

$$\left[\frac{\partial^2 W}{\partial x^2} + \frac{\partial^2 W}{\partial y^2}\right] + \left[\frac{\partial^2 W}{\partial z^2} - \mu\epsilon\frac{\partial^2 W}{\partial t^2}\right] = 0 \qquad (1)$$

The solution obtained by setting each bracket separately equal to zero is

$$W = V_1(x, y)f_1[z - (\mu\epsilon)^{-\frac{1}{2}}t] + V_2(x, y)f_2[z + (\mu\epsilon)^{-\frac{1}{2}}t] \qquad (2)$$

where V_1 and V_2 are solutions of Laplace's equation in two dimensions so that, if W_1 and W_2 are the complex potential functions of Chap. IV,

$$W_1 = U_1 + jV_1 = F_1(x + jy), \qquad W_2 = U_2 + jV_2 = F_2(x + jy) \qquad (3)$$

If $\nabla_2 V$ is defined as $\mathbf{i}(\partial V/\partial x) + \mathbf{j}(\partial V/\partial y)$ we see from 4.10 that

$$-\mathbf{k}\times\nabla_2 V = \nabla_2 U, \quad \mathbf{k}\times\nabla_2 U = \nabla_2 V, \quad \nabla\times[\nabla_2 U f(z)] = \mathbf{k}\times\nabla_2 U f'(z) \qquad (4)$$

By 10.01 (1), the vector potential for a transverse electric field is

$$\begin{aligned} A = \nabla\times\mathbf{k}W &= -\mathbf{k}\times\nabla_2 V_1 f_1[z - (\mu\epsilon)^{-\frac{1}{2}}t] - \mathbf{k}\times\nabla_2 V_2 f_2[z + (\mu\epsilon)^{-\frac{1}{2}}t] \\ &= \nabla_2 U_1(x, y) f_1[z - (\mu\epsilon)^{-\frac{1}{2}}t] + \nabla_2 U_2(x, y) f_2[z + (\mu\epsilon)^{-\frac{1}{2}}t] \end{aligned} \qquad (5)$$

Evidently the first and second terms represent, respectively, waves that travel in the positive and negative z-directions. The fields are

$$\boldsymbol{B} = \boldsymbol{\nabla} \times \boldsymbol{A} = \boldsymbol{\nabla}_2 V(x, y) f'[z \mp (\mu\epsilon)^{-\frac{1}{2}} t] \tag{6}$$

$$\boldsymbol{E} = -\frac{\partial \boldsymbol{A}}{\partial t} = \pm(\mu\epsilon)^{-\frac{1}{2}} \boldsymbol{\nabla}_2 U(x, y) f'[z \mp (\mu\epsilon)^{-\frac{1}{2}} t] \tag{7}$$

The upper sign goes with the positive wave. From (4), (6), and (7)

$$(\mu\epsilon)^{\frac{1}{2}} \boldsymbol{E} = \mp \boldsymbol{k} \times \boldsymbol{B} \tag{8}$$

Suppose that the current goes out on one set of conducting cylinders and returns on another set. It is evident from (7) that, if all members of the same set have the same potential at one value of z, then this will also be true at any other value of z. The relation between total current and charge per unit length in either set of n members can be found from (8), the magnetomotance law 7.01 (2), and Gauss's theorem 1.15 (1), thus

$$Q = \sum_{i=1}^{n} \oint \epsilon E_n \, ds_i = (\mu\epsilon)^{\frac{1}{2}} \sum_{i=1}^{n} \oint H_t \, ds_i = (\mu\epsilon)^{\frac{1}{2}} I \tag{9}$$

If L is the self-inductance per unit length, C the capacitance per unit length, and S the area outside the conductors in a plane of constant z, then with the aid of 2.07 (4), 2.19 (1), and 8.08 (1) we have

$$\frac{LI^2}{2} = \int_S \frac{B^2}{2\mu} \, dS = \int_S \frac{\epsilon E^2}{2} \, dS = \frac{Q^2}{2C} = \frac{\mu\epsilon}{2C} I^2 \tag{10}$$

Therefore, L and C are connected by the relation

$$LC = \mu\epsilon = v^{-2} \tag{11}$$

Thus the product LC equals the reciprocal of the square of the electromagnetic wave velocity in the medium outside the conductors.

If $\boldsymbol{E}$ and $\boldsymbol{B}$ are specified in the plane $z = 0$ as functions of time, it is easy from (5) to write down $\boldsymbol{A}$ for any z. Thus,

$$\boldsymbol{A} = \tfrac{1}{2} \boldsymbol{\nabla}_2 U(x, y) \{ f[t - (\mu\epsilon)^{\frac{1}{2}} z] + f[t + (\mu\epsilon)^{\frac{1}{2}} z] + g[t - (\mu\epsilon)^{\frac{1}{2}} z] - g[t + (\mu\epsilon)^{\frac{1}{2}} z] \} \tag{12}$$

If U and V are given by (3), $\boldsymbol{A}$ yields the following fields at $z = 0$:

$$\boldsymbol{E}_0 = -(\partial \boldsymbol{A}/\partial t)_0 = -\boldsymbol{\nabla}_2 U(x, y) f'(t),$$
$$\boldsymbol{B}_0 = (\boldsymbol{\nabla} \times \boldsymbol{A})_0 = -(\mu\epsilon)^{\frac{1}{2}} \boldsymbol{\nabla}_2 V(x, y) g'(t) \tag{13}$$

It is equally simple to write down $\boldsymbol{A}$ from (5) when $\boldsymbol{E}$ and $\boldsymbol{B}$ are given as functions of z at $t = 0$. Thus,

$$\boldsymbol{A} = \tfrac{1}{2} \boldsymbol{\nabla}_2 U(x, y) \{ f[(\mu\epsilon)^{\frac{1}{2}} z + t] - f[(\mu\epsilon)^{\frac{1}{2}} z - t] - g[(\mu\epsilon)^{\frac{1}{2}} z + t] - g[(\mu\epsilon)^{\frac{1}{2}} z - t] \} \tag{14}$$

If U and V are given by (3), $\boldsymbol{A}$ yields the following fields at $t = 0$:

$$\boldsymbol{E}_0 = -\nabla_2 U(x, y)f'[(\mu\epsilon)^{\frac{1}{2}}z], \qquad \boldsymbol{B}_0 = -(\mu\epsilon)^{\frac{1}{2}}\nabla_2 V(x, y)g'[(\mu\epsilon)^{\frac{1}{2}}z] \quad (15)$$

11.15. Intrinsic Impedance of a Medium.—When the dielectric of the last article is replaced by a medium of conductivity γ, 11.14 (1) becomes

$$\left[\frac{\partial^2 W}{\partial x^2} + \frac{\partial^2 W}{\partial y^2}\right] + \left[\frac{\partial^2 W}{\partial z^2} - \mu\gamma\frac{\partial W}{\partial t} - \mu\epsilon\frac{\partial^2 W}{\partial t^2}\right] = 0 \quad (1)$$

When the second bracket is equated to zero, a solution in terms of simple arbitrary functions is no longer possible so we use the Chap. V method of separation of variables which splits it into two total differential equations with a separation constant. Their solution is exponential in form and, according to whether the separation constant is real or imaginary, lead to solutions harmonic in space (transient) or to solutions harmonic in time (steady state). We shall now consider steady-state solutions in which time enters as the factor $e^{j\omega t}$ as in 11.09. The second factor in (1) now becomes $\partial^2 W/\partial z^2 - \check{\Gamma}^2 W = 0$, and in the equations of 11.14 the functions $f[z \pm (\mu\epsilon)^{-\frac{1}{2}}t]$ are now replaced by $e^{\pm\check{\Gamma}z + j\omega t}$ where

$$\check{\Gamma}^2 = j\omega\mu(\gamma + j\omega\epsilon) = (\alpha + j\beta)^2 \quad (2)$$

$$\alpha = \{(\tfrac{1}{2}\omega\mu)[(\omega^2\epsilon^2 + \gamma^2)^{\frac{1}{2}} - \omega\epsilon]\}^{\frac{1}{2}}, \qquad \beta = \{(\tfrac{1}{2}\omega\mu)[(\omega^2\epsilon^2 + \gamma^2)^{\frac{1}{2}} + \omega\epsilon]\}^{\frac{1}{2}} \quad (3)$$

Here $\check{\Gamma}$ is the propagation constant, α the attenuation constant, and β the phase constant or wave number. Thus 11.14 (6) and (7) become, for the wave in the positive or negative z-direction,

$$\boldsymbol{B} = \mp\check{\Gamma}\nabla_2 V(x, y)e^{\mp\check{\Gamma}z + j\omega t} \text{ (real part)}$$

$$= \mp\nabla_2 V(x, y)e^{\mp\alpha z}[\alpha\cos(\omega t \mp \beta z) - \beta\sin(\omega t \mp \beta z)] \quad (4)$$

$$\boldsymbol{E} = -j\omega\nabla_2 U(x, y)e^{\mp\check{\Gamma}z + j\omega t} \text{ (r.p.)} = \omega\nabla_2 U(x, y)e^{\mp\alpha z}\sin(\omega t \mp \beta z) \quad (5)$$

$$\check{\Gamma}\check{\boldsymbol{E}} = \mp j\omega(\boldsymbol{k} \times \check{\boldsymbol{B}}) \qquad \text{or} \qquad \check{\Gamma}\boldsymbol{k} \times \check{\boldsymbol{E}} = \mp j\omega\check{\boldsymbol{B}} \quad (6)$$

The differential equation and its solution for the z-directed line current are the same as for W, so that

$$\frac{\partial^2 \check{\imath}}{\partial z^2} = \check{\Gamma}^2\check{\imath}, \qquad \check{\imath} = \check{A}e^{-\check{\Gamma}z} + \check{B}e^{\check{\Gamma}z} \quad (7)$$

where $e^{j\omega t}$, which appears on both sides of the equation, is canceled out. It must be put back before taking the real part as in (4) and (5) to get the trigonometric form. In a length dz the current in the positive and negative line members is $-(\partial\check{\imath}/\partial z)\,dz$, which is partly displacement current and partly leakage conduction. This current traverses an impedance Z_c/dz so that by Ohm's law the potential across the line is

$$\check{\mathcal{E}} = -\check{Z}_c\frac{\partial\check{\imath}}{\partial z} = -\check{Y}_c^{-1}\frac{\partial\check{\imath}}{\partial z} \quad (8)$$

From 6.06 (7), the product of the resistance R per unit length between conductors by the capacitance C per unit length is $\tau\epsilon$ or ϵ/γ. Therefore the shunt admittance $\check{Y}_c$ and the series impedance $\check{Z}_L$ are, using 11.14 (1),

$$\check{Y}_C = \frac{1}{R} + j\omega C = \frac{\gamma + j\omega\epsilon}{\epsilon}\frac{\mu\epsilon}{L} = \frac{\check{\Gamma}^2}{j\omega L}, \qquad \check{Z}_L = j\omega L \tag{9}$$

For transmission in the positive z-direction, so that $\check{B} = 0$ in (7), the characteristic impedance of the line is

$$\check{Z}_k = \left(\frac{\check{\mathcal{E}}}{\check{i}}\right)_{\check{B}=0} = \check{Z}_C\check{\Gamma} = (\check{Z}_C\check{Z}_L)^{\frac{1}{2}} = \left(\frac{Z_L}{Y_C}\right)^{\frac{1}{2}} \tag{10}$$

If $\check{B}$ is not zero, the ratio of $\check{\mathcal{E}}$ to $\check{i}$ at $z = 0$, which is found from (7), (8), and (10) and called the input impedance, is

$$\check{Z}_i = \frac{(\check{A} - \check{B})\check{\Gamma}}{(\check{A} + \check{B})\check{Y}_C} = \frac{\check{A} - \check{B}}{\check{A} + \check{B}}\check{Z}_k \tag{11}$$

If the line of length l ends in a load $\check{Z}_l$ at $z = l$, then from (7) and (8)

$$\check{Z}_l = \left(\frac{\check{\mathcal{E}}}{\check{i}}\right)_{z=l} = \check{Z}_k\frac{\check{A}e^{-\check{\Gamma}l} - \check{B}e^{\check{\Gamma}l}}{\check{A}e^{-\check{\Gamma}l} + \check{B}e^{\check{\Gamma}l}} \tag{12}$$

Elimination of $\check{A}$ and $\check{B}$ from (11) and (12) gives the input impedance

$$\check{Z}_i = \check{Z}_k\frac{\check{Z}_L \cosh \check{\Gamma}l + Z_k \sinh \check{\Gamma}l}{\check{Z}_k \cosh \check{\Gamma}l + Z_L \sinh \check{\Gamma}l} \tag{13}$$

Evidently the input impedances, when the line is open or short-circuited, are $\check{Z}_k \coth \check{\Gamma}l$ and $\check{Z}_k \tanh \check{\Gamma}l$, respectively. Note that $\check{Z}_0\check{Z}_s = \check{Z}_k^2$. From (9) and (10), the characteristic impedance of the line is

$$\check{Z}_k = \frac{j\omega L}{\check{\Gamma}} = \frac{\check{V}}{\check{I}} = \frac{L\check{E}}{\check{B}} = \frac{j\omega\mu\epsilon}{\check{\Gamma}C} \tag{14}$$

Now consider the special case where the conductors are two parallel planes 1 m apart. An infinite 1-m square tubular section running in the positive z-direction has a capacitance per unit length of ϵ, so that its characteristic impedance is

$$\check{Z}_k = \left(\frac{j\omega\mu}{\gamma + j\omega\epsilon}\right)^{\frac{1}{2}} = \frac{j\omega\mu}{\alpha + j\beta} \xrightarrow[\alpha\to 0]{} \left(\frac{\mu}{\epsilon}\right)^{\frac{1}{2}} = \eta \tag{15}$$

Also $V = E_x$ and $B_y = \mu i$ where i is the current density. Note that $\check{Z}_k$ depends on the properties of the medium alone and that the configuration of the fields is identical with that in a plane polarized plane wave, so it is logical to follow Schelkunoff and call Z_k the intrinsic impedance of the medium. A comparison of (15) with 11.07 (3) shows that, in optically transparent mediums where γ is zero and μ is μ_v, the intrinsic impedance is proportional to the index of refraction. From (3) and (15), we may define the intrinsic propagation constant of the medium as

$$\check{\Gamma}_k = \alpha + j\beta = [j\omega\mu(\gamma + j\omega\epsilon)]^{\frac{1}{2}} = j\omega\mu\check{Z}_k^{-1} \quad (16)$$

We need a formula like 6.03 (2) for power absorption in an impedance. Apply an electromotance $\mathcal{E}_0 \cos \omega t$, so $\check{\mathcal{E}}$ is $\mathcal{E}_0$ and real, to an impedance $\check{Z}$ producing a current $\check{I}$ or $I_0 \cos(\omega t + \psi)$. From Ohm's law, 6.02 (1),

$$\mathcal{E}_0 = \check{I}\check{Z} = I_0 e^{j\psi}(X^2 + R^2)^{\frac{1}{2}} e^{j \tan^{-1}(X/R)} = I_0 Z \quad (17)$$

so ψ is $-j \tan^{-1}(X/R)$ and the instantaneous power absorbed is the real part of

$$P = I\mathcal{E} = \mathcal{E}_0^2 Z^{-1}(\cos^2 \omega t \cos \psi - \sin \omega t \cos \omega t \sin \psi) \quad (18)$$

11.16. Reflection at a Discontinuity. Matching Section.—Suppose the plane wave described in the last article, traveling in the positive z-direction, meets a plane boundary over which z is constant and beyond which lies a medium of capacitivity ϵ_2, permeability μ_2, and conductivity γ_2. The relations between the original wave, the reflected wave, and the transmitted wave and their respective currents are, from 11.15 (14),

$$\check{V}_1 = \check{Z}_1\check{\imath}_1, \qquad \check{V}_1' = -\check{Z}_1\check{\imath}_1', \qquad \check{V}_2 = \check{Z}_2\check{\imath}_2 \quad (1)$$

At the boundary, potential and current must be continuous so that

$$\check{V}_1 + \check{V}_1' = \check{V}_2 \qquad \text{and} \qquad \check{\imath}_1 + \check{\imath}_1' = \check{\imath}_2 \quad (2)$$

The result of solving the above equations for $\check{V}_1'$, $\check{V}_2$, $\check{\imath}_1$, and $\check{\imath}_2$ is

$$\frac{\check{V}_1'}{\check{V}_1} = -\frac{\check{\imath}_1'}{\check{\imath}_1} = \frac{\check{Z}_2 - \check{Z}_1}{\check{Z}_1 + \check{Z}_2} \qquad \frac{\check{V}_2}{\check{V}_1} = \frac{\check{\imath}_2\check{Z}_2}{\check{\imath}_1\check{Z}_1} = \frac{2\check{Z}_2}{\check{Z}_1 + \check{Z}_2} \quad (3)$$

If both mediums have zero conductivity, it will be observed from 11.15 (3), 11.15 (14), and 11.07 (3) that $\check{Z}_2/\check{Z}_1$ may be replaced by v_2/v_1, the ratio of wave velocities in the two mediums. The potential therefore obeys the same law for normal reflection as the electric field in 11.08 (18) and (19). By working with components one may use the transmission line formulas to derive the laws of reflection at any angle.

Electromagnetic waves may be passed from one nonconducting dielectric medium μ_1, ϵ_1 into a second μ_3, ϵ_3 without loss by insertion of a matching section. For perfect transmission, the input impedance of the matching layer or section must equal the characteristic impedance of the first medium. Therefore, by 11.15 (15), we must substitute in 11.15 (13)

$$\check{Z}_i = (\mu_1/\epsilon_1)^{\frac{1}{2}}, \quad \check{Z}_k = (\mu_2/\epsilon_2)^{\frac{1}{2}}, \quad \check{Z}_L = (\mu_3/\epsilon_3)^{\frac{1}{2}}, \quad \check{\Gamma}' = j\omega(\mu_2\epsilon_2)^{\frac{1}{2}} = 2\pi j\lambda_2^{-1}$$

This gives the following equation to be satisfied for a match:

$$\left(\frac{\mu_1\epsilon_2}{\mu_2\epsilon_1}\right)^{\frac{1}{2}} = \frac{(\mu_3\epsilon_2)^{\frac{1}{2}}\cos(2\pi l/\lambda_2) + j(\mu_2\epsilon_3)^{\frac{1}{2}}\sin(2\pi l/\lambda_2)}{(\mu_2\epsilon_3)^{\frac{1}{2}}\cos(2\pi l/\lambda_2) + j(\mu_3\epsilon_2)^{\frac{1}{2}}\sin(2\pi l/\lambda_2)} \tag{4}$$

The real terms on the right vanish, and the equation holds if we choose

$$l = \tfrac{1}{4}(2n+1)\lambda_2 \qquad \text{and} \qquad \mu_2\epsilon_2^{-1} = (\mu_1\epsilon_1^{-1})^{\frac{1}{2}}(\mu_3\epsilon_3^{-1})^{\frac{1}{2}} \tag{5}$$

Thus, the matching section should be an odd number of quarter wave lengths thick, and its intrinsic impedance should be the geometric mean of those of mediums 1 and 2. In terms of optics, the index of refraction of the matching layer must be the geometric mean of those on either side. These quarter wave layers are widely used to eliminate reflections from lenses.

11.17. Complex Poynting Vector.—An expression for the Poynting vector in phasor form is useful when the fields vary sinusoidally so that the operator $\partial/\partial t$ may be replaced by the factor $j\omega$. Let us write down Eq. 11.00 (2) and (1), using conjugate phasors in the latter, for which it is equally valid because they give the same two real equations. Thus

$$\nabla \times \check{E} = -j\omega\check{B} \qquad \nabla \times \hat{B} = \mu(\gamma - j\omega\epsilon)\hat{E} \tag{1}$$

Multiply the first equation by $\hat{B}$ and the second by $-\check{E}$ and add.

$$\hat{B}\cdot(\nabla \times \check{E}) - \check{E}\cdot(\nabla \times \hat{B}) = -[\mu\gamma\check{E}\cdot\hat{E} - j\omega(\mu\epsilon\check{E}\cdot\hat{E} - \check{B}\cdot\hat{B})]$$

Proceed as in 11.02, writing $\nabla\cdot(\check{E}\times\hat{B})$ for the first term, integrating over the volume v, and applying Gauss's theorem. By Ohm's law, 6.02 (3), we may replace $\check{E}$ by $\check{\imath}/\gamma$ or $\tau\check{\imath}$ where $\imath$ is the current density and so get

$$-\mu^{-1}\int_S \mathbf{n}\cdot(\check{E}\times\hat{B})\,dS = \int_V \tau i^2\,dv - j\omega\int_V(\epsilon\check{E}\cdot\hat{E} - \mu^{-1}\check{B}\cdot\hat{B})\,dv \tag{2}$$

By 10.00 (10), the real term on the right gives twice the average energy dissipated per second by eddy currents in the volume of integration. Thus, if we integrate the inward pointing normal component of the vector

$$\bar{\Pi} = \tfrac{1}{2}\mu^{-1}(\check{E}\times\hat{B})\ (\text{real part}) = \tfrac{1}{4}\mu^{-1}(\check{E}\times\hat{B} + \hat{E}\times\check{B}) \tag{3}$$

over the surface of the volume v, we get the rate at which energy is being absorbed from the wave. This formula can be used to find the energy flow across the plane $z = z_0$ of the transmission line considered in 11.15. Elimination of $\check{E}$ from (3) by 11.15 (6) and the use of 11.14 (10) and 11.15 (14) show that, for a z-directed wave, the flow is the real part of

$$\frac{-j\omega}{2\mu\check{\Gamma}}\int_S(\mathbf{k}\times\check{B})\times\hat{B}\,dS = \mathbf{k}\frac{j\omega}{\check{\Gamma}}\int_S\frac{\check{B}\hat{B}}{2\mu}\,dS = \mathbf{k}\frac{j\omega LI^2}{2\check{\Gamma}} = \mathbf{k}\frac{\check{V}\hat{I}}{2} \tag{4}$$

In 11.14 (9), a positive potential was assumed on conductors in which the current is in the positive z-direction. If either current or potential is reversed but not both, then the direction of energy flow is reversed.

11.18. Nearly Plane Waves on Imperfect Conductors. Lecher Wires. If the conductors forming the transmission line considered in 11.14 and 11.15 have finite conductivity, heat is generated in their surfaces by the eddy currents. The electric field has a z-component at the surface so that its Poynting vector is inclined toward the surface and the wave is no longer truly plane. In most practical cases, the power lost per unit length is small compared with that transmitted so that the inclination of $\mathbf{\Pi}$ to the z-axis is negligible. In this case, one uses the plane wave formula to calculate the fields in any plane normal to the z-axis. The energy loss varies as the field strength squared so it damps the fields exponentially with z and adds a real term R_i to the series impedance $\check{Z}_L$. The magnetic field penetration into the wire adds an internal inductance term $j\omega L_i$ to $\check{Z}_L$. From 11.15 (14), the characteristic impedance is

$$\check{Z}_k = \left(\frac{\check{Z}_L}{\check{Y}}\right)^{\frac{1}{2}} = \left(\frac{L[R_i + j\omega(L + L_i)]}{\mu(\gamma + j\omega\epsilon)}\right)^{\frac{1}{2}} \tag{1}$$

For many transmission lines $R_i \ll \omega L$, $L_i \ll L$, and $\gamma \ll \omega\epsilon$ so that in a vacuum the characteristic impedance and propagation constants are

$$\check{Z}_k = L(\mu_v\epsilon_v)^{-\frac{1}{2}} = 2.9979 \times 10^8 L \approx (3 \times 10^8 C)^{-1} \tag{2}$$

$$\check{\Gamma}^2 = \check{Z}_L\check{Y} = \mu L^{-1}(\gamma + j\omega\epsilon)[R_i + j\omega(L + L_i)] \tag{3}$$

It will be instructive to calculate $\check{Z}_k$ and $\check{\Gamma}$ for a Lecher wire system composed of two parallel wires of diameter d with their axes at a distance b apart. By 4.14 (2) and 11.14 (11), the external inductance is

$$L = \frac{\mu\epsilon}{C} = \frac{\mu}{\pi}\cosh^{-1}\frac{b}{d} \tag{4}$$

Note that L depends only on the configuration and not on the scale. We shall assume that the conductivity and frequency are so high that δ is small compared with d, and the formulas for the resistance per square R_i' and the inductance per square L_i', given in 10.02 (10), may be used.

$$R_iI^2 = 2R_i'\oint i_s^2\,ds \qquad L_iI^2 = 2I_i'\oint i_s^2\,ds \tag{5}$$

where i_s is the current across unit arc length in the cylindrical skin and the factor 2 is inserted to include both wires. From 11.14 (6) on the $U = U_1$ cylinder, the tangential component of B is $\partial V/\partial s$. The magnetomotance law applied to a surface element gives $\mu i_s = B$. Thus,

$$\oint i_s^2\,ds = \mu^{-2}\oint(\partial V/\partial s)^2\,ds = \mu^{-2}\oint(\partial V/\partial s)\,dV$$

From 7.09, 4.11 (2), and 4.13 (1) we have

$$\left|\frac{dW}{dz}\right| = \frac{\partial V}{\partial s} = \frac{\mu I}{\pi a} \sin \frac{\pi(U_1 + jV)}{j\mu I} \sin \frac{\pi(U_1 - jV)}{-j\mu I}$$

$$= \frac{\mu I}{\pi a}\left(\cosh^2 \frac{\pi U_1}{\mu I} - \cos^2 \frac{\pi V}{\mu I}\right)$$

$$\oint i_s^2 \, ds = \frac{I}{\mu \pi a}\int_0^{\mu I}\left(\cosh^2 \frac{\pi U_1}{\mu I} - \cos^2 \frac{\pi V}{\mu I}\right) dV = \frac{I^2}{2\pi a} \cosh \frac{2\pi U_1}{\mu I} \quad (6)$$

$$\frac{b}{d} = \cosh \frac{2\pi U_1}{\mu I}, \qquad b^2 - d^2 = 4a^2 \quad (7)$$

(7) is from 4.13 (3). Putting (6) and (7) into (5) and dividing out I^2,

$$R_i = \frac{2R_i' b}{\pi d(b^2 - d^2)^{\frac{1}{2}}} = \frac{2\tau' b}{\pi \delta d(b^2 - d^2)^{\frac{1}{2}}} \quad (8)$$

$$L_i = \frac{2L_i' b}{\pi d(b^2 - d^2)^{\frac{1}{2}}} = \frac{2\tau' b}{\pi \omega \delta d(b^2 - d^2)^{\frac{1}{2}}} \quad (9)$$

Now consider a special case where b/d is 1.5, and the line operates in a vacuum or air so that $\mu = \mu_v = 4\pi \times 10^{-7}$. Then by (4), L is 3.85×10^{-7} henry. The calculation of R_i and L_i requires, beside d/b, the frequency and the resistivity, permeability, and one dimension of the conductors. The most important conductor is copper for which γ' or $1/\tau'$ is 5.8×10^7 and $\mu = \mu_v$. Consider the frequencies 3, 300, and 30,000 megacycles which correspond to the wave lengths of 100 m, 1 m, and 1 cm for which ωL is 72.5, 725, and 7250 ohms, and the skin thickness is 3.82×10^{-5}, 3.82×10^{-6}, and 3.82×10^{-7} m. From (6), (7), and (3) we then have:

λ (m)	d (m)	R_i (ohms)	L_i (henry)	α	β
0.01	0.001	38.5	1.225×10^{-10}	5.1×10^{-2}	19
0.01	0.01	3.85	1.225×10^{-11}	5.1×10^{-3}	19
1.00	0.001	3.85	1.225×10^{-9}	5.1×10^{-3}	1.9
1.00	0.01	0.385	1.225×10^{-10}	5.1×10^{-4}	1.9
100	0.001	0.385	1.225×10^{-8}	5.1×10^{-4}	0.19
100	0.01	0.0385	1.225×10^{-9}	5.1×10^{-5}	0.19

Evidently in all cases R_i and ωL_i are very small compared with ωL so that the characteristic impedance may be obtained from (2) to be 115.5 ohms.

11.19. Group Velocity.—We have seen that any of the field amplitudes for a plane wave on a transmission line may be written in the form

$$A(\beta) = f(u_1, u_2) \cos (\omega t - \beta z) \quad (1)$$

where u_1 and u_2 are orthogonal coordinates in a plane normal to z, and β is the wave number or phase constant. When the phase velocity of a wave depends on its frequency, as in a wave guide or a dispersive medium, then it is evident that a signal consisting of a group of waves of different frequencies will change form as it advances. Let us assume that the wave numbers of the group lie between $\beta_0 - \delta$ and $\beta_0 + \delta$ and that the amplitude in the interval $d\beta$ is $A(\beta)$. The amplitude of the signal is then

$$S = \int_{\beta_0-\delta}^{\beta_0+\delta} A(\beta) e^{j(\omega t - \beta z)}\, d\beta$$

This may be split into a carrier wave factor with the velocity ω_0/β_0 and a modulation factor with a velocity $(\omega - \omega_0)/(\beta - \beta_0)$. Thus

$$S = Ce^{j(\omega_0 t - \beta_0 z)}, \qquad C = \int_{\beta_0-\delta}^{\beta_0+\delta} A(\beta) e^{j[(\omega-\omega_0)t - (\beta-\beta_0)z]}\, d\beta$$

Here $\frac{1}{2}C/\delta$ is the average amplitude of the wave group whose median phase velocity is $v_0 = \omega_0/\beta_0$. Thus we have a modulating frequency $\omega - \omega_0$ superimposed on the carrier frequency ω_0. We wish to find the velocity of a plane in which the modulation amplitude C, which constitutes the signal, is constant. By the mean value theorem for an integral, we have

$$C = 2\delta A(\beta_1) e^{j[(\omega_1-\omega_0)t - (\beta_1-\beta_0)z]}$$

where $\beta_0 - \delta \leqslant \beta_1 \leqslant \beta_0 + \delta$. If the frequency range in the group is so small that $\beta_1 - \beta_0$ and $\omega_1 - \omega_0$ are infinitesimal, the signal velocity is

$$v_s = \frac{\omega_1 - \omega_0}{\beta_1 - \beta_0} \xrightarrow[\delta \to 0]{} \left|\frac{\partial \omega}{\partial \beta}\right|_{\beta=\beta_0} \tag{2}$$

Problems

1. Two airplanes are flying at a distance d apart at a height h above a plane water surface. One sends radio signals to the other, both sending and receiving antennas being short vertical wires of length l. Show that the ratio of intensity of the signal received by water reflection to the direct signal, taking the capacitivity of water to be ϵ and taking $h \gg l$, $d \gg l$, and $\lambda \gg l$, is

$$\frac{d^6}{(d^2 + 4h^2)^3}\left\{\frac{[(\epsilon - \epsilon_v)d^2 + 4\epsilon h^2]^{\frac{1}{2}} - 2\epsilon\epsilon_v^{-\frac{1}{2}}h}{[(\epsilon - \epsilon_v)d^2 + 4\epsilon h^2]^{\frac{1}{2}} + 2\epsilon\epsilon_v^{-\frac{1}{2}}h}\right\}^2$$

2. A plane electromagnetic wave polarized at 45° to the plane of incidence is totally reflected in a prism which it enters and leaves normally. Show that the intensity of the emergent beam is $16n^2(1 + n)^{-4}I_0$, where n is the refractive index. Show that the emergent beam is elliptically polarized with a phase difference ϕ given by $\tan \frac{1}{2}\phi = \cos \theta(\sin \theta)^{-2}(\sin^2 \theta - n^2)^{\frac{1}{2}}$ where θ is the angle of incidence on the back of the prism and multiple reflections are neglected.

3. A plane electromagnetic wave of wave length λ is incident on an infinite plane plate of dielectric of thickness a. Show that the reflected beam intensity, if the beam is polarized either normal to or in the plane of incidence, is given by

$$4b^2 \sin^2 (\tfrac{1}{2}\delta)[(1 - b^2)^2 + 4b^2 \sin^2 (\tfrac{1}{2}\delta)]^{-1}I_0$$

where b^2 is the fraction of the incident light intensity reflected without entering the plate, $\delta = 4\pi na\lambda^{-1}\cos\psi$, ψ is the angle of refraction, and n is the index of refraction. Include multiple reflections and phase relations but assume no absorption.

4. Show that the law of refraction derived in 11.07 does not hold, in general, for Poynting's vector at the surface of an anisotropic medium.

5. Show that a plane normal to the optic axis where it pierces the ray surface in Fig. 11.05 is tangent to this surface on a circle that encloses the ray axis.

6. A plane wave is incident at an angle θ on the plane face of a uniaxial crystal whose normal makes an angle α with the optic axis. The angle between the plane of incidence and the plane containing the normal and the optic axis is ϕ. If $\epsilon_1 = \epsilon_2 \neq \epsilon_3$, show that the two refracted wave normal directions θ_o and θ_e are

$$\sin\theta_o'' = \frac{\epsilon_v^{\frac{1}{2}}\sin\theta}{\epsilon_1^{\frac{1}{2}}}, \qquad \sin^2\theta_e'' = \frac{2P^2 - MN \pm 2P(P^2 - M^2 - MN)^{\frac{1}{2}}}{N^2 + 4P^2}$$

$$M = \cos^2\alpha + \frac{\epsilon_1}{\epsilon_3 - \epsilon_1}, \quad P = \sin\alpha\cos\alpha\cos\phi, \; N = \frac{P^2}{\cos^2\alpha} - M^2 + \frac{\epsilon_1(\epsilon_v\sin^2\theta - \epsilon_3)}{\epsilon_v(\epsilon_3 - \epsilon_1)\sin^2\theta}$$

The minus sign is taken when $\phi < \frac{1}{2}\pi$ and the plus sign when $\frac{1}{2}\pi < \phi < \pi$.

7. Verify that possible plane wave vector potentials, when $\boldsymbol{B}$ is normal to z, are

$$\boldsymbol{A}_1 = C_1[\boldsymbol{i}j\beta' - \boldsymbol{k}(\beta'^2 - \beta_1^2)^{\frac{1}{2}}]e^{-(\beta'^2-\beta_1^2)^{\frac{1}{2}}x+j(\omega t-\beta' z)}$$
$$\boldsymbol{A}_2 = C_2\{\boldsymbol{i}j\beta'\cos[(\beta_2^2 - \beta'^2)^{\frac{1}{2}}x] - \boldsymbol{k}(\beta_2^2 - \beta'^2)^{\frac{1}{2}}\sin[(\beta_2^2 - \beta'^2)^{\frac{1}{2}}x]\}e^{j(\omega t-\beta' z)}$$

where $\beta_1^2 = \omega^2\mu_1\epsilon_1$, $\beta_2^2 = \omega^2\mu_2\epsilon_2$, and $\beta_2 > \beta_1$. With these, solve the problem of a plane surface wave moving over an infinite plane perfectly conducting surface coated with a dielectric layer μ_2, ϵ_2, of thickness a when the infinite medium above it has the constants μ_1, ϵ_1. Show that the velocity may be found from the equation

$$\epsilon_2(\beta'^2 - \beta_1^2)^{\frac{1}{2}} = \epsilon_1(\beta_2^2 - \beta'^2)^{\frac{1}{2}}\tan[(\beta_2^2 - \beta'^2)^{\frac{1}{2}}a]$$

If $\beta_1 a = 1.00$, $\mu_2 = \mu_1$, and $\epsilon_2 = 4\epsilon_1$, then $\beta_1:\beta':\beta_2 = 1:1.523:2$, roughly, so that the amplitude has dropped to 0.1 of its value at $x = a$ when $x = 3.00a$.

8. Verify that possible plane wave vector potentials, when $\boldsymbol{E}$ is normal to z, are

$$\boldsymbol{A}_1 = \boldsymbol{j}C_1e^{-(\beta'^2-\beta_1^2)^{\frac{1}{2}}x+j(\omega t-\beta' z)}, \qquad \boldsymbol{A}_2 = \boldsymbol{j}C_2\sin[(\beta_2^2 - \beta'^2)^{\frac{1}{2}}x]e^{j(\omega t-\beta' z)}$$

where $\beta_1^2 = \omega^2\mu_1\epsilon_1$, $\beta_2^2 = \omega^2\mu_2\epsilon_2$, and $\beta_2 > \beta_1$. With these, solve the problem of a plane surface wave moving over an infinite plane perfectly conducting surface coated with a dielectric layer μ_2, ϵ_2 of thickness a when the infinite medium above it has the constants μ_1, ϵ_1. Show that the velocity may be found from the equation

$$\mu_2(\beta'^2 - \beta_1^2)^{\frac{1}{2}} = -\mu_1(\beta_2^2 - \beta'^2)^{\frac{1}{2}}\cot[(\beta_2^2 - \beta'^2)^{\frac{1}{2}}a]$$

If $\beta_1 a = 1.00$, $\mu_2 = \mu_1$, and $\epsilon_2 = 4\epsilon_1$, then $\beta_1:\beta':\beta_2 = 1:1.03:2$, roughly, so that the amplitude has dropped to 0.1 of its value at $x = a$ when $x = 10.33a$.

9. Show that, if the two conductors of the section of uniform line bounded by planes of zero capacitivity and permeability at $z = 0$ and $z = l$ are given equal and opposite charges producing a difference of potential V_0 and if at $t = 0$ the surfaces of the planes are made conducting so that the resistance between conductors is R, then the potential between conductors is

$$V_s = \frac{V_0}{R + \check{Z}_k}\left(\frac{R - \check{Z}_k}{R + \check{Z}_k}\right)^s\left[R + \frac{4\check{Z}_k}{\pi}\sum_{m=0}^{\infty}\frac{(-1)^s}{2m+1}\sin\frac{(2m+1)\pi z}{l}\cos\frac{(2m+1)\pi t}{(\mu\epsilon)^{\frac{1}{2}}l}\right]$$

where $0 < z < l$ and $s(\mu\epsilon)^{\frac{1}{2}}l < t < (s+1)(\mu\epsilon)^{\frac{1}{2}}l$.

10. A uniform transmission line of length l and characteristic impedance $\check{Z}_k$ is terminated at one end by a battery of zero resistance and electromotance $\mathcal{E}$ in series with a switch and at the other by a resistance R. Neglecting field distortions at the ends, show that, if the switch is closed at $t = 0$, the current through R is

$$i = \frac{2\mathcal{E}}{R + Z_k} \sum_{s=0}^{n} \left(\frac{Z_k - R}{Z_k + R}\right)^s, \qquad (2n+1)(\mu\epsilon)^{\frac{1}{2}} l < t < (2n+3)(\mu\epsilon)^{\frac{1}{2}} l$$

11. A uniform transmission line of length l and characteristic impedance $\check{Z}_k$ is short-circuited at $z = l$ and has a switch in series with a battery of electromotance $\mathcal{E}$ and resistance R connected across it at $z = 0$. Neglecting end field distortions, show that, if the switch is closed at $t = 0$, the battery current is

$$i = \mathcal{E} \sum_{s=0}^{n} (2 - \delta_s^n) \frac{(Z_k - R)^s}{(Z_k + R)^{s+1}}, \qquad 2n(\mu\epsilon)^{\frac{1}{2}} l < t < (2n+2)(\mu\epsilon)^{\frac{1}{2}} l$$

12. If the Hertz vector has only a z-component so it satisfies the scalar wave equation 11.14 (1) for a plane wave transmission line, show that the potentials are

$$\psi = V(x, y) f[z - (\mu\epsilon)^{-\frac{1}{2}} t] \qquad \boldsymbol{A} = \boldsymbol{k}(\mu\epsilon)^{\frac{1}{2}} V(x, y) f[z - (\mu\epsilon)^{-\frac{1}{2}} t]$$

Show that when the scalar potential is eliminated by 11.01 (12), the resultant vector potential $\boldsymbol{A}'$ is identical in form with 11.14 (5).

13. A transmission line consists of two mutually external parallel cylinders of radii a and b, at an axial distance c. Show that the characteristic impedance is given by 11.18 (1) or (2) and the propagation constant by 11.18 (3), where

$$L = \frac{\mu}{2\pi} \cosh^{-1} \frac{c^2 - a^2 - b^2}{2ab}, \qquad R_i = \omega L_i = \frac{\tau'(b + a)[c^2 - (b - a)^2]}{2\pi ab\delta[(b^2 - a^2)^2 - 2c^2(b^2 + a^2) + c^4]^{\frac{1}{2}}}$$

14. A transmission line consists of two confocal elliptic cylinders with major axes $2a$ and $2b (a < b)$. If the focal distance is $2c$ and K is a complete elliptic integral, show that $\check{Z}_k$ is given by 11.18 (1) or (2) and $\check{\Gamma}$ by 11.18 (3), where

$$L = \frac{\mu}{2\pi} \ln \frac{b + (b^2 - c^2)^{\frac{1}{2}}}{a + (a^2 - c^2)^{\frac{1}{2}}}, \qquad R_i = \omega L_i = \frac{\tau'}{\pi^2 \delta}\left[\frac{1}{a} K\left(\frac{c}{a}\right) + \frac{1}{b} K\left(\frac{c}{b}\right)\right]$$

15. A transmission line consists of two parallel flat strips of width b and resistivity τ' whose faces are at a distance a apart where $a \ll b$. Show that the propagation constant and the characteristic impedance are, approximately,

$$\check{\Gamma}^2 = \frac{b}{a}(\gamma + j\omega\epsilon)\left[\frac{2\tau'}{b\delta}(1 + j) + \frac{j\omega\mu a}{b}\right], \qquad \check{Z}_k = \left\{\frac{a[2\tau'(1 + j) + j\omega\mu a\delta]}{b^2\delta(\gamma + j\omega\epsilon)}\right\}^{\frac{1}{2}}$$

16. A transmission line consists of two parallel cylinders of radii a and b, one inside the other at an axial distance c. Show that the characteristic impedance is given by 11.18 (1) or (2) and the propagation constant by 11.18 (3), where

$$L = \frac{\mu}{2\pi} \cosh^{-1} \frac{a^2 + b^2 - c^2}{2ab}, \qquad R_i = \omega L_i = \frac{\tau'(b - a)[(a + b)^2 - c^2]}{2\pi ab\delta[(b^2 - a^2)^2 - 2c^2(a^2 + b^2) + c^4]^{\frac{1}{2}}}$$

17. A lossless transmission line consists of a strip of width $2b$ centered in a slit of width $2a$ in an infinite coplanar conducting plane. Show from 4.29 (10) that the characteristic impedance given by 11.18 (2) is

$$\check{Z}_k = \tfrac{1}{4}\eta K[(1 - b^2a^{-2})^{\frac{1}{2}}][K(ba^{-1})]^{-1}$$

where η is the intrinsic impedance of 11.15 (15) and $K(k)$ is an elliptic integral.

18. Show that a transmission line of two coplanar strips, the spacing of inner edges and outer edges being $2b$ and $2a$, has, from 4.29 (9), the characteristic impedance

$$\check{Z}_k = \eta K(ba^{-1})\{K[(1 - b^2a^{-2})^{\frac{1}{2}}]\}^{-1}$$

19. A transmission line consists of a strip of width $2b$ centered in a cylinder of radius $r = a^{\frac{1}{2}}b^{\frac{1}{2}}$ or of a cylinder of this radius centered in a gap of width $2a$ in an infinite conducting plane. Show from 4.29 (10) that in either case the characteristic impedance given by 11.18 (2), if η is the intrinsic impedance, is

$$\check{Z}_k = \tfrac{1}{8}\eta K[(1 - b^2a^{-2})^{\frac{1}{2}}][K(ba^{-1})]^{-1}$$

20. A strip is coplanar with a semi-infinite plane with its edges at distances a and b from the plane edge where $a > b$. When used as a transmission line, show that, from 4.30 (2), its characteristic impedance is

$$\check{Z}_k = \tfrac{1}{2}\eta K(b^{\frac{1}{2}}a^{-\frac{1}{2}})\{K[(1 - ba^{-1})^{\frac{1}{2}}]\}^{-1}$$

where η is the intrinsic impedance and $K(k)$ is an elliptic integral.

21. A strip of width b with one edge on the axis of a cylinder of radius r, where $r > b$, forms a transmission line. Show from 4.30 that

$$\check{Z}_k = \tfrac{1}{4}\eta K[(1 - b^2r^{-2})^{\frac{1}{2}}][K(br^{-1})]^{-1}$$

22. A transmission line is formed of two unequal coplanar strips, the wider having a width w. The gap between the strips is q and the distance between their outer edges h. Show from 4.30 that the characteristic impedance is

$$\check{Z}_k = \tfrac{1}{2}\eta K(k)\{K[(1 - k^2)^{\frac{1}{2}}]\}^{-1}, \qquad k^2 = hq(w + q)^{-1}(h - w)^{-1}$$

where η is the intrinsic impedance of 11.15 (15) and $K(k)$ is an elliptic integral of modulus k.

23. A strip lies on a diameter of a cylinder of radius r, one edge being at a distance c_1 from its center and the other at a distance c_2. Show from 4.30 that t

$$\check{Z}_k = \tfrac{1}{4}\eta K[(1 - k^2)^{\frac{1}{2}}][K(k)]^{-1}, \qquad k = r(c_1 + c_2)(r^2 + c_1c_2)^{-1}$$

24. An infinite cylinder of radius c lies midway between two infinite parallel planes a distance $2b$ apart. Show from 4.31 that the characteristic impedance of this line is

$$\check{Z}_k = \tfrac{1}{4}\eta K(k)\{K[(1 - k^2)^{\frac{1}{2}}]\}^{-1}, \quad k = \cos[\tfrac{1}{2}\pi cb^{-1}(1 + \lambda)]$$

where λ is given by 4.28 (6). This is accurate to less than 1 per cent if $c \leqslant \frac{1}{2}b$. The attenuation may also be worked out by the method of 11.18 with the aid of HMF formulas.

25. An infinite strip of width $2a$ lies parallel to and halfway between two infinite parallel planes at a distance $2b$ apart. Show from problem 61 of Chap. IV that the characteristic impedance of this "strip line" is

$$\check{Z}_k = \tfrac{1}{2}\eta K(e^{-\pi ab^{-1}})\{K[(1 - e^{-2\pi ab^{-1}})^{\frac{1}{2}}]\}^{-1}$$

where $K(k)$ is a complete elliptic integral of modulus k.

26. The two halves of an infinite plane are separated by a gap of width $2h$. An infinite strip of width $2d$ lies in the gap normal to the plane so that its center line bisects the gap. Show that the characteristic impedance of this transmission line is

$$\check{Z}_k = \eta K[h(d^2 + h^2)^{-\frac{1}{2}}]\{K[d(d^2 + h^2)^{-\frac{1}{2}}]\}^{-1}$$

where η is the intrinsic impedance of the medium defined in 11.15 (15).

27. An infinite strip of width $2c$ lies midway between and normal to two infinite parallel planes a distance $2d$ apart. Show from problem 63 of Chap. IV that the characteristic impedance of this line is, if $\alpha = \frac{1}{2}c\pi/d$,

$$\check{Z}_k = \tfrac{1}{4}\eta K(\cos\alpha)[K(\sin\alpha)]^{-1}$$

28. The center line of an infinite strip of width $2h$ is coplanar with, and at a distance b from, the edge of a semi-infinite plane. The strip is normal to the plane. Show from problem 64 of Chap. IV that the characteristic impedance of this line is

$$\check{Z}_k = \tfrac{1}{2}\eta[K(k)]^{-1}K[(1-k)^2)^{\frac{1}{2}}], \qquad k = b^{-1}[(h^2 + b^2)^{\frac{1}{2}} - h]$$

29. A transmission line consists of two infinite conductors, a cross section of which forms two separate portions of the circumference of an infinite circular cylinder as specified in problem 65 of Chap. IV. Show that for this line

$$\check{Z}_k = \tfrac{1}{2}\eta K(k)\{K[(1 - k^2)^{\frac{1}{2}}]\}^{-1}$$

where the modulus k is defined in the problem referred to.

30. A transmission line consists of two parallel strips of width $2h$ in two parallel planes a distance $2d$ apart. Their center lines lie in a plane normal to the parallel planes. Show from problem 66 of Chap. IV that

$$\check{Z}_k = \eta K(ba^{-1})\{K[(1 - b^2a^{-2})^{\frac{1}{2}}]\}^{-1}$$

where ba^{-1} is found as directed in the problem referred to.

31. One side of a transmission line is a circular cylinder of radius c at the origin. The other consists of four hyperbolic cylinders given by $\pm xy = b^2$, Show that the characteristic impedance of this line is

$$\check{Z}_k = \tfrac{1}{8}\eta K(k)\{K[(1 - k^2)^{\frac{1}{2}}]\}^{-1}$$

where the modulus k is found as directed in problem 79 of Chap. IV.

32. One side of a transmission line is a circular cylinder of radius a. The other is a parabolic cylinder $y^2 = 4p(x + p)$ whose focus is the cylinder axis. Show that the characteristic impedance is

$$\check{Z}_k = \tfrac{1}{2}\eta K(k)\{K[(1 - k^2)^{\frac{1}{2}}]\}^{-1}$$

where the modulus is found as directed in problem 80 of Chap. IV.

33. One side of a transmission line is a conductor of a rectangular cross section bounded by $x' = \pm b$, $y' = \pm c$. The other side is two conducting planes at $y = \pm a$. Show that the characteristic impedance is

$$\check{Z}_k = \frac{1}{4}\eta K\left[\left(\frac{B}{A}\right)^{\frac{1}{2}}\right]\{K[(1 - A^{-1}B)^{\frac{1}{2}}]\}^{-1}$$

where A and B are determined as directed in problem 82 of Chap. IV.

34. A plane wave in a medium of permeability μ and capacitivity ϵ strikes normally a plane perfectly conducting mirror. Show that reflection is prevented by placing a thin layer of material of thickness d, capacitivity ϵ', and conductivity γ' one-quarter wave length in front of it if $\gamma' d \approx (\epsilon/\mu)^{\frac{1}{2}}$ and $\gamma' \gg \omega\epsilon'$.

References

ABRAHAM, M., AND R. BECKER: "Classical Electricity and Magnetism," Blackie, 1932. Uses vector notation and treats waves on wires.

BATEMAN, H.: "Electrical and Optical Wave Motion," Cambridge, 1915, and Dover. Discusses boundaries and wave solutions and gives extensive references.

BORN, MAX: "Optik," Springer, Berlin, 1933. A complete, illustrated treatment.

COLLIN, R. E.: "Field Theory of Guided Waves," McGraw-Hill, 1960. Chapters III and IV cover subject matter of this chapter in detail.

DRUDE, P.: "Theory of Optics," Longmans, 1920, and Dover. This is the classical book on this subject and uses the long notation.

FLUGGE, S.: "Handbuch der Physik," Vol. XVI, Springer, 1958. Several of the articles in this volume contain material on plane waves.

FORSTERLING, K.: "Lehrbuch der Optik," Hirzel, Leipzig, 1928. A complete and elegant treatment of the subject using vector notation.

FRENKEL, J.: "Lehrbuch der Elektrodynamik," 2 vols., Springer, Berlin, 1926, 1928.

GEIGER-SCHEEL: "Handbuch der Physik," Vols. XII, XV, XX, Berlin, 1927, 1928. Gives elegant treatment of electromagnetic theory in Vol. XII.

HARRINGTON, R. F.: "Time-harmonic Electromagnetic Fields," McGraw-Hill, 1961. Chapter II covers much of this material with circuit analysis.

HERTZ, H. R.: "Electric Waves," Macmillan, 1893. Drawings of radiation fields.

JACKSON, J. D.: "Classical Electrodynamics," Wiley, 1962. Chapters VI and VII are pertinent. Uses Gaussian units.

JEANS, J. H.: "The Mathematical Theory of Electricity and Magnetism," Cambridge, 1925. Gives extensive treatment using long notation.

KING, R. W. P., H. R. MIMNO, AND A. H. WING: "Transmission Lines, Antennas, and Wave Guides," McGraw-Hill, 1945. Practical detailed treatment of transmission lines, with some field theory.

MAXWELL, J. C.: "Electricity and Magnetism," 3d ed. (1891), Dover, 1954. Gives original treatment using long notation.

PANOFSKY, W. K. H., AND MELBA PHILLIPS: "Classical Electricity and Magnetism," Addison-Wesley, 1962. Chapter XI is pertinent to this chapter.

PAPAS, C. H.: "Theory of Electromagnetic Wave Propagation," McGraw-Hill, 1965. Contains a very lucid and detailed treatment of Maxwell's equations.

RAMO, S., AND J. R. WHINNERY: "Fields and Waves in Modern Radio," Wiley, 1944. Simple treatment of waves on wires.

SCHELKUNOFF, S. A.: "Electromagnetic Waves," Van Nostrand, 1943. Whole theory treated by means of the impedance concept.

STRATTON, J. A.: "Electromagnetic Theory," McGraw-Hill, 1941. Extensive and rigorous mathematical treatment of the whole subject.

TRALLI, N.: "Classical Electromagnetic Theory," McGraw-Hill, 1963. Chapters IX, X, and XI cover this chapter with additional material.

WIEN-HARMS: "Handbuch der Experimentalphysik," Vol. XI, Leipzig, 1932.

CHAPTER XII

ELECTROMAGNETIC RADIATION

12.00. The Radiation Problem.—Any system generating electromagnetic waves and incompletely enclosed by conductors loses energy because waves escape into space. If the system is not supposed to radiate, this represents a leakage loss and the only information usually required of a calculation is the power radiated. If the function of the system is to radiate, as in the case of antennas or horns, the additional information needed may include the polarization and distribution in space of the radiant energy, the nonradiative losses, the current and charge distribution in the system, and its input impedance. These will depend on the impressed frequency and electromotance, on the system's dimensions and geometrical configuration, and on the materials of which it and its surroundings are composed. When the antenna lies in empty space, a complete solution of the problem must give fields which satisfy Maxwell's equations for empty space outside the antenna and for the antenna medium inside it and which satisfy its surface boundary conditions. These fields must also lead to electric intensity and current distributions that match those of the transmission line over its junction surface with the antenna. For existing materials and configurations, such perfect solutions appear impossible.

There are, however, methods of calculating one or more of the desired quantities with considerable accuracy. One can always find, by means of retarded potentials, the fields of a given charge or current distribution. Often the distant radiation is sensitive only to the major features of this distribution, of which a rough estimate is then adequate. This also gives the ohmic losses, if they are small, with sufficient accuracy so that with the aid of Poynting's theorem the real part, but not the imaginary part, of the input impedance can be found. Information as to the actual current distribution in perfectly conducting antennas of simple geometrical forms like spheres and spheroids can be obtained by a rigorous solution of the boundary value problem in terms of orthogonal functions similar to, but more complicated than, those used in electrostatics. Skin effect formulas give the ohmic losses in terms of the magnetic fields just outside the surface. With the exception of the biconical antenna, such solutions give little information on the input reactance. The radiation from orifices is often found by assuming initial field values which are then corrected by trial to fit the boundary conditions. In the articles that follow examples of most of these methods are given.

12.01. Spherical Electromagnetic Waves. Dipole and Quadrupole Radiation.—When the strength of a multipole that appears in 1.07 or 7.10 varies in time, it generates a spherical electromagnetic wave. The electrostatic potential of the electric multipole of order n and moment $q^{(n)}$ may be written

$$\Psi_n = \sum_{i=1}^{3}\sum_{j=1}^{3}\cdots\sum_{k=1}^{3}\frac{q^{(n)}_{i,j,\ldots k}(-1)^n\,\partial^n(r^{-1})}{4\pi\epsilon n!\,\partial x_i\,\partial x_j\cdots\partial x_k} \tag{1}$$

In rectangular coordinates $x_1 = x$, $x_2 = y$, and $x_3 = z$. When $n = 1$,

$$\Psi_1 = \frac{q^{(1)}_x x}{4\pi\epsilon r^3} + \frac{q^{(1)}_y y}{4\pi\epsilon r^3} + \frac{q^{(1)}_z z}{4\pi\epsilon r^3} \tag{2}$$

This is the Ψ for a dipole of moment $[(q^{(1)}_x)^2 + (q^{(1)}_y)^2 + (q^{(1)}_z)^2]^{\frac{1}{2}}$ with direction cosines $q^{(1)}_x q^{-1}$, $q^{(1)}_y q^{-1}$, $q^{(1)}_z q^{-1}$. When $n = 2$, $q^{(2)}_{xx}$, $q^{(2)}_{yy}$ and $q^{(2)}_{zz}$ are linear quadrupoles on the x-, y-, and z-axes and $q^{(2)}_{xy} = q^{(2)}_{yx}$, $q^{(2)}_{xz} = q^{(2)}_{zx}$ and $q^{(2)}_{yz} = q^{(2)}_{zy}$ are square quadrupoles in the xy-, xz-, and yz-planes. Thus the moment is a symmetrical tensor with three different square moments and two different linear moments. One linear moment was eliminated by the relation $(\partial^2/\partial x^2 + \partial^2/\partial y^2 + \partial^2/\partial z^2)r^{-1} = 0$.

When $q^{(n)}$ oscillates, the currents transferring the charges determine the moments just as well as the charges do. Equation 7.02 (4) shows that the magnetostatic A_z of an infinitesimal current element $i_0\,dz_0$ is a function of r only. If the current $i_0 f(t)$ depends on time and the field travels with the same velocity in all directions, then the radiation A_z should be a function of r only. Thus the propagation equation and its solution are

$$\nabla^2 A_z = r^{-2}\frac{\partial}{\partial r}\left(r^2\frac{\partial A_z}{\partial r}\right) = \mu\epsilon\frac{\partial^2 A_z}{\partial t^2} = \frac{\partial^2 A_z}{v^2\,\partial t^2} \tag{3}$$

$$\mathbf{A} = \frac{\mu i_0\,\mathbf{dz}_0\,f(t \pm v^{-1}r)}{4\pi r} = \frac{\mu i_0\,dz_0\,(\mathbf{r}_1\cos\theta - \boldsymbol{\theta}\sin\theta)f(t \pm v^{-1}r)}{4\pi r} \tag{4}$$

The plus sign gives a converging wave and the minus sign a diverging one. From 1.07 (1), if the potentials of the two charges $q_0 f_1(t)$ of opposite sign at a distance dz_0 apart are superimposed, then since $\partial/\partial z_0 = -\partial/\partial z$,

$$\Psi = \frac{-q_0\,dz_0}{4\pi\epsilon}\frac{\partial}{\partial z}\left[\frac{f_1(t \pm v^{-1}r)}{r}\right] = \frac{q^{(1)}\cos\theta}{4\pi\epsilon r^2}\left[f_1(t \pm v^{-1}r) \pm \frac{rf_1'(t \pm v^{-1}r)}{v}\right] \tag{5}$$

If the current element in (4) feeds the charges in (5), then $i_0 = \partial q_0/\partial t$ and $q^{(1)}f_1(t) = i_0\,dz_0 \int f(t)\,dt$. Thus (4) and (5) satisfy 11.01 (8), and A_z and Ψ are the potentials of an electric dipole in the Lorentz gauge of 11.01.

If $f(t)$ is $\cos \omega t$, then 11.01 (1) gives the phasor fields

$$\breve{E}_r = -\frac{\partial A_z}{\partial t} - \frac{\partial \Psi}{\partial r} = \frac{\eta i_0\, dz_0}{2\pi}\left(\frac{1}{r^2} - \frac{j}{\beta r^3}\right)\cos\theta e^{-j\beta r} \tag{6}$$

$$\breve{E}_\theta = -\frac{\partial A_\theta}{\partial t} - \frac{\partial \Psi}{r\,\partial\theta} = \frac{\eta i_0\, dz_0}{4\pi}\left(\frac{j\beta}{r} + \frac{1}{r^2} - \frac{j}{\beta r^3}\right)\sin\theta e^{-j\beta r} \tag{7}$$

$$\breve{B}_\phi = \frac{\partial(rA_\theta)}{r\,\partial r} - \frac{\partial A_r}{r\,\partial\theta} = \frac{\mu i_0\, dz_0}{4\pi}\left(\frac{j\beta}{r} + \frac{1}{r^2}\right)\sin\theta e^{-j\beta r} \tag{8}$$

Replacement of $i_0\, dz_0$ by $j\omega q^{(1)}$ in (6), (7), and (8) gives the fields in terms of the moment $q^{(1)} \cos \omega t$ based on the charges. Note that

$$\beta = \omega(\mu\epsilon)^{\frac{1}{2}} = \omega v^{-1} = 2\pi\lambda^{-1} = \omega\epsilon\eta = \omega\mu\eta^{-1} \tag{9}$$

Far from the dipole the r^{-1} terms alone survive and form the radiation field, while near it the r^{-3} terms dominate and produce the quasi-static field. The r^{-2} terms give the induction field. In the Lorentz gauge the Hertz vector gives more compact formulas. From 11.01 (8), using $j\omega q^{(1)}$ for $i_0\, dz_0$ and phasor notation, (4) becomes

$$\boldsymbol{Z} = \breve{\boldsymbol{Z}} e^{j\omega t} = \boldsymbol{k}\frac{jq^{(1)}}{4\pi\epsilon r} e^{j(\omega t - \beta r)} \tag{10}$$

The linear quadrupole, $p_{zz}^{(2)}$ in (1), is found from $p_z^{(1)}$ as in the static case by differentiation of $Z_z^{(1)}$ with respect to z and division by 2!. In the wave equation $\partial/\partial z$ and ∇^2 may be interchanged only when applied to scalars or rectangular components of vectors. The formulas needed for putting $\boldsymbol{Z}$ into spherical components are, for the unit vectors,

$$\boldsymbol{i} = \boldsymbol{r}_1 \sin\theta\cos\phi + \boldsymbol{\theta}\cos\theta\cos\phi - \boldsymbol{\phi}\sin\phi \tag{11}$$

$$\boldsymbol{j} = \boldsymbol{r}_1 \sin\theta\sin\phi + \boldsymbol{\theta}\cos\theta\sin\phi + \boldsymbol{\phi}\cos\phi, \qquad \boldsymbol{k} = \boldsymbol{r}_1\cos\theta - \boldsymbol{\theta}\sin\theta \tag{12}$$

The result in phasor form with $\cos\theta$ written for z/r is

$$8\pi\epsilon r^2 \breve{\boldsymbol{Z}}_{zz} = q_{zz}^{(2)}(\beta r - j)(\boldsymbol{r}_1\cos\theta - \boldsymbol{\theta}\sin\theta)\cos\theta e^{-j\beta r} \tag{13}$$

Similar procedures for $\breve{\boldsymbol{Z}}_{xx}$ and $\breve{\boldsymbol{Z}}_{yy}$ give

$$8\pi\epsilon r^2 \breve{\boldsymbol{Z}}_{xx} = \boldsymbol{i} q_{xx}^{(2)}(\beta r - j)\sin\theta\cos\phi e^{-j\beta r} \tag{14}$$

$$8\pi\epsilon r^2 \breve{Z}_{yy} = \boldsymbol{j} q_{yy}^{(2)}(\beta r - j)\sin\theta\sin\phi e^{-j\beta r} \tag{15}$$

For a square quadrupole centered at the origin in the xy-plane with sides parallel to the x- and y-axes, $\boldsymbol{Z}$ is found by superposition of $\frac{1}{2}\,\partial Z_x^{(1)}/\partial y$ and $\frac{1}{2}\,\partial Z_y^{(1)}/\partial x$ where (10) is $Z_z^{(1)}$. This gives

$$16\pi\epsilon r^2 \breve{\boldsymbol{Z}}_{xy}^{(2)} = q_{xy}^{(2)}(\beta r - j)(\boldsymbol{i}\sin\phi + \boldsymbol{j}\cos\phi)\sin\theta e^{-j\beta r} \tag{16}$$

Likewise for square quadrupoles in the yz- and zx-planes, we get

$$16\pi\epsilon r^2\check{\boldsymbol{Z}}_{yz}^{(2)} = q_{yz}^{(2)}(\beta r - j)(\boldsymbol{j}\cos\theta + \boldsymbol{k}\sin\theta\sin\phi)e^{-j\beta r} \tag{17}$$

$$16\pi\epsilon r^2\check{\boldsymbol{Z}}_{zx}^{(2)} = q_{zx}^{(2)}(\beta r - j)(\boldsymbol{i}\cos\theta + \boldsymbol{k}\sin\theta\cos\phi)e^{-j\beta r} \tag{18}$$

From the above tensor components the most general electric quadrupole field can be built up.

From 11.01 (1) $\boldsymbol{E}$ and $\boldsymbol{B}$ satisfy the same propagation equation, so for a magnetic dipole field, from 11.03 (8) as $r \to \infty$, $v\boldsymbol{B}$ should replace the $\boldsymbol{E}$ of the electric dipole field of (6), (7), and (8), and from 11.03 (11) to propagate outward, $-\boldsymbol{E}$ should replace the $\boldsymbol{B}$. Thus writing $j\omega M$ for $i_0 dz_0$ gives

$$\check{B}_r = \frac{\mu M}{2\pi}\left(\frac{j\beta}{r^2} + \frac{1}{r^3}\right)\cos\theta e^{-j\beta r} \tag{19}$$

$$\check{B}_\theta = \frac{\mu M}{4\pi}\left(-\frac{\beta^2}{r} + \frac{j\beta}{r^2} + \frac{1}{r^3}\right)\sin\theta e^{-j\beta r} \tag{20}$$

$$\check{E}_\phi = \frac{\omega\mu M}{4\pi}\left(\frac{\beta}{r} - \frac{j}{r^2}\right)\sin\theta e^{-j\beta r} \tag{21}$$

Note that as $r \to 0$ the r^{-3} terms give exactly the curl of 7.10 (2).

The mean energy radiated from a multipole source is found by integration of the mean Poynting vector, 11.17 (3), over the sphere of large radius r. This may be put in terms of the Hertz vector by noting that in its curl the only derivative which does not disturb the r^{-1} factor is $r^{-1}\,\partial(re^{-j\beta r})/\partial r$, and when r is very large this simply replaces $e^{-j\beta r}$ by $-j\beta e^{-j\beta r}$. It acts only on the θ- and ϕ-components of $\check{\boldsymbol{Z}}$, so that if $\check{Z}_{t\infty}$ denotes the tangential component of $\check{\boldsymbol{Z}}$ as $r \to \infty$, 11.01 (12) gives, since $j\omega\mu\epsilon \cdot -j\beta$ is $\omega^2 v^{-3}$,

$$\check{\boldsymbol{B}}_{t\infty} = \omega^2 v^{-3}(-\boldsymbol{\theta}\check{Z}_{\phi\infty} + \boldsymbol{\phi}\check{Z}_{\theta\infty}) \tag{22}$$

Now from 11.03 (8), $\boldsymbol{r}_1 \times \boldsymbol{E}_t = v\boldsymbol{B}_t$, so that

$$\check{\boldsymbol{E}}_{t\infty} = \omega^2 v^{-2}(\boldsymbol{\theta}\check{Z}_{\theta\infty} + \boldsymbol{\phi}\check{Z}_{\phi\infty}) \tag{23}$$

$$\boldsymbol{\Pi} = \mu^{-1}\check{\boldsymbol{E}} \times \hat{\boldsymbol{B}} = \mu^{-1}\omega^4 v^{-5}\boldsymbol{r}_1(\check{Z}_{\theta\infty}\hat{Z}_{\theta\infty} + \check{Z}_{\phi\infty}\hat{Z}_{\phi\infty}) \tag{24}$$

For the electric dipole, from (10) and (12),

$$\check{Z}_{\theta\infty} = \frac{-jq^{(1)}\sin\theta}{4\pi\epsilon r}e^{-j\beta r}, \qquad \boldsymbol{\Pi} = \frac{\boldsymbol{r}_1\omega^4\sin^2\theta}{16\pi^2\epsilon^2\mu v^5 r^2}[q^{(1)}]^2 = 2\bar{\boldsymbol{\Pi}} \tag{25}$$

$$\bar{P} = \int\bar{\Pi}_r\,dS = \frac{\omega^4[q^{(1)}]^2}{12\pi\epsilon v^3} = \frac{4\pi^3 v[q^{(1)}]^2}{3\epsilon\lambda^4} = \frac{\omega^2(i_0\,dz_0)^2}{12\pi\epsilon v^3} = \frac{\pi(i_0\,dz_0)^2}{3\epsilon v\lambda^2} \tag{26}$$

For the magnetic dipole, from (19), (20), and (21), the radiated power is

$$\bar{P} = \int\frac{\check{E}_\phi\hat{B}_\theta}{2\mu}\,dS = \frac{\omega^4 M^2}{12\pi\epsilon v^5} = \frac{4\pi^3 M^2}{3\epsilon v\lambda^4} = \frac{\omega^4 I^2 A^2}{12\pi\epsilon v^5} = \frac{4\pi^3 I^2 A^2}{3\epsilon v\lambda^4} \tag{27}$$

In 7.10 (2) M was defined as IA for a current I in a loop of area A. For a linear electric quadrupole, from (13) and (24),

$$\bar{P} = \int \bar{\Pi}_r\, dS = \frac{\beta^2\omega^4[q_{zz}^{(2)}]^2}{128\pi^2\epsilon^2\mu v^5}\int_0^\pi \sin^3\theta\cos^2\theta\, d\theta = \frac{\omega^6[q_{zz}^{(2)}]^2}{480\pi\epsilon v^5} = \frac{2\pi^5 v[q_{zz}^{(2)}]^2}{15\epsilon\lambda^6} \quad (28)$$

For a square quadrupole, from (16), (11), and (12), then (24),

$$16\pi\epsilon r\check{\mathbf{Z}}_{t\infty} = \beta q_{xy}^{(2)}[\boldsymbol{\theta}\cos\theta\sin 2\theta + \boldsymbol{\phi}\cos 2\phi]\sin\theta e^{-j\beta r} \quad (29)$$

$$\bar{P} = \bar{\Pi}_r\, dS = \frac{\omega^6[q_{xy}^{(2)}]^2}{256\pi^2\epsilon v^5}\int_0^{2\pi}\int_0^\pi (\cos^2\theta\sin^2 2\phi + \cos^2 2\phi)\sin^3\theta\, d\theta\, d\phi$$

$$= \frac{\omega^6[q_{xy}^{(2)}]^2}{320\pi\epsilon v^5} = \frac{\pi^5 v[q_{xy}^{(2)}]^2}{5\epsilon\lambda^6} \quad (30)$$

12.02. Retarded Potentials.—A useful method of finding solutions of Maxwell's equations given in 11.00 is by means of retarded potentials. From 11.01 (6) and (7), the vector and scalar potentials propagate in a homogeneous isotropic dielectric with a velocity $(\mu\epsilon)^{-\frac{1}{2}}$. The object is to set up solutions of these equations analogous to 7.02 (5) and 3.09 (1) that give the potentials $\boldsymbol{A}$ and Ψ at the point P at the time t. Both the electric current $\boldsymbol{i}$ and the electric charge density ρ are assumed fixed in position but with magnitudes varying with time. The treatment of moving isolated charges appears in Chap. XIV. The contribution of an element dv at the point $x_1y_1z_1$, having traveled a distance r from dv to P with a velocity $(\mu\epsilon)^{-\frac{1}{2}}$, must have left dv at the time $t - (\mu\epsilon)^{\frac{1}{2}}r$ and thus must represent the conditions at dv at that time. Summing the effects of all elements gives the P potential, from 7.02 (5) and 3.09 (1), at time t,

$$\boldsymbol{A}(x,y,z,t) = \frac{\mu}{4\pi}\int\int\int \frac{\boldsymbol{i}[x_1, y_1, z_1, t - (\mu\epsilon)^{\frac{1}{2}}r]}{r}\, dx_1\, dy_1\, dz_1 \quad (1)$$

$$\Psi(x,y,z,t) = \frac{1}{4\pi\epsilon}\int\int\int \frac{\rho[x_1, y_1, z_1, t - (\mu\epsilon)^{\frac{1}{2}}r]}{r}\, dx_1\, dy_1\, dz_1 \quad (2)$$

The fields derive from these retarded potentials as in 11.01. Note that the integrands for A_x, A_y, A_z, and Ψ in (1) and (2) have exactly the form $f[t - (\mu\epsilon)^{\frac{1}{2}}]/r$ of the propagation equation solutions found in 11.03 (1).

In connection with 12.01 (4) and (5), it was shown that if the charges in the element dv in (2) are fed by the current in dv in (1), then 11.01 (8) is satisfied. Thus $\boldsymbol{A}$ and Ψ for the element dv satisfy 11.01 (8), and $\boldsymbol{A}$ is $\mu\epsilon\, \partial\boldsymbol{Z}/dt$. From this one writes down by inspection an expression for the Hertz vector that combines (1) and (2). Thus

$$\boldsymbol{Z}(x,y,z,t) = \frac{1}{4\pi\epsilon}\int\int\int\int \frac{\boldsymbol{i}[x_1, y_1, z_1, t - (\mu\epsilon)^{\frac{1}{2}}r]}{r}\, dx_1\, dy_1\, dz_1\, dt \quad (3)$$

The usefulness of (1), (2), and (3) depends on the accuracy of the estimate of the current and charge distribution in the source. If the source is a very thin perfectly conducting wire lying on a curve s, then no component of the electric field given by 11.01 (3) can lie along s and the equation of continuity must be satisfied. Thus

$$E_s = -\frac{\partial\Psi}{\partial s} - \frac{\partial A_s}{\partial t} = 0 \qquad \frac{\partial I}{\partial s} + \frac{\partial \rho}{\partial t} = 0 \tag{4}$$

But, since the wire is infinitely thin, we may go so close to it that the wave length and the radius of curvature of s is infinite compared with our distance r. Then, *if radiation is neglected*, $\mathbf{A}$ and $\mathbf{\Psi}$ are given by the formulas for a long straight wire

$$\Psi = -\frac{\rho}{2\pi}\ln r \qquad A_s = -\frac{\mu I}{2\pi}\ln r \tag{5}$$

Insert these values into (4), differentiate one with respect to t and the other with respect to s, and eliminate A_s or Ψ from the result. Thus

$$\frac{\partial^2 I}{\partial s^2} = \mu\epsilon\frac{\partial^2 I}{\partial t^2} \qquad \frac{\partial^2 \rho}{\partial s^2} = \mu\epsilon\frac{\partial^2 \rho}{\partial t^2} \tag{6}$$

If the current is harmonic in time, the solution of (6) is

$$I(s) = \check{I}_1 e^{j(\omega t-\beta s)} + \check{I}_2 e^{j(\omega t+\beta s)}, \qquad \rho(s) = \check{\rho}_1 e^{j(\omega t-\beta s)} + \check{\rho}_2 e^{j(\omega t+\beta s)} \tag{7}$$

The distribution of current and charge are therefore sinusoidal. This is the current form nearly always used in antenna theory but, as will be evident in the next article, the assumption of no radiation is not the only approximation involved.

12.03. Radiation from Linear Antenna.—Retarded potentials may be used to calculate the radiation field surrounding a linear antenna. We use the cylindrical coordinate system ρ, ϕ, z and write z' for the coordinate of a point on the axis of which the antenna occupies that portion between z_1' and z_2'. Its diameter to length ratio is so small that the current forms a sinusoidal standing wave. The origin lies inside or outside the antenna at a point where there is, or would be, a current node so that

$$I = I_0 \sin\beta z' \cos\omega t \tag{1}$$

We can use either the potentials or the Hertz vector to get the fields. The physical nature of the terminal conditions is somewhat clearer in terms of the potentials. Whether or not the current is zero at the end depends on the terminal connection. The linear charge density is, by 11.00 (5),

$$\sigma = -(\mu\epsilon)^{\frac{1}{2}} I_0 \cos\beta z' \sin\omega t \tag{2}$$

Note that σ is everywhere finite so the solution applies only when the lumped terminal charges do not radiate. It is a very simple matter to add to the solutions, when necessary, the additional terms arising from such sources. From 12.02 (1) and (2), the retarded potentials are

$$A_z = \frac{\mu I_0}{4\pi}\int_{z_1'}^{z_2'} \frac{\sin \beta z' \cos(\omega t - \beta r)}{r}\,dz' \tag{3}$$

$$\Psi = \frac{\beta I_0}{4\pi\omega\epsilon}\int_{z_1'}^{z_2'} -\frac{\cos \beta z' \sin(\omega t - \beta r)}{r}\,dz' \tag{4}$$

where $r^2 = (z' - z)^2 + \rho^2$. To get E_z from 11.01 (3), we substitute

$$u = \frac{\sin(\omega t - \beta r)}{r}, \qquad \frac{\partial u}{\partial z} = -\frac{\partial u}{\partial z'}, \qquad v = \cos \beta z', \qquad \frac{\partial v}{\partial z'} = -\beta \sin \beta z'$$

$$\frac{\partial A_z}{\partial t} = \frac{\omega\mu I_0}{4\pi\beta}\int_{z_1'}^{z_2'} u\,dv, \qquad \frac{\partial \Psi}{\partial z} = \frac{\omega\mu I_0}{4\pi\beta}\int_{z_1'}^{z_2'} v\,du$$

Thus from 11.01 (3), we have for E_z

$$E_z = -\frac{\omega\mu I_0}{4\pi\beta}\Big|uv\Big|_{z_1'}^{z_2'}$$

$$= \frac{\omega\mu I_0}{4\pi\beta}\left[\frac{\cos \beta z_1' \sin(\omega t - \beta r_1)}{r_1} - \frac{\cos \beta z_2' \sin(\omega t - \beta r_2)}{r_2}\right] \tag{5}$$

To find E_ρ and H_ϕ without the integration of (3) and (4), we substitute $u = r + z' - z$ and $v = r - z' + z$ so that $r\,du = u\,dz'$ and $r\,dv = -v\,dz'$ in the integrals of (3) and (4). With the aid of *Dw* 401.01 or *Pc* 591, these may then be written, using the upper sign for (3) and the lower for (4),

$$-\int_{u_1}^{u_2} \frac{\sin(\omega t - \beta z - \beta u)}{2u}\,du \mp \int_{v_1}^{v_2} \frac{\sin(\omega t + \beta z - \beta v)}{2v}\,dv \tag{6}$$

Note that ρ appears only in the limits of the integrals and that

$$\frac{\partial u_{1,2}}{\partial \rho} = \frac{\partial v_{1,2}}{\partial \rho} = \frac{\rho}{r_{1,2}} \tag{7}$$

so that differentiation of the integrals with respect to ρ gives

$$\frac{\rho}{2}\left[-\frac{\sin(\omega t - \beta r - \beta z')}{r(r + z' - z)} \mp \frac{\sin(\omega t - \beta r + \beta z')}{r(r - z' + z)}\right]_{z_1'}^{z_2'} \tag{8}$$

Putting over a common denominator, substituting ρ^2 for $r^2 - (z' - z)^2$, and combining the sines of the sum and difference of $\omega t - \beta r$ and $\beta z'$ give

$$E_\rho = -\frac{\partial \Psi}{\partial \rho} = -\frac{\omega\mu I_0}{4\pi\beta\rho}\left[\cos(\omega t - \beta r_2)\sin \beta z_2' - \cos(\omega t - \beta r_1)\sin \beta z_1'\right.$$
$$\left. + \frac{z_2' - z}{r_2}\sin(\omega t - \beta r_2)\cos \beta z_2' - \frac{z_1' - z}{r_1}\sin(\omega t - \beta r_1)\cos \beta z_1'\right] \tag{9}$$

$$B_\phi = -\frac{\partial A_z}{\partial \rho} = \frac{\mu I_0}{4\pi\rho}\bigg[\sin(\omega t - \beta r_2)\cos\beta z_2' - \sin(\omega t - \beta r_1)\cos\beta z_1'$$
$$+ \frac{z_2' - z}{r_2}\cos(\omega t - \beta r_2)\sin\beta z_2' - \frac{z_1' - z}{r_1}\cos(\omega t - \beta r_1)\sin\beta z_1'\bigg] \quad (10)$$

Equations (5), (9), and (10) apply to antennas whose terminal loads are not zero, provided the loads charged by them are prevented from radiating by earthed shields like the coaxial termination shown in Fig. 12.03*a*.

The cases shown in Fig. 12.03*b* and *c* can be treated by adding to the scalar potential of (4) the terminal charge retarded potential. If the load is replaced by an antenna section extending to the nearest current node, the charge on this section equals that on the load since the same

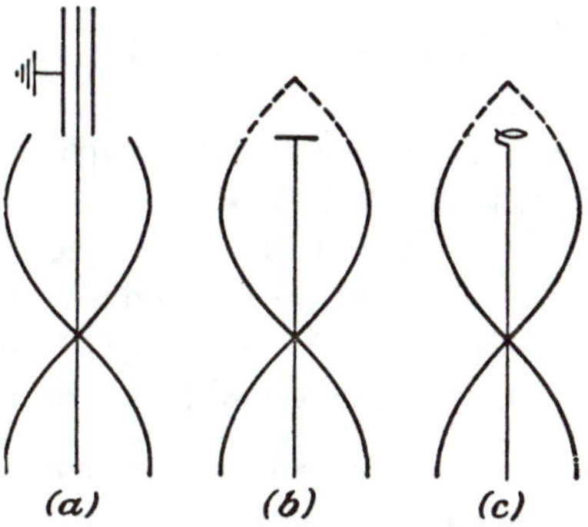

FIG. 12.03.

charging current flows into it. Thus the integrals of σ from the nearest node to z_1' and from z_2' to the nearest node give, respectively, the charge Q_1 on z_1' and Q_2 on z_2' so that, from (2),

$$Q_1 = -\omega^{-1}I_0 \sin\beta z_1' \sin\omega t, \qquad Q_2 = \omega^{-1}I_0 \sin\beta z_2' \sin\omega t$$

If the terminal loads are small enough to be treated as point charges, the corresponding retarded scalar potential to be added to (4) is

$$\Delta\Psi = \frac{I_0}{4\pi\omega\epsilon}\left[\frac{\sin\beta z_2' \sin(\omega t - \beta r_2)}{r_2} - \frac{\sin\beta z_1' \sin(\omega t - \beta r_1)}{r_1}\right] \quad (11)$$

The contributions to E_z and E_ρ are, respectively,

$$\Delta E_z = \frac{\omega\mu I_0}{4\pi\beta^2}\bigg\{-\sin\beta z_2'\frac{\partial}{\partial z}\left[\frac{\sin(\omega t - \beta r_2)}{r_2}\right]$$
$$+ \sin\beta z_1'\frac{\partial}{\partial z}\left[\frac{\sin(\omega t - \beta r_1)}{r_1}\right]\bigg\} \quad (12)$$

$$\Delta E_\rho = \frac{\omega\mu I_0}{4\pi\beta^2}\bigg\{-\sin\beta z_2'\frac{\partial}{\partial \rho}\left[\frac{\sin(\omega t - \beta r_2)}{r_2}\right]$$
$$+ \sin\beta z_1'\frac{\partial}{\partial \rho}\left[\frac{\sin(\omega t - \beta r_1)}{r_1}\right]\bigg\} \quad (13)$$

The magnetic field of (10) is unchanged. It is interesting to note that Bechmann's derivation of the linear oscillator field by use of the Hertz vector, which is quoted by Stratton, gives this load type at both ends.

If the antenna terminals are current nodes, then from (1), $\beta z_1'$ and $\beta z_2'$ must be $m_1\pi$ and $m_2\pi$, respectively, where m_1 and m_2 are positive or negative integers so that the cosine become $(-1)^{m_1}$ and $(-1)^{m_2}$. The fields given by (5), (9), and (10) then take the simplified form

$$E_z = \frac{\omega\mu I_0}{4\pi\beta}\left[\frac{(-1)^{m_1}\sin(\omega t - \beta r_1)}{r_1} - \frac{(-1)^{m_2}\sin(\omega t - \beta r_2)}{r_2}\right] \tag{14}$$

$$E_\rho = \frac{\omega\mu I_0}{4\pi\beta\rho}\left[\frac{(z_1' - z)\sin(\omega t - \beta r_1)}{(-1)^{m_1}r_1} - \frac{(z_2' - z)\sin(\omega t - \beta r_2)}{(-1)^{m_2}r_2}\right] \tag{15}$$

$$B_\phi = -\frac{\mu I_0}{4\pi\rho}[(-1)^{m_1}\sin(\omega t - \beta r_1) - (-1)^{m_2}\sin(\omega t - \beta r_2)] \tag{16}$$

Here $m_2 - m_1 = n$ is the number of current loops in the antenna.

Suppose an antenna of length $2l$ has end a at $z' = -l$, end b at $z' = l$, and is driven from the center at an arbitrary frequency so that there may or may not be a loop there, but there are nodes at the ends. Find the fields for each half separately by (5), (9), and (10) and superimpose the results. The substitutions for ends a and b are, respectively,

$$z_1' = 0,\quad z_2' = l,\quad r_1 = r_a,\quad r_2 = r,\quad z - z_1' = r_1\cos\theta_a,\quad z - z_2' = r_2\cos\theta$$
$$z_1' = 0,\quad z_2' = -l,\quad r_1 = r_b,\quad r_2 = r,\quad z - z_1' = r_1\cos\theta_b,\quad z - z_2' = r_2\cos\theta$$

The angles θ, θ_a, and θ_b are measured from the positive z-axis. Substitution in (5), (9), and (10) gives, writing $\omega = c\beta$,

$$E_z = \frac{c\mu I_0}{4\pi}\left[\frac{\sin(\omega t - \beta r_a)}{r_a} + \frac{\sin(\omega t - \beta r_b)}{r_b} - 2\frac{\cos\beta l\sin(\omega t - \beta r)}{r}\right] \tag{17}$$

$$E_\rho = \frac{c\mu I_0}{4\pi\rho}[2\cos\theta\cos\beta l\sin(\omega t - \beta r) - \cos\theta_a\sin(\omega t - \beta r_a) - \cos\theta_b\sin(\omega t - \beta r_b)] \tag{18}$$

$$B_\phi = \frac{\mu I_0}{4\pi\rho}[2\cos\beta l\sin(\omega t - \beta r) - \sin(\omega t - \beta r_a) - \sin(\omega t - \beta r_b)] \tag{19}$$

Note that I_0 is the maximum current amplitude and, if it lies on the antenna, may occur at any point except at the ends.

12.04. Distant Radiation from Linear Antenna.—We shall now obtain the radiation pattern at a great distance from a linear antenna of length $2l$ center driven at an arbitrary frequency with nodes at the ends. In 12.03 (19), write $r + l\cos\theta$ for r_a, $r - l\cos\theta$ for r_b, and $r\sin\theta$ for ρ, whence

$$B_\phi = (\mu\epsilon)^{\frac{1}{2}}E_\theta = \frac{\mu I_0[\cos\beta l - \cos(\beta l\cos\theta)]\sin(\omega t - \beta r)}{2\pi r\sin\theta} \tag{1}$$

From 11.17 (3), the mean radiation intensity in the r-direction is

$$\bar{\Pi} = \tfrac{1}{2}|E_\theta||H_\phi| = \tfrac{1}{2}\mu^{-\frac{3}{2}}\epsilon^{-\frac{1}{2}}|B_\phi|^2 \tag{2}$$

If the antenna is driven at resonance with a loop at the center so that $2\beta l = n\pi$, it is clear from these equations that, when $\bar{\Pi}$ is plotted as the radius vector in polar coordinates as a function of θ, the resultant curve has maxima and minima corresponding in number to the standing wave loops and nodes. The dashed curves in Fig. 12.04 show such a plot for $n = 3$. The total power radiated is found by the integration of $\bar{\Pi}$ over a large sphere centered at the origin. We shall calculate only the

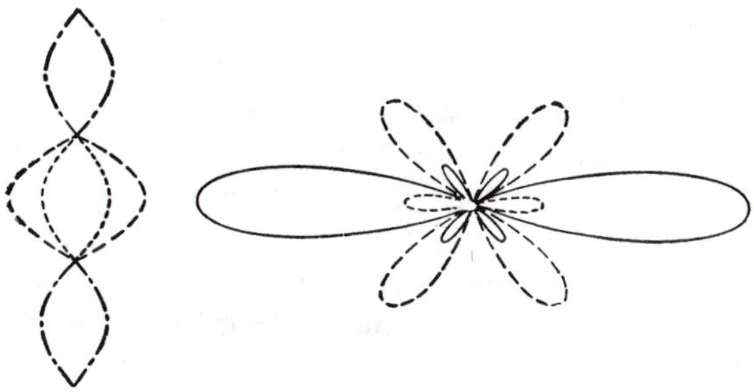

Fig. 12.04.

resonance case. When we substitute $\frac{1}{2}n\pi$ for βl and write the square of the numerator in (1) in terms of twice the angle by *Pc* 577 or *Dw* 404.22, we get

$$\bar{P} = \frac{\mu c I_0^2}{8\pi}\int_0^\pi \frac{1 - (-1)^n \cos(n\pi\cos\theta)}{1 - \cos^2\theta}\sin\theta\,d\theta \tag{3}$$

Substitution of u for $\cos\theta$ and resolution into two integrals by partial fractions show that the integrals are equal because one changes into the other when $-u$ replaces u as the variable of integration. Thus,

$$\int_{-1}^{+1}\frac{1 \pm \cos n\pi u}{2(1-u)}\,du + \int_{-1}^{+1}\frac{1 \pm \cos n\pi u}{2(1+u)}\,du = \int_{-1}^{+1}\frac{1 \pm \cos n\pi u}{1+u}\,du$$

Let $v = n\pi(1 + u)$ and (3) becomes

$$\bar{P} = \frac{\mu c I_0^2}{8\pi}\int_0^{2n\pi}\frac{1 - \cos v}{v}\,dv = \frac{\mu c I_0^2}{8\pi}[C + \ln(2n\pi) - \mathrm{Ci}(2n\pi)] \tag{4}$$

where $\text{Ci}(2n\pi)$ is the cosine integral tabulated in Jahnke and Emde and C is 0.5772. Insertion of numerical values gives for the radiation resistance

$$R_r = \frac{\bar{P}}{I_e^2} = \frac{2\bar{P}}{I_0^2} = 72.4 + 30 \ln n - 30 \text{ Ci}(6.28n) \tag{5}$$

When $n = 1$, this becomes 73.13 ohms. The radiation resistance when there is not a connected loop at the driving point is given in 12.07.

The assumption of an infinitely thin antenna is only one of the defects in the foregoing treatment. To maintain the postulated standing wave, one must supply to each element the energy lost by ohmic heating and radiation. The former is greatest at the current maximum near the driving point, and the latter is greatest near the ends of the antenna as can be shown with little difficulty from 12.03 (14) and (16). Thus it is evident not only that it is impossible to maintain the assumed current from any single driving point, but also that, to supply the losses, a damped progressive wave must be present which has no nodes and radiates in quite a different fashion from the standing wave. A rough qualitative idea of the effect on the radiation pattern of the weakening of the current near the ends of the antenna is obtained, for example, by superimposing the fields of two antennas, one oscillating in the $n = 1$ mode and one in the $n = 3$ mode. Assuming equal strengths and frequencies, this gives the radiation from a three-loop oscillator when the inner loop has twice the amplitude of the outer ones. Using (1) with $\beta l = \frac{1}{2}n\pi$ and simplifying by *Dw* 403.23 give

$$B_\phi = (\mu\epsilon)^{\frac{1}{2}}E_\theta = \frac{\mu I_0 \cos\left(\frac{1}{2}\pi \cos\theta\right)\cos\left(\pi\cos\theta\right)}{\pi r \sin\theta}\sin(\omega t - \beta r) \tag{6}$$

When substituted in (2), this gives the radiation pattern shown by the solid lines in Fig. 12.04. We note that, although the outer loops have the same amplitude as before, the outer lobes are actually cut down in magnitude, but the central lobe is quadrupled in magnitude.

12.05. Radiation from Progressive Waves.—Consider a wave of uniform amplitude that travels along a wire from $z' = -\frac{1}{2}l$ to $z' = +\frac{1}{2}l$ where it is absorbed without reflection, giving a current $I_0 \cos(\omega t - \beta z')$. From 12.02 (1), the field at x, y, z is, when $r = (\rho^2 + z^2)^{\frac{1}{2}}$ is large,

$$B_\phi = (\mu\epsilon)^{\frac{1}{2}}E_\theta = -\frac{\partial A_z}{\partial\rho} = -\frac{\mu I_0}{4\pi}\int_{-\frac{1}{2}l}^{\frac{1}{2}l}\frac{\partial}{\partial\rho}\left\{\frac{\cos[\omega t - \beta(r' + z')]}{r'}\right\}dz'$$

where $r' = [(z - z')^2 + \rho^2]^{\frac{1}{2}} \approx r - z'\cos\theta$ and $\rho = r\sin\theta$. If $r' \gg l$, we may neglect terms in r^{-2} so that, using *Dw* 401.11 or *Pc* 597,

$$B_\phi = \frac{\mu I_0}{2\pi r}\frac{\sin\theta\sin(\omega t - \beta r)\sin\left[\frac{1}{2}\beta l(1 - \cos\theta)\right]}{1 - \cos\theta} \tag{1}$$

The mean power radiated then becomes, by integration of 12.04 (2),

$$\bar{P} = \frac{\mu^{\frac{1}{2}} I_0^2}{4\pi\epsilon^{\frac{1}{2}}} \int_0^\pi \frac{\sin^2 \theta \sin^2 [\frac{1}{2}\beta l(1 - \cos \theta)]}{(1 - \cos \theta)^2} \sin \theta \, d\theta \tag{2}$$

Substitution of u for $\beta l(1 - \cos \theta)$ gives, using *Dw 403.4 or Pc* 576,

$$\bar{P} = \frac{\mu^{\frac{1}{2}} I_0^2}{4\pi\epsilon^{\frac{1}{2}}} \left[\int_0^{2\beta l} \frac{1 - \cos u}{u} \, du - \frac{1}{2\beta l} \int_0^{2\beta l} (1 - \cos u) \, du \right]$$

The first is given on page 3 of Jahnke and Emde in terms of Ci $2\beta l$ so that the substitution of numerical values and of $2\pi/\lambda$ for β give

$$\bar{P} = 30 I_0^2 \left[2.108 + \ln \frac{l}{\lambda} - \text{Ci} \frac{4\pi l}{\lambda} + \frac{\sin (4\pi l/\lambda)}{4\pi l/\lambda} \right] \tag{3}$$

12.06. Conical Transmission Lines.—The radiation fields of a linear antenna driven at the center are given by 12.03 (17) to (19). In 12.04 (5), we found the radiation resistance of such an antenna with a current loop at the driving point. No antenna reactance information is obtained by considering such infinitely thin wires because their inductance per unit length is infinite. If, to avoid this difficulty, we choose a radius not zero, then the infinite capacitance appearing across the infinitely narrow gap between input terminals again prevents reactance calculations. Schelkunoff eliminated both these difficulties by considering a biconical antenna composed of two cones whose apexes meet at the driving point. To simplify matters we should first discuss the biconical transmission line.

The propagation equation in spherical polar coordinates for a wave of angular frequency ω is, from 11.01, 3.05 (1), and 11.15 (3),

$$\frac{1}{\sin \theta} \frac{\partial}{\partial \theta}\left(\sin \theta \frac{\partial W}{\partial \theta}\right) + \frac{1}{\sin^2 \theta} \frac{\partial^2 W}{\partial \phi^2} + \frac{\partial}{\partial r}\left(r^2 \frac{\partial W}{\partial r}\right) - r^2 \check{\Gamma}^2 W = 0 \tag{1}$$

To break this into two equations, we equate the first and second pair of terms separately to zero and obtain 5.12 (6) and 12.01 (3) with $\check{\Gamma}^2$ in place of $-v^2/\omega^2$. Thus from 5.12 (7) and 12.01 (4) for an expanding wave,

$$W = r^{-1} V(\theta, \phi) e^{-\alpha r} \cos (\omega t - \beta r) \tag{2}$$

where α is the attenuation constant, β the wave number, and

$$U(\theta, \phi) + jV(\theta, \phi) = F(e^{j\phi} \tan \tfrac{1}{2}\theta) \tag{3}$$

The vector potential for a transverse electromagnetic wave is

$$\boldsymbol{A} = \boldsymbol{\nabla} \times \boldsymbol{r} W = -\boldsymbol{r}_1 \times \boldsymbol{\nabla}_2 V e^{-\alpha r} \cos (\omega t - \beta r) = \boldsymbol{\nabla}_2 U e^{-\alpha r} \cos (\omega t - \beta r) \tag{4}$$

where $\boldsymbol{\nabla}_2$ has θ- and ϕ-components only. This is identical in form with 11.14 (5) and shows that, when an alternating potential is applied between the apexes of two or more perfectly conducting cones whose surfaces are generated by radius vectors, a spherical wave is transmitted that is just equivalent to a plane wave on a cylindrical transmission line for which

$$x = \cos\phi \tan \tfrac{1}{2}\theta, \qquad y = \sin\phi \tan\tfrac{1}{2}\theta, \qquad z = r \tag{5}$$

Let us now consider the special case of two circular cones of half angle χ_1 and χ_2 whose axes intersect at an angle 2ψ, as shown in Fig. 12.06. Clearly, if the trace of these cones on a sphere of radius r is projected

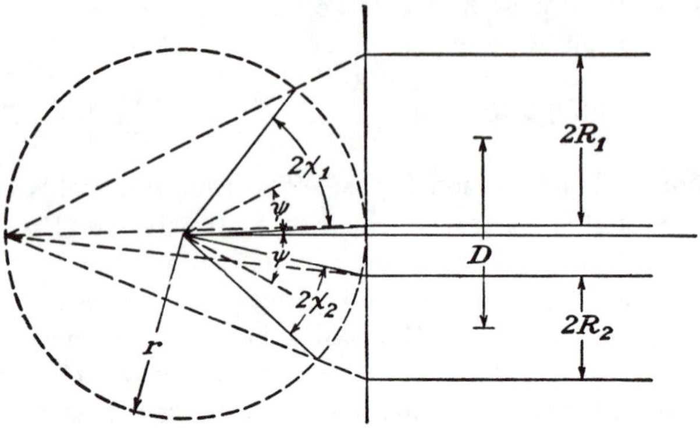

FIG. 12.06.

stereographically on the tangent plane, as in 6.20, the equivalent transmission line is the two cylinders of radii R_1 and R_2 at an axial distance D whose sections are shown in Fig. 4.13b. From Fig. 12.06,

$$\tfrac{1}{2}D + R_{1,2} = 2r\tan\tfrac{1}{2}(\psi + \chi_{1,2}), \qquad \tfrac{1}{2}D - R_{1,2} = 2r \tan\tfrac{1}{2}(\psi - \chi_{1,2})$$

Take the ratio of these equations, apply Pc 602, and solve for $\tfrac{1}{2}D/R_{1,2}$.

$$\frac{D}{2R_{1,2}} = \frac{\sin\psi}{\sin\chi_{1,2}} \tag{6}$$

The capacitance per unit length of the equivalent cylindrical line is, from 4.14 (2) or 4.13 (5), according to whether $\chi = \chi_1 = \chi_2$ or $\chi_1 \neq \chi_2$,

$$\frac{\epsilon}{C} = \frac{1}{\pi}\cosh^{-1}\frac{\sin\psi}{\sin\chi} \quad \text{or} \quad \frac{1}{2\pi}\cosh^{-1}\left(\pm\frac{4\sin^2\psi - \sin^2\chi_1 - \sin^2\chi_2}{2\sin\chi_1\sin\chi_2}\right) \tag{7}$$

The negative sign applies if one cone encloses the other, otherwise the positive sign is used. By 11.15 (10), the characteristic impedance is

$$\check{Z}_k = \frac{\epsilon}{C}\left(\frac{j\omega\mu}{\gamma + j\omega\epsilon}\right)^{\frac{1}{2}} = \frac{j\omega\mu\epsilon}{(\alpha + j\beta)C} \tag{8}$$

where μ is the permeability, ϵ the capacitivity, and γ the conductivity of the medium surrounding the perfectly conducting cones.

When $\psi = \frac{1}{2}\pi$, simpler formulas are given by taking $\theta = 0$ as the axis of one of the cones so that U is independent of ϕ. For this we choose $F(u)$ to be $C \ln u$ in (3) so that from 5.212 (4)

$$U = V_0 \frac{\ln \tan \frac{1}{2}\theta}{\ln \tan \frac{1}{2}\chi} = V_0 \frac{Q_0(\cos \theta)}{Q_0(\cos \chi)} \tag{9}$$

where the potential between the cones and the capacitance per unit length are, respectively,

$$V = 2V_0 e^{-\alpha r} \cos (\omega t - \beta r) \qquad C = \pi\epsilon(\ln \cot \tfrac{1}{2}\chi)^{-1} \tag{10}$$

12.07. The Biconical Antenna.—When both biconical transmission line cones terminate, they form a biconical antenna. If open at the ends or if closed by spherical caps, they fit into the spherical polar coordinate system to be discussed in 12.11, and the field calculation is a boundary value problem in this system. The method works best for cones whose half angles are nearly zero or 90°. In the small angle case, numerical work is simplified by using a different method.

Take perfectly conducting coaxial cones of equal length whose half angle χ is so small that by 12.02 (4) to (7) the current distribution is

$$i = I_0 \cos \omega t \sin (\beta l - \beta r) \tag{1}$$

The fields are then given by 12.03 (17) to (19) and the radiated power by an integration of Poynting's vector over the conical surfaces. Let χ_a, χ, and χ_b be the angles subtended at $z = -l$, $z = 0$, and $z = l$, respectively, by a radius ρ of the cone ending on its surface at a distance r from its apex. By Poynting's vector 11.02 (3) and the magnetomotance law 7.01 (2), the total instantaneous power radiation from both cones is

$$P = 2\int_0^l E_r H_\phi 2\pi\rho \, dr = 2\int_0^l i(E_z \cos \chi + E_\rho \sin \chi) \, dr \tag{2}$$

Inserting 12.03 (17) for E_z, 12.03 (18) for E_ρ, and (1) for i gives

$$P = \frac{c\mu I^2 \cos \omega t}{2\pi} \int_0^l \left[\left(\frac{\cos \chi}{r_a} - \frac{\cos \chi_a}{r} \right) S_a + \left(\frac{\cos \chi}{r_b} + \frac{\cos \chi_b}{r} \right) S_b \right] \sin (\beta l - \beta r) \, dr$$

where S_a and S_b have been written for $\sin (\omega t - \beta r_a)$ and $\sin (\omega t - \beta r_b)$, respectively. The terms containing $\cos \beta l$ cancel out because $\rho = r \sin \chi$. The half angle χ is chosen so small that $l + r$ and $l - r$ replace r_a and r_b and $\cos \chi_a$, $\cos \chi$, and $\cos \chi_b$ equal one. If S_a and S_b are split up so that $\sin \omega t$ and $\cos \omega t$ factor out, then with the aid of *Dw* 401.05 to 401.07

or *Pc* 595 to 597, we obtain for $2\pi P(c\mu I^2)^{-1}$

$$\frac{\sin 2\omega t}{4}\int_{\delta'}^{l-\delta}\left[\left(\frac{1}{l+r}-\frac{1}{r}\right)(\sin L-\sin R)+\left(\frac{1}{l-r}+\frac{1}{r}\right)\sin(L-R)\right]dr$$
$$-\frac{\cos^2\omega t}{2}\int_{\delta'}^{l-\delta}\left\{-\frac{l}{r(l+r)}(\cos R-\cos L)+\frac{l}{r(l-r)}[1-\cos(L-R)]\right\}dr$$

where R is written for $2\beta r$ and L for $2\beta l$. The quantities δ and δ' are to be set equal to zero except where so doing makes a term infinite. In Jahnke and Emde we find the cosine- and sine-integral formulas

$$\int_0^a \frac{\sin Ax}{x}\,dx = \mathrm{Si}(Aa), \qquad \int_0^a \frac{1-\cos Ax}{x}\,dx = C + \ln(Aa) - \mathrm{Ci}(Aa)$$

where $C = 0.5772$. With these formulas, the power expression integrates easily and $\ln\delta$ and $\ln\delta'$ cancel out so that all terms are finite. Thus

$$P = \frac{c\mu I_0^2\sin 2\omega t}{8\pi}\{2\mathrm{Si}(2\beta l) - \mathrm{Si}(4\beta l)\cos(2\beta l) + [\mathrm{Ci}(4\beta l) - C - \ln(\beta l)]\sin(2\beta l)\}$$
$$-\frac{c\mu I_0^2\cos^2\omega t}{4\pi}\{2C + 2\ln(2\beta l) - 2\mathrm{Ci}(2\beta l) + [C + \mathrm{Ci}(4\beta l) - 2\mathrm{Ci}(2\beta l) + \ln(\beta l)]\cos(2\beta l) + [\mathrm{Si}(4\beta l) - 2\mathrm{Si}(2\beta l)]\sin(2\beta l)\} \quad (3)$$

Comparison with 11.15 (18), where $\mathcal{E}^2$ is replaced by $I_0^2\check{Z}^2$, shows that the coefficient of $I_0^2\sin 2\omega t$ is $2\check{Z}_r\sin\psi$ or $2X_r$, where X_r is the radiation reactance and that of $I_0^2\cos^2\omega t$ is $\check{Z}_r\cos\psi$ or R_r the radiation resistance. If there is a current loop at the driving point so that $2\beta l = n\pi$ where n is odd, the value of R_r is identical with that of 12.04 (5). With a node at the driving point, n is even so the result is different because in (1) and (2) we assumed the current similarly directed in the two halves, whereas in 12.04 (4) they are taken in opposite directions.

Although we have now established the power loss from the antenna, we still do not know where to put it in the line to give the correct input impedance. Examination of (1) shows that H_ϕ and hence the power loss are zero at the current nodes so it cannot be put at the ends. A simple way to discover its location is to consider the case in which the perfectly conducting antenna is tuned so that the only power loss is by radiation. There is then a current loop at the driving point so that $\beta l = \frac{1}{2}(2n+1)\pi$ and the input current amplitude is I_0. Insertion of this value in 11.15 (13) after setting the attenuation constant equal to zero gives for the input impedance and the mean power expended

$$\check{Z}_i = \check{Z}_t^{-1}\check{Z}_k^2 \qquad \bar{P} = \tfrac{1}{2}\check{Z}_i I_0^2 \quad \text{(real part)} \quad (4)$$

where $\check{Z}_t$ is the terminal impedance. But (3) gives $\frac{1}{2}\check{Z}_r I_0^2$ for the mean power so for a tuned antenna $\check{Z}_r\check{Z}_t = \check{Z}_k^2$. We try this value in 11.15 (13) for the input impedance of an untuned antenna and, to justify it, show that for thin antennas the resistive power loss agrees with (3). Thus

$$\check{Z}_i = \check{Z}_k \frac{\check{Z}_k \cos \beta l + j\check{Z}_r \sin \beta l}{\check{Z}_r \cos \beta l + \check{Z}_k \sin \beta l} \tag{5}$$

Now let $\check{Z}_k$ become very large, which means a very thin cone, so that the input current becomes $I_0 \sin \beta l$ and the power input is, from (4),

$$\bar{P}_i = \tfrac{1}{2}\check{Z}_i I_0^2 \sin^2 \beta l = \tfrac{1}{2}(\check{Z}_r - \tfrac{1}{2}j\check{Z}_k \sin 2\beta l)I_0^2 \tag{6}$$

This gives a resistive power loss of $\frac{1}{2}R_r I_0^2$ as it should. We note that (5) is the same expression that would have been obtained from 11.15 (13)

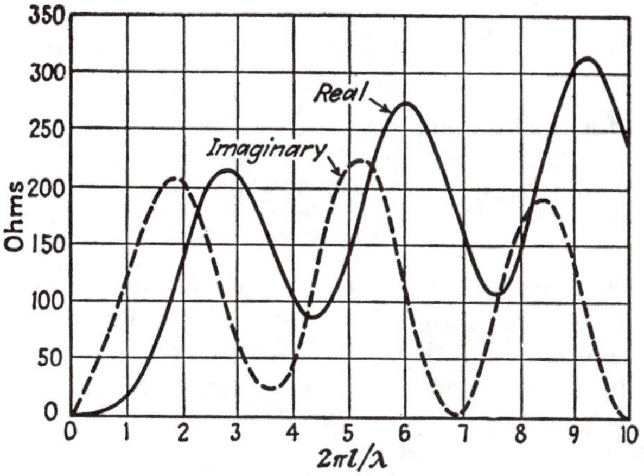

FIG. 12.07.—Real (solid) and imaginary (dotted) parts of $\check{Z}_k^2/\check{Z}_r$ plotted against $2\pi l/\lambda$.

if we had written $\beta l - \frac{1}{2}\pi$ for βl and $\check{Z}_r$ for $\check{Z}_L$. This is equivalent to cutting $\frac{1}{4}\lambda$ off the line and using $\check{Z}_r$ as the terminal impedance. Thus the effective position of the radiation impedance is one-quarter wave length from the end of the line. The inverse of the radiation impedance given by Schelkunoff in *Proc. I.R.E.*, September, 1941, pages 493–521 appears in Fig. 12.07. Clearly from (5) the resonance wave length at which $\check{Z}_i$ is real is affected by the radiation reactance, but it is still near $2\beta l = n\pi$ if $\check{Z}_k \gg X_r$.

12.08. Antenna Arrays.—The radiation pattern of a linear antenna is symmetrical about its axis. To concentrate the radiation in a single direction, a "directional array" is needed in which several radiators of length $2l$, usually identical in type and parallel to each other, are set in some regular pattern. These may differ in amplitude and phase of excitation. Let us consider radiators parallel to the z-axis and lying in the positive octant of a rectangular lattice structure, the spacing between adjacent ones being a, b, and c in the x-, y-, and z-direction, respectively. With integral u, v, and w the distance of radiator uvw from the origin is

$$\mathbf{r}_{uvw} = \mathbf{i}ua + \mathbf{j}vb + \mathbf{k}wc \tag{1}$$

Let the radius vector from the origin O to the very distant field point P be $r\mathbf{r}_1$ so that the difference in travel distance of a signal from O to P and from radiator uvw to P is $\mathbf{r}_1 \cdot \mathbf{r}_{uvw}$. Assume the dimensions of the array small compared with r so that all radiators can be considered as equidistant from P for field amplitude calculations. If oscillator uvw lags behind oscillator 000 by a phase angle ψ_{uvw} and has a current I_{uvw}, then from 12.04 (1) its contribution to B_ϕ at P is the real part of

$$\frac{\mu I_{uvw}[\cos \beta l - \cos(\beta l \cos\theta)]}{2\pi r \sin\theta} e^{j[\omega t - \psi_{uvw} - \beta(r - \mathbf{r}_1 \cdot \mathbf{r}_{uvw})]}$$

Writing F_{uvw} for the first factor gives for the whole array

$$\breve{B}_\phi = (\mu\epsilon)^{\frac{1}{2}}\breve{E}_\theta = e^{-j\beta r}\sum_u \sum_v \sum_w F_{uvw} e^{j(\beta \mathbf{r}_1 \cdot \mathbf{r}_{uvw} - \psi_{uvw})} \tag{2}$$

For identical radiators F is constant, and this formula simplifies if, when going in the positive x-, y-, or z-direction, each radiator shows a constant phase lag ξ, η, or ζ, respectively, behind the preceding one. The triple summations in (2) then breaks down into a triple product of summations.

$$F\sum_u \sum_v \sum_w e^{j[u(\beta \mathbf{r}_1 \cdot \mathbf{i}a - \xi) + v(\beta \mathbf{r}_1 \cdot \mathbf{j}b - \eta) + w(\beta \mathbf{r}_1 \cdot \mathbf{k}c - \zeta)]}$$
$$= F\sum_u e^{ju(\beta \mathbf{r}_1 \cdot \mathbf{i}a - \xi)} \sum_v e^{jv(\beta \mathbf{r}_1 \cdot \mathbf{j}b - \eta)} \sum_w e^{jw(\beta \mathbf{r}_1 \cdot \mathbf{k}c - \zeta)} \tag{3}$$

Each factor is a geometrical progression which may be summed by *Dw* 26 and gives, when m_x is the number of oscillators in the x-direction,

$$\sum_{n=0}^{m_x - 1} e^{jn\psi} = \frac{1 - e^{jm_x\psi}}{1 - e^{j\psi}} = \frac{\sin \frac{1}{2} m_x \psi}{\sin \frac{1}{2}\psi} e^{-j\frac{1}{2}(m_x - 1)\psi} \tag{4}$$

From 11.17 (4) the mean Poynting vector is $\frac{1}{2}(\mu^3\epsilon)^{-\frac{1}{2}}\breve{B}_\phi \cdot \hat{B}_\phi$ so that

$$\bar{\Pi} = \frac{F^2 \sin^2[\frac{1}{2}m_x(\beta \mathbf{r}_1 \cdot \mathbf{i}a - \xi)] \sin^2[\frac{1}{2}m_y(\beta \mathbf{r}_1 \cdot \mathbf{j}b - \eta)] \sin^2[\frac{1}{2}m_z(\beta \mathbf{r}_1 \cdot \mathbf{k}c - \zeta)]}{2\mu^{\frac{3}{2}}\epsilon^{\frac{1}{2}} \sin^2[\frac{1}{2}(\beta \mathbf{r}_1 \cdot \mathbf{i}a - \xi)] \sin^2[\frac{1}{2}(\beta \mathbf{r}_1 \cdot \mathbf{j}b - \eta)] \sin^2[\frac{1}{2}(\beta \mathbf{r}_1 \cdot \mathbf{k}c - \zeta)]} \tag{5}$$

As already stated, an array is used to concentrate radiation in special directions. The directivity or gain G is defined as the ratio of the maximum intensity Φ_M to the average intensity Φ_0 on a large sphere concentric with the array. In decibels we write it G_d. Thus

$$G = \frac{\Phi_M}{\Phi_0} \qquad G_d = 10 \log_{10} \frac{\Phi_M}{\Phi_0} \tag{6}$$

The gain function $G(\theta, \phi)$ in any direction is the ratio of $\Phi(\theta, \phi)$ to Φ_0. Thus for a half wave antenna, from 12.04 (2) and (4), the ratio of maximum to average intensity is $4[\ln (2\pi) + C - \mathrm{Ci}(2\pi)]^{-1}$ which gives a numerical value of 1.64 for G or a gain of 2.15 decibels in the equatorial plane. The transmitting pattern is the surface.

$$r = \frac{G(\theta, \phi)}{G_M} = \frac{\Phi(\theta, \phi)}{\Phi_M} \tag{7}$$

Now consider the special case of m identical, in phase, n-loop antennas, lying a half wave length apart along the x-axis. Take $m_y = m_z = 1$,

Fig. 12.08.

$\xi = 0$, $\beta a = \frac{1}{2}\beta\lambda = \pi$, and $\beta l = \frac{1}{2}n\pi$, so $\beta \mathbf{r}_1 \cdot \mathbf{i}a$ is $\pi \cos \phi \sin \theta$ in (5) which becomes

$$\bar{\Pi} = \frac{\mu^{\frac{1}{2}} I_0^2}{8\pi^2 r^2 \epsilon^{\frac{1}{2}}} \left[\frac{\cos (\frac{1}{2}n\pi \cos \theta)}{\sin \theta}\right]^2 \left[\frac{\sin (\frac{1}{2}m\pi \cos \phi \sin \theta)}{\sin (\frac{1}{2}\pi \cos \phi \sin \theta)}\right]^2 \tag{8}$$

This is called a broadside array because the second factor maximum is at right angles to the plane of the array when $\phi = \frac{1}{2}\pi$. Figure 12.08 shows the relative values of $\bar{\Pi}$ in the plane $\theta = 0$ when subsidiary maxima are omitted. Near this maximum the second factor sines are small so it becomes m^2. We cannot assume from this that, for the same total power input, the radiation in this direction is m times that of a single oscillator because the oscillators interact. To get the actual gain, we must calculate Φ_0. We note from (4) that the last factor in (8) may be written, if α is $\pi \cos \phi \sin \theta$,

$$\left|\frac{\sin \frac{1}{2}m\alpha}{\sin \frac{1}{2}\alpha} e^{\frac{1}{2}j(m-1)\alpha}\right|^2 = \left|\sum_{p=0}^{m-1} e^{jp\alpha}\right|^2 = m + 2\sum_{p=1}^{m-1}(m - p) \cos p\alpha$$

Expansion of the cosine by *Pc* 773 or *Dw* 415.02, integration over ϕ by *Pc* 483 or *Dw* 854.1, and combination of the terms independent of θ give

$$2m\pi + 2\sum_{p=1}^{m-1}(m-p)\int_0^{2\pi}\cos\,(p\pi\cos\phi\sin\theta)\,d\phi$$

$$= 2\pi\left[m^2 + \sum_{p=1}^{m-1}\sum_{s=1}^{\infty}\frac{(m-p)(jp\pi\sin\theta)^{2s}}{2^{2s-1}(s!)^2}\right]$$

To get the power radiated, multiply this by the remaining factors in (8) and by $r^2 \sin\theta\, d\theta$ and integrate from $\theta = 0$ to $\theta = \pi$. The first term was integrated in 14.05. The other integrals have the form

$$\int_0^{\pi}\cos^2\,(\tfrac{1}{2}n\pi\cos\theta)\sin^{2s-1}\theta\,d\theta = \tfrac{1}{2}\int_0^{\pi}[1+\cos\,(n\pi\cos\theta)]\sin^{2s-1}\theta\,d\theta$$

We may integrate this by *Dw* 854.1 or *Pc* 483 and 5.302 (5) which gives

$$\frac{[(s-1)!]^2 2^{2s-2}}{(2s-1)!} + \frac{(s-1)!}{2\pi^{s-1}}\left(\frac{2}{n}\right)^{s-\frac{1}{2}}J_{s-\frac{1}{2}}(n\pi)$$

The s-summation of the first term by *Dw* 442.11 or *Pc* 347 gives a result expressible as a cosine integral by page 3 of Jahnke and Emde. Thus

$$2\sum_{s=1}^{\infty}\frac{(-1)^s(p\pi)^{2s}}{2s(2s)!} = -2\int_0^{p\pi}\frac{1-\cos x}{x}\,dx = -2C - 2\ln\,(p\pi) + 2\mathrm{Ci}(p\pi)$$

Collecting terms gives for the total power radiated

$$\bar{P} = \frac{\mu I_0^2 c}{8\pi}\Bigg\{m^2C + m^2\ln\,(2n\pi) - m^2\mathrm{Ci}(2n\pi)$$
$$+ 4\sum_{p=1}^{m-1}(m-p)\left[\mathrm{Ci}(p\pi) - C - \ln\,(p\pi) + \sum_{s=1}^{\infty}\frac{(-1)^s\pi^{s+1}p^{2s}}{4ss!(2n)^{s-\frac{1}{2}}}J_{s-\frac{1}{2}}(n\pi)\right]\Bigg\} \qquad (9)$$

From (6) and (7) the directivity in decibels is 10 log $[\tfrac{1}{2}\mu cI^2m^2/(\pi\bar{P})]$. For $n = 1$, $m = 2$, the ratio Φ_M/Φ_0 is 3.81 which is more than twice that of a single half wave antenna. The directivity is 5.81 db.

12.09. Earth Effects.—Dielectrics or conductors near an antenna react on it or distort its field or both. The commonest such object is the earth's surface. To a nearly spherical wave one may apply the laws of reflection and refraction at a conducting boundary derived in 11.13. If, on the other hand, the antenna is so near the surface that the angle of incidence is a complicated function, this becomes laborious. Often the surface may be taken as flat and perfectly conducting so that the fields above it are unchanged if it is replaced by a second identical or "image" antenna located and oriented so as to make the resultant electric field normal to the former earth's surface. If all the original antennas were either normal to or parallel to the earth's surface, then evidently the resultant fields may be calculated by the formulas of the last article.

12.10. Uniqueness of Solution.—Before deriving the wave equation solutions used in boundary value problems, one should find what data are needed for a unique solution. Consider a region, bounded internally by the surfaces S_1 to S_n and externally by S_0, wherein μ, ϵ, and γ are functions of position but not of field strength, and which contains no sources. Let $\boldsymbol{E}$, $\boldsymbol{B}$ and $\boldsymbol{E}'$, $\boldsymbol{B}'$ be two solutions of Maxwell's equations which are identical throughout the region when $t = 0$. Poynting's theorem 11.02, and Ohm's law give, writing $\Delta\boldsymbol{E}$ for $\boldsymbol{E} - \boldsymbol{E}'$, $\Delta\boldsymbol{B}$ for $\boldsymbol{B} - \boldsymbol{B}'$, etc.,

$$\int_V \left\{ \frac{(\Delta i)^2}{\gamma} + \frac{\partial}{\partial t}\left[\frac{\epsilon(\Delta E)^2}{2} + \frac{(\Delta B)^2}{2\mu} \right] \right\} dv = -\sum_{i=0}^{n} \int_{S_i} \frac{\Delta \boldsymbol{E} \times \Delta \boldsymbol{B}}{\mu} \cdot \boldsymbol{n}\, dS_i$$

In order that the surface integral vanish, it is necessary that

$$\Delta\boldsymbol{E} \times \Delta\boldsymbol{B} \cdot \boldsymbol{n} = \Delta\boldsymbol{B} \times \boldsymbol{n} \cdot \Delta\boldsymbol{E} = \boldsymbol{n} \times \Delta\boldsymbol{E} \cdot \Delta\boldsymbol{B} = 0$$

Thus, if $\boldsymbol{n} \times \Delta\boldsymbol{B}$ or $\boldsymbol{n} \times \Delta\boldsymbol{E}$ is zero when $t > 0$, the surface integrals are zero. The energy term in brackets is zero or positive and was zero at $t = 0$ so, if it changes at all, it must become positive. But the first term is zero or positive, and the whole integrand is zero so that ΔE and ΔB are zero. Thus the $\boldsymbol{E}$, $\boldsymbol{B}$ and $\boldsymbol{E}'$, $\boldsymbol{B}'$ solutions are identical and are determined by the initial values of the fields in the region together with the tangential components of either $\boldsymbol{E}$ or $\boldsymbol{B}$ over its surface when $t > 0$. In practice we are usually concerned with steady-state solutions of the problem, in which case the values at any time determine all previous values.

12.11. Solutions of the Wave Equation in Spherical Coordinates.—In an isotropic insulating medium all waves are spherical at distances from the source great compared with its dimensions. In radiation problems, therefore, the most useful solutions of the propagation equation are in polar coordinates and have the form of a sum of products of orthogonal functions with coefficients which can be determined from given boundary conditions. We saw in 11.01 that in such cases the entire radiation field is derivable from a vector potential with zero divergence. Expressions for such a potential can be obtained from two solutions of the scalar propagation equation as indicated in 10.01. Thus from 10.01 (2), if W_{te} and W_{tm} are solutions of the last two of the equations,

$$\nabla^2 \boldsymbol{A} = -\mu\epsilon\omega^2 \boldsymbol{A}, \qquad \nabla^2 W_{te} = -\mu\epsilon\omega^2 W_{te}, \qquad \nabla^2 W_{tm} = -\mu\epsilon\omega^2 W_{tm} \quad (1)$$

then a solution of the first equation is obtained from the formula

$$A = \nabla \times (\mathbf{r}W_{te} + \mathbf{r} \times \nabla W_{tm}) \qquad (2)$$

Note that the solution derived from W_{te} gives a vector potential, and hence an electric field, normal to $\mathbf{r}$. This is called a transverse electric wave as the subscript suggests. From 10.01 (4), the magnetic induction is

$$B = -\nabla \times (\mu\epsilon\omega^2 \mathbf{r} W_{tm} + \mathbf{r} \times \nabla W_{te}) \qquad (3)$$

Thus the magnetic field derived from W_{tm} is normal to $\mathbf{r}$, and the wave is called transverse magnetic. We solve (1) as in 10.05 except that for more generality the factor $\Phi(\phi)$ is added. We therefore write

$$\check{W} = r^{-\frac{1}{2}}\check{R}(r)\Theta(\theta)\Phi(\phi) \qquad (4)$$

Substitution of (4) in (1) gives the differential equation of 10.05 (9) for $\check{R}$, that of 5.14 (2) for θ, and $d^2\Phi/d\phi^2 = -m^2\Phi$ for Φ. The solution for $\check{W}$ then becomes, writing β for $\omega(\mu\epsilon)^{\frac{1}{2}}$ and u for $\cos\theta$,

$$\check{W} = [AP_n^m(u) + BQ_n^m(u)][\check{C}j_n(\beta r) + \check{D}\check{k}_n(j\beta r)] \cos(m\phi + \delta_m) \qquad (5)$$

This spherical Bessel function notation is that of 5.31 and 5.37 where

$$j_n(\beta r) = \pi^{\frac{1}{2}}(2\beta r)^{-\frac{1}{2}}J_{n+\frac{1}{2}}(\beta r), \qquad \check{k}_n(j\beta r) = (\tfrac{1}{2}j\pi\beta r)^{-\frac{1}{2}}\check{K}_{n+\frac{1}{2}}(j\beta r) \qquad (6)$$

The reason for using these forms is that the first, combined with $e^{j\omega t}$, represents a standing wave and the second, from 5.37 (9), an expanding or contracting traveling wave according as v is positive or negative.

12.12. Polynomial Expansion for a Plane Wave.—A plane wave expansion in the form of 12.11 (5) is often needed to fit spherical boundary conditions. For simplicity consider a wave parallel to the $\theta = 0$ axis. It can then be referred to any other axis by 5.24. We saw in 11.15 (4) and (5) that $z = r\cos\theta$ enters plane wave formulas only in the exponent and multiplied by $\check{\Gamma}$. Thus we need the coefficient of $P_n(\cos\theta)$ in the Legendre polynomial series expansion of $e^{\check{\Gamma}r\cos\theta}$. From 5.156 (3),

$$\check{a}_n = \frac{2n+1}{2^{n+1}n!}(\check{\Gamma}r)^n \int_{-1}^{+1} e^{\check{\Gamma}ru}(1-u^2)^n\,du \qquad (1)$$

Expansion of the exponential by *Pc* 759 or *Dw* 550 gives

$$\check{a}_n = \frac{2n+1}{2^{n+1}n!}(\check{\Gamma}r)^n \sum_{s=0}^{\infty} \frac{(\check{\Gamma}r)^s}{s!} \int_{-1}^{+1} u^s(1-u^2)^n\,du \qquad (2)$$

The integral vanishes when s is odd, so write $2m$ for s and take twice the value of the integral from 0 to 1. For the $(2m)!$ in the denominator we may write $2^{2m}m!\pi^{-\frac{1}{2}}\Gamma(m+\frac{1}{2})$ by *Dw* 850.7 and 855.1 and obtain with the aid of 5.32 (2), after writing $\Gamma(n+1)$ for $n!$,

$$\check{a}_n = \frac{(2n+1)\pi^{\frac{1}{2}}}{2^{n+1}\Gamma(n+1)} \sum_{m=0}^{\infty} \frac{(\check{\Gamma}r)^{n+2m}}{2^{2m}m!\Gamma(m+\frac{1}{2})} \frac{\Gamma(m+\frac{1}{2})\Gamma(n+1)}{\Gamma(m+n+\frac{3}{2})} \tag{3}$$

$$= (2n+1)\left(\frac{\pi}{2\check{\Gamma}r}\right)^{\frac{1}{2}} \sum_{m=0}^{\infty} \frac{(\frac{1}{2}\check{\Gamma}r)^{n+2m+\frac{1}{2}}}{m!\Gamma(m+n+\frac{3}{2})} = (2n+1)\left(\frac{\pi}{2\check{\Gamma}r}\right)^{\frac{1}{2}} I_{n+\frac{1}{2}}(\check{\Gamma}r)$$

When there is no attenuation, we replace $\check{\Gamma}$ by $j\beta$ and obtain by 5.32 (2)

$$\check{a}_n = (2n+1)j^n\left(\frac{\pi}{2\beta r}\right)^{\frac{1}{2}} J_{n+\frac{1}{2}}(\beta r) = j^n(2n+1)j_n(\beta r) \tag{4}$$

The expansion is, therefore,

$$e^{j\beta r} = e^{j\beta z \cos\theta} = \sum_{n=0}^{\infty} j^n(2n+1)j_n(\beta r)P_n(\cos\theta) \tag{5}$$

For $e^{-j\beta z}$ replace β by $-\beta$ or $\cos\theta$ by $\cos(\pi-\theta)$ which is equivalent, by 5.293 (3) or 5.157, to inserting the factor $(-1)^n$ on the right. Thus

$$e^{-j\beta z} = e^{-j\beta r\cos\theta} = \sum_{n=0}^{\infty} (-j)^n(2n+1)j_n(\beta r)P_n(\cos\theta) \tag{6}$$

12.13. Radiation from Uniform Current Loop. Magnetic Dipole.—The orthogonal solutions of 12.11 usually occur in boundary value problems but may also be preferable for sources with given current distributions. For example, the radiation from a uniform current loop of radius a found, when $\lambda \gg a$ and $r \gg a$, by retarded potentials and given in 12.01 (19), (20), and (21) is more difficult by that method if $r < a$. From symmetry this source gives a transverse electric field with no free charge and hence, from 12.02 (2), no scalar potential and a vector potential with only a ϕ-component. Therefore when $r > a$ we have, from 12.11 (5),

$$\check{W}_{te} = \sum_{n=0}^{\infty} \check{A}_n P_n(\cos\theta) j_n(\beta a)\check{k}_n(j\beta r) \tag{1}$$

If $r < a$, interchange r and a. By 12.11 (2), the vector potentials are

$$r > a \qquad \check{A}_\phi = \sum_{n=0}^{\infty} \check{A}_n P_n^1(\cos\theta) j_n(\beta a)\check{k}_n(j\beta r) \tag{2}$$

$$r < a \qquad \check{A}_\phi = \sum_{n=0}^{\infty} \check{A}_n P_n^1(\cos\theta) j_n(\beta r)\check{k}_n(j\beta a) \tag{3}$$

Clearly these potentials are equal at $r = a$, and only odd values of n can appear in them for the field must be symmetrical about the $\theta = \frac{1}{2}\pi$ plane, so we replace n by $2m + 1$. The uniform current implies from 12.02 (7) that βa is very small so that we may write $(\beta r)^{2m+1}/(4m + 3)!!$ for $j_n(\beta r)$ where $r \lessgtr a$ and $(4m + 1)!!(j\beta a)^{-2m-2}$ for $\check{k}_n(j\beta a)$ from 5.31 (9) and 5.37 (11). Except for the time factor, (3) is real and identical in form with 7.13 (2) if $\alpha = \frac{1}{2}\pi$ so that the latter, multiplied by $\cos \omega t$, gives $\check{A}_\phi$ when $r \lessgtr a$. Write $(-1)^m(2m + 1)!!$ for $P^1_{2m+1}(0)$ by 5.23 (6) and 5.157 and substitute the A_m value found from (3) in (2) which gives

$$r > a, \quad \check{A}_\phi = -\frac{1}{2}\mu I \sum_{m=0}^{\infty} \frac{(2m - 1)!!(\beta a)^{2m+2}}{(2m + 2)!!(4m + 1)!!} P^1_{2m+1}(\cos \theta)\check{k}_{2m+1}(j\beta r) \quad (4)$$

To get the real part of $\check{A}_\phi e^{j\omega t}$, we may substitute, from 5.37 (13),

$$[e^{j\omega t}k_{2m+1}(j\beta r)]_{\text{real part}} = (-1)^m[n_{2m+1}(\beta r) \cos \omega t - j_{2m+1}(\beta r) \sin \omega t]$$

When r is large, all but the $m = 0$ term may be neglected because βa is small so that n_1 and j_1 are given by 5.31 (3), (4), (8), and (10). The expression for A_ϕ then becomes

$$A_\phi = \tfrac{1}{4}\mu I a^2 r^{-2} \sin \theta[\cos (\omega t - \beta r) - \beta r \sin (\omega t - \beta r)] \quad (5)$$

The electric field $-\partial A/\partial t$ agrees with 12.01 (21) provided M replaces $\pi a^2 I$ and $\nabla \times A$ gives 12.01 (19) and 12.01 (20). From 12.01 (27) the power radiated and the radiation resistance are

$$\bar{P} = 153{,}900 a^4 I^2 \lambda^{-4} \qquad R_r = 2\bar{P}I^{-2} = 307{,}800 a^4 \lambda^{-4} \quad (6)$$

Thus, in terms of antenna current, the loop radiates power inversely as the fourth power of the wave length instead of the linear antenna's square. Note from (3) that when $r < a$ there are only standing waves.

12.14. Free Oscillations of a Conducting Sphere.—The solutions of the propagation equations given in 12.11 are evidently well adapted to the study of the electromagnetic oscillations of spherical bodies. From 12.11 (2), we see that both transverse electric and transverse magnetic oscillations are possible. Examination of 12.11 (5) discloses that outside the sphere there is a standing wave if $\check{D}$ is zero, from 5.31 (7) and a diverging wave if $\check{C}$ is zero from 5.37 (12). The Q_n^m solution is excluded because it is infinite at $\theta = 0$ and $\theta = \pi$. A dielectric or imperfectly conducting sphere will have fields inside it. We then use a sum of solutions of the type of 12.11 (5) with undetermined coefficients which are different for the interior and exterior regions. The vector potentials derived from these solutions by 12.11 (2) must be matched at the boundary to satisfy the boundary conditions 7.21 (6) and (7) as was done in 10.06.

A simple and interesting case is that of a perfectly conducting sphere where, if there are initially no internal fields, there can be none at any subsequent time and energy can be lost by radiation only. The unknown frequency appearing in 12.11 (5) must be determined to fit the boundary conditions which usually make it complex. Let us consider first the transverse magnetic oscillations such as would be set up by moving the sphere in a static electric field. The simplest of these will be that for which $m = 0$ and $n = 1$ in 12.11 (5) so that by 12.11 (2)

$$\begin{aligned} \check{A} &= \nabla \times (\boldsymbol{r} \times \nabla \check{W}_{tm}) = \frac{\boldsymbol{r}_1}{r \sin\theta} \frac{\partial}{\partial\theta}\left(\sin\theta \frac{\partial \check{W}_{tm}}{\partial\theta}\right) - \frac{\boldsymbol{\theta}}{r} \frac{\partial^2(r\check{W}_{tm})}{\partial r\, \partial\theta} \\ &= \sum_p \check{A}_p\left[-\frac{2\boldsymbol{r}_1 \cos\theta}{r} \check{k}_1(j\beta_p r) + \frac{\boldsymbol{\theta} \sin\theta}{r} \frac{\partial}{\partial r}(r\check{k}_1(j\beta_p r))\right] \end{aligned} \tag{1}$$

The condition that the tangential component $-\partial\check{A}_\theta/\partial t$ of the electric field be zero at $r = a$ requires, from 5.37 (12), that

$$\left(-1 - \frac{1}{j\beta a} + \frac{1}{\beta^2 a^2}\right)e^{-j\beta a} = \left(\frac{1}{\beta a} + \frac{j + 3^{\frac{1}{2}}}{2}\right)\left(\frac{1}{\beta a} + \frac{j - 3^{\frac{1}{2}}}{2}\right)e^{-j\beta a} = 0 \tag{2}$$

Putting these two values of $\beta_p = \omega_p(\mu\epsilon)^{\frac{1}{2}}$ into 12.11 (5) yields for $\check{W}_{tm}e^{j\omega t}$

$$\sin\theta\left\{A_1\check{k}_1\left[\frac{-2jr}{(j + 3^{\frac{1}{2}})a}\right]e^{\frac{-2jt(\mu\epsilon)^{-\frac{1}{2}}}{(j+3^{\frac{1}{2}})a}} + A_2\check{k}_1\left[\frac{-2jr}{(j - 3^{\frac{1}{2}})a}\right]e^{\frac{-2jt(\mu\epsilon)^{-\frac{1}{2}}}{(j-3^{\frac{1}{2}})a}}\right\} \tag{3}$$

With the aid of 5.37 (10) and (12) and of the substitutions

$$\omega = \tfrac{1}{2}3^{\frac{1}{2}}(\mu\epsilon)^{-\frac{1}{2}}a^{-1}, \qquad \alpha = \tfrac{1}{2}(\mu\epsilon)^{-\frac{1}{2}}a^{-1}, \qquad \beta = \tfrac{1}{2}3^{\frac{1}{2}}a^{-1} \tag{4}$$

the real part of $\check{W}_{tm}e^{j\omega t}$ may be written, after collecting constants,

$$\begin{aligned} Cr^{-2}a \sin\theta[3^{\frac{1}{2}}(a - r) \sin(\omega t - \beta r - \psi) \\ - (a + r)\cos(\omega t - \beta r - \psi)]e^{-\alpha t + \frac{1}{2}a^{-1}r} \end{aligned} \tag{5}$$

From (4) and 11.09 (4), we see that the wave length of the oscillation is $7.26a$ and that the amplitude is reduced by the factor $1/e$ while the wave travels a distance equal to the diameter of the sphere. The oscillation is therefore very rapidly damped and disappears in a few cycles. The transverse electric modes can be calculated in similar fashion.

An analysis similar to that given, but more complicated, applies to a prolate spheroid. When the eccentricity of such a spheroid is large, it gives an excellent approximation to a straight wire of finite length.

12.15. Forced Oscillations of Dielectric or Conducting Sphere.—The formulas of 12.11 and 12.12 permit a rigorous calculation of the steady-state diffraction effects caused by a plane monochromatic electromagnetic wave with its electric vector in the x-direction and traveling in the

z-direction, impinging on a homogeneous sphere. We shall consider only the cases where the sphere is a perfect dielectric or a perfect conductor. From 11.15 (4), 11.15 (5), and 11.01 (2), the phasor electric field, magnectic induction, and vector potential of such a wave are

$$\check{E}_x = (\mu\epsilon)^{-\frac{1}{2}}\check{B}_y = Ee^{-j\beta z}, \qquad \check{A}_x = j\omega^{-1}Ee^{-j\beta z} \tag{1}$$

Both $\check{E}_x$ and $\check{B}_y$ have r-components, so in polar coordinates both TE and TM waves are needed. From 12.11 (2) and (3) we have

$$\begin{aligned}\check{A}_r &= \check{A}_x \sin\theta\cos\phi = [\nabla \times (\mathbf{r} \times \nabla\check{W}'_{tm})]_r,\\ \check{B}_r &= \check{B}_y \sin\theta\sin\phi = -[\nabla \times (\mathbf{r} \times \nabla\check{W}'_{te})]_r\end{aligned} \tag{2}$$

In 12.12 (6), $e^{-j\beta z}$ appears as a sum of terms like 12.11 (5). Differentiation with respect to θ brings in the sine factor of (2), so

$$j\beta r\sin\theta\, e^{-j\beta z} = -\sum_{n=1}^{\infty}(-j)^n(2n+1)j_n(\beta r)P_n^1(\cos\theta) \tag{3}$$

$$\check{A}_r = -\frac{E\cos\phi}{\omega\beta r}\sum_{n=1}^{\infty}(-j)^n(2n+1)j_n(\beta r)P_n^1(\cos\theta) \tag{4}$$

$$\check{B}_r = -\frac{E\sin\phi}{j\omega r}\sum_{n=1}^{\infty}(-j)^n(2n+1)j_n(\beta r)P_n^1(\cos\theta) \tag{5}$$

Operating on the nth term of 12.12 (5) and using 5.12 (4) give

$$\begin{aligned}[\nabla \times (\mathbf{r} \times \nabla\check{W}_n)]_r &= \left[\nabla \times \left(\boldsymbol{\phi}\frac{\partial\check{W}_n}{\partial\theta} - \frac{\boldsymbol{\theta}}{\sin\theta}\frac{\partial\check{W}_n}{\partial\phi}\right)\right]_r\\ &= \frac{j_n(\beta r)}{r}\left[\frac{1}{\sin\theta}\frac{\partial}{\partial\theta}\left(\sin\theta\frac{\partial S_n}{\partial\theta}\right) + \frac{1}{\sin^2\theta}\frac{\partial^2 S_n}{\partial\phi^2}\right] = -\frac{n(n+1)}{r}\check{W}_n\end{aligned} \tag{6}$$

Therefore those parts $\check{W}'_{tm}$ and $\check{W}'_{te}$ of $\check{W}_{tm}$ and $\check{W}_{te}$ that represent the plane incident wave are found by multiplying the nth term of (4) and (5) by $-r[n(n+1)]^{-1}$. Those parts that represent the diffracted wave must be diverging waves and so contain $\check{k}_n(j\beta r)$. Thus we have

$$\check{W}_{tm} = \frac{E\cos\phi}{\omega\beta}\sum_{n=1}^{\infty}\frac{2n+1}{n(n+1)}[(-j)^n j_n(\beta r) + \check{A}_n\check{k}_n(j\beta r)]P_n^1(\cos\theta) \tag{7}$$

$$-\check{W}_{te} = \frac{E\sin\phi}{j\omega}\sum_{n=1}^{\infty}\frac{2n+1}{n(n+1)}[(-j)^n j_n(\beta r) + \check{B}_n\check{k}_n(j\beta r)]P_n^1(\cos\theta) \tag{8}$$

Inside the sphere, fields must be finite at the origin so that

$$\check{W}_{tmi} = \frac{E\cos\phi}{\omega\beta}\sum_{n=1}^{\infty}\frac{2n+1}{n(n+1)}(-j)^n\check{A}_{ni}j_n(\beta' r)P_n^1(\cos\theta) \tag{9}$$

$$\check{W}_{tei} = \frac{E\sin\phi}{j\omega}\sum_{n=1}^{\infty}\frac{2n+1}{n(n+1)}(-j)^n\check{B}_{ni}j_n(\beta' r)P_n^1(\cos\theta) \tag{10}$$

From (6) and (2), equating normal displacements gives

$$\epsilon_i\check{A}_{ni}j_n(\beta' a) = \epsilon[j_n(\beta a) + j^n\check{A}_n\check{k}_n(j\beta a)] \tag{11}$$

From (6) and (2), equating normal magnetic inductions gives

$$-\check{B}_{ni}j_n(\beta' a) = j_n(\beta a) + j^n\check{B}_n k_n(j\beta a) \tag{12}$$

From (6) and 12.11 (2), the tangential components of $\boldsymbol{A}$ are

$$\check{A}_t = \boldsymbol{\theta}\left[\frac{1}{\sin\theta}\frac{\partial\check{W}_{te}}{\partial\phi} - \frac{1}{r}\frac{\partial^2(r\check{W}_{tm})}{\partial\theta\,\partial r}\right] - \boldsymbol{\phi}\left[\frac{\partial\check{W}_{te}}{\partial\theta} + \frac{1}{r\sin\theta}\frac{\partial^2(r\check{W}_{tm})}{\partial\phi\,\partial r}\right] \tag{13}$$

Equating tangential components of $\check{\boldsymbol{E}}$ or $\check{\boldsymbol{A}}$ involves only the r-derivatives of $\check{W}_{tm}$ because the $\check{W}_{te}$ terms are already equal by (12). The same relation is obtained from the θ- or ϕ-component.

$$\check{A}_{ni}\frac{\partial[aj_n(\beta' a)]}{\partial a} = \frac{\partial[aj_n(\beta a)]}{\partial a} + j^n\check{A}_n\frac{\partial[a\check{k}_n(j\beta a)]}{\partial a} \tag{14}$$

By comparison of 12.11 (2) and (3), the contribution of $\check{W}_{tm}$ to $\check{A}_t$ is identical with that of $\check{W}_{te}$ to $\check{B}_t$. Equating either θ- or ϕ-components of $\mu^{-1}\check{\boldsymbol{B}}$ at $r = a$ involves only the r-derivatives of $\check{W}_{te}$ because from (11), (13), and 12.11 (3) the $\check{W}_{tm}$ terms are already equal. Thus

$$-\frac{1}{\mu_i}\check{B}_{ni}\frac{\partial[aj_n(\beta' a)]}{\partial a} = \frac{1}{\mu}\left\{\frac{\partial[aj_n(\beta a)]}{\partial a} + j^n\check{B}_n\frac{\partial[a\check{k}_n(j\beta a)]}{\partial a}\right\} \tag{15}$$

We now equate values of A_{ni} from (11) and (14) and solve for $\check{A}_n$. Use of 5.37 (13), 5.31 (11), and 5.31 (12) gives $(1 - jN)^{-1}$ for $\check{A}_n$ where N is

$$\frac{\beta a\epsilon_1 j_n(\beta' a)n_{n-1}(\beta a) - \beta' a\epsilon n_n(\beta a)j_{n-1}(\beta' a) - n(\epsilon_1 - \epsilon)j_n(\beta' a)n_n(\beta a)}{\beta a\epsilon_1 j_n(\beta' a)j_{n-1}(\beta a) - \beta' a\epsilon j_n(\beta a)j_{n-1}(\beta' a) - n(\epsilon_1 - \epsilon)j_n(\beta' a)j_n(\beta a)} \tag{16}$$

The same formula is obtained for $\check{B}_n$ by solving (12) and (15) except that μ_i and μ replace ϵ_i and ϵ. When the sphere is perfectly conducting, $\check{A}_n$ and $\check{B}_n$ are obtained by equating the left sides of (12) and (14) to zero. Thus

$$\check{A}_n = \left\{1 - j\frac{\beta a n_{n-1}(\beta a) - n n_n(\beta a)}{\beta a j_{n-1}(\beta a) - n j_n(\beta a)}\right\}^{-1}, \qquad \check{B}_n = \frac{j_n(\beta a)}{j_n(\beta a) - jn_n(\beta a)} \tag{17}$$

The energy scattered by the sphere in a specified direction is, from 11.17 (3), the real part of the complex Poynting vector

$$\tfrac{1}{2}\mu^{-1}\breve{\mathbf{E}} \times \hat{\mathbf{B}} = \tfrac{1}{2}\mu^{-1}(\mu\epsilon)^{\frac{1}{2}}\breve{\mathbf{E}} \times (\mathbf{r}_1 \times \hat{\mathbf{E}}) = \tfrac{1}{2}\mathbf{r}_1\mu^{-\frac{1}{2}}\epsilon^{\frac{1}{2}}(\breve{E}_\theta\hat{E}_\theta + \breve{E}_\phi\hat{E}_\phi) \tag{18}$$

From 5.37 (12) at a great distance, neglecting terms in r^{-p} when $p > 1$,

$$\breve{k}_n(j\beta r) = \frac{1}{j\beta r}e^{-j\beta r}, \qquad \frac{\partial[r\breve{k}_n(j\beta r)]}{r\,\partial r} = -\frac{1}{r}e^{-j\beta r} \tag{19}$$

Substitution in (7), (8), and (13) gives for the tangential component of the scattered electrical intensity at a great distance from the sphere

$$\breve{\mathbf{E}}_s = -j\omega\breve{\mathbf{A}}_s = \frac{jE}{\beta r}e^{-j\beta r}\sum_{n=1}^{\infty}\frac{2n+1}{n(n+1)}\left\{\boldsymbol{\theta}\left[\breve{A}_n \sin\theta P_n^{1\prime}(\cos\theta)\right.\right.$$

$$\left.\left. - \breve{B}_n\frac{P_n^1(\cos\theta)}{\sin\theta}\right]\cos\phi - \boldsymbol{\phi}\left[\breve{A}_n\frac{P_n^1(\cos\theta)}{\sin\theta} - \breve{B}_n\sin\theta P_n^{1\prime}(\cos\theta)\right]\sin\phi\right\} \tag{20}$$

To get the total energy scattered, (18) is multiplied by $r^2 \sin\theta\, d\theta\, d\phi$ and integrated over the ranges $0 \leqslant \phi \leqslant 2\pi$ and $0 \leqslant \theta \leqslant \pi$. Substitution of (20) in (18) and integration with respect to ϕ bring in a factor π and leave, omitting the argument in the Legendre polynomials,

$$\frac{\epsilon^{\frac{1}{2}}\pi E^2}{2\mu^{\frac{1}{2}}\beta^2 r^2}\sum_{n=1}^{\infty}\sum_{m=1}^{\infty}\frac{(2n+1)(2m+1)}{n(n+1)m(m+1)}\left[(\breve{A}_n\hat{A}_m + \breve{B}_n\hat{B}_m)\left(\sin^2\theta P_n^{1\prime}P_m^{1\prime} + \frac{P_n^1P_m^1}{\sin^2\theta}\right)\right.$$

$$\left. - (\breve{A}_n\hat{B}_m + \breve{B}_n\hat{A}_m)(P_n^{1\prime}P_m^1 + P_n^1P_m^{1\prime})\right] \tag{21}$$

When multiplied by $r^2 \sin\theta\, d\theta$ and integrated from 0 to π, the first group of Legendre polynomials yields the integral following 5.231 (7) which is zero if $m \neq n$ and 5.231 (8) if $m = n$. Integration of the first term of the second group gives the negative of the second term and cancels it. The result is real so that the total scattered power is

$$\bar{P} = \frac{\pi\epsilon^{\frac{1}{2}}E^2}{\mu^{\frac{1}{2}}\beta^2}\sum_{n=1}^{\infty}(2n+1)(|A_n|^2 + |B_n|^2) \tag{22}$$

Interesting cases are discussed in the books by MacDonald and by Stratton listed at the end of the chapter.

12.16. Solution of Propagation Equation in Cylindrical Coordinates.—In 11.14 we considered the special type of cylindrical wave moving in the z-direction obtained by setting those terms in the scalar wave equation containing z and t separately equal to zero. If instead we equate the first group to $\pm\beta_{mn}^2$ and the second to $\mp\beta_{mn}^2$ and assume a sinusoidal time dependence so that $\beta^2 = \omega^2\mu\epsilon$, we obtain the equations

$$\nabla_2^2\check{U} \pm \beta_{mn}^2\check{U} = 0, \qquad \frac{d^2\check{Z}}{dz^2} + (\beta^2 \mp \beta_{mn}^2)\check{Z} = 0, \qquad \check{W} = \check{U}\check{Z} \quad (1)$$

Comparison with 5.291 (1) to (5) and 5.293 (6) shows that the form of $\check{W}$ in the ρ, ϕ, z system is, writing k_{mn}^2 for $\beta^2 - \beta_{mn}^2$, and $k_{mn}'^2$ for $\beta^2 + \beta_{mn}^2$,

$$\check{W} = (\check{A}e^{jk_{mn}z} + \check{B}e^{-jk_{mn}z})[\check{C}J_m(\beta_{mn}\rho) + \check{D}Y_m(\beta_{mn}\rho)] \cos (m\phi + \delta_m) \quad (2)$$
$$\check{W} = (\check{A}e^{jk'_{mn}z} + \check{B}e^{-jk'_{mn}z})[\check{C}I_m(\beta_{mn}\rho) + \check{D}K_m(\beta_{mn}\rho)] \cos (m\phi + \delta_m) \quad (3)$$

When $\check{C}$, $\check{D}$, k_{mn}, and k'_{mn} are real, both these equations give waves propagated only in the z-direction. If $\beta_{mn}^2 > \beta^2$ so that k_{mn} is imaginary, (2) gives a wave exponentially damped in the z-direction. If $\check{C}$ is real and $\check{D}$ is complex, we have a radial propagation component in (2). If z is absent, k_{mn} and k'_{mn} are zero, and we have cylindrical wave fronts in (2) and (3). From 5.293 (4) and 5.295 (8), $D = -jC$ gives ρ-directed waves.

If transverse electric and transverse magnetic waves are defined as those whose electric and magnetic fields, respectively, are normal to the z-axis then, putting $\boldsymbol{u} = \boldsymbol{k}$ in 10.01 (1) and (5), we obtain

$$\check{A} = \frac{\boldsymbol{\rho}_1}{\rho}\frac{\partial \check{W}_{te}}{\partial \phi} - \boldsymbol{\phi}\frac{\partial \check{W}_{te}}{\partial \rho} - \left[\boldsymbol{\rho}_1\frac{\partial^2 \check{W}_{tm}}{\partial \rho\, \partial z} + \boldsymbol{\phi}\frac{\partial^2 \check{W}_{tm}}{\rho\, \partial \phi\, \partial z} + \boldsymbol{k}\left(\beta^2\check{W}_{tm} + \frac{\partial^2 \check{W}_{tm}}{\partial z^2}\right)\right] \quad (4)$$

$$\check{B} = \boldsymbol{\rho}_1\frac{\partial^2 \check{W}_{te}}{\partial \rho\, \partial z} + \boldsymbol{\phi}\frac{\partial^2 \check{W}_{te}}{\rho\, \partial \phi\, \partial z} + \boldsymbol{k}\left(\beta^2\check{W}_{te} + \frac{\partial^2 \check{W}_{te}}{\partial z^2}\right) - \boldsymbol{\rho}_1\frac{\beta^2}{\rho}\frac{\partial \check{W}_{tm}}{\partial \phi} + \boldsymbol{\phi}\beta^2\frac{\partial \check{W}_{tm}}{\partial \rho} \quad (5)$$

12.17. Expansion in Cylindrical Harmonics for a Plane Wave.—From 11.15, the formula for a plane sinusoidal wave traveling in the direction $\boldsymbol{n}$ is

$$\check{W}e^{j\omega t} = f(u_1, u_2)e^{j(\omega t - \beta \boldsymbol{n}\cdot\boldsymbol{r})} = f(u_1, u_2)e^{j\omega t}e^{-j\beta\rho \cos (\alpha - \phi)} \quad (1)$$

where $\boldsymbol{n}$ is normal to the z-axis and makes an angle α with the plane $\phi = 0$ and u_1 and u_2 are coordinates in the plane normal to $\boldsymbol{n}$. A plane wave may be expressed in cylindrical harmonics by expanding the last exponential factor in (1) in a complex Fourier series. In order to utilize formulas already derived, it is simpler to expand real and imaginary parts separately. Let us first expand $\cos (x \sin \psi)$ in a Fourier series.

$$\cos (x \sin \psi) = \sum_{n=0}^{\infty} a_n \cos n\psi, \qquad a_n = \frac{2 - \delta_n^0}{2\pi}\int_{-\pi}^{\pi} \cos (x \sin \psi) \cos n\psi \, d\psi \quad (2)$$

Application of *Dw* 401.06 or *Pc* 592 gives two even integrands whose integral from $-\pi$ to π may be replaced by twice that from 0 to π. Thus

$$a_n = \frac{2 - \delta_n^0}{2\pi}\left[\int_0^\pi \cos(n\psi - x\sin\psi)\,d\psi + \int_0^\pi \cos(n\psi + x\sin\psi)\,d\psi\right] \quad (3)$$

From 5.302 (4), the integrals are $J_n(x)$ and $J_n(-x)$ or $(-1)^n J_n(x)$ so that a_n is zero when n is odd, and we may write $2m$ for n which gives

$$\cos(x\sin\psi) = \sum_{m=0}^{\infty}(2 - \delta_m^0)J_{2m}(x)\cos 2m\psi = \sum_{n=-\infty}^{\infty} J_n(x)\cos n\psi \quad (4)$$

We can expand $\sin(x\sin\psi)$ in a sine series in just the same way and get the difference of $J_n(x)$ and $J_n(-x)$ instead of the sum so that

$$\sin(x\sin\psi) = \sum_{m=0}^{\infty} 2J_{2m+1}(x)\sin(2m+1)\psi = \sum_{n=-\infty}^{\infty} J_n(x)\sin n\psi \quad (5)$$

If we multiply this by j and add to (4), we get the exponential form. If we then substitute $\beta\rho$ for x and $\frac{1}{2}\pi + \phi - \alpha$ for ψ, we get

$$e^{j\beta\rho\cos(\alpha-\phi)} = \sum_{n=-\infty}^{\infty} j^n J_n(\beta\rho)e^{jn(\alpha-\phi)} = J_0(\beta\rho) + 2\sum_{n=1}^{\infty} j^n J_n(\beta\rho)\cos n(\alpha - \phi) \quad (6)$$

12.18. Radiation from Apertures in Plane Conducting Screens.—The rigorous calculation of the radiation passing through an aperture is a very difficult problem. The fields must satisfy not only the propagation equations in the space outside the aperture and specified conditions at its boundaries, but also they must fit smoothly onto the fields inside the aperture. These fields are usually changed by the back radiation from the screen surrounding the aperture which practically restricts rigorous solutions to the very limited number of cases where it is mathematically feasible to treat the entire radiation space as a single volume.

It was proved in 12.10 that, if the initial field values throughout an empty volume are given together with the tangential components of either the electric or the magnetic field over its surface, then the fields at any subsequent time are uniquely determined. The second condition alone is sufficient in a steady state. Evidently with conducting screens one should use the electric field because its tangential component is known to be zero on the screen. The best first approximation to the unknown electric field over the aperture seems to be that its value is the same as with no screen. The treatment that follows is particularly applicable to plane conducting screens with no restrictions on the number or shape of the apertures nor on the form of the incident wave.

We desire a source that will give a tangential electric field $\boldsymbol{E}$ over an area S of an infinite plane and zero over the remainder. Consider a thin double current sheet with a very small distance between layers, one of which has a current density equal and opposite to the other as shown in cross section in Fig. 12.18*a*. For a sheet uniform in the direction of flow, all the current passes around the edge, but for one stronger in the center, like that shown in Fig. 12.18*b*, part will turn back before reaching the edge. If the sheet is very thin, the external magnetic field is negligible compared with that between layers so that when we apply the magnetomotance law to the rectangle *abcd*, which is normal to i and fits closely a section of the upper layer, we find B_i to be μi. As the flux $N = B_i\delta\, dl$ through the rectangle $a'b'c'd'$ changes, the electromotance $-dN/dt$ around the loop equals from symmetry $2E\, dl$ as $\delta \to 0$ so that the electric field strength $\boldsymbol{E}$ just above the sheet is $-\frac{1}{2}j\omega\mu\delta \boldsymbol{i}$. Clearly the double current sheet can be built up out of infinitesimal solenoids of cross-sectional area $\delta\, dl$, length dc, and magnetic moment $\boldsymbol{n} \times \boldsymbol{i}\delta\, dl\, dc$ which, in terms of $\boldsymbol{E}$ is $-2(j\omega\mu)^{-1}\boldsymbol{n} \times \boldsymbol{E}\, dS$. Evidently these solenoidal elements can be combined in such a way as to produce any desired variation in $\boldsymbol{E}$. It is also evident from symmetry that $\boldsymbol{E}$ is normal to the plane of the sheet outside its boundaries. From 12.13 (5) we see that the vector potential at P of the radiation from a small oscillating loop is normal to the loop axis and proportional to the sine of the angle between this axis and $\boldsymbol{r}$ the radius vector from the loop to P. Thus substitution of the moment just found for $\pi a^2 I$ gives for the diffracted vector potential

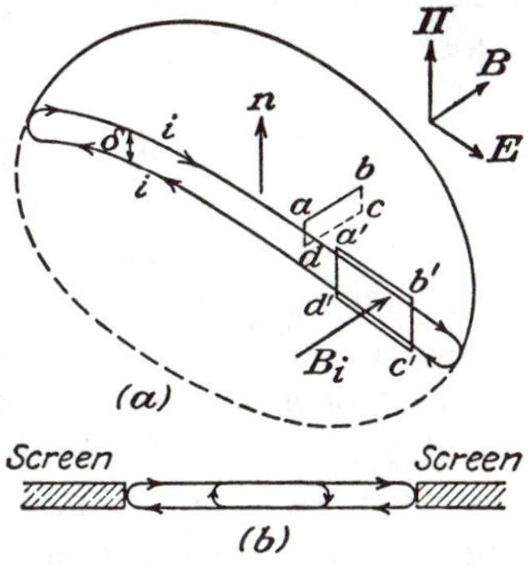

FIG. 12.18.

$$\boldsymbol{A} = \check{\boldsymbol{A}}e^{j\omega t} = \frac{1}{2\pi\omega}e^{j\omega t}\int_S \frac{(j - \beta r)(\boldsymbol{n} \times \check{\boldsymbol{E}}) \times \boldsymbol{r}_1}{r^2}e^{-j\beta r}\, dS \qquad (1)$$

where $\boldsymbol{r}_1$ is a unit vector along $\boldsymbol{r}$. At a great distance the j-term in the integrand may be neglected compared with βr. In the most general case, $\boldsymbol{E}$ will vary in magnitude, direction, and phase over the aperture and $\boldsymbol{\Pi}$ will not be parallel to $\boldsymbol{n}$, the normal to the aperture plane. Let x', y', z' refer to coordinates in the aperture and x, y, z to those of the field point. Then, since r^2 is $(x - x')^2 + (y - y')^2 + (z - z')^2$,

$$-j\omega(j - \beta r)r^{-2}\boldsymbol{r}_1 e^{-j\beta r} = \boldsymbol{\nabla}'(r^{-1}e^{-j\beta r}) = -\boldsymbol{\nabla}(r^{-1}e^{-j\beta r}) = -\boldsymbol{\nabla}\Phi$$

$$\boldsymbol{E} = -j\omega\boldsymbol{A} = \tfrac{1}{2}\pi^{-1}e^{-j\omega t}\int_S \boldsymbol{n} \times \check{\boldsymbol{E}}' \times \boldsymbol{\nabla}'\Phi\, dS \qquad \text{(real part)} \qquad (2)$$

where E' is a function of x', y', z' only. Thus writing $-\nabla\Phi$ for $\nabla'\Phi$, observing that then $\nabla \times (\Phi n \times E')$ is zero, and using the usual formula for $\nabla \times \Phi v$ give (2) the form

$$E = \tfrac{1}{2}\pi^{-1}e^{j\omega t}\nabla \times \int_S (n \times \breve{E}')r^{-1}e^{-j\beta r}\,dS \qquad \text{(real part)} \tag{3}$$

Formulas (2) and (3) are derived differently by Jackson. From the first Maxwell equation 11.00 (1), $j\omega\mu\epsilon\breve{E}$ is $\nabla \times \breve{B}$ so, except for the gradient of a scalar, canceling the curl gives

$$\breve{B}_t = j\beta(2\pi\omega)^{-1}\int_S (n \times \breve{E})r^{-1}e^{-j\beta r}\,dS \tag{4}$$

This integral equation gives the solenoidal part of the tangential $\breve{E}$ in the aperture in terms of the tangential $\breve{B}$ which is known, as will soon be shown. The diffracted field far from the aperture depends only on this part of $n \times \breve{E}$ since B is given by the curl of the A in (1) or (2) and E and B are related by 11.03 (8).

A system of sources of electromagnetic wave lies on one side of an infinite plane perfectly conducting sheet having apertures. The eddy currents in this sheet generate additional waves which combine with the original ones to produce the reflected and refracted waves. These currents, being coplanar with the apertures, can produce therein only magnetic fields normal to their plane. Thus the tangential magnetic fields in the apertures must be entirely due to the original sources unperturbed by the eddy currents.

When the incident magnetic field parallels the screen and the wave length is long compared with the relevant aperture dimension, it is often possible to calculate by potential theory the exact ratio of the normal to the tangential components of the magnetic induction in the openings and from this the tangential electric field. In Fig. 12.18 a displacement of the rectangle $a'b'c'd'$ along B_i decreases $2E\,dl$ by an amount equal to the decrease in dN/dt. This in turn equals the rate of change of the flux that escapes from both faces of the shell between the two positions. Thus, for an aperture in the xy-plane when E has only a y-component, we have

$$-\frac{\partial E_y}{\partial x} = j\omega B_z \tag{5}$$

Integration of this expression gives the tangential component of E to be used in (1). All static potential solutions will give B_z in terms of the induction tangential to the sheet far from the apertures. This standing wave value is twice the incident wave value that appears in the aperture. Examples at the end of the chapter include Bethe's small hole results.

12.19. Diffraction from Rectangular Aperture in Conducting Plane.— The formula of the last article will now be used to find the diffracted field from a rectangular aperture in a perfectly conducting sheet lying in the xy-plane. The magnetic field parallels the x-axis, and Poynting's vector makes an angle α with the z-axis as shown in Fig. 12.19a. If x_1 and y_1 are the coordinates of dS, r the radius vector from dS to P, and R that from O to P, then we have approximately, if $R \gg a$ and $R \gg b$,

$$r \approx R - x_1 \cos\phi \sin\theta - y_1 \sin\phi \sin\theta \tag{1}$$

The tangential component E_t varies in phase at $z = 0$ so that

$$\breve{E}_t e^{j\omega t} = E \cos\alpha \, e^{j(\omega t - \beta y_1 \sin\alpha)} \tag{2}$$

Because $\boldsymbol{n} \times \boldsymbol{E}$ parallels the x-axis, $(\boldsymbol{n} \times \boldsymbol{E}) \times \boldsymbol{R}$ lies parallel to the yz-plane, normal to DP and hence to $\boldsymbol{R}$ and is proportional to $\sin\theta'$. Thus

$$\breve{A}_y = -\cos\theta \csc\theta' \breve{A}, \qquad \breve{A}_z = \sin\theta \sin\phi \csc\theta' \breve{A} \tag{3}$$

$$\breve{A}_\theta = \breve{A}_y \sin\phi \cos\theta - \breve{A}_z \sin\theta, \qquad \breve{A}_\phi = \breve{A}_y \cos\phi, \qquad \breve{A}_r = 0 \tag{4}$$

When 12.18 (1) is substituted in (3), $\csc\theta'$ cancels the $\sin\theta'$. Neglect of x_1^2 and y_1^2 compared with R^2 leaves x_1 and y_1 terms only in the exponent of $\breve{\boldsymbol{A}}$. Writing in $\breve{E}_t$ from (2) and using (4), we have

$$\breve{A}_\theta = \frac{\beta E \cos\alpha \sin\phi}{2\pi\omega R} e^{-j\beta R} \int_{-\frac{1}{2}a}^{\frac{1}{2}a} \int_{-\frac{1}{2}b}^{\frac{1}{2}b} e^{j\beta[x_1 \cos\phi \sin\theta + y_1(\sin\phi \sin\theta - \sin\alpha)]} \, dx_1 \, dy_1$$

$$= \frac{2E \cos\alpha \sin\phi \sin(\frac{1}{2}\beta a \cos\phi \sin\theta) \sin[\frac{1}{2}\beta b(\sin\phi \sin\theta - \sin\alpha)]}{\pi\beta\omega R \cos\phi \sin\theta(\sin\phi \sin\theta - \sin\alpha)} e^{-j\beta R} \tag{5}$$

$$A_\phi = A_\theta \cot\phi \cos\theta \tag{6}$$

The only approximation involved in these formulas is the assumption of an unperturbed electric field over the aperture. Stratton and Chu (*Phys. Rev.*, Vol. 56, page 106) derived them by a superposition or "reflection" of two solutions of Maxwell's equations, which assume unperturbed electric and magnetic fields over the opening. This superposition was necessary to eliminate the tangential electric field over the screen and, as we see from the solution just obtained and from the uniqueness theorem, this is just equivalent to discarding all terms derived from the magnetic field. These authors check (5) and (6) at various b and α values by comparing the yz-plane intensity $E_\theta^2 + E_\phi^2$ with the rigorous two-dimensional solution for a slit ($a = \infty$) of Morse and Rubenstein (*Phys. Rev.*, Vol. 54,

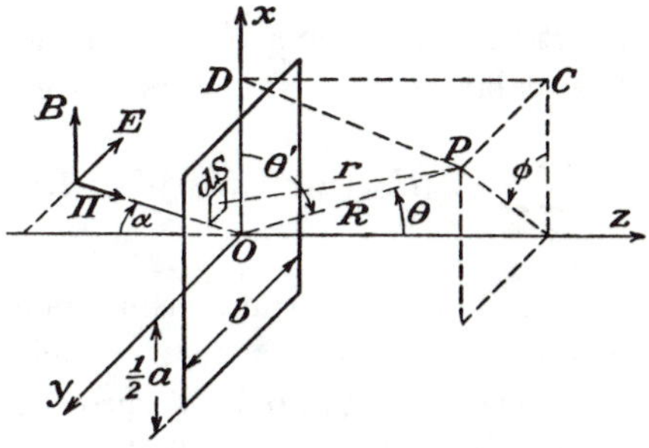

FIG. 12.19a.

page 895). The results appear in Fig. 12.19*b*, *c*, and *d*. A similar comparison of the xz-plane field with that of a slit ($b = \infty$) for $\alpha = 0$ is shown in Fig. 12.19*e*. This fairly good agreement shows the magnitude of the errors made by assuming an unperturbed electric field over the opening. Clearly the Kirchhoff scalar diffraction theory used in optics, which requires Fig. 12.19*b* and *e* to be identical, is completely wrong when applied to apertures of wave length dimensions in conducting screens. Dashed lines show rigorous values.

12.20. Orthogonal Functions in Diffraction Problems. Coaxial Line.— The method of the last two articles does not apply to curved screens, and the integral of 12.18 (1) is frequently difficult to evaluate close to the

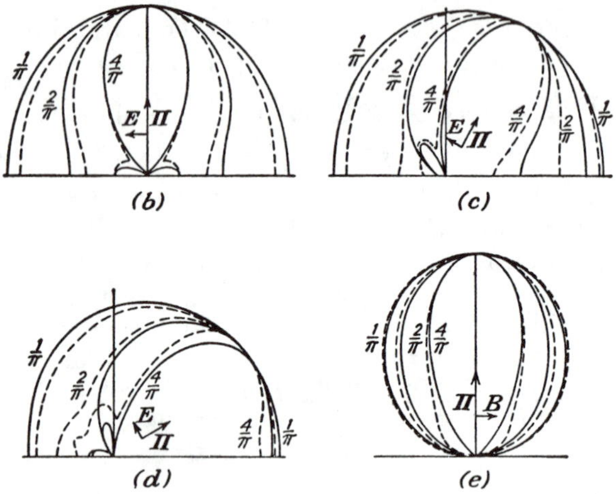

FIG. 12.19*b*, *c*, *d*, *e*.

aperture in a plane screen. In such cases we may start with solutions of the scalar propagation equation in the form

$$e^{j\omega t}\sum_n\sum_m C_{mn}U_{mn}(u_1)V_n(u_2)W_{mn}(u_3), \qquad \iint U_{nm}V_nU_{pq}V_p\,du_1\,du_2 = \delta_{pq}^{nm}D_{nm}$$

If u_3 is normal to the screen, then u_1 and u_2 are orthogonal curvilinear coordinates in its surface. The integration is over the screen and aperture surface, and δ_{pq}^{nm} is zero unless $p = n$ and $q = m$. Such solutions appear in 12.11 (5), 12.16 (2), and 12.16 (3) and others occur in wave-guide problems. For a conducting screen, we evaluate C_{mn} from the tangential component of the electric field, which is given over the aperture.

The radiation from the open end of a coaxial line can be calculated by this method. The distant field is found more easily by 12.18 (1), but the method to be used here is better near the opening. The propagation space in the line is bounded internally and externally by the cylinders $\rho = a$ and $\rho = b$, respectively, and terminates in the plane $z = 0$ which, except when $a < \rho < b$, is perfectly conducting. To find the radiation into the region of positive z, we shall assume that $b \ll \lambda$ and use electrostatic methods to calculate the field for $r < b$. This is found, for any zonal potential distribution in the aperture, by integration of that above a plane at zero potential except for a ring of radius ρ and width $d\rho$ at potential $V(\rho)$. At a point P on the axis, V_P due to the ring is got by Green's reciprocation theorem from the charge dq induced on a ring element of an earthed plane by a charge q at P. To get dq from 5.06 (3), we let $b - a = z$ and $a \to b \to \infty$. If $z > \rho$ expanding by *Pc* 757 or *Dw* 9.05 gives

$$dq = \frac{-zq2\pi\rho\, d\rho}{2\pi(\rho^2 + z^2)^{\frac{3}{2}}},$$

$$dV_P = -\frac{dq}{q}V(\rho) = V(\rho)\, d\rho \sum_{n=0}^{\infty} (-1)^n \frac{(2n + 1)!!\rho^{2n+1}}{(2n)!!z^{2n+2}} \tag{1}$$

To get dV at r, θ, we write r for z and multiply by $P_{2n+1}(\cos\theta)$. The values of the potential in the plane $z = 0$ are assumed to be

$$\rho > b, \qquad V(\rho) = 0, \qquad b > \rho > a, \qquad V(\rho) = V_0\frac{\ln(\rho/b)}{\ln(a/b)},$$
$$a > \rho > 0, \qquad V(\rho) = V_0 \tag{2}$$

When these values are inserted in (1), the result may be integrated by *Pc* 427 or *Dw* 610.9 and gives

$$V = \frac{V_0}{\ln(b/a)} \sum_{n=0}^{\infty} (-1)^n \frac{b^{2n+2} - a^{2n+2}}{(2n + 2)r^{2n+2}} \frac{(2n + 1)!!}{(2n + 2)!!} P_{2n+1}(\cos\theta) \tag{3}$$

To get the radiation fields, one puts $A = 1$ and $B = \check{C} = m = \delta = 0$ in 12.11 (5) so that it gives an expanding wave and evaluates $\check{D}_n$ by setting $r = b$ and equating the coefficient of $P_{2n+1}(\cos\theta)$ in the E_θ derived from W_{tm} to that in the E_θ given by (3). From 12.11 (5) and (2), we have

$$(\check{E}_\theta)_{r=b} = \left[\frac{j\omega}{r}\frac{\partial^2(r\check{W}_{tm})}{\partial r\, \partial\theta}\right]_{r=b} = -\frac{j\omega}{b}\sum_{n=0}^{\infty} \check{D}_n \frac{d}{db}[b\check{k}_{2n+1}(j\beta b)]P^1_{2n+1}(\cos\theta)$$

Since βb is small, we may write $-(2n + 1)(4n + 1)!!(j\beta b)^{-(2n+2)}$ for the derivative in this expression by 5.37 (11) and (12). To get E_θ at $r = b$, we write $-b^{-1}d[P_{2n+1}(\cos\theta)]/d\theta = b^{-1}P^1_{2n+1}(\cos\theta)$ for $P_{2n+1}(\cos\theta)$ in (3) and b^{2n+2} for r^{2n+2}. Equating coefficients and solving for $\check{D}_n$ give

$$\check{D}_n = \frac{jV_0}{\omega \ln (b/a)} \frac{(2n-1)!![(\beta b)^{2n+2} - (\beta a)^{2n+2}]}{(2n+2)(2n+2)!!(4n+1)!!} \tag{4}$$

The electric and magnetic fields when r is greater than b are then given by 12.11 (2), 12.11 (5), 12.15 (6), and 12.11 (3) to be

$$\check{E}_\theta = -j\omega \sum_{n=0}^{\infty} \frac{\check{D}_n}{r} \frac{d}{dr}[r\check{k}_{2n+1}(j\beta r)]P^1_{2n+1}(\cos\theta) \tag{5}$$

$$\check{E}_r = j\omega \sum_{n=0}^{\infty} r^{-1}\check{D}_n(2n+1)(2n+2)\check{k}_{2n+1}(j\beta r)P_{2n+1}(\cos\theta) \tag{6}$$

$$\check{B}_\phi = -\beta^2 \sum_{n=0}^{\infty} \check{D}_n\check{k}_{2n+1}(j\beta r)P^1_{2n+1}(\cos\theta) \tag{7}$$

These equations hold if $r > b$, and if $b \ll \lambda$ they satisfy the boundary conditions (2) rigorously. If we neglect higher powers of βb and βa, only the first term of the series survives and, if the case is further simplified by considering the field only at a great distance so that by 5.37 (12) $(j\beta r)^{-1}e^{-j\beta r}$ replaces $\check{k}_{2n+1}(j\beta r)$, then the fields become

$$B_\phi = (\mu\epsilon)^{\frac{1}{2}}E_\theta = -\frac{\beta^3(b^2-a^2)V_0}{4\omega r \ln (b/a)} \sin\theta \cos(\omega t - \beta r) \tag{8}$$

From 12.01 (8), this is identical with the distant radiation from a current element $i_0\, dz_0 \cos \omega t$ where $i_0\, dz_0$ is $\pi\omega\epsilon(b^2 - a^2)V_0/\ln (b/a)$ so that the power radiated into the upper hemisphere is half that of 12.01 (26).

$$\bar{P} = \frac{2\pi^5 c\epsilon(b^2-a^2)^2V_0^2}{3[\ln (b/a)]^2\lambda^4} \tag{9}$$

$$R_r = \frac{V_0^2}{2\bar{P}} = \frac{3[\ln (b/a)]^2\lambda^4}{4\pi^5 c\epsilon(b^2-a^2)^2} \tag{10}$$

The same method applies to other than plane screens, such, for example, as an infinite conducting cone coaxial with the line. In this case we take the same steps but start with 5.261. Other applications will be found in the problems at the end of the chapter.

Problems

1. A linear quadrupole consists of charges q, $-2q$, q at $z = -a$, 0, $+a$. Its moment $q_{zz}^{(2)}$ is $a^2q \sin \omega t$. Show from 12.01 (13) that if $a \ll r$ the fields are

$$E_r = Q(3\cos^2\theta - 1)\frac{3\beta r \cos(\omega t - \beta r) + (3 - \beta^2r^2)\sin(\omega t - \beta r)}{8\pi\epsilon r^4}$$

$$E_\theta = Q \sin 2\theta \frac{(6\beta r - \beta^3r^3)\cos(\omega t - \beta r) + (6 - 3\beta^2r^2)\sin(\omega t - \beta r)}{16\pi\epsilon r^4}$$

$$B_\phi = Q\eta\beta \sin 2\theta \frac{(3 - \beta^2r^2)\cos(\omega t - \beta r) - 3\beta r \sin(\omega t - \beta r)}{16\pi r^3}$$

2. A square quadrupole consists of charges q, $-q$, q, $-q$ at the corners of a square of area a^2 whose sides are parallel to $\phi = 0, \frac{1}{2}\pi, \pi, \frac{3}{2}\pi$, respectively. Its moment $q_{xy}^{(2)}$ is $a^2 q \sin \omega t$. Show from 12.01 (16) if $a \ll r$ that

$$E_r = 3Q[3\beta r \cos(\omega t - \beta r) - (3 - \beta^2 r^2) \sin(\omega t - \beta r)] \frac{\sin^2 \theta \sin 2\phi}{16\pi\epsilon r^4}$$

$$E_\phi = Q[(6\beta r - \beta^3 r^3) \cos(\omega t - \beta r) - (3\beta^2 r^2 - 6) \sin(\omega t - \beta r)] \frac{\sin \theta \cos 2\phi}{16\pi\epsilon r^4}$$

$$E_\theta = -E_\phi \cos \theta \tan 2\phi, \qquad B_r = 0, \qquad B_\phi = -B_\theta \cos \theta \tan 2\phi$$

$$B_\theta = \beta\eta Q[(3 - \beta^2 r^2) \cos(\omega t - \beta r) - 3\beta r \sin(\omega t - \beta r)] \frac{\sin \theta \cos 2\phi}{16\pi r^3}$$

3. It is desired to generate as pure quadrupole radiation as possible by charging the polar caps whose diameters subtend an angle 2α at the center of a sphere of radius a to potential $V_0 \cos \omega t$ while the equatorial zone is at zero potential. What is the least α value? Show that if βa is small, the value of E_θ at a great distance is

$$5\beta^2 a^2 V_0 (16r)^{-1} \sin \alpha \sin 2\alpha \sin 2\theta \cos(\omega t - \beta r)$$

4. Two coaxial uniform current loops, $I \cos \omega t$ and $-I \cos \omega t$, are a distance b apart. (a) Find the smallest value of b in wave lengths such that at a great distance the radiation at 60° is zero. (b) Verify that at a great distance there will be a maximum of radiation at 60° if $\beta b = 5.296$ or $b = 0.843\lambda$, approximately.

5. An electric dipole lies in the xy-plane at the origin and rotates about the z-axis with an angular velocity ω pointing in the positive x-direction when $t = 0$. Show that the components of the magnetic induction are

$$B_r = 0, \qquad B_\theta = -\frac{\omega\mu M}{4\pi r^2}[r\beta \sin(\omega t - \beta r - \phi) - \cos(\omega t - \beta r - \phi)]$$

$$B_\phi = \frac{\omega\mu M \cos \theta}{4\pi r^2}[r\beta \cos(\omega t - \beta r - \phi) + \sin(\omega t - \beta r - \phi)]$$

6. An antenna of length $2l$ supports a standing wave of $2n + 1$ loops and current amplitude I_0. If there are nodes at the ends, show that the mean energy radiated per second from an element dz at a distance z from the center is

$$\frac{\mu c l I_0^2 \, dz}{4\pi(l^2 - z^2)} \cos^2 \frac{(2n + 1)\pi z}{2l}$$

7. Two antennas, each $\frac{1}{2}(2n + 1)\lambda$ long, lie parallel to the z-axis with their centers at $x = 0$ and $x = a$. The one at $x = a$ lags 90° in phase behind the one at the origin. Show that the radiation intensity of this "end-fire" array is

$$\bar{\Pi}_1 = \frac{\mu c I_0^2 \cos^2 [\frac{1}{2}(2n + 1)\pi \cos \theta] \cos^2 \{\frac{1}{4}\pi[1 - (4a/\lambda) \sin \theta \cos \phi]\}}{2\pi^2 r^2 \sin^2 \theta}$$

Draw a rough sketch of the heart-shaped intensity pattern when $\theta = \frac{1}{2}\omega$ and $a = \frac{1}{4}\lambda$.

8. Show that, when $a = \frac{1}{4}\lambda$, the directivity of the double antenna of the last problem is twice that of a single antenna resonating in the same mode.

9. A set of p end-fire in-phase pairs like that of the last problem are set at half wave intervals along the y-axis. Show that the radiation intensity pattern is

$$\bar{\Pi} = \bar{\Pi}_1 \frac{\sin^2 (\frac{1}{2}p\pi \sin \theta \sin \phi)}{\sin^2 (\frac{1}{2}\pi \sin \theta \sin \phi)}$$

where $\bar{\Pi}_1$ is given in problem 7 with $a = \frac{1}{4}\lambda$. Show that the directivity is twice that of p single antennas spaced at half wave length intervals along the y-axis.

10. A number p of half wave in-phase antennas are placed end to end along the z-axis. Show that the radiation intensity at a great distance is

$$\bar{\Pi} = \frac{\mu c I_0^2 \cos^2(\frac{1}{2}\pi\cos\theta)\sin^2(\frac{1}{2}p\pi\cos\theta)}{8\pi^2 r^2\sin^2\theta\sin^2(\frac{1}{2}\pi\cos\theta)}$$

Note that this is the same situation as when a long linear antenna is loaded at half wave length intervals to suppress alternate phases.

11. The current in a circular loop of radius a of wire is $I_0 \sin n\phi \cos \omega t$. Show by the use of retarded potentials and with the aid of the formulas of 12.16 that at r, θ, ϕ, when r is very large, the potential A_ϕ is

$$\tfrac{1}{2}\mu a r^{-1} I_0 J_n'(\beta a\sin\theta)\sin n\phi\sin(\tfrac{1}{2}n\pi + \omega t - \beta r)$$

12. Electromagnetic waves are scattered from a conducting sphere whose radius is small compared with a wave length. Show that, at a distance from the sphere large compared with its radius, the radiation scattered at an angle $\frac{1}{3}\pi$ with the direction of the incident beam is plane polarized.

13. A plane polarized electromagnetic wave falls upon a nonconducting sphere of capacitivity ϵ_1 and permeability μ_1 whose radius is very small compared with the wave length. Show that the scattered radiation is the same as if there were at the origin an electric dipole parallel to the incident electric vector and a magnetic loop dipole normal to it, their moments being, respectively,

$$M = \frac{4\pi a^3(\epsilon_1 - \epsilon)\epsilon E}{\epsilon_1 + 2\epsilon}\cos\omega t \qquad \text{and} \qquad \pi r^2 I = \frac{4\pi a^3(\mu_1 - \mu)E}{\mu c(\mu_1 + 2\mu)}\cos\omega t$$

14. A plane electromagnetic wave, whose y-directed E-vector is $E\cos(\omega t - \beta x)$, impinges on a perfectly conducting cylinder of radius a whose axis is the z-axis. Show that the total fields are given by the real parts of

$$E_\rho = -\frac{j\omega}{\rho}\frac{\partial \check{W}_{te}}{\partial\phi}e^{j\omega t}, \qquad E_\phi = j\omega\frac{\partial\check{W}_{te}}{\partial\rho}e^{j\omega t}, \qquad B_z = \beta^2\check{W}_{te}e^{j\omega t}$$

$$\check{W}_{te} = \frac{jE}{\omega\beta}\sum_{n=0}^{\infty}\frac{(-j)^n(2-\delta_n^0)[J_n'(\beta a)Y_n(\beta\rho) - J_n(\beta\rho)Y_n'(\beta a)]}{J_n'(\beta a) - jY_n'(\beta a)}\cos n\phi$$

15. In a medium $\mu\epsilon$ a plane electromagnetic wave whose y-directed electric vector is $E\cos(\omega t - \beta x)$ impinges on a cylinder $\mu'\epsilon'$ of radius a, whose axis is the z-axis. Show that the scattered radiation is given by the real part of

$$B_z = \beta^2\check{W}_{te}e^{j\omega t}, \qquad E_\rho = -\frac{j\omega}{\rho}\frac{\partial\check{W}_{te}}{\partial\phi}e^{j\omega t}, \qquad E_\phi = j\omega\frac{\partial\check{W}_{te}}{\partial\rho}e^{j\omega t}$$

$$\check{W}_{te} = (\omega\beta)^{-1}E\sum_{n=0}^{\infty}\check{B}_n[J_n(\beta\rho) - jY_n(\beta\rho)]\cos n\phi$$

$$\check{B}_n = \frac{(-j)^n(2-\delta_n^0)[(\mu'\epsilon)^{\frac{1}{2}}J_n(\beta a)J_n'(\beta' a) - (\mu\epsilon')^{\frac{1}{2}}J_n'(\beta a)J_n(\beta' a)]}{(\mu\epsilon')^{\frac{1}{2}}J_n(\beta' a)[J_n'(\beta a) - jY_n'(\beta a)] - (\mu'\epsilon)^{\frac{1}{2}}J_n'(\beta' a)[J_n(\beta a) - jY_n(\beta a)]}$$

16. A plane electromagnetic wave, whose z-directed E-vector is $E\cos(\omega t - \beta x)$, impinges on a perfectly conducting cylinder of radius a whose axis is the z-axis. Show that the total fields are given by the real parts of

$$E_z = j\omega\beta^2 \check{W}_{tm} e^{j\omega t}, \qquad B_\rho = -\frac{\beta^2}{\rho}\frac{\partial \check{W}_{tm}}{\partial \phi} e^{j\omega t}, \qquad \beta_\phi = \beta^2 \frac{\partial \check{W}_{tm}}{\partial \rho} e^{j\omega t}$$

$$\check{W}_{tm} = \frac{E}{\omega\beta^2}\sum_{n=0}^{\infty}\frac{(-j)^n(2-\delta_n^0)[J_n(\beta a)Y_n(\beta\rho) - J_n(\beta\rho)Y_n(\beta a)]}{J_n(\beta a) - jY_n(\beta a)}\cos n\phi$$

17. In a medium $\mu\epsilon$ a plane electromagnetic wave whose z-directed electric vector is $E \cos(\omega t - \beta x)$ impinges on a cylinder $\mu'\epsilon'$ of radius a whose axis is the z-axis. Show that the scattered radiation is given by the real part of

$$E_z = j\omega\beta^2 \check{W}_{tm} e^{j\omega t}, \qquad B_\rho = -\frac{\beta^2}{\rho}\frac{\partial \check{W}_{tm}}{\partial \phi} e^{j\omega t}, \qquad B_\phi = \beta^2 \frac{\partial \check{W}_{tm}}{\partial \rho} e^{j\omega t}$$

$$\check{W}_{tm} = \frac{E}{j\omega\beta^2}\sum_{n=0}^{\infty}\check{C}_n[J_n(\beta\rho) - jY_n(\beta\rho)]\cos n\phi$$

$$\check{C}_n = \frac{(-j)^n(2-\delta_n^0)[(\mu\epsilon')^{\frac{1}{2}}J_n(\beta a)J_n'(\beta' a) - (\mu'\epsilon)^{\frac{1}{2}}J_n'(\beta a)J_n(\beta' a)]}{(\mu'\epsilon)^{\frac{1}{2}}J_n(\beta' a)[J_n'(\beta a) - jY_n'(\beta a)] - (\mu\epsilon')^{\frac{1}{2}}J_n'(\beta' a)[J_n(\beta a) - jY_n(\beta a)]}$$

18. The portion of a perfectly conducting sphere of radius a between $\theta = \gamma$ and $\theta = \pi - \gamma$ is removed, and a potential difference $2V_0 \cos \omega t$ is maintained between the cones at $r = a$. Assuming that the potential in the gap at $r = a$ is given by 12.06 (9) show that the radiation fields are the real parts of

$$E_\theta = V_0 r^{-1} e^{j\omega t}\sum_{n=0}^{\infty}\check{C}_n[(2n+2)\check{k}_{2n+1}(j\beta r) - j\beta r\check{k}_{2n+2}(j\beta r)]P_{2n+1}^1(\cos\theta)$$

$$E_r = -V_0 r^{-1} e^{j\omega t}\sum_{n=0}^{\infty}(2n+1)(2n+2)\check{C}_n\check{k}_{2n+1}(j\beta r)P_{2n+1}(\cos\theta)$$

$$B_\phi = -V_0 j\beta^2\omega^{-1}e^{j\omega t}\sum_{n=0}^{\infty}\check{C}_n\check{k}_{2n+1}(j\beta r)P_{2n+1}^1(\cos\theta)$$

$$\check{C}_n = \frac{\omega\mu(4n+3)P_{2n+1}(\cos\gamma)}{\pi\beta\check{Z}_k(2n+1)(2n+2)[(2n+2)\check{k}_{2n+1}(j\beta a) - j\beta a\check{k}_{2n+2}(j\beta a)]}$$

where $\check{Z}_k$ is given by 12.06 (8) and (10).

19. Calculate the input impedance of the arrangement of the last problem by proving its equivalence to a conical transmission line terminated at $r = a$ by a load comprised of an infinite number of impedances $\check{Z}_0, \check{Z}_1, \ldots$ in parallel where

$$\frac{1}{\check{Z}_n} = \frac{\pi a j\beta^2 \sin\gamma}{\mu\omega}\check{C}_n\check{k}_{2n+1}(j\beta a)P_{2n+1}^1(\cos\gamma)$$

20. A medium $\mu\epsilon$ fills the space between an infinite dielectric cylinder $\mu'\epsilon'$ of radius a and a perfectly conducting cylinder $\rho = b$. Show that a plane transverse electric wave propagates along the cylinder for each value of u_n that satisfies

$$\frac{\mu' J_1(u_n)}{u_n J_0(u_n)} = \frac{\mu[J_1(v_n)Y_1(v_n b/a) - Y_1(v_n)J_1(v_n b/a)]}{v_n[J_0(v_n)Y_1(v_n b/a) - Y_0(v_n)J_1(v_n b/a)]} = \frac{\mu R_1(v_n)}{v_n R_0(v_n)}$$

provided that $\omega^2a^2\mu'\epsilon' > u_n^2$ and $\omega^2a^2\mu\epsilon > v_n^2$. Show that the velocity of the nth wave is $\omega a(\omega^2a^2\mu'\epsilon' - u_n^2)^{-\frac{1}{2}}$ and that the fields in $\mu'\epsilon'$ are

$$-k_n\breve{E}_\phi = \omega\breve{B}_\rho = k_n\breve{C}_n e^{-jk_nz}J_1(u_n\rho/a), \qquad j\omega a\breve{B}_z = u_n\breve{C}_n e^{-jk_nz}J_0(u_n\rho/a)$$

where $\omega^2a^2\mu'\epsilon' - u_n^2 = \omega^2a^2\mu\epsilon - v_n^2 = k_n^2a^2$, and $a < b$.

21. If in the last problem there is no reflector at $\rho = b$, show that with the same materials and frequencies a plane wave is impossible but that otherwise the results of problem 20 hold if the Hankel functions $J_0 - jY_0$ and $J_1 - jY_1$ are substituted for R_0 and R_1, respectively. This requires that $\rho = a$ be an equipotential.

22. An infinite medium $\mu\epsilon$ surrounds an infinite dielectric cylinder $\mu'\epsilon'$ of radius a. Show that a plane transverse electric wave can be propagated along the cylinder for each value of u_n that satisfies the equations

$$\mu' v_n K_0(v_n)J_1(u_n) + \mu u_n K_1(v_n)J_0(u_n) = 0, \qquad \omega^2a^2\mu'\epsilon' - u_n^2 = \omega^2a^2\mu\epsilon + v_n^2 = k_n^2a^2$$

provided that $\omega^2a^2\mu'\epsilon'$ is greater than u_n^2 and ϵ' is greater than ϵ. Show that the velocity of the wave is $\omega a(\omega^2a^2\mu'\epsilon' - u_n^2)^{-\frac{1}{2}}$ and that the external fields are

$$k_n\breve{E}_\phi = -\omega\breve{B}_\rho = k_n\breve{C}e^{-jk_nz}K_1(v_na^{-1}\rho), \qquad -j\omega a\breve{B}_z = v_n\breve{C}e^{-jk_nz}K_0(v_na^{-1}\rho)$$

23. If $Y_1(v_nb/a)$ is zero, verify that the conditions of problem 20 are satisfied approximately by $\epsilon = 1.99\epsilon' = 1.99\epsilon_v$, $\mu = \mu' = \mu_v$, $\omega a = 10^9$, $u = 1$, $v = 3.46$, and show that the phase velocity is approximately 3.15×10^8 m/sec.

24. Verify that the conditions of problem 22 are satisfied approximately by $\epsilon' = 1.72\epsilon = 1.72\epsilon_v$, $\mu = \mu' = \mu_v$, $\omega a = 10^9$, $u = 2.65$, $v = 1$, and show that the phase velocity in the z-direction is approximately 2.87×10^8 m/sec.

25. A medium $\mu\epsilon$ fills the space between an infinite dielectric cylinder $\mu'\epsilon'$ of radius a and a perfectly conducting cylinder $\rho = b$. Show that a plane transverse magnetic wave propagates along the cylinder for each value of u_n that satisfies

$$\frac{u_nJ_0(u_n)}{\epsilon' J_1(u_n)} = \frac{v_n[J_0(v_n)Y_0(v_nb/a) - Y_0(v_n)J_0(v_nb/a)]}{\epsilon[J_1(v_n)Y_0(v_nb/a) - Y_1(v_n)J_0(v_nb/a)]} = \frac{v_nR_0(v_n)}{\epsilon R_1(v_n)}$$

provided that $\omega^2a^2\mu'\epsilon' > u_n^2$ and $\omega^2a^2\mu\epsilon > v_n^2$. Show that the velocity of the nth wave is $\omega a(\omega^2a^2\mu'\epsilon' - u_n^2)^{-\frac{1}{2}}$ and that the fields in $\mu'\epsilon'$ are

$$\breve{E}_\rho = k_n(\omega\mu'\epsilon')^{-1}\breve{B}_\phi = j\breve{C}e^{-jk_nz}J_1(u_n\rho/a), \qquad k_na\breve{E}_z = u_n\breve{C}_ne^{-jk_nz}J_0(u_n\rho/a)$$

where $\omega^2a^2\mu'\epsilon' - u_n^2 = \omega^2a^2\mu\epsilon - v_n^2 = k_n^2a^2$ and $a < b$.

26. If there is no reflector at $\rho = b$ in the last problem, show that with the same materials and frequencies a plane wave is impossible but that otherwise the results of the last problem hold if the Hankel functions $J_0 - jY_0$ and $J_1 - jY_1$ are substituted for R_0 and R_1, respectively.

27. An infinite medium $\mu\epsilon$ surrounds an infinite dielectric cylinder $\mu'\epsilon'$ of radius a. Show that a plane transverse magnetic wave can be propagated along the cylinder for each value of u_n that satisfies the equations

$$v_n\epsilon' J_1(u_n)K_0(v_n) + u_n\epsilon J_0(u_n)K_1(v_n) = 0, \qquad \omega^2a^2\mu'\epsilon' - u_n^2 = \omega^2a^2\mu\epsilon + v_n^2 = k_n^2a^2$$

provided that $\omega^2a^2\mu'\epsilon'$ is greater than u_n^2. Show that the velocity of the nth wave is $\omega a(\omega^2a^2\mu'\epsilon' - u_n^2)^{-\frac{1}{2}}$ and that the fields outside the cylinder are

$$\breve{E}_\rho = k_n(\omega\mu\epsilon)^{-1}\breve{B}_\phi = \breve{C}_ne^{-jk_nz}K_1(v_na^{-1}\rho), \qquad k_na\breve{E}_z = jv_n\breve{C}_ne^{-jk_nz}K_0(v_na^{-1}\rho)$$

28. If $Y_0(v_n b/a)$ is zero, verify that the conditions of problem 25 are satisfied approximately by $\epsilon = 2\epsilon' = 2\epsilon_v$, $\mu = \mu' = \mu_v$, $\omega a = .85 \times 10^9$, $u_n = 1$, $v_n = 3$, and show that the phase velocity in the z-direction is approximately 3.21×10^8 m/sec.

29. Verify that the conditions of problem 27 are satisfied approximately by taking $\epsilon' = 1.81\epsilon = 1.81\epsilon_v$ $\mu = \mu' = \mu_v$, $\omega a = 0.99 \times 10^9$, $u_n = 2.8$, $v_n = 1$ and show that the phase velocity is approximately 2.88×10^8 m/sec.

30. An infinite plane perfectly conducting surface is coated with a dielectric layer $\mu_2\epsilon_2$ of thickness a and underlies an infinite dielectric medium $\mu_1\epsilon_1$. By 12.16, show a diverging surface TM wave is possible whose vector potentials are the real parts of

$$A_1 = C_1[\varrho_1(\beta'^2 - \beta_1^2)^{\frac{1}{2}}H_0^{(2)\prime}(\beta'\rho') - k\beta'H_0^{(2)}(\beta'\rho)]e^{-(\beta'^2-\beta_1{}^2)^{\frac{1}{2}}z+j\omega t}$$
$$A_2 = C_2\{\varrho_1(\beta_2^2 - \beta'^2)^{\frac{1}{2}} \sin [(\beta_2^2 - \beta'^2)z]H_0^{(2)\prime}(\beta'\rho) - k\beta' \cos [(\beta_2^2 - \beta'^2)^{\frac{1}{2}}z]H_0^{(2)}(\beta'\rho)\}e^{j\omega t}$$

where $\beta_1^2 = \omega^2\mu_1\epsilon_1$, $\beta_2^2 = \omega^2\mu_2\epsilon_2$, and $\beta_2 > \beta' > \beta_1$. Show that its velocity is the same as that of the plane wave of problem 7, Chap. XI.

31. An infinite plane perfectly conducting surface is coated with a dielectric layer $\mu_2\epsilon_2$ of thickness a and underlies an infinite dielectric medium $\mu_1\epsilon_1$. By 12.16, show a diverging surface TE wave is possible whose vector potentials are the real parts of

$$A_1 = \phi C_1 e^{-(\beta'^2-\beta_1{}^2)^{\frac{1}{2}}z}H_1^{(2)}(\beta'\rho)e^{j\omega t}, \qquad A_2 = \phi C_2 \sin [(\beta_2^2 - \beta'^2]^{\frac{1}{2}}z]H_1^{(2)}(\beta'\rho)e^{j\omega t}$$

where $\beta_1^2 = \omega^2\mu_1\epsilon_1$, $\beta_2^2 = \omega^2\mu_2\epsilon_2$, and $\beta_2 > \beta > \beta_1$. Show that its velocity is the same as that of the plane wave of problem 8, Chap. XI.

32. An infinite perfectly conducting cylinder of radius a which is coated with a dielectric layer $\mu_2\epsilon_2$ of outer radius b passes through an infinite medium $\mu_1\epsilon_1$. Show by 12.16 that possible vector potentials for a surface TM wave are the real parts of

$$A_1 = C_1[-\varrho_1 j\beta'K_1(p_1\rho) + kp_1K_0(p_1\rho)]e^{j(\omega t-\beta'z)}$$
$$A_2 = C_2[-\varrho_1 j\beta'R_1(p_2\rho) - kp_2R_0(p_2\rho)]e^{j(\omega t-\beta'z)}$$

where $\beta_1^2 = \omega^2\mu_1\epsilon_1$, $\beta_2^2 = \omega^2\mu_2\epsilon_2$, $\beta_2 > \beta' > \beta_1$, $p_1^2 = \beta'^2 - \beta_1^2$, $p_2^2 = \beta_2^2 - \beta'^2$, and

$$R_0(p_2\rho) = Y_0(p_2a)J_0(p_2\rho) - J_0(p_2a)Y_0(p_2\rho)$$
$$R_1(p_2\rho) = Y_0(p_2a)J_1(p_2\rho) - J_0(p_2a)Y_1(p_2\rho)$$

Show that the velocity may be found from the equation

$$-\epsilon_2p_1K_0(p_1b)R_1(p_2b) = \epsilon_1p_2K_1(p_1b)R_0(p_2b)$$

If $a = 1$, $b = 2$, $\mu_2 = \mu_1$, $\epsilon_2 = 4\epsilon_1$, and $\beta_1 b = 1.174$, show that $\beta_1:\beta':\beta_2 = 1:1.048:2$ and that, when $\rho = 4.350b$, then B_ϕ has one-tenth of its value at $\rho = b$.

33. Obtain the result in 12.20 (8) by using 12.18 (1).

34. The plane of polarization of the incident wave in Fig. 12.19a is rotated so that $\boldsymbol{E}$ parallels the x-axis. Show that the diffracted vector potential at a great distance is, if A_θ' is given by 12.19 (5),

$$\breve{A}_\theta = \frac{-\cos\phi}{\cos\alpha\sin\phi}\breve{A}_\theta', \qquad \breve{A}_\phi = \frac{\cos\theta}{\cos\alpha}\breve{A}_\theta'$$

35. The opening of 12.19 is not rectangular but annular of internal and external radii a and b. Show that the diffracted vector potential at a great distance is

$$\breve{A}_\theta = \frac{\tan\phi}{\cos\theta}\breve{A}_\phi = \frac{-E\cos\alpha\sin\phi}{\omega RQ}[aJ_1(\beta Qa) - bJ_1(\beta Qb)]e^{-j\beta R}$$

where $Q = (\sin^2\alpha + \sin^2\theta - 2\sin\alpha\sin\theta\sin\phi)^{\frac{1}{2}}$.

36. The polarization plane of the incident beam in the last problem is rotated so that $\boldsymbol{E}$ parallels the x-axis. Show that the vector potential of the diffracted field at a great distance is

$$\check{A}_\theta = -\frac{\cot\phi}{\cos\theta}\check{A}_\phi = \frac{-E\cos\phi}{\omega RQ}[bJ_1(\beta Qb) - aJ_1(\beta Qa)]e^{-j\beta R}$$

37. Show that, if in 12.19 or in the last three problems, there are not one but two identical openings centered at $y = c$ and $y = -c$ then, if $R \gg c$ the single opening potential must be multiplied by $2\cos[\beta c(-\sin\alpha + y/R)]$.

38. If in the last problem the openings are at $x = c$ and $x = -c$, show that the multiplication factor is $2\cos(\beta cx/R)$.

39. A spherical shell of radius a is perfectly absorbing inside and perfectly conducting outside. An electric dipole of moment $M\cos\omega t$ is placed at the origin pointed in the $\theta = 0$ direction, and the part of the shell between $\theta = \alpha'$ and $\theta = \alpha''$ is removed. Show that 12.11 gives the external diffracted vector potential where

$$\check{W}_{tm} = M(1 - \beta^2a^2 + j\beta a)\sum_{n=1}^{\infty}\check{A}_nP_n(\cos\theta)\check{k}_n(j\beta r)$$

$$\check{A}_n = e^{-j\beta a}\left[\frac{(2n+1)P_n(u) - (n+2)uP_{n-1}(u) - (n-1)uP_{n+1}(u)}{8\pi j\omega\epsilon a^2(n-1)(n+2)\partial[a\check{k}_n(j\beta a)]/\partial a}\right]_{u=\cos\alpha'}^{u=\cos\alpha''}$$

40. The dipole in the shell of the last problem is replaced by a small coaxial loop of wire of radius b carrying a current $Ie^{j\omega t}$. Show that 12.11 (2) gives the vector potential of the diffracted field outside where

$$\check{W}_{te} = \mu Ib^2(1 + j\beta a)\sum_{n=1}^{\infty}\check{A}_nP_n(\cos\theta)\check{k}_n(j\beta r)$$

$$\check{A}_n = e^{-j\beta a}\left[\frac{(n+2)uP_{n-1}(u) + (n-1)uP_{n+1}(u) - (2n+1)P_n(u)}{8a^2\check{k}_n(j\beta a)(n-1)(n+2)}\right]_{u=\cos\alpha'}^{u=\cos\alpha''}$$

41. The boundary of a circular hole in an infinite thin plane conducting sheet is $\rho = a$. The potential of an impinging wave at the hole surface is $A_\phi = f(\rho)$. If the wave in the hole is unperturbed by the boundary, show that the vector potential of the radiated field at a great distance where $R \gg \lambda$ is the real part of

$$A_\phi = \beta R^{-1}\cos\theta e^{j(\omega t - \beta R)}\int_0^a \rho f(\rho)J_1(\beta\rho\sin\theta)\,d\rho$$

If the source is a very small loop which carries a current $\check{I}e^{j\omega t}$ and is coaxial with, and at a distance c from the hole and if $a \ll \lambda$ and $c \ll \lambda$, show that

$$\check{A}_\phi e^{j\omega t} = -\mu\beta^2\check{I}b^2(16R)^{-1}[2c - (2c^2 + a^2)(c^2 + a^2)^{-\frac{1}{2}}]\sin 2\theta e^{j(\omega t - \beta R)}$$

42. A slit of width $2a$ in an infinite plane conducting surface is bisected by the z-axis. The tangential electric field over its surface is $\boldsymbol{j}f_1(y_0) + \boldsymbol{k}f_2(y_0)$. Show with the aid of 5.331 (3) that the vector potential of the radiation field is

$$\boldsymbol{A} = \tfrac{1}{2}(\mu\epsilon)^{\frac{1}{2}}e^{j\omega t}\int_{-a}^{+a}[\boldsymbol{\phi}f_1(y_0) + \boldsymbol{k}\cos\phi' f_2(y_0)]H_1^{(2)}(\beta R')dy_0$$

where R' is the shortest distance from the strip of width dy_0 at y_0 to the field point and ϕ' is the angle between R' and the normal to the surface.

43. A plane wave whose electric vector is $\boldsymbol{k}E_0e^{j(\omega t-\beta x)}$ impinges normally on the slotted plane of the last problem. Solid lines in Fig. 4.23 show the magnetic field near the slit. Get E_z as in 12.18 (3) and then show from the last problem that if the phase is constant over the slit, the diffracted field is

$$E_z = -j\omega A_z = \tfrac{1}{4}\pi\beta^2a^2E_0 \cos\phi H_1^{(2)}(\beta\rho)e^{j\omega t}$$

where ρ is measured from, and ϕ around, the z-axis, $\rho \gg a$ and $\beta\rho$ is unrestricted. In 4.23 B, the standing wave amplitude, is twice the incident amplitude.

44. An electric field $iE_0e^{j\omega t}$ is bounded by the infinite plane conducting sheet $x = 0$ in which there is a slit of width $2a$ bisected lengthwise by the z-axis. The dotted lines in Fig. 4.23 show $\boldsymbol{E}$ near the slit. From Art. 4.23 and problem 42 show that if the phase is constant over the slit the radiated field is

$$E_\phi = -\tfrac{1}{8}j\pi\beta^2a^2E_0 \sin\phi H_1^{(2)\prime}(\beta\rho)$$

where ρ is measured from, and ϕ around, the z-axis, $\rho \gg a$, and $\beta\rho$ is unrestricted.

45. A plane wave whose magnetic vector is $\boldsymbol{k}Be^{j(\omega t-\beta x)}$ impinges normally on the slotted plane of problem 42. This creates an electric field across the slit whose form, for slowly varying fields, is shown by the ellipses in Fig. 4.22. Find $Cf_1(y_0)$ in problem 42 from 4.22 (3) and so, assuming constant phase over the slit, obtain A_ϕ for the radiated field. Find C approximately by equating the curl of A_ϕ when ρ is zero to the incident B. Thus show that, neglecting $(\beta a)^2$ terms,

$$B_z = \frac{\pi B_0e^{j\omega t}}{2[\pi - j\ln(\frac{1}{2}\gamma\beta a)]}H_0^{(2)}(\beta\rho)$$

where $\ln\gamma = -.11593$, ρ is measured from the z-axis, $\rho \gg a$, and $\beta\rho$ is unrestricted.

46. A plane polarized electromagnetic wave whose wave front makes an angle α with the yz-plane impinges on the slotted plane of problem 42. If its magnetic vector parallels the slit, show from problems 44 and 45 that the diffracted intensity at a great distance R from the slit is, if βa is so small that phase differences over the slit are negligible and $-\ln(\beta a) \gg 1$,

$$\bar{\Pi} = \frac{\pi}{2\beta R}\left\{\frac{\cos\alpha}{\ln(\beta a)} - \frac{\beta^2a^2\sin\alpha\sin\phi}{4}\right\}^2\bar{\Pi}_0$$

47. The wave of the last problem has its electric vector parallel to the slit. Show that, if βa is very small so that phase differences over the slit may be neglected, then from problem 43 the diffracted intensity is

$$\bar{\Pi} = \frac{\pi\beta^3a^4\cos^2\phi\cos^2\alpha}{8R}\bar{\Pi}_0$$

48. A hole of radius a is cut in an infinite plane perfectly conducting sheet which bounds a uniform electric standing wave field $E_0e^{j\omega t}$. By neglecting phase differences over the hole, show from 5.272 (6) and 12.18 (1) that the vector potential at a great distance from the center of the hole is

$$\check{A}_\theta = \frac{-jE_0[\sin(\beta a\sin\theta) - \beta a\sin\theta\cos(\beta a\sin\theta)]}{\pi\omega\beta R\sin^2\theta}e^{-j\beta R}$$

49. Show by the methods of 5.272 that, if a uniform standing wave magnetic field of induction $B_0e^{j\omega t}$ exists in the x-direction above and parallel to an infinite thin plane perfectly conducting sheet at $z = 0$ having a hole of radius a in it, the z-component of

B in the hole is $2B(x/\pi)(a^2 - x^2 - y^2)^{-\frac{1}{2}}$. Hence by 12.18, neglecting phase differences, show that at a great distance from the center of the hole below the sheet, where the azimuth angle from the x-axis is ϕ and the colatitude angle from the hole axis is θ, the vector potential is

$$\breve{A}_\theta = \frac{2jB_0 \sin\phi\,[\sin(\beta a \sin\theta) - \beta a \sin\theta \cos(\beta a \sin\theta)]}{\pi\beta^2 R \sin^3\theta} e^{-j\beta R}, \qquad \breve{A}_\phi = \breve{A}_\theta \cos\theta \cot\phi$$

50. A plane polarized wave strikes a thin plane perfectly conducting sheet with a hole of radius a in it at an angle α from the normal. If E_0 parallels the sheet, show, when βa is very small in problem 49, that the intensity of the diffracted radiation is

$$\bar{\Pi} = \frac{4\beta^4 a^6 \cos^2\alpha(1 - \sin^2\theta\cos^2\phi)}{\pi^2 R^2}\bar{\Pi}_0$$

Note that the standing wave B_0 of problem 49 is twice the incident running wave induction.

51. Show that the average power radiated from the hole in the last problem is

$$\bar{P} = \frac{16\beta^4 a^6 \cos^2}{3\pi}\bar{\Pi}_0$$

52. In problem 50 let the electric vector lie in the plane of incidence and show, when βa is very small in problems 48 and 49, that the intensity of the diffracted radiation is

$$\bar{\Pi} = \frac{\beta^4 a^6}{\pi^2 R^2}[4\sin^2\phi\cos^2\theta + (2\cos\phi - \sin\alpha)^2]\bar{\Pi}_0$$

Note that the standing wave amplitudes in problems 48 and 49 are twice those in Π_0.

53. Show that the average power radiated from the hole in the last problem is

$$\bar{P} = \frac{4\beta^4 a^6(4 + \sin^2\alpha)}{3\pi}\bar{\Pi}_0$$

54. An electromotance $\mathcal{E}$ is maintained across the center of a very narrow slot bounded by $x = \pm\frac{1}{2}\delta$, $z = \pm l$ in the infinite plane conducting face $y = 0$. Assuming that the field in the slot in this plane is $\boldsymbol{E}_0 = \boldsymbol{i}\delta^{-1}\mathcal{E}\sin[\beta(l - z)]$, show by 12.18 (1) that the electrical intensity outside the slot, where $y > 0$, is

$$\boldsymbol{E} = \tfrac{1}{2}\pi^{-1}\mathcal{E}(\boldsymbol{i}y - \boldsymbol{j}x)\sum_{\pm}\int_{\frac{1}{2}(l\mp l)}^{\frac{1}{2}(l\pm l)} R^{-3}(1 + j\beta R)\sin[\beta(l \mp z_1)]e^{j(\omega t - \beta R)}\,dz_1$$

where $R^2 = x^2 + y^2 + (z - z_1)^2$. Show by comparison with the curl of 12.03 (3) that

$$[\boldsymbol{E}]_{\text{slot}} = -2\mathcal{E}(\mu I_0)^{-1}[\boldsymbol{B}]_{\text{antenna}}$$

so that the fields are identical in form with those of an antenna driven at the center but with $\boldsymbol{B}$ and $\boldsymbol{E}$ interchanged so that from 12.03 (18)

$$E_\phi = -\tfrac{1}{2}\mathcal{E}(\pi\rho)^{-1}[2\cos\beta l\sin(\omega t - \beta r) - \sin(\omega t - \beta r_a) - \sin(\omega t - \beta r_b)]$$

where r_a and r_b are the distances between the ends of the slot and the field point.

55. Apply the results of the last problem to the resonant slot $l = \frac{1}{4}\lambda$ and thus show by 12.04 (4) that the radiation resistance of the slot is 363 ohms if it radiates from both faces.

NOTE: The results of the following problems are based on the Kirchhoff theory of diffraction. They are useful in optics but should be used with caution elsewhere.

56. Let $U = \psi(x, y, z)e^{j\omega t}$ be a component of the Hertz vector in an insulating medium so that it satisfies 11.01 (10) when $r = \infty$. Let

$$V = r^{-1}e^{j(\omega t - \beta r)} = \phi(x, y, z)e^{j\omega t}$$

be a similar component for a spherical wave originating at a point P. Insert ψ and ϕ in Green's theorem, 3.06 (4), and, taking the volume of integration to lie between a very small sphere surrounding P and some larger surface S, enclosing the sphere, show that

$$4\pi\psi_p = \int_S [r^{-1}e^{-j\beta r}\nabla\psi - \psi\nabla(r^{-1}e^{-j\beta r})] \cdot \boldsymbol{n}\, dS$$

where ψ_p is the value of ψ at P. This formula, the basis of Kirchhoff's diffraction theory, gives the effect at P in terms of its integral over a surface surrounding P.

57. Let S have apertures, and let U be a spherical wave originating at Q outside S. Assuming that U has the same value over the apertures as if S were absent and is zero over the remainder of S, show that

$$\psi_p = \frac{j\beta}{4\pi}\int_{S'} \frac{1}{rr_1}\left(\frac{\boldsymbol{n}\cdot\boldsymbol{r}}{r} - \frac{\boldsymbol{n}\cdot\boldsymbol{r}_1}{r_1}\right)e^{-j\beta(r+r_1)}\, dS'$$

where $\boldsymbol{r}_1$ and $\boldsymbol{r}$ are the radius vectors from Q and P to the apertures and both are much larger than β^{-1}. S' is the area of the openings.

58. Let there be a single aperture, let R and R_1 be the mean distances from its center O to P and Q, and let the coordinates of any point in it, referred to O, be x' and y'. Expand r and r_1 in powers of x' and y' and, since $\beta = 2\pi\lambda^{-1}$, show that

$$\psi_p = \frac{j}{2\lambda RR_1}\left(\frac{\boldsymbol{n}\cdot\boldsymbol{R}}{R} - \frac{\boldsymbol{n}\cdot\boldsymbol{R}_1}{R_1}\right)e^{-\frac{2\pi j(R+R_1)}{\lambda}}\int_{S'} e^{-\frac{2\pi jF(x',y',1/R,1/R_1)}{\lambda}}\, dS'$$

59. Consider a plane wave of intensity I_0 incident normally on a circular aperture of radius a so that $R_1 = \infty$, and take $R \gg a$ and P at $x = x$, $y = 0$ so that

$$F(x', y', R^{-1}, R_1^{-1}) \approx -xx'R^{-1} = \rho' \cos\theta' \sin\alpha$$

where α is the angle subtended by x at 0. Evaluate the integral by 5.302 (3) and (2), and show that the diffracted intensity is

$$I_d = \tfrac{1}{4}I_0R^{-2}\cot^2(\tfrac{1}{2}\alpha)[J_1(2\pi a\lambda^{-1}\sin\alpha)]^2a^2$$

60. Take Q and P on the axis of a circular aperture, of radius a and take R_1 and R much larger than a. Show that, if the incident intensity is I_0, that at P is

$$4R_1^2I_0(R_1 + R)^{-2}\sin^2[\pi a^2(2\lambda)^{-1}(R^{-1} + R_1^{-1})]$$

References

Most recent references give extensive bibliographies.

FLUGGE, S.: "Handbuch der Physik," Vol. XVI, Springer, 1958.

GEIGER-SCHEEL: "Handbuch der Physik," Vols. XII, XV, XX, Berlin, 1927, 1928. Gives elegant treatment of electromagnetic theory in Vol. XII.

HARRINGTON, R. F.: "Time-harmonic Electromagnetic Fields," McGraw-Hill, 1963. Discusses multipoles, Babinet's principle, and pertinent topics.

Hertz, H. R.: "Electric Waves," Macmillan, 1893. Radiation field drawings.

Jackson, J. D.: "Classical Electrodynamics," Wiley, 1962. Good treatment of multipoles, Babinet's principle and diffraction, and cgs units.

King, R. W. P.: "Theory of Linear Antennas," Harvard, 1956. A complete treatment with extensive bibliography.

MacDonald, H. M.: "Electromagnetism," Bell, 1934. Solves diffraction problems.

Mason, M., and W. Weaver: "The Electromagnetic Field," University of Chicago Press, 1929, and Dover. Gives excellent treatment of retarded potentials.

Maxwell, J. C.: "Electricity and Magnetism," 3d ed. (1891), Dover, 1954. The original treatment of the subject.

Panofsky, W. K. H., and Melba Phillips: "Classical Electricity and Magnetism," Addison-Wesley, 1962. Covers multipoles and related topics.

Papas, C. H.: "Theory of Electromagnetic Wave Propagation," McGraw-Hill, 1965. Covers multipoles and antennas with radiation patterns and applications.

Schelkunoff, S. A.: "Advanced Antenna Theory," Wiley, 1952.

Schelkunoff, S. A., and T. Friis: "Antennas; Theory and Practice," Wiley, 1952. Formulas for multipoles and antennas with specific cases in problems.

Silver, S.: "Microwave Antenna Theory and Design," McGraw-Hill, 1949. Treats general antenna theory and diffraction extensively.

Stratton, J. A.: "Electromagnetic Theory," McGraw-Hill, 1941. Extensive and rigorous mathematical treatment of the whole subject.

Tralli, N.: "Classical Electromagnetic Theory," McGraw-Hill, 1963. Treats concisely much of the material of this chapter.

Van Bladel, J.: "Electromagnetic Fields," McGraw-Hill, 1964. Expands topics of this chapter to include stratified mediums, pulses, and other cases.

Wien-Harms: "Handbuch der Experimentalphysik," Vol. XI, Leipzig, 1932.

CHAPTER XIII

WAVE GUIDES AND CAVITY RESONATORS

13.00. Waves in Hollow Cylindrical Tubes.—That plane waves may be guided by two or more mutually external perfectly conducting cylinders was shown in 11.14 to 11.18. If the propagation space is enclosed by a perfectly conducting cylinder, other unattenuated wave types will propagate at very high frequencies, and if this is the only boundary, a simple plane wave is no longer possible. Each external and internal (if any) boundary of a cylindrical wave guide is generated by moving a straight line transversely keeping it parallel to the z-axis. It is often considered closed by the $z = 0$ plane and extends from $z = 0$ to $z = \infty$. In the text following 11.01 (5) it was pointed out that in the Coulomb gauge the vector potential $\boldsymbol{A}$ is derivable from a solution W of the scalar propagation equation in two ways, one giving transverse electric and one transverse magnetic waves. The differential equations are

$$\nabla^2 \boldsymbol{A} = \mu\epsilon \frac{\partial^2 \boldsymbol{A}}{\partial t^2}, \qquad \nabla^2 W_{te} = \mu\epsilon \frac{\partial^2 W_{te}}{\partial t^2}, \qquad \nabla^2 W_{tm} = \mu\epsilon \frac{\partial^2 W_{tm}}{\partial t^2} \tag{1}$$

We are interested in a steady-state solution of angular frequency ω. In this case from 10.01 a vector potential satisfying the first equation and its curl may be obtained from solutions of the scalar equations by

$$\check{\boldsymbol{A}} = \boldsymbol{\nabla} \times (\boldsymbol{k}\check{W}_{te} + \boldsymbol{k} \times \boldsymbol{\nabla}\check{W}_{tm}), \qquad \check{\boldsymbol{B}} = -\boldsymbol{\nabla} \times (\boldsymbol{k}\beta^2\check{W}_{tm} + \boldsymbol{k} \times \boldsymbol{\nabla}\check{W}_{te}) \tag{2}$$

where $\beta^2 = \omega^2\mu\epsilon$. These equations are written out in cylindrical polar coordinates in 12.16 (4) and (5). As in 12.16, we may write $\check{W}$ as a product of a function of z by a function of the transverse coordinates so that either scalar equation in (1) splits into two. Thus

$$\check{W} = U\check{Z}, \qquad \nabla_2^2 U \pm \beta_{mn}^2 U = 0, \qquad \frac{d^2\check{Z}}{dz^2} + (\beta^2 \mp \beta_{mn}^2)\check{Z} = 0 \tag{3}$$

Here U depends only on the transverse coordinates u_1 and u_2, and boundary and symmetry conditions fix β_{mn}. If $\boldsymbol{n}$ is a unit vector which like $\check{\boldsymbol{A}}$ is normal to the curved surface, then from (2) the conditions there are

$$0 = \boldsymbol{n} \times \check{\boldsymbol{A}}_{te} = -\boldsymbol{n} \cdot \boldsymbol{\nabla}(\boldsymbol{k}\check{W}_{te}) \qquad \text{or} \qquad \frac{\partial U_{te}}{\partial n} = 0 \tag{4}$$

$$\begin{aligned} 0 &= \boldsymbol{n} \times [\boldsymbol{\nabla} \times (\boldsymbol{k} \times \boldsymbol{\nabla}\check{W}_{tm})] = \boldsymbol{n} \times [\boldsymbol{k}\nabla^2\check{W}_{tm} - \boldsymbol{k} \cdot \boldsymbol{\nabla}(\boldsymbol{\nabla}\check{W}_{tm})] \\ &= -\beta^2\boldsymbol{n} \times \boldsymbol{k}\check{W}_{tm} - \frac{\partial}{\partial z}(\boldsymbol{n} \times \boldsymbol{\nabla}\check{W}_{tm}) = \boldsymbol{s}_1\left[\beta^2\check{W}_{tm} + \frac{\partial^2\check{W}_{tm}}{\partial z^2}\right] - \boldsymbol{k}\frac{\partial^2\check{W}_{tm}}{\partial s\,\partial z} \\ &= \boldsymbol{s}_1\beta_{mn}^2\check{W}_{tm} - \boldsymbol{k}\frac{\partial^2\check{W}_{tm}}{\partial s\,\partial z} \end{aligned}$$

where s is a coordinate measured around a cylindrical surface along its curve of intersection with a plane normal to the z-axis so that $\mathbf{s}_1 = \mathbf{k} \times \mathbf{n}$. The conditions imposed on U at the boundary are therefore

$$U_{tm} = 0 \qquad \text{or} \qquad \beta_{mn} = 0 \qquad \text{and} \qquad \frac{\partial U_{tem}}{\partial s} = 0 \tag{5}$$

The second pair of conditions in (5) are those met by the principal waves described in 11.14 to 11.18 whose velocity is independent of frequency. These are possible only with two or more conducting cylinders.

The differential equations of (3) are of the second order so each has two solutions, and the general solution is of the form

$$\breve{W} = [\breve{A}U_1(u_1, u_2) + \breve{B}U_2(u_1, u_2)](\breve{C}e^{-\breve{\Gamma}_{mn}z} + \breve{D}e^{\breve{\Gamma}_{mn}z}) \tag{6}$$

$$\breve{\Gamma}_{mn} = j(\beta^2 - \beta^2_{mn})^{\frac{1}{2}} = \alpha_{mn} + j\beta'_{mn} \tag{7}$$

When $\beta > \beta_{mn}$, the propagation constant $\breve{\Gamma}_{mn}$ is a pure imaginary so $\alpha = 0$ and the first and second exponential terms in (6) indicate unattenuated waves in the positive and negative z-direction, respectively. If $\beta < \beta_{mn}$, we put $\breve{C} = 0$ when z is positive or $\breve{D} = 0$ when z is negative since the fields must remain finite. For the remaining term, $\breve{\Gamma}_{mn}$ is real so $\beta'_{mn} = 0$ which gives an exponentially damped term. The cutoff frequency ν_{mn} and the cutoff wave length for the mnth mode are

$$\nu_{mn} = \frac{\beta_{mn}}{2\pi}(\mu\epsilon)^{-\frac{1}{2}} = \frac{\omega_{mn}}{2\pi} \qquad \lambda_{mn} = \frac{2\pi}{\beta_{mn}} \tag{8}$$

To extend the concept of characteristic impedance to wave guides, the definition of 11.15 (15) is put in terms of $\breve{\mathbf{E}}$ and $\breve{\mathbf{B}}$ by 11.15 (6). Thus

$$|\breve{Z}_k|_{te} = \mp\frac{\mu\breve{E}_{te}}{\mathbf{k} \times \breve{\mathbf{B}}_{te}} = \frac{\pm j\omega\mu|\nabla \times \mathbf{k}\breve{W}_{te}|}{|\mathbf{k} \times [\nabla \times (\nabla \times \mathbf{k}\breve{W}_{te})]|} = \frac{\pm j\omega\mu|\nabla \times \mathbf{k}\breve{W}_{te}|}{|\partial(\nabla \times \mathbf{k}\breve{W}_{te})/\partial z|} = \pm\frac{j\omega\mu}{\breve{\Gamma}_{mn}} \tag{9}$$

$$|\breve{Z}_k|_{tm} = \pm\frac{\mu|\mathbf{k} \times \breve{\mathbf{E}}_{tm}|}{|\breve{\mathbf{B}}_{tm}|} = \mp\frac{j|\mathbf{k} \times [\nabla \times (\mathbf{k} \times \nabla\breve{W}_{tm})]|}{\omega\epsilon|\nabla \times \mathbf{k}\breve{W}_{tm}|} = \mp\frac{j|\partial(\nabla \times \mathbf{k}\breve{W}_{tm})/\partial z|}{\omega\epsilon|\nabla \times \mathbf{k}\breve{W}_{tm}|}$$

$$= \mp\frac{j\breve{\Gamma}_{mn}}{\omega\epsilon} \tag{10}$$

In terms of the cutoff frequency ν_{mn}, these are

$$|\breve{Z}_k|_{te} = \frac{\pm\mu^{\frac{1}{2}}\epsilon^{-\frac{1}{2}}}{[1 - (\nu_{mn}/\nu)^2]^{\frac{1}{2}}}, \qquad |\breve{Z}_k|_{tm} = \pm\mu^{\frac{1}{2}}\epsilon^{-\frac{1}{2}}[1 - (\nu_{mn}/\nu)^2]^{\frac{1}{2}} \tag{11}$$

From 11.15 (6), the upper or lower sign is to be used according as the wave is in the negative or positive z-direction, respectively. Note that both wave types give a real characteristic impedance, indicating power propagation above the cutoff frequency and a reactive one below it. Observe that this differs from the transmission line definition in 11.16.

From (6) and (7), when $\beta > \beta_{mn}$, waves propagate with a phase velocity

$$v_{mn} = \frac{\omega}{\beta'_{mn}} = \frac{\omega}{(\beta^2 - \beta^2_{mn})^{\frac{1}{2}}} = \frac{v}{[1 - (\nu_{mn}/\nu)^2]^{\frac{1}{2}}} = \frac{v}{[1 - (\lambda/\lambda_{mn})^2]^{\frac{1}{2}}} \quad (12)$$

where v is the velocity of a free wave in the medium filling the tube. The signal velocity is given by 11.19 (2) and (12) to be

$$(v_s)_{mn} = \frac{\partial\omega}{\partial\beta'_{mn}} = \frac{1}{(\mu\epsilon)^{\frac{1}{2}}}\frac{\partial\beta}{\partial\beta'_{mn}} = \frac{\beta'_{mn}}{\beta(\mu\epsilon)^{\frac{1}{2}}} = \frac{1}{\mu\epsilon v_{mn}} = \frac{v^2}{v_{mn}} \quad (13)$$

Thus $v_{mn} > v > (v_s)_{mn}$. At sufficiently high frequencies, both phase and signal velocities approach the velocity of the free wave.

From (2) and (3), the electric and magnetic fields in terms of U are

$$\breve{E}_{te} = -j\omega\breve{A}_{te} = j\omega k \times \nabla U_{te}e^{-j\beta'_{mn}z} \quad (14)$$

$$\breve{B}_{te} = \nabla \times \breve{A}_{te} = (-j\beta'_{mn}\nabla_2 U_{te} + k\beta^2_{mn}U_{te})e^{-j\beta'_{mn}z} \quad (15)$$

$$\breve{E}_{tm} = -j\omega\breve{A}_{tm} = (\omega\beta'_{mn}\nabla_2 U_{tm} + kj\omega\beta^2_{mn}U_{tm})e^{-j\beta'_{mn}z} \quad (16)$$

$$\breve{B}_{tm} = \nabla \times \breve{A}_{tm} = \beta^2 k \times \nabla U_{tm}e^{-j\beta'_{mn}z} \quad (17)$$

where U_{te} and U_{tm} satisfy (3) and from (2) have different dimensions.

13.01. Attenuation in Hollow Wave Guides.—Wave guides are used to transmit high-frequency power from one point to another so transmission losses are important. If an imperfect dielectric fills the tube, energy will be dissipated there, but in the common air-filled guide such losses are seldom important and their calculation will be left to the problems at the end of the chapter. The eddy current energy loss in the walls is, however, unavoidable and must be considered. In all practical cases, the wall conductivity is so high that field solutions for perfectly conducting tubes are excellent approximations and can be used to calculate P. In any case where losses in a quantity are proportional to the quantity itself, there is exponential attenuation. Equation 10.02 (8) shows that the power loss in a conducting surface is proportional to the square of the tangential magnetic field at the surface. Therefore the fields are damped exponentially as they pass down the tube and, if their attenuation factor is α, that of the Poynting vector is 2α. The periodic factors disappear in the time average $\bar{P}$ of the transmitted power leaving $\bar{P}e^{-2\alpha z}$. The z-derivative of $\bar{P}$, evaluated by 10.02 (8) and divided by $\bar{P}$, is

$$2\alpha = -\frac{1}{\bar{P}}\frac{\partial P}{\partial z} = \frac{\tau}{2\mu^2\bar{P}\delta}\oint \breve{B}\cdot\hat{B}\,ds \quad (1)$$

where $\partial\bar{P}/\partial z$ is the average power loss per unit length, $\breve{B}$ the magnetic induction next the wall, τ the wall resistivity, and δ the skin thickness. At the wall $\nabla_2 U_{te} = s_1\partial U_{te}/\partial s$ so that, from 13.00 (15) and (17),

$$\breve{B}_{te}\cdot\hat{B}_{te} = (\beta^2 - \beta^2_{mn})\left(\frac{\partial U_{te}}{\partial s}\right)^2 + \beta^4_{mn}U^2_{te}, \quad \breve{B}_{tm}\cdot\hat{B}_{tm} = \beta^4\nabla_2 U_{tm}\cdot\nabla_2 U_{tm} \quad (2)$$

By 11.17 (3), the average z-directed energy passing through unit area per second is the real part of $\frac{1}{2}\mu^{-1}\breve{\boldsymbol{E}} \times \hat{\boldsymbol{B}}$ so that 13.00 (14) to (17) give

$$|\Pi_{te}|_z = -\tfrac{1}{2}\mu^{-1}\omega\beta'_{mn}[(\boldsymbol{k} \times \nabla U_{te}) \times \nabla_2 U_{te}]_z = \tfrac{1}{2}\mu^{-1}\omega(\beta^2 - \beta^2_{mn})^{\frac{1}{2}}\nabla_2 U_{te} \cdot \nabla_2 U_{te} \tag{3}$$

$$\begin{aligned}|\Pi_{tm}|_z &= \tfrac{1}{2}\mu^{-1}\omega\beta'_{mn}\beta^2[\nabla_2 U_{tm} \times (\boldsymbol{k} \times \nabla U_{tm})]_z \\ &= \tfrac{1}{2}\mu^{-1}\omega\beta^2(\beta^2 - \beta^2_{mn})^{\frac{1}{2}}\nabla_2 U_{tm} \cdot \nabla_2 U_{tm}\end{aligned} \tag{4}$$

To simplify the surface integral of the scalar product in (3) and (4), we expand and note that the first term of the surface integral vanishes when transformed into a line integral around the boundary because, by 13.00 (4) or (5), either U or $\boldsymbol{n} \cdot \nabla_2 U$ is zero there. Thus, using 13.00 (3),

$$\int_S \nabla_2 U \cdot \nabla_2 U \, dS = \int_S [\nabla_2 \cdot (U\nabla_2 U) - U\nabla_2^2 U] \, dS = \beta^2_{mn}\int_S U^2 \, dS \tag{5}$$

From (1), (2), and (3), the attenuation of transverse electric waves is

$$\alpha_{te} = \frac{\tau}{2\mu v\delta}\,\frac{\beta^{-2}_{mn}[1 - (\nu_{mn}/\nu)^2]\oint(\partial U_{te}/\partial s)^2\,ds + (\nu_{mn}/\nu)^2\oint U^2_{te}\,ds}{[1 - (\nu_{mn}/\nu)^2]^{\frac{1}{2}}\int_S U^2_{te}\,dS} \tag{6}$$

Here δ is the skin depth $(\frac{1}{2}\omega\mu'\gamma)^{-\frac{1}{2}}$, μ' the permeability of the wall, ν the frequency, ν_{mn} the cutoff frequency, v the free wave velocity in $\mu\epsilon$, and β_{mn} is $2\pi v^{-1}\nu_{mn}$. From (1), (2), and (4), since $\nabla_2 U$ is normal to the wall,

$$\alpha_{tm} = \frac{\tau}{2\mu v\delta}\,\frac{\oint(\partial U_{tm}/\partial n)^2\,ds}{\beta^2_{mn}[1 - (\nu_{mn}/\nu)^2]^{\frac{1}{2}}\int_S U^2_{tm}\,dS} \tag{7}$$

This equation shows that the attenuation of transverse magnetic waves increases with frequency in the same way for all forms of cross section. From Eq. (6) the attenuation of transverse electric waves depends on the form of the tube cross section and increases with frequency unless the first numerator term vanishes, as in some circular pipe modes.

13.02. The Rectangular Wave Guide.—Consider a tube bounded by the planes $x = 0$, $x = a$, $y = 0$, $y = b$. By inspection, solutions $U\ (x,y)$ of 13.00 (3), which satisfy boundary conditions 13.00 (4) and (5), are

$$U_{te} = C_{mn}\cos\frac{m\pi x}{a}\cos\frac{n\pi y}{b} \qquad U_{tm} = C_{mn}\sin\frac{m\pi x}{a}\sin\frac{n\pi y}{b} \tag{1}$$

$$\beta^2_{mn} = \pi^2\left(\frac{m^2}{a^2} + \frac{n^2}{b^2}\right) \quad \text{or} \quad \nu^2_{mn} = \tfrac{1}{4}v^2\left(\frac{m^2}{a^2} + \frac{n^2}{b^2}\right) = \frac{v^2}{\lambda^2_{mn}} \tag{2}$$

if m and n are integers. From 13.00 (14) and (15) the TE fields are

$$\breve{\boldsymbol{E}}_{te} = j\omega\pi\breve{C}\left(\boldsymbol{i}\frac{n}{b}\cos\frac{m\pi x}{a}\sin\frac{n\pi y}{b} - \boldsymbol{j}\frac{m}{a}\sin\frac{m\pi x}{a}\cos\frac{n\pi y}{b}\right)e^{-j\beta'_{mn}z} \tag{3}$$

$$\begin{aligned}\breve{\boldsymbol{B}}_{te} = \breve{C}\Big\{j\pi\beta'_{mn}\Big[\boldsymbol{i}\frac{m}{a}\sin\frac{m\pi x}{a}\cos\frac{n\pi y}{b} + \boldsymbol{j}\frac{n}{b}\cos\frac{m\pi x}{a}\sin\frac{n\pi y}{b}\Big] \\ + \boldsymbol{k}\beta^2_{mn}\cos\frac{m\pi x}{a}\cos\frac{n\pi y}{b}\Big\}e^{-j\beta'_{mn}z}\end{aligned} \tag{4}$$

From (1), (2), and 13.01 (6), the attenuation of these waves is

$$|\alpha_{te}|_{mo} = \frac{\tau}{\mu v \delta} \frac{a + 2b(\nu_{mo}/\nu)^2}{ab[1 - (\nu_{mo}/\nu)^2]^{\frac{1}{2}}} \tag{5}$$

$$|\alpha_{te}|_{mn} = \frac{2\tau}{\mu v \delta}\left\{\frac{(m^2b + n^2a)[1 - (\nu_{mn}/\nu)^2]^{\frac{1}{2}}}{m^2b^2 + n^2a^2} + \frac{(a + b)(\nu_{mn}/\nu)^2}{ab[1 - (\nu_{mn}/\nu)^2]^{\frac{1}{2}}}\right\} \tag{6}$$

From (2), the longest cutoff wave length for the TE waves in a guide for which $a = 2$ cm and $b = 4$ cm occurs when $m = 0$, $n = 1$, for then $\lambda_{01} = 8$ cm. This is independent of a. The next longest at $m = 1$, $n = 0$ is $\lambda_{10} = 4$ cm and is independent of b. As we shall see later, the

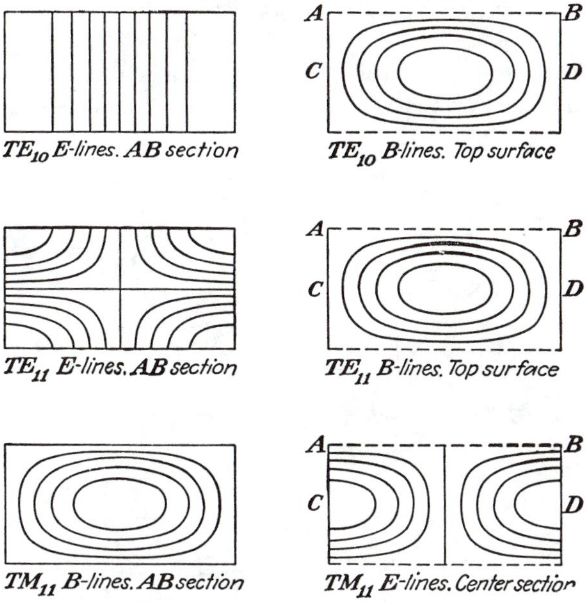

FIG. 13.02a.

guide coupling can be arranged to excite some modes and not others. For a 5-cm wave transmitted only in the TE_{01} mode, $\delta \approx 8.5 \times 10^{-7}$ m in copper. Thus, from (5), the attenuation is $\alpha_{te} \approx 0.005$ so by *Pc* 757 or *Dw* 550.2, the field strength drops about $\frac{1}{2}$ per cent and the energy 1 per cent in 1 m of guide. From 13.00 (12), the λ_{10} attenuation is

$$\alpha_{10} = j\beta'_{10} = 0.3\pi = 0.94$$

or about 200 times that of λ_{01}. From 13.00 (12) and (13), the signal velocity is 78 per cent that of a free wave. Figure 13.02a shows field maps of the TE_{10} and TE_{11} waves. From 13.00 (16) and (17), the TM

wave fields are

$$\check{\boldsymbol{E}}_{tm} = \check{C}\left\{\omega\pi\beta'_{mn}\left[\boldsymbol{i}\frac{m}{a}\cos\frac{m\pi x}{a}\sin\frac{n\pi y}{b} + \boldsymbol{j}\frac{n}{b}\sin\frac{m\pi x}{a}\cos\frac{n\pi y}{b}\right]\right.$$
$$\left. + \boldsymbol{k}j\omega\beta^2_{mn}\sin\frac{m\pi x}{a}\sin\frac{n\pi y}{b}\right\}e^{-j\beta'_{mn}z} \quad (7)$$

$$\check{\boldsymbol{B}}_{tm} = \pi\beta^2\check{C}\left(-\boldsymbol{i}\frac{n}{b}\sin\frac{m\pi x}{a}\cos\frac{n\pi y}{b} + \boldsymbol{j}\frac{m}{a}\cos\frac{m\pi x}{a}\sin\frac{n\pi y}{b}\right)e^{-j\beta'_{mn}z} \quad (8)$$

From (1), (2), and 13.01 (7), the attenuation of these waves is

$$\alpha_{tm} = \frac{2\tau}{\mu v\delta}\,\frac{m^2b^3 + n^2a^3}{ab(m^2b^2 + n^2a^2)[1 - (\nu_{mn}/\nu)^2]^{\frac{1}{2}}} \quad (9)$$

From (2), the longest TM cutoff wave length in a guide for which $a = 2$ cm and $b = 4$ cm occurs when $m = n = 1$ at $\lambda_{11} = 3.58$ cm, and the next is

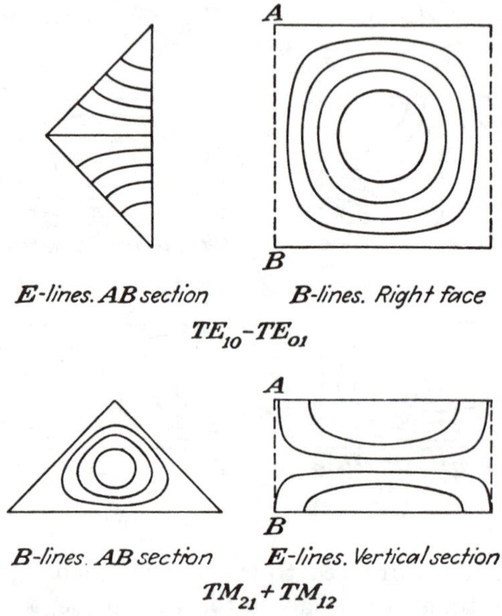

FIG. 13.02*b*.

at $\lambda_{12} = 2.83$ cm. A field map of the TM_{11} wave is shown in Fig. 13.02*a*.

Superposition of two square wave-guide fields gives that of the isosceles right-triangular guide of Fig. 13.02*b* with the same cutoff frequency.

13.03. Green's Function for Rectangular Guide.—A wave guide is often excited by wire loops or stubs carrying sinusoidal currents. The resultant fields can always be found by integration if the Green's function for a current element is known. The most general current element is

$$\check{\boldsymbol{I}}(x,y,z)\,ds = \boldsymbol{i}\check{I}_x(x,y,z)\,dx + \boldsymbol{j}\check{I}_y(x,y,z)\,dy + \boldsymbol{k}\check{I}_z(x,y,z)\,dz \tag{1}$$

It is convenient to work in the Coulomb gauge of 11.01 (4) and 11.01 (5) in which the vector potential comes entirely from the solenoidal part of $\check{\boldsymbol{I}}$ and satisfies the wave equation and Ψ comes from the lamellar part of $\check{\boldsymbol{I}}$ and satisfies Poisson's equation being everywhere in phase. If $|\check{\boldsymbol{I}}|$ is not a function of x, y, z as in a uniform current loop, then only $\check{\boldsymbol{A}}$ is needed, but if it depends on x, y, z as in stub excitation, then both $\check{\boldsymbol{A}}$ and Ψ are needed for the whole field. In all cases the propagated field as well as the magnetic field comes entirely from $\check{\boldsymbol{A}}$. The Green's function must satisfy the boundary conditions that the electric field is normal to the guide walls and the magnetic field tangential to them. The procedure will be to find Green's function for $\check{I}_x$, $\check{I}_y$, and $\check{I}_z$ and form the general case by (1). The calculations are made for elements at $z = z_0$ in a guide running from $z = -\infty$ to $z = +\infty$. The function for a guide closed at $z = 0$ is then obtained by superimposing an image Green's function of suitable sign and orientation in which $-z_0$ replaces z_0.

First consider a y-directed element. It is evident that the guide walls could be replaced by an infinite set of images in the $z = z_0$ plane, each image being parallel or antiparallel to the element and so having a magnetic field normal to the y-axis. Such a field is given by replacement of $\boldsymbol{k}$ by $\boldsymbol{j}$ in 13.00 (2), so that

$$\check{\boldsymbol{A}} = \nabla \times \boldsymbol{j} \times \nabla \check{W}, \qquad \check{\boldsymbol{B}} = \nabla \times \boldsymbol{j}\beta^2\check{W} \tag{2}$$

$$\check{W} = -\check{C}_{mn} \sin\frac{m\pi x}{a} \cos\frac{n\pi y}{b} e^{-j\beta'_{mn}|z-z_0|} \tag{3}$$

Ampère's circuital law, 7.01 (2), is used to get $\check{C}_{mn}$ for which $\check{B}_x$ is needed. From (2) and (3),

$$\check{B}_x = j\beta^2 \sum_{m=1}^{\infty}\sum_{n=0}^{\infty} \beta'_{mn}\check{C}_{mn(y)} \sin\frac{m\pi x}{a} \cos\frac{n\pi y}{b} e^{-j\beta'_{mn}|z-z_0|} \tag{4}$$

Now integrate $\check{B}_x$ across the guide at constant y in the $z = z_0$ plane except at $x = x_0$, which the path avoids by an infinitesimal detour as shown in Fig. 13.03a. On this path, from symmetry, $\check{B}_x$ is zero except between y_0 and $y_0 + dy_0$ where it passes around the current element and equals $\frac{1}{2}\mu\check{I}$ from 7.01 (2). Multiplication of (4) by $\sin(m\pi x/a)\cos(n\pi y/b)\,dx\,dy$ and integration over the ranges $0 < x < a$ and $0 < y < b$ yield

$$\tfrac{1}{2}\mu\check{I}_y\,dy_0 \sin\frac{m\pi x_0}{a} \cos\frac{n\pi y_0}{b} = j\beta'_{mn}\beta^2\tfrac{1}{4}ab(1 + \delta_n^0)\check{C}_{mn(y)} \tag{5}$$

This determines $\check{C}_{mn(y)}$ for $\check{A}_{(y)}$ which takes the form

$$\check{\mathbf{A}}_{(y)} = \sum_{m=1}^{\infty}\sum_{n=0}^{\infty} j\check{C}_{mn(y)}\left[-\mathbf{i}\frac{mn\pi^2}{ab}\cos\frac{m\pi x}{a}\sin\frac{n\pi y}{b} + \mathbf{j}\left(\beta^2 - \frac{n^2\pi^2}{b^2}\right)\sin\frac{m\pi x}{a}\cos\frac{n\pi y}{b} - \mathbf{k}\frac{j\beta'_{mn}n\pi}{b}\sin\frac{m\pi x}{a}\sin\frac{n\pi y}{b}\right]e^{-j\beta'_{mn}|z-z_0|} \qquad (6)$$

$$(\beta'_{mn})^2 = \omega^2\mu\epsilon - \left(\frac{m\pi}{a}\right)^2 - \left(\frac{n\pi}{b}\right)^2 = \beta^2 - \beta^2_{mn} \qquad (7)$$

There remains the local field due to the lamellar part of $\check{I}_y\,dy_0$. This is found from problem 113, Chap. V, by interchanging m and n, using the

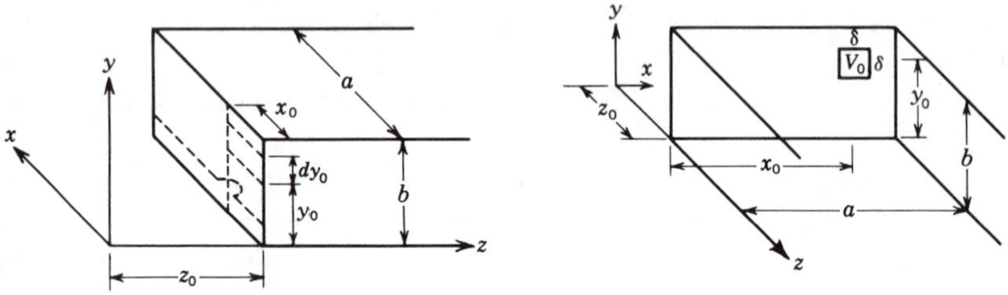

(a) FIG. 13.03a, b. (b)

β_{mn} notation of (7), writing $\check{I}_y/(j\omega)$ for q, differentiation with respect to y_0 as in 1.07 (1), and replacement of $(\partial V/\partial y_0)\,dy_0$ by Ψ. Thus

$$\check{\Psi}_{(y)} = \frac{2\check{I}_y\,dy_0}{j\omega\epsilon ab^2}\sum_{m=1}^{\infty}\sum_{n=1}^{\infty}\frac{n}{\beta_{mn}}\sin\frac{m\pi x_0}{a}\cos\frac{n\pi y_0}{b}\sin\frac{m\pi x}{a}\sin\frac{n\pi y}{b}e^{-\beta_{mn}|z-z_0|} \qquad (8)$$

The potentials of the x-component in (1) can be formed from (5), (6), and (8) by interchanging x_0 and y_0, a and b, m and n, and x and y as well as permuting $\mathbf{i}$, $\mathbf{j}$, and $\mathbf{k}$ in $\check{\mathbf{A}}_{(y)}$. Thus

$$\check{\mathbf{A}}_{(x)} = \sum_{m=0}^{\infty}\sum_{n=1}^{\infty} j\check{C}_{mn(x)}\left[\mathbf{i}\left(\beta^2 - \frac{m^2\pi^2}{a^2}\right)\sin\frac{n\pi y}{b}\cos\frac{m\pi x}{a} - \mathbf{j}\frac{mn\pi^2}{ab}\cos\frac{n\pi y}{b}\sin\frac{m\pi x}{a} + \mathbf{k}j\beta'_{mn}\frac{m\pi}{a}\sin\frac{m\pi x}{a}\sin\frac{n\pi y}{b}\right]e^{-j\beta'_{mn}|z-z_0|} \qquad (9)$$

$$\check{\Psi}_{(x)} = \frac{2\check{I}_x\,dx_0}{j\omega\epsilon a^2 b}\sum_{m=1}^{\infty}\sum_{n=1}^{\infty}\frac{m}{\beta_{mn}}\cos\frac{m\pi x_0}{a}\sin\frac{n\pi y_0}{b}\sin\frac{m\pi x}{a}\sin\frac{n\pi y}{b}e^{-\beta_{mn}|z-z_0|} \qquad (10)$$

As there are no orthogonal functions in the z-direction, a description of the dipole moment in terms of x and y, such as that worked out following 12.21 (8), is needed. This suggests producing a dipole field by the arrangement of Fig. 13.03b in which the $z = z_0$ plane is earthed except for the δ^2 area at $x = x_0$, $y = y_0$, which is raised to potential $V_0 e^{j\omega t}$. To get V_0 in terms of $\check{I}_z\, dz_0$, start with 12.21 (8).

$$\check{I}_z\, dz_0 = \frac{j\omega\pi\epsilon(b^2 - a^2)V_0}{\ln(b/a)} \xrightarrow[a\to b]{} 2j\omega\epsilon V_0 \pi a^2 = 2j\omega\epsilon V_0\, dS \tag{11}$$

Clearly the magnetic fields of $i_z\, dz_0$ and all its images are normal to z and given by 13.02 (7) and 13.02 (8). Thus the tangential component of $\check{\boldsymbol{E}}$ in the plane $z = z_0$ is

$$\check{\boldsymbol{E}}_t = \omega\pi \sum_{m=1}^{\infty}\sum_{n=1}^{\infty} \check{C}_{mn(z)}\beta'_{mn}\left[\boldsymbol{i}\frac{m}{a}\cos\frac{m\pi x}{a}\sin\frac{n\pi y}{b} + \boldsymbol{j}\frac{n}{b}\sin\frac{m\pi x}{a}\cos\frac{n\pi y}{b}\right] \tag{12}$$

Denote the bracket by [ ], take the scalar product of (12) by [ ] $dx\,dy$, and integrate over the z_0-plane. The integral of [ ]·[ ] $dx\,dy$ on the right gives $\frac{1}{4}ab\beta_{mn}^2$. On the left side the tangential $\check{\boldsymbol{E}}$ is zero except on the boundaries of the V_0 square, and its integral across the boundary is V_0 or $-V_0$ depending on the direction of crossing. Therefore the integral of the left side over the vertical edges of the square is

$$V_0\,\delta\frac{m}{a}\left\{\cos\left[\frac{m\pi}{a}\left(x_0 + \frac{1}{2}\delta\right)\right] - \cos\left[\frac{m\pi}{a}\left(x_0 - \frac{1}{2}\delta\right)\right]\right\}\sin\frac{n\pi y_0}{b}$$

The y-integral is similar with m, n, y_0, x_0, a, b, replacing n, m, x_0, y_0, b, a. Add the two integrals, combine terms and note that δ is small, and obtain for the left side $V_0\,\delta^2\pi\beta_{mn}^2 \sin(m\pi x_0/a)\sin(n\pi y_0/b)$. Thus

$$\check{C}_{mn(z)} = \frac{-2I_z\, dz_0}{\omega^2\epsilon ab\beta'_{mn}}\sin\frac{m\pi x_0}{a}\sin\frac{n\pi y_0}{b} \tag{13}$$

If $\check{C}$ in 13.02 (7) is replaced by $\check{C}_{mn(z)}$ and it is summed over m and n and multiplied by j/ω, it becomes $\check{\boldsymbol{A}}_{(z)}$.

$$\check{\boldsymbol{A}}_{(z)} = \sum_{m=1}^{\infty}\sum_{n=1}^{\infty}\check{C}_{mn(z)}\left\{j\pi\beta'_{mn}\left[\boldsymbol{i}\frac{m}{a}\cos\frac{m\pi x}{a}\sin\frac{n\pi y}{b} + \boldsymbol{j}\frac{n}{b}\sin\frac{m\pi x}{a}\cos\frac{n\pi y}{b}\right]\right.$$
$$\left. - \boldsymbol{k}\beta_{mn}^2\sin\frac{m\pi x}{a}\sin\frac{n\pi y}{b}\right\}e^{-j\beta'_{mn}|z-z_0|} \tag{14}$$

The scalar potential for the z-directed dipole can also be written down from problem 113, Chap. V, by calculating $(\partial V/\partial z_0)\, dz_0$ and making the same substitutions as for (8).

$$\Psi_{(z)} = \frac{2I_z\, dz_0}{j\omega\pi\epsilon ab}\sum_{m=0}^{\infty}\sum_{n=0}^{\infty} \sin\frac{m\pi x_0}{a}\sin\frac{n\pi y_0}{b}\sin\frac{m\pi x}{a}\sin\frac{n\pi y}{b}e^{-\beta_{mn}|z-z_0|} \quad (15)$$

The integral of the Green's function over a wave-guide antenna plays the same role as the retarded potential integrals in free-space antennas, and both require a knowledge of the current distribution, which is difficult to calculate. A plausible sinusoidal distribution is usually assumed, and the transmitted wave is rarely very sensitive to errors in this assumption. If the wire in loop or stub is not too thick, then the fields outside it are essentially the same as for an infinitely thin antenna along its axis. In input impedance calculations the fields between axis and surface must be excluded to avoid infinite field energies.

Each mode acts as an independent circuit shunted across the antenna element. A given current element excites every mode that has an electric field along it. The antenna impedance is minus the sum of the line integrals of the mode electric fields along the wire surface, in the direction of the current. Add to this the capacitive contribution of the scalar potential (if any) field integrated over the surface of the wire. It will be noted in the rigorous inductance formulas of Chap. VIII that the per unit length inductance of a wire becomes infinite if the radius becomes zero. If the input reactance of a thin wire-guide antenna is found by integration of the mode fields along its surface, no single term approaches infinity, and the convergence of the series becomes slower and vanishes at zero radius. Thus in the rectangular-guide case where the walls can be replaced by image antennas, a much better method is to add to the free-space antenna reactance, which can go to infinity and is often in closed form, the mutual reactances of the image antennas which are usually independent of the wire radius. The resultant series is then rapidly convergent. Examples of such calculations appear among the problems.

13.04. Aperture Excitation of Wave Guides. Coaxial Opening.—Wave-guide fields set up by waves entering through orifices in the sides or end give many of the same calculation problems encountered with waves diffracted through apertures and discussed in 12.18, 12.19, and 12.20. Few exact solutions are possible but in many cases the assumption that the incident wave in the opening is unperturbed as in 12.19 gives useful results. If the aperture dimensions are sufficiently small compared with a wave length, quasi-static methods give correct tangential fields. Thus the double current sheet method of 12.18 yields accurate solutions.

An example of aperture excitation which uses the double current sheet of 12.18 and the Green's function of 13.03 is the coaxial opening shown in Fig. 13.04*a* centered at $x = d$, $y = 0$, $z = c$ in the bottom of a rectangular guide closed at $z = 0$. The frequency is such that only the TE_{10} mode is transmitted, and it is absorbed at some positive value of z. The value of the output resistance of the coaxial line is desired. To make E_y zero at the closed end $z = 0$ requires two double current sheets of opposite sign, one at $z = c$ and one at $z = -c$ as shown in Fig. 13.04*b*. The electric field of the TE_{10} mode everywhere parallels the y-axis, so the infinitesimal thickness dy_0 of the toroidal double layer is exaggerated in Fig. 13.04*c*. The exact value of E over the opening is unknown, but the

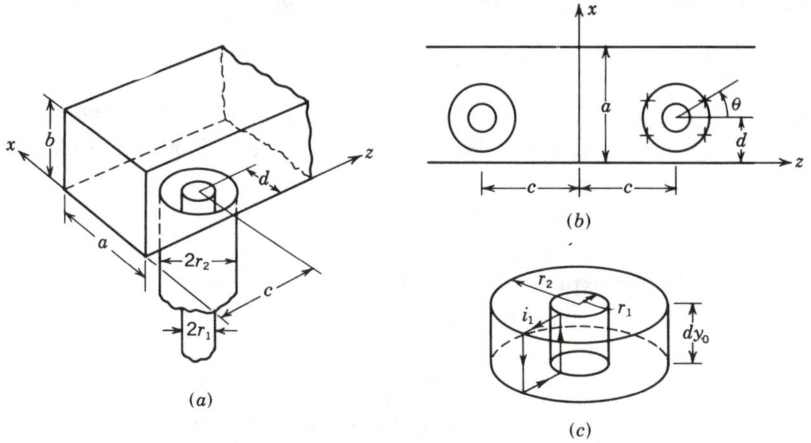

FIG. 13.04*a*, *b*, *c*.

field of the principal TEM mode in the coaxial line which transports the power and is the only field far from the opening is inversely proportional to r. If the potential of the central conductor is V and the current density in the double layer in amperes per radian is i, then from the text below Fig. 12.18,

$$E = \frac{\frac{1}{2}j\omega\mu i\, dy_0}{r} = \frac{V}{r \ln (r_2/r_1)} \tag{1}$$

The currents encircle the entire section of the toroidal current sheet of Fig. 13.04*c*. Thus they are solenoidal and none of the scalar potentials in 13.03 can appear. The amplitude of the TE_{10} mode excited by a vertical element of width $d\theta$ may be written down from 13.03 (6) by placing the torus in the $z = 0$ plane in a guide running from $z = -\infty$ to $z = \infty$ and bounded by $x = 0$, $x = a$, $y = -b$, and $y = b$. From symmetry the $y = 0$ plane is then an equipotential except for the coaxial aperture. Putting $m = 1$, $n = 0$, $I = i\,d\theta$, $y = 0$, and $2b$ for b in 13.03 (5) gives

$$d\check{C}_{10} = \frac{\mu i\, dy_0 \sin(\pi x_0/a)\, d\theta}{2jab\beta'_{10}\beta^2} = \frac{-\check{V}\sin(\pi x_0/a)\,d\theta}{\omega ab\beta^2\beta'_{10}\ln(r_2/r_1)} \tag{2}$$

For the total amplitude this must be integrated over the vertical walls. It is simpler to combine the four elements at $d \pm r_2 \sin\theta$ and $c \pm r_2 \cos\theta$ shown by crosses in Fig. 13.04b so that for the outer wall the integral is

$$\frac{1}{2}\int_0^{\frac{1}{2}\pi} \cos\left(\frac{\pi r_2}{a}\sin\theta\right)\cos(\beta'_{10}r_2\cos\theta)\,d\theta \tag{3}$$

In the notation of 13.03 (7) with $\tan\phi$ for $\pi/(\beta'_{10}a)$ this may be written

$$\tfrac{1}{2}\int_0^{\frac{1}{2}\pi}\{\cos[\beta r_2\cos(\theta-\phi)] + \cos[\beta r_2\cos(\theta+\phi)]\}\,d\theta \tag{4}$$

Both terms in the integrand have a period of $\frac{1}{2}\pi$, so that the value of the integral is independent of ϕ. With $\phi = 0$ it becomes 5.302 (3) so

$$\frac{1}{2}\int_0^{\pi}\cos(\beta r_2\cos\theta)\,d\theta = \frac{1}{2}\pi J_0(\beta r_2) = \frac{1}{2}J_0\left(\frac{2\pi r_2}{\lambda}\right) \tag{5}$$

Addition of similar expressions for the current at the inner edge of the torus and for the image torus in Fig. 13.04b gives, with $c + r_2 < z$,

$$\check{A}_{10} = \check{C}_{10}\sin\frac{\pi x}{a}\,e^{j(\omega t-\beta'_{10}z)} \tag{6}$$

$$\check{C}_{10} = 4\pi\check{V}\sin\frac{\pi d}{c}\,\frac{J_0(2\pi r_2/\lambda) - J_0(2\pi r_1/\lambda)}{j\omega\beta'_{10}ab\ln(r_2/r_1)} \tag{7}$$

The TE_{10} electric field and the magnetic induction are

$$\check{B}_x = \frac{-\partial\check{A}_{10}}{\partial z} = j\beta'_{10}\check{A}_{10}, \qquad \check{E}_y = -j\omega\check{A}_{10} \tag{8}$$

The power transmitted is the mean Poynting vector

$$\frac{\frac{1}{2}\check{E}\hat{B}}{\mu} = \frac{\frac{1}{2}\omega\beta'_{10}}{\mu}|C_{10}|^2\sin^2\frac{\pi x}{a} \tag{9}$$

Integration of this over the range $0 < x < a$, $0 < y < b$ and setting the result equal to $\frac{1}{2}V^2/R$ give for the radiation resistance

$$R = \frac{\mu\omega\beta'_{10}ab\ln^2(r_2/r_1)}{8\pi^2\sin^2(\pi d/a)\sin^2(\beta'_{10}c)[J_0(2\pi r_2/\lambda) - J_0(2\pi r_1/\lambda)]^2} \tag{10}$$

The same result is obtained by the integration of the Poynting vector over the coaxial opening, using $\boldsymbol{r}_1 V[r\ln(r_2/r_1)]^{-1}$ for $\boldsymbol{E}$ and $\nabla \times \boldsymbol{A}_{10}$ for $\boldsymbol{B}$, but the integration is more difficult.

13.05. Rectangular Guide Filled by Two Mediums.—Suppose the guide is filled from $x = 0$ to $x = c$ with a lossless medium $\mu_1\epsilon_1$ and from c to a with $\mu_2\epsilon_2$, where $a - c = d$. The wave velocity or wave number and the cutoff frequency are desired. To satisfy the boundary conditions at all times and all y values, the waves in both mediums must have the same velocity and y dependence. This is impossible for most waves purely TM or TE with respect to z but works out well for waves TE or TM with respect to x. Such waves derive from 13.00 (2) if $\boldsymbol{i}$ and W_{tex} replace $\boldsymbol{k}$ and W_{te}. The $(\pi m/a)^2$ term in 13.02 (2) is still valid but the $(\pi n/b)^2$ term no longer holds and will be replaced by p_1^2 in $\mu_1\epsilon_1$ and p_2^2 in $\mu_2\epsilon_2$. Thus the W_{tex} forms in mediums 1 or 2 that satisfy the boundary conditions at the guide walls are

$$\breve{A}_1 \sin p_1 x \cos\frac{m\pi y}{b} e^{-j\beta'_{mp}z} \qquad \text{or} \qquad \breve{A}_2 \sin p_2(a - x) \cos\frac{m\pi y}{b} e^{-j\beta'_{mp}z} \tag{1}$$

From the relations in 13.00 (12), when $\beta_1^2 = \omega^2\mu_1\epsilon_1$ and $\beta_2^2 = \omega^2\mu_2\epsilon_2$,

$$(\beta'_{mp})^2 = \beta_1^2 - \left(\frac{m\pi}{b}\right)^2 - p_1^2 = \beta_2^2 - \left(\frac{m\pi}{b}\right)^2 - p_2^2 \tag{2}$$

For continuity of E_t and H_t at $x = c$, $(E_z)_1 = (E_z)_2$ and $\mu_2(B_z)_1 = \mu_1(B_z)_2$. Thus

$$A_2 \sin p_1 c = A_2 \sin p_2 d \qquad \text{since } E_z = \frac{j\omega\, \partial W_{tex}}{\partial y} \tag{3}$$

$$\mu_2 p_1 A_2 \cos p_1 c = -\mu_1 p_2 A_2 \cos p_2 d \qquad \text{since } B_z = \frac{\partial^2 W_{tex}}{(\partial x\, \partial z)} \tag{4}$$

The ratio of (3) to (4) gives the transcendental equation

$$\mu_1 p_2 \tan p_1 c = -\mu_2 p_1 \tan p_2 d \tag{5}$$

The same equation results from $(E_y)_1/(B_y)_1 = (E_y)_2/(B_y)_2$ at $z = c$, so all tangential components of $\boldsymbol{E}$ and $\boldsymbol{H}$ are continuous if (5) holds. Equating the normal components of $\boldsymbol{B}$ gives (3). For a given frequency and medium, p_1 and p_2 as functions of β'_{mn} from (2) may be inserted in (5), the left and right sides plotted as a function of β'_{mp}, and its correct value found by their intersection. For the cutoff frequency β'_{mp} is zero, so from (2)

$$(p_1^2)_c = \omega_c^2\mu_1\epsilon_1 - \left(\frac{m\pi}{b}\right)^2, \qquad (p_2^2)_c = \omega_c^2\mu_2\epsilon_2 - \left(\frac{m\pi}{b}\right)^2 \tag{6}$$

Insertion of these values in (5) and solution of the resultant transcendental equation as before yield ω_c. Note that if p_1 is real in (6), then p_2 is imaginary and vice versa. The cutoff frequencies will lie between those of the equivalent modes in $\mu_1\epsilon_1$ and $\mu_2\epsilon_2$. There are an infinite number of values of p_1, p_2, and ω_c.

The boundary conditions for this guide can also be satisfied by waves TM to x by a perfectly analogous procedure. The chief results appear in problem 18 at the end of this chapter.

13.06. Thin Iris in Rectangular Guide.—The junction of two rectangular wave guides, which are affixed normally to opposite sides of the perfectly conducting $z = d$ plane in which holes have been cut to connect their interiors, forms a plane discontinuity. If in both guides only the TE_{10} wave is propagated, then standing waves can be set up so that there is an electric node at $z = d$. There is then no transverse electric field

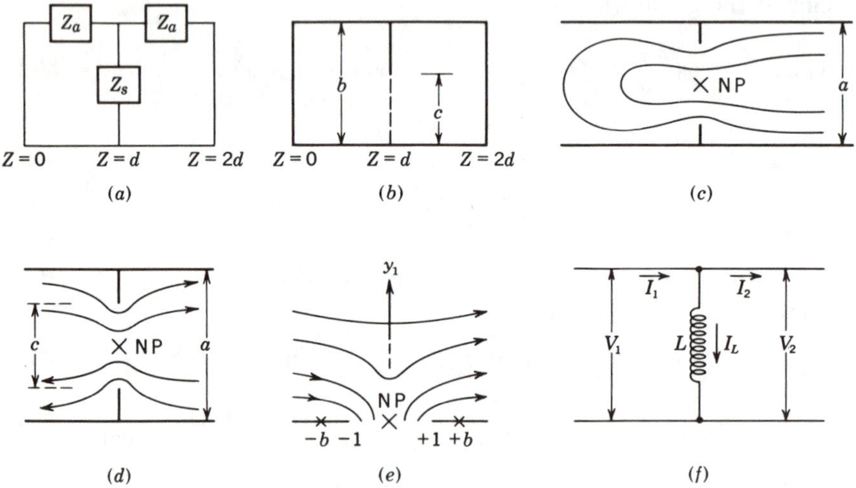

FIG. 13.06*a*, *b*, *c*, *d*, *e*, *f*.

there, and no magnetic field links any element of $z = d$. In a transmission line only a pure shunt element Z_S may be inserted at an electric node without upsetting the pattern. Now consider the special case in which a thin conducting iris at $z = d$ in a rectangular guide partially blocks TE_{10} standing waves set up with nearest nodes at $z = 0$ and $z = 2d$. From 13.02 (3), the TE_{10} electric field when $0 < z < d$ is

$$E_y = E_0 \sin\left(\frac{\pi x}{a}\right) \sin \beta'_{10} z \cos \omega t \qquad (1)$$

Higher mode fields must be present to cancel this field where it is tangent to the iris. By 13.02 (3), these contain the factor $e^{-|\beta'_{mn}(z-d)|}$ which, when λ_{10} is large, will confine them to a z-interval so short that all local fields are in phase and may be found by static field methods. This is called the "quasi-static" approach.

The wave-guide section $0 < z < 2d$ is equivalent to the short-circuited T section of Fig. 13.06a in which the currents are directed as shown. It is usual in treating wave guides to use "normalized" impedances which we designate by a superior ^{0}. These are found by dividing each impedance by the characteristic guide impedance which normalizes to 1. For a guide transmitting only the TE_{10} mode with no attenuation, 11.15 (13) is

$$\check{Z}_i^0 = \frac{\check{Z}_i}{\check{Z}_k} = \frac{\check{Z}_L^0 \cos \beta'_{10}l + j \sin \beta'_{10}l}{\cos \beta'_{10}l + j\check{Z}_L^0 \sin \beta'_{10}l} \tag{2}$$

For the short-circuited T section, $\check{Z}_L^0 = 0$ and $l = d$, so for resonance

$$\check{Z}_s^0 = -\tfrac{1}{2}\check{Z}_i^0 = -\tfrac{1}{2}j \tan \beta'_{10}d \tag{3}$$

Now consider the special case where all sections parallel to the $x = 0$ plane, and hence to the electric field, look alike. The electric fields parallel the $x = 0$ plane so that, in the region near $z = d$ where phase differences may be neglected, $\mathbf{E}$ at any value of x must obey the equation of continuity in y and z and can be derived from a potential that satisfies Laplace's equation in y and z. Identical currents cross all electric nodes in a guide so that the total charge on top and bottom between nodes is always the same. Let the charge per unit width in the interval $0 < z < 2d$ at some value x_0 of x be Q_{10} for the TE_{10} mode and Q_0 for all modes. Let the static discontinuity capacitance per unit width be C_1 and the potential across the guide at $x = x_0$, $z = d$ be $V_d = bE_d$. Then from (1),

$$C_1 = \frac{Q_0 - Q_c}{bE_d} = \frac{2}{bE_d}\left(\int_0^{\frac{1}{4}\lambda_{10}} \epsilon E_y\,dz - \int_0^d \epsilon E_y\,dz\right) = \frac{2\epsilon \cot \beta'_{10}d}{b\beta'_{10}} \tag{4}$$

Insertion of this value in (3) gives

$$Z_s^0 = \frac{\epsilon}{jb\beta'_{10}C_1} = \frac{1}{j\omega C^0} \qquad \text{or} \qquad C^0 = \frac{b\beta'_{10}}{\omega\epsilon}C_1 \tag{5}$$

As a specific example, consider an iris which leaves a window of height c in a guide of height b, as shown in Fig. 13.06b and Fig. 4.22b where the equipotentials are dotted except at the boundaries $U = 0$ and $U = \frac{1}{2}\pi$. Insertion of $a = c$, $W = jV$, and $z = jd$ in 4.22 (6) gives for V, as V and d approach infinity,

$$V = \frac{\frac{1}{2}\pi d}{b} - \ln \sin \frac{\frac{1}{2}\pi c}{b} = \frac{\frac{1}{2}\pi d}{b} + \ln \csc \frac{\frac{1}{2}\pi c}{b} \tag{6}$$

The charge per unit length between 0 and d is ϵV. The last term times ϵ equals half the additional charge on the guide due to the presence of the iris for a potential difference of $\frac{1}{2}\pi$. Therefore twice the second term multiplied by $2\epsilon/\pi$ gives C_1. Insertion of C_1 into (5) and substitution of $2\pi\lambda_g^{-1}$ for β'_{10} give for the normalized susceptance B^0

$$B^0 = \omega C^0 = \frac{8b}{\lambda_g} \ln \csc \frac{\frac{1}{2}\pi c}{b} \tag{7}$$

For the symmetrical iris bounded by $U = -\frac{1}{2}\pi$ and $U = \frac{1}{2}\pi$ in Fig. 4.22b, the only change is the potential difference, which is now π instead of $\frac{1}{2}\pi$, so that (7) must be divided by 2. The least attenuated TE_{11} mode is now out by symmetry; thus the result is more precise. From Fig. 4.22b by images the center strip behaves exactly the same way as the symmetrical iris. The errors in the above value of B^0 are about 2 and 9 per cent at b/λ_g values 0.1 and 0.2. These can be improved by a variational method given later.

Another interesting case is when all sections parallel to the $y = 0$ plane, and hence to the magnetic field, look alike. Suppose the iris is at $z = d$ and the frequency is such that only the TE_{10} wave propagates. All fields are independent of y, so n is zero in 13.03 (1) to (4) and the betas are

$$\beta'_1 = \frac{2\pi}{\lambda_g} \qquad \beta^1_m = (\beta^2 - \beta^2_m)^{\frac{1}{2}}, \qquad \beta_m = \frac{m\pi}{a} \qquad \beta = \frac{\omega}{(\mu\epsilon)^{\frac{1}{2}}} = \frac{2\pi}{\lambda} \tag{8}$$

This case, unlike the previous one, has charges only on the guide walls. Thus in the Coulomb gauge of 11.01 the entire field derives from the vector potential. Take the origin at one end of the short-circuited T section of Fig. 13.06a so that the $\boldsymbol{B}$ component amplitudes at $z = d$ are

$$B_x = A_1\beta'_1 \cos(\beta'_1 d) \sin(\beta_1 x) + \sum_{m=2}^{\infty} mA_m|\beta'_m| \cosh|\beta'_m d| \sin(\beta_m x) \tag{9}$$

$$B_z = -A_1\beta_1 \sin(\beta'_1 d) \cos(\beta_1 x) - \sum_{m=2}^{\infty} mA_m\beta_m \sinh|\beta'_m d| \cos(\beta_m x) \tag{10}$$

Multiply (9) by $\sin\beta_1 x\,dx$ and (10) by $\cos\beta_1 x\,dx$, integrate from $x = 0$ to $x = a$, eliminate A_1, and substitute from (3) to give

$$\tan(\beta'_1 d) = -\frac{\beta'_1 a \int_0^a B_z \cos\beta_1 x\,dx}{\pi \int_0^a B_x \sin\beta_1 x\,dx} \tag{11}$$

where the shunt impedance is shown in Fig. 13.06a. This requires the static approximation for the higher modes only at $z = d$.

As a specific example, take the symmetrical iris with a gap of width c as shown in Fig. 13.06d. To visualize the equivalent magnetostatic case, let $\lambda \gg a$ so the magnetic loops in Fig. 13.02 are much elongated in the z-direction. As the impermeable iris edges push in from the walls, they press the lines together as shown in Fig. 13.06c. No matter where this occurs in the loop, B_z in the window is positive at one edge, negative at the other, zero at the center, and vanishes on the iris outside the window while B_x is nearly zero in the window and tangent to the iris walls. If the guide walls in Fig. 13.06*d* are folded outward at the iris junction to the coplanar position, Fig. 13.06*e* results. These two maps bear the same relation to each other as those in Fig. 4.22*b*. The field in (*b*) comes from cutting a slit 2 units wide, shown in Fig. 13.06*e* in an infinite sheet of permeability zero that separates two equal and opposite uniform magnetic fields. Figure 4.23 shows the result with only one field present. Superposition of a similar field rotated 180° gives Fig. 13.06*e*, and application of the transformation below Fig. 4.22*b*, but with the origin at the iris base, is Fig. 13.06*d*.

$$W = C'(z_1'^2 - 1)^{\frac{1}{2}}, \qquad z_1' = b_1 \sin\left[\frac{\pi}{a}(z' - \tfrac{1}{2}a)\right] \tag{12}$$

Eliminate z_1' and make $z' = \frac{1}{2}(a - c)$ when $z_1' = 1$ to give

$$W = C\left(\cos^2\frac{\pi z'}{a} - \sin^2\frac{\frac{1}{2}\pi c}{a}\right)^{\frac{1}{2}} \tag{13}$$

For the $z = d$ plane of the guide, $y' = 0$ so this yields, by 4.11 (1),

$$B_x + jB_z = \pm C\frac{\partial[\cos^2(\pi x/a) - \sin^2(\frac{1}{2}\pi c/a)]^{\frac{1}{2}}}{\partial x} \tag{14}$$

Thus with respect to the guide axis, B_x is an even function that is zero when $\frac{1}{2}(a - c) < x < \frac{1}{2}(a + c)$ and B_z is an odd function that is zero if $0 < x < \frac{1}{2}(a - c)$ and if $\frac{1}{2}(a + c) < x < a$. The integration in (4) can be done by parts, letting $dv = B_x\,dx$ and $u = \sin(\pi x/a)$ in one and $dv = B_z\,dx$ and $u = \cos(\pi x/a)$ in the other. The uv products vanish, and if $\sin(\pi x/a)$ is replaced by $t\cos(\frac{1}{2}\pi c/a)$ in one and $\cos(\pi x/a)$ by $t\sin(\pi c/a)$ in the other, they lead to *Dw* 350.01. Substitution in (11) gives

$$\check{Z}_s^0 = \frac{a}{\lambda_g}\tan^2\frac{\frac{1}{2}\pi c}{a} = j\omega L_s^0 \tag{15}$$

Thus the iris is inductive, and the normalized shunt susceptance is

$$B^0 = \frac{B}{\check{Y}_k} = -\frac{1}{\omega L} = -\frac{\lambda_g}{a}\cot^2\left(\frac{\pi c}{2a}\right) \tag{16}$$

An exact value could be found by solving an integral equation given in 13.08, but this is impractical. A better way is to improve the above solutions by a variational method. This method, also applicable to some static problems, will now be discussed, as in Borgnis and Papas.

13.07. A Variational Method for Improving Approximate Solutions.—Let some physical quantity such as the iris susceptance in 13.06 be expressed in the form of the integral equation

$$Q = \int_S h(r)E(r)\,dS \tag{1}$$

where $h(r)$ is known but $E(r)$ is an unknown function, such as the electric field, of the coordinate r in the surface S, which can be the window area in 13.06, over which the integration extends. Now consider the integral

$$h(r) = \int_{S'} G(r,r')E(r')\,dS' \tag{2}$$

where r' relates to area S' and $G(r,r') = G(r',r)$. Put (2) into (1) so

$$Q = \int_S\int_{S'} E(r')G(r,r')E(r)\,dS\,dS' \tag{3}$$

Division of (3) by the square of (1) yields

$$\frac{1}{Q} = \frac{\int_{S'}\int_S E(r')G(r,r')E(r)\,dS\,dS'}{\left[\int_S h(r)E(r)\,dS\right]^2} \tag{4}$$

This expression has the useful property that, if trial values of $E(r)$ are inserted, the Q value is stationary about the correct value of $E(r)$ so that a first-order variation in $E(r)$ gives a second-order Q variation. To prove this multiply both sides of (4) by the denominators giving

$$Q\int_S\int_{S'} E(r')G(r,r')E(r)\,dS\,dS' = \left[\int_S h(r)E(r)\,dS\right]^2 \tag{5}$$

Take the variation, which follows differential rules, thus

$$Q\int_S\int_{S'} \delta E(r')G(r,r')E(r)\,dS\,dS' + Q\int_S\int_{S'} E(r')G(r,r')\,\delta E(r)\,dS\,dS'$$
$$= -\delta Q\left[\int_S h(r)E(r)\,dS\right]^2 + 2\int_S h(r)\,\delta E(r)\,dS\int_S h(r)E(r)\,dS \tag{6}$$

Since $G(r,r') = G(r',r)$, r and r' may be interchanged in one integral on the left so it adds to the other. In the last right-side term put (2) for $h(r)$ in the first integral and the Q of (1) for the second. Double integrals cancel leaving $\delta Q = 0$. The value of (4) is independent of the $E(r)$ units. Borgnis and Papas, Collin, Harrington, Marcuvitz, and others treat the subject fully. Schwinger first used this method in this way.

13.08. Inductive Iris Susceptance by Variational Method.—The technique of the last article, where Q is B^0 or $(B^0)^{-1}$, may be used to improve the quasi-static solutions of Art. 13.06. The accuracy of the result depends on how close the trial value of $E(r)$ is to the true value of the tangential $\boldsymbol{E}$ in the window or the tangential $\boldsymbol{B}$ (current density) on the iris. The conformal transformations of 13.06 will provide good trial values in both cases if the static E_t and B_t have the same directions as the actual E_t and B_t. To find the latter in the capacitive case, both vector and scalar potentials are needed in the Coulomb gauge because there is charge inside the walls and the divergence of the iris current is not zero. Thus it is easier to use the magnetic Hertz vector of 11.01 (13) for the fields. This shows that E_t is y-directed, but B_t has both x- and y-components and thus is not compatible with the quasi-static value. This case will therefore be relegated to the problems at the end of the chapter. There is no charge on the inductive iris; therefore only the vector potential of the Coulomb gauge is needed. This gives both E_t and B_t independent of y in both magnitude and direction and provides an ideal case for using the quasi-static fields as trial values in the variational integrals. From 13.02 (3) and 13.02 (4), after the absorption of some common factors into C_m, the guide fields become

$$\check{E}_y(x,z) = j\omega \sum_{m=1}^{\infty} \check{C}_m \sin \beta_m x e^{-j\beta'_m z} \tag{1}$$

$$\check{\boldsymbol{B}}(x,z) = \sum_{m=1}^{\infty} \check{C}_m(-\boldsymbol{i}\beta'_m \sin \beta_m x + \boldsymbol{k}j\beta_m \cos \beta_m x)e^{-j\beta'_m z} \tag{2}$$

where m is odd since the fields are symmetrical about $x = \frac{1}{2}a$. In the iris plane the electric field is zero on the iris surface and E_a in the window. Multiplication of (1) by $\sin \beta_m x\, dx$ and integration over the guide section give C_m which, put into (1), gives

$$\check{E}(x) = \frac{2}{a} \sum_{m=1}^{\infty} \sin \beta_m x \int_a E_a(x') \sin \beta_m x'\, dx'\, e^{-j\beta'_m z} \tag{3}$$

But in the window, as proved in the paragraph following 12.18 (4), $\check{B}(x)$ is that of the incident wave, which is normalized to $\sin(\beta_1 x)$. From 11.00 (1), $B^i_x(x,0)$ is $\omega^{-1}\, \partial E^i(x,z)/\partial z$. Thus transfer of the $m = 1$ term, which is the TE_{10} amplitude, to the left side gives at $z = 0$

$$\beta'_1 \sin \beta_1 x\left[\frac{a}{2} - \int_a \check{E}_a(x') \sin \beta_1 x' dx'\right] = \sum_{m=3}^{\infty} \beta'_m \sin \beta_m a \int_a \check{E}_a(x') \sin \beta_m x'\, dx'$$

Multiplication of both sides by $E_a(x)$, integration over the aperture, and division by $\beta_1'[\int E_a(x) \sin \beta_1 x\, dx]^2$ give

$$\frac{a}{2\int E_a \sin \beta_1 x\, dx} - 1 = \sum_{m=3}^{\infty} \frac{\beta_m'[\int E_a(x) \sin \beta_m x\, dx]^2}{\beta_1'[\int E_a(x) \sin \beta_1 x\, dx]^2} \tag{4}$$

The left side of this equation is expressible in terms of the iris susceptance. Let the total potential and current be V_1 and I_1 when $z < 0$ and V_2 and I_2 when $z > 0$ as in Fig. 13.06*f*. Both the incident wave V_i and the reflected wave V_r contribute to V_1, so that

$$V_1 = V_i + V_r, \qquad Z_k I_1 = V_i - V_r, \qquad Z_k I_2 = V_2 \tag{5}$$

Combine with this the relations in (6), valid at $z = 0$, to get (7).

$$V_1 = V_2 = j\omega L I_L, \qquad I_1 = I_2 + I_L \tag{6}$$

$$2V_i = V_1 + Z_k I_1 = 2V_2 + \frac{Z_k V_2}{j\omega L} \qquad \text{so} \qquad \frac{Z_k}{j\omega L} = 2\left(\frac{V_i}{V_2} - 1\right) \tag{7}$$

From (3) the ratio of normalized TE_{10} incident to TE_{10} transmitted is

$$\frac{E_i}{E_2} = \frac{V_i}{V_2} = \left[\frac{2}{a}\int E_a(x) \sin \beta_1 x\, dx\right]^{-1} \tag{8}$$

Substitution of this into (4) and replacement of $j\beta_m'$ by $(\beta_m^2 - \beta^2)^{\frac{1}{2}}$ give

$$\frac{Z_k}{\omega L} = \frac{-B}{Y_k} = -B_0 = \sum_{m=3}^{\infty} \frac{2(\beta_m^2 - \beta^2)^{\frac{1}{2}}[\int E_a(x) \sin \beta_m x\, dx]^2}{\beta_1'[\int E_a(x) \sin \beta_1 x\, dx]^2} \tag{9}$$

Integration of both integrals in (9) by parts replaces the sines with cosines and $E_a(x)$ by $\partial E_a(x)/\partial x$ which, by Maxwell's equations, can be replaced by $-j\omega B_z(x)$ since $E_a(x)$ is y-directed. Writing $kB_a(x)$ for $j\omega E_a'(x)$ and replacement of some of the betas in (9) by (1) give

$$aB^0 = -\lambda_g \sum_{m=3}^{\infty} (1 - 4a^2 m^{-2} \lambda^{-2})^{\frac{1}{2}} \left[\frac{\int B_a(x) \cos \beta_m x\, dx}{\int B_a(x) \cos \beta_1 x\, dx}\right]^2 \tag{10}$$

The imaginary part of 13.06 (14) gives an excellent trial value of $B_a(x)$. Integration limits are $x = \frac{1}{2}(a - c)$ and $x = \frac{1}{2}(a + c)$, but the integrands are even about $x = \frac{1}{2}a$ and thus the limits $\frac{1}{2}a$ and $\frac{1}{2}(a + c)$ may be used. Since m is odd, replace it by $2n + 1$ so that the sum goes from $n = 1$ to $n = \infty$. In 13.06 (14) and in (10), substitute θ for $\beta_1 c$ and $\frac{1}{2}(2n + 1)$ $(\pi + \phi)$ for $\beta_m x$. The integrals then lead to Mehler's Legendre function integrals found in MO, page 52, WW, page 315, HMF 22.5.36 and

22.10.11, and elsewhere. Thus

$$C_1 \int_0^\theta \frac{\sin\phi \sin(n+\frac{1}{2})\phi\, d\phi}{(\cos\phi - \cos\theta)^{\frac{1}{2}}} = C_2 \int_0^\theta \frac{[\cos(n-\frac{1}{2})\phi - \cos(n+\frac{3}{2})\phi]\, d\phi}{(\cos\phi - \cos\theta)^{\frac{1}{2}}}$$
$$= C_3[P_{n-1}(\cos\theta) - P_{n+1}(\cos\theta)] = C_3 \frac{2n+1}{n(n+1)} \sin\theta P_n^1(\cos\theta) \qquad (11)$$

When $n = 0$, the denominator integral is $C_3(1 - \cos\theta)$. Substitution of these values into (10) and application of *Dw* 406.3 give

$$B^0 = -\frac{\lambda_g}{a} \cot^2\left(\frac{\pi c}{2a}\right) \sum_{n=1}^{\infty} \left[(2n+1)^2 - \left(\frac{2a}{\lambda}\right)^2\right]^{\frac{1}{2}} \left[\frac{P_n^1 \cos(\pi c/a)}{n(n+1)}\right]^2 \qquad (12)$$

The convergence of this series can be improved by taking the difference between it and a summable series having identical terms when n is large. Note that if $a = 1$ and $\alpha = \theta$ in 7.13 (2), then the r integral of A_ϕ from 0 to ∞, if r^{-n-1} replaces r^n when $r > 1$, is

$$2(\mu I \sin\alpha)^{-1} \int_0^\infty A_\phi\, dr = \sum_{n=1}^{\infty} (2n+1) n^{-2} (n+1)^{-2} [P_n^1(\cos\alpha)]^2 \qquad (13)$$

Here A_ϕ is the vector potential of a circular loop of radius $\sin\alpha$ lying on a unit sphere. Doing the same integration on the first integral in 7.10 with $a = \sin\alpha$, $\rho = r\sin\alpha$, and $z = (r-1)\cos\alpha$ gives

$$\sum_{n=1}^{\infty} (2n+1) n^{-2} (n+1)^{-2} [P_n^1(\cos\alpha)]^2 = \pi^{-1} \int_0^\pi \cos\phi \ln(1 - \cos\phi)\, d\phi = 1 \qquad (14)$$

This might have been foreseen because $\lambda = \infty$ was a tacit assumption used to get 13.06 (16). Thus the series in (12) may be replaced by

$$1 + \sum_{n=1}^{\infty} \{[(2n+1)^2 - 4a^2\lambda^{-2}]^{\frac{1}{2}} - 2n - 1\} \left[n^{-1}(n+1)^{-1} P_n^1\left(\cos\frac{\pi c}{a}\right)\right]^2 \qquad (15)$$

in which the series converges so rapidly that in the cases checked five terms gave three digits correct in B^0 and ten terms gave four digits. Jahnke and Emde tabulate $dP_n/d\theta$ and $|P_n^1(\cos\theta)| = |dP_n/d\theta|$.

The integrals in (12) are positive definite and are an absolute minimum for the true fields so (12) provides an upper limit for B^0. A short proof of this appears in Borgnis and Papas, page 122. A lower limit is found by use of a variational expression, based on the obstacle current, which bears a strong resemblance to (9), and reads

$$-\frac{Y_k}{B} = -\frac{1}{B^0} = \frac{\beta_1' \sum_{m=3}^{\infty} (\beta_m^2 - \beta^2)^{-\frac{1}{2}} \left[\int_0^{\frac{1}{2}(a-c)} B_x(x) \sin \beta_m x \, dx \right]^2}{2\left[\int_0^{\frac{1}{2}(a-c)} B_x(x) \sin \beta_1 x \, dx \right]^2} \tag{16}$$

where the obstacle current density $i_y(x)$ has been replaced by $B_x(x)$ to which it is proportional when $z = 0$. The integration is over the iris surface since B_x is zero in the window. Note that all terms of (9) except the integrals have been inverted. The real part of 13.06 (14) is a good trial value for $B_x(x)$. Put this value into the numerator integral of (16), and substitute $\pi - \phi$ for $2\pi x/a$, θ for $\pi c/a$, and $2n + 1$ for m, which is odd from symmetry. The integral then becomes, disregarding the sign,

$$C_1 \int_\theta^\pi \frac{\sin \phi \cos (n + \frac{1}{2})\phi \, d\phi}{(\cos \theta - \cos \phi)^{\frac{1}{2}}} = C_2 \int_\theta^\pi \frac{[\sin (n + \frac{3}{2})\phi - \sin(n - \frac{1}{2})\phi] \, d\phi}{(\cos \theta - \cos \phi)^{\frac{1}{2}}}$$

$$= C_3[P_{n-1}(\cos \theta) - P_{n+1}(\cos \theta)] = C_3 \frac{2n+1}{n(n+1)} \sin \theta P_n^1(\cos \theta) \tag{17}$$

The second form of Mehler's integral, already referred to, was used here. When $n = 0$, the denominator integral is $C_3(1 - \cos \theta)$. Substitution of these values into (16) and use of *Dw* 406.2 give

$$\frac{1}{B^0} = -\frac{a}{\lambda_g} \tan^2 \left(\frac{\pi c}{2a}\right) \sum_{n=1}^{\infty} \frac{(2n+1)\{P_n^1[\cos (\pi c/a)]\}^2}{[1 + 4a^2\lambda^{-2}(2n+1)^{-2}]^{\frac{1}{2}} n^2(n+1)^2} \tag{18}$$

Exactly as with (12) the series may be put in the rapidly convergent form

$$1 + \sum_{n=1}^{\infty} \left\{ \left[1 - \frac{4a^2}{(2n+1)^2\lambda^2} \right]^{-\frac{1}{2}} - 1 \right\} (2n+1) \left\{ \frac{P_n^1[\cos (\pi c/a)]}{n(n+1)} \right\}^2 \tag{19}$$

This provides a lower limit for the susceptance. The following table gives values of the correction factor by which 13.06 (16) should be multi-

Correction factor table

c/a	λ = 0.5		λ = 1.0		λ = 1.3	
	0.2,0.8	0.4,0.6	0.2,0.8	0.4,0.6	0.2,0.8	0.4,0.6
Upper limit	0.9768	0.9582	0.8995	0.8149	0.8102	0.6385
Lower limit	0.9758	0.9558	0.8745	0.7550	0.6760	0.2940
Arithmetic mean	0.9763	0.9570	0.8867	0.7849	0.7434	0.4662
Handbook	0.9765	0.9578	0.8913	0.8045	0.7594	0.5881

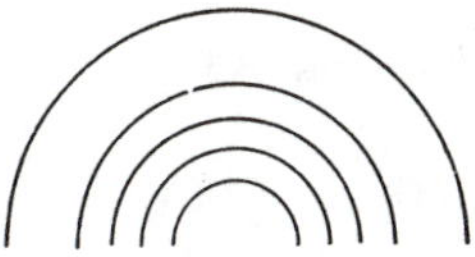

TE_{01} mode. *E*-lines in *AB* section spaced to give equal intensity increments.

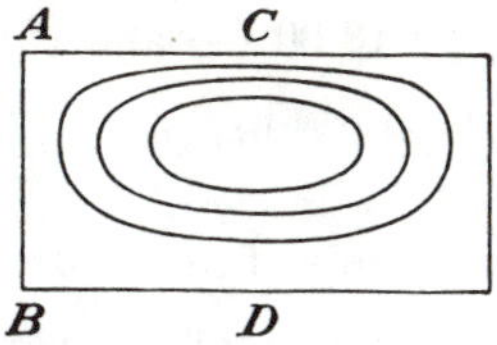

TE_{01} mode. Principal section of equiflux tubes of magnetic induction.

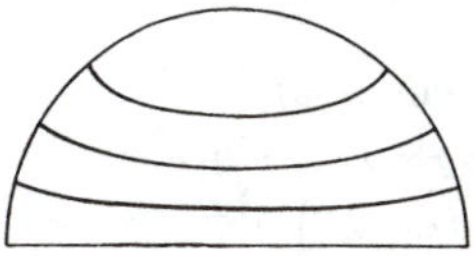

TE_{11} mode. *E*-lines in *AB* section spaced to give equal intensity increments on center line.

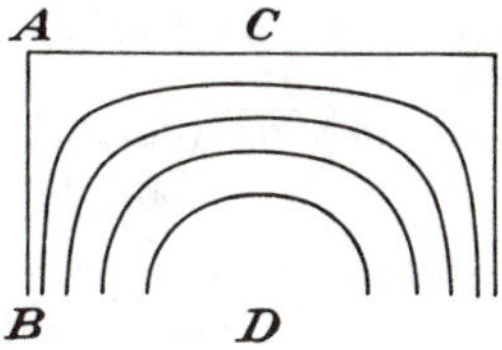

TE_{11} mode. *B*-lines in principal section spaced to give equal induction increments on center line.

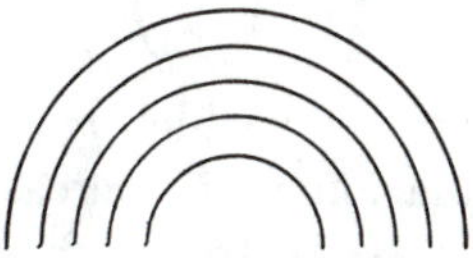

TM_{01} mode. *B*-lines in *AB* section spaced to give equal induction increments.

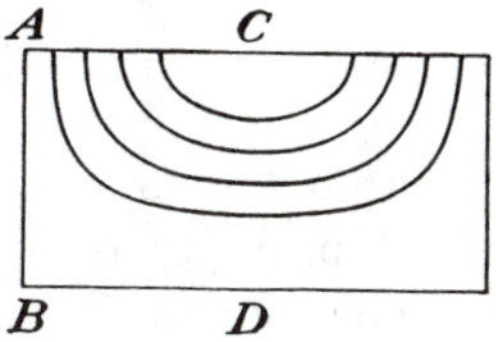

TM_{01} mode. Principal section of equiflux tubes of displacement.

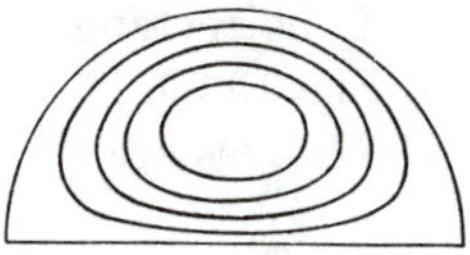

TM_{11} mode. *B*-lines in *AB* section spaced to give equal induction increments on center line.

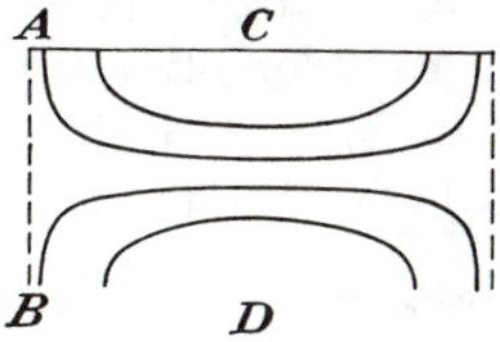

TM_{11} mode. *E*-lines in principal section spaced to give equal intensity increments on center line.

Fig. 13.09.

plied according to (12) and (18) and their average compared with the accurate value of the $1a$ formula on page 221 of the "Waveguide Handbook" over the ranges plotted on pages 222 and 223. It is clear that the variational treatment has made a great improvement in the quasi-static value and that the upper limit alone does pretty well.

13.09. The Circular Wave Guide.—Consider a wave guide of radius a. From 12.16, a solution of 13.00 (3) which is finite on the axis is

$$\check{U}(\rho,\phi) = J_m(\beta_{mn}\rho)(\check{C}_{mn} \cos m\phi + \check{D}_{mn} \sin m\phi) \tag{1}$$

where, by 13.00 (4) and (5), β_{mn} must be chosen so that

$$|\partial J_m(\beta_{mn}a)/\partial a|_{te} = 0 \qquad |J_m(\beta_{mn}a)|_{tm} = 0 \tag{2}$$

The most important zeros of these functions are

$$J_0'(3.832) = J_0'(7.016) = J_1'(1.841) = J_1'(5.331) = 0 \tag{3}$$
$$J_0(2.405) = J_0(5.520) = J_1(3.832) = J_1(7.016) = 0 \tag{4}$$

From 13.00 (14) and (15), the fields of the TE waves are

$$\begin{aligned}[\check{\mathbf{E}}]_{te} = {} & j\omega[\mathbf{\varrho}_1 m\rho^{-1}J_m(\beta_{mn}\rho)(\check{C}_{mn}\sin m\phi - \check{D}_{mn}\cos m\phi) \\ & + \boldsymbol{\phi}\beta_{mn}J_m'(\beta_{mn}\rho)(\check{C}_{mn}\cos m\phi + \check{D}_{mn}\sin m\phi)]e^{-j\beta'_{mn}z} \quad (5) \\ [\check{\mathbf{B}}]_{te} = {} & \{j\beta'_{mn}[-\mathbf{\varrho}_1\beta_{mn}J_m'(\beta_{mn}\rho)(\check{C}_{mn}\cos m\phi + \check{D}_{mn}\sin m\phi) \\ & + \boldsymbol{\phi}m\rho^{-1}J_m(\beta_{mn}\rho)(\check{C}_{mn}\sin m\phi - \check{D}_{mn}\cos m\phi)] \\ & + \mathbf{k}\beta_{mn}^2J_m(\beta_{mn}\rho)(\check{C}_{mn}\cos m\phi + \check{D}_{mn}\sin m\phi)\}e^{-j\beta'_{mn}z} \quad (6)\end{aligned}$$

From (1), (2), 13.01 (6), and 5.295 (6), the TE wave attenuation is

$$\alpha_{te} = \frac{\tau}{\mu v a\delta}\left[1 - \left(\frac{\nu_{mn}}{\nu}\right)^2\right]^{-\frac{1}{2}}\left[\frac{m^2}{\beta_{mn}^2a^2 - m^2} + \left(\frac{\nu_{mn}}{\nu}\right)^2\right] \tag{7}$$

We notice that the circularly symmetrical waves for which $m = 0$ have the unusual property of a decreasing attenuation with increasing frequency. The longest TE cutoff wave length λ_{11} is, from (2), (3), and 13.00 (8), at $3.42a$ and the next, λ_{01}, is at $1.64a$. Take $a = 2.34$ cm so that λ_{11} is 8 cm and compare with the 2- by 4-cm guide of the last article at the same cutoff. The 5-cm free wave length attenuation is 0.0028 m^{-1} which is considerably less than the 0.005 m^{-1} of the rectangular guide. Field maps of some of the TE modes appear in Fig. 13.09.

From 13.00 (16) and (17), the fields of the TM waves are

$$\begin{aligned}\check{\mathbf{E}}_{tm} \doteq {} & \{\omega\beta'_{mn}[\mathbf{\varrho}_1\beta_{mn}J_m'(\beta_{mn}\rho)(\check{C}_{mn}\cos m\phi + \check{D}_{mn}\sin m\phi) \\ & - \boldsymbol{\phi}m\rho^{-1}J_m(\beta_{mn}\rho)(\check{C}_{mn}\sin m\phi - \check{D}_{mn}\cos m\phi)] \\ & + \mathbf{k}j\omega\beta_{mn}^2J_m(\beta_{mn}\rho)(\check{C}_{mn}\cos m\phi + \check{D}_{mn}\sin m\phi)\}e^{-j\beta'_{mn}z} \quad (8) \\ \check{\mathbf{B}}_{tm} = {} & [\mathbf{\varrho}_1\beta^2 m\rho^{-1}J_m(\beta_{mn}\rho)(\check{C}_{mn}\sin m\phi - \check{D}_{mn}\cos m\phi) \\ & + \boldsymbol{\phi}\beta^2\beta_{mn}J_m'(\beta_{mn}\rho)(\check{C}_{mn}\cos m\phi + \check{D}_{mn}\sin m\phi)]e^{-j\beta'_{mn}z} \quad (9)\end{aligned}$$

From (1), (2), 13.01 (7), and 5.295 (6), the TM wave attenuation is

$$\alpha_{tm} = \frac{\tau}{\mu a v\delta}\left[1 - \left(\frac{\nu_{mn}}{\nu}\right)^2\right]^{-\frac{1}{2}} \tag{10}$$

The longest λ_{mn} values are $\lambda_{01} = 2.61a$, $\lambda_{11} = 1.64a$, and $\lambda_{02} = 1.14a$.

13.10. Green's Function for a Circular Guide.—The wave patterns set up by a wire loop or stub carrying sinusoidal currents can be found by integration of the Green's function of a current element. Instead of the x-, y-, z-components used for the rectangular guide in 13.03 (1), the current element will be resolved into ρ-, ϕ-, z-components, thus

$$\check{I}(\rho,\phi,z)\,ds = \boldsymbol{\rho}_1\check{I}_\rho(\rho,\phi,z)\,d\rho + \boldsymbol{\phi}_1\check{I}_\phi(\rho,\phi,z)\rho\,d\phi + \boldsymbol{k}\check{I}_z(\rho,\phi,z)\,dz \tag{1}$$

As in 13.03 the Coulomb guage, in which $\boldsymbol{A}$ arises entirely from the solenoidal part of $\check{I}(\rho,\phi,z)$ and Ψ entirely from the lamellar part, is convenient. The comments of 13.03 also apply here. The calculations and results for radial ρ- and longitudinal z-components follow, but the ϕ-component results, whose calculation resembles that for the ρ-component, are given in problem 27 at the end of the chapter.

Consider a current element of strength $\boldsymbol{\rho}_1 Ie^{j\omega t}$ and of length $d\rho_0$ at ρ_0, 0 in the $z = 0$ plane. If the element is at $\phi = \phi_0$ instead of at $\phi = 0$, $\phi - \phi_0$ replaces ϕ in the equation of the last article. The complete TE solution is a double sum over m and n of terms like 13.09 (5) and (6). To find the amplitude of the pq-term, one sets $t = 0$, $z = z_0$, and $C_{mn} = 0$ in 13.09 (6), $\check{E}_\rho$ being an even function of ϕ, takes its scalar product by

$$[\boldsymbol{\rho}_1\beta_{pq}J'_p(\beta_{pq}\rho)\sin p\phi + \boldsymbol{\phi}_1 p\rho^{-1}J_p(\beta_{pq}\rho)\cos p\phi]\rho\,d\rho\,d\phi$$

and integrates from $\phi = -\pi$ to $\phi = \pi$ and from $\rho = 0$ to $\rho = a$. The first integration eliminates all but the pth term in the m-summation and puts π in place of $\sin^2 p\phi$ and $\cos^2 p\phi$. If v_1 is written for $\beta_{pn}\rho$ and v_2 for $\beta_{pq}\rho$, the remaining ρ-integrals are

$$\pi\beta_{pq}\beta_{pn}^{-1}\int_0^{\beta_{pn}a}\left[J'_p(v_1)J'_p(v_2) + \frac{p^2}{v_1v_2}J_p(v_1)J_p(v_2)\right]v_1\,dv_1 \tag{2}$$

By 5.296 (11), the integral is zero unless $n = q$ when, by 5.296 (12), its value is $\frac{1}{2}(\beta_{pq}^2a^2 - p^2)J_p^2(\beta_{pq}a)$. In the z_0-plane, B_ρ is zero everywhere and B_ϕ is zero except over the element $d\rho_0$ at ρ_0, 0. But by 7.01 (2), at $\rho_0 \oint B_\phi\rho\,d\phi$ is $-\frac{1}{2}\mu I$. Solving for D_{pq} gives

$$[\check{D}_{pq}]_{te} = \frac{p\mu I\,d\rho_0\,J_p(\beta_{pq}\rho_0)}{\pi\rho_0 j\beta'_{pq}(\beta_{pq}^2a^2 - p^2)[J_p(\beta_{pq}a)]^2}e^{j\beta'_{pq}z_0} \tag{3}$$

For the TM modes excited by the radial element, set $t = 0$, $z = z_0$, and $\check{D}_{mn} = 0$ in 13.09 (9) and take its scalar product by

$$[\boldsymbol{\rho}_1 p\rho^{-1}J_p(\beta_{pq}\rho)\sin p\phi + \boldsymbol{\phi}_1\beta_{pq}J'_p(\beta_{pq}\rho)\cos p\phi]\rho\,d\rho\,d\phi$$

And integrate from $\phi = -\pi$ to $\phi = \pi$ and from $\rho = 0$ to $\rho = a$. The integrals are identical with those found before except for a $2 - \delta_p^0$ factor, and the integral in (1) is $\frac{1}{2}[\beta_{pq}aJ'_p(\beta_{pq}a)]^2$ because $J_p(\beta_{pq}a)$ is zero, so

$$[\check{C}_{pq}]_{tm} = -\frac{\mu I\,d\rho_0\,J'_p(\beta_{pq}\rho_0)(2 - \delta_p^0)}{2\pi\beta_{pq}[\beta aJ'_p(\beta_{pq}a)]^2}e^{j\beta'_{pq}z_0} \tag{4}$$

For the longitudinal component, use the source of 13.03 (11), set $z = z_0$, $\check{D}_{mn} = 0$ in 13.03 (8), and take the scalar product of both sides by

$$[\varrho_1 \rho \beta_{pq} J'_p(\beta_{pq}\rho) \cos p\phi - \phi p J_p(\beta_{pq}\rho) \sin p\phi]\, d\rho\, d\phi = \boldsymbol{R}_{pq}\, d\rho\, d\phi$$

Integrate over the ranges $0 < \rho < a$ and $0 < \phi < 2\pi$. After integration of the right side with respect to ϕ, we obtain with the aid of 5.296 (11) and (12), keeping in mind that $J_p(\beta_{pq}a)$ is zero,

$$\int_0^a \int_0^{2\pi} \boldsymbol{E} \cdot \boldsymbol{R}_{pq}\, d\rho\, d\phi = \frac{\pi\omega\beta'_{pq}\check{C}_{pq}}{2 - \delta_p^0} \beta_{pq}^2 a^2 [J'_p(\beta_{pq}a)]^2 e^{-j\beta'_{pq}z_0} \tag{5}$$

The arcs $c + \frac{1}{2}\delta$ and $c - \frac{1}{2}\delta$ and radial lines at $\frac{1}{2}\Delta\phi$ and $-\frac{1}{2}\Delta\phi$ bound dS_0. The integral over the arcs, since the integrand contains no ϕ, is

$$\beta_{pq}\{(c + \tfrac{1}{2}\delta)J'_p[\beta_{pq}(c + \tfrac{1}{2}\delta)] - (c - \tfrac{1}{2}\delta)J'_p[\beta_{pq}(c - \tfrac{1}{2}\delta)]\}V_0\,\Delta\phi \xrightarrow[\delta\to 0]{} \beta_{pq}[\beta_{pq}cJ''_p(\beta_{pq}c) + J'_p(\beta_{pq}c)]V_0\,\delta\,\Delta\phi \tag{6}$$

On the radial sides δ is so small that we may set $\rho = c$, which gives

$$\delta p c^{-1} J_p(\beta_{pq}c)[\sin(\tfrac{1}{2}p\,\Delta\phi) - \sin(-\tfrac{1}{2}p\,\Delta\phi)]V_0 = p^2\,\Delta\phi\,\delta c^{-1} V_0 J_p(\beta_{pq}c) \tag{7}$$

The sum of (6) and (7) gives the left side of (5). When we write dS_0 for $c\delta\,\Delta\phi$, substitute for $V_0\,dS_0$ from 13.03 (11), combine the Bessel functions by Bessel's equation 5.293 (3), and solve for $\check{C}_{pq}$, we obtain

$$[\check{C}_{pq}]_{tm} = \frac{j(2 - \delta_p^0)I\,dz_0\,J_p(\beta_{pq}c)}{2\pi\beta'_{pq}\omega^2\epsilon[aJ'_p(\beta_{pq}a)]^2} e^{j\beta'_{pq}z_0} \tag{8}$$

It should be noted that, in (4) and (7), $-J_{p+1}(\beta_{pq}a)$ may be substituted for $J'_p(\beta_{pq}a)$ by 5.294 (2) because $J_p(\beta_{pq}a)$ is zero.

In the Coulomb gauge the scalar potentials Ψ_ρ and Ψ_z of the radial and longitudinal components of the current element in (1) may be derived from the point charge Green's function of 5.297 (4). For the radial component write $-\beta_{mn}\,d\rho_0\,J'_s(\beta_{mn}\rho_0)$ for $J_s(\mu_r b)$, where the second equation in 13.09 (2) defines β_{mn}, and $I/(j\omega)$ for q. Thus the element at ρ_0, 0, z_0 yields

$$\check{\Psi}_\rho = \frac{jI\,d\rho_0}{2\pi\omega\epsilon a^2} \sum_{n=1}^{\infty} \sum_{m=0}^{\infty} (2 - \delta_m^0) e^{-\beta_{mn}|z-z_0|} \frac{J'_m(\beta_{mn}\rho_0)J_m(\beta_{mn}\rho)}{[J_{s+1}(\beta_{mn}a)]^2} \cos m\phi \tag{9}$$

For the longitudinal element replace q by $I/(j\omega)$ and the exponential factor by $-\beta_{mn}\,dz_0\,e^{-\beta_{mn}|z-z_0|}$ giving the result

$$\check{\Psi}_z = \frac{jI\,dz_0}{2\pi\omega\epsilon a^2} \sum_{n=1}^{\infty} \sum_{m=0}^{\infty} (2 - \delta_m^0) e^{-\beta_{mn}|z-z_0|} \frac{J_m(\beta_{mn}\rho_0)J_m(\beta_{mn}\rho)}{[J_{s+1}(\beta_{mn}a)]^2} \cos m\phi \tag{10}$$

The fields are everywhere in phase.

13.11. Loop Coupling with Circular Guide.—The formulas of the last article may be applied to a square plane loop at $\phi = 0$ in a wave guide of radius a closed by a conducting wall at $z = 0$ where the boundary condition is satisfied by using an image loop, as shown in Fig. 13.11. The current is assumed uniform so that no scalar potentials are needed, although the integrals can be set up for any given distribution. Only the radial ends of the loops generate TE waves. Addition of the contributions to $\check{D}_{mn}$ as given by 13.10 (3) for elements $I\,d\rho_0$ at ρ_0, d; $-I\,d\rho_0$ at ρ_0, b; $I\,d\rho_0$ at ρ_0, $-b$; and $-I\,d\rho_0$ at ρ_0, $-d$, and integration from $\rho_0 = c$ to $\rho_0 = a$ give

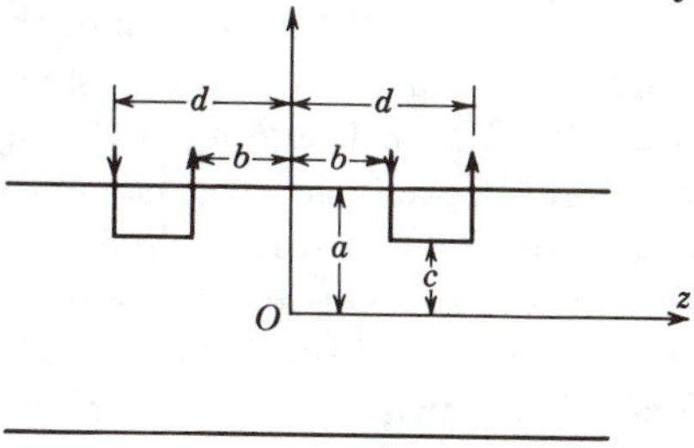

Fig. 13.11.

$$[\check{D}_{mn}]_{te} = \frac{-2m\mu I[\sin(\beta'_{mn}d) - \sin(\beta'_{mn}b)]}{\pi\beta'_{mn}(\beta^2_{mn}a^2 - m^2)[J_m(\beta_{mn}a)]^2}\int_{\beta_{mn}c}^{\beta_{mn}a}\frac{J_m(v)}{v}\,dv \tag{1}$$

where β_{mn} is given by 13.09 (2), β'^2_{mn} is $\beta^2 - \beta^2_{mn}$, and z is greater than d. The integral can be evaluated by using for $J_m(v)$ the series 5.293 (3) of which only a few terms are usually needed. The TE fields are given by putting this value of $\check{D}_{mn}$ into 13.09 (5) and (6) and letting $\check{C}_{mn}$ be zero.

The last article shows that TM waves will be generated by all sides of the loop. Proceeding as for the TE waves but using 13.10 (4) instead of 13.10 (3), we find for the sides of the loop when $z > d$,

$$[\check{C}'_{mn}]_{tm} = \frac{j\mu I_0(2 - \delta^0_m)[\sin(\beta'_{mn}d) - \sin(\beta'_{mn}b)]J_m(\beta_{mn}c)}{\pi[\beta\beta_{mn}aJ'_m(\beta_{mn}a)]^2} \tag{2}$$

From Fig. 13.11 for the longitudinal wires, we add by 13.10 (8) effects of elements at z_0 and $-z_0$, integrate from $z_0 = b$ to $z_0 = d$, and get

$$z > d, \qquad [\check{C}''_{mn}]_{tm} = \frac{-(2 - \delta^0_m)\mu I_0[\sin(\beta'_{mn}d) - \sin(\beta'_{mn}b)]J_m(\beta_{mn}c)}{j\pi[\beta\beta'_{mn}aJ'_m(\beta_{mn}a)]^2} \tag{3}$$

Addition of (2) and (3) gives, since β^2 equals $\beta'^2_{mn} + \beta^2_{mn}$ by 13.00 (7), for all TM modes generated by the loop

$$[\check{C}_{mn}]_{tm} = \frac{j(2 - \delta^0_m)\mu I_0[\sin(\beta'_{mn}d) - \sin(\beta'_{mn}b)]J_m(\beta_{mn}c)}{\pi[\beta'_{mn}\beta_{mn}aJ'_m(\beta_{mn}a)]^2} \tag{4}$$

This value, when substituted in 13.09 (8) and (9), gives the fields. For transmitted waves β'_{mn} is real, so that the trigonometric terms in the numerators may be written, by *Dw* 401.09,

$$\sin(\beta'_{mn}d) - \sin(\beta'_{mn}b) = 2\sin[\pi(d - b)\lambda'^{-1}_{mn}]\cos[\pi(d + b)\lambda'^{-1}_{mn}] \tag{5}$$

It is at once evident that if the length $d - b$ of the loop is an integral number of guide wave lengths or if the distance $\frac{1}{2}(d + b)$ of its center from the closed end is odd number of half wave lengths, then this mode is not excited. If the bottom of the loop is on the guide axis, only axially symmetric TM waves appear since $J_m(0) = 0$ unless $m = 0$.

13.12. Orifice Coupling with Circular Guide.—In the last article the wave-guide excitation was calculated for an assumed current distribution in the antenna or current loop. We shall now calculate it by assuming the tangential electric field component over the orifice. The propagation space of a coaxial line is bounded by the cylinders $\rho = b$ and $\rho = c$ where $c > b$ and terminates in the plane $z = 0$ which is perfectly conducting save for the annular opening. This plane closes the end of a circular wave guide bounded by the cylinder $\rho = a$ extending from $z = 0$ to $z = \infty$. From symmetry it is clear that only the TM waves independent of ϕ can be excited, and from 13.09 (2) and (8), we must have

$$J_0(\beta_n a) = 0, \qquad \check{E}_\rho = \omega \sum_{n=1}^{\infty} C_n \beta'_n \beta_n J_1(\beta_n \rho) e^{-j\beta'_n z} \tag{1}$$

As in 12.21 (2), the assumed boundary conditions at $z = 0$ are

$$\rho < b, \quad c < \rho < a, \quad E_\rho = 0; \quad b < \rho < c, \quad \check{E}_\rho = \check{V}_0 \left(\rho \ln \frac{b}{c}\right)^{-1} \tag{2}$$

Insert these values for E_ρ into (1), multiply both sides by $\rho J_1(\beta_n \rho)\, d\rho$, integrate from $\rho = 0$ to $\rho = a$, and solve for C_n:

$$C_n = \frac{2\check{V}_0[J_0(\beta_n b) - J_0(\beta_n c)]}{\omega \beta'_n \beta_n^2 a^2 \ln (b/c)[J_1(\beta_n a)]^2} \tag{3}$$

A simpler way to derive (3) is to replace the opening by the double current sheet of Fig. 13.04c. By 13.04 (1) the angular current density is

$$i\, dz_0 = \frac{2V_0}{j\omega\mu \ln (c/b)} \tag{4}$$

and in the Green's function of 13.10 (8) $p = m$, $q = n$, and I is replaced by $i\, d\phi_0$ at $\rho = c$ and by $-i\, d\phi_0$ at $\rho = b$. In 13.09 (8) and (9), when ϕ is replaced by $\phi - \phi_0$ and integrated around the opening, the result is zero except when $m = 0$ where its value is 2π. Thus 13.10 (8) gives (3). To estimate the input impedance of the guide when coupled to the coaxial line, integrate Poynting's vector over the opening. From 11.17 (3) by using the ratio of $\check{E}_\rho$ to $\hat{B}_\phi$ obtained from 13.09 (8) and (9), we have

$$\overline{\Pi} = \frac{1}{2\mu} \int \check{E}_\rho \hat{B}_\phi\, dS = \frac{\beta^2}{2\omega\mu\beta'_n} \int_b^c E^2 2\pi\rho\, d\rho = \frac{V_0^2}{2\hat{Z}_i} \tag{5}$$

The square of (1) is integrated by 5.297 (3).

$$\frac{1}{\hat{Z}_i} = \frac{4\pi\omega\epsilon}{a^4 \ln^2 (b/c)} \sum_{n=1}^{\infty} \frac{[J_0(\beta_n b) - J_0(\beta_n c)]^2\{[cJ_1(\beta_n c)]^2 - [bJ_1(\beta_n b)]^2\}}{\beta_n^2 \beta_n' [J_1(\beta_n a)]^2} \tag{6}$$

Thus the guide presents a set of parallel impedances, one for each mode. For transmitted modes, β_n' is real giving resistive impedance; for others, β_n' is imaginary giving reactive impedance. An exact solution requires a matching of fields across the opening, using 13.13 (3) on the coaxial side.

13.13. The Coaxial Wave Guide.—Transmission lines whose propagation space is bounded internally by $\rho = a$ and externally by $\rho = b$ fall in the class treated in 11.14 which have principal modes that can transmit waves at any frequency. From 11.18 (5), the resistance per unit length is

$$R_i I^2 = R_i\left(\oint_a i^2\, ds + \oint_b i^2\, ds\right) = \frac{I^2 R_i}{2\pi}\left(\frac{1}{b} + \frac{1}{a}\right) = \frac{rI^2(a+b)}{2\pi ab\delta} = \omega L_i I^2 \tag{1}$$

The characteristic impedance is then given by 11.18 (1) to be

$$\check{Z}_k = \left\{\frac{L[R_i + j\omega(L + L_i)]}{\mu(\gamma + j\omega\epsilon)}\right\}^{\frac{1}{2}} \qquad \text{where } L = \frac{\mu}{2\pi} \ln \frac{b}{a} \tag{2}$$

The inductance per unit length was written down from $LC = \mu\epsilon$ and 2.04 (2). If the skin inductance L_i and dielectric conductivity γ are small, there is a flat minimum in the attenuation when $\partial(R_i/L)/\partial(b/a)$ is zero. This value, which occurs at $b = 3.6a$ is only 5 per cent below that at $2.5a$ and $5.0a$.

At short enough wave lengths, higher order waves of the type discussed in the last article are possible in a coaxial line. These waves appear chiefly in boundary conditions at discontinuities. The axial region is now excluded so both Bessel functions can occur. Thus from 12.17 (2),

$$\check{U} = (\check{C}_{mn} \cos m\phi + \check{D}_{mn} \sin m\phi) R_m(\beta_{mn}\rho) \tag{3}$$

$$[R_m(\beta_{mn}\rho)]_{te} = J_m(\beta_{mn}\rho) Y_m'(\beta_{mn}a) - Y_m(\beta_{mn}\rho) J_m'(\beta_{mn}a) \tag{4}$$

$$[R_m(\beta_{mn}\rho)]_{tm} = J_m(\beta_{mn}\rho) Y_m(\beta_{mn}a) - Y_m(\beta_{mn}\rho) J_m(\beta_{mn}a) \tag{5}$$

where, from 13.00 (4) and (5), β_{mn} must be chosen so that

$$\frac{\partial[R_m(\beta_{mn})b]_{te}}{\partial b} = 0 \qquad [R_m(\beta_{mn}b)]_{tm} = 0 \tag{6}$$

From (4), (5), and (6), the boundary conditions of 13.00 (4) and (5) are satisfied at $\rho = a$ and $\rho = b$. The fields, attenuations, etc., can be worked out exactly as in the last article but are of little interest.

13.14. Plane Discontinuities in Coaxial Lines.—Electrostatic methods yield nearly exact solutions of certain electromagnetic wave problems. To show this, write 13.00 (1) in terms of wave length with $\boldsymbol{E}$ for $\boldsymbol{A}$.

$$\nabla^2 \boldsymbol{E} = -\omega^2\mu\epsilon\boldsymbol{E} = -4\pi^2\lambda^{-2}\boldsymbol{E} \xrightarrow[\lambda\to\infty]{} 0 \tag{1}$$

Thus for wave lengths very long compared with the dimensions of the region involved, the instantaneous field is indistinguishable from a solution of Laplace's equation that satisfies the same boundary conditions.

Consider two coaxial lines coaxially joined to the $z = 0$ plane which is perfectly conducting save where holes connect the propagation spaces. The wave lengths are so long that only the principal mode is propagated. Clearly, if standing waves are set up so that there is an electric node at $z = 0$, the discontinuity has no effect for there is no radial electric field there and the magnetic induction links no element of this plane. In any transmission line, only a pure shunt element can be inserted at an electric node in a standing wave pattern without upsetting it. Therefore this discontinuity acts as a pure shunt element across the line at $z = 0$.

Now take the special case of coaxial annular openings and suppose that standing waves are set up so that $z = 0$ lies halfway between two adjacent electric nodes. Obviously the radial field of the principal mode alone is not normal to the plane face at $z = 0$, so local higher mode fields of the type of 13.04 (4) or (5) are needed there. These contain the factor $e^{-|\beta'_{0n}z|}$ as in 13.03 (5) which confines them to line sections so short compared with the wave length that all the local fields are in phase. By (1), the instantaneous electric field in this section is identical with the static field between inner and outer conductors, and the capacitance across the line at $z = 0$ is the excess of the actual static capacitance of the section over the sum of the capacitances given by the concentric cylindrical capacitor formula for each half of it.

The simplest case occurs when the ratio of the external to the internal radius is nearly one. A small sector of a cylindrical capacitor then looks like a narrow strip of parallel plate capacitor which is solvable by a conformal transformation. The case of a parallel plate capacitor with a step in one wall, as shown by the lines $V = 0$ and $V = \frac{1}{2}\pi$ in Fig. 6.07*b*, is treated in 6.07. Half the increase of resistance per unit length is given by 6.08 (5), so by 6.08 (4) the increase in capacitance per unit length is $2\epsilon/(\tau\Delta R)$. The discontinuity capacitance is the product of this by the circumference $2\pi r$ of the propagation space. Thus, from 6.07 (5),

$$C = 2\pi r C_1 = 2\epsilon r\left(\frac{h^2 + k^2}{hk}\ln\frac{k + h}{k - h} + 2\ln\frac{k^2 - h^2}{4hk}\right) \tag{2}$$

Whinnery and Jamison (*Proc. I.R.E.* Vol. 32, page 98) have shown empirically that, when a discontinuity is in one wall, the circumference of the other wall should be used. If the ratio of internal to external radius is 5, this rule still gives an error of only about 10 per cent. Reflections from such obstacles are calculated by 11.16 (3) where, if Z_{k1} and Z_{k2} are the line characteristic impedances on the two sides of the obstacle,

$$Z_1 = Z_{k1}, \qquad \frac{1}{Z_2} = j\omega C + \frac{1}{Z_{k2}} = \frac{2\pi jC}{\lambda} + \frac{1}{Z_{k2}} \tag{3}$$

13.15. Guides of Arbitrary Section. Elliptic Guides.—A most important bit of information about a wave guide is its dominant mode cutoff frequency. From 13.00 (8) this lowest transmitted frequency ω_{mn} is $\beta_{mn}(\mu\epsilon)^{-\frac{1}{2}}$ or for short $\beta_c(\mu\epsilon)^{-\frac{1}{2}}$. It always refers to a TE mode and for the rectangular guide, from 13.02 (2) and (3), is TE_{10} and for the circular guide, from 13.09 (2), (3), and (5), is TE_{11}. From 13.03 (3) and (4), β_c and the transverse coordinates u_1 and u_2 are connected by

$$\nabla^2 U(u_1,u_2) + \beta_c^2 U(u_1,u_2) = 0, \qquad \text{and} \qquad \frac{\partial U}{\partial n} = 0 \qquad \text{on } s \tag{1}$$

where s refers to the curve bounding a cross section of the guide of area S. The fields derive from U by 13.00 (14) and (15). In only a few simple sections can U be written as a product of orthogonal functions of u_1 and u_2 as in 13.02 (1) and 13.09 (1). However, as in 13.07, stationary variational formulas exist for β_c in terms of trial U_t values which may be improved by iteration and which satisfy the equations

$$\nabla^2 U_t^{(n+1)} + U_t^{(n)} = 0 \qquad \text{and} \qquad \int U_t^{(n)}\, dS = 0 \tag{2}$$

where $U_t^{(n)}$ is the nth trial function. The integral ensures zero net flux through a guide section because magnetic flux lines must close, and 13.00 (15) shows that U_{te} is proportional to B_z. Borgnis and Papas, Chap. XIV, give an excellent detailed treatment which covers both TE and TM modes.

At any z, (1) holds for all u_1 and u_2 values. Thus its product by U, integrated over the cross section, is

$$\int (U\nabla^2 U + \beta_c^2 U^2)\, dS = 0 \tag{3}$$

Now replace the first term by means of the vector formula (4) to get (5).

$$\nabla \cdot (\phi\nabla\psi) = \phi\nabla^2\psi + \nabla\phi \cdot \nabla\psi \tag{4}$$

$$\int \nabla \cdot (U\nabla U)\, dS - \int (\nabla U)^2\, dS + \beta_c^2 \int U^2\, dS = 0 \tag{5}$$

From the two-dimensional Gauss theorem 3.00 (2), in which dS, ds, and $U\nabla U$ replace dV, dS, and $\mathbf{A}$, the first integral is $\int U(\partial U/\partial n)\, ds$, which is zero by (1). Thus solving (5) for β_c^2 gives

$$\beta_c^2 = \frac{\int(\nabla U)^2\, dS}{\int U^2\, dS} \tag{6}$$

This has the familiar variational form of 13.07 (4). To prove that it is stationary, take the variation of the last two terms of (5) and obtain

$$2\beta_c \delta\beta_c \int U^2\, dS + 2\beta_c^2 \int U\, \delta U\, dS = 2\int \nabla U \cdot \nabla(\delta U)\, dS \tag{7}$$

By (3) and the Gauss theorem, the right side may be written

$$-2\oint \delta U \frac{\partial U}{\partial n}\, ds + 2\int \delta U\, \nabla^2 U\, dS = -2\int \nabla U \cdot \nabla(\delta U)\, dS \tag{8}$$

Add (7) and (8). The line integral is zero on s by (1) and two others add to give $\int \delta U(\nabla^2 U + \beta_c^2 U)\, dS$, also zero by (1), so $\delta\beta_c$ is zero. Thus in (6) a first-order variation in U gives no variation in β_c. It can also be shown that, if the trial U is sufficiently near the correct form for β_c^2, the right side of (6) gives a result greater than β_c^2, just as it did in 13.08 (10) and (16). The square of (1), being positive, is greater than zero if an incorrect U form is used. Substitution of (6) for β_c^2 in this square and rearrangement give

$$\frac{\int(\nabla^2 U)^2\, dS}{\int(\nabla U)^2\, dS} \geqslant \frac{\int(\nabla U)^2\, dS}{\int U^2\, dS} \tag{9}$$

with correct U values both equal β_c^2. Now take the gradient of (1), square, substitute the left side of (9) for β_c^2, and rearrange to get

$$\frac{\int[\nabla(\nabla^2 U)]^2\, dS}{\int(\nabla^2 U)^2\, dS} \geqslant \frac{\int(\nabla^2 U)^2\, dS}{\int(\nabla U)^2\, dS} \tag{10}$$

The same remarks apply to (9) and (10). Combine (6), (9), and (10) into

$$\frac{\int[\nabla(\nabla^2 U)]^2\, dS}{\int(\nabla^2 U)^2\, dS} \geqslant \frac{\int(\nabla^2 U)^2\, dS}{\int(\nabla U)^2\, dS} \geqslant \frac{\int(\nabla U)^2\, dS}{\int U^2\, dS} \geqslant \beta_c^2 \tag{11}$$

To show how the iterative process improves a trial function, assume that a suitable function $U_t^{(n)}$ for which $\partial U/\partial n$ is zero on s has been found and that a better one $U_t^{(n+1)}$ is desired. Substitute $U_t^{(n+1)}$ for U in (11) and then replace $\nabla^2 U_t^{(n+1)}$ by $-U^{(n)}$ from (2) to give

$$\frac{\int[\nabla U_t^{(n)}]^2\, dS}{\int[U_t^{(n)}]^2\, dS} \geqslant \frac{\int[U_t^{(n)}]^2\, dS}{\int[\nabla U_t^{(n+1)}]^2\, dS} \geqslant \frac{\int[\nabla U_t^{(n+1)}]^2\, dS}{\int[U_t^{(n+1)}]^2\, dS} \geqslant \beta_c^2 \tag{12}$$

The first and third terms have the same form and the third lies closer to β_c^2. Therefore the function $U_t^{(n+1)}$ obtained by solving (2) for $U_t^{(n+1)}$ gives an improved value of β_c^2. Any desired precision can be obtained by repetition of the process.

The foregoing technique works much better for the dominant mode cutoff frequency of an elliptical guide than that used for the special case of a circular guide in 13.09. Take the origin on the guide axis, so that the equations for the boundary and its slope are

$$\left(\frac{x'}{a}\right)^2 + \left(\frac{y'}{b}\right)^2 = 1, \qquad \frac{dy'}{dx'} = \frac{-a^2y'}{b^2x'} \tag{13}$$

where boundary coordinates are primed. To visualize a suitable trial function, imagine the ellipse formed by rounding the corners of the rectangular guide in the TE_{10} E-line map in Fig. 13.02*a*. The E-lines bend to stay normal to the boundary, and the TE_{10} B-line map is little changed in the $y = 0$ plane, but the right and left boundary close in as $|y| \to b$. Thus from symmetry the B_z to which U is proportional by 13.00 (15) is oppositely directed in the right and left halves and so is odd in x. The B-line maps in the $+y$- and $-y$-plane are identical if $|y| < b$ so B_z is even in y. A simple dimensionless trial function meeting all these conditions is

$$U^{(0)} = \frac{x}{a}(1 + Bx^2 + Cy^2) \tag{14}$$

From (1) $\nabla U^{(0)}$ must be tangent to the boundary ellipse, and thus from (13)

$$\frac{\partial U^{(0)}/\partial x'}{\partial U^{(0)}/\partial y'} = \frac{1 + 3B(x')^2 + C(y')^2}{2Cx'y'} = \frac{-a^2y'}{b^2x'} \tag{15}$$

Collect all the terms of the last equation on one side, put over a common denominator, factor out x', and write $(x')^2$ in terms of $(y')^2$ by (13). The resultant equation must hold for all y' values so the constant term and the $(y')^2$ terms are separately equal to zero. The constant term gives B, which is then put into the coefficient of $(y')^2$ to give C. Thus (14) is

$$U^{(0)} = \frac{x}{a}\left(1 - \frac{x^2}{3a^2} - \frac{y^2}{2a^2 + b^2}\right) \tag{16}$$

The integrals of (6) then become

$$\frac{\int(\nabla U)^2\,dS}{\int U^2\,dS} = \frac{[\pi b(5a^2 + 2b^2)(3a^2 + b^2)]/[6a(2a^2 + b^2)^2]}{[\pi ab(101a^4 + 80a^2b^2 + 17b^2)]/[144(2a^2 + b^2)^2]} \tag{17}$$

If p is written for b/a, the last inequality in (11) is

$$\beta_c a \leqslant \left[\frac{24(5 + 2p^2)(3 + p^2)}{101 + 80p^2 + 17p^4}\right]^{\frac{1}{2}} \tag{18}$$

For the circle $a = b$, so $p = 1$ and $\beta_c a = 1.842$, which agrees with 13.09 (3) to better than 0.1 per cent. This is a more precise value than can be read from the graph on page 83 of the "Waveguide Handbook," in which $\pi/[2\beta_c aE(e)]$ is plotted against $e = [1 - (b/a)^2]^{\frac{1}{2}}$. On the next page appears a map of the fields of the fundamental or ${}_eH_{11}$ mode for $e = 0.75$. A table of the values of the complete elliptic integral $E(e)$ is given in HMF, page 609.

13.16. Cavity Resonators. Normal Modes.—When an electromagnetic disturbance occurs in a closed cavity with perfectly conducting walls so the energy cannot be dissipated, waves reflect back and forth between the walls indefinitely. Because the tangential electric field vanishes at each wall, a wave of arbitrary form must be built up of elementary standing waves of frequencies such as to give electric nodes at the walls. These standing waves constitute the normal oscillation modes of the cavity. Clearly, when two continuous waves having the same frequency, polarization, and intensity travel in opposite directions on a transmission line or a wave guide, there are transverse electric nodes at half wave length intervals. If we form a cavity by inserting at any two nodes perfectly conducting plane sheets normal to the axis of the guide, then that portion of the trapped waves between the sheets is a normal mode of the cavity. It is evident that, if the spacing of the sheets is arbitrary, then the normal frequencies will be those for which the guide wave length is an integral number of half wave lengths. This fact can be used to find the normal or natural oscillation frequencies of cavities formed in the manner just described from any section of wave guide. The two oppositely directed waves in 13.00 (6) must have equal amplitudes so $\check{C} = \pm\check{D}$. If the cavity ends are at $z = 0$ and $z = d$, then for a nondissipative cavity the real part of 13.00 (6) becomes, for one of the pth transverse electric modes,

$$W_{te} = [AU_1(u_1, u_2) + BU_2(u_1, u_2)]_{te} \sin\frac{p\pi z}{d} \cos(\omega_p t + \psi_p)_{te} \tag{1}$$

and for one of the pth transverse magnetic modes,

$$W_{tm} = [AU_1(u_1, u_2) + BU_2(u_1, u_2)]_{tm} \cos\frac{p\pi z}{d} \cos(\omega_p t + \psi_p)_{tm} \tag{2}$$

where β'_{mn} is chosen equal to $p\pi/d$ to make the tangential electric field vanish over the ends. The boundary conditions of 13.00 (4), (5), and (7) used to determine β_{mn} still hold. Thus, writing β_{mnp} for β and $p\pi/d$ for β'_{mn}, we may solve 13.00 (7) for the normal frequency of the mnpth mode in an evacuated cavity.

$$\omega_{mnp} = (\mu_v\epsilon_v)^{-\frac{1}{2}}\beta_{mnp} = v\beta_{mnp} = v(\beta_{mn}^2 + \beta_{mn}'^2)^{\frac{1}{2}} = v[\beta_{mn}^2 + p^2\pi^2 d^{-2}]^{\frac{1}{2}} \tag{3}$$

For rectangular wave guides β_{mn} is given by 13.02 (2), for circular wave guides by 13.09 (2), and for the higher modes of the coaxial wave guide by 13.13 (6). For the principal coaxial modes, β_{mn} is zero by 13.00 (5).

13.17. Independent Oscillation Modes of a Cavity.—An electromagnetic disturbance in a cavity will usually excite several modes simultaneously. We shall now prove that the instantaneous total energy is the sum of the instantaneous energies of each of the constituent modes. To prove that

$$\int_V E^2\,dv = \sum_{i=1}^{n}\int_V E_i^2\,dv \qquad \text{and} \qquad \int_V B^2\,dv = \sum_{i=1}^{n}\int_V B_i^2\,dv \tag{1}$$

we substitute $\boldsymbol{E}_i$ or $\boldsymbol{B}_i$ for $\boldsymbol{\Psi}$ and $\boldsymbol{E}_j$ or $\boldsymbol{B}_j$ for $\boldsymbol{\Phi}$ in the vector analogue, 3.06 (6), of Green's theorem, and for the double curl we write $\omega^2\mu\epsilon$ from 11.01 (1) or (2). Thus, we obtain

$$\mu\epsilon(\omega_j^2 - \omega_i^2)\int_V \boldsymbol{E}_i\cdot\boldsymbol{E}_j\,dv = \int_S[\boldsymbol{E}_j\times(\nabla\times\boldsymbol{E}_i) - \boldsymbol{E}_i\times(\nabla\times\boldsymbol{E}_j)]\cdot\boldsymbol{n}\,dS \tag{2}$$

But $\boldsymbol{E}_i$ and $\boldsymbol{E}_j$ are normal to the boundary, so both vectors in the bracket are tangential to it and their scalar product with $\boldsymbol{n}$ is zero. Thus the surface integral is zero, and we have, since $\boldsymbol{E} = -j\omega\boldsymbol{A}$,

$$\omega_i \neq \omega_j, \qquad \int_V \boldsymbol{E}_i\cdot\boldsymbol{E}_j\,dv = \int_V \boldsymbol{B}_i\cdot\boldsymbol{B}_j\,dv = \int_V \boldsymbol{A}_i\cdot\boldsymbol{A}_j\,dv = 0 \tag{3}$$

Thus, in any perfectly conducting cavity filled with a nondissipative dielectric, those oscillation modes having different frequencies are quite independent. If cavity losses are so large that the resonance curves of two modes overlap, then such conclusions are not valid because $\omega_j^2 - \omega_i^2$ on the left side of (2) may be zero part of the time.

The vector potential used here is always derived by taking the curl of $\boldsymbol{W}$ so that its divergence is zero. As shown in 11.01, all fields that satisfy Maxwell's equations are derivable from such potentials if charges are absent. If the cavity contains an electrode carrying an oscillating charge, then by 11.01 (1) a scalar potential satisfying Poisson's equation is needed whose electric field throughout the cavity oscillates in phase with the charge. That this field $-\nabla\Phi$ has an energy independent of that of the field $-j\omega\boldsymbol{A}$ or $-j\omega\nabla\times\boldsymbol{W}$ is proved from 3.00 (2) by writing

$$\int_V(\nabla_w\times\boldsymbol{W})\cdot\nabla\Phi\,dv = \int_V\nabla_w\cdot(\boldsymbol{W}\times\nabla\Phi)\,dv = \int_S\boldsymbol{n}\cdot(\boldsymbol{W}\times\nabla\Phi)\,dS = 0 \tag{4}$$

The surface integral is zero because $\nabla\Phi$ is parallel to $\boldsymbol{n}$.

13.18. Energy and Damping of Normal Modes. Quality.—In a lossless, charge-free cavity, the Coulomb gauge gives, by 11.01 (2), (3), and (4),

$$\boldsymbol{B} = \nabla\times\boldsymbol{A} \qquad \boldsymbol{E} = -j\omega\boldsymbol{A} \tag{1}$$

The instantaneous magnetic energy W_m and the electric energy W_e for field magnitudes $\boldsymbol{H}$ and $\boldsymbol{E}$ are then, from 8.02 (2), 11.00 (1), and 1.14 (6),

$$W_m = \tfrac{1}{2}\mu^{-1}\int B^2\,dv = \tfrac{1}{2}\int \nabla \times \boldsymbol{H}\cdot\boldsymbol{A}\,dv = \tfrac{1}{2}\sum_i \omega_i^2\epsilon\int \boldsymbol{A}_i\cdot\boldsymbol{A}_i\,dv \tag{2}$$

$$W_e = \tfrac{1}{2}\epsilon\int E^2\,dv = \tfrac{1}{2}j^2\sum_i \omega_i^2\epsilon\int \boldsymbol{A}_i\cdot\boldsymbol{A}_i\,dv \tag{3}$$

This shows that the time average W_m and W_e are equal so that

$$\int(\mu|H^2| - \epsilon|E^2|)\,dv = \int(\mu\,\check{\boldsymbol{H}}\cdot\hat{\boldsymbol{H}} - \epsilon\check{\boldsymbol{E}}\cdot\hat{\boldsymbol{E}})\,dv = 0 \tag{4}$$

The metals of most cavity walls conduct well, so that an observable loss in amplitude appears only after many free oscillation cycles. Therefore damping is a small perturbation. Thus the skin effect formulas of 10.02 applied to the undamped field formulas give the energy losses. So, from 10.02 (8),

$$\bar{P} = (2\mu^2\gamma\delta)^{-1}\int\boldsymbol{B}_t\cdot\boldsymbol{B}_t\,dS \tag{5}$$

where γ and δ are the conductivity and the skin thickness of the wall and $\boldsymbol{B}_t$ is the magnetic induction tangent to the wall. Cavity resonator damping is given in terms of its "quality" or "Q" value defined by

$$W_t = W_0e^{-\omega t/Q} \tag{6}$$

where W_t is the energy at a time t after W_0. Division of this equation by its derivative with respect to t gives

$$Q = -\frac{\omega W}{\partial W/\partial t} = \frac{\omega W}{\bar{P}} \tag{7}$$

Note from (2) and (3) that the total energy swings back and forth between purely magnetic and purely electric, but the sum is constant except for the damping factor. Thus, from (2),

$$Q = \frac{\omega^3\mu^2\epsilon\gamma\delta\int\boldsymbol{A}\cdot\boldsymbol{A}\,dv}{\int\boldsymbol{B}_t\cdot\boldsymbol{B}_t\,dS} = \frac{\omega\mu\gamma\delta\int\boldsymbol{B}\cdot\boldsymbol{B}\,dv}{\int\boldsymbol{B}_t\cdot\boldsymbol{B}_t\,dS} = \frac{2\mu\int B^2\,dv}{\mu'\delta\int B_t^2\,dS} \tag{8}$$

where μ' is the wall permeability.

In the last article it was noted that the energy transfers from one resonant mode to another if the resonance curve overlaps, so it is well to know how this curve depends on Q. Suppose the oscillation starts at $t = 0$ when the energy is all magnetic. Then from (6),

$$B = B_0e^{-\frac{1}{2}\omega_pt/Q}\cos\omega_pt \tag{9}$$

where ω_p is the undamped resonant frequency of the pth mode. By Fourier's integral theorem, B as a function of ω is found by multiplication by $\pi^{-1}e^{j\omega t}$ and integration from 0 to ∞. Write the cosine in exponential form by *Dw* 408.02 for easy integration, and assume a very large Q to make the $\omega + \omega_p$ term negligible when ω is near ω_p, so that

$$\breve{B}(\omega) = \frac{\frac{1}{2}B_0/\pi}{\frac{1}{2}(\omega_p/Q_p) - j(\omega - \omega_p)}$$

and

$$|B(\omega)| = \frac{Q_pB_0/(\omega_p\pi)}{[1 + 4Q_p^2(\omega - \omega_p)^2/\omega_p^2]^{\frac{1}{2}}} \tag{10}$$

Thus $|B(\omega)|$ is peaked at $\omega = \omega_p$, and from (2) and (10), W_m and B^2 have half their peak value when $\omega_{\frac{1}{2}} = \omega_p \pm \frac{1}{2}\omega_p/Q_p$. The peak width there is the "band width" $\Delta\omega$. Thus

$$\Delta\omega = \frac{\omega_p}{Q} \tag{11}$$

13.19. Normal Modes of a Cylindrical Cavity.—The shape and materials used in a cylindrical resonant cavity fix the relative field strengths and Q-factors. From 13.00 (2) and 13.16 (1) and (2), the fields are

$$\boldsymbol{E}_{te} = -\frac{\partial \boldsymbol{A}_{te}}{\partial t} = \omega_p(\boldsymbol{k} \times \nabla U_{te}) \sin \frac{p\pi z}{d} \sin(\omega_p t + \phi_p) \tag{1}$$

$$\boldsymbol{B}_{te} = \left(\frac{p\pi}{d}\nabla U_{te} \cos\frac{p\pi z}{d} - \boldsymbol{k}\beta_{mn}^2 U_{te} \sin\frac{p\pi z}{d}\right)\cos(\omega_p t + \phi_p) \tag{2}$$

$$\boldsymbol{E}_{tm} = \omega_p\left(\frac{p\pi}{d}\nabla U_{tm} \sin\frac{p\pi z}{d} - \boldsymbol{k}\beta_{mn}^2 U_{tm} \cos\frac{p\pi z}{d}\right)\sin(\omega_p t + \phi_p) \tag{3}$$

$$\boldsymbol{B}_{tm} = \beta_{mnp}^2(\boldsymbol{k} \times \nabla U_{tm}) \cos\frac{p\pi z}{d} \cos(\omega_p t + C_p) \tag{4}$$

From 13.00 (2) and 13.16 (1), the energy integral over the volume and the loss integrals over the two ends and the side walls are, for TE waves,

$$\int A^2\,dv = \int (\boldsymbol{k} \times \nabla U_{te})^2\,dS \int_0^d \sin^2\frac{p\pi z}{d}\,dz = \frac{d\int(\nabla U)^2\,dS}{2 - \delta_p^0} \tag{5}$$

$$2\int B_t^2\,dS = 2\left(\frac{p\pi}{d}\right)^2\int(\boldsymbol{k} \times \nabla U)^2\,dS = 2\left(\frac{p\pi}{d}\right)^2\int(\nabla U)^2\,dS \tag{6}$$

$$\int B_t^2\,dS = \oint\left[\left(\frac{p\pi}{d}\right)^2(\boldsymbol{n} \times \nabla U)^2 + \beta_{mn}^4 U^2\right]ds\int_0^d \sin^2\frac{p\pi z}{d}\,dz$$

$$= [d(2 - \delta_p^0)]^{-1}\oint\left[\left(\frac{p\pi\,\partial U}{\partial s}\right)^2 + d^2\beta_{mn}^4 U^2\right]ds \tag{7}$$

Thus from 13.18 (8) the cavity quality Q_{te} is

$$\frac{\beta^2\mu d^3\int(\nabla U)^2\,dS}{4p^2\pi^2(2 - \delta_p^0)\mu_e'\delta_e\int(\nabla U)^2\,dS + 2d\mu_s'\delta_s\oint[(p\pi\,\partial U/\partial s)^2 + d^2\beta_{mn}^4U^2]\,ds} \tag{8}$$

where s is the curve bounding a section S normal to z and μ_e', μ_s' and δ_e, δ_s are the conductivities and skin depths of the end and side walls. From 13.00 (2) and 13.16 (2), the volume integral of B_{tm}^2 is the same as (5) multiplied by β^4 and the surface integrals and Q are

$$2\int B_t^2\,dS = 2\beta_{mnp}^4\int (\mathbf{k}\times\nabla U)^2\,dS = 2\beta_{mnp}^4\int(\nabla U)^2\,dS \tag{9}$$

$$\int B_t^2\,dS = \beta_{mnp}^4[\mathbf{n}\times(\mathbf{k}\times\nabla U)]^2\,ds\int_0^d \cos^2\left(\frac{p\pi z}{d}\right)dz$$
$$= \beta_{mnp}^4 d(2-\delta_p^0)^{-1}\oint\left(\frac{\partial U}{\partial s}\right)^2 ds \tag{10}$$

$$Q_{tm} = \frac{\mu d\int(\nabla U)^2\,dS}{4(2-\delta_p^0)\mu_e'\delta_e\int(\nabla U)^2\,dS + 2\mu_s'\delta_s\,d\oint(\partial U/\partial s)^2\,ds} \tag{11}$$

In the articles that follow the detailed properties of the rectangular, right circular, and coaxial cylindrical cavities are worked out. Results for other simple forms appear in the problems at the end of the chapter.

13.20. Properties of a Rectangular Cavity.—From 13.16 (3) and 13.02 (2), the resonant frequencies of a rectangular cavity are

$$\nu_{mnp}^2 = \frac{v^2}{\lambda_{mnp}^2} = \frac{v^2}{4}\left(\frac{m^2}{a^2}+\frac{n^2}{b^2}+\frac{p^2}{d^2}\right) = \frac{v^2}{4\pi^2}\beta_{mnp}^2 = \frac{\omega_{mnp}^2}{4\pi^2} \tag{1}$$

If M, N, and P stand for $m\pi/a$, $n\pi/b$, and $p\pi/d$ and T_{mnp} for $\omega_{mnp}t + \psi_{mnp}$, then from 13.02 (1) and 13.19 (1) to (4), the fields are

$$\mathbf{E}_{te} = \omega_{mnp}C[\mathbf{i}N\cos Mx\sin Ny - \mathbf{j}M\sin Mx\cos Ny]\sin Pz\sin T_{mnp} \tag{2}$$

$$\mathbf{B}_{te} = C\{P[\mathbf{i}M\sin Mx\cos Ny + \mathbf{j}N\cos Mx\sin Ny]\cos Pz - \mathbf{k}(M^2+N^2)\cos Mx\cos Ny\sin Pz\}\cos T_{mnp} \tag{3}$$

$$\mathbf{E}_{tm} = \omega_{mnp}C'\{P[\mathbf{i}M\cos Mx\sin Ny + \mathbf{j}N\sin Mx\cos Ny]\sin Pz - \mathbf{k}(M^2+N^2)\sin Mx\sin Ny\cos Pz\}\sin T_{mnp} \tag{4}$$

$$\mathbf{B}_{tm} = \beta_{mnp}^2C'[\mathbf{i}N\sin Mx\cos Ny - \mathbf{j}M\cos Mx\sin Ny]\cos Pz\cos T_{mnp} \tag{5}$$

When m, n, or p is zero, $\mathbf{E}$ lies along the corresponding axis and ν is then independent of its length. In this most common case, the fields become, taking $p = 0$ and writing C_0 for $\pi^2C'(m^2a^{-2}+n^2b^{-2})$,

$$E_z = -\omega_{mn}C_0\sin\frac{m\pi x}{a}\sin\frac{n\pi y}{b}\sin(\omega t+\psi)_{mn} \tag{6}$$

$$\mathbf{B} = \pi C_0(\mathbf{i}N\sin Mx\cos Ny - \mathbf{j}M\cos Mx\sin Ny)\cos(\omega t+\psi)_{mn} \tag{7}$$

If the electric field is along any other axis, the fields are given by a cyclic permutation of x, y, z, a, b, c and m, n, p.

If none of the integers m, n, p is zero and (1) is used for ω_{mnp} and (3) for $\boldsymbol{B}$, then 13.18 (4) gives for Q_{te} at resonance

$$Q_{te} = \frac{\frac{1}{4}v\mu\gamma\delta\pi(m^2b^2 + n^2a^2)(m^2b^2d^2 + n^2a^2d^2 + p^2a^2b^2)^{\frac{3}{2}}}{p^2a^3b^3[n^2a(a + d) + m^2b(b + d)] + d^3(a + b)(m^2b^2 + n^2a^2)^2} \tag{8}$$

If (1) is used for ω_{mnp}, (5) for B_{tm}, and neither m nor n is zero, then

$$Q_{tm} = \frac{v\mu\gamma\delta\pi(m^2b^2 + n^2a^2)(m^2b^2d^2 + n^2a^2d^2 + p^2a^2b^2)^{\frac{1}{2}}}{2[ab(2 - \delta_p^0)(m^2b^2 + n^2a^2) + 2d(n^2a^3 + m^2b^3)]} \tag{9}$$

When $a > b > d$ and $m = n = 1$, the lowest resonance Q, at $p = 0$, is

$$Q_{tm} = \frac{v\mu\gamma\delta\pi d(a^2 + b^2)^{\frac{3}{2}}}{2[ab(a^2 + b^2) + 2d(a^3 + b^3)]} \tag{10}$$

For $a = b$ and the cube $a = b = c$, this Q becomes

$$Q_{tm} = \frac{v\mu\gamma\delta\pi d}{2^{\frac{1}{2}}(a + 2d)}, \qquad Q_{tm} = \frac{v\mu\gamma\delta\pi}{2^{\frac{1}{2}}3} \tag{11}$$

where v is the free wave velocity in the dielectric of permeability μ in the cavity, γ is the wall conductivity, and δ is the skin depth.

13.21. Properties of a Right Circular Cylindrical Cavity.—From 13.11 the resonant frequencies of this cavity are, in terms of the velocity v and wave length λ_{mnp} of a free wave in the medium that fills it,

$$\nu_{mnp}^2 = \frac{v^2}{\lambda_{mnp}^2} = \frac{v^2}{4\pi^2}\left(\beta_{mn}^2 + \frac{p^2\pi^2}{d^2}\right) = \frac{v^2}{4\pi^2}\beta_{mnp}^2 = \frac{\omega_{mnp}^2}{4\pi^2} \tag{1}$$

where β_{mnp} is chosen to satisfy 13.09 (2). From 13.09 (1) and 13.19 (4) the fields are, if $s(\phi)$ stands for $\sin(m\phi + \psi_m')$, $c(\phi)$ for $\cos(m\phi + \psi_m)$, J_m for $J_m(\beta_{mn}\rho)$, and P for $p\pi/d$,

$$\boldsymbol{E}_{te} = \omega_{mnp}C\left[\boldsymbol{\rho}_1\left(\frac{m}{\rho}\right)J_m s(\phi) + \boldsymbol{\phi}\beta_{mn}J_m'c(\phi)\right]\sin(Pz)\sin(\omega t + \alpha)_{mnp} \tag{2}$$

$$\boldsymbol{B}_{te} = C\{P[\boldsymbol{\rho}_1\beta_{mn}J_m'c(\phi) - \boldsymbol{\phi}\left(\frac{m}{\rho}\right)J_m s(\phi)]\cos(Pz) + \boldsymbol{k}\beta_{mn}^2J_m c(\phi)\sin(Pz)\}\cos(\omega t + \alpha)_{mnp} \tag{3}$$

$$\boldsymbol{E}_{tm} = \omega_{mnp}C'\left\{P\left[\boldsymbol{\rho}_1\beta_{mn}J_m'c(\phi) - \boldsymbol{\phi}\left(\frac{m}{\rho}\right)J_m s(\phi)\right]\sin(Pz) - \boldsymbol{k}\beta_{mn}^2J_m c(\phi)\cos(Pz)\right\}\sin(\omega t + \alpha)_{mnp} \tag{4}$$

$$\boldsymbol{B}_{tm} = \beta_{mnp}^2C'[\boldsymbol{\rho}_1\left(\frac{m}{\rho}\right)J_m s(\phi) + \boldsymbol{\phi}\beta_{mn}J_m'c(\phi)]\cos(Pz)\cos(\omega t + \alpha)_{mnp} \tag{5}$$

For a TE wave, $p \neq 0$. Obtain Q from 13.19 (8) and evaluate ρ-integrals by 5.296 (12) with U from 13.09 (1) and (2) and $\mu_e'\delta_e = \mu_s'\delta_s$. Thus

$$Q_{te} = \frac{v\mu\gamma\delta a d^3 \beta_{mnp}^3(\beta_{mn}^2 a^2 - m^2)}{2\{p^2\pi^2[m^2(d - 2a) + 2\beta_{mn}^2 a^3] + \beta_{mn}^4 a^2 d^3\}} \tag{6}$$

$$Q_{tm} = \frac{v\mu\gamma\delta a d\beta_{mnp}}{2[a(2 - \delta_p^0) + d]} = \frac{\mu a d}{4\mu'\delta[(2 - \delta_p^0)a + d]} \tag{7}$$

13.22. Multiply Connected Cylindrical Cavities.—In a cavity which is multiply connected, it is possible to draw a closed curve which cannot be contracted to a point without passing outside the cavity. If the cavity is bounded externally by two parallel planes and by a cylindrical surface normal to them and internally by one or more closed cylindrical surfaces also normal to them, modes of oscillation are possible whose frequency depends only on the plane spacing. These are called the principal modes. They satisfy the second set of boundary conditions in 13.00 (5) in which β_{mn} is zero and represent standing waves of the transmission-line type treated in 11.14 to 11.18. From 13.16 (3), the resonance frequencies of the principal modes of such a multiply connected cavity are

$$\nu_p = \frac{v}{\lambda_p} = \frac{\frac{1}{2}v\beta_p}{\pi} = \frac{\frac{1}{2}\omega}{\pi} = \frac{\frac{1}{2}pv}{d} \tag{1}$$

The fields can be written down from 11.14 (6) and (7). Thus

$$\boldsymbol{E} = C_p \nabla U(x,y) \sin Pz \sin \omega_p t, \qquad \boldsymbol{B} = C_p(\mu\epsilon)^{\frac{1}{2}} \nabla V(x,y) \cos Pz \cos \omega_p t \tag{2}$$

where $U(x,y)$ and $V(x,y)$ are the conjugate functions of Chap. IV and $U(x,y)$ has the value U_1 on one set of cylindrical surfaces and U_2 on the other. From 4.10 (1), the line integral of ∇U or ∇V around any path of constant U is the increment $[V]$ in V. This is usually 2π as with the elliptic cylinders (write U for V) of Fig. 4.22*a*.

The integrals of 13.18 (4) must be found to calculate Q. The potential drop between the two surfaces is $U_2 - U_1$, so the instantaneous electric energy in a cavity length dz is, from 4.11 (4) with U and V interchanged,

$$W_e\, dz = \tfrac{1}{2}C(U_2 - U_1)^2\, dz \sin^2 \omega_p t = \tfrac{1}{2}\epsilon[V](U_2 - U_1)\, dz \sin^2 \omega_p t \tag{3}$$

Integration over z gives the total electric energy which equals, but is 90° out of phase with, the magnetic energy. Thus

$$\int \boldsymbol{B}\cdot\boldsymbol{B}\, dv = \mu\epsilon d[V](U_2 - U_1)C_p^2 \cos^2 \omega_p t = \tfrac{1}{2}d\int_{S_0} \boldsymbol{B}\cdot\boldsymbol{B}\, dS \tag{4}$$

where S_0 covers the two end surfaces. Since ∇V, $\partial U/\partial s$, and $|dW/dz_1|$

are identical, the surface integral of B_t^2 over side surfaces U_1 and U_2 is

$$\tfrac{1}{2}\mu\epsilon d C_p^2 \oint_{1,2} (\nabla V)^2 \, ds \cos^2 \omega_p t = \tfrac{1}{2}\mu\epsilon d C_p^2 \oint_{1,2} |dW/dz_1| \, dV \cos^2 \omega_p t \quad (5)$$

Substitution of (4) and (5) into 13.18 (4) gives

$$Q_p = \frac{\mu d}{4\mu'\delta_p}\left[1 + \frac{d}{4[V](U_2 - U_1)}\left(\oint_1 \left|\frac{dz_1}{dW}\right|^{-1} dV + \oint_2 \left|\frac{dz_1}{dW}\right|^{-1}\right) dV\right]^{-1} \quad (6)$$

To evaluate the line integrals, z_1 is expressed as a function of W as in 4.22 (3) so that dz_1/dW is a function of U and V. For the first integral U is set equal to U_1, for the second to U_2, and both are integrated over the range of V necessary to cover U.

13.23. Coaxial Cable Resonators.—In the simplest kind of multiply connected cavity the curved walls are concentric cylinders. The principal modes are found as in the last article. From 4.12, we have

$$W = \ln z_1, \qquad z_1 = e^W, \qquad U = \ln \rho, \qquad V = \theta \quad (1)$$

The fields given by 13.22 (1) and (2) are, therefore,

$$E_\rho = \frac{C_p}{\rho} \sin \frac{p\pi z}{d} \sin \omega_p t, \qquad B_\theta = \frac{C_p}{\rho v} \cos \frac{p\pi z}{d} \cos \omega_p t \quad (2)$$

The range $[V]$ of V is 2π so, if the outer radius is b and the inner a,

$$(U_2 - U_1) = \ln \frac{b}{a}, \qquad \oint_1 \left|\frac{\partial z_1}{\partial W}\right|^{-1} dV = e^{-U_1} \int_0^{2\pi} dV = 2\pi e^{-U_1} = \frac{2\pi}{a} \quad (3)$$

In similar fashion the other line integral gives $2\pi/b$. Substitution of these results in 13.22 (10) gives for the pth principal mode

$$Q_p = \frac{\mu d}{2\mu'\delta}\left[1 + \frac{(a + b)d}{4ab \ln (b/a)}\right]^{-1} \quad (4)$$

In addition to the principal modes, there are the types considered in 13.09 (3) to (6), which involve two kinds of Bessel functions. If d is large coıpared with $b - a$, these modes have much higher frequencies than the principal mode $p = 1$. The methods of 13.19 apply to these.

13.24. Normal Modes of a Spherical Cavity.—The solution of the propagation equation in 12.11 involves products of spherical harmonics and spherical Bessel functions. Of the θ and R terms given in 12.11 (5), only the $P_n(\cos \theta)$ and $j_n(\beta r)$ terms survive because $Q_n(\cos \theta)$ is infinite on the axis and $k_n(j\beta r)$ represents a progressive wave, whereas in the perfectly conducting oscillating cavity only standing waves can exist. The electric field and the magnetic induction are, from 12.11 (5), (2), and (3) and 12.13 (6) and (13), when we write $s(\phi)$ for $\sin (m\phi + \psi_m)$,

$c(\phi)$ for $\cos(m\phi + \psi_m)$, j_n for $j_n(\beta_{pn}r)$, and u for $\cos\theta$,

$$\boldsymbol{E}_{te} = \omega C[\boldsymbol{\theta} m \csc\theta P_n^m(u)s(\phi) - \boldsymbol{\phi}\sin\theta P_n^{m\prime}(u)c(\phi)]j_n \sin(\omega t + \alpha) \quad (1)$$

$$\boldsymbol{B}_{te} = \frac{C}{r}\left[\boldsymbol{r}_1 n(n+1)P_n^m(u)c(\phi)j_n - \boldsymbol{\theta}\sin\theta P_n^{m\prime}(u)c(\phi)\frac{\partial(rj_n)}{\partial r} - \boldsymbol{\phi} m\csc\theta P_n^m(u)s(\phi)\frac{\partial(rj_n)}{\partial r}\right]\cos(\omega t + \alpha) \quad (2)$$

$$\boldsymbol{E}_{tm} = \frac{\omega C'}{r}\left[\boldsymbol{r}_1 n(n+1)P_n^m(u)c(\phi)j_n - \boldsymbol{\theta}\sin\theta P_n^{m\prime}(u)c(\phi)\frac{\partial(rj_n)}{\partial r} - \boldsymbol{\phi} m\csc\theta P_n^m(u)s(\phi)\frac{\partial(rj_n)}{\partial r}\right]\sin(\omega t + \alpha) \quad (3)$$

$$\boldsymbol{B}_{tm} = \omega^2\mu\epsilon C'[\boldsymbol{\theta} m\csc\theta P_n^m(u)s(\phi) - \boldsymbol{\phi}\sin\theta P_n^{m\prime}(u)c(\phi)]j_n\cos(\omega t + \alpha) \quad (4)$$

At the cavity surface $r = a$, the tangential component of $\boldsymbol{E}$ vanishes, so that β_{pn} must be chosen to satisfy the boundary conditions

$$[j_n(\beta_{pn}a)]_{te} = 0, \qquad \frac{\partial[aj_n(\beta_{pn}a)]_{tm}}{\partial a} = 0 \quad (5)$$

The first form of 13.18 (8), with $\boldsymbol{E}$ for $-j\omega A$ and $(2\mu'\delta^2)$ for $\omega\gamma$ from 10.02, is best for Q_{te}. By (1) and (2) this gives

$$Q_{te} = \frac{2\mu^2\epsilon\int_{-1}^{+1}\int_0^{2\pi} f_1(u,\phi)\,du\,d\phi\int_0^a r[j_n(\beta_{pn}r)]^2\,dr}{\mu'\delta\int_{-1}^{+1}\int_0^{2\pi} f_2(u,\phi)\,du\,d\phi\,\{\partial[aj_n(\beta_{pn}a)]/\partial a\}^2} \quad (6)$$

After the ϕ integration the u-integrals in the numerator and denominator are identical by 5.231 (8) and cancel out so that by means of 5.31 (10) and integration by 5.296 (6), which is valid for nonintegral orders by HTF II (48), page 70, there results

$$Q_{te} = \frac{2\omega^2\mu\epsilon(\frac{1}{2}\pi/\beta_{pn})\int_0^a r[J_{n+\frac{1}{2}}(\beta_{pn}r)]^2\,dr}{\mu'\delta[\frac{1}{2}\pi/(\beta_{pn}a)][J'_{n+\frac{1}{2}}(\beta_{pn}a)]^2} = \frac{\omega^2\mu^2\epsilon a}{\mu'\,\delta\beta_{pn}^2} \quad (7)$$

Use of the last form of 13.18 (8) for Q_{tm} gives in analogous fashion

$$Q_{tm} = \frac{\mu a}{\mu'\delta}\left[1 + \frac{(2n+1)^2}{4\beta_{pn}^2a^2}\right] \quad (8)$$

13.25. Normal Modes of an Imperfect Cavity.—So far all cavities have been treated as completely enclosed by perfectly conducting walls. Wall loss effects were found in 13.18 (11). In practice there must be openings in the walls through which waves enter or electrodes are inserted. These are usually made as few and small as possible so as to modify the fields only slightly. Let the vector potential and resonance frequency of the actual cavity be $\boldsymbol{A}$ and ω and of the perfect cavity $\boldsymbol{A}_0$ and ω_0. Insertion of $\boldsymbol{A}$ for $\boldsymbol{\Psi}$, $\boldsymbol{A}_0$ for $\boldsymbol{\Phi}$, and $\omega^2\mu\epsilon$ for the double curl in 3.06 (6) gives

$$\mu\epsilon(\omega^2 - \omega_0^2)\int_V \boldsymbol{A}\cdot\boldsymbol{A}_0\,dv = \int_S[\boldsymbol{A}\times(\nabla\times\boldsymbol{A}_0) - \boldsymbol{A}_0\times(\nabla\times\boldsymbol{A})]\cdot\boldsymbol{n}\,dS \quad (1)$$

Let v be the volume of the perfect cavity and S its surface to which $\boldsymbol{A}_0$ is normal so that the second term in the surface integral vanishes. If the hole is small, $\omega \approx \omega_0$, $\omega + \omega_0 \approx 2\omega_0$, and in most of v $\boldsymbol{A} \approx \boldsymbol{A}_0$ so that

$$\omega_0 - \omega = \frac{\int \boldsymbol{E}\times\boldsymbol{B}_0\cdot\boldsymbol{n}\,dS}{2j\omega_0^2\mu\epsilon\int\boldsymbol{A}\cdot\boldsymbol{A}\,dv} = \frac{-j\int\boldsymbol{E}_t\times\boldsymbol{B}_0\cdot\boldsymbol{n}\,dS}{4\mu W_m} \quad (2)$$

where W_m, $\boldsymbol{E}_t$, and $\boldsymbol{B}_0$ are instantaneous values of the magnetic energy in the cavity and the tangential components of the actual $\boldsymbol{E}$ and of the $\boldsymbol{B}_0$ with no hole. The part of $\boldsymbol{E}_t$ due to $\boldsymbol{B}_0$ is found by solving the magnetostatic problem of an infinite plane face of zero permeability pierced by the given hole which bounds a field of induction $\boldsymbol{B}_0$ that is tangential to it and uniform except for distortion near the hole. This expresses the normal $\boldsymbol{B}_n$ over the hole surface in terms of $\boldsymbol{B}_0$. The effect of the hole is wiped out by covering it with a double current sheet like that in 12.18 whose radiation exactly cancels its absorption. The current sheet $\boldsymbol{B}_n$ and $\boldsymbol{E}_t$ are exactly equal and opposite to those of the hole, but the $\boldsymbol{B}_t$ adds to restore the standing wave $\boldsymbol{B}_0$. For an aperture in the xy-plane and an x-directed $\boldsymbol{B}_0$, $(E_t)_y$ can be obtained from $\boldsymbol{B}_n$ by integration of 12.18 (5) Problem 49 of Chap. XII gives $\boldsymbol{B}_n$ for a circular aperture in a thin wall. If, with no hole, the normal electric field is $\boldsymbol{E}_0$, then its contribution to $\boldsymbol{E}_t$ is given by the electrostatic solution of the problem of a plane conducting face that bounds a field of intensity $\boldsymbol{E}_0$ which is uniform far from the hole. Article 5.272 treats the case of a circular hole in a thin wall. The hole losses must be added to the wall losses to get Q.

13.26. Effects of Small Foreign Body on Cavity Mode Frequency.—The electric field distribution in a complicated cavity is sometimes found by noting the frequency change produced when a small dielectric sphere is placed at various points therein. The perturbing effect on $\boldsymbol{E}_0$ and $\boldsymbol{H}_0$ of a small body may be found by starting with the vector identities

$$\nabla\cdot\left[\frac{\partial\check{\boldsymbol{E}}}{\partial\omega}\times\hat{\boldsymbol{H}}_0\right] = \hat{\boldsymbol{H}}_0\cdot\left(\frac{\nabla\times\partial\check{\boldsymbol{E}}}{\partial\omega}\right) - \frac{\partial\check{\boldsymbol{E}}}{\partial\omega}\cdot(\nabla\times\hat{\boldsymbol{H}}_0) \quad (1)$$

$$\nabla\cdot\left(\hat{\boldsymbol{E}}_0\times\frac{\partial\check{\boldsymbol{H}}}{\partial\omega}\right) = \frac{\partial\check{\boldsymbol{H}}}{\partial\omega}\cdot(\nabla\times\hat{\boldsymbol{E}}_0) - \hat{\boldsymbol{E}}_0\cdot\left(\frac{\nabla\times\partial\check{\boldsymbol{H}}}{\partial\omega}\right) \quad (2)$$

The integrals of the right sides of (1) and (2) over the cavity volume are zero because those on the left side transform by Gauss's theorem, 3.00 (2), into surface integrals which vanish on S since $\boldsymbol{n}\cdot\boldsymbol{E}\times\boldsymbol{H}$ is zero if $\boldsymbol{n}\times\boldsymbol{E}$ is. Now transform the sum of the right sides of (1) and (2) by elimination of all vector products by Maxwell's equations, 11.00 (1) and (2), which apply to $\check{E}$ and $\check{H}$. For the conjugates $\hat{E}$ and $\hat{H}$, j must be replaced by $-j$ in 11.00. The result after multiplication by j is

$$0 = \int \left[\mu \check{H} \cdot \hat{H}_0 + \epsilon \check{E} \cdot \hat{E}_0 + \omega \left(\check{H} \cdot \hat{H}_0 \frac{\partial \mu}{\partial \omega} + \check{E} \cdot \hat{E}_0 \frac{\partial \epsilon}{\partial \omega} \right) \right] dv \qquad (3)$$

Replace the differential ∂ by the variation δ and solve for $\delta\omega/\omega$. Thus

$$\frac{\delta\omega}{\omega} = -\frac{\int (\delta\mu \check{H} \cdot \hat{H}_0 + \delta\epsilon \check{E} \cdot \hat{E}_0)\, dv}{\int (\mu \check{H} \cdot \hat{H}_0 + \epsilon \check{E} \cdot \hat{E}_0)\, dv} \qquad (4)$$

Clearly an increase in μ or ϵ or both will decrease the resonance frequency. From 11.06 (2), the field products in (4) may be interpreted in terms of energy density, so that for small perturbations

$$\frac{\delta\omega}{\omega} = -\frac{1}{W} \int \left(\frac{\delta\epsilon}{\epsilon} \frac{dW_e}{dv} + \frac{\delta\mu}{\mu} \frac{dW_m}{dv} \right) dv = \frac{\delta W}{W} \qquad (5)$$

If the foreign object is so small that current and charge distributions on the walls are essentially unaltered, then usually the phase differences over the volume it occupies may be neglected. Thus a static solution is valid. Even if the object greatly distorts currents and charges, but only on one wall so near that phase differences between it and adjacent wall areas are small, a static solution may work. This is especially true of the plane walls in rectangular and cylindrical cavities where image methods apply. One then replaces the wall by an image cavity and image object and works with that mode of the composite cavity which gives the same fields in the region of the actual cavity. Some situations like this appear in problems at the end of the chapter.

As an example consider a small sphere of radius r_0, permeability μ_1, and capacitivity ϵ_1 inserted in an empty cavity. From 5.18 (1) and (2) with $C = D = 0$, and from the analogous magnetic case, the fields inside are

$$E_1 = E_0 \frac{3\epsilon_v}{\epsilon_1 + 2\epsilon_v}, \qquad H_1 = H_0 \frac{3\mu_v}{\mu_1 + 2\mu_v} \qquad (6)$$

Assume a nearly uniform E_0 and H_0 in the volume $4\pi r_0^3/3$ of the sphere. With $\epsilon_1 - \epsilon_v$ for $\delta\epsilon$ and $\mu_1 - \mu_v$ for $\delta\mu$, (4) gives

$$\frac{\delta\omega}{\omega} = \frac{4\pi r_0^3 \{[\epsilon_v(\epsilon_1 - \epsilon_v)/(\epsilon_1 + 2\epsilon_v)]|E_0|^2 + [\mu_v(\mu_1 - \mu_v)/(\mu_1 + 2\mu_v)]|H_0|^2\}}{\epsilon_v \int |E_0|^2\, dv + \mu_v \int |H_0|^2\, dv} \qquad (7)$$

For a small, perfectly conducting sphere set $\epsilon_1 = \infty$, $\mu_1 = 0$, so that

$$\frac{\delta\omega}{\omega} = \frac{4\pi r_0^3 (\epsilon_v |E_0|^2 - \frac{1}{2}\mu_v |H_0|^2)}{\epsilon_v \int |E_0|^2\, dv + \mu_v \int |H_0|^2\, dv} \qquad (8)$$

The values of E_0 and H_0 to be used in the numerators of (7) and (8) are the unperturbed values at the centers of the sphere.

13.27. Wall Deformation Effect on Mode Frequencies.—Anywhere inside a conducting cavity where the magnetic field touches the wall it exerts a pressure $\frac{1}{2}\mu H^2$ by 8.14 (2). Where an electric field touches the wall, there is a pull or tension of $\frac{1}{2}\epsilon E^2$ from 1.14 (2). Thus an inward deformation of the wall puts in energy by compression of the magnetic lines and absorbs energy from the contraction of the electric lines. If one assumes that the deformation is smooth and shallow, so the original fields are little altered in the process, then

$$\delta W \approx -\tfrac{1}{2}\int_{\Delta \mathrm{v}} (\mu \check{\boldsymbol{H}}_0 \cdot \hat{\boldsymbol{H}}_0 - \epsilon \check{\boldsymbol{E}}_0 \cdot \hat{\boldsymbol{E}}_0)\, d\mathrm{v} \tag{1}$$

Insertion of this into 13.26 (5) gives

$$\frac{\delta\omega}{\omega} = -\frac{\int_{\Delta \mathrm{v}} (\mu|H_0|^2 - \epsilon|E_0|^2)\, d\mathrm{v}}{\int_{\mathrm{v}} (\mu|H_0|^2 + \epsilon|E_0|^2)\, d\mathrm{v}} \tag{2}$$

Thus an inward perturbation will raise the resonance frequency if made at a point of large $\boldsymbol{H}$ and will lower it if made at a point of large $\boldsymbol{E}$.

13.28. Stationary Formulas for Cavity Resonance Frequency.—It is useful to have a calculation method for the resonance frequency of a cavity whose boundaries do not fit a coordinate system in which the wave equation is separable. Two such formulas come from 13.18 (4) which state that the time average electric and magnetic energies are equal, or

$$\int (\mu|H|^2 - \epsilon|E|^2)\, d\mathrm{v} = 0 \tag{1}$$

The first comes from putting (1) in terms of $\boldsymbol{H}$ by 11.00 (1) and the second by putting it in terms of $\boldsymbol{E}$ by 11.00 (2). The results are

$$\int [\omega^2\mu\epsilon|H|^2 - |\nabla \times \boldsymbol{H}|^2]\, d\mathrm{v} = 0 = \int [\omega^2\mu\epsilon|E|^2 - |\nabla \times \boldsymbol{E}|^2]\, d\mathrm{v} \tag{2}$$

$$\omega_p^2\mu\epsilon = \beta_p^2 = \frac{\int |\nabla \times \boldsymbol{H}_p|^2\, d\mathrm{v}}{\int |H_p|^2\, d\mathrm{v}} = \frac{\int |\nabla \times \boldsymbol{E}_p|^2\, d\mathrm{v}}{\int |E_p|^2\, d\mathrm{v}} \tag{3}$$

These may be proved stationary exactly as 13.07 (5) was by multiplying up the denominator and taking the variation of β_p and H_p in one case and of β_p and E_p in the other. Clearly the integral $\int |H_p|^2\, d\mathrm{v}$ or $\int |E_p|^2\, d\mathrm{v}$ which multiplies $\delta\beta_p$ does not vanish. After some vector manipulation, including the use of Gauss's theorem, 3.00 (2), the two other volume integrands combine to give the left side of 11.01 (1) or (2), and thus vanish. The surface integral over the boundary is zero if

$$\boldsymbol{n} \times \boldsymbol{E}_p = 0 \text{ for } H_p, \qquad \boldsymbol{n} \times \delta\boldsymbol{E}_p = 0 \text{ for } E_p \tag{4}$$

Thus the $\boldsymbol{H}$ integrals for $\delta\beta_p$ are stationary regardless of the choice of $\delta\boldsymbol{E}$ but the $\boldsymbol{E}$ integral only if the trial field $\boldsymbol{E} + \delta\boldsymbol{E}$ is zero on S.

As a specific example consider the spheroidal cavity

$$\left(\frac{\rho}{a}\right)^2 + \left(\frac{z}{b}\right)^2 = 1 \tag{5}$$

From symmetry only a ϕ independent H_ϕ occurs which has zero divergence and is tangent to the spheroid surface. To vanish on the z-axis the factor ρ is indicated and to be even in z a suitable form is

$$H_\phi = \rho(1 - B\rho^2 - Cz^2), \qquad \nabla \times H_\phi = \boldsymbol{\rho}_1 2C\rho z + \boldsymbol{k}2(1 - 2B\rho^2 - Cz^2) \tag{6}$$

The constants B and C can be adjusted to make $\boldsymbol{E}$ normal to the spheroid surface. From (6) and Maxwell's equations,

$$\frac{E_\rho}{E_z} = \frac{Cz\rho}{1 - 2B\rho^2 - Cz^2} = -\frac{\partial z}{\partial \rho} = \frac{b^2\rho}{a^2 z} \tag{7}$$

Eliminate ρ by (5) and require that (7) hold for all z-values, so that

$$2a^2B = 1, \qquad (a^2 + b^2)C = 1 \tag{8}$$

Put these values into (6), carry out the integrations indicated in (3), where $dv = 2\pi\rho\, d\rho\, dz$ and obtain

$$\beta^2 a^2 = \frac{18(5a^2 + 4b^2)(3a^2 + 2b^2)}{33a^4 + 52a^2b^2 + 22b^4} \tag{9}$$

This result may be checked, for $a = b$, against the exact value for a spherical TM mode given in 13.24 (5). Thus

$$\beta a = 2.751 \quad (9), \qquad \beta a = 2.744 \quad 13.24\ (5)$$

For the general ellipsoid see Borgnis and Papas, page 180.

13.29. Complex Cavities.—Some of the most important cavities are the simply connected ones whose boundaries coincide only in part with single

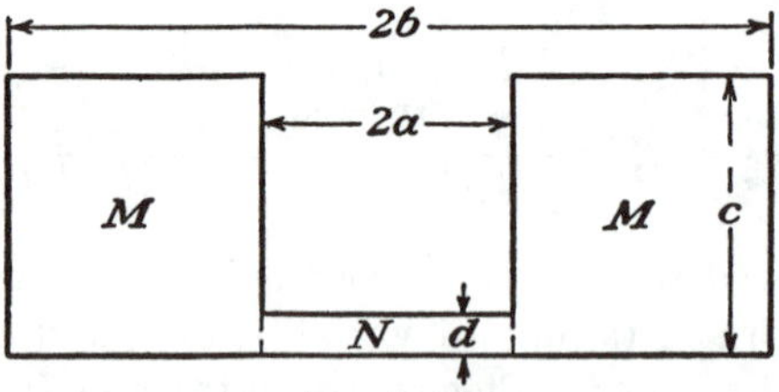

FIG. 13.29.

coordinate surfaces. Usually a special type of solution is required for each case. Consider, for example, a right circular cylindrical cavity of radius b and length c, to one end of the interior of which a shorter closed right circular cylinder of radius a has been fastened coaxially leaving a gap d at the other end. This cavity, shown in section in Fig. 13.29, falls outside the categories so far treated. We shall indicate the nature of the exact solution by Hahn's method (*J. Appl. Phys.*, Vol. 12, page 62) then give a nearly exact method that applies in the important case where d is much less than a, b, or λ, and finally present a rough approximation that uses the transmission line analogy. We desire the cavity fields and the lowest normal frequency. From 13.19 and 13.21, TM field components in the M-region which give E_t zero at $\rho = b$, $z = 0$, and $z = c$ are

$$\check{E}_z = \check{C}_0 \frac{R_0(\beta_0\rho)}{R_0(\beta_0 a)} + \sum_{m=1}^{\infty} \check{C}_m \frac{R_0(\beta_m\rho)}{R_0(\beta_m a)} \cos\frac{m\pi z}{c} \tag{1}$$

$$\check{B}_\phi = \frac{\beta_0^2}{j\omega}\left[\frac{\check{C}_0 R_0'(\beta_0\rho)}{\beta_0 R_0(\beta_0 a)} + \sum_{m=1}^{\infty} \frac{\check{C}_m R_0'(\beta_m\rho)}{\beta_m R_0(\beta_m a)} \cos\frac{m\pi z}{c}\right] \tag{2}$$

$$\begin{aligned} R_0(\beta_0\rho) &= J_0(\beta_0 b)Y_0(\beta_0\rho) - Y_0(\beta_0 b)J_0(\beta_0\rho) \\ R_0(\beta_m\rho) &= I_0(\beta_m b)K_0(\beta_m\rho) - K_0(\beta_m b)I_0(\beta_m\rho) \end{aligned} \tag{3}$$

$$\beta_m^2 = \beta_0^2 - \frac{m^2\pi^2}{c^2}, \qquad \beta_n^2 = \beta_0^2 - \frac{n^2\pi^2}{d^2}, \qquad \beta_0^2 = \beta^2 = \omega^2\mu\epsilon = \frac{4\pi^2}{\lambda^2} \tag{4}$$

In the N-region, fields finite at $\rho = 0$ giving $E = 0$ at $z = 0$ and $z = d$ are

$$\check{E}_z' = \check{A}_0 \frac{J_0(\beta_0\rho)}{J_0(\beta_0 a)} + \sum_{n=1}^{\infty} \check{A}_n \frac{I_0(\beta_n\rho)}{I_0(\beta_n a)} \cos\frac{n\pi z}{d} \tag{5}$$

$$\check{B}_\phi' = \frac{\beta_0^2}{j\omega}\left[\frac{\check{A}_0 J_0'(\beta_0\rho)}{\beta_0 J_0(\beta_0 a)} + \sum_{n=1}^{\infty} \frac{\check{A}_n I_0'(\beta_n\rho)}{\beta_n I_0(\beta_n a)} \cos\frac{n\pi z}{d}\right] \tag{6}$$

A formula for $\check{A}_p$ in terms of all the $\check{C}_m$'s is found by equating $\check{B}_\phi$ to $\check{B}_\phi'$, setting $\rho = a$, multiplying through by $\cos(p\pi z/d)\,dz$, and integrating from $z = -d$ to $z = d$. Similarly $\check{C}_q$ in terms of all the $\check{A}_n$'s is obtained by equating $\check{E}_z$ to $\check{E}_z'$ when $0 < |z| < d$ and to zero when $d < |z| < c$, multiplying by $\cos(q\pi z/c)\,dz$, and integrating from $z = -c$ to $z = c$. Elimination of the $\check{A}_n$'s from these equations and division by $\check{C}_0$ give $\check{C}_q/\check{C}_0$ in terms of the $\check{C}_m/\check{C}_0$ ratios. There are an infinite number of these equations, one for each q, so that to obtain $\check{C}_m/\check{C}_0$ an infinite determinant must be solved. Insertion of these values of $\check{C}_q/\check{C}_0$ into the $q = 0$ equation gives

$$1 = \frac{J_0(\beta_0 a)}{J_0'(\beta_0 a)}\left[\check{C}_0 \frac{R_0'(\beta_0 a)}{R_0(\beta_0 a)} + \frac{\beta_0 c}{\pi d} \sum_{m=1}^{\infty} \frac{C_m}{m\beta_m} \frac{R_0'(\beta_m a)}{R_0(\beta_m a)} \sin\frac{m\pi d}{c}\right] \tag{7}$$

The lowest value of β_0 that satisfies this equation gives β_0 by (4). The amplitude of excitation has been chosen to make $A_0 = 1$.

The calculation just outlined is evidently a major undertaking which is exact in theory but not in practice because of the finite amount of time available. However Hahn has prepared tables that make it feasible. If d is much less than a, b, and λ, then the following method gives λ with greater accuracy than that with which a, b, and c are usually measurable and enables us to calculate Q. The tangential electric field between M and N at $\rho = a$ is independent of a and b if d is small enough and depends only on d, c, and the potential across the gap. We may then take a and b infinite, which makes it the two-dimensional problem solved in 6.07 where h and k replace d and c. The choice of $c \gg d$ simplifies the work by deleting the bends at $x_1 = \pm 1$ in Fig. 6.07a. The procedure of 6.07 gives

$$z = \frac{2d}{\pi}\left[(z_1^2 - 1)^{\frac{1}{2}} + \sin^{-1}\frac{1}{z_1}\right] \tag{8}$$

$$E_x = \left|\frac{\partial V}{\partial x}\right| = \left|\frac{\partial W}{\partial z}\right|_{i.p.} = \left|\frac{V_0}{d(z_1^2 - 1)^{\frac{1}{2}}}\right|_{i.p.} \tag{9}$$

where z and z_1 are complex variables and the potential difference across the gap is $2V_0$. To get E_x on the x-axis in terms of x, we find complex values of z_1 which give $y = 0$ in (8). Substitution of these values in (8) and (9) gives x and E_x. To within 1 part in 5000, these fit the formula

$$E_x = V_0 d^{-1}\{A + B\cos(\pi x d^{-1}) + C(xd^{-1})^2 + D[\sec(\tfrac{1}{2}\pi x d^{-1})]^{.355}\} \tag{10}$$

where $A = .34091$, $B = -.00656$, $C = -.07785$, and $D = .49910$. Integration of (5) from $z = 0$ to $z = d$ with $\rho = a$ shows that $V_0 = A_0$. Insert this value of V_0 into (10), substitute z for x, equate the result to (1) when $-d < z < d$, and equate (1) to zero when $d < |z| < c$, multiply through by $\cos(p\pi z/c)\,dz$ and integrate from $z = -c$ to $z = c$. Solve the resultant equation for C_p and insert in (7) which is then solved by trial for β. Moreno (see references) gives curves for such a cavity on pages 230 to 238. It is evident that insertion of A_n and C_m in (2) and (6) gives the magnetic inductions so that Q can be calculated from 13.18 (4).

The transmission line concept furnishes an approximate value of the resonance frequency. If $d \ll c$, the region M of the cavity resembles a section of coaxial line short-circuited at $z = c$ whose input impedance is $\frac{1}{2}\ln(b/a)j\beta(\pi\epsilon\omega)^{-1}\tan\beta c$ from 11.15 (13) and 13.13 (2). The region N represents a capacitor of impedance $(j\omega C_0)^{-1}$ shunted across the line. At resonance, these impedances are equal and opposite so β satisfies, approximately,

$$\beta\tan\beta c = \frac{2\pi\epsilon}{C_0\ln(b/a)} \tag{11}$$

We can only estimate C_0. Treatment of the N-region as a parallel plate capacitor where $C_0 \approx \epsilon\pi a^2/d$ gives a lower limit because it neglects the fringing field. In the special case where $d \ll a$, $d \ll c$, and $a \approx b$, C_0 can be found quite accurately as will be seen in problem 63.

13.30. Excitation of Cavities. Inductive Coupling.—The most general cavity oscillation state is a superposition of modes already considered. Not only does each mode oscillate as an independent circuit, but also the exciting to cavity current ratio in any mode depends on the geometrical configuration only, as with linear circuits. E. U. Condon (*Rev. Mod. Phys.*, Vol. 14, No. 4) simplified this mutual inductance calculation by breaking up the vector potential $\breve{\mathbf{A}}_i$ of the ith mode into two factors $\breve{C}_i$ and $\mathbf{A}_i^0$ where $\breve{C}_i$ is adjusted to make the integral of $\mathbf{A}_i^0 \cdot \mathbf{A}_i^0$ over v, the cavity volume, equal to v. These "normalized," dimensionless, and by 13.17, orthogonal vector potentials depend only on the geometry and are related to the steady-state cavity vector potential by the equations

$$\breve{\mathbf{A}} = \sum_i \breve{C}_i \mathbf{A}_1^0, \qquad \int \breve{\mathbf{A}}_i \cdot \hat{\mathbf{A}}_i \, d\mathrm{v} = C_1^2 \int \mathbf{A}_i^0 \cdot \mathbf{A}_i^0 \, d\mathrm{v} = \mathrm{v} C_i^2 \tag{1}$$

Assume that the Q-values of the ith mode and its neighbors are so high that the resonance curves discussed in 13.17 and 13.18 do not overlap and that the time-dependence of $\breve{\mathbf{A}}_i$ is $\exp\,[(j - \frac{1}{2}Q^{-1})\omega_i t]$. Thus 11.01 (6) becomes

$$\nabla^2 \breve{\mathbf{A}}_i + \omega_i^2 \mu\epsilon(1 + jQ^{-1})\breve{\mathbf{A}}_i = 0 \tag{2}$$

Let $\breve{\mathbf{A}}$ be the total cavity vector potential excited by the solenoidal part of the loop current density $\mathbf{i}(\mathbf{r},t)$ of frequency ω. Then 11.01 (5) is

$$\nabla^2 \breve{\mathbf{A}} + \omega^2 \mu\epsilon \breve{\mathbf{A}} = -\mu \breve{\mathbf{i}} = \sum_i (\nabla^2 \breve{\mathbf{A}}_i + \omega^2 \mu\epsilon \breve{\mathbf{A}}_i) \tag{3}$$

Elimination of $\nabla^2 \breve{\mathbf{A}}_i$ by (2) gives

$$\epsilon \sum_i [\omega^2 - \omega_i^2(1 + jQ^{-1})]\breve{\mathbf{A}}_i = -\breve{\mathbf{i}} \tag{4}$$

Multiply through by $\hat{\mathbf{A}}_i$, integrate over the cavity volume, and solve for $\breve{C}_i$. For a thin wire loop $\mathbf{i}\,d\mathrm{v}$ is $\mathbf{i}\,dS\,ds$ or $I\,\mathbf{ds}$, so that

$$\breve{C}_i = \frac{-IM_i^0}{(\omega^2 - \omega_i^2 - j\omega_i^2 Q^{-1})\epsilon \mathrm{v}} \qquad \text{where } M_i^0 = \int \mathbf{A}_i^0 \cdot \mathbf{ds} \tag{5}$$

Here, by analogy with 8.03 (2), M_i^0 is the normalized mutual inductance between the ith mode and the wire. The electric field along the wire must be compensated by the electromotance $\mathcal{E}$ applied to the loop, so that

$$\mathcal{E} = -\int \breve{\mathbf{E}} \cdot \mathbf{ds} = j\omega \int \breve{\mathbf{A}} \cdot \mathbf{ds} \tag{6}$$

Put (5) into (1), multiply by $\mathbf{A}_i^0$, integrate around the loop, add the IR

drop due to loop resistance and divide by I to get the input impedance.

$$\check{Z} = \frac{\check{\mathcal{E}}}{I} = R + \sum_i \frac{-j\omega(M_i^0)^2}{\epsilon \mathrm{v}(\omega^2 - \omega_i^2 - j\omega_i^2 Q^{-1})} \tag{7}$$

Since Q_i is usually a large number, the first terms in the denominator dominate unless $\omega \approx \omega_i$. Modes for which $\omega_i > \omega$ give inductive reactance, modes for which $\omega_i < \omega$ give a capacitative reactance. This formula is useful in showing the qualitative behavior of driving loop reactance, but M_i^0 must be modified to give quantitative results. The origin of the difficulty lies in the assumption of an infinitely thin wire near which $\boldsymbol{B}$ and hence the inductance L are infinite. In practice, if the wire is thin enough compared with wave length, the cavity fields will be the same as if the exciting current were all on its center line, and its surface will coincide with the boundary of a tube of induction. The true flux linkage can therefore by found by integrating $\check{\boldsymbol{A}}$ from one end of the loop to the other along any curve lying in its surface. The definition of M_i^0 in (5) should be modified accordingly. One should note that the path of integration in (5) can be closed by going back to the starting point along the cavity boundary which contributes nothing to M_i^0 because $\check{\boldsymbol{A}}$ is normal to it. The integral of B_i^0 over the loop area would do as well.

13.31. Inductive Coupling to a Circular Cylindrical Cavity.—The theory just developed will be clarified by considering the case of a circular cylindrical cavity excited in its lowest natural frequency, the TE_{11} mode shown in Fig. 13.09, so that $\beta_{11}a$ is 1.841. From 13.21 (1), this is

$$\nu_{11} = \frac{v}{2}\left(\frac{\beta_{11}^2}{\pi^2} + \frac{1}{d^2}\right)^{\frac{1}{2}} = 1.5 \times 10^8\left(\frac{0.343}{a^2} + \frac{1}{d^2}\right)^{\frac{1}{2}} = \frac{\omega_{11}}{2\pi} \tag{1}$$

From 13.21 (2) and 5.296 (8) and (12), $\boldsymbol{A}_i^0$, which contains all geometric factors and is normalized so that $\int \boldsymbol{A}_i^0 \cdot \boldsymbol{A}_i^0 \, d\mathrm{v} = \mathrm{v}$, is

$$\boldsymbol{A}_i^0 = \frac{2a[\boldsymbol{\varrho}_1 J_1(\beta_{11}\rho) \sin \phi + \boldsymbol{\phi}\beta_{11}\rho J_1'(\beta_{11}\rho) \cos \phi] \sin (\pi z/d)}{\rho(\beta_{11}^2 a^2 - 1)^{\frac{1}{2}} J_1(\beta_{11}a)} \tag{2}$$

A small loop inserted through the wall at some point excites this mode. Evidently from (2), one encircles the maximum TE_{11} flux for the closest coupling on the flat ends by linking B_ρ near the center or B_ϕ near the edge or on the curved surface by linking B_ϕ near the ends or B_z near the center. Weak corner fields bar the B_ϕ choice. The ratio of B_z where $\rho \approx a$, $\phi = 0$, $z \approx \frac{1}{2}d$ to B_ρ at $\rho \approx 0$, $\phi = 0$, $z \approx 0$ is

$$\frac{M_z^0}{M_\rho^0} = \frac{|B_z|_{\max}}{|B_{\rho\max}} = \frac{\beta\, dJ_1(\beta a)}{\pi J_1'(0)} = \frac{0.682d}{a} \tag{3}$$

Thus for $0.682d > a$, the tightest coupling for a given size loop will be near the curved wall halfway between the ends. Let us assume that the small loop of area S on the $z = \frac{1}{2}d$ plane at $\phi = 0$ is so near the wall that the value of B_z at the wall may be used over its whole area. To justify this, we observe that, if ρ is $0.95a$, then $J_1(\beta\rho)$ is $0.997\, J_1(\beta a)$. Thus at $\phi = 0$, $z = \frac{1}{2}d$, $\rho \approx a$, the mutual inductance is, from 13.21 (3),

$$M^0 = |\nabla \times A_1^0|_z S = \frac{2\beta_{11}^2 aS}{(\beta_{11}^2 a^2 - 1)^{\frac{1}{2}}} = \frac{4.39S}{a} \tag{4}$$

From 13.21 (6) and 13.30 (7), in a copper cavity where $\gamma = 5.7 \times 10^7$, the Q-factor and the additional loop resistance at resonance are

$$Q = \frac{9.92(ad)^{\frac{1}{2}}(d^2 + 2.92a^2)^{\frac{3}{2}}}{4.11a^3 + 0.86a^2 d + d^3} \times 10^4, \qquad R' = \frac{1255S^2 Q}{a^3(d^2 + 2.92a^2)^{\frac{1}{2}}} \tag{5}$$

The maximum potential across the cavity, at $z = \frac{1}{2}d$, $\phi = \frac{1}{2}\pi$, is $\left|2\int_0^a E_\rho\, d\rho\right|$ or $\left|2\omega\int_0^a A_\rho\, d\rho\right|$. The potential magnification is the ratio of this to the loop electromotance and from 13.30 (1), (6), and (2) is

$$\frac{V}{\mathcal{E}} = \frac{2\omega C_{11}\int_0^a A_\rho^0\, d\rho}{\omega C_{11} M^0} = \frac{4a\int_0^a \rho^{-1} J_1(\beta_{11}\rho)\, d\rho}{M^0(\beta_{11}^2 a^2 - 1)^{\frac{1}{2}} J_1(\beta_{11}a)} = \frac{0.812a^2}{S} \tag{6}$$

The integral, found by integration of the series of 5.293 (3), is 0.80. The M^0 comes from (4) and $J_1(\beta_{11}a)$ is 0.582 from Jahnke and Emde.

It is of interest to take some numerical values at resonance. Let the area S of the loop be 1 cm², the radius a of the cavity 10 cm, the length d 20 cm, and the medium inside a vacuum, then

$$\omega \approx 7.25 \times 10^9 \text{ rad/sec}, \qquad \nu \approx 1.15 \times 10^9 \text{ cyc/sec}, \qquad \lambda \approx 26.0 \text{ cm}$$
$$M^0 = 0.00039 \text{ m}, \qquad Q = 36{,}000, \qquad R' \approx 1715 \text{ ohms}$$

Note that M^0, the *normalized* mutual inductance, is in meters, not henrys, from (4). If the peak loop electromotance in phase with $\boldsymbol{E}$ is 1000 volts, then $\mathcal{E}_0 \approx 1000$ volts, $I = 0.583$ amp, $V_{max} = 81{,}200$ volts, and $P = 292$ watts. Some other excitation methods appear in the problems.

13.32. Excitation of Cavity by Internal Electrode.—When a cavity is excited by a nonuniform current as when an electrode terminates inside it, the divergence of the current density $\boldsymbol{i}$ is no longer zero, and thus there is a charge density σ. In such cases, by 11.01 (4), the fields may be derived in the Coulomb gauge from a vector potential and a scalar potential that satisfies Poisson's equation. The mutual inductance with each normal mode is still given by 13.31 (4) because, as proved in 13.17 (4), the lamellar part contributes nothing to the volume integral. The field of the scalar potential contributes an additional electric energy in phase with the electrode charge that represents a purely capacitive reactance. As in the antenna problem, the difficulty of finding the current and charge distribution appears. A very thin wire stub has negligible capacitance, and its inductance is fixed almost entirely by its radius, so that the 12.02 reasoning applies giving the 12.02 (7) sinusoidal distribution. A fairly accurate result is obtainable when the electrode is charged by a thin wire in which the current may be taken uniform, so that the charge is entirely on the electrode and may be calculated as in electrostatics. Equations (1) to (6) of 13.30 are unchanged, but q/C_0 adds to (7) where $I = j\omega q$ and C_0 is the electrode capacitance. Thus

$$\check{Z} = R - \frac{j}{\omega C_0} + \sum_i \frac{j\omega M_i^2}{\epsilon v(\omega_i^2 - \omega^2 + j\omega_i^2/Q_i)} \tag{1}$$

When $\omega = \omega_0$, $\check{Z}$ is no longer real because of C_0, but if $\omega \approx \omega_i$, only the ith term of the sum is important. Therefore equating the reactance to zero gives

$$\epsilon v Q^2(\omega_i^2 - \omega^2)^2 + \omega^2 C_0 M_i^2 Q^2(\omega_i^2 - \omega^2) + \omega_i^4 \epsilon v = 0 \tag{2}$$

Often the coefficient of $\omega_i^2 - \omega^2$ is much larger than that of $(\omega_i^2 - \omega^2)^2$, so that from the last two terms alone, writing $2\omega_i$ for $\omega + \omega_i$, we get

$$\frac{\omega - \omega_i}{\omega_i} = \frac{\epsilon v}{2C_0 M_i^2 Q_i^2} = \frac{1}{2\omega_i R' C_0 Q_i} \tag{3}$$

As an example take a sphere of radius r supported with its center a distance l from the wall by a thin wire, and use this to excite the TE_{11} mode of the last article. For tight coupling the wire should coincide with a strong vector potential component. From 13.21 (2), this position is at $z = \frac{1}{2}d$, $\phi = \frac{1}{2}\pi$. As with 13.31 (4), $J_1(ka)$ may be used for $J_1(k\rho)$ for some distance from the wall. Thus 13.30 (5) and 13.31 (2) yield

$$M^0 = \int_{a-1}^{a} \mathbf{A}^0 \cdot d\mathbf{s} = \frac{2a\int \rho^{-1}\,d\rho}{(\beta_{11}^2 a^2 - 1)^{\frac{1}{2}}} = \frac{2a}{(\beta_{11}^2 a^2 - 1)^{\frac{1}{2}}} \ln \frac{a}{a - l} \tag{4}$$

From (7), the resistance contribution to the coupling circuit is

$$R' = 109aQ(d^2 + 2.92a^2)^{-\frac{1}{2}}\{\ln[a(a - l)^{-1}]\}^2 \tag{5}$$

If the ball is sufficiently small and near the wall, the capacitance C_0 in (1) approximates that of sphere and plane which from 5.082 (1) is

$$C_0 = 4\pi\epsilon[r + r^2(2l)^{-1} + r^2(4l^2 - r^2)^{-1} + \cdots] \tag{6}$$

The condition that (3) be valid is strongly satisfied for any reasonable choice of values, so that ω is very close to ω_1. From (1) and 13.31 (6), the ratio of the maximum potential across the cavity to that across the wire due to R' and C_0 is

$$\frac{V}{\mathcal{E}} = \frac{1.664 \times 10^{11} C_0 Q}{d(1 + \omega^2 R'^2 C_0^2)} \ln \frac{a}{a - l} \tag{7}$$

In the calculation of $\mathcal{E}$ the reactance of the higher modes was neglected. When this is not negligible, it can be found from (1). Consider the mode of the last article, so that ω, ν, λ, and Q have the values given there, and take l and r to be 1 cm and 1 mm, respectively.

$$C_0 \approx 1.17 \times 10^{-13} \text{ farad}, \qquad M^0 \approx 0.136 \text{ m}$$

$$R' \approx 16{,}560 \text{ ohms}, \qquad \frac{\omega_i - \omega}{\omega_i} = 0.99 \times 10^{-6}$$

A peak stub potential of 1000 volts gives, neglecting higher mode terms,

$$V_{\max} \approx 27{,}100 \text{ volts}, \qquad I \approx 0.0604 \text{ amp}, \qquad \bar{P} \approx 30.2 \text{ watts}$$

13.33. Cavity Excitation through Orifice.—When a wave guide ends in a cavity, a rigorous calculation involves matching the vector potential across the junction surface. On either side this is represented by an infinite series of terms, so that the problem is like that encountered in 13.29 where Hahn's method was used. Often a fair approximation is found by assuming the field over the opening as in 13.12. Examples of such solutions will be found among the problems that follow.

Problems

1. From 11.15 and 13.00 (7), show that when the dielectric conductivity in a wave guide is γ, the wave number and the dielectric attenuation are, respectively,

$$\beta' = 2^{-\frac{1}{2}}\{[(\beta^2 - \beta_m^2)^2 + \omega^2\mu^2\gamma^2]^{\frac{1}{2}} + \beta^2 - \beta_m^2\}^{\frac{1}{2}}$$
$$\alpha' = 2^{-\frac{1}{2}}\{[(\beta^2 - \beta_m^2)^2 + \omega^2\mu^2\gamma^2]^{\frac{1}{2}} - \beta^2 + \beta_m^2\}^{\frac{1}{2}}$$

Note that these formulas hold both above and below cutoff.

2. If the conductivity to capacitivity ratio is small compared with the difference between the frequency ν transmitted in the mth mode and the cutoff frequency ν_m, show that the attenuation α' and the phase velocity v'_m are, approximately,

$$\alpha' = \frac{1}{2}\mu v \gamma\left[1 - \left(\frac{\nu_m}{\nu}\right)^2\right]^{-\frac{1}{2}}, \qquad v'_m = v_m\left[1 - \frac{1}{8}\left(\frac{\mu\gamma v_m^2}{\omega}\right)^2\right]$$

where v is the free wave velocity and v_m the phase velocity for zero conductivity.

3. A rectangular wave guide is excited by a loop lying in the plane $x = d$. Show by images that all vector potential components are parallel to this plane, and thus the waves generated by this uniform current loop may be derived as in 13.00, where

$$\breve{W} = \breve{C}_{mn} \sin\frac{m\pi x}{a}\cos\frac{n\pi y}{b}e^{-j\beta'_{mn}z}, \qquad \breve{\mathbf{A}} = \nabla \times \mathbf{i}\breve{W}$$

$$\breve{\mathbf{E}} = -\omega\breve{C}_{mn}\left[\mathbf{j}\beta'_{mn}\cos\frac{n\pi y}{b} + \mathbf{k}j\frac{n\pi}{b}\sin\frac{n\pi y}{b}\right]\sin\frac{m\pi x}{a}e^{-j\beta'_{mn}z}$$

$$\breve{\mathbf{B}} = \breve{C}_{mn}\left\{\mathbf{i}(\beta'_{mo})^2\sin\frac{m\pi x}{a}\cos\frac{n\pi y}{b} - \frac{m\pi}{a}\left[\mathbf{j}\frac{n\pi}{b}\sin\frac{n\pi y}{b} + \mathbf{k}j\beta'_{mn}\cos\frac{n\pi y}{b}\right]\cos\frac{m\pi x}{a}\right\}e^{-j\beta'_{mn}z}$$

where $(\beta'_{mn})^2 = \beta^2 - \beta^2_{mn}$. Note that this combines TE and TM waves.

4. Show from 13.03 (6) and (14) that the vector potential of a small loop of wire with area dS, current $Ie^{j\omega t}$, and x-directed magnetic moment which lies in the x_0-plane at $y = y_0$ and $z = z_0$ in a rectangular wave guide is, if $z > z_0$,

$$\breve{\mathbf{A}} = -\frac{\mu I\,dS}{ab}\sum_{m=1}^{\infty}\sum_{n=0}^{\infty}(2 - \delta_n^0)\sin\frac{m\pi x_0}{a}\cos\frac{n\pi y_0}{b}\left[\mathbf{j}\cos\frac{n\pi y}{b} + \mathbf{k}\frac{jn\pi}{\beta'_{mn}b}\sin\frac{n\pi y}{b}\right]\sin\frac{m\pi x}{a}e^{-j\beta'_{mn}(z-z_0)}$$

5. Show from 13.03 (6) and (9) that the vector potential of a small loop of wire of area dS carrying a current $Ie^{j\omega t}$, having a z-directed magnetic moment, and lying in the z_0-plane at x_0, y_0 in a rectangular wave guide is, if $z > z_0$,

$$\breve{\mathbf{A}} = \frac{j\mu I\,dS}{ab}\sum_{m=0}^{\infty}\sum_{n=0}^{\infty}\frac{2 - \delta_m^0 - \delta_n^0}{\beta'_{mn}}\cos\frac{m\pi x_0}{a}\sin\frac{n\pi y_0}{b}\left[\mathbf{i}\frac{n\pi}{b}\sin\frac{n\pi y}{b}\cos\frac{m\pi x}{a} - \mathbf{j}\frac{m\pi}{a}\sin\frac{m\pi x}{a}\cos\frac{n\pi y}{b}\right]e^{-j\beta'_{mn}(z-z_0)}$$

6. A rectangular wave guide, closed at $z = 0$, is driven by half a rectangular loop of wire with its legs normal to the y-axis and its side, of length c, at x_0, z_0. If only the TE_{10} wave is transmitted, show that the radiation resistance is

$$R_r = \frac{2\omega\mu c^2}{\beta'_{10}ab}\sin^2\frac{\pi x_0}{a}\sin^2\beta'_{10}z_0 \qquad \text{where } \beta'_{10} = \frac{2\pi}{\lambda}\left(1 - \frac{\lambda^2}{4a^2}\right)^{\frac{1}{2}}$$

7. A rectangular wave guide, closed at $z = 0$, is driven by a semicircular loop of radius c of thin wire that carries a uniform current. The loop lies in the $x = d$ plane and its center and ends are in the $z = 0$ plane. The frequency is such that only the TE_{10} wave is transmitted. Show from 13.03 (6) that the radiation resistance is

$$R_r = \frac{2\mu\omega\pi^2c^2}{\beta'_{10}ab}[J_1(\beta'_{10}c)]^2\sin^2\frac{\pi d}{a} \qquad \text{where } \beta'_{10} = \frac{2\pi}{\lambda}\left(1 - \frac{\lambda^2}{4a^2}\right)^{\frac{1}{2}}$$

8. Show, by application of the method of images to the result of problem 44 and with the aid of 5.298 (7) and 5.35 (4), that the radiation reactance in problem 7 is

$$X_0 + \frac{2\pi\mu\omega c^2}{a}\sum_{n=1}^{\infty}\left\{\pi Y_0(\beta_1 nb)\left[J_1(\beta_1 c)\sin\frac{\pi d}{a}\right]^2 + 2\sum_{p=2}^{\infty}K_0(\beta_p nb)\left[I_1(\beta_p c)\sin\frac{p\pi d}{a}\right]^2\right\}$$

The problem 45 reactance with $\alpha = \pi$ is X_0. The loop center is at $y = \frac{1}{2}b$.

$$\beta_1 = \frac{2\pi}{\lambda}\left(1 - \frac{\lambda^2}{4a^2}\right)^{\frac{1}{2}}, \qquad \beta_n = \frac{2\pi}{\lambda}\left(\frac{n^2\lambda^2}{4a^2} - 1\right)^{\frac{1}{2}}$$

Except for a large loop, the first term is much larger than the sum of the image terms.

9. The rectangular wave guide of 13.02, closed at $z = 0$ by a perfectly conducting plane, is excited by a wire of radius r parallel to the y-axis at $x = d$, $z = c$ that carries a uniformly distributed current $jI \cos \omega t$. If $r \ll a$, $r \ll \lambda$, $r \ll c$, and $r \ll d$, show by superimposing solutions of the type in problem 32 that the reactance is

$$-\frac{\omega\mu b}{4}\left\{\frac{2}{\pi}(\ln \tfrac{1}{2}\beta r + C) + 2\sum_{n=1}^{\infty} Y_0(2n\beta a) - \sum_{n=-\infty}^{+\infty} Y_0(2\beta|na - d|)\right.$$
$$\left. - \sum_{n=0}^{\infty}(2 - \delta_n^0)Y_0[2\beta(n^2a^2 + c^2)^{\frac{1}{2}}] + \sum_{n=-\infty}^{+\infty} Y_0(2\beta[(na - d)^2 + c^2]^{\frac{1}{2}}\right\}$$

10. If in the preceding problem c is of the same order of magnitude as r but the wire does not actually touch the end wall, show that if terms in β^4c^4, β^6c^6, etc., are neglected, pairs of terms may be combined to give the reactance

$$X = \frac{\omega\mu b}{2\pi}\left\{\cosh^{-1}\frac{c}{r} - \pi\beta c^2\sum_{n=1}^{\infty}\frac{Y_1(2\beta na)}{na} + \pi\beta c^2\sum_{n=-\infty}^{+\infty}\frac{Y_1(2\beta|na - d|)}{2|na - d|}\right\}$$

which, if $d = \frac{1}{2}a$, takes the form

$$X = \frac{\omega\mu b}{2\pi}\left\{\cosh^{-1}\frac{c}{r} + 2\pi\beta c^2\sum_{n=1}^{\infty}(-1)^{n+1}\frac{Y_1(n\beta a)}{na}\right\}$$

11. The rectangular guide of 13.02, closed at $z = 0$ by a perfectly conducting plane, is excited by a wire of radius r carrying a current $jI \cos \omega t$ parallel to the y-axis at $x = d$, $z = c$. If r is small, show that the fields when $\beta_m' = (\omega^2\mu\epsilon - \pi^2m^2a^{-2})^{\frac{1}{2}}$ are

$$\breve{\boldsymbol{E}} = -\frac{2\omega\mu I}{a}\sum_{m=1}^{\infty} j\frac{1}{\beta_m'}\sin\frac{m\pi d}{a}\sinh j\beta_m'c \sin\frac{m\pi x}{a}e^{-j\beta_m' z}$$

$$\breve{\boldsymbol{B}} = \frac{2\mu I}{a}\sum_{m=1}^{\infty}\sin\frac{m\pi d}{a}\sinh j\beta_m'c\left[\boldsymbol{i}\sin\frac{m\pi x}{a} - \boldsymbol{k}\frac{jm\pi}{\beta_m' a}\cos\frac{m\pi x}{a}\right]e^{-j\beta_m' z}$$

if $z > c$. When $z < c$, interchange z and c in $\boldsymbol{E}$ and for $\boldsymbol{B}$ write

$$\breve{\boldsymbol{B}} = \frac{2\mu I}{a}\sum_{m=1}^{\infty}\sin\frac{m\pi d}{a}\left[-\boldsymbol{i}\cosh j\beta_m' z\sin\frac{m\pi x}{a} - \boldsymbol{k}\frac{jm\pi}{\beta_m' a}\sinh j\beta_m' z\cos\frac{m\pi x}{a}\right]e^{-j\beta_m' c}$$

12. In the last problem take the exciting frequency ν so that $nv < 2a\nu < (n + 1)v$ where v is the free wave velocity in the tube dielectric. Show by integration of the Poynting vector across the tube section at large z that the radiation resistance is

$$R_r = \frac{2\omega\mu b}{a}\sum_{m=1}^{n}(\beta_m')^{-1}\sin^2\beta_m' c\sin^2\frac{m\pi d}{a}$$

13. A rectangular wave guide, $a = 10$ cm, $b = 2$ cm, and closed at $z = 0$, transmits waves whose free wave length is 15 cm. It is driven by a 0.5-mm radius wire across the guide at $x = 5$ cm, $z = c$. Show that, to make its resistance equal 100 ohms, c should be 2.62 cm, 14.0 cm, etc. Show that the phase and signal velocities are $1.52v$ and $0.66v$, respectively, and that the inductive reactance is 274 ohms.

14. The wave guide of problems 11 and 12 is short-circuited at $x = d_2$, $z = c_2$ by a perfectly conducting wire. The impedances $\check{Z}_1$ and $\check{Z}_2$ of each wire alone in terms of its position and size may be calculated from 11 and 12. Show that, when both are present, the impedance $\check{Z}_1'$ of the first wire is $\check{Z}_1' = [\check{Z}_1\check{Z}_2 + \omega^2(M^0)^2]/\check{Z}_2$ where

$$M^0 = \frac{2\mu b}{a}\sum_{m=1}^{\infty}(\beta_m')^{-1}\sin\frac{m\pi d}{a}\sin\frac{m\pi d_2}{a}\sin\beta_m' c e^{-i\beta_m' c_2}$$

where β_m' is real for transmitted modes.

15. It is desired to eliminate the reactance in problem 13. Show from 14 that this can be done without changing the resistance by insertion of a short-circuiting wire on the center line at $c_2 = 11.39$ cm from the end, so that M^0 is real, provided its radius is 4.5 mm and assuming that our formulas hold for such a radius.

16. A rectangular wave guide, $a = 1$, $b = 2$, and closed at $z = 0$, is excited by the application of equal and opposite potentials to the ends of a rectangular loop in the $x = \frac{1}{2}$ plane. The legs of the loop are at $y = \frac{1}{2}$ and $y = \frac{3}{2}$, and their length is not given. The loop is driven at the frequency of half unit length free waves. Show that only the following waves can be transmitted: TM_{12}, TM_{32}, TM_{16}, (TE_{10}), TE_{12}, TE_{16}, (TE_{30}), TE_{32}; give reasons for excluding other waves.

17. A rectangular wave guide is filled with a medium $\mu_1\epsilon_1$ when $0 < x < c$ and $\mu_2\epsilon_2$ when $c < x < a$. Starting with 13.00 (2) and choosing

$$(W_{tmx})_1 = A_1\cos p_1x\sin\frac{n\pi y}{b}e^{-i\beta'_{pn}z}$$

$$(W_{tmx})_2 = A_2\cos[p_2(a - x)]\sin\frac{n\pi y}{b}e^{-i\beta'_{pn}z}$$

show that β_{pn}' may be found by solving the transcendental equation

$$p_2\epsilon_1\cot p_1c = -p_1\epsilon_2\cot p_2(a - c)$$

as in 13.05. Indicate how the cutoff frequency may be found.

18. If the divisions between the mediums in the wave guides in 13.05 and the last problem are at $y = c$, show that the equations to be solved for β_{pn}' are

$$-\mu_1q_2\tan q_1c = \mu_2q_1\tan q_2(b - c), \qquad \text{for } TEX$$
$$-\epsilon_1q_2\cot q_1c = \epsilon_2q_1\cot q_2(b - c), \qquad \text{for } TMX$$

19. Show that the variational expression for the normalized susceptance of 13.06 (7) is

$$B^0 = \frac{8b}{\lambda_g}\sum_{n=1}^{\infty}\frac{\left[\int_0^c E(y)\cos(n\pi y/b)\,dy\right]^2}{[n^2 - (2b/\lambda_g)^2]^{\frac{1}{2}}\left[\int_0^c E(y)\,dy\right]^2}$$

Use the trial E in the quasi-static expression of 4.22, namely,

$$E_y = \frac{\cos(\frac{1}{2}\pi y/b)}{[\sin^2(\frac{1}{2}\pi c/b) - \sin^2(\frac{1}{2}\pi y/b)]^{\frac{1}{2}}} = \frac{2^{\frac{1}{2}}\cos(\frac{1}{2}\phi)}{(\cos\phi - \cos\theta)^{\frac{1}{2}}}$$

where $\phi = \pi y/b$ and $\theta = \pi c/b$. With these coordinates show that

$$\int_0^c E(y)\cos\left(\frac{n\pi y}{b}\right)dy = \frac{1}{2}b\left[P_n\left(\cos\frac{\pi c}{b}\right) + P_{n-1}\left(\cos\frac{\pi c}{b}\right)\right] = \frac{1}{2}bL_n\frac{\pi c}{b}$$

Using the fact that the trial E_y is exact at cutoff where $\lambda_g = \infty$ and B^0 is given by 13.06 (7), write the variational upper limit in the form

$$B^0 = \frac{2b}{\lambda_g}\left\{4\ln\left(\csc\frac{\frac{1}{2}\pi c}{b}\right) + \sum_{n=1}^{\infty}[(n^2 - 4b^2\lambda_g^{-2})^{-\frac{1}{2}} - n^{-1}]\left(L_n\frac{\pi c}{b}\right)^2\right\}$$

Note that the static value provides a lower limit.

20. The rectangular wave guide in 13.02 is obstructed by a thin conducting iris in which a slit of width d with its center at $y = c$ runs from $x = 0$ to $x = a$. Calculate the normalized shunt susceptance as in 13.06. Verify that the transformation

$$W = \sin^{-1}\frac{[-\cos(\pi z'/b) + \cos(\pi c/b)\cos(\frac{1}{2}\pi d/b)]}{[\sin(\pi c/b)\sin(\frac{1}{2}\pi d/b)]}$$

gives $U = -\frac{1}{2}\pi$ when $x' = 0$, $0 \leqslant y' \leqslant \infty$ and when $y' = 0$, $0 \leqslant x' \leqslant (c - \frac{1}{2}d)$; $V = 0$ when $y' = 0$, $(c - \frac{1}{2}d) \leqslant x' \leqslant (c + \frac{1}{2}d)$; $U = \frac{1}{2}\pi$ when $y = 0$, $(c + \frac{1}{2}d) \leqslant x' \leqslant b$ and when $x' = b$, $0 \leqslant y' \leqslant \infty$ and that when $y' \to \infty$, instead of 13.06 (6), the stream function V becomes

$$V = \frac{\pi y}{b} - \ln\left(\sin\frac{\pi c}{b}\sin\frac{\frac{1}{2}\pi d}{b}\right)$$

The potential difference is $U_2 - U_1 = \pi$. Thus as in 13.06

$$B^0 = -\frac{4b}{\lambda_g}\ln\left(\sin\frac{\pi c}{b}\sin\frac{\frac{1}{2}\pi d}{b}\right)$$

This is the quasi-static approximation to the "Waveguide Handbook," 1a, page 218.

21. One-half of the iris in Fig. 13.06c is removed leaving a gap of width d. Put the right-angle bends at $+1$ and $-b$ in Fig. 13.06e and take the origin at the base of the iris, so that the transformation and magnetic induction in the iris plane are

$$W = C\cos\frac{\frac{1}{2}\pi z}{a}\left(\sin^2\frac{\frac{1}{2}\pi c}{a} - \sin^2\frac{\frac{1}{2}\pi z}{a}\right)^{\frac{1}{2}}$$

$$B_x + jB_y = C_1\frac{\partial\{\cos(\frac{1}{2}\pi x/a)[\sin^2(\frac{1}{2}\pi c/a) - \sin^2(\frac{1}{2}\pi x/a)]^{\frac{1}{2}}\}}{\partial x}$$

Proceed as with 13.06 (13) to show that the quasi-static normalized susceptance is

$$B^0 = -\frac{\lambda_g}{a}\cot^2\frac{\frac{1}{2}\pi d}{a}\left(1 + \csc^2\frac{\frac{1}{2}\pi d}{a}\right)$$

22. Show that the upper and lower limits for the B^0 of the last problem found by insertion of its B_z and B_x into the variational formulas 13.08 (10) and 13.08 (16) are

$$B^0 = -\frac{\lambda_g}{a}\sum_{m=2}^{\infty}\left[1-\left(\frac{2a}{m\lambda}\right)^2\right]^{\frac{1}{2}}\left[\frac{P_m(\cos 2\psi)+P_{m-1}(\cos 2\psi)}{2\cos^2\psi}\right]^2$$

$$\frac{1}{B^0} = -\frac{a}{\lambda_g}\sum_{m=2}^{\infty}\left[m^2-\left(\frac{2a}{\lambda}\right)^2\right]^{-\frac{1}{2}}\left[\frac{P_m(\cos 2\gamma)-P_{m-1}(\cos 2\gamma)}{2\sin^2\gamma}\right]^2$$

where ψ is $\frac{1}{2}\pi d/a$ and γ is $\frac{1}{2}\pi c/a$. Then assume, as proved for the symmetrical inductive iris by 13.06 (14), that when $\lambda = \infty$, the above formulas give the quasi-static values of the last problem. Thus write the above limits in the more rapidly convergent form

$$B^0 = \frac{\lambda_g}{a}\left\{\cot^2\psi[1+\csc^2\psi]+\frac{1}{4}\sec^4\psi\sum_{m=2}^{\infty}\left\{\left[1-\left(\frac{2a}{m\lambda}\right)^2\right]^{\frac{1}{2}}-1\right\}[P_m(\cos 2\psi) + P_{m-1}(\cos 2\psi)]^2\right\}$$

$$\frac{1}{B^0} = \frac{-a}{\lambda_g}\left\{\cot^2\gamma[1+\sec^2\gamma]^{-1}+\frac{1}{4}\csc^4\gamma\sum_{m=2}^{\infty}\left\{\left[m-\left(\frac{2a}{\lambda}\right)^2\right]^{-\frac{1}{2}}-m^{-1}\right\}[P_m(\cos 2\gamma) - P_{m-1}(\cos 2\gamma)]^2\right\}$$

23. The inductive iris in a rectangular wave guide is a centered strip of width c. Note from the symmetry that identical areas on the top and bottom of the guide must charge via the strip so no current circulates around the windows, and the stream function has the same value, $V = 0$, on strip and wall. Show that in the plane of the strip

$$B_x + jB_z = C\left(\cos^2\frac{\pi x}{a}-\frac{E}{K}\right)\left(\sin^2\frac{\frac{1}{2}\pi c}{a}-\cos^2\frac{\pi x}{a}\right)^{-\frac{1}{2}}$$

where E and K are complete elliptic integrals of modulus $k = \cos(\frac{1}{2}\pi c/a)$. Hence show that the quasi-static normalized shunt susceptance is

$$B^0 = -\frac{\lambda_g}{a}\frac{E-\frac{1}{2}(1-k^2)K}{E-(1-\frac{1}{2}k^2)K}$$

24. Show that the upper and lower limits for the B^0 of the last problem, found by insertion of its B_z and B_x into the variational formulas 13.08 (10) and (16) are

$$B^0 = -\frac{\lambda_g}{a}\sum_{n=2}^{\infty}\left[1-\frac{4a^2}{(2n+1)^2\lambda^2}\right]^{\frac{1}{2}}f(\theta),\qquad \frac{1}{B^0} = \frac{-a}{\lambda_g}\sum_{n=2}^{\infty}\left[(2n+1)^2-\frac{4a^2}{\lambda^2}\right]^{-\frac{1}{2}}f(\theta)$$

$$f(\theta) = \left\{\frac{P_{n+1}[\cos(\pi d/a)]+P_{n-1}[\cos(\pi d/a)]+2[1-2(E/K)]P_n[\cos(\pi d/a)]}{\cos(\pi d/a)+3-4(E/K)}\right\}^2$$

Assume, as proved for the symmetric inductive iris by 13.06 (14), that when $\lambda = \infty$ the above formulas give the quasi-static solutions of the last problem. Thus write the above limits in the more rapidly convergent forms

$$B^0 = \frac{-\lambda_g}{a}\left\{\frac{E - \frac{1}{2}(1 - k^2)K}{E - (1 - \frac{1}{2}k^2)K} + \sum_{n=2}^{\infty}\left(\left[1 - \frac{4a^2}{(2n+1)^2\lambda^2}\right]^{\frac{1}{2}} - 1\right)f(\theta)\right\}$$

$$\frac{1}{B^0} = \frac{-a}{\lambda_g}\left\{\frac{E - (1 - \frac{1}{2}k^2)K}{E - \frac{1}{2}(1 - k^2)K} + \sum_{n=2}^{\infty}\left(\left[(2n+1)^2 - \frac{4a^2}{\lambda^2}\right]^{-\frac{1}{2}} - \frac{1}{2n+1}\right)f(\theta)\right\}$$

25. The wave-guide section is a right isosceles triangle with legs of length a. Show that for the simplest TE wave the cutoff frequency ν_c is $\frac{1}{2}v/a$ and that

$$\alpha = \frac{1}{\mu\gamma v a\delta}\left[\left(1 - \frac{\nu_c^2}{\nu^2}\right)(1 + 2^{\frac{1}{2}}) + (3 + 2^{\frac{1}{2}}2)\frac{\nu_c^2}{\nu^2}\right]\left(1 - \frac{\nu_c^2}{\nu^2}\right)^{-\frac{1}{2}}$$

where v is the velocity of a free wave in the medium that fills the guide.

26. The section of a wave guide is a right triangle with two sides of length a. Show that the simplest TM mode cutoff frequency ν_c is $\frac{1}{2}\,5^{\frac{1}{2}}v/a$ and that

$$\alpha = \frac{(2 + 2^{\frac{1}{2}})\tau}{\mu v a\delta}\left[1 - \left(\frac{\nu_c}{\nu}\right)^2\right]^{-\frac{1}{2}}$$

where v is the velocity of a free wave in the medium that fills the guide.

27. A circular wave guide of radius a that extends from $z = -\infty$ to $z = \infty$ is excited by a current element of strength $\boldsymbol{\phi} I_0\rho_0\, d\phi e^{j\omega t}$ at ρ_0, 0, z_0. Show that the amplitude of the TE modes in 13.09 (5) and (6) is

$$[\check{C}_{mn}]_{te} = \frac{j(2 - \delta_m^0)\mu I_0\, d\phi\beta_{mn}\rho_0 J_m'(\beta_{mn}\rho_0)}{2\pi\beta_{mn}'(\beta_{mn}^2a^2 - m^2)[J_m(\beta_{mn}a)]^2}e^{j\beta'_{mn}z_0} \qquad [\check{D}_{mn}]_{te} = 0$$

where $J_m'(\beta_{mn}a)$ is zero, and that for the TM modes in 13.03 (8) and (9) it is

$$[\check{D}_{mn}]_{tm} = -\frac{\mu I_0\, d\phi_0 m J_m(\beta_{mn}\rho_0)}{\pi[\beta\beta_{mn}aJ_m'(\beta_{mn}a)]^2}e^{j\beta'_{mn}z_0} \qquad [\check{C}_{mn}]_{tm} = 0$$

where $J_m(\beta_{mn}a)$ is zero.

28. A circular wave guide of radius a that extends from $z = -\infty$ to $z = \infty$ is excited by a small current loop of magnetic moment $\boldsymbol{k}I\, dSe^{j\omega t}$ that lies in the plane $z = z_0$ at ρ_0, 0. Show that only TE modes appear whose amplitudes in 13.09 (5) and (6) are

$$[\check{C}_{mn}]_{te} = -\frac{j(2 - \delta_m^0)\mu I\, dS\beta_{mn}^2 J_m(\beta_{mn}\rho_0)}{2\pi\beta_{mn}'(\beta_{mn}^2a^2 - m^2)[J_m(\beta_{mn}a)]^2}e^{j\beta'_{mn}z_0} \qquad [\check{D}_{mn}]_{te} = 0$$

29. A circular wave guide of radius a that extends from $z = -\infty$ to $z = \infty$ is excited by a small current loop of magnetic moment $\boldsymbol{\phi}I\, dSe^{j\omega t}$ that lies in the plane $\phi = 0$ at $z = z_0$, $\rho = \rho_0$. Show that, if $z > z_0$, the amplitudes of the TE modes in 13.09 (5) and (6) and of the TM modes in 13.09 (8) and (9) are, respectively,

$$[\check{D}_{mn}]_{te} = \frac{-m\mu I\, dSJ_m(\beta_{mn}\rho_0)}{\pi\rho_0(\beta_{mn}^2a^2 - m^2)[J_m(\beta_{mn}a)]^2}e^{j\beta'_{mn}z_0}, \qquad [\check{C}_{mn}]_{te} = 0$$

$$[\check{C}_{mn}]_{tm} = -\frac{j\mu I\, dS(2 - \delta_m^0)J_m'(\beta_{mn}\rho_0)}{2\pi\beta_{mn}\beta_{mn}'[aJ_m'(\beta_{mn}a)]^2}e^{j\beta'_{mn}z_0} \qquad [\check{D}_{mn}]_{tm} = 0$$

30. The plane $\phi = 0$ in a circular cylindrical wave guide is an infinitely thin conducting sheet. Show that the lowest TE cutoff frequency is given by

$$\beta_{\frac{1}{2}} a = 1.1656 \qquad \text{or} \qquad \tan(2\pi\nu_c a v^{-1}) = 4\pi\nu_c a v^{-1}$$

and that, if the sheet has finite conductivity, the attenuation is infinite.

31. Show that the lowest TM cutoff frequency in problem 30 is $\frac{1}{2}v/a$ and that the attenuation is infinite if the sheet has finite conductivity.

32. A uniform current $I \cos \omega t$ flows in that portion of the perfectly conducting cylinder $\rho = a$ which lies between two infinite parallel perfectly conducting planes normal to its surface. Show that the phasor vector potential of the outward moving electromagnetic radiation between the planes is

$$\breve{A}_z = \frac{\mu I[J_0(\beta\rho) - jY_0(\beta\rho)]}{2\pi a\beta[J_1(\beta a) - jY_1(\beta a)]}$$

where $\beta^2 = \omega^2\mu\epsilon$, and from 5.293 (5) $Y_1(\beta a)$ approaches $2/(\pi\beta a)$ as βa approaches zero. Note, from the Jahnke and Emde tables, that if $a < 0.0181\lambda$ then

$$[J_1(\beta a)] < 0.01[Y_1(\beta a)]$$

33. Show that in the last problem the wire's input resistance and reactance are

$$R_i = \frac{\mu^{\frac{1}{2}} l}{\pi^2 \beta a^2 M \epsilon^{\frac{1}{2}}}, \qquad X_i = \frac{\mu^{\frac{1}{2}} l[J_0(\beta a)J_1(\beta a) + Y_0(\beta a)Y_1(\beta a)]}{2\pi a M \epsilon^{\frac{1}{2}}}$$

where M is $[J_1(\beta a)]^2 + [Y_1(\beta a)]^2$, and l is the distance between the planes.

34. The planes in problem 32 are connected at $\rho = b$ by a perfectly conducting cylinder. Show that the phasor vector potential between the planes is

$$\breve{A}_z = \frac{\mu I[Y_0(\beta b)J_0(\beta\rho) - J_0(\beta b)Y_0(\beta\rho)]}{2\pi a\beta[J_0(\beta b)Y_1(\beta a) - Y_0(\beta b)J_1(\beta a)]}$$

and calculate the input reactance.

35. Electromagnetic waves move radially between two parallel perfectly conducting planes normal to the z-axis. If the electric field has only a z-component and is independent of ϕ and if the ratio of $\breve{E}_z$ to $\breve{H}_\phi$ at $\rho = a$ is $\breve{Z}_a$, show with the aid of 12.16 (2), (4), and (5) that at $\rho = b$ the ratio is

$$\left|\frac{\breve{E}_z}{\breve{H}_\phi}\right|_b = \breve{Z}_b = \frac{\mu v[J_0(\beta a)Y_0(\beta b) - J_0(\beta b)Y_0(\beta a)] - j\breve{Z}_a[J_1(\beta a)Y_0(\beta b) - J_0(\beta b)Y_1(\beta a)]}{\epsilon v\breve{Z}_a[J_1(\beta a)Y_1(\beta b) - J_1(\beta b)Y_1(\beta a)] + j[J_0(\beta a)Y_1(\beta b) - J_1(\beta b)Y_0(\beta a)]}$$

36. A wave guide of radius b transmits the TM_{01} wave of Fig. 13.09 when $\nu > \nu_{01}$. For this wave to pass a gap in the walls, a field must exist across the gap to transfer the induced charge. Thus show from the previous problem that the transmission of the frequency $\nu_c > \nu_{01}$ is blocked by covering the small gap with a coaxial cylindrical box whose plane ends terminate at its edges, if the box radius satisfies the relation

$$J_1[2\pi(\mu\epsilon)^{\frac{1}{2}}\nu_c b]Y_0[2\pi(\mu\epsilon)^{\frac{1}{2}}\nu_c a] = J_0[2\pi(\mu\epsilon)^{\frac{1}{2}}\nu_c a]Y_1[2\pi(\mu\epsilon)^{\frac{1}{2}}\nu_c b]$$

37. It is desired to insulate two sections of the outer walls of radius b of a coaxial guide for d-c potentials. Show from problem 35 that this can be done without interfering with the transmission of the principal mode at frequency ν_c by leaving a small gap between sections, provided each terminates in a plane flange whose outer edge of radius a is coaxial with the line and satisfies the relation

$$J_0[2\pi(\mu\epsilon)^{\frac{1}{2}}\nu_c b]Y_1[2\pi(\mu\epsilon)^{\frac{1}{2}}\nu_c a] = J_1[2\pi(\mu\epsilon)^{\frac{1}{2}}\nu_c a]Y_0[2\pi(\mu\epsilon)^{\frac{1}{2}}\nu_c b]$$

Assume there is a magnetic or current node at the flange's edge.

38. If I is the coaxial line current in the last problem, t is the gap length, and v is $(\mu\epsilon)^{-\frac{1}{2}}$, show that the potential between the outer flange edges is

$$V_0 = \frac{It}{\pi^2\omega\epsilon a b[J_1(\omega a/v)Y_1(\omega b/v) - J_1(\omega b/v)Y_1(\omega a/v)]}$$

39. A sectorial horn formed by the planes $z = 0$, $z = b$ and $\phi = 0$, $\phi = \alpha'$ is excited by a uniform current $\mathbf{k}I \cos \omega t$ in a thin wire at $\rho = c$, $\phi = \alpha$. The wave is completely absorbed at $\rho = \infty$. From 12.16 (2) and (4), find suitable forms for A'_z when $\rho < c$ and A_z when $c < \rho$. Set up the Fourier series by equating A'_z to A_z at $\rho = c$, integrating $B_\phi - B'_\phi$ over the surface $\rho = c$, and applying the magnetomotance law as in 7.31 (9).

$$0 < \rho < c, \qquad \breve{A}'_z = \frac{\pi\mu I}{j\alpha'}\sum_{m=1}^{\infty}\left[J_{\frac{m\pi}{\alpha'}}(\beta c) - jY_{\frac{m\pi}{\alpha'}}(\beta c)\right]J_{\frac{m\pi}{\alpha'}}(\beta\rho)\sin\frac{m\pi\alpha}{\alpha'}\sin\frac{m\pi\phi}{\alpha'}$$

$$c < \rho < \infty, \qquad \breve{A}_z = \frac{\pi\mu I}{j\alpha'}\sum_{m=1}^{\infty}\left[J_{\frac{m\pi}{\alpha'}}(\beta\rho) - jY_{\frac{m\pi}{\alpha'}}(\beta\rho)\right]J_{\frac{m\pi}{\alpha'}}(\beta c)\sin\frac{m\pi a}{\alpha'}\sin\frac{m\pi\phi}{\alpha'}$$

40. Using the last problem, integrate Poynting's vector 11.17 (4) over a section of the horn when ρ is large, and thus show that the radiation resistance is

$$R_r = \frac{\omega\mu\pi b}{\alpha'}\sum_{m=1}^{\infty}\left[J_{\frac{m\pi}{\alpha'}}(\beta c)\right]^2\sin^2\frac{m\pi\alpha}{\alpha'}$$

41. The horn angle in problem 39 is $\alpha' = \pi/n$ where n is an integer and the wire radius is a. If $a \ll \alpha c$ and $a \ll [(\pi/n) - \alpha]c$, show that the boundary conditions are satisfied by superimposed fields of the type of problem 32 according to the principle of images of Fig. 4.06. If the wave is completely absorbed at $\rho = \infty$, show that the input impedance is, approximately, if $a \ll \lambda$ and R is the ohmic resistance of the wire,

$$R + \frac{\omega\mu b}{4}\left\{1 - \frac{2j}{\pi}(\ln \tfrac{1}{2}\beta a + C) + \sum_{s=1}^{n-1}\left[J_0\left(2\beta c\sin\frac{s\pi}{n}\right) - jY_0\left(2\beta c\sin\frac{s\pi}{n}\right)\right]\right.$$
$$\left. - \sum_{s=0}^{n-1}\left[J_0\left(2\beta c\sin\frac{(s+1)\pi - n\alpha}{n}\right) - jY_0\left(2\beta c\sin\frac{(s+1)\pi - n\alpha}{n}\right)\right]\right\}$$

The sum of the Y_0 terms will usually be negligible compared with the logarithm term.

42. Show that, if an electromagnetic wave having n electric field peaks in the z-direction is to be propagated radially in the horn bounded by the perfectly conducting planes $z = 0$, $z = b$, $\phi = 0$, and $\phi = \alpha$, it is necessary that $\nu > \frac{1}{2}n(\mu\epsilon)^{-\frac{1}{2}}b^{-1}$.

43. A uniform current $\boldsymbol{\phi}Ie^{j\omega t}$ flows in a circular loop $\rho = c$ that lies in the plane $z = d$ between two infinite conducting planes $z = 0$ and $z = a$. With the aid of 12.16 show that there is no propagation of energy if $\lambda > 2a$ and that, in this case, the vector potential between the planes, when $\rho > c$, is, from 12.16

$$\check{A}_\phi = \frac{2c\mu I}{a}\sum_{n=1}^{\infty} I_1(\beta_n c)K_1(\beta_n \rho)\sin\frac{n\pi d}{a}\sin\frac{n\pi z}{a} \qquad \text{where } \beta_n = \frac{2\pi}{\lambda}\left[\left(\frac{n\lambda}{2a}\right)^2 - 1\right]^{\frac{1}{2}}$$

44. If in the last problem $a < \lambda < 2a$ so that a single mode is propagated, show that, when $\rho > c$, the phasor vector potential is

$$-\frac{jc\mu I}{a}\left\{\pi H_1^{(2)}(\beta_1\rho)J_1(\beta_1 c)\sin\frac{\pi d}{a}\sin\frac{\pi z}{a} + 2j\sum_{n=2}^{\infty} K_1(\beta_n\rho)I_1(\beta_n c)\sin\frac{n\pi d}{a}\sin\frac{\pi n z}{a}\right\}$$

where $H_n^{(2)}(\beta_1\rho)$ is the Hankel function of 5.293 (10), $\beta_1 = (2\pi/\lambda)[1 - (\frac{1}{2}\lambda/a)^2]^{\frac{1}{2}}$, and $\beta_n = (2\pi/\lambda)[(\frac{1}{2}n\lambda/a)^2 - 1]^{\frac{1}{2}}$.

45. A sectional horn is bounded by the planes $z = 0$, $z = a$ and $\phi = 0$, $\phi = \alpha$ and is excited by a thin curved wire at $\rho = c$, $z = d$ carrying a uniform current $\boldsymbol{\phi} I\cos\omega t$. If $a < \lambda < 2a$ and the radius r of the wire is much less than c, and if the radiation is completely absorbed at $\rho = \infty$, show from problem 44 that

$$R_r = \frac{\pi\omega\alpha c^2\mu}{a}\left[J_1(\beta_1 c)\sin\frac{\pi d}{a}\right]^2$$

$$X_r = \frac{2\omega\alpha c^2\mu}{a}\left\{-\left[\frac{1}{2}\pi J_1(\beta_1 c)Y_1(\beta_1 c) + \frac{a}{2\pi c}\right]\sin^2\frac{\pi d}{a}\right.$$
$$\left. + \sum_{n=2}^{\infty}\left[K_1(\beta_n c)I_1(\beta_n c) - \frac{a}{2\pi n c}\right]\sin^2\frac{n\pi d}{a} + \frac{a}{4\pi c}\left[\ln\sin\frac{\pi d}{a} + \ln\frac{2a}{\pi r}\right]\right\}$$

Note that the series converges very rapidly.

46. If in the last problem $\lambda = 1.5a$, $c = a$, $\alpha = 18°$, $d = \frac{1}{2}a$, show that to make the reactance vanish the radius of the wire should be $0.0114a$ and that the radiation resistance in this case is 274.5 ohms.

47. The planes $x = 0$, $y = 0$, $x + y = a$, $z = 0$, and $z = d$ bound a cavity. Show that the natural frequency of the simplest TE mode is $\nu = v(a^2 + d^2)^{\frac{1}{2}}/(2ad)$ and that

$$Q = \frac{\mu a d(a^2 + d^2)}{\mu'\delta[2a^3 + (1 + 2^{\frac{1}{2}})a^2 d + (3 + 2^{\frac{1}{2}}2)d^3]}$$

where v is the velocity of a free wave in the dielectric filling the cavity.

48. The planes $x = 0$, $y = 0$, $x + y = a$, $z = 0$, and $z = d$ bound a cavity. Show that the natural frequency of the simplest TM mode is $\nu = 5^{\frac{1}{2}}v/(2a)$ and that

$$Q = \frac{\mu a d}{\mu'\delta[(2 + 2^{\frac{1}{2}})d + a]}$$

where v is the velocity of a free wave in the dielectric filling the cavity.

49. A cavity is bounded by the planes $x = 0$ and $x = d$ and by the prism $y = 0$, $y = b$, $y = z$, $z - y = 2b$. If the electric field is entirely x-directed and if one assumes that the only periodic solution that gives $A_x = 0$ at $y = 0$ and $z = y$ is

$$\check{A}_x = \check{C}[\sin p(z - y)\sin q(z + y) - \sin q(z - y)\sin p(z + y)]$$

where p and q are different arbitrary constants, show that the lowest resonance wave length of the cavity is $2b/5^{\frac{1}{2}}$ given by $p = \frac{3}{2}\pi/b$ and $q = \frac{1}{2}\pi/b$.

50. A check of the nodal lines shows that the lowest purely TM mode in the last problem consists of four right isosceles triangular cells. Hence show by problem 48, adding up losses on four side, two diagonal, and eight end faces, that

$$Q = 2\mu(\delta\mu')^{-1}bd[(2 + 2^{\frac{1}{2}})d + 2b]^{-1}$$

This exceeds 13.20 (10) and 13.21 (7) using the same lowest resonance ω.

51. A cavity is formed by short-circuiting the biconical transmission line bounded by the cones $\theta = \alpha$ and $\theta = \pi - \alpha$ by the spherical surface $r = d$. Show that the wave lengths of the principal mode resonances are $4d/(2p + 1)$ and that

$$Q = \frac{2\mu d \sin\alpha \ln\cot\frac{1}{2}\alpha}{\mu'\delta\{C + \ln[(2p+1)\pi] - \text{Ci}[(2p+1)\pi] + 2\sin\alpha\ln\cot\frac{1}{2}\alpha\}}$$

where C is Euler's constant 0.5772.

52. A cavity resonator is formed by planes $z = 0$, $z = d$ and two confocal elliptic cylinders orthogonal to them whose major and minor semi-axes are M_1, m_1 and M_2, m_2. Show with the aid of 4.22 and 13.22 that for the pth principal mode of this cavity

$$Q_p = \frac{\mu d}{2\mu'\delta_p}\left\{1 + \frac{d(M_1K_2 + M_2K_1)}{2\pi M_1M_2[\ln(M_2 + m_2) - \ln(M_1 + m_1)]}\right\}^{-1}$$

where $M_2 > M_1$ and K_1 and K_2 are complete elliptic integrals of modulus $(M_1^2 - m_1^2)^{\frac{1}{2}}/M_1$ and $(M_2^2 - m_2^2)^{\frac{1}{2}}/M_2$, respectively.

53. A cavity is bounded by the planes $z = 0$, $z = d$, and by two circular cylinders orthogonal to them of radii a and b whose axes are a distance c apart where $a > b + c$. Show with the aid of 4.13 and 13.22 that for the pth principal mode

$$Q = \frac{\mu d}{2\mu'\delta_p}\left\{1 + \frac{d(a-b)[(a+b)^2 - c^2]}{4ab[(a^2-b^2)^2 - 2(a^2+b^2)c^2 + c^4]^{\frac{1}{2}}\cosh^{-1}[(a^2+b^2-c^2)/(2ab)]}\right\}^{-1}$$

54. A rectangular cavity of width a, height b, and length $2d \geqslant 2a$ is partly closed at the center from $y = c$ to $y = b$ by a thin conducting iris and excited in the TE_{10} mode. Consider it as a transmission line T section short-circuited at the ends with the capacitance of 13.06 forming the leg of the T, and thus show that the lowest free space resonance wave length is given by

$$\lambda = \frac{2a\lambda_1}{(4a^2 + \lambda_1^2)^{\frac{1}{2}}} \qquad 1 = \frac{4b}{\lambda_1}\left(\ln\csc\frac{\pi c}{2b}\right)\left(\tan\frac{2\pi d}{\lambda_1}\right)$$

55. If $a = 2$, $b = 1$, $c = \frac{1}{2}$, and $d = 2$ in the last problem, show that $\lambda = 3.692$.

56. Obtain an approximate solution for the fields in problem 54 as follows: Use fields TE to x from problem 3 but with a $\cos\beta'_{1n}z\cos\omega t$ dependence on z and t in W_{tex}. Take $z = 0$ at the left end of the cavity and assume the quasi-static result of 4.22 (6) for $\boldsymbol{E}$ at $z = d$, so that

$$\frac{E_y}{\sin(\pi x/a)} = \frac{2^{\frac{1}{2}}V_0\cos(\frac{1}{2}\pi y/b)}{b[\cos(\pi y/b) - \cos(\pi c/b)]^{\frac{1}{2}}} \qquad \text{or} \qquad \frac{E_y}{\sin(\pi x/a)} = 0$$

according as $0 < y < c$ or $c < y < b$, where V_0 is the electrostatic potential across the iris. Thus get the coefficients in problem 3 by using the Mehler integral, see 13.08 (11), and so obtain for the electric field in the cavity, when $0 < z < d$ and

$u = \cos(\pi c/b)$,

$$E_y = \frac{V_0}{b}\sin\frac{\pi x}{a}\left\{\frac{\sin\beta'_{10}z}{2\sin\beta'_{10}d} + \sum_{n=1}^{\infty}[P_n(u) + P_{n-1}(u)]\frac{\sinh|\beta'_{1n}z|}{\sinh|\beta'_{1n}d|}\cos\frac{n\pi y}{b}\right\}\cos\omega t$$

$$E_z = \frac{V_0}{b}\sin\frac{\pi x}{a}\sum_{n=1}^{\infty}[P_n(u) + P_{n-1}(u)]\frac{n\pi\cosh|\beta'_{1n}z|}{|\beta'_{1n}a|\sinh|\beta'_{1n}d|}\sin\frac{n\pi y}{b}\cos\omega t$$

where $|\beta'_{1n}|^2 = |\omega^2\mu\epsilon - (n\pi/b)^2 - (\pi/a)^2|$ and ω is given roughly by problem 54. Note that in the window E_z is small compared with E_y but does not actually vanish, so that the electric field line which leaves the edge in the $z = d$ plane strikes the $y = 0$ plane normally but slightly displaced from the $z = d$ plane. The boundary condition on the iris is rigorously met, thus the result is best for small windows. This also gives one-quarter of the field in a cavity of height $2b$ with a centered window of width $2c$ and in a cavity with height $2b$ and a centered strip of width $2b - 2c$.

57. Use the method and notation of the last problem but assume

$$\frac{E_y}{\sin(\pi x/a)} = 0 \qquad \text{or} \qquad \frac{E_z}{\sin(\pi x/a)} = \frac{2^{\frac{1}{2}}V_0\cos(\frac{1}{2}\pi y/b)}{b[\cos(\pi c/b) + \cos(\pi y/b)]^{\frac{1}{2}}}$$

depending on whether $a < y < c$ or $c < y < b$. Thus get the coefficients in problem 3 by using the second Mehler integral in 13.08 and obtain

$$E_y = \frac{V_0}{b}\sin\frac{\pi x}{a}\left\{\frac{\sin|\beta'_{10}z|}{2\sin|\beta'_{10}d|} + \sum_{n=1}^{\infty}\frac{|\beta'_{1n}b|}{n\pi}[P_n(u) + P_{n-1}(u)]\frac{\sinh|\beta'_{1n}z|}{\cosh|\beta'_{1n}d|}\cos\frac{n\pi y}{b}\right\}\cos\omega t$$

$$E_z = \frac{V_0}{b}\sin\frac{\pi x}{a}\sum_{n=1}^{\infty}[P_n(u) + P_{n-1}(u)]\frac{\cosh|\beta'_{1n}z|}{\cosh|\beta'_{1n}d|}\sin\frac{n\pi y}{b}\cos\omega t$$

Here E_y is small compared with E_z on the iris but does not actually vanish as it should so the iris surface is not exactly plane. The boundary condition $E_z = 0$ is rigorously met in the window. Thus these expressions are best for a small iris and a large window.

58. A rectangular cavity of width a, height b, and length $2d > 2a$ is partly closed at the center by a thin iris having a slit running from $y = 0$ to $y = b$ with edges at $x = \frac{1}{2}(a - c)$ and $x = \frac{1}{2}(a + c)$. Proceed as in problem 54 to show, with the aid of 13.06 (16), that the lowest resonance frequency corresponds to a free space wave length λ given by

$$\lambda = \frac{2a\lambda_1}{(4a^2 + \lambda_1^2)^{\frac{1}{2}}}, \qquad 1 = \frac{2a}{\lambda_1}\cot\frac{2\pi d}{\lambda_1}\tan^2\frac{\pi c}{2a}$$

59. If $a = 2$, $b = 1$, $c = \frac{1}{2}$, and $d = 2$ in the last problem, show that $\lambda = 2.7488$.

60. Obtain an approximate solution for the fields in problem 58 as follows: Use the fields of 13.06 (9) and (10) but with $\cos\omega t$ time dependence. Take $z = 0$ at the left end of the cavity as in Fig. 13.06c and assume the quasi-static B_x at $z = d$ given by 13.06 (14). Thus obtain for the magnetic fields in the cavity, using relations in 13.08,

$$B_x = \frac{\cos|\beta'_1 z|}{\cos|\beta'_1 d|}\sin\frac{\pi x}{a} + \sum_{m=3}^{\infty}\frac{A_m\cosh|\beta'_m z|}{A_1\cosh|\beta'_m d|}\sin\frac{m\pi x}{a}$$

$$B_z = -\frac{\pi\sin|\beta'_1 z|}{|\beta'_1 a|\cos|\beta'_1 d|}\cos\frac{\pi x}{a} - \sum_{m=3}^{\infty}\frac{A_m m\pi\sinh|\beta'_m z|}{A_1|\beta'_m a|\cosh|\beta'_m d|}\cos\frac{m\pi x}{a}$$

where $|\beta_m'^2| = |\omega^2\mu\epsilon - (m\pi/a)^2|$, is given by problem 58, m is odd, and

$$A_{2n+1} = (-1)^n\left[\frac{2n+1}{n(n+1)}\right]\sin\frac{\pi c}{a}P_n^1\left[\cos\frac{\pi c}{a}\right]$$

Note that on the iris B_z is small but does not actually vanish, so that the iris surface deviates slightly from a magnetic flux line. The boundary condition is rigorously met in the window. Thus this result is best for wide windows.

61. Obtain a second approximate solution of the last problem by assuming the quasi-static B_z at $z = d$. Thus obtain for the magnetic fields in the cavity

$$B_x = \frac{|\beta_1'a|\cos|\beta_1'z|}{\pi\sin|\beta_1'd|}\sin\frac{\pi x}{a} + \sum_{m=3}^{\infty}\frac{A_m|\beta_m'a|\cosh|\beta_m'z|}{A_1 m\pi\sinh|\beta_m'd|}\sin\frac{m\pi x}{a}$$

$$B_z = -\frac{\sin|\beta_1'z|}{\sin|\beta_1'd|}\cos\frac{\pi x}{a} - \sum_{m=3}^{\infty}\frac{A_m\sinh|\beta_m'z|}{A_1\sinh|\beta_m'd|}\cos\frac{m\pi x}{a}$$

where the notation is that of the last problem. Note that in the window the B_x is small but not actually zero, so that if the image field is drawn in the region $d < z < 2d$, there will be a discontinuity in the slope of the flux lines in the plane of the window. On the iris the boundary conditions are rigorously met. Thus this result is best for narrow windows.

62. Use the second form of the stationary variational formula 13.28 (3) to get the resonance frequency of the spheroid $(\rho/a)^2 + (z/b)^2 = 1$. Take for the components of the electric field the divergenceless form, which meets the symmetry conditions,

$$E_z = A - B\rho^2 - Cz^2, \qquad E_\rho = 2C\rho z$$

Determine A, B, and C so that $\boldsymbol{E}$ is normal to the spheroid, substitute into 13.28 (3), and show that

$$\beta A \approx \left[\frac{14(3a^2+2b^2)}{a^2(5a^2+4b^2)}\right]^{\frac{1}{2}} \xrightarrow[a\to b]{} 2.788$$

63. If $d \ll a$, $d \ll c$, and $a \approx b$ in the reentrant cavity of 13.29, show with the aid of the results in problem 60, Chap. IV, that C_0 in 13.29 (11) is, approximately,

$$C_0 = \epsilon\left[\frac{\pi a^2}{d} + 4a\ln\frac{d^2+k^2}{4kd} + \frac{4ak}{d}\tan^{-1}\frac{d}{k} + \frac{4ad}{k}\tan^{-1}\frac{k}{d} - \frac{2\pi ad}{k}\right]$$

where k is $b - a$. If $a = 0.01$, $b = 0.015$, $d = 0.001$, $c = 0.02$, this gives $\beta = 40.6$. Moreno's graphs give $\beta = 41.0$. The accuracy improves as a approaches b.

64. A rectangular cavity of dimensions a, b, d in the x-, y-, z-directions is to be excited in the TM_{110} mode by a wire parallel to the z-axis at $x = a_1$, $y = b_1$. The wire is terminated to give uniform current in the section of resistance R inside the cavity. If d is much less than a or b, show that the input impedance is

$$R' = R + \frac{2\gamma\delta\mu^2 v^2 d^2(a^2+b^2)}{ab(a^2+b^2)+2d(a^3+b^3)}\sin^2\frac{\pi a_1}{a}\sin^2\frac{\pi b_1}{b}$$

Show that the electromotance multiplication is $\csc(\pi a_1/a)\csc(\pi b_1/b)$.

65. Show by the image method and with the aid of problem 32 that the reactance load on the driving circuit of the last problem when the radius of the wire is c is

$$-\frac{\omega\mu d}{2\pi}\ln c + \frac{\omega\mu d}{4}\sum_{-\infty}^{\infty}\sum_{-\infty}^{\infty}((1-\delta_n^0\delta_m^0)Y_0[2\beta(n^2a^2+m^2b^2)^{\frac{1}{2}}] - Y_0\{2\beta[(na-a_1)^2 + m^2b^2]^{\frac{1}{2}}\} - Y_0\{2\beta[n^2a^2+(mb-b_1)^2]^{\frac{1}{2}}\} + Y_0\{2\beta[(na-a_1)^2+(mb-b_1)^2]^{\frac{1}{2}}\})$$

Observe that the first term is much larger than the rest, if c is small, and that in the remainder the $m = n = 0$ terms are most important.

66. The cavity in problem 64 is of copper, $\gamma = 5.8 \times 10^7$, and its a-, b-, and d-dimensions are 10, 20, and 2 cm, respectively. Show that its TM_{110} resonance frequency is 1.676×10^9 cycles per second, its Q is roughly 9100, and that for an input resistance of 100 ohms the driving wire is at $x = 0.0512$ cm, $y = 10$ cm; at $x = 5$ cm, $y = 0.1024$ cm, or anywhere on the curve $\sin(\pi a_1/10)\sin(\pi b_1/20) = 0.0160$. Show that for 1 watt input, the rms potential across the center is 625 volts.

67. A rectangular cavity is driven by a semicircular loop of radius r in the $y = b_1$ plane centered at $z = d_1$, $x = 0$. From 13.20 (2) and 13.30 (5) A^0_{mnp} is

$$\pi 8^{\frac{1}{2}}\left[\mathbf{i}\left(\frac{n}{\beta_{mn}b}\right)\right]\cos\frac{m\pi x}{a}\sin\frac{n\pi y}{b} - \mathbf{j}\left(\frac{m}{\beta_{mn}a}\right)\sin\frac{m\pi x}{a}\cos\frac{n\pi y}{b}\Bigg]\sin\frac{p\pi z}{d}$$

Show that the normalized mutual inductance between loop and the TE_{mnp} mode is

$$M^0 = \frac{8^{\frac{1}{2}}\pi^3 npr}{\beta_{mn}\beta_{mp}bd}\cos\frac{n\pi b_1}{b}\sin\frac{p\pi d_1}{d}J_1(\beta_{mp}r)$$

The integration is simplified by using $\int(\nabla \times A^0)_y\, dS$ in place of $\int A^0 \cdot ds$.

68. Solve the preceding problem for the TM_{mnp} mode using 13.20 (4) and 13.30 (5). Show that the normalized vector potential A^0_{mnp} is, in the notation of 13.20,

$$8^{\frac{1}{2}}(\beta_{mnp}\beta_{ma})^{-1}[P(\mathbf{i}M\cos Mx\sin Ny + \mathbf{j}N\sin Mx\cos Ny)\sin Pz - \mathbf{k}\beta^2_{mn}\sin Mx\sin Ny\cos Pz]$$

where $\beta^2_{mn} = M^2 + N^2$ and $\beta^2_{mnp} = M^2 + N^2 + P^2$. Show, using the same integrals as in problem 67, that the normalized mutual inductance between the loop and the TM_{mnp} mode is

$$M^0 = \frac{8^{\frac{1}{2}}\pi^4 mnpr}{\beta_{mnp}\beta_{mn}\beta_{mp}ad^2}\cos\frac{n\pi b_1}{b}\sin\frac{p\pi d_1}{d}J_1(\beta_{mp}r)$$

69. A spherical copper cavity of radius a is excited by a small plane loop of area S so close to the wall that the value of $\boldsymbol{B}$ at the wall may be used over S. Show that, when driven at the TM_{01} resonance frequency, the input resistance of the loop is

$$R_i + R + \frac{3\omega^2\mu^2\gamma\delta(4\beta^2_{01}a^2 - 9)S^2}{4\pi a^2(4\beta^2_{01}a^2 - 1)} = R + \frac{1.854 \times 10^5\gamma\delta S^2}{a^4}$$

and that, if $\beta_{01}a$ is 2.744 and R is the loop resistance, the potential multiplication is

$$\frac{V}{\mathcal{E}} = \frac{2}{\beta^2_{01}S}\left[\frac{\mathrm{Si}(\beta_{01}a)}{j_1(\beta_{01}a)} - 1\right] = \frac{7.42}{\beta^2_{01}S}$$

70. A rectangular cavity is filled with material $\mu_1\epsilon_1$ when $0 < y < \frac{1}{2}b$ and material $\mu_2\epsilon_2$ when $\frac{1}{2}b < y < b$. If $\mu_1\epsilon_1 = \mu_2\epsilon_2$ but $\mu_1 \neq \mu_2$ and $\epsilon_1 \neq \epsilon_2$, find the ratio of the energy stored in $\mu_1\epsilon_1$ to that in $\mu_2\epsilon_2$ when the cavity is excited in the TE_{101} mode.

71. A cubical cavity is oscillating at the frequency $\nu^2 = (2\mu\epsilon a^2)^{-1}$. Show that it is possible to have equal average pressures on all six walls, and state what directions the electric field at the center can have in this case. Show that the rms force on the wall is then $\epsilon E_c^2 a^2/24$.

72. The surface of a rectangular cavity, given by $x = 0$, $x = a$; $y = 0$, $y = \frac{1}{4}a$; and $z = 0$, $z = a$, is formed of material of conductivity γ and skin depth δ. It is excited by a straight wire carrying a uniform current $I_0 \cos \omega t$ that lies in the plane $z = \frac{1}{2}a$ and passes across the corner from $y = \frac{1}{8}a$ to $z = \frac{1}{8}a$ at 45°. When ω is the lowest resonance frequency, show that the power dissipation is $P = \frac{2}{3}(\mu\gamma\delta\epsilon^{-1}\pi^{-2})(1 - \cos \frac{1}{8}\pi)I_0^2$.

73. A cavity is bounded by $y = 0$, $y = 0.5$; $x = 0$, $x = 2$ if $0 < z < 2$; $x = 1$, $x = 2$ if $2 < z < 4$. Show that if it oscillates with the configuration of the TE_{202} mode of a $2 \times 0.5 \times 4$ rectangular cavity, then $Q = \pi v \mu \gamma \delta 5^{\frac{3}{2}}/32$.

References

Borgnis, F. E., and C. H. Papas: "Randwertprobleme der Mikrowellenphysik," Springer, 1955. Excellent detailed discussion of wave guides and cavities including integral equation and variational methods.

Collin, R. E.: "Field Theory of Guided Waves," McGraw-Hill, 1960. Gives complete exposition, with examples, of the mathematical techniques for wave guides of all types.

Corson, D. R., and Paul Lorrain: "Introduction to Electromagnetic Fields and Waves," Freeman, 1962. Illustrated introductory treatment.

Flugge, S.: "Encyclopedia of Physics," Vol. XVI, Springer, 1958. Borgnis and Papas give a very comprehensive and readable treatment of "Electromagnetic Wave Guides and Resonators."

Gray, D. E.: "American Institute of Physics Handbook," 2d ed., McGraw-Hill, 1963. Section 5b gives useful collection of formulas and pictures.

Harrington, R. F.: "Time-harmonic Electromagnetic Fields," McGraw-Hill, 1961. Very complete, detailed, and practical treatment with problems.

Jackson, J. D.: "Classical Electrodynamics," Wiley, 1962. Thirty pages on these subjects. All formulas in both cgs and mks units.

Jones, D. S.: "The Theory of Electromagnetism," Macmillan, 1964. Treats cavities from a mathematician's viewpoint.

Johnson, C. C.: "Field and Wave Electrodynamics," McGraw-Hill, 1965. Well-illustrated and readable treatment.

King, R. W. P., H. R. Mimno, and A. H. Wing: "Transmission Lines, Antennas, and Wave Guides," McGraw-Hill, 1945. Well-illustrated descriptions of wave guides and cavities.

Marcuvitz, N.: "Waveguide Handbook," McGraw-Hill, 1951. Most complete collection of formulas and graphs with some theory.

Montgomery, C. G., R. H. Dicke, and E. M. Purcell: "Principles of Microwave Circuits," McGraw-Hill, 1948. Covers most of the material in this chapter.

Moreno, T.: "Microwave Transmission Design Data," McGraw-Hill, 1948. Good collection of graphs and tables for microwave work with some theory.

Panofsky, W. K. H., and Melba Phillips: "Classical Electricity and Magnetism," 2d ed., Addison-Wesley, 1962. Brief but lucid treatment.

Ramo, S., and J. R. Whinnery: "Fields and Waves in Modern Radio," 2d ed., Wiley, 1953. Clear, well-illustrated, practical discussion.

Schelkunoff, S. A.: "Electromagnetic Waves," Van Nostrand, 1943. Extended discussion of theory with many graphs.

Slater, J. C.: "Microwave Electronics," Van Nostrand, 1950. Complete theoretical treatment of wave guides and cavities.

Tralli, N.: "Classical Electromagnetic Theory," McGraw-Hill, 1963. Very brief but lucid discussion.

Van Bladel, J.: "Electromagnetic Fields," McGraw-Hill, 1964. Discusses additional topics connected with wave guides and cavities.

CHAPTER XIV

SPECIAL RELATIVITY AND THE MOTION OF CHARGED PARTICLES

14.00. The Postulates of Special Relativity.—In previous chapters, we considered electric and magnetic interaction of charges and currents whose magnitudes may vary with time but whose positions are fixed relative to each other and the observer. The special theory of relativity seems to furnish the most firmly established experimental basis for the discussion of the interaction of systems moving in a straight line relative to each other or to the observer.

The two fundamental postulates of this theory are:

1. Physical laws, or the equations defining them, have the same form, *in vacuo,* in all cartesian coordinate systems that have a uniform translatory motion relative to each other.

2. The velocity of light, *in vacuo,* is the same for all observers, independent of the relative velocity of the source of light and the observer.

A great number of observable phenomena, mechanical, optical, and electrical, predicted by these postulates have been investigated, and in all cases the results of experiments made on a terrestrial scale confirm them.

14.01. The Lorentz Transformation Equations.—Consider the two coordinate systems shown in Fig. 14.01 in which the x and x' axes coincide

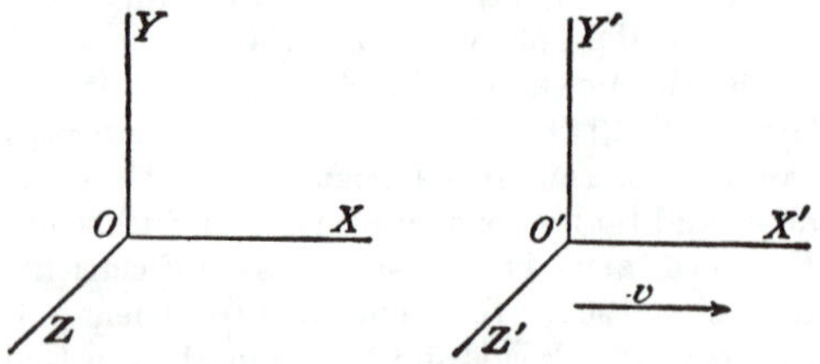

FIG. 14.01.

and the system S' is moving with a uniform velocity v in the x-direction with respect to S. To find the equations connecting x, y, z and t with x', y', z' and t' which will satisfy the two postulates of special relativity, we shall consider a simple experiment. Suppose that at the instant $t = t' = 0$ the two origins, O and O', are in coincidence and that at this instant a light pulse leaves the common origin. Our postulates require that S and S' each observe a spherical wave spreading out from his origin with a velocity c. Thus, at any future instant, the equation of the wave front must be for S

$$x^2 + y^2 + z^2 = c^2t^2 \tag{1}$$

For S' it must be

$$x'^2 + y'^2 + z'^2 = c^2t'^2 \tag{2}$$

From the symmetry of the situation as viewed by S and S', we are led to try setting $z = z'$ and $y = y'$ so that, to satisfy these equations, we must have

$$x^2 - c^2t^2 = x'^2 - c^2t'^2 \tag{3}$$

For S, at the time t, O' has the coordinate vt; and for S', at the time t', O has the coordinate $-vt'$. The simplest relations which will give this result are

$$x' = \kappa(x - vt) \qquad x = \kappa'(x' + vt') \tag{4}$$

Eliminating x' from these equations gives

$$t' = \kappa\left[t - \frac{x}{v}\left(1 - \frac{1}{\kappa\kappa'}\right)\right] \tag{5}$$

Substituting for x' and t' in (3) from (4) and (5), we obtain an equation in which the variables appear as x^2, xt, and t^2, and since this equation must be satisfied for any positive values of t and any values of x, this requires that the coefficients of x^2, xt, and t^2 vanish separately. Setting these equal to zero and solving for κ and κ' we obtain

$$\kappa = \kappa' = \left[1 - \left(\frac{v}{c}\right)^2\right]^{-\frac{1}{2}} \tag{6}$$

Thus, from (4) and (5), the transformation formulas become

$$\begin{aligned} x' &= \kappa(x - vt) & x &= \kappa(x' + vt') \\ y' &= y, \quad z' = z & y &= y', \quad z = z' \\ t' &= \kappa\left(t - \frac{xv}{c^2}\right) & t &= \kappa\left(t' + \frac{x'v}{c^2}\right) \end{aligned} \tag{7}$$

Suppose that the signals are emitted from a point fixed in the moving system S' at times t_1' and t_2' and received by an observer in the S-system at times t_1 and t_2. Then from (7)

$$\Delta t' = t_1' - t_2', \qquad \Delta t = t_1 - t_2 = \kappa(t_1' - t_2') = \kappa\Delta t' \tag{8}$$

This "time dilation" has been observed experimentally with fast mesons produced by cosmic rays whose average apparent lifetime, as estimated from their velocity and travel distance, may be increased by as much as ten times its rest value by the high speed.

14.02. Transformation Equations for Velocity and Acceleration.—We shall designate velocities relative to the S coordinate system by

$$u_x = \frac{dx}{dt}, \qquad u_y = \frac{dy}{dt}, \qquad u_z = \frac{dz}{dt} \tag{1}$$

and those relative to the S'-system by

$$u'_x = \frac{dx'}{dt'}, \qquad u'_y = \frac{dy'}{dt'}, \qquad u'_z = \frac{dz'}{dt'} \tag{2}$$

By differentiating the first three equations of the first group in 14.01 (7) with respect to t' and those in the second group with respect to t and then eliminating dt/dt' and dt'/dt, respectively, from the right sides by differentiating the fourth equation of the opposite group, we get

$$u'_x = \frac{u_x - v}{1 - \dfrac{u_x v}{c^2}}, \qquad u_x = \frac{u'_x + v}{1 + \dfrac{u'_x v}{c^2}} \tag{3}$$

$$u'_{y,z} = \frac{u_{y,z}}{\kappa\left(1 - \dfrac{u_x v}{c^2}\right)}, \qquad u_{y,z} = \frac{u'_{y,z}}{\kappa\left(1 + \dfrac{u'_x v}{c^2}\right)} \tag{4}$$

These are the transformation equations for velocity. It is interesting to note that even if we give the two systems the relative velocity $v = c$ and a point in the S'-system has a velocity $u'_x = c$, the value of u given by (3) is still c. Thus, the velocity of light may be regarded as the upper limit of possible velocities.

By a very similar process, differentiating (3) and (4), we obtain for the accelerations

$$\frac{du_x}{dt} = \left[\kappa\left(1 + \frac{u'_x v}{c^2}\right)\right]^{-3} \frac{du'_x}{dt'} \tag{5}$$

$$\frac{du_{y,z}}{dt} = \left[\kappa\left(1 + \frac{u'_x v}{c^2}\right)\right]^{-2} \frac{du'_{y,z}}{dt'} - \frac{u'_{y,z} v}{\kappa^2 c^2}\left(1 + \frac{u'_x v}{c^2}\right)^{-3} \frac{du'_x}{dt'} \tag{6}$$

We get the corresponding equation for du'_x/dt' and $du'_{y,z}/dt'$ by interchanging primed and unprimed quantities and substituting $+v$ for $-v$. We note that a constant acceleration in the S'-system does not, in general, imply a constant acceleration in the S-system.

14.03. Variation of Mass with Velocity.—The first postulate requires that the laws of conservation of energy and of momentum be satisfied for all observers. To see the result of this requirement, let us consider a simple hypothetical experiment devised by Tolman and illustrated in

Fig. 14.03. At the instant when the two origins are coincident, let S' project a sphere with a velocity u from B' toward O', and let S project one from A parallel to y with a velocity u. Let OA and OB be chosen in such a way that the spheres collide when their centers are aligned in the Y-direction. The collisions as observed by S and S' are shown. We shall choose the mass of the spheres so that, when at rest with respect

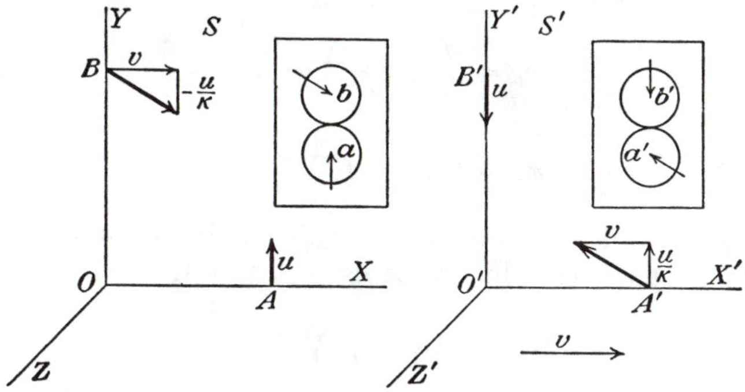

FIG. 14.03.

to any observer, both have, for him, the mass m_0. We shall assume that the mass is a function of the magnitude of the velocity so that

$$m = f(u^2) \qquad \text{and} \qquad m_0 = f(0) \tag{1}$$

By the formulas of the last articles, the initial velocities are

$$\begin{array}{llll} \text{For } S & & \text{For } S' & \\ u_{ax} = 0, & u_{ay} = u & u'_{ax} = -v, & u'_{ay} = \dfrac{u}{\kappa} \\ u_{bx} = v, & u_{by} = -\dfrac{u}{\kappa} & u'_{bx} = 0, & u'_{by} = -u \end{array} \tag{2}$$

Let us indicate velocities after the impact by bars. From the conditions of impact of smooth spheres, neither observer will see any transfer of the x-component of momentum. Thus, S observes that $m_b v = \bar{m}_b v$ or

$$vf\left(v^2 + \frac{u^2}{\kappa^2}\right) = vf\left(v^2 + \frac{\bar{u}_b^2}{\kappa^2}\right)$$

This requires that

$$\bar{u}_b = -u \qquad \text{and hence} \qquad \bar{m}_b = m_b \tag{3}$$

The negative sign is chosen since we know it to be correct for small velocities, where $\kappa = 1$. For the y-component, S observes that

$$uf(u^2) - \frac{m_b u}{\kappa} = \bar{u}_a f(\bar{u}_a^2) - \frac{\bar{m}_b \bar{u}_b}{\kappa}$$

In view of (3) and the known result when $v = 0$, this can be satisfied only if $\bar{u}_a = -u$ so that this equation becomes, by dividing out u,

$$f(u^2) = m_a = \frac{m_b}{\kappa} \tag{4}$$

If the velocity of projection u becomes very small, then

$$m_b = f\left(v^2 + \frac{u^2}{\kappa^2}\right) \to f(v^2)$$

and

$$m_a = f(u^2) \to f(0) = m_0 \tag{5}$$

Thus, the mass of an object in motion with a velocity v with respect to a given observer appears to him to be increased by the factor

$$\kappa = \left[1 - \left(\frac{v}{c}\right)^2\right]^{-\frac{1}{2}} \tag{6}$$

over its mass when at rest.

We shall find it convenient to introduce here two new symbols κ_1 and κ_1' defined by

$$\kappa_1 = \left(1 - \frac{u_x^2 + u_y^2 + u_z^2}{c^2}\right)^{-\frac{1}{2}} \tag{7}$$

$$\kappa_1' = \left(1 - \frac{u_x'^2 + u_y'^2 + u_z'^2}{c^2}\right)^{-\frac{1}{2}} \tag{8}$$

where u_x, u_y, and u_z are the components of velocity of a particle as observed by S and u_x', u_y', and u_z' the components of velocity of the same particle as observed by S'. Using the relations 14.02 (3) and (4), we obtain the relations

$$\kappa_1 = \kappa\kappa_1'\left(1 + \frac{u_x'v}{c^2}\right) \tag{9}$$

$$\kappa_1' = \kappa\kappa_1\left(1 - \frac{u_xv}{c^2}\right) \tag{10}$$

14.04. The Transformation Equations for Force.—The nature of the equations connecting the forces observed by S and S' depends on whether force is defined as the product of mass by acceleration or as the rate of change of momentum. We shall take the latter definition so that, from 14.03 (5), we have

$$\boldsymbol{F} = \frac{d(m\boldsymbol{u})}{dt} = m_0\frac{d(\kappa_1\boldsymbol{u})}{dt} \tag{1}$$

By writing out the components of this equation, we see that with this definition force and acceleration are not, in general, in the same direction. Carrying out the differentiation, we have

$$\boldsymbol{F} = m\frac{d\boldsymbol{u}}{dt} + \boldsymbol{u}\frac{dm}{dt} = m_0\kappa_1\frac{d\boldsymbol{u}}{dt} + m_0\kappa_1^3\boldsymbol{u}\frac{u}{c^2}\frac{du}{dt}$$
$$= m_0\kappa_1^3\left[\left(1 - \frac{u^2}{c^2}\right)\frac{d\boldsymbol{u}}{dt} + \frac{u\boldsymbol{u}}{c^2}\frac{du}{dt}\right] \qquad (2)$$

If the force is applied in the direction $\boldsymbol{u}_1$ of u, then $\boldsymbol{u} = \boldsymbol{u}_1 u$ and

$$\frac{d\boldsymbol{u}}{dt} = \frac{\boldsymbol{u}_1\,du}{dt}$$

so that we have

$$\boldsymbol{F}_l = m_0\kappa_1^3\frac{d\boldsymbol{u}}{dt} = m_l\frac{d\boldsymbol{u}}{dt} \qquad (3)$$

If du/dt is zero, *i.e.*, if the velocity changes in direction but not in magnitude, then (2) shows that the force is again in the direction of the acceleration which is at right angles to $\boldsymbol{u}$, and we have

$$\boldsymbol{F}_t = m_0\kappa_1\frac{d\boldsymbol{u}}{dt} = m_t\frac{d\boldsymbol{u}}{dt} \qquad (4)$$

The quantities $m_l = m_0\kappa_1^3$ and $m_t = m_0\kappa_1$ are usually called the longitudinal and transverse masses, respectively, of the particle.

The forces observed by S' will be of the same form as those observed by S; thus,

$$\boldsymbol{F}' = m_0\frac{d(\kappa_1'\boldsymbol{u}')}{dt'}$$

Let us substitute in the x-component of (1) for $\kappa_1 u_x$ from 14.02 (3) 14.03 (9) and for dt'/dt from differentiating 14.01 (7), and we have

$$F_x = m_0\frac{d(\kappa_1 u_x)}{dt} = m_0\left(1 + \frac{u_x' v}{c^2}\right)^{-1}\frac{d[\kappa_1'(u_x' + v)]}{dt'}$$
$$= F_x' + m_0 v\left(1 + \frac{u_x' v}{c^2}\right)^{-1}\left[-\frac{u_x'}{c^2}\frac{d(\kappa_1' u_x')}{dt'} + \frac{d\kappa_1'}{dt'}\right]$$
$$= F_x' + m_0 v\left(1 + \frac{u_x' v}{c^2}\right)^{-1}\left[\left(1 - \frac{u_x'^2}{c^2}\right)\frac{d\kappa_1'}{dt'} - \frac{\kappa_1' u_x'}{c^2}\frac{du_x'}{dt'}\right]$$

But

$$1 - \frac{u_x'^2}{c^2} = \frac{u_y'^2 + u_z'^2}{c^2} + \frac{1}{\kappa_1'^2} \quad \text{and} \quad -\frac{\kappa_1' u_x'\,du_x'}{c^2\,dt'} = \frac{\kappa_1' u_y'\,du_y'}{c^2\,dt'} + \frac{\kappa_1' u_z'\,du_z'}{c^2\,dt'} - \frac{1}{\kappa_1'^2}\frac{d\kappa_1'}{dt'}$$

so that, making the substitution, we have

$$F_x = F'_x + \frac{m_0 v}{c^2 + u'_x v}\left[u'_y\left(\kappa'_1\frac{du'_y}{dt'} + u'_y\frac{d\kappa'_1}{dt'}\right) + u'_z\left(\kappa'_1\frac{du'_z}{dt'} + u'_z\frac{d\kappa'_1}{dt'}\right)\right]$$

$$= F'_x + \frac{u'_y v}{c^2 + u'_x v}F'_y + \frac{u'_z v}{c^2 + u'_x v}F'_z \tag{5}$$

Similar operations for F_y and F_z give

$$F_{y,z} = \frac{c^2}{\kappa(c^2 + u'_x v)}F'_{y,z} \tag{6}$$

These are the transformation equations for force.

14.05. Force on Charge Moving in Magnetic Field.—The first postulate of special relativity requires that the laws of electrostatics shall be identical for all observers. If we assume, in addition, that the magnitude of charges is the same for all observers, we shall see that it follows that the forces observed by S between charges moving in his system differ from electrostatic forces. We shall see from 14.16 that the invariance of charge follows from the invariance of the Maxwell equations. One might call these additional forces electrokinetic forces but, as we shall see, they are identical with those we have already called magnetic forces.

Let us suppose that there are point charges q and q_1 fixed in the $x'y'$-plane at $x' = 0$, $y' = y'_1$ and $x' = x'$, $y' = 0$, respectively, in the S'-system. From Coulomb's law, the components of the force observed by S' to act on q are

$$F'_x = \frac{-qq_1x'}{4\pi\epsilon_v(x'^2 + y_1'^2)^{\frac{3}{2}}}, \qquad F'_y = \frac{qq_1y'_1}{4\pi\epsilon_v(x'^2 + y_1'^2)^{\frac{3}{2}}}, \qquad F'_z = 0 \tag{1}$$

To S, these charges appear to be moving in the positive x-direction with a velocity v, and the forces he observes may be obtained by substituting in 14.04 (5) from 14.01 (7) and (1). To both S and S', the force appears independent of the time; so let us take it at $t = 0$, $t' = -x'vc^{-2}$. This gives, since $u'_x = u'_y = u'_z = 0$, $x' = \kappa x$, etc., the result

$$F_x = \frac{-qq_1\kappa x}{4\pi\epsilon_v(\kappa^2x^2 + y_1^2)^{\frac{3}{2}}}, \qquad F_y = \frac{qq_1y_1}{4\pi\epsilon_v\kappa(\kappa^2x^2 + y_1^2)^{\frac{3}{2}}}, \qquad F_z = 0 \tag{2}$$

Now let us see what forces S observes if q_1 is replaced by an infinite line of charges uniformly spaced along the x-axis and moving with a velocity v. The charge on any element of axis is, by hypothesis, the same for both observers so that the force due to any element is given, by substituting in (2), the value

$$q_1 = \sigma\,dx = \sigma'\,dx' \tag{3}$$

In passing, we should note that the assumption that q_1 is the same for both observers requires, since from 14.01 (7) $dx' = \kappa\, dx$, that

$$\sigma = \kappa\sigma' \tag{4}$$

so that the charge densities observed by S and S' are different. By substituting (3) in (2) and integrating from $x = -\infty$ to $x = +\infty$, we get for the forces on q due to the moving line charge, by *Pc* 138 or *Dw* 200.03 and 201.03, if we write β for v/c,

$$F_x = 0, \qquad F_z = 0, \qquad F_y = \frac{q(1-\beta^2)\sigma}{2\pi\epsilon_v y_1} \tag{5}$$

By putting $\beta = 0$, we find the electrostatic force observed by S to be $q\sigma/(2\pi\epsilon_v y_1)$ so that the additional force, due to the motion, is seen to be

$$\Delta F_y = -\frac{qv^2\sigma}{2\pi\epsilon_v c^2 y_1} = -\frac{\mu_v \sigma v}{2\pi y_1}\cdot qv \tag{6}$$

But the current measured by S is $i = \sigma v$. Furthermore, we have seen in 7.14 (3) that S attributes to this current a magnetic induction in the z-direction at $x = 0$, $y = y_1$ of the amount $B_z = \mu_v i/(2\pi y_1)$. Substituting in (6), we see that the additional force on q due to its motion is

$$\Delta F_y = -B_z q v \tag{7}$$

Therefore, we say that a point charge q, moving with a uniform velocity $\boldsymbol{v}$ at right angles to a uniform field of induction $\boldsymbol{B}$, is acted on by a force at right angles to both $\boldsymbol{v}$ and $\boldsymbol{B}$ given by

$$\boldsymbol{F}_m = q[\boldsymbol{v} \times \boldsymbol{B}] \tag{8}$$

In rectangular coordinates, this is

$$\boldsymbol{F}_m = q[\boldsymbol{i}(v_y B_z - v_z B_y) + \boldsymbol{j}(v_z B_x - v_x B_z) + \boldsymbol{k}(v_x B_y - v_y B_x)] \tag{9}$$

In 14.16, we shall see that (8) also follows from the invariance of Maxwell's equations. If we let $\boldsymbol{v}q = I\, d\boldsymbol{s}$, this is exactly the law of force for an element $d\boldsymbol{s}$ of a circuit carrying a current I in a field of magnetic induction $\boldsymbol{B}$. This law as derived from Ampère's experiments in 7.18 was valid only when integrated around a closed circuit. We have now extended it to include isolated moving charges. We should note that (8) should apply quite accurately when $\boldsymbol{B}$ is variable as regards both time and position provided that the variation is sufficiently slow so that $\boldsymbol{B}$ may be considered uniform and constant over the region occupied by the actual charge involved. We may expect (8) to break down for any actual charged particles if we go to subatomic distances and frequencies.

14.06. Motion of Charge in Uniform Magnetic Field.—Suppose a particle having a charge q and a mass m has an initial velocity $\boldsymbol{v}$ in a magnetic field $\boldsymbol{B}$. Let $\boldsymbol{s}_1$ be a unit vector making an angle α with $\boldsymbol{B}$ such that $\boldsymbol{v} = v\boldsymbol{s}_1$. Then, from 14.05 (8), we have

$$\frac{d(m\boldsymbol{v})}{dt} = \boldsymbol{s}_1\frac{d(mv)}{dt} + mv\frac{d\boldsymbol{s}_1}{dt} = qv[\boldsymbol{s}_1 \times \boldsymbol{B}] \tag{1}$$

Writing out components, we have

$$\boldsymbol{s}_1\frac{d(mv)}{dt} = 0 \tag{2}$$

showing that v is constant, and

$$m\frac{d\boldsymbol{s}_1}{dt} = q[\boldsymbol{s}_1 \times \boldsymbol{B}] \tag{3}$$

showing that $d\boldsymbol{s}_1$ is normal to $\boldsymbol{s}_1$ and $\boldsymbol{B}$ so that we must use the transverse mass 14.04 (4) for m. Let $d\phi$ be the angle, measured in the plane normal to $\boldsymbol{B}$, through which $\boldsymbol{s}_1$ has been turned by the change $d\boldsymbol{s}_1$, and let ρ be the radius of curvature of the path corresponding to $d\phi$. Then, carrying out the multiplication in (3) and dividing through by $s_1 \sin\alpha$, we have

$$\frac{m_t\, ds_1}{s_1 \sin\alpha\, dt} = \frac{m_t\, d\phi}{dt} = \frac{m_t v \sin\alpha}{\rho} = qB \tag{4}$$

Thus, the path is a helix of constant angular pitch $\frac{1}{2}\pi - \alpha$, lying on the circular cylinder whose radius is

$$\rho = \left|\frac{m_0 v \sin\alpha}{qB[1 - (v/c)^2]^{\frac{1}{2}}}\right| \tag{5}$$

and whose axis is parallel to $\boldsymbol{B}$. In the special case, where $\boldsymbol{v}$ is normal to $\boldsymbol{B}$, α is $\frac{1}{2}\pi$ so that the charge travels in a circular path.

14.07. Energy of a Charged Moving Particle.—We define the energy given to a charged particle by the action of a force to be the work done on it. In an infinitesimal interval of time dt, a particle will move a distance $d\boldsymbol{r} = \boldsymbol{u}\, dt$. Thus, from 14.04 (3), the work done on it is

$$dW = \boldsymbol{F} \cdot d\boldsymbol{r} = F_l\, dr = m_0\kappa_1^3\frac{du}{dt}\, dr = m_0\kappa_1^3 u\, du \tag{1}$$

Substitution of 14.03 (7) for κ_1 and integration from 0 to u give

$$W = (m_t - m_0)c^2 = (\kappa_1 - 1)m_0c^2 \tag{2}$$

This proves that an energy increase is equivalent to a mass increase, which has become one of the most accurately and completely verified laws of physics. Confirmation has come chiefly from the vast amount of data on nuclear disintegrations. Thus we have excellent experimental grounds for the statement that a quantity of energy ΔW always has associated with it a mass Δm given by the equation

$$\Delta W = c^2\,\Delta m \tag{3}$$

From (1) and 14.05 (8) a charged particle gets no energy from the magnetostatic part of a static electromagnetic field. Thus it cannot enter regions where the electrostatic potential exceeds the sum of the starting point potential and that needed to give it its initial velocity.

14.08. Magnetic Cutoff of Thermionic Rectifier.—As an example of a calculation of the motion of a charge when both electric and magnetic fields are present, let us consider a device known as the magnetron. The space between a pair of concentric circular conducting cylinders is evacuated, and charged particles, usually electrons, are set free at the surface of the inner cylinder with negligible initial velocity. The cylinders are maintained at different potentials so that the charges are accelerated from the inner, of radius a, toward the outer, of radius b. A magnetic field is applied parallel to the axis of the cylinders. The induction $\boldsymbol{B}$ is a function only of ρ, the distance from the axis. We wish to find the induction $\boldsymbol{B}$ that is just sufficient to prevent particles from reaching the outer cylinder when the potential is V. If ϕ is the longitude angle, then the charges will just reach the plate when $d\rho/dt = \dot{\rho} = 0$ there, so that we have $b\,d\phi/dt = b\dot{\phi} = v_m$. Setting the rate of change of angular momentum equal to the torque acting, we have

$$\frac{d}{dt}(m\rho^2\dot{\phi}) = -q\rho B\dot{\rho} \tag{1}$$

At the beginning of the path, $\dot{\phi} = 0$, and at the end, $\rho\dot{\phi} = v$, $\rho = b$, and $m = \kappa_1 m_0$. Integrating (1) over this path, we have

$$\kappa_1 m_0 b v = -q\int_a^b \rho B\,d\rho \tag{2}$$

If N is the total magnetic flux passing between the cylinders, then $N = 2\pi\int_a^b \rho B\,d\rho$ so that this can be written

$$\kappa_1 m_0 b v = -\frac{Nq}{2\pi} \tag{3}$$

Since the charge has acquired all the energy by falling through the potential V, we have, from 14.07 (2),

$$Vq = (\kappa_1 - 1)m_0 c^2 \tag{4}$$

Eliminating κ_1 and v from (3) by means of (4) gives

$$N = -2\pi bc\left[\frac{V}{c^2}\left(\frac{2m_0}{q} + \frac{V}{c^2}\right)\right]^{\frac{1}{2}} \tag{5}$$

If the potential is sufficiently small, the second term on the right can be neglected. In terms of potential, (5) becomes

$$V = \frac{m_0c^2}{q}\left[\left(1 + \frac{N^2q^2}{4\pi^2c^2b^2m_0^2}\right)^{\frac{1}{2}} - 1\right] \tag{6}$$

14.09. Path of Cosmic Particle in Uniform Field.—As an example of the calculation of a path of a charged particle influenced by both electrical and mechanical forces, we shall consider a particle, traveling in a plane normal to a uniform field ***B***, in a medium that opposes the motion of the particle with a force proportional to its velocity. It is found experimentally that the high-speed charged particles of cosmic origin dissipate their energy nearly uniformly along their paths so that the retarding force is that assumed. Let us write down the components of the equation of motion from 14.01 (1) and 14.05 (8), giving

$$\frac{d}{dt}(m_0\kappa_1\dot{y}) = -qB\dot{x} - K\dot{y} \tag{1}$$

$$\frac{d}{dt}(m_0\kappa_1\dot{x}) = qB\dot{y} - K\dot{x} \tag{2}$$

We now integrate these two equations from any point on the path to the origin, where $\dot{x} = \dot{y} = 0$ and $x = y = 0$ and take the ratio of the results, obtaining

$$\frac{\dot{y}}{\dot{x}} = \frac{dy}{dx} = \frac{-qBx - Ky}{qBy - Kx} \tag{3}$$

Setting $y = r \sin \theta$ and $x = r \cos \theta$, this becomes

$$\frac{dr}{r} = \frac{K}{qB}d\theta \tag{4}$$

Choosing $\theta = 0$ when $r = 1$ and integrating from this point to any point on the path, we have

$$\ln r = \frac{K}{qB}\theta \qquad \text{or} \qquad r = e^{\frac{K\theta}{qB}} \tag{5}$$

Thus, the path is an equiangular spiral. For a high-speed cosmic electron, K roughly equals $3qP \times 10^{-4}$ where P is the air pressure in atmospheres.

14.10. Magnetic Field of Moving Charge.—Let us consider again the force between two moving charges determined in 14.05 (2). In view of the conclusions of 14.05, we may say that the force that S observes to act on the charge crossing the y-axis due to the charge moving along the x-axis he ascribes to the combined electric and magnetic fields of the latter. The x-component which acts in the direction of motion of q must be considered entirely electrostatic, but the y-component includes both electric and magnetic parts. Thus, we have

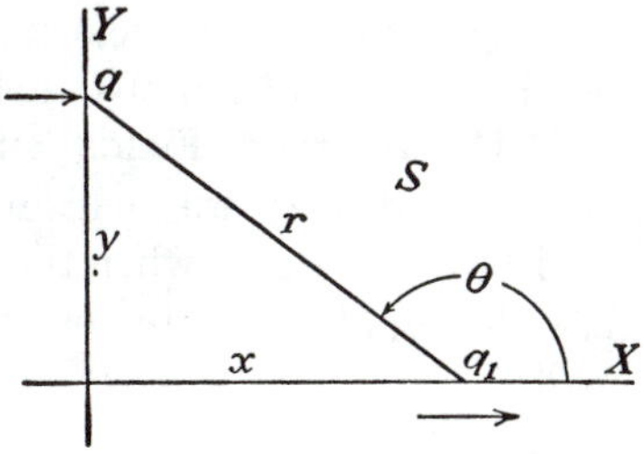

FIG. 14.10.

$$F_x = F_e \cos\theta, \qquad F_y = F_e \sin\theta - F_m \tag{1}$$

$$F_m = F_x \tan\theta - F_y = -F_x \frac{y}{x} - F_y \tag{2}$$

We use the angle θ in Fig. 14.10 since, as we shall see in 14.16, the aberration of a signal received by q from q_1 is zero. Substituting in the values of the forces from 14.05 (2), we obtain

$$F_m = \frac{qq_1 y(\kappa^2 - 1)}{4\pi\epsilon_v \kappa(\kappa^2 x^2 + y^2)^{\frac{3}{2}}} = \frac{qq_1\kappa v^2 y}{4\pi\epsilon_v c^2(\kappa^2 x^2 + y^2)^{\frac{3}{2}}} \tag{3}$$

S observes that this force does not vary with time and concludes that, since q is a point charge, it is effectively moving in a uniform field so that it is legitimate to compute the magnetic field in which it moves by means of 14.05 (8) and (3). Thus, the magnetic field due to q_1 is

$$B_z = \frac{\mu_v q_1 v y \kappa}{4\pi(\kappa^2 x^2 + y^2)^{\frac{3}{2}}} \tag{4}$$

By definition, the electric field acting along r is, substituting from (1),

$$E = \frac{F_e}{q} = -\frac{F_x r}{qx} = \frac{q_1 \kappa r}{4\pi\epsilon_v(\kappa^2 x^2 + y^2)^{\frac{3}{2}}} \tag{5}$$

By applying in the formulas of 11.01,

$$\boldsymbol{B}_z = (\boldsymbol{\nabla} \times \boldsymbol{A}_x)_z = -\boldsymbol{k}\frac{\partial A_x}{\partial y}$$

$$\boldsymbol{E} = -\boldsymbol{\nabla}\phi - \frac{d\boldsymbol{A}}{dt}$$

$$E = \left[\left(-\frac{\partial\phi}{\partial x} + v\frac{\partial A_x}{\partial x}\right)^2 + \left(\frac{\partial\phi}{\partial y}\right)^2\right]^{\frac{1}{2}}$$

we can easily verify that the vector and scalar potentials corresponding to (4) and (5) are

$$A_x = \frac{\mu_v q_1 \kappa v_x}{4\pi(\kappa^2 x^2 + y^2)^{\frac{1}{2}}} \tag{6}$$

$$\phi = \frac{q_1 \kappa}{4\pi\epsilon_v(\kappa^2 x^2 + y^2)^{\frac{1}{2}}} \tag{7}$$

It should be noted that (4), (5), (6), and (7) apply to a point charge moving with uniform velocity in a straight line and are stated in terms of the actual position of the charge at the instant of measurement.

14.11. Retarded Fields and Potentials of Moving Charge.—Since electric and magnetic fields are propagated with a finite velocity c, the field at Q at a time t when the charge that produced it is at P will actually have left q_1 at some previous point on its path indicated by $[P]$. If, *after leaving* $[P]$, the charge had changed its motion, the field at Q at the time t would still be the same. Thus, in the case of nonuniform motion, it is better to describe the field at a time t in terms of its motion at a time $t - ([r]/c)$, called the retarded time, where $[r]$ is the radius vector drawn from the retarded position $[P]$ to the point of measurement Q. Since the field at Q is independent of the path taken by q_1 after leaving $[P]$, we shall assume that its state of motion at $[P]$ continues unchanged so that it moves in a straight line with a uniform velocity $[v]$ to a fictitious position P'. We now calculate the field at Q for this uniform motion by the method of the last article. From Fig. 14.11, we can write down the necessary relations between the fictitious and the retarded magnitudes

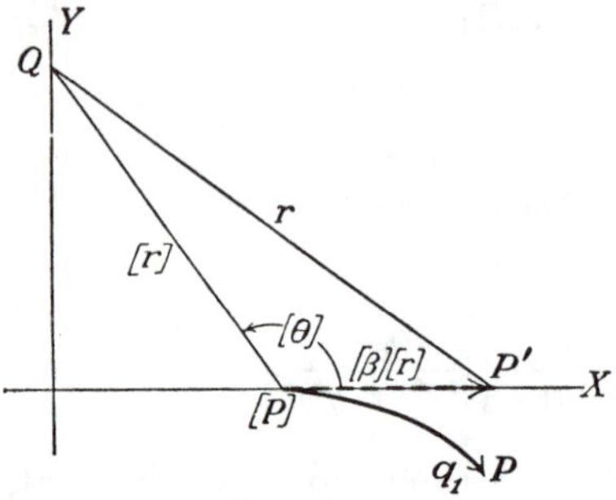

FIG. 14.11.

$$r^2 = [r]^2(1 + [\beta]^2 - 2[\beta] \cos [\theta]) \tag{1}$$
$$y^2 = [y]^2 = [r]^2 \sin^2 [\theta] \tag{2}$$

where $[\beta]$ is $[v]/c$ and $[\boldsymbol{\beta}]$ will be $[\mathbf{v}]/c$. The binomial denominator factor of 14.10 (3) becomes, when $[\kappa]$ is written for $(1 - [\beta]^2)^{-\frac{1}{2}}$,

$$\kappa^2 x^2 + y^2 = [\kappa]^2(r^2 - [\beta]^2 y^2) = [\kappa]^2[r]^2(1 - [\beta] \cos [\theta])^2 \tag{3}$$

Substitution of (2) and (3) into 14.10 (4) and (5) gives $\boldsymbol{B}$ and $\boldsymbol{E}$ for a moving charge in terms of retarded time. Thus

$$\boldsymbol{k}B_z = \frac{\boldsymbol{k}\mu_v q_1[v][y]}{4\pi[\kappa]^2[r]^3(1 - [\beta] \cos [\theta])^3} = \frac{\mu_v q_1[\mathbf{v}] \times [r]}{4\pi[\kappa]^2([r] - [r] \cdot [\boldsymbol{\beta}])^3} = \frac{[\mathbf{r}] \times \boldsymbol{E}}{[r]c} \tag{4}$$

This also applies to a slowly varying velocity provided we use the values $[\beta]$ and $[\boldsymbol{v}]$ at the position $[P]$. The electric field of 14.10 acts along $\boldsymbol{r}$, and Fig. 14.11 gives the vector relation

$$\boldsymbol{r} = [\boldsymbol{r}] - [r][\boldsymbol{\beta}] = [r]([\boldsymbol{r}_1] - [\boldsymbol{\beta}])$$

where $[\boldsymbol{r}_1]$ is a unit vector along $[\boldsymbol{r}]$. Thus 14.10 (5) may be written in terms of the retarded values $[\beta]$, $[\boldsymbol{\beta}]$, $[\boldsymbol{r}_1]$, and $[\boldsymbol{v}]$.

$$\boldsymbol{E} = \frac{q_1(1 - [\beta]^2)([\boldsymbol{r}_1] - [\boldsymbol{\beta}])}{4\pi\epsilon_v[r]^2(1 - [\beta]\cos[\theta])^3} = \frac{q_1([\boldsymbol{r}_1] - [\boldsymbol{\beta}])}{4\pi\epsilon_v[\kappa]^2[r]^2[p]^3} \tag{5}$$

$$[p] = 1 - [\beta]\cos\theta = 1 - [\boldsymbol{r}_1]\cdot[\boldsymbol{\beta}] \tag{6}$$

Substitution of (1) and (2) into 14.10 (6) and (7) gives for the retarded or Lienard-Wiechert potentials due to a moving point charge

$$\boldsymbol{A} = \frac{\mu_v q_1[\boldsymbol{v}]}{4\pi[r][p]} = \frac{q_1[\boldsymbol{\beta}]}{4\pi\epsilon_v c[r][p]}, \qquad \phi = \frac{q_1}{4\pi\epsilon_v[r][p]} \tag{7}$$

When $[\beta]$ is small, these expressions obviously are identical with those given by 12.02 (1) and (2) for small charge. When $[\beta]$ is large, we can obtain (7) from 12.02 (1) and (2) by taking account of the variation in the retarded time within the infinitesimal volume of the charge. This variation cannot be neglected even when the charge shrinks to a point. Mason and Weaver give a full discussion of this point.

14.12. Radiation from an Accelerated Charge.—The results of the last article may be applied to the accelerated motion of an actual charge such as an electron provided that the acceleration is so small that the variation of retarded time throughout the charge may be neglected. If this is not the case, it is necessary to know the configuration of the latter in order to find the field and this, at our present state of knowledge, is not possible. Let us suppose that the charge is sufficiently small so 14.11 (7) may be used and compute the fields from these potentials.

Assume that the retarded position of the charge q_1 is known as a function of the retarded time $[t]$. Thus the data given are

$$[\boldsymbol{v}] = -\frac{\partial[\boldsymbol{r}]}{\partial[t]}, \qquad \frac{\partial[\boldsymbol{v}]}{\partial[t]} = \frac{\partial^2[\boldsymbol{r}]}{\partial[t]^2} \tag{1}$$

The field of the accelerated charge will be found from 14.11 (7) by the familiar formulas 11.01 (2) and (3), so that

$$\boldsymbol{E} = -\frac{\partial \boldsymbol{A}}{\partial t} - \nabla\phi, \qquad \boldsymbol{B} = \nabla \times \boldsymbol{A} \tag{2}$$

Here the differentiations are carried out at the field point, so $\partial/\partial t$ is evaluated with $[r]$ constant and ∇ with $[t]$ constant. By definition $[r]$

is $c(t - [t])$. Therefore it follows that, using 14.11 (6),

$$\frac{\partial[r]}{c\,\partial t} = 1 - \frac{\partial[t]}{\partial t} = \frac{\partial[r]}{\partial[t]}\frac{\partial[t]}{c\,\partial t} = -[\mathbf{r}_1]\cdot[\boldsymbol{\beta}]\frac{\partial[t]}{\partial t},$$

$$\frac{\partial[t]}{\partial t} = \frac{1}{[p]}, \qquad \frac{\partial[r]}{\partial[t]} = -[\mathbf{r}_1]\cdot[\boldsymbol{\beta}]c \tag{3}$$

Let ∇' be the operator with $[t]$ constant. Then

$$\nabla[r] = -c\nabla[t] = \nabla'[r] + \nabla[t]\frac{\partial[r]}{\partial[t]} = [\mathbf{r}_1] - [\mathbf{r}_1]\cdot[\boldsymbol{\beta}]c\nabla[t]$$

$$\nabla[t] = -\frac{[\mathbf{r}_1]}{c[p]}, \qquad \nabla = \nabla' - \frac{[\mathbf{r}_1]}{[p]}\frac{\partial}{c\,\partial[t]} \tag{4}$$

Application of the operators (3) and (4) to (2) gives

$$\mathbf{E} = -\frac{1}{[p]}\frac{\partial \mathbf{A}}{\partial[t]} - \nabla'\phi + \frac{[\mathbf{r}_1]}{[p]c}\frac{\partial\phi}{\partial[t]} \tag{5}$$

$$\mathbf{B} = \nabla' \times \mathbf{A} + \frac{[\mathbf{r}_1]}{c[p]} \times \frac{\partial \mathbf{A}}{\partial[t]} \tag{6}$$

The component of $[\mathbf{v}]$ normal to $[\mathbf{r}_1]$ which is the vector difference of $[\mathbf{v}]$ and the component, $[\mathbf{r}_1][\mathbf{r}_1]\cdot[\mathbf{v}]$, of $[\mathbf{v}]$ along $[\mathbf{r}_1]$, divided by $[r]$ is the time rate of change of $[\mathbf{r}_1]$, so that

$$\frac{\partial[\mathbf{r}_1]}{c\,\partial[t]} = \frac{[\mathbf{r}_1][\mathbf{r}_1]\cdot[\boldsymbol{\beta}] - [\boldsymbol{\beta}]}{[r]} \tag{7}$$

$$\frac{\partial([r][p])}{c\,\partial[t]} = -[\mathbf{r}_1]\cdot[\boldsymbol{\beta}] + [\beta]^2 - [\mathbf{r}]\cdot\frac{[\dot{\boldsymbol{\beta}}]}{c}, \qquad [\dot{\boldsymbol{\beta}}] = \frac{\partial[\boldsymbol{\beta}]}{\partial[t]} \tag{8}$$

$$\nabla'([r][p]) = \nabla'([r] - [\mathbf{r}]\cdot[\boldsymbol{\beta}]) = [\mathbf{r}_1] - [\boldsymbol{\beta}] \tag{9}$$

Substitution for $\mathbf{A}$ and ϕ from 14.11 (7) and for the derivative from (8) and (9) in (5) gives for the electrical intensity

$$\mathbf{E} = \frac{q_1}{4\pi\epsilon_v}\left[\frac{-[\dot{\boldsymbol{\beta}}]}{c[r][p]^2} + \frac{[\mathbf{r}_1] - [\boldsymbol{\beta}]}{[r]^2[p]^3}\left\{[p] - \frac{\partial([r][p])}{c\,\partial[t]}\right\}\right]$$

$$= \frac{q_1}{4\pi\epsilon_v}\left\{\frac{([\mathbf{r}_1] - [\boldsymbol{\beta}])(1 - [\beta]^2)}{[r]^2[p]^3} + \frac{[\mathbf{r}_1]\cdot[\dot{\boldsymbol{\beta}}]([\mathbf{r}_1] - [\boldsymbol{\beta}]) - [p][\dot{\boldsymbol{\beta}}]}{c[r][p]^3}\right\} \tag{10}$$

$$\mathbf{E} = \frac{q_1}{4\pi\epsilon_v}\left\{\frac{([\mathbf{r}_1] - [\boldsymbol{\beta}])(1 - [\beta]^2)}{[r]^2[p]^3} + \frac{[\mathbf{r}_1] \times \{([\mathbf{r}_1] - [\boldsymbol{\beta}]) \times [\dot{\boldsymbol{\beta}}]\}}{c[r][p]^3}\right\} \tag{11}$$

As before $[\mathbf{r}]$ is $[\mathbf{r}_1][r]$, $[\boldsymbol{\beta}]$ is $[\mathbf{v}]/c$, $[p]$ is $1 - [\mathbf{r}_1]\cdot[\boldsymbol{\beta}]$, and $[\mathbf{r}_1]$, $[\boldsymbol{\beta}]$, and $[p]$ are dimensionless. Similar substitution in (6) gives

$$\mathbf{B} = \frac{q_1}{4\pi\epsilon_v}\left\{\frac{([\boldsymbol{\beta}] \times [\mathbf{r}_1])(1 - [\beta]^2)}{c[r]^2[p]^3} + \frac{[\mathbf{r}_1] \times ([\mathbf{r}_1] \times \{([\mathbf{r}_1] - [\boldsymbol{\beta}]) \times [\dot{\boldsymbol{\beta}}]\})}{c^2[r][p]^3}\right\} \tag{12}$$

Note that $\boldsymbol{B}$ is normal to $\boldsymbol{E}$ and $[\boldsymbol{r}_1]$ because

$$c\boldsymbol{B} = [\boldsymbol{r}_1] \times \boldsymbol{E} \tag{13}$$

Observe that the $[\dot{\boldsymbol{\beta}}]$ term in (10) or (11) is proportional to $[r]^{-1}$, so that the Poynting vector integral over an infinite sphere does not vanish and represents energy radiated by the accelerated charge. The remaining term in (11) is identical with 14.11 (5) and thus, together with the remaining term in (12), represents the fields of a charge moving with uniform velocity in a straight line.

14.13. Charge Radiation When Acceleration Parallels Velocity.—When $[\dot{\boldsymbol{v}}]$ is parallel or antiparallel to $[\boldsymbol{v}]$, the radiation part of $\boldsymbol{B}$ by 14.12 (12) is

$$\boldsymbol{B} = \frac{q[\boldsymbol{r}_1] \times \{[\boldsymbol{r}_1] \times ([\boldsymbol{r}_1] \times [\dot{\boldsymbol{\beta}}])\}}{4\pi c^2\epsilon_v[r][p]^3} = \frac{-q[\boldsymbol{r}_1] \times [\dot{\boldsymbol{\beta}}]}{4\pi c^2\epsilon_v[r][p]^3} \tag{1}$$

since $[\boldsymbol{\beta}] \times [\dot{\boldsymbol{\beta}}] = 0$. In terms of polar coordinates r, θ, ϕ, as shown in Fig. 14.13*a*, with $[\dot{\boldsymbol{v}}]$ taken antiparallel, this becomes

$$B_\phi = \frac{-q[\dot{v}] \sin [\theta]}{4\pi\epsilon_v c^3[r](1 - [\beta] \cos [\theta])^3} \tag{2}$$

When an electron strikes the target of an X-ray tube, $[\boldsymbol{v}]$ and $[\dot{\boldsymbol{v}}]$ are antiparallel and the target penetration depth is very small compared with

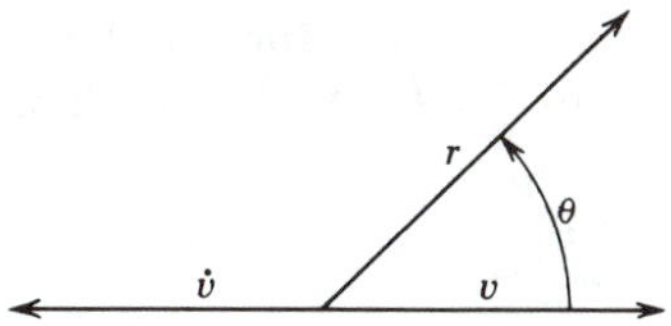

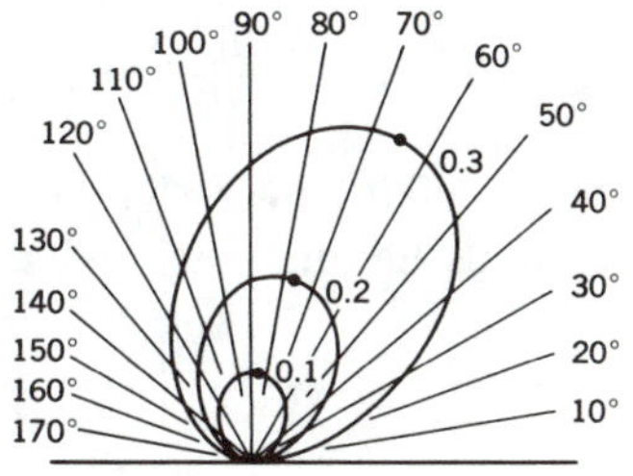

FIG. 14.13*a*, *b*.

the observation distance. Therefore the wave is spherical, $[r] \to r$, $[\theta] \to \theta$, and cB_ϕ equals E_θ. This formula will now be used to find the total radiated energy density for a given θ when the electron is stopped. The instantaneous Poynting vector 11.02 (3) becomes, since $(\mu_v\epsilon_v)^{-1}$ is c^2,

$$\Pi = \frac{\boldsymbol{E} \times \boldsymbol{B}}{\mu_v} = \frac{q^2[\dot{v}]^2 \sin^2\theta[\boldsymbol{r}_1]}{16\pi^2c^3\epsilon_v r^2(1 - [\beta] \cos \theta)^6} \tag{3}$$

The total energy per pulse $\boldsymbol{T}(\theta,r)$ passing radially through unit area at $\boldsymbol{r}$, θ is found by integration of Π over the transit time t_0, corresponding to a deceleration time $[t_0]$. Thus from 14.12 (3)

$$T = \int_0^{t_0} \mathbf{\Pi}\, dt = \int_0^{t_0} \mathbf{\Pi}[p]d[t] = \int_0^{t_0} \mathbf{\Pi}(1 - [\beta] \cos \theta)d[t] \tag{4}$$

Write $d[v]$ for $[\dot{v}]d[t]$ and integrate by *Dw* 90.5. Thus

$$\begin{aligned} T_r &= \frac{q^2[\dot{v}] \sin^2 \theta}{16\pi^2 c^3 \epsilon_v r^2} \int_0^{[v]} \frac{d[v]}{\{1 - ([v]/c) \cos \theta\}^5} \\ &= \frac{q^2[\dot{v}] \sin^2 \theta}{64\pi^2 c^2 \epsilon_v r^2 \cos \theta} \left[\frac{1}{(1 - [\beta] \cos \theta)^4} - 1 \right] \end{aligned} \tag{5}$$

The assumption that the deceleration of the electron is constant, used to derive (5), is certainly not rigorously true at the target of an X-ray tube, but the result agrees roughly with the intensity measurements of the properly polarized part of the continuous X-rays or bremsstrahlung. Figure 14.13*b* shows a polar plot of T_r as a function of θ as calculated from (5) by Sommerfeld. Curves for three values of β (0.1, 0.2, and 0.3) for the same $[\dot{v}]$ are shown, and the angle for maximum T_r is indicated by a dot on the curve.

Another application of (3) is to the calculation of the power lost due to radiation by electrons in a linear accelerator. The power, measured at the charge, radiated per unit area is $d\Pi/d[t]$, and when multiplied by $r^2\, d\Omega$ this gives the power radiated through the solid angle $d\Omega$ at an angle θ.

$$\frac{d[P]}{d\Omega} = \frac{q^2[\dot{v}]^2 \sin^2 \theta}{16\pi^2 \epsilon_v c^3 (1 - [\beta] \cos \theta)^5} \tag{6}$$

Integration of (6) over the whole solid angle, using x for $\cos \theta$ so $d\Omega$ is $-2\pi\, dx$ and the limits are $+1$ and -1, gives, by *Dw* 90.5 and 92.5,

$$[P] = \frac{q^2[\dot{v}]^2}{6\pi \epsilon_v c^3 (1 - [\beta]^2)^3} \tag{7}$$

One is usually interested in (6) when $[\beta]$ is nearly 1 so $\sin \theta \approx \theta$ is near zero, and from 14.01 (6), $(1 - [\beta])^{-1} = \kappa^2(1 + [\beta])$ is very close to $2\kappa^2$. Writing $1 - \frac{1}{2}\theta^2$ for $\cos \theta$ gives $1 - [\beta] \cos \theta$ the form $\frac{1}{2}\kappa^{-2}(1 + \kappa^2\theta^2)$ so

$$\frac{d[P]}{d\Omega} \xrightarrow[{[\beta] \to 1}]{} \frac{2q^2[\dot{v}]^2 \kappa^2 (\kappa\theta)^2}{\pi^2 \epsilon_v c^3 (1 + \kappa^2 \theta^2)^5} \tag{8}$$

This has its maximum at $\kappa\theta = \frac{1}{2}$, where from 14.01 (6) its value is

$$\left[\frac{d[P]}{d\Omega}\right]_{\max} \xrightarrow[{[\beta] \to 1}]{} \frac{512 q^2 [\dot{v}]^2}{3125\pi^2 \epsilon_v c^2 (1 - [\beta]^2)^4} \tag{9}$$

14.14. Charge Radiation with Acceleration at Right Angles to Velocity. Radiation losses may be very important when high-speed charged particles are bent in magnetic fields as is the case in cyclotrons and synchrotrons. As in the last article one first finds Poynting's vector by squaring the radiation term in 14.12 (10), noting that $[\boldsymbol{\beta}] \cdot [\dot{\boldsymbol{\beta}}]$ is now zero, that $B = E/c$, and that $(\mu_v \epsilon_v)^{-1} = c^2$. The result is

$$\Pi = \frac{E^2}{c\mu_v} = \frac{q^2\{[p]^2[\dot{v}]^2 - ([\boldsymbol{r}_1] \cdot [\dot{\boldsymbol{v}}])^2(1 - [\beta]^2)\}}{16\pi^2\epsilon_v c^3[r]^2[p]^6} \tag{1}$$

Let the orbit and hence $[\dot{\boldsymbol{v}}]$ and $[\boldsymbol{v}]$ lie in the xy-plane of Fig. 14.14*a*, and let the observer be in the yz-plane so $[\boldsymbol{r}_1]$ makes an angle α with the z-axis. The component of $[\boldsymbol{r}_1]$ in the xy-plane is of length $\sin \alpha$ and parallels the y-axis, so that $[\boldsymbol{r}_1] \cdot [\dot{\boldsymbol{\beta}}]$ is $-[\dot{\beta}] \sin \alpha \sin \phi$ and $[\boldsymbol{r}_1] \cdot [\boldsymbol{\beta}]$ is $[v] \sin \alpha \cos \phi$

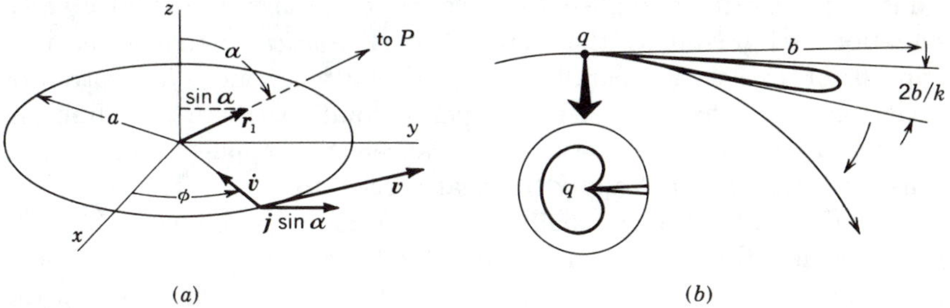

Fig. 14.14*a*, *b*.

where ϕ is the angle between the orbit radius to the electron and the x-axis. Then $[p]$ is $1 - [\beta] \sin \alpha \cos \phi$ from 14.11 (6), and $d[P]/d\Omega$ is calculated from Π as 14.13 (6) is from 14.13 (3) and yields

$$\frac{d[P]}{d\Omega} = \frac{q^2[\dot{v}]^2}{16\pi^2\epsilon_v c^3(1 - [\beta] \sin \alpha \cos \phi)^3}\left\{1 - \frac{(1 - [\beta]^2) \sin^2 \alpha \sin^2 \phi}{(1 - [\beta] \sin \alpha \cos \phi)^2}\right\} \tag{2}$$

For the radiated power integrate (2) over Ω, with $0 < \alpha < \pi$ and $0 < \phi < 2\pi$. The following coordinate change facilitates integration

$$\sin \alpha \cos \phi = \frac{x}{r} \rightarrow \frac{z'}{r} = \cos \theta'$$

$$\sin \alpha \sin \phi = \frac{y}{r} \rightarrow \frac{x'}{r} = \sin \theta' \cos \phi'$$

$$\cos \alpha = \frac{z}{r} \rightarrow \frac{y'}{r} = \sin \theta' \sin \phi'$$

$$\frac{d[P]}{d\Omega} = \frac{q^2[\dot{v}]^2}{16\pi^2\epsilon_v c^3(1 - [\beta] \cos \theta')^3}\left\{1 - \frac{(1 - [\beta]^2) \sin^2 \theta' \cos^2 \phi'}{(1 - [\beta] \cos \theta')^2}\right\} \tag{3}$$

The integral of the second term gives 14.13 (7) multiplied by $-\frac{1}{2}(1 - [\beta]^2)$, and the first term integral is 14.13 (7) multiplied by $\frac{3}{2}(1 - [\beta]^2)$, so that

$$[P] = \frac{q^2[\dot{v}]^2}{6\pi\epsilon_v c^3(1 - [\beta]^2)} \tag{4}$$

It is interesting to compare the radiation loss when a force F acts on a charged particle in the direction of its velocity to that when the same force acts at right angles to it. When $[\dot{v}]^2$ is eliminated from 14.13 (7) by means of 14.04 (3), there results

$$P_l = \frac{q^2\kappa^6}{6\pi\epsilon_v c^3}\left(\frac{F}{m_0\kappa^3}\right)^2 = \frac{q^2F^2}{6\pi\epsilon_v m_0^2c^3} \tag{5}$$

When $[\dot{v}]^2$ is eliminated from (4) by 14.04 (4), the result is

$$P_t = \frac{q^2\kappa^4}{6\pi\epsilon_v c^3}\left(\frac{F}{m_0\kappa}\right)^2 = \frac{q^2\kappa^2F^2}{6\pi\epsilon_v m_0^2c^3} \tag{6}$$

Thus the same force applied for acceleration transverse to $[\mathbf{v}]$ gives a radiation loss greater by the factor κ^2 or $(1 - [v/c]^2)^{-1}$ than when it is applied for acceleration parallel to $[\boldsymbol{v}]$, so that when any synchrotron or cyclotron has a fixed acceleration per revolution, there is a limiting velocity at which the radiation loss cancels this applied acceleration. This limitation does not apply to linear accelerators.

As with 14.13 (8), one usually wants the radiation intensity near its peak, which in this case lies on a tangent to the circle in Fig. 14.14*a*. For a distant observer in the yz-plane, when $[\beta] \to 1$, this flash leaves the electron when $\alpha = \frac{1}{2}\pi$, $\phi = \omega[t] = 0$, or from (3) at $\theta' = 0$. Putting into (3) the relations connecting 14.13 (7) and (8) gives

$$\frac{d[P]}{d\Omega}\underset{[\beta]\to 1}{\longrightarrow} \frac{q^2[\dot{v}]^2\kappa^6}{2\pi^2\epsilon_v c^3(1+\kappa^2\theta'^2)}\left\{1 - \left(\frac{2\kappa\theta'\cos\phi'}{1+\kappa^2\theta'^2}\right)^2\right\} \tag{7}$$

Evidently the intensity is a maximum when the second term in parentheses is zero at $\theta' = 0$ and is zero when this term is 1. Thus

$$\left(\frac{d[P]}{d\Omega}\right)_{\max} = \frac{q^2[\dot{v}]^2}{2\pi^2\epsilon_v c^3(1-[\beta]^2)^3} \tag{8}$$

In the plane of the loop from (3), $z = 0$, $\phi' = 0$, $\alpha = \frac{1}{2}\pi$. Thus, if $\kappa\theta' = \pm 1$,

$$\frac{d[P]}{d\Omega} = 0 \qquad \text{and} \qquad \theta'_{\min} = \pm(1-[\beta]^2)^{\frac{1}{2}} = \phi_{\min} = \omega_0[t]_{\min} \tag{9}$$

From (8) and (9), at high velocities where $[\beta] \to 1$, the beam in the forward direction becomes extremely intense and narrow, as shown in Fig. 14.14*b*.

This beam, like a rotating searchlight, sweeps across an observer in the plane of the orbit once per revolution. The interval between the two $[t]_{\min}$ values bounding the peak is, from (9),

$$\Delta[t] = \frac{2(1-[\beta]^2)^{\frac{1}{2}}}{\omega_0} = \frac{d[t]}{dt}\Delta t = \frac{\Delta t}{1-[\beta]} \tag{10}$$

where $d[t]/dt$ from 14.12 (3) has been inserted. For high velocities, $[\beta] \to 1$ and $1 - [\beta]$ is written $\frac{1}{2}(1 + [\beta])(1 - [\beta]) = \frac{1}{2}(1 - [\beta]^2)$, so that the interval Δt in which the beam sweeps across the observer is

$$\Delta t = \frac{(1 - [\beta]^2)^{\frac{3}{2}}}{\omega_0} = \frac{1}{\kappa^3 \omega_0} \tag{11}$$

There is a general rule that when a pulse of duration Δt is expressed as a Fourier integral in ω, the frequency range needed is ω_0 to $\omega_0 + \Delta\omega$ where $\Delta\omega\,\Delta t$ is approximately 1. Thus the highest frequency seen will be, roughly, neglecting ω_0 compared with $\Delta\omega$,

$$\omega_{\max} \sim \Delta\omega \sim (1 - [\beta]^2)^{-\frac{3}{2}}\omega_0 = \kappa^3\omega_0 \tag{12}$$

Many different approximations may be used in the calculation of the intensity as a function of ω. Much depends on the distribution of electrons in the orbit and whether they radiate coherently or not. If the spacing of N electrons around the orbit is perfectly random and they radiate incoherently, then the loss is N times that of a single electron. Julian Schwinger in *Phys. Rev.*, Vol. 75, pages 1912 to 1925, 1949, gives a detailed treatment of this case.

14.15. Cherenkov Radiation.—When a particle moves through a medium with a velocity $[v]$ exceeding $(\mu\epsilon)^{-\frac{1}{2}}$, then even with $[v]$ constant,

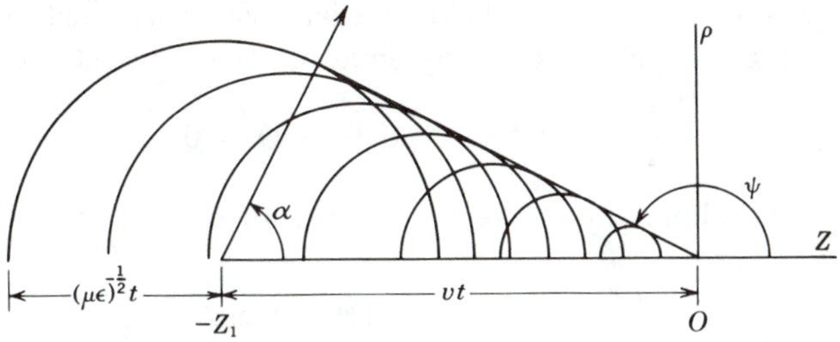

FIG. 14.15.

it loses energy by radiation. Figure 14.15 gives a physical picture of what occurs when $[v] > (\mu\epsilon)^{-\frac{1}{2}}$, so that the particle outruns its radiation and shows a section of the half plane $\phi = 0$. The particle path with constant $[v]$ is shown from $-z_1$ to the origin O. Seven semicircles with equidistant centers indicate the locus at $t = 0$ of the pulses generated at these points by the particle. It is evident from the figure that there is one direction α, normal to the cone of half angle $\frac{1}{2}\pi - \alpha$, in which all pulses are in phase. Each pulse can be resolved into a band of frequencies or Fourier components. The field without dispersion would be infinite at the conical sheet of thickness zero, but actually dispersion will make the fields finite and of thin, but not zero, thickness. The normal direction of this shock wave is

$$\mu\epsilon[v]^2 = \cos^2 \alpha \tag{1}$$

The potentials of 14.11 (11) can be modified to give the Cherenkov field when the particle moves in a straight line with uniform velocity $[v]$. A cylindrical coordinate system ρ, ϕ, z is used with the origin at the particle when $t = 0$. The equation for $[r]$ is

$$[r]^2 = (\mu\epsilon)^{-1}(t - [t])^2 = (z - [v][t])^2 + \rho^2 \tag{2}$$

Solve the last two members of this equation for $[t]$, subtract $[t]$ from t, and then use the first two members to get

$$[r] = \frac{(\mu\epsilon)^{\frac{1}{2}}[v](z - [v]t) \pm \{(z - [v]t)^2 + \rho^2(1 - \mu\epsilon[v]^2)\}^{\frac{1}{2}}}{1 - \mu\epsilon[v]^2} \tag{3}$$

If the velocity of light $(\mu\epsilon)^{-\frac{1}{2}}$ in the medium exceeds $[v]$, then $\mu\epsilon[v]^2$ is less than 1 and both terms under the radical are positive. Thus its value exceeds that of the first term and its sign must be positive to make $[r]$ positive. This is the case treated in 14.11. If $\mu\epsilon[v]^2$ is greater than 1, then the second term in the radical and the denominator in (3) are negative. If the square root is real, its magnitude is less than that of the first term. Therefore $[r]$ is positive only if $[v]t < z$, and thus z must be negative in Fig. 14.15 where t is zero, for a field to exist. Since $[r]$ must be real, the curve bounding the field is found by equating the radical to zero. Thus

$$(z - [v]t)^2 - \rho^2(\mu\epsilon[v]^2 - 1) = 0 \tag{4}$$

The slope of this bounding curve is constant for

$$\frac{dz}{d\rho} = -(\mu\epsilon[v]^2 - 1)^{\frac{1}{2}} = \tan \psi \tag{5}$$

for all values of t, since the cone moves with the particle without change of slope. Note that $dz/d\rho = z/\rho$ when $t = 0$, so that when $t = 0$ the particle is at the origin. Also (5) agrees with (1) because $\psi - \frac{1}{2}\pi = \alpha$.

It can be shown that the $[r][p]$ of 14.11 (6) and (7) becomes

$$[r][p] = \pm\{(z - [v]t)^2 + \rho^2(1 - \mu\epsilon[v]^2)\}^{\frac{1}{2}} \tag{6}$$

Thus there are two potentials for every point inside the cone. This is also evident from Fig. 14.15, where two wavelets intersect at every point. The absolute value of $[r][p]$ must be used, so the total potential ϕ is

$$2\pi\epsilon\phi = q\{(z - [v]t)^2 + \rho^2(1 - \mu\epsilon[v]^2)\}^{-\frac{1}{2}} \tag{7}$$

which is infinite on the cone with no dispersion. The vector potential is

$$A_z = \mu\epsilon[v]\phi \tag{8}$$

The electric field components are given by

$$2\pi E_z = \mu_v q(z - [v]t)\{(\mu\epsilon)^{-1} - [v]^2\}|[r][p]|^{-\frac{3}{2}} \tag{9}$$
$$2\pi E_\rho = \mu_v q\rho\{(\mu\epsilon)^{-1} - [v]^2\}|[r][p]|^{-\frac{3}{2}} \tag{10}$$

These formulas show that $\boldsymbol{E}$ lies in the cone in Fig. 14.15 where $t = 0$. From symmetry the magnetic induction is B_ϕ which also lies in its surface.

Equation (1) shows that measurement of the angle α at which light of known wave length and velocity is emitted when a high-speed particle traverses a medium; $\mu\epsilon$ gives the velocity of the particle. The weak radiation from single particles can be amplified for particle-counting purposes. A very complete discussion of the theoretical and experimental aspects of this subject is given by J. V. Jelley (see references).

14.16. Transformation of Maxwell's Equations.—The first postulate of special relativity requires that the Maxwell equations have the same form for both S and S'. We have now extended these to include moving isolated charges; thus we put $\rho\boldsymbol{u}$ in place of $\boldsymbol{i}$ in 11.00 (1), giving for the equations that must be satisfied in free space

$$\begin{aligned}
&\nabla \times \frac{\boldsymbol{B}}{\mu_v} = \rho\boldsymbol{u} + \epsilon_v\frac{\partial \boldsymbol{E}}{\partial t} && \nabla \times \frac{\boldsymbol{B}'}{\mu_v} = \rho'\boldsymbol{u}' + \epsilon_v\frac{\partial \boldsymbol{E}'}{\partial t'}\\
&\nabla \times \boldsymbol{E} = -\frac{\partial \boldsymbol{B}}{\partial t} && \nabla \times \boldsymbol{E}' = -\frac{\partial \boldsymbol{B}'}{\partial t'}\\
&\nabla \cdot \boldsymbol{E} = \rho/\epsilon_v && \nabla \cdot \boldsymbol{E}' = \rho'/\epsilon_v\\
&\nabla \cdot \boldsymbol{B} = 0 && \nabla \cdot \boldsymbol{B}' = 0
\end{aligned} \tag{1}$$

From 14.01 (7), we have the differential relations

$$\frac{\partial x'}{\partial x} = \kappa, \qquad \frac{\partial t'}{\partial x} = -\kappa\frac{v}{c^2}, \qquad \frac{\partial x'}{\partial t} = -\kappa v, \qquad \frac{\partial t'}{\partial t} = \kappa \tag{2}$$

Thus, we get the relations

$$\begin{aligned}
\frac{\partial}{\partial x} &= \frac{\partial}{\partial x'}\frac{\partial x'}{\partial x} + \frac{\partial}{\partial t'}\frac{\partial t'}{\partial x} = \kappa\left(\frac{\partial}{\partial x'} - \frac{v}{c^2}\frac{\partial}{\partial t'}\right)\\
\frac{\partial}{\partial t} &= \frac{\partial}{\partial x'}\frac{\partial x'}{\partial t} + \frac{\partial}{\partial t'}\frac{\partial t'}{\partial t} = \kappa\left(\frac{\partial}{\partial t'} - v\frac{\partial}{\partial x'}\right)\\
\frac{\partial}{\partial y} &= \frac{\partial}{\partial y'}, \qquad \frac{\partial}{\partial z} = \frac{\partial}{\partial z'}
\end{aligned} \tag{3}$$

Changing the independent variables in the first group of (1) from the S- to the S'-system by means of (3) and using the velocity relations from 14.02, we find that the substitution of $\boldsymbol{E}$, $\boldsymbol{B}$, and ρ from the following equations gives the second group in (1) and they are therefore the desired transformation equations.

$$E_x = E'_x \qquad E_{y,z} = \kappa(E'_{y,z} \pm vB'_{z,y}) \tag{4}$$

$$B_x = B'_x \qquad B_{y,z} = \kappa(B'_{y,z} \mp c^{-1}\beta E'_{z,y}) \tag{5}$$

$$\rho = \rho'\kappa\left(1 + \frac{u'_x v}{c^2}\right) \tag{6}$$

where the lower sign goes with the second subscript. These equations show that the way in which an electromagnetic field is separated into electric and magnetic parts depends on the state of motion of the observer. To get $\boldsymbol{E}'$, $\boldsymbol{B}'$, and ρ' in terms of $\boldsymbol{E}$, $\boldsymbol{B}$, and ρ, we need interchange only primed and unprimed quantities and change the sign of β and v.

We can show that (6) implies the invariance of electric charge assumed in 14.05. Elimination of the last factor in Eq. (6) by 14.03 (9) gives $\rho\kappa'_1 = \rho'\kappa_1$ where κ_1 and κ'_1 are defined by 14.03 (7) and (8). Thus, if S_0 is an observer stationary with respect to the charge,

$$\rho_0 = \rho\left[1 - \left(\frac{u}{c}\right)^2\right]^{\frac{1}{2}} = \rho'\left[1 - \left(\frac{u'}{c}\right)^2\right]^{\frac{1}{2}} \tag{7}$$

The volume elements seen by S and S' are contracted by (2), so that

$$ds_0 = \left[1 - \left(\frac{u}{c}\right)^2\right]^{-\frac{1}{2}} ds = \left[1 - \left(\frac{u'}{c}\right)^2\right]^{-\frac{1}{2}} ds' \tag{8}$$

Combining (7) and (8) shows that S and S' observe the same charge, for

$$\rho_0\, ds_0 = \rho\, ds = \rho'\, ds' = q \tag{9}$$

We can also show that (4) and (5) lead to the expression for the force acting on a charge moving in a magnetic field given in 14.05 (8). Suppose that S' observes a charge at rest in an electrostatic field in his system, giving

$$F'_{x,y,z} = q'E'_{x,y,z} \tag{10}$$

Using the transformation equation for force [14.05 (5)], setting $u_x = v$, $u_y = 0$, and $u_z = 0$, and using (4) and (5), we have, since $q = q'$,

$$F_x = qE_x, \qquad F_{y,z} = q(E_{y,z} \mp u_x B_{z,y}) \tag{11}$$

Writing this in vector form gives

$$\boldsymbol{F} = q[\boldsymbol{E} + (\boldsymbol{u} \times \boldsymbol{B})] \tag{12}$$

which agrees with 14.05 (8).

14.17. Ground Speed of an Airplane.—One of the most difficult experimental problems in aviation is that faced by an aviator in determining his speed relative to the ground when conditions are such that he can see nothing outside his airplane. Since an airplane flying horizontally is cutting the vertical component of the earth's field, an electromotive force, proportional to its speed, must be induced across any lateral conductor on the airplane. The idea has been suggested and patented that ground speed can be determined by measuring this electromotive force. Several methods of doing this suggest themselves, one of the most obvious of which is to rotate an elongated conductor in a horizontal plane about its center. For a uniform speed of rotation, the induced charges are constant if the airplane is stationary. If, however, the latter is in motion, the induced electromotive force is always, in the Northern Hemisphere, from right to left. Thus, there is a transfer of charge from one end of the conductor to the other, producing at its center an alternating current whose frequency is that of the rotation and whose magnitude is proportional to the ground speed of the airplane.

This current is very small but could be measured by sensitive instruments in a well-shielded laboratory. The question arises as to whether the effect disappears if the apparatus is electrically shielded in the airplane. We know that the magnetic field will penetrate nonmagnetic metallic conductors, but we also know that the induced electromotive forces in the shields will set up electric fields tending to counteract the fields induced inside. So far, we have looked at this problem from the point of view of a stationary observer S on the ground and found it quite complicated. Let us now see how it looks to the observer S' in the airplane who is to make the measurements. Let z be the vertical direction and x the direction of motion and let the components of the earth's field be B_x, B_y, and B_z. There is also usually a vertical electric field present which we shall call $\bar{E}_z$. These are the fields observed by S, from which we determine by 14.11 (4), the magnetic field observed by S' to be

$$B'_x = B_x, \qquad B'_{y,z} = \kappa(B_{y,z} \pm c^{-1}\beta\bar{E}_{z,y}) \approx B_{y,z} \tag{1}$$

Since the speed of an airplane is small compared with that of light, $\beta \approx 0$, and $\kappa \approx 1$ so that the magnetic fields observed by S' are unchanged. For the electric fields, we have

$$E'_x = 0, \qquad E'_y = -\kappa v B_z \approx -vB_z, \qquad E'_z = \kappa\bar{E}_z + \kappa v B_y \approx \bar{E}_z \tag{2}$$

Thus, S' finds his airplane in a transverse electric field vB_z. He cannot therefore use any metallic shields about his apparatus. Furthermore, this field is extremely minute compared with $\bar{E}_z$, the normal atmospheric electric field, so that the slightest lateral tilt of the airplane will completely nullify the measurements.

14.18. Motion of Charged Particle in Crossed Electric and Magnetic Fields.—In the last article, we had an example of the simplification of a problem by using the transformation equations 14.16 (4), (5), and (6) to introduce a field. We shall now use these same equations to eliminate a field. Consider a charged particle starting from the origin with initial velocity components v_x, v_y, and v_z under the influence of a uniform electric field in the y-direction and a uniform magnetic field in the z-direction. For an observer moving along the x-axis with a uniform velocity v, the fields will appear to be, from 14.16 (4) and (5),

$$E'_x = 0, \qquad E'_y = \kappa E_y - \kappa v B_z, \qquad E'_z = 0 \tag{1}$$

$$B'_x = 0, \qquad B'_y = 0, \qquad B'_z = \kappa(B_z - c^{-1}\beta E_y) \tag{2}$$

Thus if $E_y < cB_z$, we choose the velocity with which S' moves such that

$$E_y = vB_z \tag{3}$$

so he observes only a single uniform magnetic field in the z-direction, given by (2), to be present. If the charge is released when the origins are coincident, then initially $t = t' = x = x' = y = y' = z = z' = 0$. For S', the components of the initial velocity are v'_x, v'_y, and v'_z. From 14.06 (5), he observes the particle to move in a helical path of pitch γ' on a circular cylinder of radius a, with an angular velocity ω', where

$$a = \left|\frac{\kappa'_1 m_0 (v'^2_x + v'^2_y)^{\frac{1}{2}}}{qB'_z}\right| \qquad \text{and} \qquad \tan\gamma' = \frac{v'_z}{(v'^2_x + v'^2_y)^{\frac{1}{2}}} \tag{4}$$

If we make use of 14.03 (7), (8), and (10), κ'_1 may be written

$$\kappa'_1 = \left(1 - \frac{v'^2_x + v'^2_y + v'^2_z}{c^2}\right)^{-\frac{1}{2}} = (1-\beta^2)^{-\frac{1}{2}}\left(1 - \frac{v^2_x + v^2_y + v^2_z}{c^2}\right)^{-\frac{1}{2}}\left(1 - \beta\frac{v_x}{c}\right)$$

$$= \kappa\kappa_1\left(1 - \beta\frac{v_x}{c}\right) = \frac{\kappa\kappa_1(c^2B - v_xE)}{c^2B} \tag{5}$$

Substituting from (2) and (5) in (4), we obtain

$$a = \frac{\kappa_1 m_0 c[c^2(v_xB - E)^2 + (c^2B^2 - E^2)v_y^2]^{\frac{1}{2}}}{q(c^2B^2 - E^2)} \tag{6}$$

The angular velocity about the axis of the cylinder is

$$\omega' = \frac{(v'^2_x + v'^2_y)^{\frac{1}{2}}}{a} = \frac{q(c^2B^2 - E^2)}{\kappa_1 m_0 (c^2B - v_xE)} \tag{7}$$

where, looking in the positive z-direction, ω' is counterclockwise if q is positive.

In most cases, $E > v_xB$ so that v'_x is negative. In this case, a line normal to the z-axis and tangent to the cylinder at the origin makes an acute angle ψ_0 with the negative x'-axis, where, from 14.02 (3) and (4),

$$\tan\psi_0 = -\frac{v'_y}{v'_x} = \frac{v_y}{\kappa(v - v_x)} = \frac{v_y(c^2B^2 - E^2)^{\frac{1}{2}}}{c(E - v_xB)} \tag{8}$$

We can write down the position of the charge at any time in the S'-system.

$$x' = a[-\sin\omega't'\cos\psi_0 + (1 - \cos\omega't')\sin\psi_0] = a[\sin\psi_0 - \sin(\omega't' + \psi_0)]$$
$$y' = a[\sin\omega't'\sin\psi_0 + (1 - \cos\omega't')\cos\psi_0] = a[\cos\psi_0 - \cos(\omega't' + \psi_0)]$$

Let us now introduce new symbols ψ and b which are defined by

$$\psi = \omega't' + \psi_0 = \omega'\kappa\left(t - \frac{xE}{c^2B}\right) + \psi_0 \tag{9}$$

$$b = \frac{v}{\omega'} = \frac{\kappa_1 m_0 E(c^2B - v_xE)}{qB(c^2B^2 - E^2)} \tag{10}$$

Writing x and y in terms of x', y', and t' by 14.01 (7) and writing t' in terms of ψ and b, we have

$$x\left(1 - \frac{E^2}{c^2B^2}\right)^{\frac{1}{2}} = b\psi - b\psi_0 + a\sin\psi_0 - a\sin\psi \tag{11}$$

$$y = a\cos\psi_0 - a\cos\psi \tag{12}$$

These are the rigorous equations of the path of the particle in terms of the parameter ψ. We notice that when $\psi = 2n\pi + \psi_0$, the position of the particle is

$$x_n = 2n\pi b\left(1 - \frac{E^2}{c^2B^2}\right)^{-\frac{1}{2}}, \qquad y_n = 0$$

Thus, the paths of all particles starting from the origin with the same value of b intersect periodically at the same points along the z-axis.

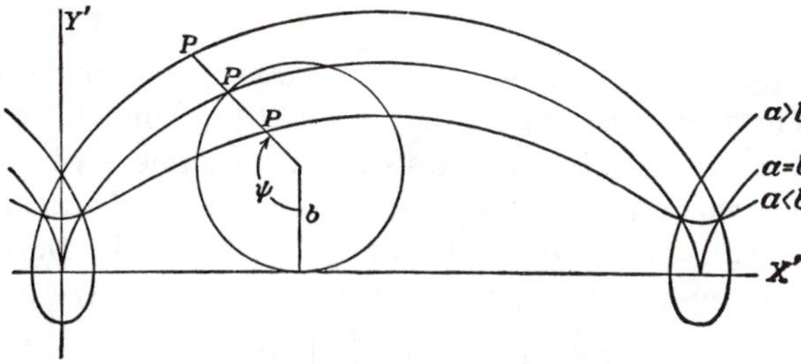

FIG. 14.18.

Furthermore, we notice from (10) that if the initial velocities are small compared with that of light, the distance between these points of intersection depends only on the field strengths and the ratio of the charge to the mass of the particles. This principle is utilized in one type of mass spectrometer.

With an electric field of 100,000 volts per meter and with an induction of 0.1 weber per square meter, the quantity $E^2/(cB)^2$ occurring in (11) is 1/90,000. In such a case, (11) and (12) become the parametric equations of a trochoid, which is the curve traced by a point on the radius of a rolling circle. To make the circle roll on the x'-axis symmetrically, we choose a new origin so that

$$x' = x + b\psi_0 - a\sin\psi_0 \qquad \text{and} \qquad y' = y - a\cos\psi_0 + b$$

We thus obtain the paths shown in Fig. 14.18 in which we see that b is the radius of this rolling circle and a is the distance along this radius from its center to P, the point tracing the curve. The cycloidal path, traced when $a = b$, was used by J. J. Thomson, in 1899, in finding e/m_0 for photoelectrons.

14.19. Aberration and Doppler Effect.—In 11.01, we saw that all the properties of an electromagnetic wave can be obtained from the Hertzian vector $\boldsymbol{Z}$ which, in free space, satisfies the propagation equation 11.01 (10).

$$\nabla^2\boldsymbol{Z} = \frac{1}{c^2}\frac{d^2\boldsymbol{Z}}{dt^2} \tag{1}$$

A general solution of this equation for a plane wave was given in 11.03 (1). In the case of a monochromatic wave of frequency ν, we may write this, using 11.09 (4), since $v = c$,

$$\boldsymbol{Z} = \boldsymbol{A}\cos\frac{2\pi\nu}{c}(\boldsymbol{n}\cdot\boldsymbol{r} - ct) \tag{2}$$

This represents the ray as observed by S. The same ray will appear to S' to be of frequency ν' and to be propagated in some direction $\boldsymbol{n}'$, so that

$$\boldsymbol{Z}' = \boldsymbol{A}'\cos\frac{2\pi\nu'}{c}(\boldsymbol{n}'\cdot\boldsymbol{r}' - ct') \tag{3}$$

the velocity, by the second postulate of 14.00, appearing the same to S and S'. The transformation equations 14.01 (7) must transform (3) into (2), and, as we see, this requires that the argument of the cosine in (3) when expressed in terms of ν, $\boldsymbol{n}$, $\boldsymbol{r}$, and t must be identical with that of the cosine in (2). To hold for all values of x, y, z, and t, the coefficients of each of these must be the same in both arguments. We may write

$$\boldsymbol{n}\cdot\boldsymbol{r} - ct = lx + my + nz - ct \tag{4}$$

Using the transformation equation 14.01 (7), we have, by rearranging terms,

$$\begin{aligned}\boldsymbol{n}'\cdot\boldsymbol{r}' - ct' &= l'x' + m'y' + n'z' - ct' \\ &= \kappa\left(l' + \frac{v}{c}\right)x + m'y + n'z - \kappa(c + l'v)t\end{aligned} \tag{5}$$

Equating the coefficients mentioned, we have

$$\frac{\kappa\left(l' + \frac{v}{c}\right)}{l} = \frac{m'}{m} = \frac{n'}{n} = \frac{\kappa(c + l'v)}{c} = \frac{\nu}{\nu'} \tag{6}$$

These equations enable us to find the effect of the motion of a source of electromagnetic radiation relative to an observer upon the measurements of its radiation made by him. The directional effect is known as aberration and the frequency effect as the Doppler effect. Let the source be at rest in the S'-system and the observer be at rest in the S-system. Then the light that appears to S to come from the l,m,n direction will appear to S' to travel in the l',m',n' direction, where the relations between l,m,n and l',m',n' are given by (6). Equate the first and fourth fractions in (6), put $l = \cos\theta$ and $l' = \cos\theta'$, and we have

$$\cos\theta = \frac{\cos\theta' + \beta}{1 + \beta\cos\theta'} \tag{7}$$

where $\beta = v/c$. This is the rigorous formula for aberration. Equating the last two fractions, we obtain

$$\nu = \frac{1 + \beta\cos\theta'}{(1 - \beta^2)^{\frac{1}{2}}}\nu' \tag{8}$$

This is the rigorous expression for the change in the observed frequency of electromagnetic radiation when the source is moving relative to the observer. If the source is approaching the observer, $\cos\theta'$ is positive, and if it is receding, $\cos\theta'$ is negative so that in the first case the frequency is apparently increased and in the second decreased. From the fundamental postulates, the aberration and Doppler effect appear only when there is a relative motion of source and observer. Thus (7) and (8) are equally valid if the observer is moving and the source is stationary. We might expect to find deviations from these formulas at astronomical distances since the special theory of relativity has been completely verified only on a terrestrial scale, but they should be precise on this scale.

Problems

1. A long straight filament of radius a carries a current I and emits electrons with negligible initial velocity which are accelerated by a potential V toward a long concentric cylindrical anode of radius b. Show that the relation between I, a, b, and V such that the electrons just fail to reach the anode is $V = [\mu^2 I^2 q/(8\pi^2 m)] \ln^2 (b/a)$, where the dependence of m on the velocity is neglected.

2. Treat the preceding problem rigorously when $E_\rho > cB_\phi$ by transforming to a system of coordinates moving parallel to the axis of the cylinder with such a velocity that the magnetic field disappears. Show by this means that it is impossible for an electron, once leaving the filament, to return.

3. Treat the preceding problem rigorously when $E_\rho < cB_\phi$ by transforming to a system of axes moving parallel to the axes of the cylinder with such a velocity that the electric field disappears. Show that the relation such that electrons just fail to reach the anode is

$$2\pi m_0 c(A^{\frac{1}{2}} - v_1') = \mu I q(c^2 - A)^{\frac{1}{2}}(1 - \beta_1^2)^{\frac{1}{2}} \ln\frac{b}{a}$$

where $A = v_1'^2 + v_2'^2$ and v_1' and v_2' are related to the initial velocities v_1 and v_2 in the x and ρ directions by the relations

$$v_1' = \frac{v_1 - c\beta_1}{1 - \dfrac{v_1\beta_1}{c}}, \qquad v_2' = \frac{v_2(1-\beta_1^2)^{\frac{1}{2}}}{1 - \dfrac{v_1\beta_1}{c}}$$

where $v_1 = E_\rho/B_\phi$ and we neglect the initial v_ϕ.

4. A vacuum tube contains a cylindrical cathode of radius r_0 which is surrounded by a coaxial anode of radius r_1. This tube is placed with its axis on the axis of rotation of the confocal hyperbolic pole pieces of an electromagnet. Show that the potential which must be applied in order that electrons, in the plane of symmetry, reach the anode, if the magnetic field in this plane is given by $B_0 b(r^2 + b^2)^{-\frac{1}{2}}$, is rigorously

$$\frac{m_0c^2}{q}\left(\left\{1 + \frac{B^2q^2b^2}{m_0^2c^2r_1^2}[(r_1^2 + b^2)^{\frac{1}{2}} - (r_0^2 + b^2)^{\frac{1}{2}}]^2\right\}^{\frac{1}{2}} - 1\right)$$

5. A parallel beam of electrons which have been accelerated through a potential V carries a current I. If the section of the beam is circular of radius a, show that the acceleration, normal to the beam, given to an electron at its surface due to electric and magnetic forces is

$$\dot{v}_n = \frac{I}{4\pi\epsilon_v a}\left(\frac{2q}{mV}\right)^{\frac{1}{2}}\left[\left(1 + \frac{Vq}{2mc^2}\right)^{\frac{1}{2}}\left(1 + \frac{Vq}{mc^2}\right)^2\right]^{-1}$$

6. A point charge q moves in the field of a stationary point charge Q. Using the conservation of angular momentum p and of energy and the expression 14.07 (2) for kinetic energy, show that the equation of its path is $r^{-1} = A + B\cos\gamma\phi$, where

$$\gamma^2 = 1 - \left(\frac{qQ}{4\pi\epsilon_v cp}\right)^2$$

7. A line normal to two plane sheets at a distance a apart in a vacuum is concentric with a hole of diameter b in each sheet. A beam of charged particles, all having the same energy, but having a maximum divergence angle α from the normal, emerges from the first hole. Neglecting α compared with unity, find the set of values of the magnetic field which, when applied normal to the sheets, will make these particles pass through the second hole. Show that the maximum diameter of the beam between the holes is then $[2a/(n\pi)]\sin\alpha + b$, where n is an integer.

8. If, in the last problem, all particles comprising the cone originate at the same point on the axis and the magnetic field includes this point, show that the minimum value of the field which will bring these particles to a focus at the image point beyond the second hole is $2\pi mv\alpha[q(a\alpha + b)]^{-1}$, where α is the angle subtended at the source by any radius of the first aperture. Show also that this field increases the number of ions passing the first aperture by the factor $\{\pi b/[2(a\alpha + b)]\}^2\{\sin[\frac{1}{2}\pi b/(a\alpha + b)]\}^{-2}$.

9. Two similar circular parallel conducting cylindrical shells carry charges $+Q$ and $-Q$ per unit length. Inside each, a wire is so situated that when a current I flows through them in opposite directions each shell coincides with a magnetic line of force. Show how, if $cQ < I$, it is possible to transform to a set of moving axes such that only a magnetic field appears.

10. In the last problem, show that, if a charged particle starts from rest on one cylinder, the component of its velocity parallel to the cylinder at r_1, r_2, is given by

$$v_z = I \ln \frac{r_1 r_{20}}{r_2 r_{10}} \left(Q \ln \frac{r_1 r_{20}}{r_2 r_{10}} + \frac{2\pi m_0}{\mu q} \right)^{-1}$$

here r_{10} and r_{20} are the distances from the starting point to the two wires.

11. In a two-dimensional magnetostatic field where the vector potential has only a z-component, let the total velocity of an ion of charge q and rest mass m_0 be v and let its z-component be v_z, at a point where the vector potential is A_1. Show that, when it is at a point where the vector potential is A_2, the z-component of its velocity is given by $v_z - q(m_0 c)^{-1}(c^2 - v^2)^{\frac{1}{2}}(A_2 - A_1)$.

12. Show that, in any two-dimensional case in which the electric and magnetic fields are orthogonal, it is possible to choose a set of moving axes such as to eliminate either $\boldsymbol{E}$ or $c\boldsymbol{B}$ whichever is the smaller.

13. The position of a charge q is given by $s = a \cos \omega t$. If $\omega a c^{-1} \ll 1$ and $a \ll r$, show, by using 14.11 (7), that the periodic parts of $\boldsymbol{E}$ and $\boldsymbol{B}$ due to q are identical with 12.01 (6), (7), and (8) where r is the distance from its mean position to the field point, θ is the angle between r and s, and $M = qa$.

14. The positions s_1 and s_2 of two identical charges q are $s_1 = -s_2 = 2a \cos(\frac{1}{2}\omega t)$. Proceeding as in the last problem, taking $ar^{-1} \ll \omega a c^{-1} \ll 1$, and letting $Q_0 = qa^2$, show that the periodic part of the field at a great distance is

$$cB_\phi = E_\theta = -\frac{\omega^3 Q_0 \sin 2\theta}{8\pi\epsilon_v c^3 r} \sin \omega\left(t - \frac{r}{c}\right)$$

Compare with the linear quadrupole in problem 1, Chap. XII.

15. An electron moves in an axially symmetrical magnetostatic field such that

$$A_\phi(\rho, z) \neq 0, \qquad A_\rho = 0, \qquad A_z = 0$$

Show that if its angular velocity about the axis at the point $\rho = \rho_1$ and $z = z_1$ is ω_1 and if at the point $\rho = \rho_2$ and $z = z_2$ it is ω_2, then

$$m\rho_1^2\omega_1 - m\rho_2^2\omega_2 = e[\rho_1 A_\phi(\rho_1, z_1) - \rho_2 A_\phi(\rho_2, z_2)]$$

16. At $t = t_0$ and x-directed particle of charge q, mass m_0, and velocity v_0 enters at $x = 0$, $y = \frac{1}{2}b$, $z = 0$ an evacuated retangular wave guide transmitting a TE_{01} wave. Show that its equation of motion for an observer moving with the signal velocity is

$$\frac{d(mv')}{dt'} = \frac{c\beta_{01} q E_0}{\omega} \sin(c\beta_{01} t')$$

where E_0 is the maximum electric field intensity in the stationary coordinate system. Show that, if $v_x \ll c$, its position at any time t in this system is

$$x = qE_0(\omega^2 m_0)^{-1}[\omega(t - t_0)\cos \omega t_0 - \sin \omega t + \sin \omega t_0] + v_0(t - t_0), \qquad y = \tfrac{1}{2}b, \qquad z = 0$$

References

A. Electrical

Abraham, M., and R. Becker: "The Classical Theory of Electricity and Magnetism," (trans. 14th German ed.), Hafner, 1950.

Jackson, J. D.: "Classical Electrodynamics," Wiley, 1962. More extensive and detailed treatment than that given here.

JEANS, J. H.: "The Mathematical Theory of Electricity and Magnetism," Cambridge, 1925. Gives fairly extensive treatment.

JELLEY, J. V.: "Čerenkov Radiation," Pergamon, 1958. Most extensive coverage of theory and experiment.

JONES, D. S.: "The Theory of Electromagnetism," Macmillan, 1964. Treatment somewhat different from most in places.

LORENTZ, H. A.: "The Theory of Electrons," 2d ed., Dover, 1952. Written by a participant in the classical development.

MASON, M., and W. WEAVER: "The Electromagnetic Field," University of Chicago Press, 1929, and Dover. A nonrelativistic treatment of the fields of moving charges.

PANOFSKY, W. K. H., and MELBA PHILLIPS: "Classical Electricity and Magnetism," 2d ed., Addison-Wesley, 1962. More detailed and extensive treatment than that given here.

SOMMERFELD, A.: "Electrodynamics: Lectures on Theoretical Physics," Vol. III, Academic, 1952. A lucid treatment.

STRATTON, J. A.: "Electromagnetic Theory," McGraw-Hill, 1941. Treats the vector transformations in first chapter.

TOLMAN, R. C.: "Relativity, Thermodynamics and Cosmology," Oxford, 1934. Gives clear and detailed treatment of special relativity.

TRALLI, N.: "Classical Electromagnetic Theory," McGraw-Hill, 1963. Very concise and lucid treatment.

ZWORYKIN, V. K., G. A. MORTON, E. G. RAMBERG, J. HILLIER, and A. W. VANCE: "Electron Paths and the Electron Microscope," Wiley, 1945. Gives details of path tracing of charged particles in electrostatic and magnetostatic fields. Extensive references to periodic literature.

B. MATHEMATICAL

See Chap. V.

APPENDIX

SYSTEMS OF ELECTRICAL UNITS

There are three ways in which absolute systems of electrical units may be grouped. According to the mechanical units used, they belong either to the centimeter-gram-second (cgs) system or to the meter-kilogram-second (mks) system. According to the magnitudes of the units, they belong to either the classical or the practical system. According to the way in which a 4π factor enters, they belong to either the rationalized or the unrationalized system. The mks units used in this book belong to the practical rationalized mks group. The tables in this Appendix give the interrelations between these units, practical unrationalized cgs units, and classical unrationalized cgs units.

Equivalent formulas in all the rationalized systems are very similar. But the way in which the 4π is inserted in defining the units in the Georgi mks system differs from the method used in the corresponding classical systems in that it is so arranged that the rationalization does not affect the common practical units such as the coulomb and volt. This is done by inserting a 4π in the capacitivity and the permeability. Since all numerical tables listing the electric and magnetic properties of materials give the relative capacitivity $K = \epsilon/\epsilon_v$ and the relative permeability $K_m = \mu/\mu_v$, it is often the practice, in the practical system, to write for these quantities $K\epsilon_v$ and $K_m\mu_v$, where ϵ_v and μ_v are the values, for a vacuum, of the capacitivity, dielectric constant, or permittivity and of the permeability.

If we are given a formula in one system of units, we can always write it in any other system by substituting in the given formula the values for the same quantities expressed in the desired units. To make this clear, let us, as an example, express Coulomb's law for the force between two charges in classical unrationalized cgs units. A given physical quantity, when expressed in cgs esu, will be written without subscripts and the same quantity, measured in Georgi mks units, with subscript 1. From Table II,

$$q_1 = 3^{-1}10^{-9}q \qquad \text{and} \qquad \epsilon_1 = (36\pi)^{-1}10^{-1}\epsilon \tag{1}$$

From Table I, for mechanical units, we see that

$$F_1 = 10^{-5}F \qquad \text{and} \qquad r_1 = 10^{-2}r \tag{2}$$

The formula 1.01 (1) for Coulomb's law, which is in mks units, is

$$F_1 = \frac{q_1q_1'}{4\pi\epsilon_1r_1^2} = \frac{q_1q_1'}{4\pi K\epsilon_vr_1^2} \tag{3}$$

Substitution of F_1, q_1, ϵ_1, and r_1 from (1) and (2) and simplification give

$$F = \frac{qq'}{\epsilon r^2} \tag{4}$$

The same procedure can, of course, be used for the conversion of magnetic formulas. An elaborate discussion of the latter is given by Kennelly in *Am. Phil. Soc. Proc.*, Vol. 76, page 343 (1936). We notice that, if we put ϵ equal to 1 in (3) and (4), the formulas are still different. Thus it is important, even if the result is to be applied to a vacuum, that the formula to be transposed, if in esu or emu, be written for a medium in which μ and ϵ are not unity.

TABLE I

A. A formula given in cgs units is expressed in mks units by replacing each symbol by its value in the right-hand column.

Energy............	W joule	10^7W ergs
Force............	F newton	10^5F dynes
Length............	l meter	10^2l centimeters
Mass............	m kilogram	10^3m grams
Power............	P watt	10^7P ergs/sec

B. A formula given in rationalized mks units is expressed in cgs units by replacing each symbol by its value in the extreme right-hand column.

Energy..........	W erg	$10^{-7}W$ joule
Force..........	F dyne	$10^{-5}F$ newton
Length..........	l centimeter	$10^{-2}l$ meter
Mass..........	m gram	$10^{-3}m$ kilogram
Power..........	P erg/sec	$10^{-7}P$ watt

TABLE II

A formula given in rationalized mks units is expressed in cgs esu or Gaussian units by replacing each symbol by its value in the extreme right-hand column. $c \approx 3 \times 10^{10}$. In very precise work each factor 3 is replaced by 2.9979.

Quantity	Esu	Emu	Practical	
Capacitance............	C	$c^{-2}C$	$9^{-1}10^{-11}C$ farad	
Capacitivity............	ϵ	$c^{-2}\epsilon$	$9^{-1}10^{-11}\epsilon$ 4π-farad/cm	$(36\pi)^{-1}10^{-9}\epsilon$ farad/meter
Charge, quantity.......	q	$c^{-1}q$	$3^{-1}10^{-9}q$ coulomb	
Conductance...........	G	$c^{-2}G$	$9^{-1}10^{-11}G$ mho	
Conductivity, surface...	γ'	$c^{-2}\gamma'$	$9^{-1}10^{-11}\gamma$ mho	
Conductivity, volume...	γ	$c^{-2}\gamma$	$9^{-1}10^{-11}\gamma$ mho/cm	$9^{-1}10^{-9}\gamma$ mho/meter
Current density, surface.	i'	$c^{-1}i'$	$3^{-1}10^{-9}i'$ amp/cm	$3^{-1}10^{-7}i'$ amp/meter
Current density, volume	i	$c^{-1}i$	$3^{-1}10^{-9}i$ amp/cm²	$3^{-1}10^{-5}i$ amp/meter²
Current...............	I	$c^{-1}I$	$3^{-1}10^{-9}I$ ampere	
Displacement..........	D	$c^{-1}D$	$3^{-1}10^{-9}D$ 4π-coul/cm²	$(12\pi)^{-1}10^{-5}D$ coul/m²
Elastance.............	S	c^2S	$9 \times 10^{11}S$ darafs	
Electromotance.........	$\mathcal{E}$	$c\mathcal{E}$	$300\mathcal{E}$ volts	
Field intensity, electric..	E	cE	$300E$ volts/cm	$30{,}000E$ volts/meter
Impedance............	Z	c^2Z	$9 \times 10^{11}Z$ ohms	
Inductance............	L	c^2L	$9 \times 10^{11}L$ henrys	
Potential, electric.......	V	cV	$300V$ volts	
Reactance.............	X	c^2X	$9 \times 10^{11}X$ ohms	
Resistance.............	R	c^2R	$9 \times 10^{11}R$ ohms	
Resistivity, surface.....	s	c^2s	$9 \times 10^{11}s$ ohms	
Resistivity, volume.....	τ	$c^2\tau$	$9 \times 10^{11}\tau$ ohm-cms	$9 \times 10^9\tau$ ohm-meters

Table III

A formula given in rationalized mks units is expressed in cgs emu or Gaussian (starred) units by replacing each symbol with its value in the extreme right-hand column. $c = 2.9979 \times 10^{10} \approx 3 \times 10^{10}$.

Quantity	Emu	Esu	Practical	
Capacitance	C	c^2C^*		10^9C farads
Charge, quantity	q	cq^*		$10q$ coulombs
Conductance	G	c^2G^*		10^9G mhos
Conductivity, surface	γ'	$c^2\gamma'^*$		$10^9\gamma'$ mhos
Conductivity, volume	γ	$c^2\gamma^*$	$10^9\gamma$ mhos/cm	$10^{11}\gamma$ mhos/meter
Current density, surface	i'	ci'^*	$10i'$ amps/cm	$10^3i'$ amps/meter
Current density, volume	i	ci^*	$10i$ amps/cm^2	10^5i amps/meter^2
Current	I	cI^*		$10I$ amperes
Elastance	S	$c^{-2}S^*$		$10^{-9}S$ daraf
Electromotance	$\mathcal{E}$	$c^{-1}\mathcal{E}^*$		$10^{-8}\mathcal{E}$ volt
Field intensity, electric	E	$c^{-1}E^*$	$10^{-8}E$ volt/cm	$10^{-6}E$ volt/meter
Field intensity, magnetic	H^*	cH	H oersteds	$(10^3/4\pi)H$ amp-turns/m
Flux, magnetic	N^*	$c^{-1}N$	N maxwells	$10^{-8}N$ weber
Impedance	Z	$c^{-2}Z^*$		$10^{-9}Z$ ohm
Inductance	L	$c^{-2}L^*$		$10^{-9}L$ henry
Induction, magnetic	B^*	$c^{-1}B$	B gauss	$10^{-4}B$ weber/meter^2
Magnetic moment (dipole)	M'^*	$c^{-1}M'$	$4\pi M'$ maxwell-cm	$4\pi 10^{-10}M'$ weber-m
Magnetic moment (loop)	M^*	cM	$10M$ ampere-cm^2	$10^{-3}M$ ampere-m^2
Magnetization (dipole)	M'^*	$c^{-1}M'$	$4\pi M'$ maxwells/cm^2	$4\pi 10^{-4}M'$ webers/m^2
Magnetization (loop)	M^*	cM	$10M$ amperes/cm	$1000M$ amperes/m
Magnetomotance	Ω^*	$c\Omega$	Ω gilberts	$(10/4\pi)\Omega$ amp-turns
Permeability	μ^*	$c^{-2}\mu$	μ gauss/oersteds	$4\pi 10^{-7}\mu$ henry/m
Pole strength, magnetic	m^*	$c^{-1}m$	$4\pi m$ maxwells	$4\pi 10^{-8}m$ weber
Potential, electric scalar	V	$c^{-1}V^*$		$10^{-8}V$ volt
Potential, magnetic vector	A^*	$c^{-1}A$	A gauss-cm	$10^{-6}A$ weber/meter
Reactance	X	$c^{-2}X^*$		$10^{-9}X$ ohm
Reluctance	R'^*	c^2R'	R' gilb/max	$(10^9/4\pi)R'$ amp-turns/web
Resistance	R	$c^{-2}R^*$		$10^{-9}R$ ohm
Resistivity, surface	s	$c^{-2}s^*$		$10^{-9}s$ ohm
Resistivity, volume	τ	$c^{-2}\tau^*$	$10^{-9}\tau$ ohm-cm	$10^{-11}\tau$ ohm-meter

TABLE IV

A formula given in cgs emu, cgs esu, or Gaussian units may be expressed in (*a*) unrationalized cgs practical units or (*b*) rationalized mks units by replacing each symbol with its value in the emu, esu, or starred column, respectively.

Quantity	(*a*) Practical cgs units (*b*) Rationalized mks units	Emu	Esu
Capacitance	C farad	$10^{-9}C$	$9 \times 10^{11}C^*$
Capacitivity	(*a*) ϵ 4π-farad/cm	$10^{-9}\epsilon$	$9 \times 10^{11}\epsilon^*$
	(*b*) ϵ farad/meter	$4\pi 10^{-11}\epsilon$	$36\pi 10^{9}\epsilon^*$
Charge, quantity	q coulomb	$10^{-1}q$	$3 \times 10^{9}q^*$
Conductance	G mho	$10^{-9}G$	$9 \times 10^{11}G^*$
Conductivity, surface	γ' mho	$10^{-9}\gamma'$	$9 \times 10^{11}\gamma'^*$
Conductivity, volume	(*a*) γ mho/cm	$10^{-9}\gamma$	$9 \times 10^{11}\gamma^*$
	(*b*) γ mho/meter	$10^{-11}\gamma$	$9 \times 10^{9}\gamma^*$
Current	I ampere	$10^{-1}I$	$3 \times 10^{9}I^*$
Current density, surface	(*a*) i' ampere/cm	$10^{-1}i'$	$3 \times 10^{9}i'^*$
	(*b*) i' amp/meter	$10^{-3}i'$	$3 \times 10^{7}i'^*$
Current density, volume	(*a*) i ampere/cm^2	$10^{-1}i$	$3 \times 10^{9}i^*$
	(*b*) i amp/meter^2	$10^{-5}i$	$3 \times 10^{5}i^*$
Displacement, electric	(*a*) D 4π-coul/cm^2	$10^{-1}D$	$3 \times 10^{9}D^*$
	(*b*) D coulomb/m^2	$4\pi 10^{-5}D$	$12\pi 10^{5}D^*$
Elastance	S daraf	$10^{9}S$	$9^{-1}10^{-11}S^*$
Electromotance	$\mathcal{E}$ volt	$10^{8}\mathcal{E}$	$(300)^{-1}\mathcal{E}^*$
Field intensity, electric	(*a*) E volt/cm	$10^{8}E$	$(300)^{-1}E$
	(*b*) E volt/meter	$10^{6}E$	$3^{-1}10^{-4}E$
Field intensity, magnetic	(*a*) H oersted	H^*	$3 \times 10^{10}H$
	(*b*) H amp-turn/m	$4\pi 10^{-3}H^*$	$12\pi 10^{7}H$
Flux, magnetic	(*a*) N maxwell	N^*	$3^{-1}10^{-10}N$
	(*b*) N weber	$10^{8}N^*$	$(300)^{-1}N$
Impedance	Z ohm	$10^{9}Z$	$9^{-1}10^{-11}Z^*$
Inductance	L henry	$10^{9}L$	$9^{-1}10^{-11}L^*$
Induction, magnetic	(*a*) B gauss	B^*	$3^{-1}10^{-10}B$
	(*b*) B weber/meter^2	$10^{4}B^*$	$3^{-1}10^{-6}B$
Magnetic moment (dipole)	(*a*) M' maxwell-cm	$(4\pi)^{-1}M'^*$	$(12\pi)^{-1}10^{-10}M'$
	(*b*) M' weber-meter	$(4\pi)^{-1}10^{10}M'^*$	$(12\pi)^{-1}M'$
Magnetic moment (loop)	(*a*) M ampere-cm^2	$10^{-1}M^*$	$3 \times 10^{9}M$
	(*b*) M amp-meter^2	$10^{3}M^*$	$3 \times 10^{13}M$
Magnetization (dipole)	(*a*) M maxwell/cm^2	$(4\pi)^{-1}M^*$	$(12\pi)^{-1}10^{-10}M$
	(*b*) M weber/meter^2	$(4\pi)^{-1}10^{4}M^*$	$(12\pi)^{-1}10^{-6}M$
Magnetization (loop)	(*a*) M ampere/cm	$10^{-1}M^*$	$3 \times 10^{9}M$
	(*b*) M amp/meter	$10^{-3}M^*$	$3 \times 10^{7}M$
Magnetomotance	(*a*) Ω gilbert	Ω^*	$3 \times 10^{10}\Omega$
	(*b*) Ω ampere-turn	$4\pi 10^{-1}\Omega^*$	$12\pi \times 10^{9}\Omega$
Permeability	(*a*) μ gauss/oersted	μ^*	$9^{-1}10^{-20}\mu$
	(*b*) μ henry/meter	$(4\pi)^{-1}10^{7}\mu^*$	$(36\pi)^{-1}10^{-13}\mu$
Pole strength	(*a*) m maxwell	$(4\pi)^{-1}m^*$	$(12\pi)^{-1}10^{-10}m$
	(*b*) m weber	$(4\pi)^{-1}10^{8}m^*$	$(1200\pi)^{-1}m$
Potential, electric scalar	V volt	$10^{8}V$	$(300)^{-1}V^*$

TABLE IV.—(*Continued*)

Quantity	(a) Practical cgs units (b) Rationalized mks units	Emu	Esu
Potential, magnetic vector	(*a*) A gauss-cm	A*	$3^{-1}10^{-10}A$
	(*b*) A weber/meter	10^6A*	$3^{-1}10^{-4}A$
Reactance	X ohm	10^9X	$9^{-1}10^{-11}X$*
Reluctance	(*a*) R' gilbert/maxwell	R'*	$9 \times 10^{20}R'$
	(*b*) R' amp-t/weber	$4\pi 10^{-9}R'$*	$36\pi \times 10^{11}R'$
Resistance	R ohm	10^9R	$9^{-1}10^{-11}R$*
Resistivity, surface	s ohm	10^9s	$9^{-1}10^{-11}s$*
Resistivity, volume	(*a*) τ ohm-cm	$10^9\tau$	$9^{-1}10^{-11}\tau$*
	(*b*) τ ohm-meter	$10^{11}\tau$	$9^{-1}10^{-9}\tau$*

TABLE V

This table gives the physical dimensions of electrical units. Length is l, mass is m, time is t, charge is q, capacitivity is ϵ, and permeability is μ. Any column may be used in any system of units, but usually the first with $\epsilon = 1$ is used for esu, the second with $\mu = 1$ for emu, and the third for mks.

Quantity		Esu ($\epsilon = 1$)	Emu ($\mu = 1$)	Mks
Capacitance	C	ϵl	$\mu^{-1}l^{-1}t^2$	$m^{-1}l^{-2}t^2q^2$
Capacitivity	ϵ	ϵ	$\mu^{-1}l^{-2}t^2$	$m^{-1}l^{-3}t^2q^2$
Charge, quantity	q, Q	$\epsilon^{\frac{1}{2}}m^{\frac{1}{2}}l^{\frac{3}{2}}t^{-1}$	$\mu^{-\frac{1}{2}}m^{\frac{1}{2}}l^{\frac{1}{2}}$	q
Conductance	G	ϵlt^{-1}	$\mu^{-1}l^{-1}t$	$m^{-1}l^{-2}tq^2$
Conductivity, surface	γ'	ϵlt^{-1}	$\mu^{-1}l^{-1}t$	$m^{-1}l^{-2}tq^2$
Conductivity, volume	γ	ϵt^{-1}	$\mu^{-1}l^{-2}t$	$m^{-1}l^{-3}tq^2$
Current	I	$\epsilon^{\frac{1}{2}}m^{\frac{1}{2}}l^{\frac{3}{2}}t^{-2}$	$\mu^{-\frac{1}{2}}m^{\frac{1}{2}}l^{\frac{1}{2}}t^{-1}$	$t^{-1}q$
Current density, surface	i'	$\epsilon^{\frac{1}{2}}m^{\frac{1}{2}}l^{\frac{1}{2}}t^{-2}$	$\mu^{-\frac{1}{2}}m^{\frac{1}{2}}l^{-\frac{1}{2}}t^{-1}$	$l^{-1}t^{-1}q$
Current density, volume	i	$\epsilon^{\frac{1}{2}}m^{\frac{1}{2}}l^{-\frac{1}{2}}t^{-2}$	$\mu^{-\frac{1}{2}}m^{\frac{1}{2}}l^{-\frac{3}{2}}t^{-1}$	$l^{-2}t^{-1}q$
Displacement, electric	D	$\epsilon^{\frac{1}{2}}m^{\frac{1}{2}}l^{-\frac{1}{2}}t^{-1}$	$\mu^{-\frac{1}{2}}m^{\frac{1}{2}}l^{-\frac{3}{2}}$	$l^{-2}q$
Elastance	S	$\epsilon^{-1}l^{-1}$	μlt^{-2}	$ml^2t^{-2}q^{-2}$
Electromotance	$\mathcal{E}$	$\epsilon^{-\frac{1}{2}}m^{\frac{1}{2}}l^{\frac{1}{2}}t^{-1}$	$\mu^{\frac{1}{2}}m^{\frac{1}{2}}l^{\frac{3}{2}}t^{-2}$	$ml^2t^{-2}q^{-1}$
Field intensity, electric	E	$\epsilon^{-\frac{1}{2}}m^{\frac{1}{2}}l^{-\frac{1}{2}}t^{-1}$	$\mu^{\frac{1}{2}}m^{\frac{1}{2}}l^{\frac{1}{2}}t^{-2}$	$mlt^{-2}q^{-1}$
Field intensity, magnetic	H	$\epsilon^{\frac{1}{2}}m^{\frac{1}{2}}l^{\frac{1}{2}}t^{-2}$	$\mu^{-\frac{1}{2}}m^{\frac{1}{2}}l^{-\frac{1}{2}}t^{-1}$	$l^{-1}t^{-1}q$
Flux, magnetic	N	$\epsilon^{-\frac{1}{2}}m^{\frac{1}{2}}l^{\frac{1}{2}}$	$\mu^{\frac{1}{2}}m^{\frac{1}{2}}l^{\frac{3}{2}}t^{-1}$	$ml^2t^{-1}q^{-1}$
Impedance	Z	$\epsilon^{-1}l^{-1}t$	μlt^{-1}	$ml^2t^{-1}q^{-2}$
Inductance	L	$\epsilon^{-1}l^{-1}t^2$	μl	ml^2q^{-2}
Induction, magnetic	B	$\epsilon^{-\frac{1}{2}}m^{\frac{1}{2}}l^{-\frac{3}{2}}$	$\mu^{\frac{1}{2}}m^{\frac{1}{2}}l^{-\frac{1}{2}}t^{-1}$	$mt^{-1}q^{-1}$
Magnetic moment (dipole)	M'	$\epsilon^{-\frac{1}{2}}m^{\frac{1}{2}}l^{\frac{3}{2}}$	$\mu^{\frac{1}{2}}m^{\frac{1}{2}}l^{\frac{5}{2}}t^{-1}$	$ml^3t^{-1}q^{-1}$
Magnetic moment (loop)	M	$\epsilon^{\frac{1}{2}}m^{\frac{1}{2}}l^{\frac{7}{2}}t^{-2}$	$\mu^{-\frac{1}{2}}m^{\frac{1}{2}}l^{\frac{5}{2}}t^{-1}$	$l^2t^{-1}q$
Magnetization (dipole)	M'	$\epsilon^{-\frac{1}{2}}m^{\frac{1}{2}}l^{-\frac{3}{2}}$	$\mu^{\frac{1}{2}}m^{\frac{1}{2}}l^{-\frac{1}{2}}t^{-1}$	$mt^{-1}q^{-1}$
Magnetization (loop)	M	$\epsilon^{\frac{1}{2}}m^{\frac{1}{2}}l^{\frac{1}{2}}t^{-2}$	$\mu^{-\frac{1}{2}}m^{\frac{1}{2}}l^{-\frac{1}{2}}t^{-1}$	$l^{-1}t^{-1}q$
Magnetomotance	Ω	$\epsilon^{\frac{1}{2}}m^{\frac{1}{2}}l^{\frac{3}{2}}t^{-2}$	$\mu^{-\frac{1}{2}}m^{\frac{1}{2}}l^{\frac{1}{2}}t^{-1}$	$t^{-1}q$
Permeability	μ	$\epsilon^{-1}l^{-2}t^2$	μ	mlq^{-2}
Pole strength, magnetic	m	$\epsilon^{-\frac{1}{2}}m^{\frac{1}{2}}l^{\frac{1}{2}}$	$\mu^{\frac{1}{2}}m^{\frac{1}{2}}l^{\frac{3}{2}}t^{-1}$	$ml^2t^{-1}q^{-1}$
Potential, electric scalar	V	$\epsilon^{-\frac{1}{2}}m^{\frac{1}{2}}l^{\frac{1}{2}}t^{-1}$	$\mu^{\frac{1}{2}}m^{\frac{1}{2}}l^{\frac{3}{2}}t^{-2}$	$ml^2t^{-2}q^{-1}$
Potential, magnetic vector	A	$\epsilon^{-\frac{1}{2}}m^{\frac{1}{2}}l^{-\frac{1}{2}}$	$\mu^{\frac{1}{2}}m^{\frac{1}{2}}l^{\frac{1}{2}}t^{-1}$	$mlt^{-1}q^{-1}$
Reactance	X	$\epsilon^{-1}l^{-1}t$	μlt^{-1}	$ml^2t^{-1}q^{-2}$
Reluctance	R'	ϵlt^{-2}	$\mu^{-1}l^{-1}$	$m^{-1}l^{-2}q^2$
Resistance	R	$\epsilon^{-1}l^{-1}t$	μlt^{-1}	$ml^2t^{-1}q^{-2}$
Resistivity, surface	s	$\epsilon^{-1}l^{-1}t$	μlt^{-1}	$ml^2t^{-1}q^{-2}$
Resistivity, volume	τ	$\epsilon^{-1}t$	μl^2t^{-1}	$ml^3t^{-1}q^{-2}$

TABLE VI

Numerical values for empty space in the rationalized mks system

Velocity of light	c	$(\mu_v \epsilon_v)^{-\frac{1}{2}} = 2.99793 \times 10^8$ meters/second
Capacitivity	ϵ_v	$8.85419 \times 10^{-12} \approx (36\pi \times 10^9)^{-1}$ farad/meter
Permeability	μ_v	$4\pi \times 10^{-7} \approx 1.25664 \times 10^{-6}$ henry/meter
Intrinsic impedance	$(\mu_v/\epsilon_v)^{\frac{1}{2}}$	376.730 ohms
Intrinsic admittance	$(\epsilon_v/\mu_v)^{\frac{1}{2}}$	2.65442×10^{-3} mho

Numerical values of atomic constants in the rationalized mks system *

Faraday constant ($0^{16} = 16$)	F	$= 96{,}487 \pm 2$ coulombs/gm equivalent
Avogadro number ($0^{16} = 16$)	N	$= (6.0225 \pm 0.0002) \times 10^{23}$ 1/mole
Planck constant	h	$= (6.6256 \pm 0.0005) \times 10^{-34}$ joule sec
Electronic charge	e	$= (1.60210 \pm 0.00007) \times 10^{-19}$ coulomb
Specific electronic charge	e/m	$= (1.75880 \pm 0.00002) \times 10^{11}$ coulombs/kg

* COHEN, E. R., and J. W. M. DUMOND, *NBS Tech. News Bul.* October, 1963, page 175.

INDEX

(Problem numbers are italicized and enclosed in parentheses with the number of the page on which they occur)

D

J

K

L

Q